Pearson Australia
(a division of Pearson Australia Group Pty Ltd)
459–471 Church Street, Level 1, Building B,
Richmond, Victoria 3121
PO Box 23360, Melbourne, Victoria 8012

www.pearson.com.au

First published 2018 by Pearson Australia
2024 2023 2022 2021
10 9 8 7 6 5 4 3 2 1

Publisher: Hannah Gorman
Project Managers: Shelly Wang, Beth Zeme
Production Managers: Anji Bignell, Natalie Lincoln
Development Editor: Zoe Hamilton
Series Designer/Cover Designer: Anne Donald
Rights & Permissions Editor: Amirah Fatin
Editor: Marcia Bascombe
Proofreader: Helen Koehne
Indexer: Jon Jermey
Illustrator/s: DiacriTech
Printed in Malaysia (CTP-VVP)

National Library of Australia Cataloguing-in-Publication entry

Pearson Chemistry 11 : Western Australia / Geoff Quinton (coordinating author).

ISBN: 9781488617720 (pbk.)

Target Audience: For secondary school age

Subjects: Chemistry—Study and teaching (Secondary)—Western Australia.

Chemistry—Textbooks.

Pearson Australia Group Pty Ltd ABN 40 004 245 943

Attributions
WACE Chemistry ATAR Course Year 11 Syllabus

This book is dedicated to Pat and Mike Quinton

Disclaimer
The selection of internet addresses (URLs) provided for this book was valid at the time of publication and was chosen as being appropriate for use as a secondary education research tool. However, due to the dynamic nature of the internet, some addresses may have changed, may have ceased to exist since publication, or may inadvertently link to sites with content that could be considered offensive or inappropriate. While the authors and publisher regret any inconvenience this may cause readers, no responsibility for any such changes or unforeseeable errors can be accepted by either the authors or the publisher.

Some of the images used in *Pearson Chemistry 11 Western Australia Student Book* might have associations with deceased Indigenous Australians. Please be aware that these images might cause sadness or distress in Aboriginal or Torres Strait Islander communities.

The Chemistry Education Association was formed in 1977 by a group of teachers from secondary and tertiary institutions. It aims to promote the teaching of chemistry, particularly in secondary schools. The CEA has established a tradition of providing up to date text and electronic material and support resources for both students and teachers and professional development opportunities for teachers.

CEA offers scholarship and bursaries to students and teachers to further their interest in chemistry. CEA supports STAV with sponsorship for the Chemistry, Science Drama Awards and The Science Talent Search.

PEARSON

CHEMISTRY

WESTERN AUSTRALIA

Writing and Development Team

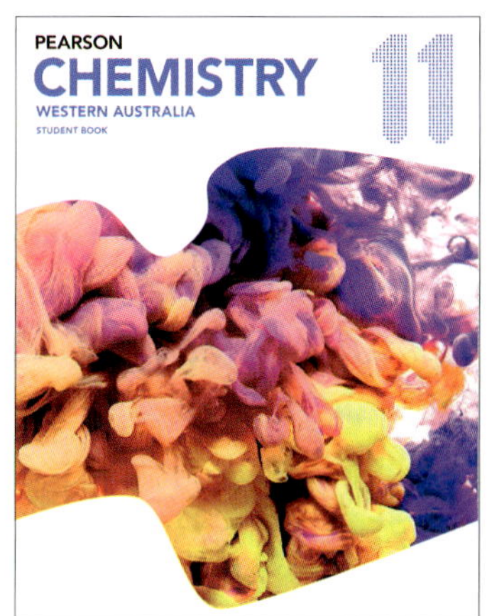

Geoff Quinton
President of ASTA
Head of science
Coordinating author

Erin Bruns
Senior science teacher
Author

Simon Carrello
Chemistry teacher
Author

John Clarke
Chief executive officer STAWA
Author

Chris Commons
Chemistry lecturer
Author and reviewer

Penny Commons
Chemistry lecturer
Author

Lanna Derry
Teacher
Contributing author

Bob Hogendoorn
Senior consultant
Contributing author

Elissa Huddart
Teacher
Contributing author

Phil Jones
Educator
Author

Allan Knight
Science education consultant
Author

Claire Molinari
Chemistry teacher and Director of the Centre for Excellence, Christ Church Grammar School
Author

Bill Offer
Former head of science and science teacher
Author and reviewer

Pat O'Shea
Teacher
Contributing author

Maria Porter
Teacher
Contributing author

Bob Ross
Teacher
Contributing author

Patrick Sanders
Teacher
Contributing author

Robert Sanders
Teacher
Contributing author

Sophie Selby-Pham
Master of Science
Answer checker

Lyndon Smith
Former member of the SCSA curriculum development team
Reviewer and answer checker

Marguerite van der Klashorst
Secondary teacher
Author

Contents

UNIT 1: Chemical fundamentals: structure, properties and reactions

UNIT 2: Molecular interactions and reactions

AREA OF STUDY 5: INTERACTING MOLECULES

How can opposite charges affect chemicals?

AREA OF STUDY 6 CHEMICALS IN THE ENVIRONMENT

How do our atmosphere and hydrosphere behave?

AREA OF STUDY 7 CHEMISTRY IN OUR WATER

What? How much? How do we know?

AREA OF STUDY 8 RATES OF CHEMICAL REACTIONS

How can chemicals be forced to change?

AREA OF STUDY 9 SCIENCE INQUIRY SKILLS IN CHEMISTRY

How to use this book

Pearson Chemistry 11 Western Australia

Pearson Chemistry 11 Western Australia has been written to the *WACE Chemistry ATAR Course, Year 11 Syllabus 2015*. Each chapter is clearly divided into manageable sections of work. Best practice literacy and instructional design are combined with high-quality, relevant photos and illustrations. Explore how to use this book below.

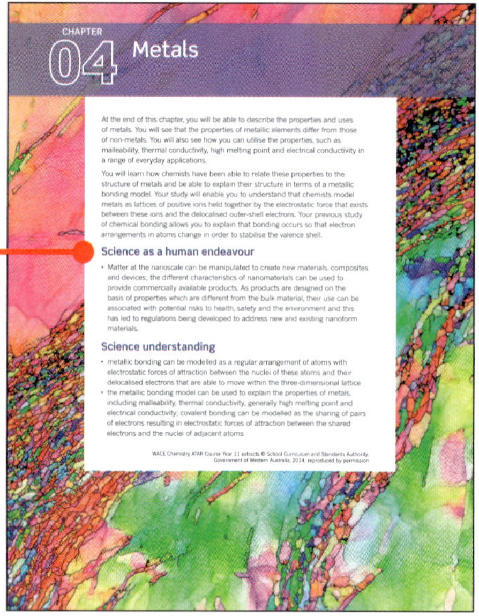

CHAPTER 04 Metals

Science as a human endeavour

Science understanding

Chapter opening page

The chapter opening page links the syllabus to the chapter content. Science understanding and Science as a human endeavour addressed in the chapter are clearly listed.

4.3 Reactivity of metals

DETERMINING THE REACTIVITY OF METALS

Reactivity with water

CHEMFILE Reactivity of group 1 metals

Reactivity with acids

CHEMFILE

ChemFiles include a range of interesting information and real-world examples.

EXTENSION

Extension boxes include material that goes beyond the core content of the syllabus. They are intended for students who wish to expand their depth of understanding in a particular area.

Reasons for different reactivities of metals

EXTENSION Extracting iron from iron ore

Highlight box

Focuses students' attention on important information such as key definitions, formulae and summary points.

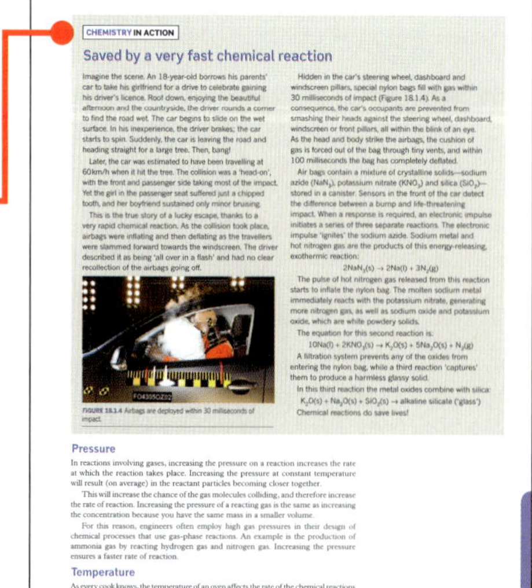

CHEMISTRY IN ACTION

Saved by a very fast chemical reaction

Pressure

Temperature

CHEMISTRY IN ACTION

Chemistry in Action boxes place Chemistry in an applied situation or relevant context and encourages students to think about the development of chemistry and its use and influence of chemistry in society.

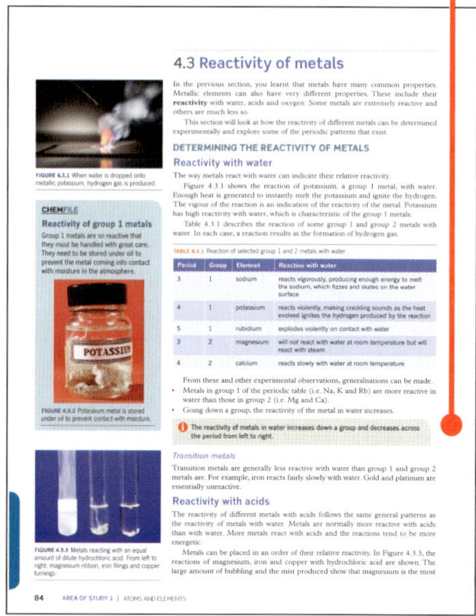

4.3 Reactivity of metals

DETERMINING THE REACTIVITY OF METALS

Reactivity with water

Reactivity with acids

Worked example

Worked examples are set out in steps that show both thinking and working. This enhances student understanding by clearly linking underlying logic to the relevant calculations.

Each Worked example is followed by a Worked example: Try yourself. This mirror problem allows students to immediately test their understanding.

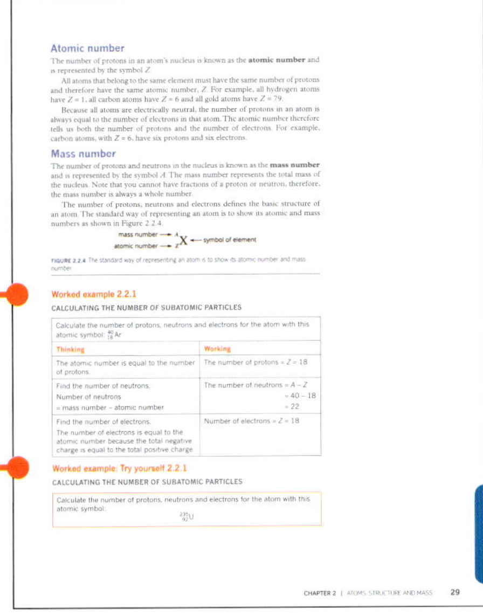

Unit review

Each unit finishes with a comprehensive set of exam-style questions, which assist students to draw together their knowledge and understanding and apply it to this style of questions.

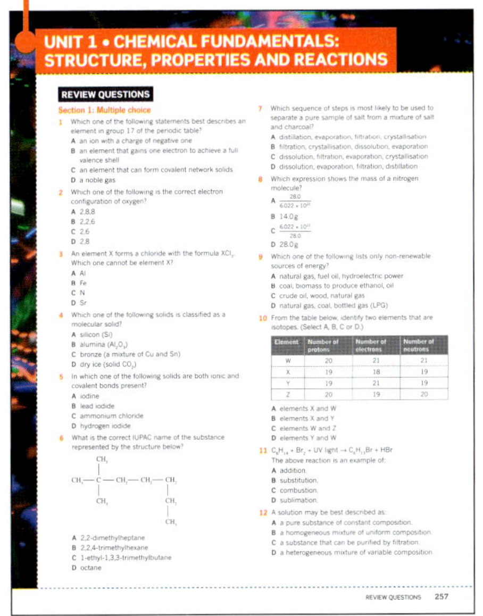

Section SUMMARY

Each section includes a summary to assist students consolidate key points and concepts.

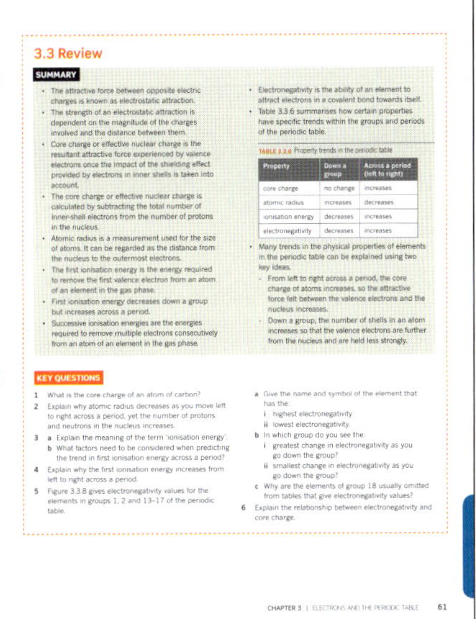

Section review Key questions

Each section finishes with questions to test students' understanding and ability to recall the key concepts of the section.

Key terms and glossary

Key terms are shown in bold and listed at the end of each chapter. A comprehensive glossary at the end of the book includes and defines all the key terms.

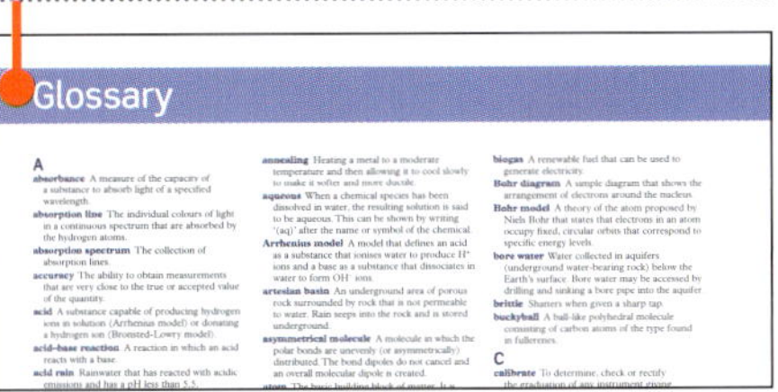

Chapter review

Each chapter finishes with a set of higher order questions to test students' ability to apply the knowledge gained from the chapter.

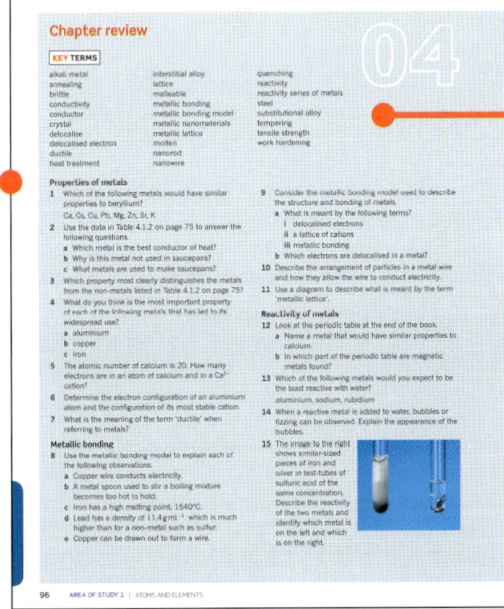

Answers

Numerical answers and key short response answers are included at the back of the book. Comprehensive answers and fully worked solutions for all section review questions, Worked example: Try yourself exercises, chapter review questions and unit review questions are provided via *Pearson Chemistry 11 Western Australia Teacher Reader+*.

Pearson Chemistry 11 Western Australia

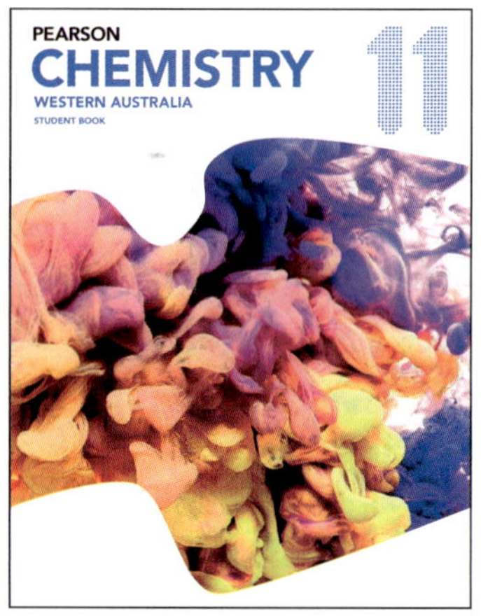

Student Book

Pearson Chemistry 11 Western Australia has been written to fully align with the *WACE Chemistry ATAR Course, Year 11 Syllabus 2015*. The series includes the very latest developments and applications of Chemistry and incorporates best practice literacy and instructional design to ensure the content and concepts are fully accessible to all students.

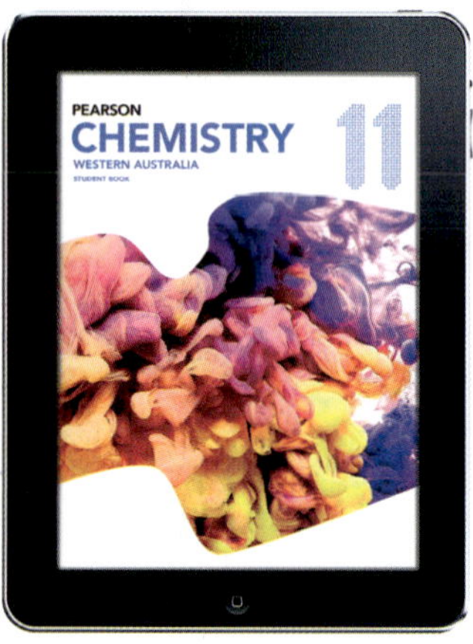

Reader+

Pearson Reader+ lets you use the Student Book online or offline on any device. Pearson Reader+ includes interactive activities and videos to enhance learning and test understanding. The teacher version includes support material to assist planning and implementing the syllabus.

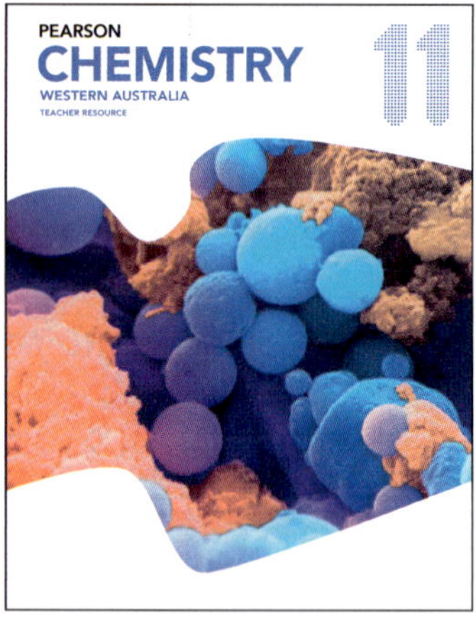

Teacher Resource

The *Pearson Chemistry 11 Western Australia Teacher Resource* consists of a print book and online resources that provide comprehensive teacher support, including teaching programs, fully worked solutions to all questions, chapter tests and practice exams.

Access digital resources at **pearsonplaces.com.au**
Browse and buy at **pearson.com.au**

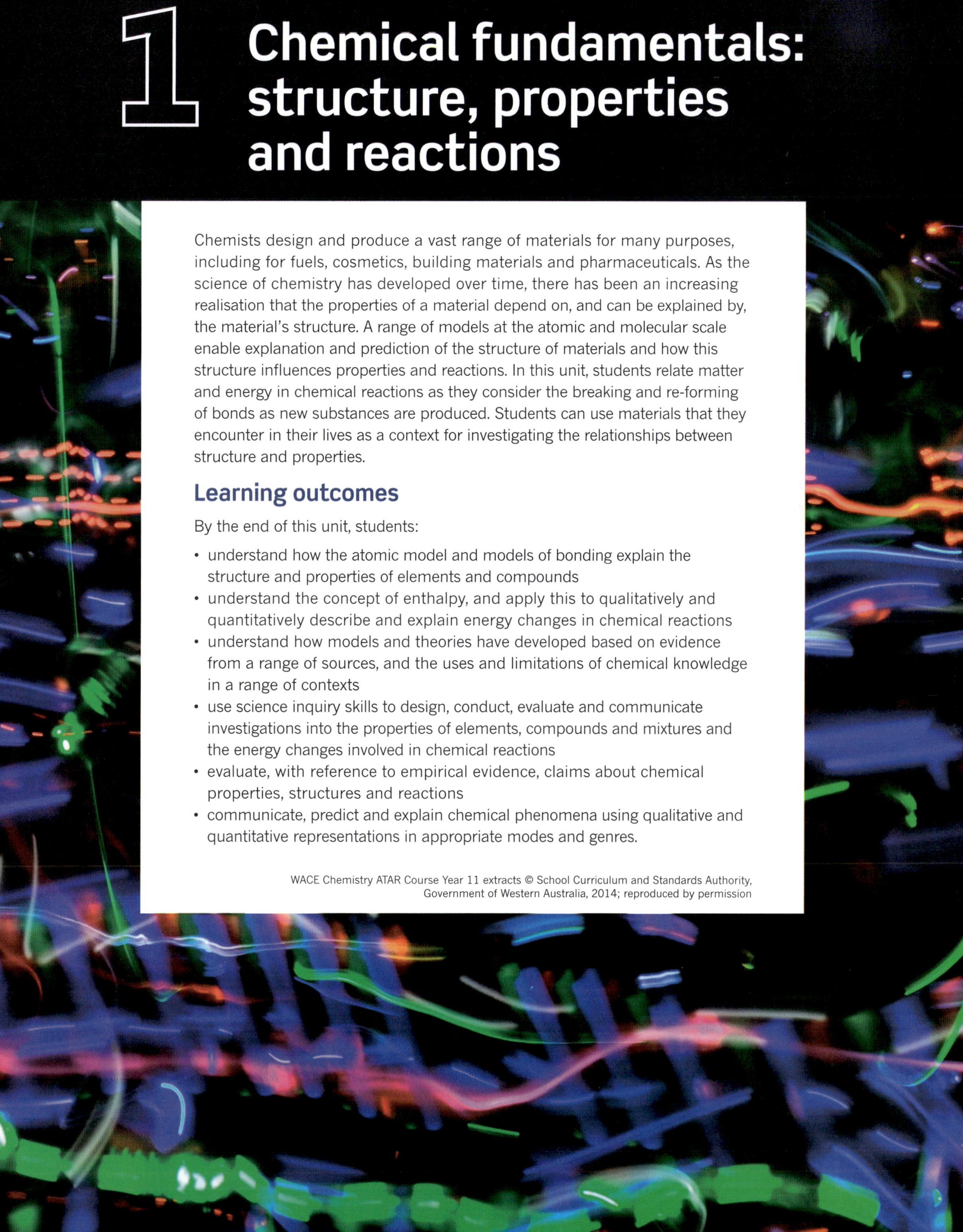

UNIT 1

Chemical fundamentals: structure, properties and reactions

Chemists design and produce a vast range of materials for many purposes, including for fuels, cosmetics, building materials and pharmaceuticals. As the science of chemistry has developed over time, there has been an increasing realisation that the properties of a material depend on, and can be explained by, the material's structure. A range of models at the atomic and molecular scale enable explanation and prediction of the structure of materials and how this structure influences properties and reactions. In this unit, students relate matter and energy in chemical reactions as they consider the breaking and re-forming of bonds as new substances are produced. Students can use materials that they encounter in their lives as a context for investigating the relationships between structure and properties.

Learning outcomes

By the end of this unit, students:

- understand how the atomic model and models of bonding explain the structure and properties of elements and compounds
- understand the concept of enthalpy, and apply this to qualitatively and quantitatively describe and explain energy changes in chemical reactions
- understand how models and theories have developed based on evidence from a range of sources, and the uses and limitations of chemical knowledge in a range of contexts
- use science inquiry skills to design, conduct, evaluate and communicate investigations into the properties of elements, compounds and mixtures and the energy changes involved in chemical reactions
- evaluate, with reference to empirical evidence, claims about chemical properties, structures and reactions
- communicate, predict and explain chemical phenomena using qualitative and quantitative representations in appropriate modes and genres.

CHAPTER

01 Materials in our world

Many of the most significant scientific advancements of civilisation are related to the discovery, development, production or application of new materials. New materials are so fundamental to the advancement of technology that entire periods of history are known by the materials that proliferated in that age. And just as technological progress is highly dependent on the structure and properties of new materials, many of the scientific problems facing society today are the result of the limitations of existing materials.

At the end of this chapter, you will be able to identify how the structure of a material affects the material's properties and, therefore, its uses. Mixing substances can produce new materials with different properties. These mixtures include alloys, composites and materials containing nanoparticles.

In addition, you will learn how separating and purifying materials can lead to useful products. A variety of separation techniques use the different properties of mixed substances to separate and purify those substances.

Science as a human endeavour

- Matter at the nanoscale can be manipulated to create new materials, composites and devices; the different characteristics of nanomaterials can be used to provide commercially available products. As products are designed on the basis of properties which are different from the bulk material, their use can be associated with potential risks to health, safety and the environment and this has led to regulations being developed to address new and existing nanoform materials.

Science understanding

- materials are pure substances with distinct measurable properties, including melting and boiling points, reactivity, hardness and density; or mixtures with properties dependent on the identity and relative amounts of the substances that make up the mixture
- pure substances may be elements or compounds which consist of atoms of two or more elements chemically combined; the formulae of compounds indicate the relative numbers of atoms of each element in the compound
- differences in the physical properties of substances in a mixture, including particle size, solubility, density, and boiling point, can be used to separate them
- nanomaterials are substances that contain particles in the size range 1–100 nm and have specific properties relating to the size of these particles which may differ from those of the bulk material

1.1 Materials science

Materials science is one of the most rapidly advancing areas of science in the world today. This interdisciplinary field uses chemistry to control the structure of materials down to the atomic (microscopic) level while utilising physics and engineering to control the structure of materials at a practical and visible (macroscopic) level. This gives materials scientists the ultimate control over the properties of materials.

Having control over the structure and properties of materials has led to the development of new and useful technologies. These technologies can be found in things you use every day, such as smartphones, LCD screens and even sunscreen. Today, materials science is also being used to create technologies that were once thought to be science fiction, such as 3D printers and artificial skin.

MATERIALS

The term **material** usually describes substances that are used to make objects. For example, wood, paper and nylon are all classified as materials because they can be used to create houses, books and clothes, respectively. Substances that are not considered materials include chemicals such as hydrochloric acid, chlorophyll and carbon dioxide. While these are all extremely useful substances, other objects are not made out of these substances. Therefore these substances are not usually classified as materials.

Materials are often mixtures of many substances—for example, cement or bitumen. However, materials can also be pure elements or compounds.

Elements are substances that are made up of just one type of atom. This means they consist of atoms with the same atomic number (the number of protons in the nucleus). Pure metals such as gold or silver are examples of elements that are also materials. Carbon is an example of a non-metallic element that forms a variety of materials such as charcoal, graphite, diamond and even nanotubes.

Compounds also make up a huge variety of materials. Compounds are pure substances made of more than one type of atom. They consist of more than one element in fixed proportions.

The formula of a compound indicates the relative numbers of atoms of each element in the compound. For example, silica (SiO_2) is a compound made up of silicon and oxygen atoms and is the main component of beach sand. The compound contains twice as many oxygen atoms as silicon atoms. Silica is a component of glass, quartz and gemstones. Calcium carbonate ($CaCO_3$) is another compound that makes up several different materials. These materials include chalk, limestone and marble.

The way in which a material is used is determined by the material's physical and chemical properties. These properties are special features of the material, such as its colour, hardness, melting and boiling points, whether it conducts electricity or heat and how easily the material reacts with other chemicals.

Elements and compounds are pure substances, and therefore their properties, such as melting point, boiling point and conductivity, cannot be altered. However, the properties of elements and compounds depend on the arrangement of their atoms and molecules. The properties of elements and compounds are distinct and measurable for a given arrangement of atoms or molecules.

Unlike elements and compounds, the properties of mixtures can be changed depending on how much of each component is added to the mixture. The ability to control the properties of mixtures makes mixtures very useful materials.

Although there are thousands of different pure materials and countless more materials that are mixtures, many that we use fall into one of three classifications: metals, polymers or ceramics.

Metals

Metals account for around 80% of all known elements and around 24% of the total mass of the Earth. Only a few metals are naturally found in their elemental metallic form. These metals, known as native metals, include gold and copper. Most metals are found as compounds, known as minerals, which make up the ores mined from the Earth's crust.

Metals (Figure 1.1.1) are valuable materials due to their useful set of properties including high tensile strength, ductility, malleability, shiny lustre, high melting points, and thermal and electrical conductivity.

The characteristic of a metal required for a particular application can often be improved, or its weaknesses mitigated, by using an alloy of that metal. An **alloy** is a mixture of a metal with other metals or small amounts of non-metals. For example, iron is an abundant and easily worked metal, which in its pure form is relatively soft and prone to corrosion. If iron is alloyed with other elements, primarily carbon, the result is a much stronger and corrosion-resistant alloy known as steel.

The key reason for the attractive set of physical properties demonstrated by metals and their alloys is the nature of the bonding that exists within metals. This will be examined in greater detail in Chapter 4.

FIGURE 1.1.1 An example of different metals (clockwise from the bottom left): copper, aluminium, zinc, iron and, in the centre, lead

Polymers

A **polymer**, from Greek *poly* meaning 'many' and *mer* meaning 'parts', is a material with a molecular structure that is composed of many repeating smaller units bonded together. Polymeric materials include plastics, such as polyethylene (polyethene) and nylon, and rubbers, such as latex. There are both natural polymers, such as wool, silk and paper (Figure 1.1.2), and synthetic polymers, such as polystyrene.

Polymers present a vastly different set of physical properties compared to metals: they are less dense and corrosion-resistant, offer excellent electrical resistance and polymers of a biological nature offer good compatibility with human tissue.

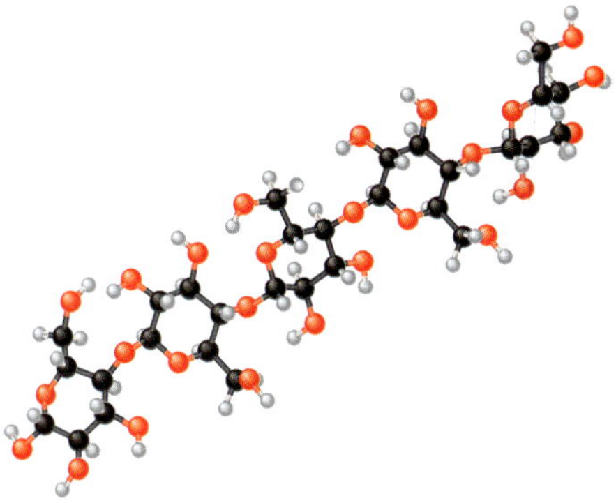

FIGURE 1.1.2 The pages of this book, the cotton used in clothes and the cell walls of green plants are all made of the naturally occurring polymer cellulose.

CHEMISTRY IN ACTION

3D printing

3D printing is a rapidly growing area of technology that started development in the early 1980s. It involves the manufacture of three-dimensional objects by the deposition or 'printing' of successive layers of a material controlled by a computer.

Initially 3D-printing materials were limited to various plastics (Figure 1.1.3), but this has since improved to include metals, ceramics and even biological polymers in the manufacture of replacement human tissue. 3D printing is becoming commonplace with many consumer-level 3D-printing machines available, and even a number of public libraries offering 3D-printing services.

FIGURE 1.1.3 This 3D printer uses an acrylonitrile–butadiene–styrene polymer to make an object.

Ceramics

FIGURE 1.1.4 A small cylindrical magnet being levitated above a liquid nitrogen–cooled, superconducting ceramic by the Meissner effect

A **ceramic** is an inorganic, non-metallic solid. Ceramics contain metal, non-metal and metalloid elements held together by ionic and covalent bonds (these types of chemical bonding are covered in Chapters 5 and 6). The degree of order within ceramic materials can range from highly ordered (crystalline) to highly irregular (amorphous). As with polymers, some ceramics such as kaolinite are naturally occurring and used to make porcelain, and others are synthetic, such as silicon carbide used as an abrasive.

As such a wide range of elements and bonding can occur in ceramics, they demonstrate a wide range of properties, but generally they are hard, have high compressive strength and are able to withstand high temperatures. Most ceramics are good insulators while others demonstrate semiconductor and superconducting properties (Figure 1.1.4).

COMPOSITES

A **composite material** is a combination of two or more distinct materials with significantly different physical and chemical properties. The resultant material demonstrates a range of properties that would be unobtainable by using one of the individual materials on its own.

Reinforced concrete (Figure 1.1.5) is a common example of a composite material. It consists of a concrete matrix with embedded steel bars (rebar). In this arrangement, the relatively low tensile strength of the concrete, which is a ceramic, is counteracted by the high tensile strength of the steel, an alloy, while maintaining the high compressive strength of the concrete.

FIGURE 1.1.5 The steel rebar can be seen here embedded in the concrete structure.

CHEMFILE

Damascus steel

Damascus steel was a type of steel used for the manufacture of bladed weapons in the Middle East up until the 17th century. It is characterised by distinctive banding patterns reminiscent of flowing water (Figure 1.1.6). The performance of Damascus steel blades became almost legendary; they were reputed to be tough, resistant to shattering and able to hold a sharp edge far better than steel produced in other regions.

In 2006, a research team in Germany published an investigation into Damascus steel revealing a network of nanowires and carbon nanotube reinforcing, formed in the steel during the forging process. The presence of these reinforcing fibres to form a composite material is the likely reason for Damascus steel's legendary durability.

The knowledge of the forging of Damascus steel was lost in the mid-1700s. Many have tried to rediscover the original forging methods, but they remain a mystery.

FIGURE 1.1.6 The blade pattern typical of Damascus steel blades

1.1 Review

SUMMARY

- Materials are either pure substances or mixtures.
- The properties of materials determine how they are used.
- Materials can be mixed to produce new materials with different properties.
- Alloys are materials made by mixing metals with other metals or small quantities of non-metals.
- Polymers are large molecules made of smaller repeating units.
- Ceramics are inorganic, non-metallic solids formed from a mixture of metal and non-metal elements.
- Composite materials are materials made from two or more different materials, e.g. glass and plastic to make fibreglass.

KEY QUESTIONS

1. What is a material?
2. What are the typical physical properties of a metal?
3. What is an alloy?
4. What are the typical physical properties of a polymer?
5. What are the typical physical properties of a ceramic?

1.2 Nanotechnology

Chemistry is the study of the structure and behavior of matter, and traditionally chemists focused on the structure of atoms and the properties of bulk materials composed of countless molecules. However, there are particles and structures that fall between these two extreme sizes. These are particles that are larger than individual atoms, but still smaller than the wavelength of visible light and thus cannot be viewed even with the most powerful optical microscopes.

At these scales, properties such as the surface area to volume ratio of particles changes, allowing surface area effects, also known as quantum effects, to become more pronounced. The impact of these quantum effects can dramatically alter the properties of a material; for example, turning electrical insulators into superconductors or making visible substances seemingly invisible.

Nanotechnology is a branch of materials science that investigates the design, properties and applications of materials produced on this scale.

NANOSCALE

The term **nanoscale** refers to structures that are between 1 and 100 nanometres across. A **nanometre** (nm) is one billionth of a metre (10^{-9} m). To illustrate this scale, silicon atoms are around 0.2 nanometres across and a human hair is up to 100 000 nanometres in diameter (Figure 1.2.1).

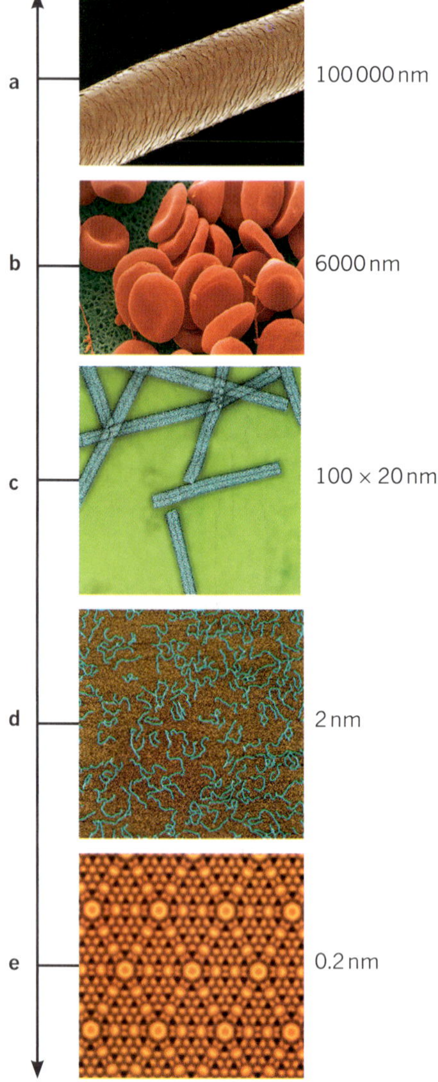

FIGURE 1.2.1 This scale compares the sizes of objects down to the nanometre size. (a) The width of a human hair is approximately 100 000 nm. (b) The diameter of red blood cells is 6000 nm. (c) Tobacco mosaic virus is 100 nm long and 20 nm wide. (d) DNA strands are about 2 nm wide. (e) Silicon atoms are 0.2 nm across.

The problem for scientists in the early days of nanotechnology was that the tools and techniques designed for working at this scale did not exist. In the early 1980s, the development of atomic force microscopes and scanning tunnelling microscopes allowed scientists to both view and manipulate individual atoms.

NANOMATERIALS

Nanomaterials are substances, both natural and synthetic, that are composed of single units that exist in the nanoscale. Although there is much discussion about the progress of synthetic nanomaterials, there are numerous examples of naturally occurring nanomaterials. For example, spider silk, butterfly wings and the bottom of gecko feet all have fascinating properties as a result of their nanostructure.

CHEMFILE

What do opals and butterflies have in common?

Opal is a nanomaterial made of tiny spheres of silica. These spheres diffract light to produce spectacular flashes of red, green and blue. Butterflies also get their coloured patterns from nanostructures on the surfaces of their wings.

FIGURE 1.2.2 (Top) An opal is made of tiny spherical nanoparticles of silica. (Bottom) The colour of a butterfly's wing is due to nanostructures in the wing.

One naturally occurring nanomaterial that has been a hot topic of research for scientists over the past two decades is a family of carbon molecules known as the fullerenes. **Fullerenes** are three-dimensional structures formed by networks of carbon atoms. The fullerene that promises a wide range of applications is a cylindrical tube known as a **carbon nanotube** (Figure 1.2.3). These nanotubes are formed from a flat, two-dimensional layer of carbon atoms arranged in hexagons, known as **graphene**.

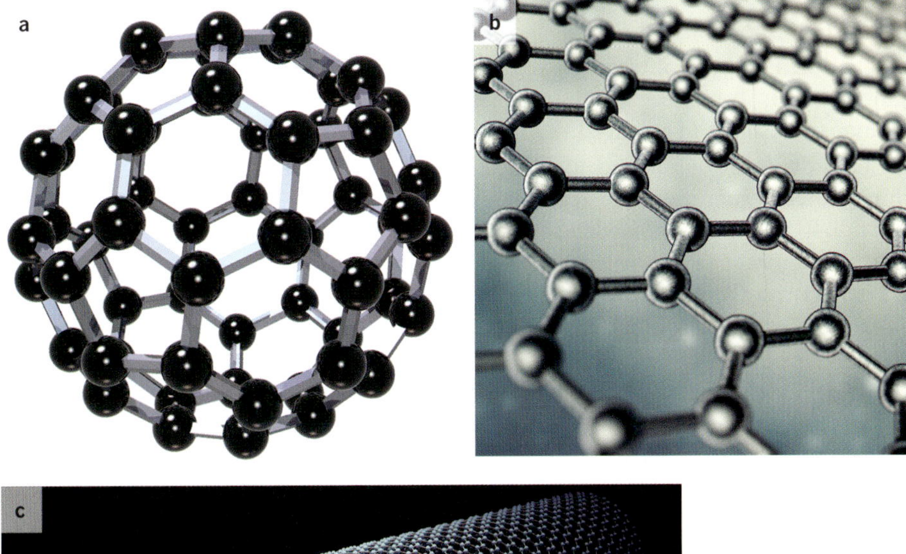

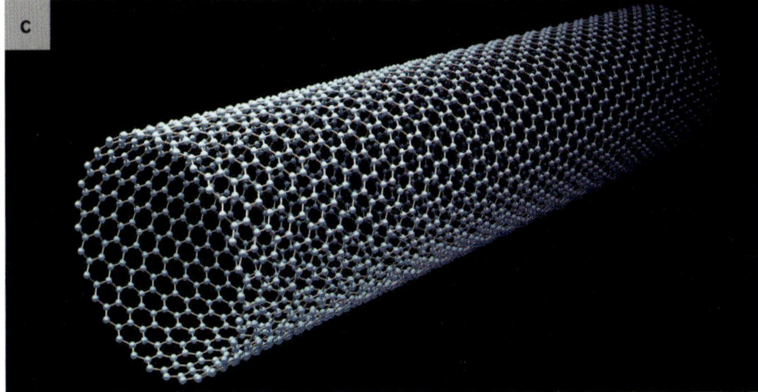

FIGURE 1.2.3 Example of different fullerenes: (a) the soccerball-shaped buckyball, (b) the flat chickenwire, like graphene and (c) the rolled, cylindrical nanotube

Carbon nanotubes have demonstrated exceptional strength and stiffness, finding use as reinforcement in composite materials. They also demonstrate interesting electrical properties and are the basis for a range of superconductor research. You will learn more about graphene and nanotubes in Chapter 7.

Materials on the nanoscale can be fabricated either using top-down or bottom-up techniques. Each technique is suited to particular applications and presents a different set of strengths.

Top-down fabrication

Top-down fabrication starts with material of a larger scale than desired. Material is then either selectively removed or the size of the material is progressively reduced, by grinding, until the required size and shape is achieved. An analogy for top-down fabrication would be a sculptor starting with a large block of marble and carving away pieces until the statue is formed.

Top-down techniques are currently the most commonly utilised methods and are used to make everything from computer chips to sunscreen. Some of the benefits of top-down fabrication are that large quantities of material can be produced relatively cheaply and the product demonstrates a good level of uniformity. Some disadvantages of top-down fabrication techniques are that they are limited to relatively simple structures and are limited by the scale of the tools used to remove material from the starting medium.

Bottom-up fabrication

Bottom-up fabrication is still in its early stages of development and involves physically building, or growing, the required material atom by atom or molecule by molecule. Individually selected atoms or molecules are successively built up until the required shape and size of material is formed. An analogy for bottom-up fabrication would be a sculptor joining individual calcium, carbon and oxygen atoms one at a time until the statue is formed. In this way, it is very much like existing chemical syntheses, but on a nanoscale.

One area of keen interest to scientists is the development of 'self-assembling' nanomaterials. It has already been demonstrated that molecules can be designed to arrange themselves in required shapes and structures, given appropriate conditions.

The advantage of bottom-up fabrication methods is they can be used for far more complicated structures than top-down methods due to the ability to manipulate individual atoms or molecules at the nanoscale. A disadvantage of bottom-up fabrication methods is they do not scale up to commercial levels efficiently and thus are only currently economical for research and niche applications.

CHEMISTRY IN ACTION

Superhydrophobic coatings

One developing application of nanomaterials is superhydrophobic coatings. These are synthetic, nanoscopic, surface coatings that mimic the behaviour of leaves from the lotus plant, which trap an imperceptible layer of air on the surface, preventing contact with liquids.

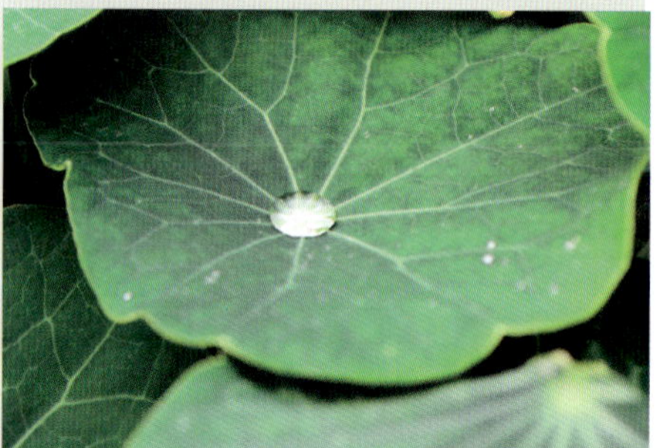

FIGURE 1.2.4 The liquid repelling effect of a lotus leaf is due to naturally occurring hydrophobic waxes and microscopic hairs on its surface.

The coatings can be applied with a simple aerosol spray and once treated, demonstrate remarkable anti-wetting characteristics, repelling water and dirt and essentially becoming self-cleaning. Applications for such treatments include materials that never need cleaning, surgical equipment that never carries pathogens and tomato sauce bottles that always allow you to get the last bit out. Unfortunately, current coatings demonstrate extremely poor durability, making them unsuitable for most applications, but research is ongoing.

NANOPARTICLES

Nanoparticles are a specific type of nanomaterial, which are currently the topic of intense scientific investigation. **Nanoparticles** are usually spherical particles with diameters of about 1–100 nm. It is in this size range that the properties of materials begin to change from those that are normally observed for bulk material to those that result from the greater contribution of surface effects. Nanoparticles have potential applications in medicine, physics, optics and electronics.

Although nanoparticles are a hot topic of cutting-edge scientific research at the moment, nanoparticles have been used throughout history. One of the curious properties of nanoparticles is how their optical properties are markedly different from the bulk material. For example, gold as a bulk material demonstrates a characteristic shiny yellow lustre valued for its appearance (Figure 1.2.5), yet gold nanoparticles can take on a range of different colours (Figure 1.2.6). The variable colours granted by gold nanoparticles of different sizes were used by artisans as far back as Rome in the 4th century for colouring glass and ceramics (Figure 1.2.7).

FIGURE 1.2.5 Bulk gold has a characteristic colour and typical metallic properties.

FIGURE 1.2.6 These vials contain gold particles of various sizes. The different-sized particles are different colours because they interact with light differently.

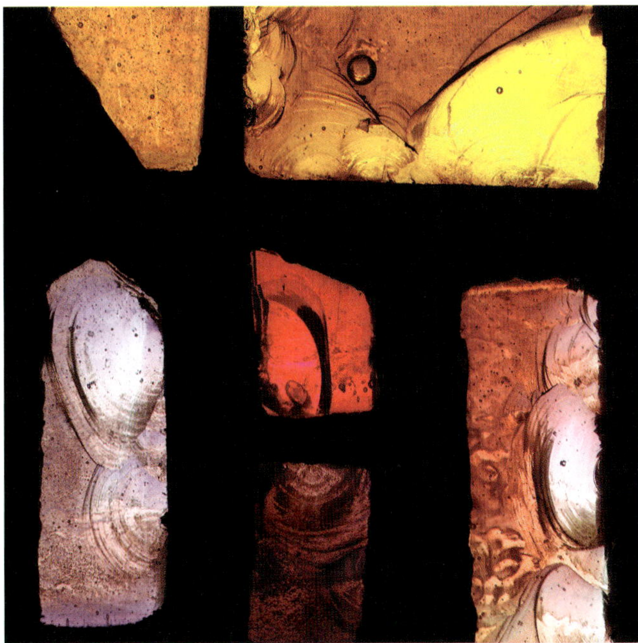

FIGURE 1.2.7 The red colour in this stained glass window is caused by gold nanoparticles trapped in the glass. The deep yellow colour is caused by silver nanoparticles.

One interesting application of nanoparticles that demonstrates different optical properties from the bulk is in sunscreens. Normally sunscreens contain metal oxides, such as zinc oxide, which appear white when applied to the skin (Figure 1.2.8). However, when zinc nanoparticles are used, they interact with light differently and appear clear.

FIGURE 1.2.8 White zinc on faces is a common sight on Australian beaches during summer.

CHEMFILE

Nanoparticles: risks to health and the environment

Nanoparticles have opened up a range of technological possibilities. However, the development of nanoparticles has also raised concerns about the possible dangers to humans and the environment.

In the past, other useful materials such as asbestos have been found to have devastating side effects. Therefore, the CSIRO (Commonwealth Scientific and Industrial Research Organisation) and other scientists around the world are studying the potential dangers associated with nanoparticles and their applications.

THE PROBLEM WITH NANOPARTICLES

Nanoparticles are so small that they can travel through the air, through skin and into your bloodstream and even into cells. Inside the body, the particles may interact with biomolecules to cause unwanted chemical reactions. This makes nanoparticles potentially dangerous if breathed in or if they are in contact with the skin (for example, in sunscreen, fabrics or cosmetics).

NANOPARTICLES IN SUNSCREENS

CSIRO scientists in Australia are looking at the zinc oxide nanoparticles used in sunscreens to determine whether they are safe. This research focuses on whether the nanoparticles can penetrate skin, their long-term health effects and how they might affect the environment.

Initial studies suggest that small amounts of zinc oxide from sunscreens are absorbed into the body and can be detected in the blood and urine. It is still not clear whether the absorbed zinc oxide has any harmful effects on the human body. The most recent research indicates that the cells of the immune system can break down the nanoparticles.

CHEMFILE

Why so blue?

A **colloid** is a mixture where one insoluble substance is dispersed through another. It is distinguished from a **suspension** in that colloids contain particles on the nanoscale that will not settle out of the mixture over time. For example, milk is a colloid of droplets of fat dispersed in water.

A common alternative medicine available at most pharmacies is colloidal silver. Colloidal silver consists of silver metal nanoparticles dispersed throughout a liquid, normally water. Although silver and silver compounds are toxic for bacteria, colloidal silver has been marketed as a cure-all for many conditions, with claims ranging from it being an essential dietary supplement to curing cancer.

One of the side effects of excessive silver consumption is a condition known as argyria, in which silver compounds accumulate in the skin, react to sunlight and turn the skin blue-grey (Figure 1.2.9). Although the condition is not life threatening, the effects are permanent and irreversible.

FIGURE 1.2.9 The blue-grey complexion of an individual suffering from argyria as a result of excessive colloidal silver consumption.

1.2 Review

SUMMARY

- Nanotechnology is the study of materials on the nanoscale.
- A nanometre (nm) is a billionth of a metre (10^{-9} m).
- Nanomaterials are materials with nanoscale features that give the material useful and important properties.
- Nanomaterials can be found in nature, manufactured in the laboratory or engineered from other materials.
- Nanoparticles are particles with diameters of about 1–100 nm.
- The properties of nanoparticles can be different from the bulk material that they are made from.
- Nanoparticles can travel through the air, through skin and even though cells, making them both potentially useful and dangerous for humans.

KEY QUESTIONS

1 Convert the following lengths into nanometres. Use scientific notation in your answers.
   a 8.35 cm
   b 1.35 mm
   c 4.2 mm

2 Explain why nanoparticles could play an important role in transporting medicines to certain regions of the body in treatments such as chemotherapy.

3 Why are the properties of nanoparticles so radically different from those of the bulk material?

1.3 Purifying materials

Alloys, composites and some nanomaterials are examples of how chemists can combine substances to create materials with different properties and for new applications. However, scientists can also develop new technologies by extracting, separating and purifying materials. For example:

- Separating crude oil: Crude oil is made up of a large number of useful compounds that can be separated into fuels, lubricants and chemicals to make plastics.
- Purifying silicon: 99.999% pure silicon can be used to make solar panels, but 99.9999999% pure silicon is required to make silicon microchips!
- Extracting DNA: The ability to extract and purify DNA from cells has fundamentally changed medicine, forensics and even agriculture.

Separation and purification techniques play an important role in many industries and laboratories. In each case, the separation technique uses differences in the physical and chemical properties of the mixed substances to separate the components of the mixture. Such differences in properties include particle size, density, solubility, boiling point and electric charge.

SEPARATION BY PARTICLE SIZE

One of the simplest separation techniques is to separate the substances on the basis of particle size. This is commonly done through sieving or filtration.

Sieving

Sieving is used to separate a mixture of solids with different particle sizes. The technique involves passing the mixture through a mesh. Particles that are smaller than the holes in the mesh pass through, leaving larger particles behind.

This simple technique is commonly used during baking to separate lumps from powders such as flour or cocoa. The technique is also an important tool in industries like mining. Figure 1.3.1 shows metal **ore** being poured into a hopper crusher, where it is crushed and passed through a sieve. The small particles that pass through the sieve move onto the next stage of processing. The larger particles remain in the hopper crusher to be crushed further.

FIGURE 1.3.1 After ore is mined, it is placed into a hopper crusher like this one. The ore is then pulverised. The smallest pieces fall through a sieve, while the larger pieces are crushed further.

Filtration

Filtration is used to separate solid particles from a liquid or gas. Air filters in vacuum cleaners, extraction fans or industrial chimneys are examples of filters that separate particles from air. Pool filters, coffee plungers and tea bags are domestic examples of filters that separate solid particles from liquids.

In the laboratory, scientists use filter paper to separate liquids and solids. This is normally done in one of two ways.

Gravitational filtration

Gravitational filtration uses the weight of the solid–liquid mixture to push the mixture through filter paper. The filter paper is first folded to produce a cone-shaped filter that fits into a funnel. The funnel is then placed into a conical flask, as shown in Figure 1.3.2. The solid–liquid mixture is then poured into the filter paper and left to run through the funnel. The purified liquid that collects in the flask is known as the **filtrate**. The solid collected in the filter paper is known as the **residue**.

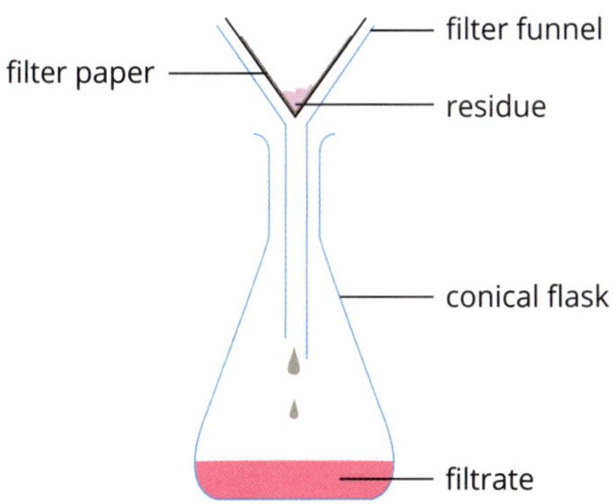

FIGURE 1.3.2 A gravitational filtration apparatus

Vacuum filtration

Vacuum filtration is faster than gravitational filtration and also helps to dry the residue more quickly. The apparatus for this technique is shown in Figure 1.3.3. The funnel, called a Büchner funnel, is usually ceramic, with a flat base and vertical sides. The flat base has many holes for the liquid to flow through. A flat piece of filter paper is placed at the bottom of the funnel. The funnel is then placed in a vacuum flask, which is a conical flask with a rubber seal and a side arm for attaching a vacuum tube. The solid–liquid mixture poured into the funnel is sucked into the flask by the vacuum. The solid residue is trapped by the filter paper.

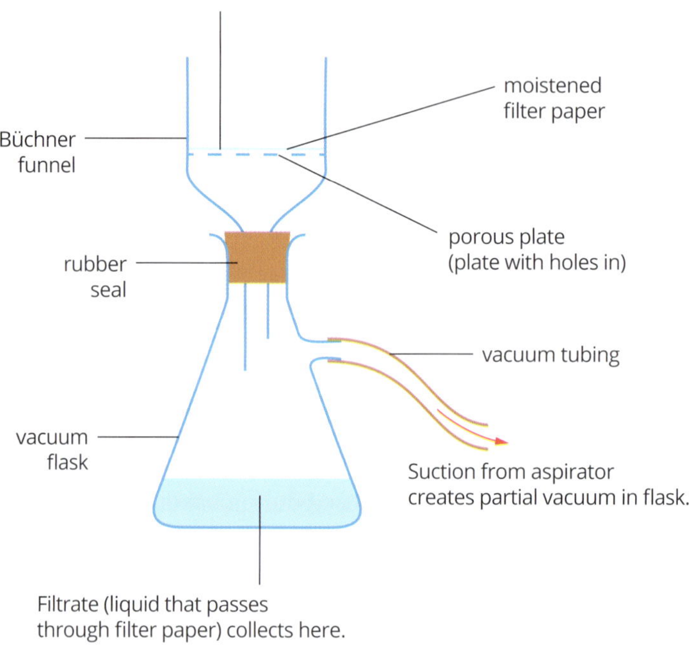

FIGURE 1.3.3 A vacuum filtration apparatus

SEPARATION BY DENSITY

Density is a measure of the mass per unit of volume of a substance. The difference in density between substances in a solid–liquid mixture determines whether each substance floats or sinks. The denser substance will sink and the less dense substance will float. Separating materials based on their density is relatively easy and cheap. However, the size of the particles in the materials also play a role. Even dense materials can stay suspended in a liquid if the particle size is very small, making it harder to separate very fine particles.

FIGURE 1.3.4 In this mixture of clay and water, while most of the clay has settled to the bottom, the smallest clay particles are still suspended in the liquid.

Sedimentation and decantation

If you drop a handful of sand, rocks and clay into a clear container of water, you will observe a few things. Firstly, the larger rocks will sink to the bottom, followed by the finer grains of sand. If you leave the mixture for days or weeks, you may also see that even finer particles start to settle. However, the finest particles will remain suspended in the liquid (Figure 1.3.4).

This process is known as **sedimentation**, or settling, and is a form of gravitational separation. The liquid can then be separated from the **sediment** by carefully pouring the liquid into another container. This process of pouring liquid from the sediment is known as **decantation**.

While not highly efficient, sedimentation is a cheap method of separating solid particles from liquids in large volumes.

- In water treatment plants, the sewage is first stored in large sedimentation tanks so that the solid waste can be separated from the liquid waste after it settles to the bottom of the tank (Figure 1.3.5).
- In wine production, the grape juices are stored in large settling tanks before being fermented. The solid particles sink to the bottom, allowing their separation from the grape juices.

FIGURE 1.3.5 Sedimentation tanks are used in water treatment plants to separate solid waste from liquid waste.

Separation funnels

If two liquids have different densities and are immiscible (don't mix), then the liquids can be separated with a separation funnel. The **separation funnel** (Figure 1.3.6) is a glass flask connected to a thin outlet tube with a small tap. When immiscible liquids are placed in the flask, two layers will form. The less dense liquid floats to the top and the denser liquid sinks to the bottom. When the tap is opened, the denser liquid flows out first and is separated from the less dense liquid.

Separation funnels are used in a process known as liquid–liquid extraction. This technique can be used to extract compounds like fragrances from plant oils. The plant oil containing the fragrance is mixed with water and a liquid known as ether. The water and the ether are immiscible but the fragrance will dissolve in the ether. The fragrance can be separated out of the oil and into the ether by gently shaking the water and ether in the separation funnel. After being left to stand for some time, the water and ether will separate. The ether, containing the fragrance, can then be poured out of the separating funnel and into a beaker.

FIGURE 1.3.6 Separation funnels are often used in organic chemistry where many organic solvents display significantly different densities from aqueous solutions.

Centrifugation

Spinning a mixture rapidly can speed up the sedimentation process and extract finer particles that may not settle out naturally. This process is known as **centrifugation**. Separation occurs because spinning the mixture results in the denser particles being pushed to the outside of the container by centrifugal force.

Centrifugation is used extensively in research and in medical and forensics laboratories since it is particularly useful for separating mixtures in small quantities. In Figure 1.3.7, vials of blood are placed in a centrifuge. This machine spins the vials around at up to 20 000 revolutions per minute.

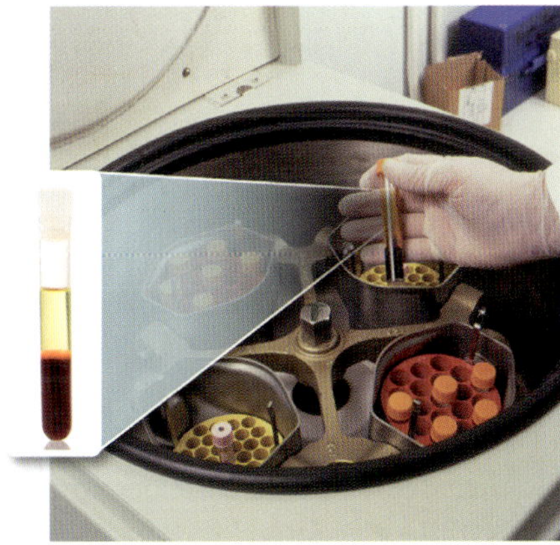

FIGURE 1.3.7 Centrifuges like this one are used to separate the components of blood in small quantities. Blood that has been centrifuged separates into three distinct layers. The top layer is the plasma, the middle layer is the buffy coat and the bottom layer is composed of red blood cells.

The centrifuge separates the blood into three components—plasma, the buffy coat (a mixture of white blood cells and platelets) and a layer of red blood cells. The plasma and red blood cells can be used for blood transfusions. The buffy coat is used to extract DNA or detect some diseases.

SEPARATION BY BOILING POINT

For mixtures containing compounds that do not breakdown under heating, separation can be based on the difference in boiling points.

Evaporation

Some solids dissolve in liquids to form a solution. The liquid is known as the solvent and the dissolved solid is called the solute. Dissolved solids cannot be separated by filtration, sedimentation or centrifugation. Instead, the liquid solvent needs to be evaporated or boiled off in order to recover the solid solute. This technique is used in the Dampier salt mine in Western Australia and in the salt pans in Cusco Region, Peru (Figure 1.3.8). During the evaporation process, the water is lost to the atmosphere as water vapour. However, it is possible to collect both the solute and the solvent using the process of distillation.

FIGURE 1.3.8 Salt pans of Cusco Region, Peru

CHEMISTRY IN ACTION

Gas centrifuges

For the separation of substances in a mixture, such as isotopes where the differences in mass between the components are so minute, a gas centrifuge is employed. Gas centrifuges work on the same principle as traditional centrifuges except gases are used as the feed material and the process is conducted in multiple stages.

The most prominent use of gas centrifuges is the enrichment of uranium for nuclear power and weapons. This requires the separation of radioactive ^{235}U isotopes from the majority ^{238}U isotopes. The ^{235}U isotope represents only 0.72% of the atoms present in natural samples of uranium.

The difficulty of this process is the major hurdle for many nations who desire nuclear weapons.

FIGURE 1.3.9 The interior view of a gas centrifuge used for enriching uranium for nuclear power stations

Distillation

Distillation uses the same principle as evaporation but is performed in an apparatus (Figure 1.3.10) in such a way that the evaporated liquid can be recovered. The solution is heated in a flask known as the distillation flask to vaporise the liquid. The vapour passes through a condenser, which is a tube cooled with running water. The condenser cools the vapour, causing it to condense back into a liquid and form droplets along the inside of the condenser. The condensed liquid drips out of the condenser into a second flask, known as the receiving flask. The liquid collected is called the distillate.

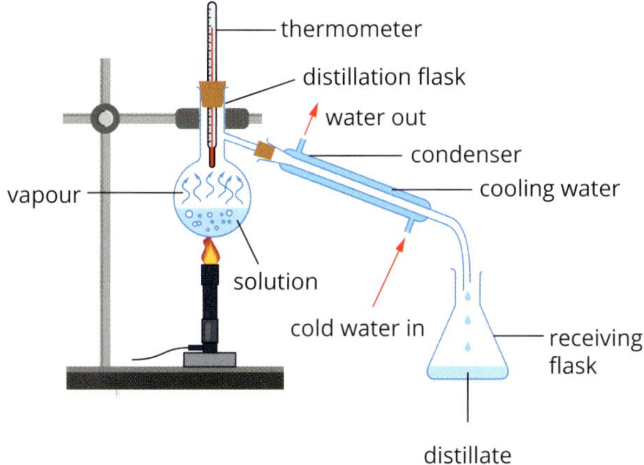

FIGURE 1.3.10 A set-up of a distillation apparatus

Fractional distillation

The distillation method can also be used to separate miscible (can mix) liquids that have boiling points that are only slightly different from each other. This technique is known as **fractional distillation** and uses the same apparatus as simple distillation with one important addition: a fractionating column is placed between the flask containing the mixture and the condenser. A fractionating column is a hollow column packed with high surface area material (Figure 1.3.11).

The fractionating column allows increased contact between rising vapour and falling condensate. This results in the more volatile component exiting the column as a vapour at the top and the less volatile component condensing back into the distillation flask.

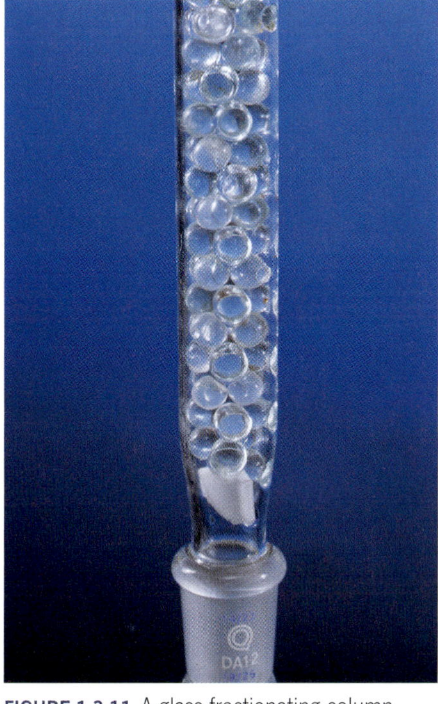

FIGURE 1.3.11 A glass fractionating column filled with glass beads

CHEMISTRY IN ACTION

Distillation towers

One of the key structures in an oil refinery is the distillation tower. These towers are distillation columns that are up to 100 metres tall. They are heated from the bottom and allow the separation of crude oil into different fractions based on their different boiling points.

The highest boiling point fractions, such as asphalt and fuel oil, are collected at the bottom whereas the lowest boiling point fractions, such as petrol for cars and other gases, are collected at the top of the column.

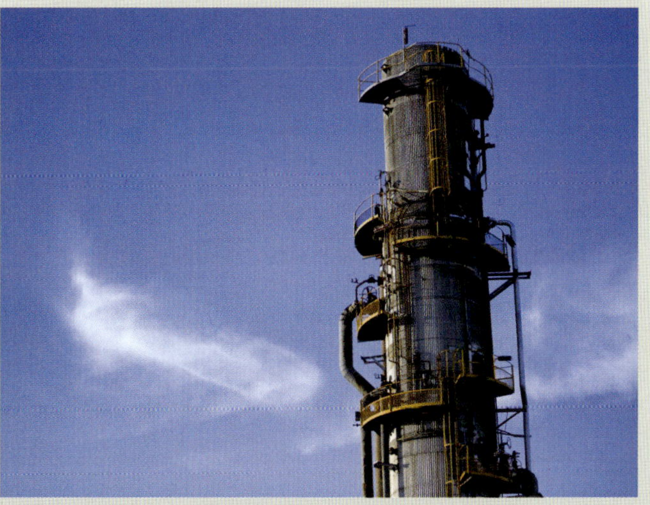

FIGURE 1.3.12 The distillation tower at an oil refinery

SEPARATION BY ELECTRIC CHARGE

Electrostatic separation

Objects that have opposite electric charges attract each other. The force of attraction between opposite charges is known as the electrostatic force and can be used to separate charged particles from uncharged particles.

Electrostatic filters use the electrostatic force to separate solid particles from a gas. The technique is used in mineral processing and industrial chimneys to remove smoke particles from waste gases before releasing the gases into the atmosphere. The smoke is first passed through a negatively charged grid, which gives the smoke particles a negative charge. The smoke is then passed between positively charged plates or electrodes that attract and collect the particles.

Chromatography

Other important techniques are the various forms of chromatography, including thin layer chromatography (TLC), gas chromatography (GC), paper chromatography and high performance liquid chromatography (HPLC). Chromatography separates liquids or gases based on their differing affinity for various materials present in the chromatography apparatus.

All chromatography techniques employ a stationary and mobile phase. If a compound from the mixture introduced into a chromatography apparatus demonstrates a high affinity for the mobile phase and a low affinity for the stationary phase, it will be moved quickly through the system as the mobile phase moves. If a different component in the mixture demonstrates a high affinity for the stationary phase and a low affinity for the mobile phase, it will remain in the system longer, thus effecting a separation of the compounds. Chromatography will be covered in greater detail in Chapter 13.

EXTENSION

Electrophoresis

Electrophoresis is an electrostatic-based separation technique that relies on the different migration rates of particles dispersed in a fluid when exposed to an electric field.

Particles suspended in a fluid obtain a surface charge as a result of processes such as deprotonation (losing a hydrogen ion), protonation (gaining a hydrogen ion) or adsorption of other ions. This surface charge responds to an applied electric charge to the fluid in the same way as other charged particles interacting electrostatically; positively charged particles are attracted to the negatively charged terminal and vice-versa for negatively charged particles. This electrostatic attraction causes the particles to migrate, or move through the fluid, the rate of which depends on their charge and size. Over a period of time, particles of different sizes (and therefore roughly molecular weights) migrate different distances (Figure 1.3.13), affecting the separation of the components.

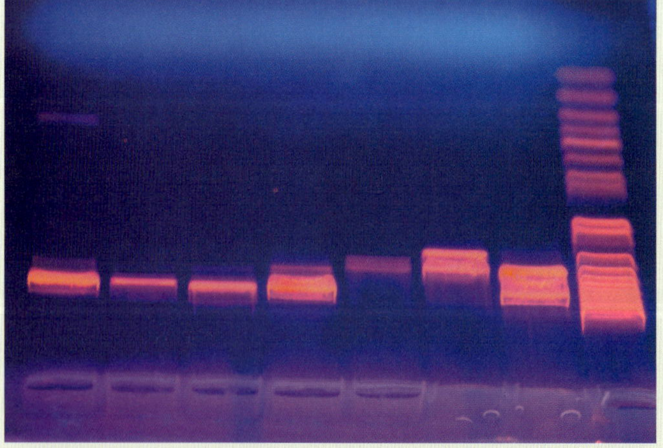

FIGURE 1.3.13 A type of electrophoresis known as gel electrophoresis is commonly used to analyse DNA and proteins.

1.3 Review

SUMMARY

- Useful materials can be created by extracting and purifying substances.
- Separation techniques use differences in the materials' physical and chemical properties in order to separate them. Such differences include particle size, density, boiling point and electric charge.
- Sieving can be used to separate mixtures of solids based on particle size.
- Filtration can be used to separate mixtures of solids and liquids based on particle size.
- Sedimentation and decantation can separate mixtures of solids and liquids based on density.
- Mixtures of immiscible liquids of different densities can be separated using a separation funnel.
- A mixture of components where the differences in density are very small can be separated by centrifugation.
- Evaporation can separate the solute from the solvent in solutions but the solvent is not retained.
- Distillation can separate the solute from the solvent in solutions, or even two miscible liquids, and all components are retained.
- Fractional distillation is effective at separating mixtures of miscible liquids with similar boiling points.
- A mixture of substances that respond differently to external electric fields can be separated by electrostatic means.

KEY QUESTIONS

1 Why would sieving a mixture of salt and pepper not be an effective separation technique?

2 Can you use sedimentation and decantation to separate the solvent from the solute of a solution? Why/why not?

3 Is it possible to use a separation technique outlined in this chapter to separate the hydrogen from the oxygen in water? Why/why not?

4 Draw a fully labelled scientific diagram of a distillation apparatus.

5 For each of the following sets of mixtures, outline what technique(s) could be employed to separate the components.

- **a** water and oil
- **b** water and sugar
- **c** sand and gravel
- **d** sugar and pebbles
- **e** pebbles and woodchips

6 A mixture of four different hydrocarbons is to be separated by fractional distillation; their details are in the table below.

Hydrocarbon	Boiling point (°C)
pentane	36.1
heptane	98.4
octane	125.0
benzene	80.1

In what order will the hydrocarbons be collected in the receiving flask?

Chapter review

01

KEY TERMS

alloy
carbon nanotube
centrifugation
ceramic
colloid
composite material
compound
decantation
density
distillation
element
filtrate
filtration
fractional distillation
fullerene
graphene
material
mineral
nanomaterial
nanometre
nanoparticle
nanoscale
nanotechnology
ore
polymer
residue
sediment
sedimentation
separation funnel
sieving
suspension

Materials science

1 What is a composite material?

2 Papier mâché is a commonly used composite material composed of what two materials?

3 Classify each of the following materials as a metal, ceramic or polymer: aluminium, cotton, nylon, porcelain, stainless steel.

4 Gold is a precious, soft, yellow metal valued for its rarity and appearance. When used in jewellery, gold is often alloyed with other metals. Suggest a reason why.

5 Diamond, graphite and carbon nanotubes are all elemental forms of carbon. Suggest a reason why their physical properties differ so significantly when their chemical composition is identical.

6 In what way is the composite material known as reinforced concrete superior to normal concrete for construction?

7 Why are kitchen sinks made from stainless steel and not iron?

8 Conduct research to find out what are the major components of the following alloys.

- **a** bronze
- **b** pewter
- **c** white gold

Nanotechnology

9 Convert the following lengths into nanometres (express your answers in scientific notation).

- **a** 5 cm
- **b** 12 mm
- **c** 0.02 mm

10 Zinc oxide powder and zinc oxide nanoparticles both absorb UV light. What property of zinc oxide nanoparticles makes them more suitable than zinc oxide powder for use in sunscreen?

11 a Describe the difference between top-down and bottom-up fabrication methods of nanomaterials.

b What are the advantages and disadvantages of each method?

12 What is a colloid and how is it different from a suspension?

Purifying materials

13 What is the main advantage of distillation over evaporation?

14 A common survival technique to obtain drinking water when lost in the outdoors is to dig a hole in the ground, fill it with waste or salt water, place a clean container in the centre of the hole and cover the hole with a piece of plastic weighed down with a small rock in the centre. Explain in terms of separation techniques how this apparatus can produce drinkable water.

15 Crystallisation is the evaporation of the solvent from a solution to form crystals of the solute. Explain why this method is not suitable for purifying mixtures of multiple soluble salts from solution.

16 Briefly outline the steps you would take in a laboratory to separate the following mixtures while retaining each component.

- **a** sawdust, sand and sugar
- **b** oil, water and ethanol
- **c** instant coffee and tea leaves

Connecting the main ideas

17 Research the composite materials produced with carbon-fibre reinforcing, stating their advantages over traditional materials.

18 Describe some of the concerns regarding the use of nanoparticles in consumer products.

19 One of the concerns raised regarding nanotechnology is the scenario suggested by early nanotechnology pioneers involving 'grey goo'. Research the notion of 'grey goo', summarising the scenario.

CHAPTER 02

Atoms: structure and mass

Our understanding of the structure of the atom has evolved over centuries and continues to change in the light of new evidence. When the existence of atoms was proposed by John Dalton in 1808, it resurrected an idea that originated around two thousand years earlier in ancient Greece and India. The philosophy that matter was composed of tiny, solid spheres that could not be divided into smaller parts has developed into an evidence-based theory that allows us to predict and explain the properties and behaviour of matter.

Science as a human endeavour

- Findings from a range of scientific experiments contributed to the understanding of the atom, enabling scientists, including Dalton, Thomson, Rutherford, Bohr and Chadwick to develop models of atomic structure and make reliable predictions about the mass, charge and location of the sub-atomic particles.

Science understanding

- the relative atomic mass (atomic weight), A_r is the ratio of the average mass of the atom to 1/12 the mass of an atom of ^{12}C; relative atomic masses of the elements are calculated from their isotopic composition
- isotopes are atoms of an element with the same number of protons but different numbers of neutrons and are represented in the form ^{A}X (IUPAC) or X-A
- isotopes of an element have the same electron configuration and possess similar chemical properties but have different physical properties
- mass spectrometry involves the ionisation of substances and the separation and detection of the resulting ions; the spectra which are generated can be analysed to determine the isotopic composition of elements and interpreted to determine relative atomic mass

2.1 Atomic theory

FIGURE 2.1.1 John Dalton (1766–1844) proposed that matter was composed of atoms.

Around 450 BCE, the Greek philosopher Leucippus proposed that, if a piece of **matter** were subdivided over and over again, eventually it would not be able to be divided any further. A follower of Leucippus called Democritus expanded on this idea and named the final piece of matter *atomos*, meaning 'indivisible'. He believed different substances had *atomos* of different sizes and shapes, which gave substances different properties. Unfortunately, Democritus' ideas were not widely accepted; the preferred teachings were those of Aristotle, who believed that all matter was composed of the four elements: earth, air, fire and water.

The notion of *atomos* was not revisited until 1808, when English chemist John Dalton (Figure 2.1.1) published *A New System of Chemical Philosophy*, in which he explained his observations of chemical reactions in terms of tiny particles that he called atoms. Dalton thought that all matter was made up of these indivisible atoms and that the atoms of different elements had different weights. He further proposed that compounds are formed from the combination of the atoms of two or more elements in simple numerical ratios. He stated that atoms of a given element are identical in size and mass and when atoms combine to form compounds, they do so only in the simplest ratio. Not all of these points are true, however, his innovative ideas became the foundation of modern atomic theory.

THE STRUCTURE OF ATOMS

Most of the points of Dalton's atomic theory were well received and could be supported with experimental data, but by the end of the 1800s, there was increasing evidence to suggest that atoms are made up of smaller, **subatomic particles**. This shifted the focus of scientists of the time from the existence of atoms to the subatomic composition and structure of atoms.

FIGURE 2.1.2 British physicist J. J. Thomson

One such scientist was Joseph John Thomson (Figure 2.1.2) who carried out experiments with an apparatus known as a cathode ray tube, which would later be utilised in the first televisions. A cathode ray tube consists of a sealed glass tube with metal terminals at each end that act as electrodes. When electricity passes through the tube, a stream of particles is emitted and the tube glows with a coloured light.

In 1897, Thomson discovered that, regardless of the gas in the tube or the type of metal electrodes used, the particles flow from the negative electrode to the positive electrode and the stream of particles is repelled by another negative electrode. He concluded that the particles must be negatively charged and be present in all atoms. Since atoms are electrically neutral, he also proposed that they must contain positive particles as well so that together the charges cancel out.

To explain the existence of these two subatomic particles, Thomson developed the 'plum pudding' **model** of the atom where the negatively charged particles were spread throughout the positively charged sphere of the atom like the raisins of a plum pudding. These negatively charged particles became known as **electrons**.

FIGURE 2.1.3 New Zealand–born physicist Ernest Rutherford

As scientists gathered more experimental evidence, it became apparent that Thomson's model of the atom was not correct. In 1910, New Zealand–born Ernest Rutherford (Figure 2.1.3) led a group of scientists who performed an experiment that changed the accepted view of the atom.

Rutherford fired a beam of positively charged **alpha particles** at a very thin sheet of gold foil. Most of the alpha particles passed straight through the foil and produced flashes on a fluorescent screen on the other side. Rutherford was surprised to discover, however, that some particles were deflected as they passed through the foil, and some even bounced back. This experiment is described in Figure 2.1.4.

Rutherford concluded that most of an atom must be empty space to allow the majority of the alpha particles to pass through it. Most of the mass of an atom and its positive charge must be located in a tiny, central region. This region must have deflected the alpha particles. He called this region the **nucleus** of the atom and surrounded it with the negative electrons similar to planets orbiting the Sun.

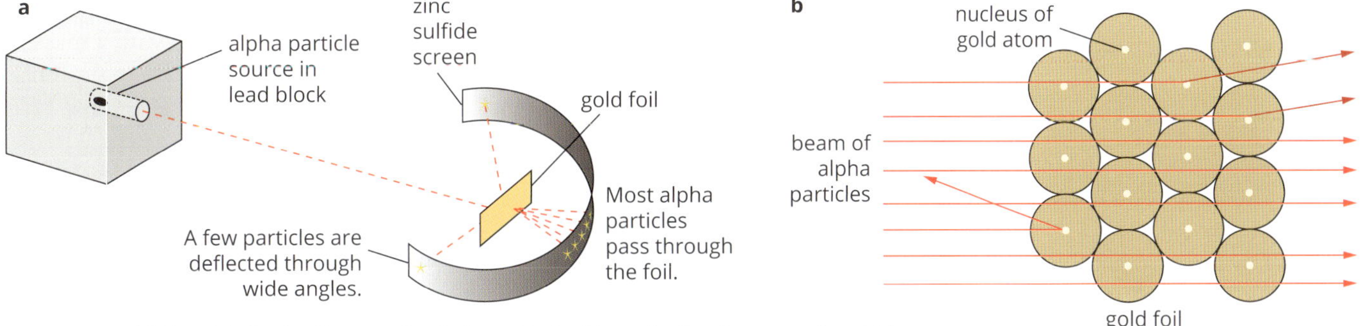

FIGURE 2.1.4 (a) Ernest Rutherford's apparatus that provided evidence for the discovery of nuclei in atoms. (b) Only those alpha particles that closely approach the nuclei in the gold foil are deflected significantly. Most particles pass almost directly through the foil.

Rutherford's model of orbiting electrons would be further improved by Niels Bohr. You will learn more about the Bohr model of the atom in Chapter 3.

In further experiments by Rutherford in 1920, he found that firing a stream of alpha particles into nitrogen gas produced a signal typical of hydrogen nuclei. He went on to determine that there are hydrogen nuclei in all other nuclei. He called this elementary particle present in all nuclei the **proton**.

Rutherford also had reason to believe the nucleus contained a yet undiscovered neutral particle. He believed it to be a proton with a closely orbiting electron, as his experimental data indicated that a neutral particle with a mass similar to a proton should exist.

A student, then colleague, of Rutherford, the English physicist James Chadwick, later confirmed the existence of Rutherford's proposed neutral particle. Chadwick showed it is a distinct particle and not the proton–electron pairing previously suggested. He called this new elementary particle the **neutron**.

Electrons

Electrons are negatively charged particles that behave like a cloud of negative charge around the nucleus. This cloud gives the atom its size and volume. An electron has a mass approximately 1800 times smaller than that of a proton or neutron. Therefore, electrons contribute very little to the total mass of an atom. However, the volume of space occupied by the cloud of electrons is 10 000–100 000 times larger than that of the nucleus.

Negative particles attract positive particles. This is called **electrostatic attraction**. The electrostatic attraction of the atom's electrons to the protons within the nucleus keeps the atom bound together. The charge on an electron is equal but opposite to the charge on a proton. Electrons are said to have a charge of –1 whereas protons have a charge of +1.

In some circumstances, electrons can be easily removed from an atom. For example, when you rub a rubber balloon on a woollen jumper or dry hair, electrons are transferred to the balloon and the balloon develops a negative charge. The negative charge is observed as an electrostatic force that can stick the balloon to a wall. You will learn more about the removal of electrons from atoms when looking at ionic compounds in Chapter 5.

The electricity that powers lights and appliances is the result of electrons moving in a current through wires. Sparks and lightning are also caused by electrons moving through air (Figure 2.1.5).

FIGURE 2.1.5 Lightning is a sudden electrostatic discharge during a thunderstorm.

The nucleus

The nucleus of an atom is approximately 10 000–100 000 times smaller than the size of the atom. To put this in perspective, if an atom were the size of Perth Stadium, then the nucleus would be about the size of a pea (Figure 2.1.6). Nonetheless, the nucleus contributes around 99.97% of the atom's mass. This means that atomic nuclei are extremely dense.

FIGURE 2.1.6 If an atom were the size of Perth stadium (built in 2017), then the nucleus would be about the size of a pea.

The subatomic particles in the nucleus—the protons and neutrons—are referred to collectively as **nucleons**. Protons are positively charged particles with a mass of approximately 1.673×10^{-27} kg. Neutrons are almost identical in mass to protons.

Table 2.1.1 summarises the properties of protons, neutrons and electrons.

TABLE 2.1.1 Properties of the subatomic particles

Particle	Symbol	Charge	Mass relative to a proton	Mass (kg)
proton	p^+	+1	1	1.673×10^{-27}
neutron	n^0	0	1	1.675×10^{-27}
electron	e^-	−1	$\frac{1}{1800}$	9.109×10^{-31}

EXTENSION

Quarks

The proton, neutron and electron are not the only particles that have been discovered and characterised. In the early 1960s, there were so many particles discovered that they were collectively referred to as the 'particle zoo'. They were all believed to be elementary particles, that is, particles that are not in turn made of other particles.

In the early 1960s, two physicists, Murray Gell-Man and George Zweig, separately suggested these particles, including the protons and neutrons, weren't elementary particles but were composed of a subset of particles known as quarks. The six types, or flavours, of quarks are up, down, strange, charm, top and bottom.

The quarks are now included in a list of 17 elementary particles with the most recently discovered, the Higgs Boson, confirmed in 2012.

2.1 Review

SUMMARY

- All substances are made up of atoms.
- Atoms are composed of a small, positively charged nucleus surrounded by a negatively charged cloud of electrons.
- The nucleus is made up of two subatomic particles—protons and neutrons. These particles are referred to as nucleons.
- Protons and neutrons are similar in size and mass while electrons are approximately 1800 times smaller.
- The charges on protons and electrons are equal but opposite.
- Neutrons are uncharged.
- Positive and negative particles experience an attractive force known as electrostatic attraction.

KEY QUESTIONS

1 Name the three subatomic particles that atoms are composed of.

2 What subatomic particles make up the most mass of an atom and where are they found?

3 How are electrons held within the cloud surrounding the nucleus?

2.2 Describing atoms

Even before the structure of the atom was well understood, scientists had begun classifying and categorising them using qualitative and quantitative data obtained through experimentation. Thanks to the contributions of Dalton and other scientists, objective measurements of the composition of atoms can now be made and they can be described using a number of derived quantities.

DESCRIBING ATOMS

Fundamentally, the characteristics of atoms are derived from the numbers of the subatomic particles they contain—the protons, the neutrons and the electrons.

Elements

As Dalton predicted, elements are made of just one type of atom and these atoms are identical. Most non-metallic elements, such as sulfur, form molecules with a definite number of sulfur atoms (8), however, some non-metals, such as carbon, can form extensive networks numbering in the thousands of atoms (Figure 2.2.1). So far, scientists have discovered 118 different elements. About 98 of these occur in nature; the other elements have only been observed in the laboratory.

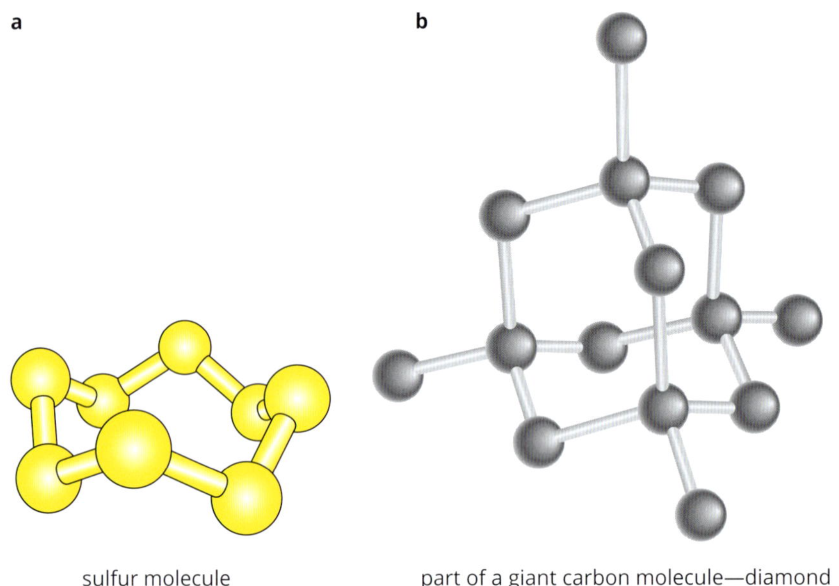

FIGURE 2.2.1 (a) Most non-metal elements, such as sulfur, form molecules. (b) Some other elements, such as carbon, form covalent network lattices or giant molecules, shown here by diamond.

Each element has a unique name and **chemical symbol**. Table 2.2.1 lists the chemical symbols of some of the most common elements.

TABLE 2.2.1 Chemical symbols and names of some of the most common elements

Element	Symbol	Element	Symbol
aluminium	Al	mercury	Hg
argon	Ar	nitrogen	N
carbon	C	oxygen	O
chlorine	Cl	potassium	K
copper	Cu	silver	Ag
hydrogen	H	sodium	Na
iron	Fe	uranium	U

The chemical symbol is usually made up of one or two letters. The first letter is always capitalised and subsequent letters are always lower case.

In many cases, the chemical symbol corresponds to the name of the element. For example, nitrogen has the chemical symbol N, chlorine has the chemical symbol Cl and uranium has the chemical symbol U.

However, some chemical symbols do not correspond to the name of the element. For example, sodium has the chemical symbol Na, potassium has the chemical symbol K and iron has the chemical symbol Fe. This is because the chemical symbols have been derived from the Latin or Greek names of the elements. In Latin, sodium is known as *natrium*, potassium, *kalium* and iron, *ferrum*.

The atomic symbols are usually displayed in a periodic table as shown in Figure 2.2.2. The **periodic table** helps to group elements that have similar chemical properties. You will learn more about the periodic table in Chapter 3.

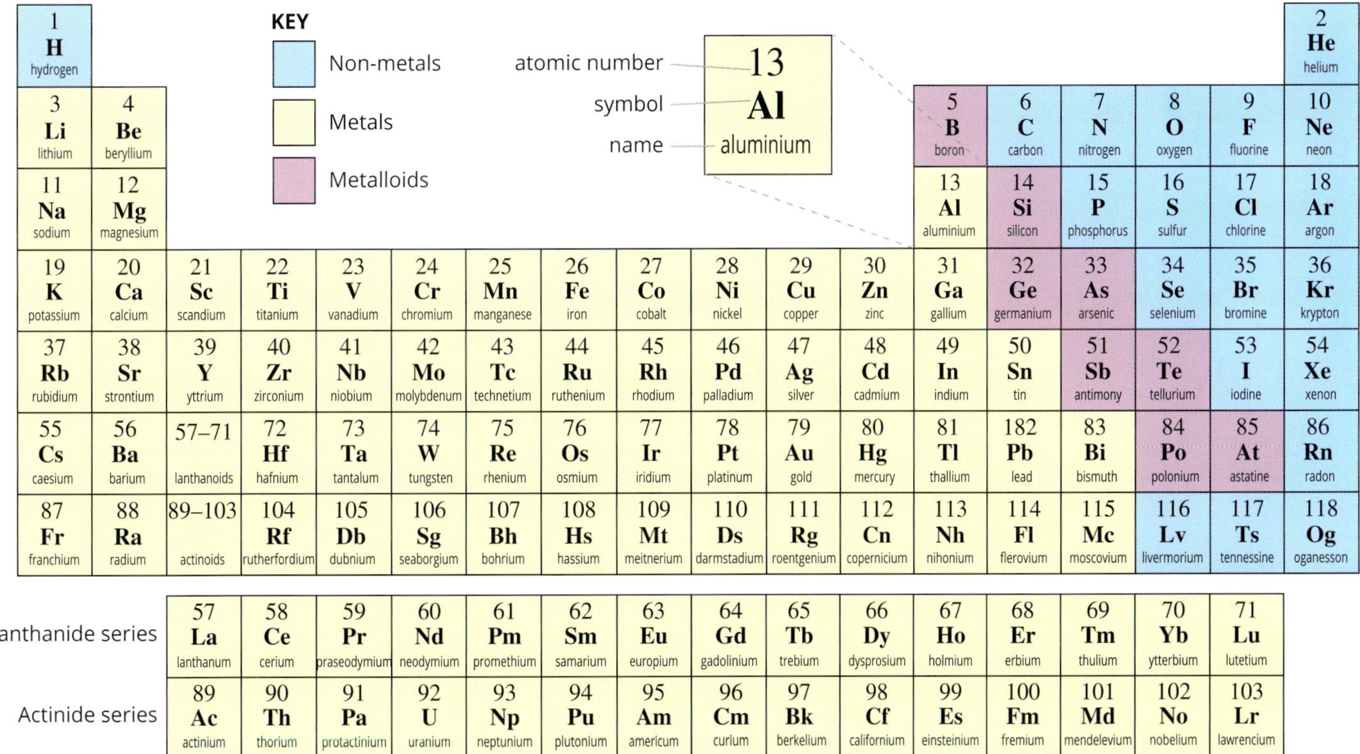

FIGURE 2.2.2 The periodic table groups the chemical elements according to their atomic structure.

CHEMFILE

Viewing atoms

Dalton's **atomic theory of matter** assumed that atoms are spherical. Atoms cannot be seen with conventional microscopes. So, there was no way to confirm the shape of atoms. It wasn't until 1981 that a microscope capable of viewing atoms was developed by IBM researchers Gerd Binning and Heinrich Rohrer. This type of microscope is known as a **scanning tunnelling microscope (STM)**. Using STMs, scientists confirmed that atoms are indeed spherical.

STMs use an extremely sharp metal tip to detect atoms. The tip is scanned, line by line, across the surface of a crystal. As the tip moves, it measures minute height differences in the crystal's surface due to the individual atoms. This is similar to the way sight-impaired people use their finger to sense braille on a page. The data from the tip is then sent to a computer that constructs an image of the atoms. An STM image of silicon atoms on the surface of a silicon wafer is shown in Figure 2.2.3.

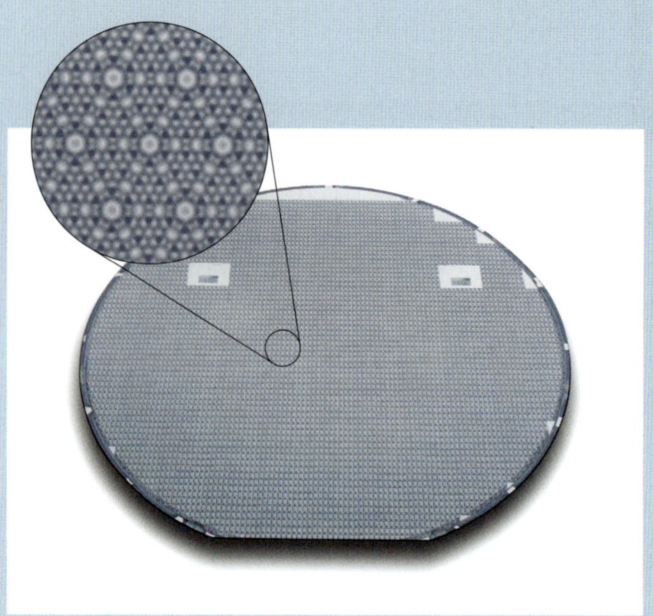

FIGURE 2.2.3 This image of silicon atoms in a silicon wafer was taken with a scanning tunnelling microscope (STM).

EXTENSION

The periodic table: updated periodically

IUPAC (International Union of Pure and Applied Chemistry) is the worldwide representative body for chemists. In 2016, IUPAC announced the addition of four new elements to the periodic table. The four elements, which were discovered between 2004 to 2010, are:

- nihonium – symbol Nh, element 113
- moscovium – symbol Mc, element 115
- tennessine – symbol Ts, element 117
- oganesson – symbol Og, element 118.

Nihonium, tennessine and oganesson were prepared by bombarding previously known heavy atoms with other atoms, while moscovium is a radioactive decay product of tennessine.

Scientists are currently searching for the first element in the eighth row of the periodic table.

Atomic number

The number of protons in an atom's nucleus is known as the **atomic number** and is represented by the symbol Z.

All atoms that belong to the same element must have the same number of protons and therefore have the same atomic number, Z. For example, all hydrogen atoms have $Z = 1$, all carbon atoms have $Z = 6$ and all gold atoms have $Z = 79$.

Because all atoms are electrically neutral, the number of protons in an atom is always equal to the number of electrons in that atom. The atomic number therefore tells us both the number of protons and the number of electrons. For example, carbon atoms, with $Z = 6$, have six protons and six electrons.

Mass number

The number of protons and neutrons in the nucleus is known as the **mass number** and is represented by the symbol A. The mass number represents the total mass of the nucleus. Note that you cannot have fractions of a proton or neutron, therefore, the mass number is always a whole number.

The number of protons, neutrons and electrons defines the basic structure of an atom. The standard way of representing an atom is to show its atomic and mass numbers as shown in Figure 2.2.4.

mass number → $^{A}_{Z}\text{X}$ ← symbol of element
atomic number →

FIGURE 2.2.4 The standard way of representing an atom is to show its atomic number and mass number.

Worked example 2.2.1

CALCULATING THE NUMBER OF SUBATOMIC PARTICLES

Calculate the number of protons, neutrons and electrons for the atom with this atomic symbol: $^{40}_{18}\text{Ar}$

Thinking	Working
The atomic number is equal to the number of protons.	The number of protons = $Z = 18$
Find the number of neutrons. Number of neutrons = mass number – atomic number	The number of neutrons = $A - Z$ $= 40 - 18$ $= 22$
Find the number of electrons. The number of electrons is equal to the atomic number because the total negative charge is equal to the total positive charge.	Number of electrons = $Z = 18$

Worked example: Try yourself 2.2.1

CALCULATING THE NUMBER OF SUBATOMIC PARTICLES

Calculate the number of protons, neutrons and electrons for the atom with this atomic symbol:

$^{235}_{92}\text{U}$

2.2 Review

SUMMARY

- All the atoms of an element have the same number of protons. This number is called the atomic number (Z).
- In a neutral atom, the number of protons and electrons is equal.
- The mass number (A) of an atom is the total number of protons plus neutrons in the nucleus.

KEY QUESTIONS

1 Give the chemical symbol for the following elements.
  a sodium
  b copper
  c neon
  d potassium
  e beryllium
  f calcium

2 What term is given to the number of protons in the nucleus of an atom?

3 What term is given to the total number of protons and neutrons in the nucleus of an atom?

4 For any uncharged atom, which two subatomic particles are equal in number?

5 Calculate the numbers of protons, neutrons and electrons in the atom $^{31}_{15}P$.

6 What is the correct name of the element that has an atom with seven protons and eight neutrons?

2.3 Isotopes

One of the ideas Dalton put forward when he was developing modern atomic theory was that the atoms of an element are all identical; this is not strictly true.

All atoms that belong to the same element have the same number of protons in the nucleus and therefore the same atomic number, Z. However, not all atoms that belong to the same element have the same mass number, A. For example, hydrogen atoms can have mass numbers of 1, 2 or 3. In other words, hydrogen atoms may contain just a single proton, a proton and a neutron, or a proton and two neutrons as shown in Figure 2.3.1. Atoms that have the same number of protons (atomic number) but different numbers of neutrons (and therefore different mass numbers) are known as **isotopes**.

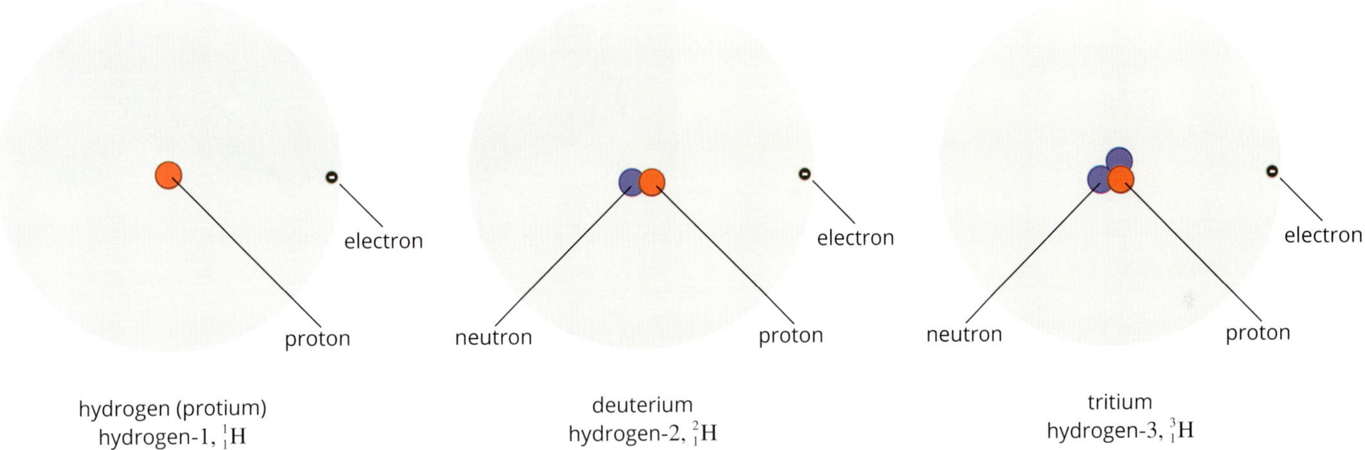

hydrogen (protium)
hydrogen-1, ${}^{1}_{1}H$

deuterium
hydrogen-2, ${}^{2}_{1}H$

tritium
hydrogen-3, ${}^{3}_{1}H$

FIGURE 2.3.1 The three isotopes of hydrogen are given special names. A hydrogen atom with just one proton in its nucleus is known as hydrogen (or protium). A hydrogen atom with one proton and one neutron is known as deuterium. A hydrogen atom with one proton and two neutrons is known as tritium.

ISOTOPES

Carbon also has three naturally occurring isotopes. These three isotopes are known as carbon-12, carbon-13 and carbon-14. Carbon-12 atoms have a mass number of 12, carbon-13 atoms have a mass number of 13 and carbon-14 atoms have a mass number of 14. In the 1950s and 1960s, nuclear weapons testing caused a spike in carbon-14 in the atmosphere. This has been declining in the last 50 years. These three carbon isotopes can be represented as shown in Figure 2.3.2.

${}^{12}_{6}C$	${}^{13}_{6}C$	${}^{14}_{6}C$
carbon-12	carbon-13	carbon-14

FIGURE 2.3.2 Carbon has three naturally occurring isotopes.

Isotopes have identical chemical properties but different physical properties such as mass and density. In particular, some isotopes that have a number of neutrons significantly different from the number of protons are unstable and hence **radioactive**. These isotopes undergo radioactive decay, emitting various forms of radiation and turn into lighter, more stable nuclei. Carbon-14, having eight neutrons compared to six protons, is unstable and radioactive.

CHEMISTRY IN ACTION

Lucas Heights

Australia has 33% of the world's uranium deposits but no nuclear facilities generating electricity.

Australia does, however, have a working nuclear reactor in the Sydney suburb of Lucas Heights and has had since the late 1950s. The Open-Pool Australian Lightwater reactor (OPAL) is used for research, mineral analysis and the production of radioisotopes for medical and industrial applications.

In 1996, the residential area of Lucas Heights was renamed Barden Ridge, as the association with the Lucas Heights nuclear reactor facility had adversely impacted the value of nearby real estate.

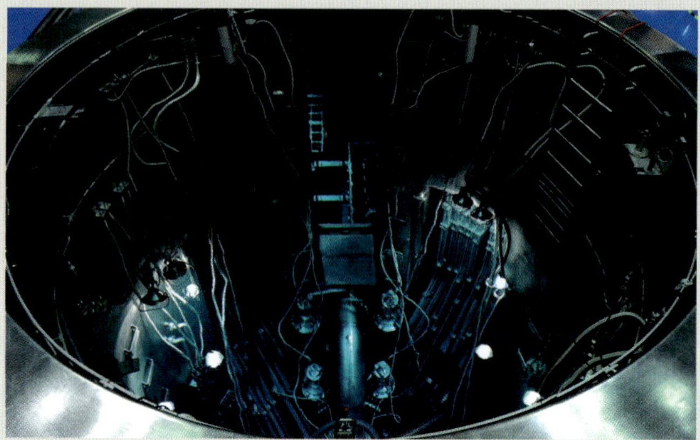

FIGURE 2.3.3 The OPAL nuclear research reactor at Lucas Heights in Sydney

Relative atomic masses

Chemists have devised a standard for measuring the masses of individual atoms. The standard chosen is the mass of an atom of the most common isotope of carbon, ${}^{12}_{6}C$ or carbon-12, which is given the value of 12 units. One unit of mass on this scale is therefore one-twelfth of the mass of a ${}^{12}_{6}C$ isotope. The mass of an individual isotope of each element compared to this standard is known as its **relative isotopic mass**.

There are two different isotopes of the element chlorine: ${}^{35}_{17}Cl$ and ${}^{37}_{17}Cl$. The isotopes have different mass numbers because of the different numbers of neutrons. Remember that the mass of an atom is mainly determined by the number of protons and neutrons in their nucleus, since the mass of the electrons is relatively small.

The relative isotopic masses of the two chlorine isotopes are experimentally determined to be 34.969 (chlorine-35) and 36.966 (chlorine-37). Since the masses of a proton and a neutron are similar and close to 1 on the ${}^{12}C = 12$ scale, the relative isotopic mass of an isotope is almost, but not exactly, equal to the number of protons plus neutrons in the nucleus.

Table 2.3.1 below shows the relative isotopic mass and the **relative isotopic abundances** of some elements. The relative abundance of each isotope can be expressed as a percentage abundance and is used to indicate the proportion of each isotope in a sample of the element.

TABLE 2.3.1 The relative mass and abundance of hydrogen, oxygen and silver isotopes

Element	Isotopes	Relative isotopic mass	Relative isotopic abundance (%)
hydrogen	${}^{1}H$	1.008	99.986
	${}^{2}H$	2.014	0.014
	${}^{3}H$	3.016	0.0001
oxygen	${}^{16}O$	15.995	99.76
	${}^{17}O$	16.999	0.04
	${}^{18}O$	17.999	0.20
silver	${}^{107}Ag$	106.9	51.8
	${}^{109}Ag$	108.9	48.2

Relative isotopic masses do not have a unit because they are a comparison to the standard, carbon-12, which is 12 units exactly.

Most elements consist of a mixture of isotopes. For the purpose of the calculations you will be doing in later chapters, it is convenient to calculate the average relative mass of an atom in this naturally occurring mixture of isotopes. This average is known as the **relative atomic mass** of an element. It is sometimes referred to as atomic weight and given the symbol A_r.

The relative atomic mass of an element is a weighted average of the relative isotopic masses because it takes into account the relative abundances of each isotope in natural samples of the element.

One way of calculating the relative atomic mass of a sample of an element with a mixture of isotopes is to calculate the contribution of each isotope to the total mass, and then add these contributions. Another way is to use a mathematical expression to calculate the weighted mean. Both these methods are shown in Worked example 2.3.1.

The relative atomic mass of an element is the weighted average of the relative masses of the isotopes of the element relative to ^{12}C.

Worked example 2.3.1

CALCULATING THE RELATIVE ATOMIC MASS OF CHLORINE: METHOD 1

Chlorine is made up of two isotopes, ^{35}Cl and ^{37}Cl. ^{35}Cl has a relative isotopic mass of 34.969 and a relative abundance of 75.80%. ^{37}Cl has a relative isotopic mass of 36.966 and a relative abundance of 24.20%. Calculate the relative atomic mass of chlorine.

Thinking	Working
Calculate the relative atomic mass contribution by the chlorine-35 isotope, which is the relative isotopic mass of the Cl-35 isotope, times the percentage abundance, divided by 100.	$34.969 \times \frac{75.80}{100} = 26.50$
Calculate the relative atomic mass contribution by the chlorine-37 isotope.	$36.966 \times \frac{24.20}{100} = 8.945$
Add the relative atomic mass contributions of all the isotopes.	$26.50 + 8.945 = 35.45$

Worked example 2.3.1

CALCULATING THE RELATIVE ATOMIC MASS OF CHLORINE: METHOD 2

Chlorine is made up of two isotopes, ^{35}Cl and ^{37}Cl. ^{35}Cl has a relative isotopic mass of 34.969 and a relative abundance of 75.80%. ^{37}Cl has a relative isotopic mass of 36.966 and a relative abundance of 24.20%. Calculate the relative atomic mass of chlorine.

Thinking	Working
Identify the relative isotopic masses and the abundance of each isotope.	^{35}Cl relative isotopic mass = 34.969 relative abundance = 75.80% ^{37}Cl relative isotopic mass = 36.966 relative abundance = 24.20%
Use the expression for calculating the weighted mean (A_r) from the percentage abundance and the relative isotopic masses: $A_r = \frac{(\% \times \text{relative isotopic mass}) + (\% \times \text{relative isotopic mass})}{100}$	$A_r = \frac{(75.80 \times 34.969) + (24.20 \times 36.966)}{100}$
Calculate the relative atomic mass. (Give answer to 4 significant figures.)	$A_r = 35.45$

Worked example: Try yourself 2.3.1

CALCULATING THE RELATIVE ATOMIC MASS OF SILVER

Use the data from Table 2.3.1 to calculate the relative atomic mass of silver.

If you are given the relative atomic mass of a sample of an element, and you know that only two isotopes are present, it is also possible to work backwards and calculate the relative abundances of the isotopes. The technique for doing this is shown in Worked example 2.3.2.

Worked example: 2.3.2

CALCULATING THE ABUNDANCE OF ISOTOPES FROM RELATIVE ATOMIC MASSES AND ISOTOPIC MASSES

The relative atomic mass of rubidium is 85.47. The relative isotopic masses of its two isotopes, ^{85}Rb and ^{87}Rb, are 84.95 and 86.94. Calculate the relative abundances of the isotopes.

Thinking	Working
Use ***a*** to represent the percentage abundance of the lighter isotope. The percentage abundance of the heavier isotope must therefore equal (100 – ***a***).	Abundance of 85Rb = ***a*** Abundance of 87Rb = (100 – ***a***)
Substitute the values into the following expression (used in method 2 above): $A_r = \frac{(\% \times \text{relative isotopic mass}) + (\% \times \text{relative isotopic mass})}{100}$	$85.47 = \frac{(84.95 \times \mathbf{a}) + (86.94 \times (100 - \mathbf{a}))}{100}$
Expand the top line of the equation.	$85.47 = \frac{84.95\mathbf{a} + 8694 - 86.94\mathbf{a}}{100}$
Solve the equation to find ***a*** (the abundance of the lighter isotope).	$8547 = 84.95\mathbf{a} + 8694 - 86.94\mathbf{a}$ $8547 - 8694 = 84.95\mathbf{a} - 86.94\mathbf{a}$ $-147 = -1.99\mathbf{a}$ $\mathbf{a} = \frac{147}{1.99} = 73.87$ Abundance of ^{85}Rb = 73.87%
Determine the abundance of the heavier isotope using (100 – ***a***).	Abundance of ^{87}Rb = 100 – 73.87 = 26.13%

CHEMFILE

New standard atomic weights for 19 elements

In 2013, IUPAC voted to change the standard atomic weights of 19 elements. Of the 19 elements, four of them have had renewed isotopic abundance values leading to a change in standard atomic weights. For the remaining 15, the new weights are a result of improvements to the precision of analytical devices.

With improvements to the accuracy of these values, all other sciences and research that involves these values will also see an improvement in accuracy.

Worked example: Try yourself 2.3.2

CALCULATING THE ABUNDANCE OF ISOTOPES FROM RELATIVE ATOMIC MASSES AND ISOTOPIC MASSES

The relative atomic mass of copper is 63.54. The relative isotopic masses of its two isotopes, ^{63}Cu and ^{65}Cu, are 62.95 and 64.95. Calculate the relative abundances of the isotopes.

For comparison purposes, the standard atomic weight can also be used. The **standard atomic weight** is the relative atomic mass of the element based on the agreed proportions of isotopes in a 'normal' sample of the element on Earth.

2.3 Review

SUMMARY

- The isotopes of an element have the same atomic number but different mass numbers. They have the same number of protons but different numbers of neutrons.
- Isotopes of an element have the same chemical properties but different physical properties.
- Isotopes with a significantly different number of neutrons to protons tend to be unstable and therefore radioactive.
- Radioactive isotopes undergo radioactive decay, emitting various forms of radiation to become lighter, more stable nuclei.
- Relative atomic masses are the weighted average of the relative isotopic masses taking into account their relative abundance in nature.

KEY QUESTIONS

1. What is an isotope?
2. Explain the similarities and differences between isotopes of the same element.
3. How many more neutrons does an atom of carbon-14 have than an atom of the carbon-12 isotope?
4. Use the data in Table 2.3.1 on page 32 to calculate the relative atomic mass of:
   a oxygen
   b hydrogen.
5. The element lithium has two isotopes:
   - ^{6}Li with a relative isotopic mass of 6.02
   - ^{7}Li with a relative isotopic mass of 7.02

   The relative atomic mass of lithium is 6.94. Calculate the percentage abundance of the lighter isotope.

2.4 Mass spectrometry

Mass spectrometry is a technique used to measure the mass of atoms or molecules. In the context of atomic structure, it can be used to identify the presence and relative abundance of isotopes in a sample of an element. It other situations, it can also be used to measure the mass of molecules, which provides evidence to help determine their structure.

A MASS SPECTROMETER

Relative isotopic masses of elements and their isotopic abundances are determined by using an instrument called a **mass spectrometer**, which was invented by Francis Aston in 1919. This instrument:

- separates the individual isotopes in a sample of the element
- determines the mass of each isotope, relative to the carbon-12 standard
- calculates the relative abundances of the isotopes in the sample.

A mass spectrometer, as shown in Figure 2.4.1, operates as follows:

1. The sample is vaporised and then ionised using high-energy electrons.
2. The ions are separated and accelerated according to their mass-to-charge $\frac{m}{z}$ ratios in an electric/magnetic field.
3. The ions that have a particular mass-to-charge ratio are detected by a device that counts the number of ions that strike it.

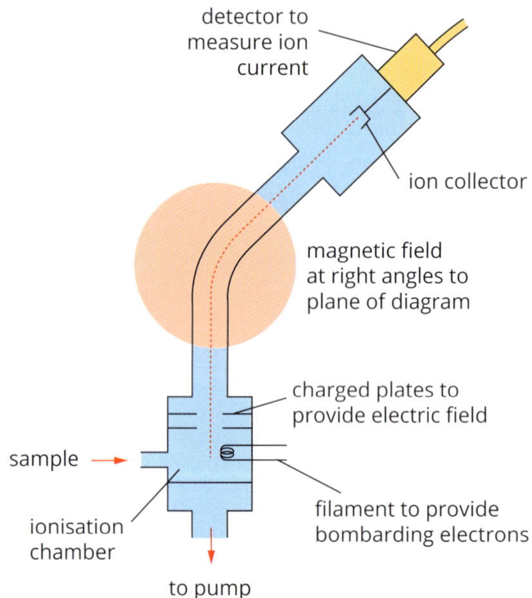

FIGURE 2.4.1 A mass spectrometer is an apparatus used for separating particles, such as isotopes, according to their mass and charge.

Mass spectra

The information obtained from a mass spectrometer is presented graphically as a **mass spectrum**, which plots the relative abundance of each ion against its mass-to-charge ratio, $\frac{m}{z}$.

For simple mass spectra, most of the ions produced in the instrument have a +1 charge, so the value of the $\frac{m}{z}$ ratio is equal to the mass of the ion.

A mass spectrum of aluminium is shown in Figure 2.4.2. The isotopic composition of an element can be seen from its mass spectrum. Aluminium exists as only one natural isotope and so there is only one peak (line) in the mass spectrum of this element. The mass-to-charge ratio corresponds with the aluminium's relative isotopic mass.

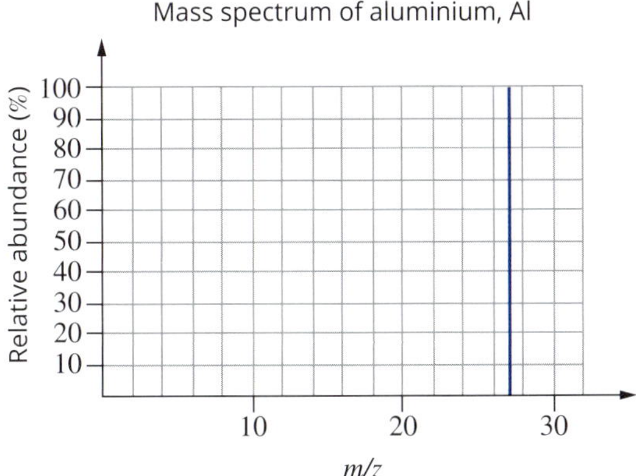

FIGURE 2.4.2 The mass spectrum of aluminium. There is only one peak because aluminium consists of only one isotope.

Many elements occur as a mixture of two or more isotopes. For most elements, the proportion of its isotope is approximately the same in every sample, regardless of from where on Earth the sample has been taken. The percentage amount of an isotope in the natural element is called the relative isotopic abundance of the isotope in that element.

The mass spectrum of magnesium is shown in Figure 2.4.3. Magnesium consists of three isotopes, ${}^{24}_{12}Mg$, ${}^{25}_{12}Mg$ and ${}^{26}_{12}Mg$ with relative abundances of 79%, 10% and 11%, respectively, and so its mass spectrum contains three peaks with the height of each peak representing 79%, 10% and 11% respectively of the total peak height.

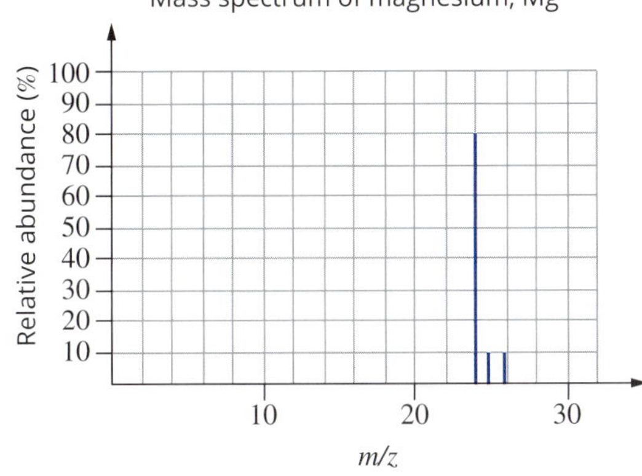

FIGURE 2.4.3 A mass spectrum of magnesium showing three peaks for its three different isotopes

Worked example 2.4.1

CALCULATING PERCENTAGE ABUNDANCE OF EACH ISOTOPE FROM THE MASS SPECTRUM

Figure 2.4.4 shows a simplified mass spectrum of magnesium. From this mass spectrum, calculate the percentage abundance of each of its three isotopes.

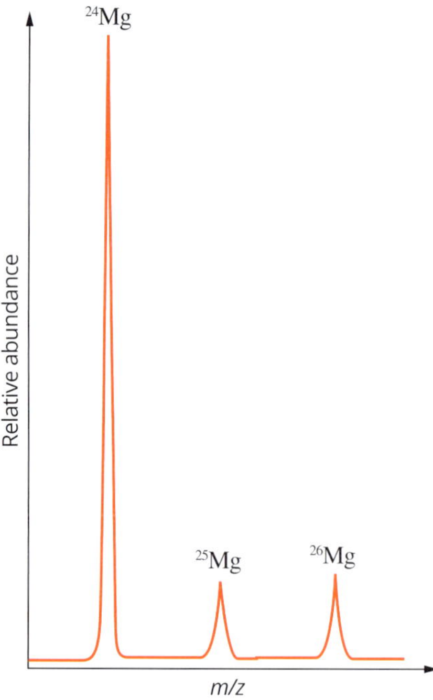

FIGURE 2.4.4 Mass spectrum of magnesium

Thinking	Working
Measure the peak height for each isotope using a ruler.	From the spectrum, the height of each peak is: $^{24}Mg = 7.9\,cm$ $^{25}Mg = 1.0\,cm$ $^{26}Mg = 1.1\,cm$
Calculate the total peak height for the three isotopes by adding the individual peak heights.	Total peak height = 7.9 + 1.0 + 1.1 = 10.0 cm
Substitute the peak height for each isotope into the formula: $\%\text{ abundance} = \frac{\text{peak height}}{\text{total peak height}} \times 100\%$	$\%\text{ abundance } ^{24}Mg = \frac{7.9}{10.0} \times 100$ $= 79\%$ $\%\text{ abundance } ^{25}Mg = \frac{1.0}{10.0} \times 100$ $= 10\%$ $\%\text{ abundance } ^{26}Mg = \frac{1.1}{10.0} \times 100$ $= 11\%$

Worked example: Try yourself 2.4.1

CALCULATING PERCENTAGE ABUNDANCE OF EACH ISOTOPE FROM THE MASS SPECTRUM

Figure 2.4.5 shows a simplified mass spectrum of lead. From this mass spectrum, calculate the percentage abundance of each of its three isotopes.

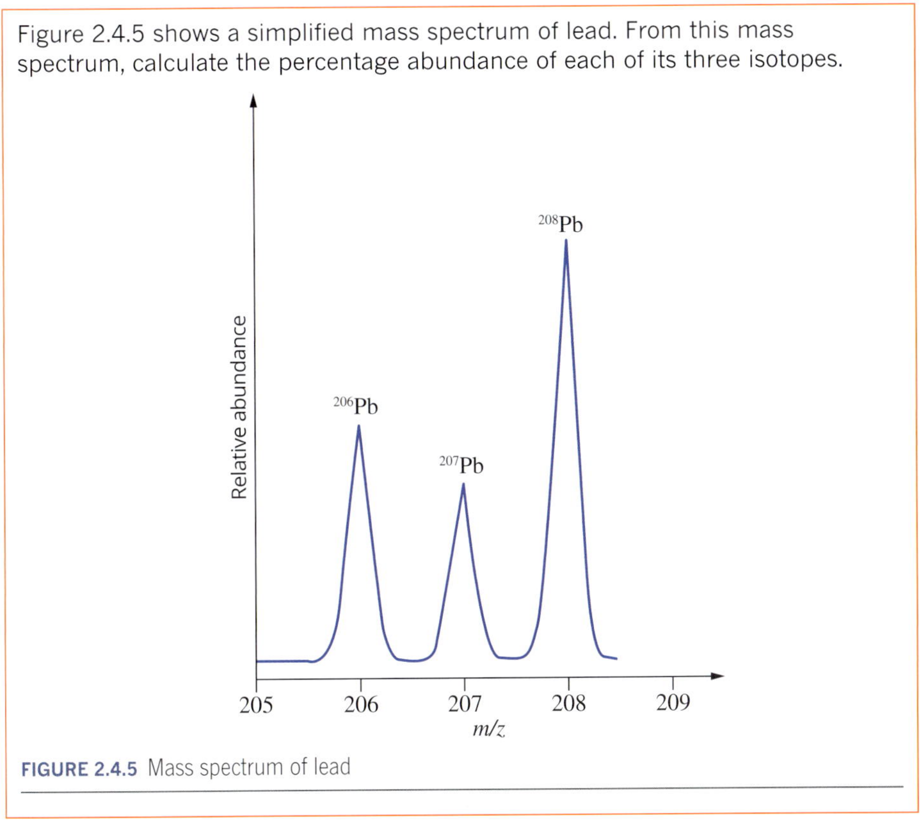

FIGURE 2.4.5 Mass spectrum of lead

2.4 Review

SUMMARY

- Isotopic abundances can be measured by mass spectrometry.
- A mass spectrometer:
  - separates the individual isotopes in a sample of the element
  - determines the mass of each isotope, relative to the carbon-12 standard
  - calculates the relative abundances of the isotopes in the sample.
- A mass spectrum is a plot of the measured abundance of each isotope against the mass-to-charge ratio.
- In a mass spectrum of an element:
  - the number of peaks indicates the number of isotopes
  - the position of each peak on the horizontal axis indicates the relative isotopic mass
  - the relative heights of the peaks correspond to the relative abundance of the isotopes.

KEY QUESTIONS

1 What are the three key operating steps of a mass spectrometer?

2 What does the number of peaks in a mass spectrum represent?

3 What does the position of the peaks along the *x*-axis represent?

4 What does the height of peaks represent?

5 The mass spectrum of zirconium is shown in the graph below.

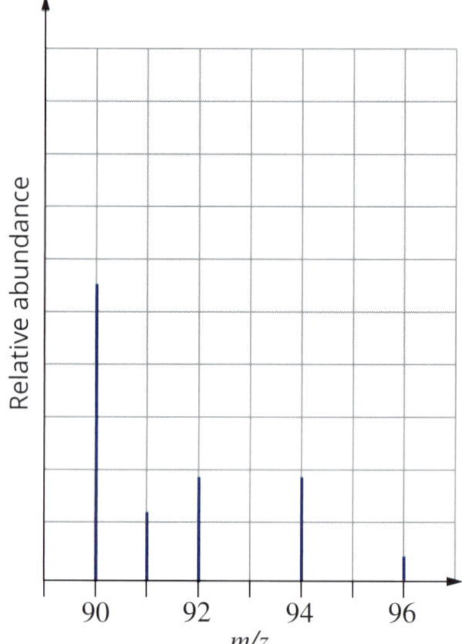

a Measure the peak heights to calculate the percentage abundance of each zirconium isotope.

b Use the percentage abundances calculated in part **a** to determine the relative atomic mass of zirconium.

Chapter review

KEY TERMS

alpha particle
atom
atomic number
atomic theory of matter
chemical symbol
electron
electrostatic attraction
isotope
mass number
mass spectrometer
mass spectrum
matter
model
neutron
nucleon
nucleus
periodic table
proton
radioactive
relative atomic mass
relative isotopic abundance
relative isotopic mass
scanning tunnelling microscope (STM)
standard atomic weight
subatomic particle

Atomic theory

1 Which of Dalton's predictions about the nature of atoms was later proven to be incorrect?

2 Where would you find 99.97% of the mass of an atom?

3 How are protons, neutrons and electrons arranged in an atom?

4 Compare the mass and charge of protons, neutrons and electrons.

5 Who is credited with discovering each of the subatomic particles?

Describing atoms

6 Is the periodic table organised by atomic number or mass number or both?

7 An atom of chromium can be represented by the symbol $^{52}_{24}Cr$.

a Determine its atomic number and mass number.

b Determine the number of electrons, protons and neutrons in the chromium atom.

8 An atom of manganese can be represented by the symbol $^{55}_{25}Mn$.

a Determine its atomic number and mass number.

b Determine the number of electrons, protons and neutrons in the manganese atom.

9 Explain why the number of electrons in an atom equals the number of protons.

10 Using the element bromine as an example, explain why elements are best identified by their atomic number and not their mass number.

Isotopes

11 Two atoms both have 20 neutrons in their nucleus. The first has 19 protons and the other has 20 protons. Are they isotopes? Why or why not?

12 The standard on which all relative masses are based is the ^{12}C isotope, which is given a mass of 12 exactly. Explain why, in the table of relative atomic masses in the Appendix at the end of the book, the relative atomic mass of carbon is listed as 12.01.

13 Determine the percentage abundance of the lighter isotope of each of the following elements.

a Gallium: relative isotopic masses 68.95 and 70.95 respectively; $A_r = 69.72$

b Boron: relative isotopic masses 10.02 and 11.01 respectively; $A_r = 10.81$

14 The separation of certain isotopes from the bulk of an element is a highly technical process. Why can this process not be carried out by chemical means?

15 The two isotopes argon-40 and calcium-40 both have the same mass number. In terms of the numbers of protons, neutrons and electrons, what do these two isotopes share the same number of?

Mass spectrometry

16 When a sample of palladium is placed in a mass spectrometer, the peaks are recorded as the relative isotopic masses and corresponding percentage abundances given in the table below.

Relative isotopic mass	Abundance (%)
101.9049	0.9600
103.9036	10.97
104.9046	22.23
105.9032	27.33
107.9039	26.71
109.9044	11.80

Calculate the relative atomic mass of palladium.

17 The table below gives isotopic composition data for argon and potassium.

Element	Atomic number	Relative isotopic mass	Relative abundance (%)
argon	18	35.978	0.307
		37.974	0.060
		39.974	99.633
potassium	19	38.975	93.3
		39.976	0.011
		40.974	6.69

a Determine the relative atomic masses of argon and potassium.

b Explain why the relative atomic mass of argon is greater than that of potassium, even though potassium has a larger atomic number.

18 The mass spectrum of chromium is shown in the graph below.

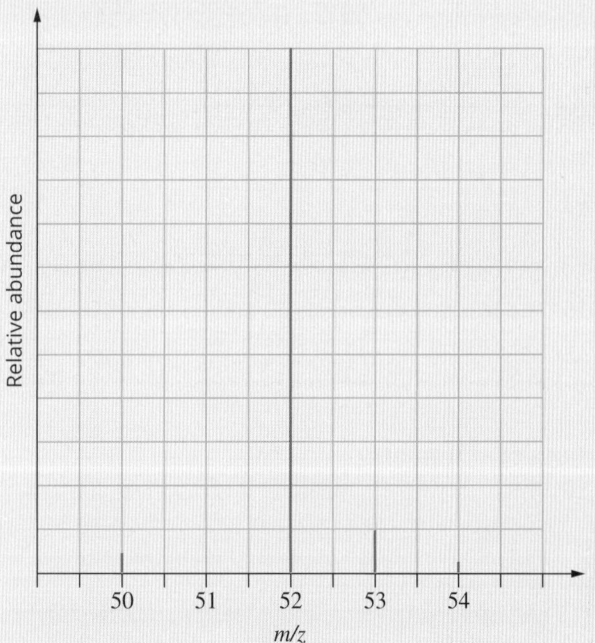

a Measure the peak heights to calculate the percentage abundance of each chromium isotope.

b Use the percentage abundances calculated in part **a** to determine the relative atomic mass for chromium.

19 The relative atomic mass of europium is 151.96. The relative isotopic masses of its two isotopes are 150.92 and 152.92. Calculate the relative abundances of the isotopes in naturally occurring europium.

20 Based on the mass spectrum of neon shown below, give an estimate of the relative atomic mass of neon to the closest whole number.

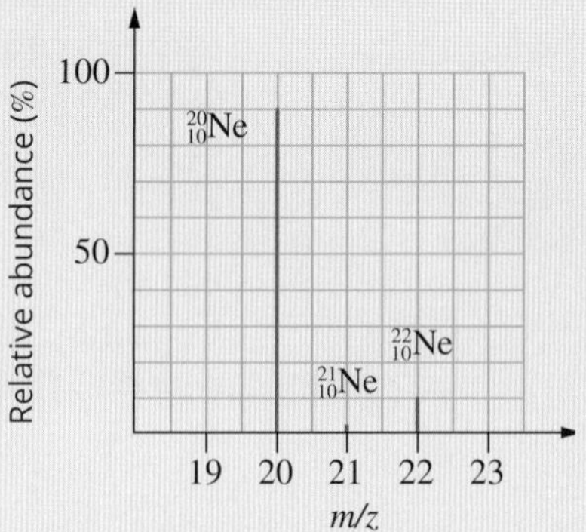

Connecting the main ideas

21 Using suitable examples, clearly distinguish between:

a mass number

b relative isotopic mass

c relative atomic mass.

22 The Banana Equivalent Dose is an informal expression of ionising radiation exposure a person would be exposed to by eating one average-sized banana. The major source of radioactivity in bananas is the naturally occurring radioactive isotope potassium-40.

A sample of bananas was homogenised, dissolved and the potassium extracted. The resultant solution was analysed by mass spectrometry producing the following results.

Relative isotopic mass	Peak area
38.96	454 212.6
39.96	56.984 73
40.96	32 779.37

a Calculate the percentage abundance of the radioactive potassium-40 isotope.

b Using the results, determine the relative atomic mass of potassium to 2 decimal places.

CHAPTER

03 Electrons and the periodic table

In 1911, Ernest Rutherford put forward the planetary model of the atom, consisting of the nucleus surrounded by orbiting electrons, just like the planets orbit the Sun in our solar system. Rutherford's model made a number of significant improvements over the preceding plum-pudding model but was unable to explain the well-known phenomenon of each element emitting a characteristically unique set of radiation (spectra) when heated. To account for this, a new model of the atom needed to be developed. This chapter outlines this new model of the atom—one that explains the behaviour of elements and the types of compounds that they create.

Science understanding

- atoms can be modelled as a nucleus, surrounded by electrons in distinct energy levels, held together by electrostatic forces of attraction between the nucleus and electrons; the location of electrons within atoms can be represented using electron configurations
- flame tests and atomic absorption spectroscopy (AAS) are analytical techniques that can be used to identify elements; these methods rely on electron transfer between atomic energy levels and are shown by line spectra
- the structure of the periodic table is based on the atomic number and the properties of the elements
- the elements of the periodic table show trends across periods and down main groups, including in atomic radii, valencies, 1st ionisation energy and electronegativity as exemplified by groups 1, 2, 13–18 and period 3

3.1 Electronic structure of atoms

FIGURE 3.1.1 The spectacular colours in this Australia Day fireworks display are emitted by metal atoms that have been heated to very high temperatures.

When fireworks explode, they create a spectacular show of coloured lights (Figure 3.1.1). The light is produced by metal atoms that have been heated by the explosion. These coloured lights posed a significant problem for early scientists. The planetary model for the atom put forward by Rutherford could not explain why each metal produced a colour of light unique to that particular metal. However, the light was a clue that ultimately led to a better understanding of the arrangement of electrons in atoms.

BOHR MODEL OF THE ATOM

In 1913, Niels Bohr developed a new model of the hydrogen atom that explained the production of the light characteristic of hydrogen. The **Bohr model** proposed the following.

- Electrons revolve around the nucleus in fixed, circular orbits.
- The electrons' orbits correspond to specific energy levels or shells in the atom.
- Electrons can only occupy these fixed energy levels and cannot exist between two energy levels.
- Electron orbits of larger radii correspond to energy levels of higher energy.

In the Bohr model, the electron orbits the nucleus at a fixed distance from the nucleus. This region of space the electron occupies is known as an **electron shell**. The shell corresponds with a fixed quantity of energy and any electrons occupying that shell possess the same quantity of energy, hence the shell is also referred to as an **energy level**.

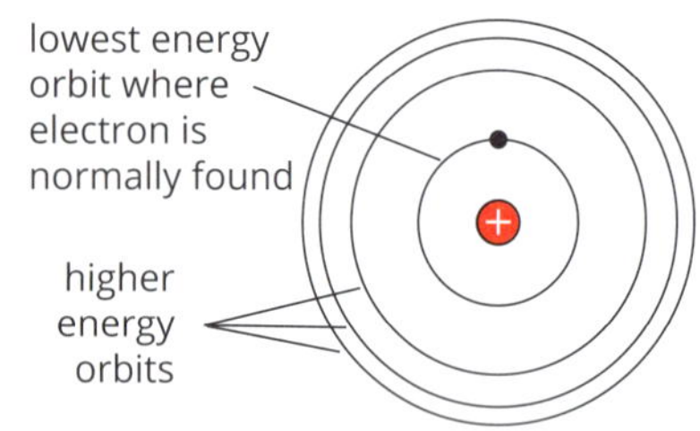

FIGURE 3.1.2 The Bohr model of a hydrogen atom. Bohr suggested that electrons move in orbits of particular energies.

Bohr also suggested that it is possible for electrons to move between the energy levels by absorbing or emitting energy in the form of light. Bohr's model (Figure 3.1.2) provided close agreement between the calculated theoretical frequencies, or colours, of light emitted by hydrogen atoms and the experimentally observed values.

Scientists quickly extended Bohr's model of the hydrogen atom to other atoms. Scientists proposed that electrons are grouped in different energy levels. The lowest energy levels contain electrons closest to the nucleus (Figure 3.1.3).

The orbit in which an electron moves depends on the energy of the electron; electrons with low energy are in orbits closest to the nucleus while high-energy electrons are in outer orbits.

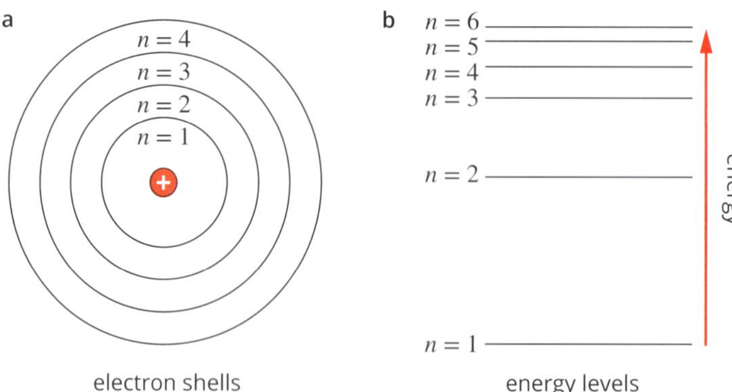

FIGURE 3.1.3 (a) The electron shells of an atom (n) are labelled using integers. The first shell is the shell closest to the nucleus and the radius of each shell increases as the shell number increases. (b) Each shell corresponds to an energy level that electrons can occupy. The first shell has the lowest energy and the energy increases as the shell number increases.

The lowest energy shell is the shell closest to the nucleus and is labelled $n = 1$. Shells with higher values of n correspond to higher energy levels. As the values of n increase, the energy levels get closer together.

The **outermost shell** of an atom is known as the atom's **valence shell**. The electrons in this outer shell are called **valence electrons**. These electrons require the least amount of energy to be removed from the atom. The valence electrons are involved in chemical reactions. Consequently, if you know the number of valence electrons in the atoms of an element, then you can predict the chemical properties of the element.

Later in this chapter, and in Chapters 4–6 you will learn about how atoms tend to lose, gain or share valence electrons in order to achieve eight electrons in their outer shell when they are involved in chemical reactions. This is known as the **octet rule**.

The Bohr model of the atom was revolutionary when it was proposed. Before Bohr, Rutherford and other scientists believed that electrons could orbit the nucleus at any distance from the nucleus. This picture of the electrons was based on how scientists observed the world around them. For example, planets can revolve at varying distances around the Sun. However, Bohr's theory stated that electrons only occupy specific, circular orbits. This was the first suggestion that the physics inside atoms might be very different from the physics we experience in our daily lives.

The shell model of the atom that developed from the work of Niels Bohr could explain why light is emitted when atoms are heated, however, there were some things that the model could not explain. Despite these limitations, the Bohr model is still a valuable predictive tool for the behaviour of some electrons.

EXTENSION

Quantum mechanical model of the atom

The word *quantum* means 'a specific amount'. In the Bohr model, the electrons can only have specific amounts of energy depending on which shell they are in. The energy of the electrons is said to be quantised.

In 1926, the German scientist Erwin Schrödinger proposed that electrons behaved as waves around the nucleus. This was fundamentally different from Bohr, who treated electrons as tiny, hard particles.

Using a mathematical approach and his wave theory, Schrödinger developed a model of the atom called quantum mechanics. This is the model of the atom that most scientists use today.

The equations used in quantum mechanics are complex. The electron is now seen as a vague, elusive object that behaves like a cloud of negative charge (Figure 3.1.4). Unlike the Bohr model, there are no definite orbits for electrons in this model of the atom. Electrons are thought to move in regions of space surrounding the nucleus called orbitals.

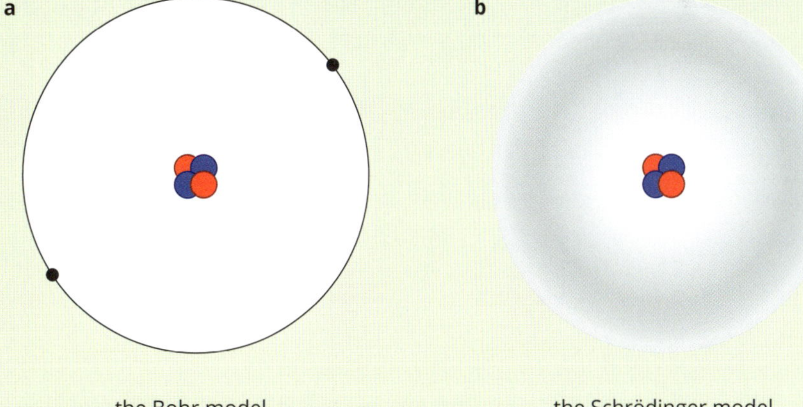

FIGURE 3.1.4 (a) Bohr regarded electrons as particles that travel along a defined path in circular orbits. (b) In Schrödinger's quantum mechanical approach, the electrons behave as waves and occupy a three-dimensional space around the nucleus. The region occupied by the electrons is known as an orbital.

3.1 Review

SUMMARY

- When atoms are heated, they emit light that is characteristic of that element.
- The Bohr model of the atom was the first atomic model to explain the origin of emission spectra.
- The Bohr model assumes that electrons can only exist in fixed, circular orbits of specific energies. These orbits later came to be known as energy levels or shells.

KEY QUESTIONS

1 Explain how emission spectra are evidence for electron shells in the Bohr model.

2 What four assumptions did Bohr make about the electronic structure of the atom?

3 What form of energy is emitted when an electron moves from a higher energy shell to a lower energy shell?

3.2 Electron arrangement in the periodic table

Chapter 2 looked at research into elements and the nature of atoms, the existence of subatomic particles, their charge and mass, the way these particles are arranged in an atom, and the way they behave. As scientists' understanding of the atom improved, and more elements were discovered, a way of organising this knowledge was needed.

THE PERIODIC TABLE

In the 19th century, chemists wanted to use information they had gathered about the elements to organise the elements into a useful and practical format. The work of these scientists led to what is now known as the periodic table. The early forms of the periodic table were very different from the one in use today; many elements had not been discovered and scientists only had limited information about some other elements. An early form of the periodic table was created by Russian chemist Dimitri Mendeleev (Figure 3.2.1).

At the time the early periodic table was being constructed, the existence of subatomic particles was unknown, and so the elements were placed in order of the increasing mass of the atoms. Mendeleev recognised that the chemical properties of the elements varied periodically with increasing atomic mass. He arranged elements with similar properties in vertical columns. Mendeleev proposed his periodic law—the properties of elements vary periodically with their atomic weights.

It is now known that the atomic number (number of protons) is what makes one element fundamentally different from another element. The elements in the modern periodic table are therefore arranged in rows in order of increasing atomic number.

Chemists use the number of electrons in the outer shell (valence electrons) to organise the elements into columns. The form of the periodic table in common use is shown in Figure 3.2.2.

FIGURE 3.2.1 Russian chemist Dimitri Mendeleev is commonly acknowledged as the creator of the basis for the modern periodic table.

CHEMFILE

Relative atomic mass

The property of an element that Mendeleev called atomic weight is now more commonly called relative atomic mass. Relative atomic mass should not be confused with the mass number. Scientists at the time of Mendeleev did not know about subatomic particles. They only knew the relative masses of atoms of each element.

1	2	3	4	5	6	7	8	9	10	11	12	13	14	15	16	17	18
1 **H** hydrogen 1.008																	2 **He** helium 4.003
3 **Li** lithium 6.968	4 **Be** beryllium 9.012											5 **B** boron 10.82	6 **C** carbon 12.01	7 **N** nitrogen 14.01	8 **O** oxygen 16.00	9 **F** fluorine 19.00	10 **Ne** neon 20.18
11 **Na** sodium 22.99	12 **Mg** magnesium 24.31											13 **Al** aluminium 26.98	14 **Si** silicon 28.09	15 **P** phosphorus 30.97	16 **S** sulfur 32.07	17 **Cl** chlorine 35.45	18 **Ar** argon 39.95
19 **K** potassium 39.10	20 **Ca** calcium 40.08	21 **Sc** scandium 44.96	22 **Ti** titanium 47.88	23 **V** vanadium 50.94	24 **Cr** chromium 52.00	25 **Mn** manganese 54.94	26 **Fe** iron 55.85	27 **Co** cobalt 58.93	28 **Ni** nickel 58.69	29 **Cu** copper 63.55	30 **Zn** Zinc 65.38	31 **Ga** gallium 69.72	32 **Ge** germanium 72.59	33 **As** arsenic 74.92	34 **Se** selenium 78.96	35 **Br** bromine 79.90	36 **Kr** krypton 83.80
37 **Rb** rubidium 85.47	38 **Sr** strontium 87.62	39 **Y** yttrium 88.91	40 **Zr** zinconium 91.22	41 **Nb** niobium 92.91	42 **Mo** molybdenum 95.94	43 **Tc** technetium	44 **Ru** ruthenium 101.1	45 **Rh** rhodium 102.9	46 **Pd** palladium 106.4	47 **Ag** silver 107.9	48 **Cd** cadmium 112.4	49 **In** indium 114.8	50 **Sn** tin 118.7	51 **Sb** antimony 121.8	52 **Te** tellurium 127.6	53 **I** iodine 126.9	54 **Xe** xexon 131.3
55 **Cs** caesium 132.9	56 **Ba** barium 137.3	57–71 ***La** lanthanum 138.9	72 **Hf** hafnium 178.5	73 **Ta** tantalum 180.9	74 **W** tungsten 183.9	75 **Re** rhenium 186.2	76 **Os** osmium 190.2	77 **Ir** iridium 192.2	78 **Pt** platinum 195.1	79 **Au** gold 197.0	80 **Hg** mercury 200.6	81 **Tl** thallium 204.4	82 **Pb** lead 207.2	83 **Bi** bismuth 209.0	84 **Po** polonium	85 **At** astatine	86 **Rn** radon
87 **Fr** francium	88 **Ra** radium 226.0	89–103 ****Ac** actinium	104 **Rf** rutherfordium	105 **Db** dubnium	106 **Sg** seaborgium	107 **Bh** bohrium	108 **Hs** hassium	109 **Mt** meitnerium	110 **Ds** darmstadtium	111 **Rg** roentgenium	112 **Cn** copernicium	113 **Uut** ununtrium	114 **Fl** flerovium	115 **Uup** ununpentium	116 **Lv** livermorium	117 **Uus** ununseptium	118 **Uuo** ununoctium
key:		* Lanthanide series		58 **Ce** cerium 140.1	59 **Pr** praseodymium 140.9	60 **Nd** neodymium 144.2	61 **Pm** promethium	62 **Sm** samarium 150.4	63 **Eu** europium 152.0	64 **Gd** gadolinium 157.3	65 **Tb** terbium 158.9	66 **Dy** dysprosium 162.5	67 **Ho** holmium 164.9	68 **Er** erbium 167.3	69 **Tm** thulium 168.9	70 **Yb** ytterbium 173.0	71 **Lu** lutetium 175.0
Atomic number **Symbol** Name Standard atomic weight		** Actinide series		90 **Th** thorium 232.0	91 **Pa** protactinium	92 **U** uranium 238.0	93 **Np** neptunium	94 **Pu** plutonium	95 **Am** americium	96 **Cm** curium	97 **Bk** berkelium	98 **Cf** californium	99 **Es** einsteinium	100 **Fm** fermium	101 **Md** mendelevium	102 **No** nobelium	103 **Lr** lawrencium

FIGURE 3.2.2 This is the most common form of the periodic table in use.

CHEMFILE

Element 117 and the Canberra scientists

Four atoms of a super-heavy element known as number 117 have been created in a German laboratory as part of a collaboration with Professor David Hinde, director of the heavy ion accelerator facility at the Department of Nuclear Physics at the Australian National University.

Because the element is radioactive, the atoms disappeared within one-tenth of a second, but the scientists were able to confirm the element's first observation in 2010. This new element was given the temporary name of ununseptium ('one-one-seven' in Latin), and in 2015 was given the name tennessine by the International Union of Pure and Applied Chemistry (IUPAC). The name recognises the contribution of a team from Oak Ridge National Laboratory in Tennessee, USA, one of the teams that confirmed its existence. The atoms of element 117 are the second-heaviest atoms ever observed, being 40% heavier than an atom of lead (element 118 is now the heaviest).

Professor Hinde said that because such an extremely small amount of the element was synthesised, it will not be practical to use it. But the Australian researchers will continue to concentrate on the quantum science behind the element's synthesis and attempt to produce another super-heavy element.

FIGURE 3.2.3 Professor David Hinde is director of the heavy ion accelerator facility at the Department of Nuclear Physics at the Australian National University in Canberra.

The modern periodic table has several key features.

- The periodic table is arranged in order of increasing atomic number.
- The atomic number is located above each chemical symbol.
- The relative atomic mass is located under each chemical symbol.
- The horizontal rows are known as **periods** and are labelled 1–7.
- The vertical columns are known as **groups** and are labelled 1–18.
- **Main group elements** are elements in groups 1, 2 and 13–18.
- The elements in groups 3–12 are known as **transition metals**.
- The lanthanoids and actinoids in periods 6 and 7 are shown separately from the main part of the table simply to make the table itself smaller and easier to manage.

Some periodic tables you will see also indicate other properties of the elements such as boiling point or whether the element is a solid, liquid or gas at room temperature.

Groups

Elements in the periodic table are arranged into vertical columns called groups. For main group elements, the group number can be used to determine the number of valence electrons (outer-shell electrons) in an atom of the element.

In groups 1 and 2, the number of valence electrons is equal to the group number. For example, magnesium is in group 2 and therefore has two valence electrons.

In groups 13–18, the number of valence electrons is equal to the group number minus 10. For example, oxygen is in group 16 so oxygen has six outer-shell electrons. Similarly, neon is in group 18 so neon has eight valence electrons. Helium is an important exception. It is located in group 18 but only has two valence electrons. Helium is placed in group 18 because it is unreactive, like other group 18 elements. This information is summarised in Table 3.2.1.

TABLE 3.2.1 The number of valence electrons in elements belonging to each group

Group	Number of valence electrons
1	1
2	2
13	3
14	4
15	5
16	6
17	7
18	8*

*Exception: helium has only two valence electrons.

The valence electrons are the electrons that are involved in chemical reactions. As a consequence, the number of valence electrons determines many of the chemical properties that an element exhibits.

Table 3.2.2 shows the names of some groups in the periodic table.

TABLE 3.2.2 Names of different groups in the periodic table

Group	Name
1	alkali metals
2	alkaline earth metals
17	halogens
18	noble gases

Because elements in the same group have the same number of valence electrons, elements in the same group have similar chemical properties. For example, the alkali metals are elements in group 1 (with the exception of hydrogen). They all have one valence electron (Figure 3.2.4) and are all relatively soft metals and are highly reactive with water and oxygen.

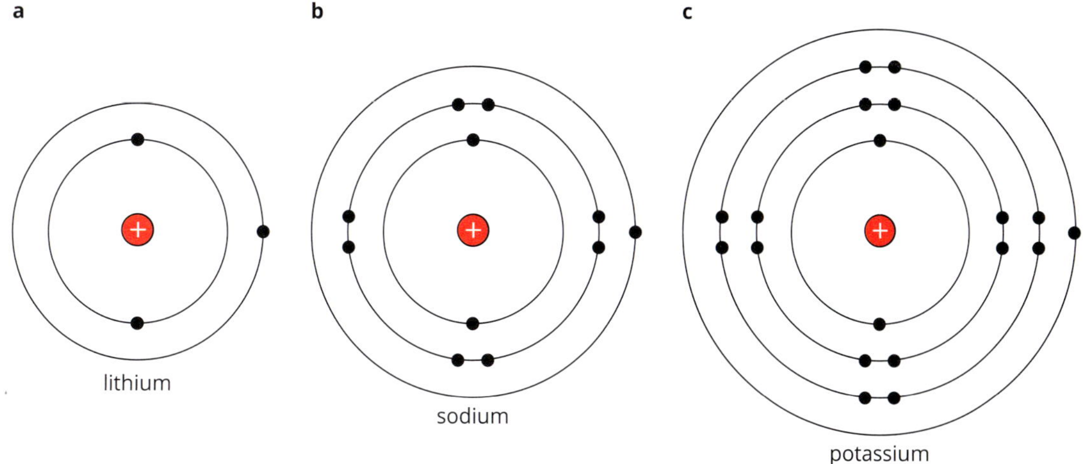

FIGURE 3.2.4 The arrangement of electrons in the first three alkali metals (a) lithium (b) sodium and (c) potassium

Fluorine, chlorine, bromine and iodine are halogens (group 17). They all have seven valence electrons and are all coloured and highly reactive.

The **noble gases** (group 18) are a particularly interesting group. The noble gases have a very stable electron arrangement: helium has a full outer shell of two electrons and the other members of this group have a full outer shell of eight electrons. Chemical reactions involve the rearrangement of valence electrons to achieve a stable outer shell. Noble gases have a stable electron configuration, so they do not tend to lose or gain electrons. This means that the noble gases have low reactivity or are **inert**.

Periods

The horizontal rows in the periodic table are called periods. Periods are numbered 1–7. The number of a period gives information about the arrangement of electrons in each atom. The period an element is located within is equal to the number of occupied electron shells in the element's atoms. For example, the outer shell of magnesium and chlorine (Figure 3.2.5) is the third shell and both of these elements are in period 3.

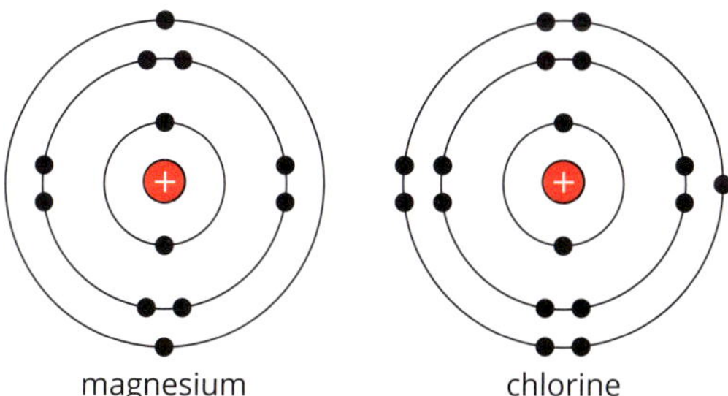

FIGURE 3.2.5 The arrangement of electrons in magnesium and chlorine. The valence electrons of both atoms occupy the third shell as they are both in period 3.

Electron configuration

In all atoms, the electrons are as close to the nucleus as possible. This means that electrons will generally occupy inner shells before outer shells. For example, an atom of lithium has three electrons. Two electrons will occupy the first shell as that is all it can hold. The third electron is in the second shell.

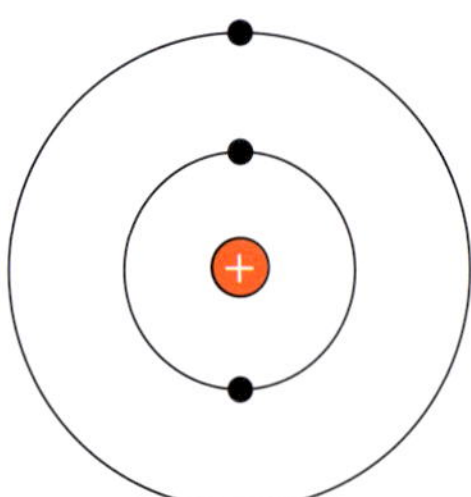

FIGURE 3.2.6 The Bohr diagram for the lithium atom with three electrons. Two electrons occupy the inner shell and the remaining one occupies the outer or valence shell.

A **Bohr diagram** is a simple diagram that shows the arrangement of electrons around the nucleus. In such diagrams, only the shells that are occupied are drawn. The Bohr diagram for lithium is shown in Figure 3.2.6.

A more succinct and concise way of communicating the arrangement of electrons in an atom is the **electron configuration**. The electron configuration is the number of electrons in each shell separated by commas, starting at the shell with the lowest energy (closest to the nucleus).

For example, in the previous example, you saw that a lithium atom has two electrons in the inner, lowest energy shell, followed by one electron in the outer, valence shell.

Thus the electron configuration for lithium is 2,1.

To determine the electron configuration for any of the first 20 elements, you can simply use an element's location on the periodic table to indicate the total number of electrons present, the number of shells it has and how many electrons occupy each shell.

A maximum of two electrons can fit into the first shell, eight in the second shell and eight in the third shell. Using this information it is possible to work out the electron configurations of atoms as shown in Worked example 3.2.1.

Worked example 3.2.1

DETERMINING THE ELECTRON CONFIGURATION OF AN ATOM

Determine the electron configuration of sulfur.	
Thinking	**Working**
Locate sulfur (S) on the periodic table and find its atomic number. This is the number of protons in the nucleus.	16 protons
As atoms are uncharged, they have an equal number of positive charges (protons) as negative charges (electrons).	16 electrons
Start allocating electrons from lowest energy to highest, ensuring each shell is complete before progressing to the next shell. The first shell can accommodate a maximum of two electrons. There are 14 electrons remaining to be allocated.	2
The second shell can accommodate a maximum of eight electrons. There are six electrons remaining to be allocated.	2,8
The remaining six electrons (less than a complete third shell) are allocated to the third (valence) shell.	2,8,6

Worked example: Try yourself 3.2.1

DETERMINING THE ELECTRON CONFIGURATION OF AN ATOM

Determine the electron configuration of argon.

Although this is an easy pattern to use to determine the electron configuration of any of the first 20 elements, this pattern does not continue from scandium onward. Scandium is the first of the transition metals, a region of the periodic table also known as the *d*-block. To determine the electron configuration of the transition metals onward, a greater understanding of the Schrödinger (quantum mechanical) model of the atom is required, which is beyond the scope of this course.

> In the Bohr model, the maximum number of electrons in the first shell is two, the second shell is eight and the third shell is eight.

Ions

Atoms have an equal number of positive charges (protons) and negative charges (electrons). This gives them a neutral or zero overall charge. The number of electrons an atom has, however, is not fixed. Through chemical reactions or the addition of energy, the total number of electrons an atom possesses can be increased or decreased. This in turn also changes the electron configuration of the atom.

The electrons are gained to or lost from the valence shell. This shell contains the outermost and most weakly held electrons. When the number of positive charges (protons) does not equal the number of negative charges (electrons), a charged particle, known as an **ion**, is the result.

As electrons are negatively charged, if an atom obtains additional electron(s) and the number of positive charges (protons) remains constant; the atom will find itself with an overall negative charge equal to the number of electrons gained. Negatively charged ions are known as **anions**.

> An ion is any atom of an element with more or fewer electrons than protons. If an atom loses electrons, it becomes positively charged as there are now more protons than electrons. A positively charged ion is called a cation. If an atom gains electrons, it becomes negatively charged and is called an anion.

For example, if a chlorine atom with an electron configuration of 2,8,7 gains one valence electron, its electron configuration becomes 2,8,8. Having a total of 17 protons and now 18 electrons, the chlorine atom acquires a negative charge and becomes a chloride ion, denoted as Cl^-.

Conversely, if an atom loses electrons and the number of positive charges (protons) remains constant, the atom will have an overall positive charge equal to number of electrons lost. Positively charged ions are known as **cations**.

For example, if a sodium atom with an electron configuration of 2,8,1 loses its one valence electron, its electron configuration becomes 2,8. Having a total of 11 protons and only 10 electrons, the sodium atom now acquires a positive charge and becomes a sodium ion, denoted as Na^+.

Table 3.2.3 shows the electron configuration of the atom and ion of some of the first 20 elements.

TABLE 3.2.3 Atom and ion electron configurations

Atom		Ion	
Symbol	Electron configuration	Symbol	Electron configuration
Li	2,1	Li^+	2
F	2,7	F^-	2,8
Na	2,8,1	Na^+	2,8
Mg	2,8,2	Mg^{2+}	2,8
Al	2,8,3	Al^{3+}	2,8
S	2,8,6	S^{2-}	2,8,8
Ca	2,8,8,2	Ca^{2+}	2,8,8

Notice that the ions share the same electron configuration as three of the noble gases—helium (2), neon (2,8) or argon (2,8,8).

The formation of ions and the associated compounds will be covered in greater depth in Chapter 5.

3.2 Review

SUMMARY

- The periodic table is a tool for organising elements according to their chemical and physical properties.
- The elements of the periodic table are arranged in order of increasing atomic number.
- Columns in the periodic table are known as groups and are numbered 1–18.
- The number of valence electrons in an atom of an element can be determined by the group in which it is located (Table 3.2.1 on page 48).
- Certain groups in the periodic table have specific names (Table 3.2.2 on page 48).
- The main group elements are in groups 1, 2 and 13–18 in the periodic table.
- Elements in between the main group elements (groups 3–12) are known as transition metals.
- Rows in the periodic table are known as periods and are numbered 1–7.
- The number of occupied electron shells of an atom of an element is equal to the number of the element's period.
- Bohr diagrams can be used to show the number and location of electrons surrounding a nucleus.
- The electron configuration of an atom indicates the arrangement of electrons in each shell.
- A maximum of two electrons can fit into the first shell, eight in the second shell and eight in the third shell.

KEY QUESTIONS

1 Construct the electron configurations for atoms with:
  a five electrons
  b 12 electrons
  c 20 electrons.

2 Write the electron configuration of the following elements.
  a Be
  b S
  c Ar
  d Mg
  e Ne

3 Write the name and symbol of the element with the following electron configurations.
  a 2
  b 2,7
  c 2,8,3
  d 2,5
  e 2,8,7

3.3 Trends in the periodic table

The periodic table does not just provide information about an element's electron configuration. It can also be used as a tool for summarising the relative properties of elements and explaining the trends observed in those properties.

You have already seen how the group number of an element determines how many valence electrons an atom of that element has and how the period indicates how many electron shells are occupied in an atom of an element. Properties such as atomic radius, ionisation energy and electronegativity show common trends in the periodic table and can be predicted and explained by examining the underlying forces present in each atom.

ELECTROSTATIC ATTRACTION

The behaviour of magnets when brought in close proximity of each other is a familiar phenomenon. If two north or south poles are brought together, they repel, whereas if a north and south pole are brought together, they attract. Indeed, the saying 'opposites attract' is a commonly used phrase and it is generally understood that positive and negative charges attract just as the north and south poles of two magnets do.

Although magnetic and electrostatic attraction can seem to be similar, the nature of the force that is present is different. The electrostatic attraction that exists between positive and negative charges is the force that, in addition to holding individual atoms together, holds atoms together in molecules and then these molecules together in all forms of matter.

The force of attraction between two charges is described mathematically by Coulomb's law:

$$F = k\frac{q_1 q_2}{r^2}$$

where q_1 and q_2 are the magnitudes of the charges involved, r is the distance between them and k is a constant. Although the use of this formula in calculations is not part of this course, understanding how changes to the quantities involved in the formula affect the strength of the electrostatic attraction allows us to explain many trends in the periodic table.

From this equation two important points can be discerned.

- The strength of any electrostatic attraction is directly proportional to the magnitude of the charges involved. That is, as the size of the charges increases, the force of attraction between the charges also increases.
- The strength of any electrostatic attraction is inversely proportional to the distance between the charges squared. That is, as the distance between the charges increases, the force of attraction decreases. Note that as the distance, r, is squared, any change to the distance between the charges will have a more significant effect on the strength of the attraction than a change of similar magnitude to the size of the charges.

It is these two factors, the charges (on ions, electrons or protons) and the distance (between these charged particles) involved in each electrostatic attraction that allow the other trends present to be explained as you move through the periodic table.

Core charge

The **core charge** (or **effective nuclear charge**) of an atom is a measure of the attractive force felt by the valence electrons towards the nucleus. Core charge, although not a physically measurable quantity, is a useful concept to help explain patterns in the periodic table.

Consider an atom of lithium, which has an atomic number of three. It has three protons in its nucleus, two electrons in the first shell and one electron in the second shell (Figure 3.3.1).

- The valence electron is attracted to the three positive charges in the nucleus.
- This electron is also repelled by the two electrons in the inner shell, reducing the attraction to the nucleus.

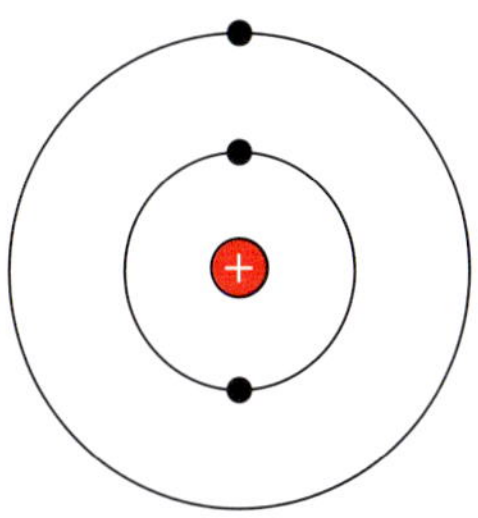

FIGURE 3.3.1 A lithium atom with one valence electron and two electrons in the inner shell. The atom has a core charge of +1.

- The electrons in the inner shell are said to shield the valence electron from the full attraction of the nucleus.
- The valence electron is therefore effectively attracted to the nucleus as if there were a +1 nuclear charge.
- This atom is therefore said to have a core charge of +1.

In atoms with two or more shells filled with electrons, the attraction between the nucleus and valence electrons is reduced by repulsion between the inner-shell electrons and the valence-shell electrons.

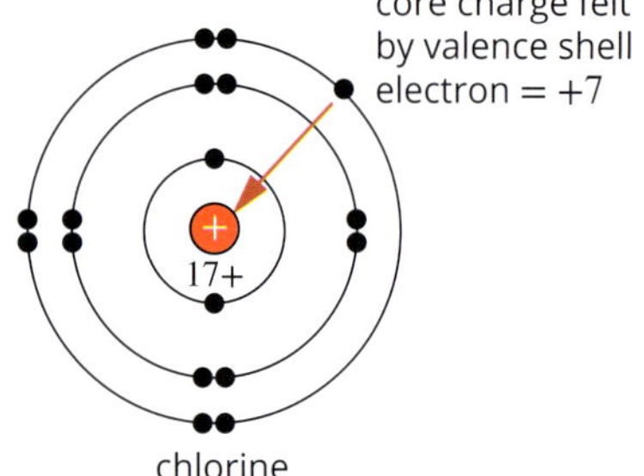

FIGURE 3.3.2 A Bohr diagram for the chlorine atom showing the valence-shell electron shielded by 10 electrons in shells $n = 1$ and $n = 2$.

For example, an atom of chlorine (Figure 3.3.2) has 17 protons and seven valence electrons; the number of electrons in the inner shells is 10. The core charge of a chlorine atom is therefore 17 − 10 = +7.

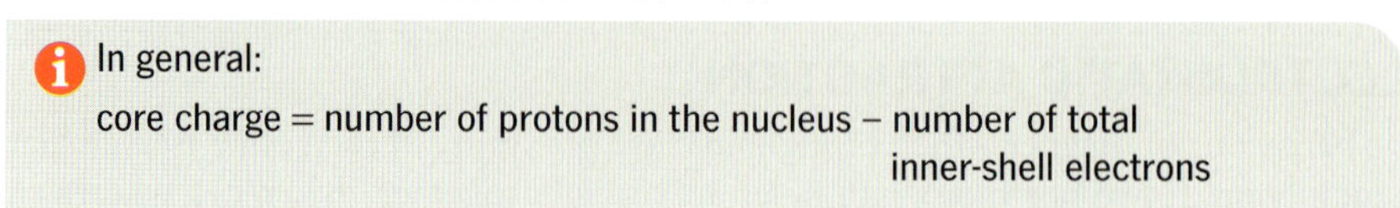

In general:

core charge = number of protons in the nucleus − number of total inner-shell electrons

Consider the atoms of two different elements in group 1, lithium and sodium, shown in Figure 3.3.3.

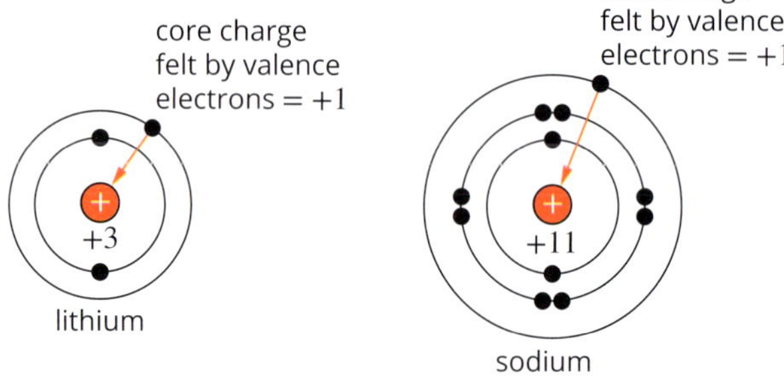

FIGURE 3.3.3 Lithium and sodium atoms both have a core charge of +1.

As for all group 1 elements, the valence electron of a lithium atom and a sodium atom experience a core charge of +1.

Moving down a group you can see that the core charge remains constant. Thus, any differences in the strength of the attraction between the valence electron in lithium and sodium and their respective nuclei cannot be explained in terms of differences in charge. However, the number of electron shells increases.

Hence, the valence electrons are further from the nucleus. This increase in distance has a marked effect on the force of attraction the valence electrons experience. Consequently, they will be pulled less strongly towards the nucleus.

Now consider sodium and chlorine. They are both in period 3 in the periodic table (Figure 3.3.4).

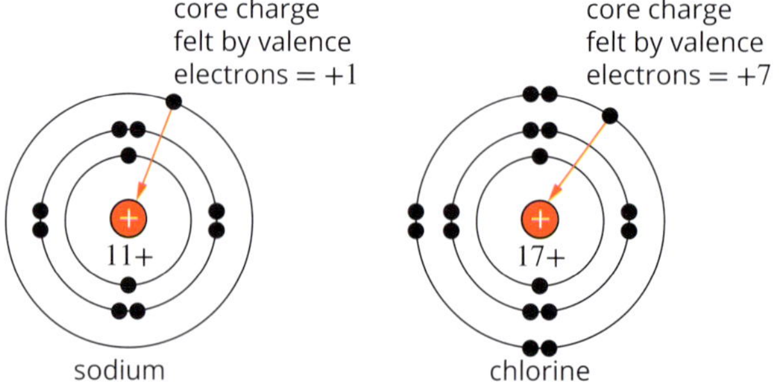

FIGURE 3.3.4 The core charges of two period 3 elements, sodium and chlorine, are +1 and +7 respectively.

Moving across a period, you can see that the number of electron shells remains constant. Thus, any differences in the strength of the attraction between the valence electron(s) in sodium and chlorine and their respective nuclei cannot be explained in terms of differences in distance. However, the core charge increases.

Hence, the valence electrons are attracted by an increasing core charge. This increase in charge affects the force of attraction the valence electrons experience. Consequently, they will be pulled more strongly towards the nucleus.

Table 3.3.1 summarises how the attraction between the nucleus and valence electrons changes in the periodic table.

TABLE 3.3.1 The changes in attraction between the nucleus and valence electrons within groups and periods of the periodic table

	Trend in core charge	Trend in nuclei–valence electron distance	Trend in attraction between the nucleus and valence electrons
down a group	remains constant	increases	The valence electrons are more weakly held moving down a group as, although the core charge stays constant, the valence electrons are further from the nucleus (there are more shells in the atom) and thus experience a weaker electrostatic attraction.
left to right across a period	increases	remains constant	The valence electrons are more strongly held moving across a period as, although the distance between the valence electrons and the nucleus distance remains essentially constant (same electron shell), the valence electrons are subject to a higher core charge and thus experience a stronger electrostatic attraction.

The trends of atomic radii, ionisation energy and electronegativity are easily explained in terms of the strength of attraction experienced by the valence electrons.

ATOMIC RADIUS

Atomic radius is a measurement used for the size of atoms. It can be regarded as the distance from the nucleus to the valence-shell electrons. Figure 3.3.5 depicts the atomic radii of many of the main group elements.

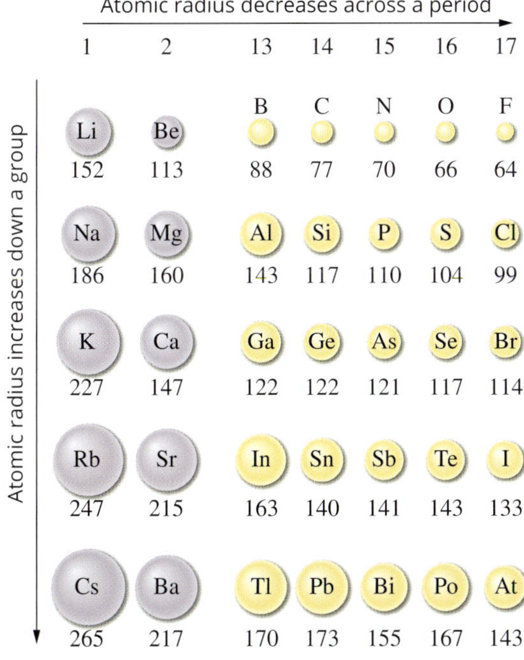

FIGURE 3.3.5 The relative sizes of atoms of selected main group elements. The atomic radii are given in picometres (pm). A picometre is 10^{-12} m.

Atoms do not have sharply defined boundaries and so it is not possible to measure their radii directly. One method of obtaining atomic radii, and therefore an indication of atomic size, is to measure the distance between nuclei of atoms of the same element in molecules. For example, in a hydrogen molecule (H_2) the two nuclei are 74 picometres (pm) apart. The radius of each hydrogen atom is assumed to be half of that distance, i.e. 37 pm.

Table 3.3.2 explains the trends in atomic radius in the periodic table.

TABLE 3.3.2 Trends in atomic radius in the periodic table

	Trend in atomic radius	Explanation
down a group	increases	Core charge stays constant and the number of shells increases as you move down a group. As a result, atomic radius increases down a group.
left to right across a period	decreases	As you move across a period, the number of occupied shells in the atoms remains constant but the core charge increases. The valence electrons become more strongly attracted to the nucleus, so atomic radius decreases across a period.

IONISATION ENERGY

The valence electrons are held to a nucleus due to the electrostatic attraction between the two. If an atom is given sufficient energy, this electrostatic attraction can be overcome and an electron can be completely removed from the atom. The process of removing an electron from an atom and forming an ion is called **ionisation**.

The valence electrons are removed first because they are the furthest electrons from the nucleus and are the least strongly held. When ionisation occurs, the atom will have one less electron than the number of protons in the nucleus, and becomes a positively charged ion.

EXTENSION

Ionising radiation

Radiation is a term that can have negative connotations in wider society but, scientifically, radiation is simply the transmission of energy via waves or particles across space. Electromagnetic radiation is in the form of waves, whereas alpha and beta radiation, generated from nuclear decay, is in the form of particles.

Certain high-energy forms of radiation are classified as ionising radiation. This is radiation that carries with it sufficient energy to remove electrons from atoms or molecules, thereby ionising them. Exposure to ionising radiation poses significant health risks as the molecules within cells, such as DNA, can be damaged by ionising radiation, possibly leading to cancer.

Of the electromagnetic spectrum, gamma rays, X-rays and some ultraviolet light is considered ionising. Other forms of electromagnetic radiation, such as microwaves used in cooking appliances and radio waves used in mobile phones, are not energetic enough to cause the ionisation of atoms or molecules. These forms of radiation can only cause electrons to become 'excited', as they move from lower to higher energy levels (shells).

First ionisation energy

The energy required to remove one electron from an atom of an element in the gas phase is called the **ionisation energy**. For example, the ionisation energy of sodium is 494 kJ per mole of sodium atoms.

The **first ionisation energy** is the energy required to remove the first valence electron from an atom in the gas phase. Figure 3.3.6 shows the first ionisation energies of most main group elements.

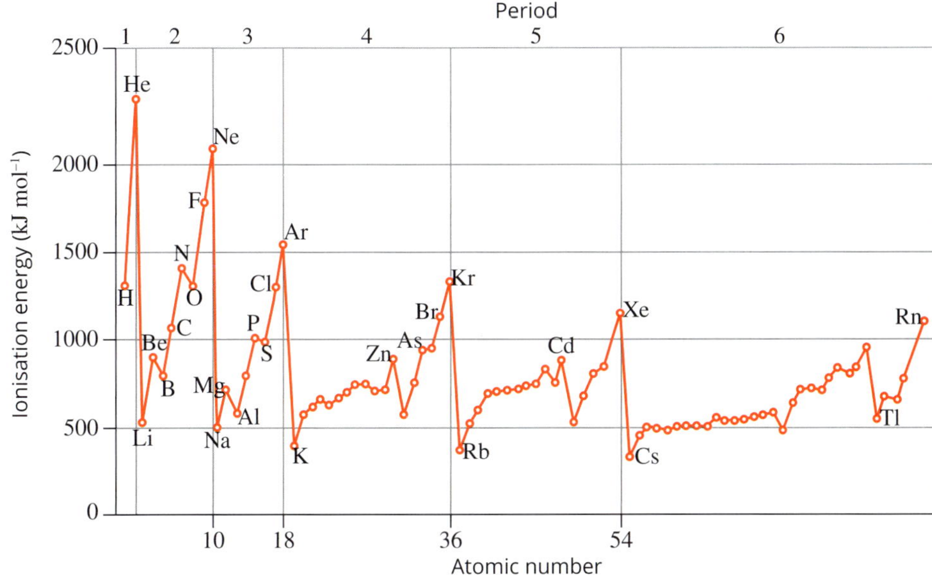

FIGURE 3.3.6 The first ionisation energy of a range of main group elements

The magnitude of the first ionisation energy reflects how strongly the valence electron is attracted to the nucleus of the atom. The more strongly the valence electron is attracted to the nucleus, the more energy is required to remove it from the atom and the higher the first ionisation energy.

The main group elements (groups 1, 2, 13–18) of the periodic table follow these general patterns. As the core charge increases across a period, so too does the first ionisation energy. As atomic radius increases down a group (due to the additional electron shell), the electrons are further from the nucleus. Therefore, the valence electrons in elements lower in a group can be removed more easily, meaning the first ionisation energy decreases. This is summarised in Table 3.3.3.

TABLE 3.3.3 The trend in first ionisation properties in groups and periods of the periodic table

	Trend in ionisation energy	Explanation
down a group	decreases	Core charge stays constant and the number of shells increases as you move down a group. Therefore, the valence electrons are less attracted to the nucleus as they are further from the nucleus. As a result, the energy required to overcome the attraction between the nucleus and the valence electron is less, and the first ionisation energy decreases down a group.
left to right across a period	increases	Core charge increases and the number of occupied shells remains constant as you move across a period. As a result, the valence electrons become more strongly attracted to the nucleus, and more energy is required to remove an electron. Therefore, the first ionisation energy increases across a period.

Successive ionisation energy

After the removal of the first valence electron through ionisation, it is possible to then remove the remaining valence electrons and also the electrons that occupy the inner shells if sufficient energy is supplied. The energy required to achieve the sequential removal of electrons from the atom is called the **successive ionisation energy**.

Consider a sodium atom with a total of 11 electrons. As each successive electron is removed, these ionisation energies can be measured. Figure 3.3.7 shows that the first electron to be removed has the lowest ionisation energy and is therefore the easiest to remove. The following eight electrons are slightly more difficult, and finally, the last two require substantially more energy to remove.

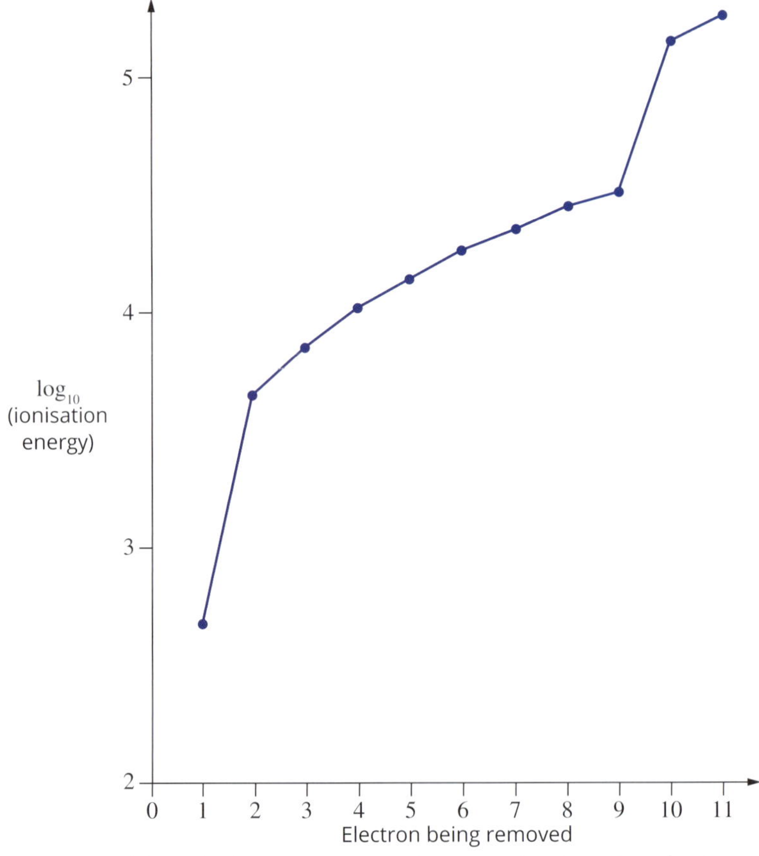

FIGURE 3.3.7 Graph of the ionisation energies of a sodium atom. (The logarithm of ionisation energy is used to provide a more convenient vertical scale.)

A sodium atom has the electron configuration of 2,8,1, which means there is only one valence electron in the third shell. As this electron is further away from the positive nucleus, and the electrostatic attraction to the nucleus it experiences is reduced by the shielding provided by the two filled inner shells of electrons, it is relatively weakly held in the atom compared to the other electrons and thus has the smallest ionisation energy.

The following eight electrons are all removed from the second shell of the sodium atom. Notice that there is a significant increase in ionisation energy from the first valence electron to the next eight electrons. As these eight electrons occupy the second shell, which is closer to the nucleus, they experience a greater electrostatic attraction than the first valence electron in the third shell and thus require a greater amount of energy to overcome this stronger attraction. Remember that, according to Coulomb's law, distance has a significant effect on the force of attraction between two charged particles. These electrons also receive less shielding from filled inner shells than the original valence electron in the third shell. This pattern is continued for the two electrons in the innermost shell. Being closest to the nucleus, they are the most tightly held electrons, requiring a much greater quantity of energy to remove them from the atom.

There is a gradual increase in the energy required to remove the second through to the eighth electrons. As these electrons all occupy the second shell, they all exist at similar distances from the nucleus, so the increasing energy requirement to remove them cannot be attributed to the changes in the distance to the nucleus. Each successive removal of an electron does, however, result in an increasing positively charged ion, as the protons in the nucleus no longer have sufficient electrons to balance their positive charge.

The removal of subsequent electrons from an increasingly positively charged ion (Figure 3.3.7) will require a greater quantity of energy to overcome the greater electrostatic attraction experienced by each electron (Table 3.3.4).

TABLE 3.3.4 The ionisation equations for the successive ionisations of sodium

Electron being removed	Ionisation equation	Ionisation energy ($kJ\,mol^{-1}$)
1	$Na \rightarrow Na^{+} + e^{-}$	496
2	$Na^{+} \rightarrow Na^{2+} + e^{-}$	4562
3	$Na^{2+} \rightarrow Na^{3+} + e^{-}$	6910
4	$Na^{3+} \rightarrow Na^{4+} + e^{-}$	9543
5	$Na^{4+} \rightarrow Na^{5+} + e^{-}$	13354
6	$Na^{5+} \rightarrow Na^{6+} + e^{-}$	16613
7	$Na^{6+} \rightarrow Na^{7+} + e^{-}$	20117
8	$Na^{7+} \rightarrow Na^{8+} + e^{-}$	25496
9	$Na^{8+} \rightarrow Na^{9+} + e^{-}$	28932
10	$Na^{9+} \rightarrow Na^{10+} + e^{-}$	141362
11	$Na^{10+} \rightarrow Na^{11+} + e^{-}$	159076

Ionisation energy is the energy required to completely remove an electron from an atom. This energy increases as more electrons are removed from the atom.

METALLIC CHARACTER

Metals conduct electricity and are usually solids at room temperature. Conversely, non-metallic elements usually do not conduct electricity and many are gases at room temperature.

The differences between the properties of metals and non-metals are related to the number of electrons in the outer shell of their atoms. In general, elements with atoms containing one, two or three valence electrons tend to behave as metals, whereas those with four or more valence electrons behave as non-metals.

Metals are located on the left side of the periodic table and non-metals are on the right. Elements known as the **metalloids** are located between the metals and non-metals. The metalloids exhibit both metallic and non-metallic properties. Silicon is one of the most abundant metalloids. It is a brittle solid, which is a common property of non-metals. However, it is also a semiconductor, meaning it can exhibit **electrical conductivity**, making it useful in many electronic devices such as computers and calculators.

The classification of an element as metallic or non-metallic, however, is not always clear. Across the periodic table, the change from metals to non-metals can be considered as a continuum. **Metallic character** is the name given to describe how closely an element exhibits the properties commonly associated with metals, namely, that it readily loses an electron to form a cation. This is closely related to ionisation energy.

As seen previously, the first ionisation energy decreases as you move down a group due to the increasing distance from the nucleus (due to additional electron shells) and from the right to left of a period (due to the reducing core charge). The elements with the most weakly held valence electrons, having the smallest first ionisation energy and the greatest metallic character, are towards the bottom left-hand side of the periodic table.

ELECTRONEGATIVITY

Electronegativity is the ability of an atom to attract electrons in a covalent bond towards itself. The more strongly the valence electrons of an atom are attracted to the nucleus of the atom, the greater the electronegativity. The strength of this attraction, as for all electrostatic attraction, depends on the magnitude of the charges involved and the distance between them. Figure 3.3.8 shows the electronegativity of many of the main group elements.

Electronegativity increases across a period.

Electronegativity decreases down a group.

1	2	13	14	15	16	17
Li	Be	B	C	N	O	F
1.0	1.6	2.0	2.6	3.0	3.4	4.0
Na	Mg	Al	Si	P	S	Cl
0.9	1.3	1.6	1.9	2.2	2.6	3.2
K	Ca	Ga	Ge	As	Se	Br
0.8	1.0	1.8	2.0	2.2	2.6	3.0
Rb	Sr	In	Sn	Sb	Te	I
0.8	1.0	1.8	2.0	2.1	2.1	2.7
Cs	Ba	Tl	Pb	Bi	Po	At
0.8	0.9	2.0	2.3	2.0	2.0	2.2
Fr	Ra					
0.7	0.9					

FIGURE 3.3.8 The electronegativity of elements generally decreases down a group and increases across a period, from left to right.

The values given for electronegativity are dimensionless quantities, that is, they have no unit and are calculated theoretically rather than measured experimentally.

The trends observed in the electronegativity of the elements are summarised in Table 3.3.5.

TABLE 3.3.5 Trends in electronegativity in groups and periods of the periodic table

	Trend in electronegativity	Explanation
down a group	decreases	The core charge stays constant and the number of shells increases down a group. Therefore, valence electrons are less strongly attracted to the nucleus as they are further from the nucleus. As a result, electronegativity decreases.
left to right across a period	increases	The number of occupied shells in the atoms remains constant but the core charge increases across a period. Therefore, the valence electrons become more strongly attracted to the nucleus. As a result, electronegativity increases.

Although electronegativity is related to the attraction of electrons within a covalent bond, the electronegativity values are useful in predicting the types of bonding in a chemical and are therefore an important concept in Chemistry. This will be covered in greater depth in Chapters 5 and 6.

3.3 Review

SUMMARY

- The attractive force between opposite electric charges is known as electrostatic attraction.
- The strength of an electrostatic attraction is dependent on the magnitude of the charges involved and the distance between them.
- Core charge or effective nuclear charge is the resultant attractive force experienced by valence electrons once the impact of the shielding effect provided by electrons in inner shells is taken into account.
- The core charge or effective nuclear charge is calculated by subtracting the total number of inner-shell electrons from the number of protons in the nucleus.
- Atomic radius is a measurement used for the size of atoms. It can be regarded as the distance from the nucleus to the outermost electrons.
- The first ionisation energy is the energy required to remove the first valence electron from an atom of an element in the gas phase.
- First ionisation energy decreases down a group but increases across a period.
- Successive ionisation energies are the energies required to remove multiple electrons consecutively from an atom of an element in the gas phase.
- Electronegativity is the ability of an element to attract electrons in a covalent bond towards itself.
- Table 3.3.6 summarises how certain properties have specific trends within the groups and periods of the periodic table.

TABLE 3.3.6 Property trends in the periodic table

Property	Down a group	Across a period (left to right)
core charge	no change	increases
atomic radius	increases	decreases
ionisation energy	decreases	increases
electronegativity	decreases	increases

- Many trends in the physical properties of elements in the periodic table can be explained using two key ideas.
  - From left to right across a period, the core charge of atoms increases, so the attractive force felt between the valence electrons and the nucleus increases.
  - Down a group, the number of shells in an atom increases so that the valence electrons are further from the nucleus and are held less strongly.

KEY QUESTIONS

1 What is the core charge of an atom of carbon?

2 Explain why atomic radius decreases as you move left to right across a period, yet the number of protons and neutrons in the nucleus increases.

3 a Explain the meaning of the term 'ionisation energy'.
  b What factors need to be considered when predicting the trend in first ionisation energy across a period?

4 Explain why the first ionisation energy increases from left to right across a period.

5 Figure 3.3.8 gives electronegativity values for the elements in groups 1, 2 and 13–17 of the periodic table.
  a Give the name and symbol of the element that has the:
    i highest electronegativity
    ii lowest electronegativity.
  b In which group do you see the:
    i greatest change in electronegativity as you go down the group?
    ii smallest change in electronegativity as you go down the group?
  c Why are the elements of group 18 usually omitted from tables that give electronegativity values?

6 Explain the relationship between electronegativity and core charge.

3.4 Quantisation of energy

The fact that metals, when heated strongly, emit light of a specific colour that is characteristic of the given element was a well-known phenomenon throughout the early 1800s. In fact, scientists were using this practice to confirm the presence of certain metals in substances long before Bohr proposed a model of the atom that enabled its explanation.

QUANTISED ENERGY OF ELECTRONS

What allowed Bohr's model to explain these light emissions was that it was able to identify that the electrons in atoms can only be found at certain discrete energies. These are the energies associated with the shells in which the electrons can be found. The energy levels of the electrons are described as being **quantised**.

Bohr's model allows electrons to jump between these discrete energy levels or shells. They accomplish this by either absorbing or emitting energy.

Consider a hydrogen atom with one proton and one electron. Usually, the electron exists in the $n = 1$ shell. This is the lowest energy state of the atom and is called the **ground state**.

When the hydrogen atom is heated strongly, the electron absorbs the energy and jumps to a higher energy level. This higher energy state is known as the **excited state**.

The higher energy excited state is inherently unstable and remains for only fractions of a second. The excited atom then emits the recently absorbed energy as the electron returns to a lower energy level. The energy that is emitted by the atom is in the form of electromagnetic radiation.

Electromagnetic radiation

Visible light is a form of energy and a type of **electromagnetic radiation**. Other forms of electromagnetic radiation include radio waves and X-rays. The range of different forms of electromagnetic radiation is referred to as the **electromagnetic spectrum**, as seen in Figure 3.4.1.

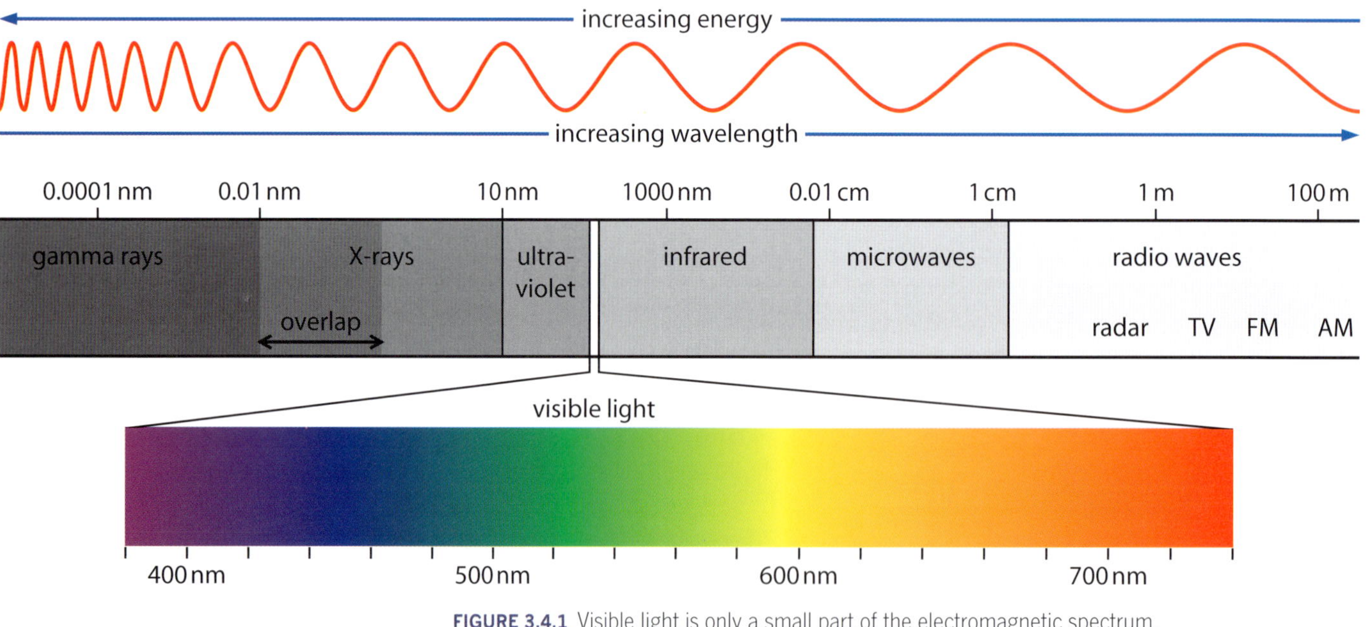

FIGURE 3.4.1 Visible light is only a small part of the electromagnetic spectrum.

Radiation from each portion of the electromagnetic spectrum can be described in terms of its frequency, wavelength and energy. Different colours of light can also be described in terms of different frequencies, wavelengths and energies. For example, compared to red light, violet light has a higher energy and higher frequency, but a shorter wavelength.

While excited atoms returning to their ground state emit electromagnetic radiation, atoms in their ground state can also absorb certain amounts of electromagnetic radiation (**absorbance**).

By observing and measuring either the electromagnetic radiation absorbed by atoms in their ground state or that is emitted by excited atoms, scientists can build up a picture of the energy levels present in the atom. As the arrangement and interactions of the electric charges in an atom are unique to that particular element, so is the energy of the absorbed or emitted electromagnetic radiation. This electromagnetic radiation can be observed as either absorption spectra or emission spectra.

CHEMFILE

Uranium glass

Uranium is a heavy, dense, grey metal, which is commonly used as a fuel for nuclear power due to its radioactive U-238 isotope.

From around the late 1800s to early 1900s, uranium oxide was popular as a glass colourant, producing a range of yellow and green decorative glass. Most pieces usually only contain trace levels of uranium, around 2%, but some pieces contain up to 25% uranium. Uranium glass is slightly radioactive but is still considered safe.

A peculiar property of uranium glass is it fluoresces bright green when exposed to ultraviolet light (Figure 3.4.2). The process of fluorescence is similar to that of atoms emitting radiation as their electrons return from an excited state to a lower energy state, except that in the case of fluorescence, the emitted radiation is of a different colour from the incident radiation.

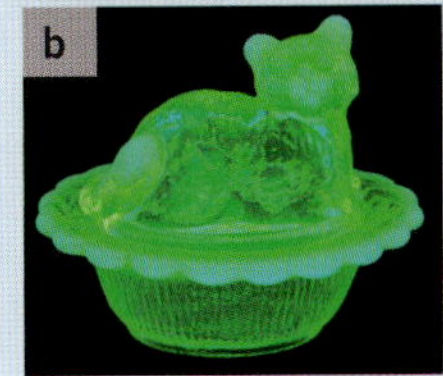

FIGURE 3.4.2 A piece of uranium glass viewed (a) under visible light and (b) under ultraviolet light

ABSORPTION SPECTRA

Electrons in energy levels close to the nucleus have the lowest energies and experience the strongest attraction to the nucleus. An electron can jump up to a higher energy level if it absorbs energy that corresponds exactly to the difference in energy between the lower energy level and the higher energy level (Figure 3.4.3).

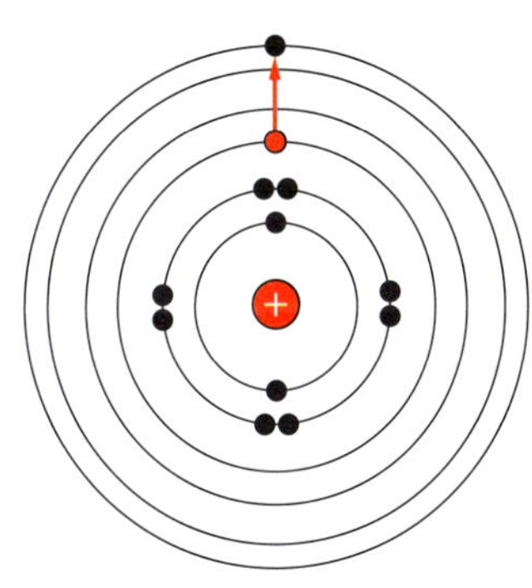

FIGURE 3.4.3 Absorbing energy can cause an electron to jump to a higher energy level.

When an electron jumps to a higher energy level, it is not necessarily the level directly above it. For example, in the case of a hydrogen atom with a single valence electron, if it is exposed to sufficient energy, the valence electron can jump to the second, third or even fourth energy level. As these energy levels represent distinct, quantised energies, the energy required to promote a ground-state electron to these higher levels are all different, but identical for any hydrogen atom.

The different quantities of energy required to promote these ground-state electrons are supplied by electromagnetic radiation of varying wavelengths and frequencies, which are observed as different colours of light (Figure 3.4.4).

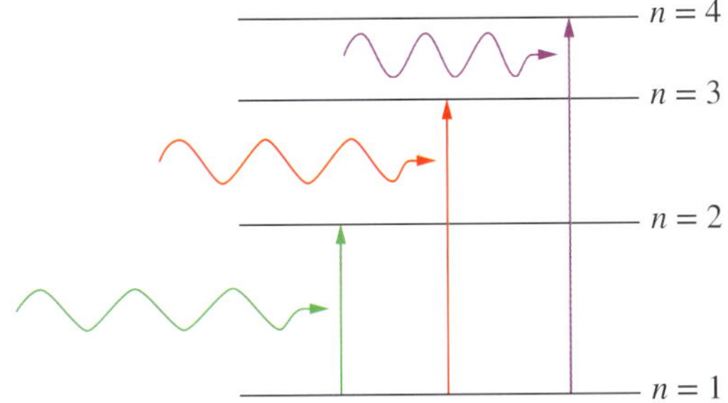

FIGURE 3.4.4 Energy of a specific wavelength and frequency corresponds to jumps between specific energy levels (electron shells) in the atom.

FIGURE 3.4.5 The absorption spectrum of hydrogen

> The black lines in an absorption spectrum represent electrons gaining energy as they move to higher energy levels in the atom.

A sample of hydrogen atoms in their ground state exposed to a continuous spectrum of visible light consisting of every possible wavelength will only absorb certain colours. These colours have energies that perfectly correspond to the energy needed to promote an electron to a higher energy level. The individual colours of light in this continuous spectrum that are absorbed by the hydrogen atoms are known as **absorption lines**. Collectively, these absorption lines form an **absorption spectrum**. The absorption spectrum for hydrogen is shown in Figure 3.4.5.

The dark regions in the spectrum correspond to the colours of light that provide the exact quantity of energy required to promote its valence electrons to a higher energy level. Even though there is only one electron in each hydrogen atom, there are several absorptions lines because each individual line is caused by the promotion of an electron to the second, third or fourth energy level or higher. In any sample of hydrogen there will be many hydrogen atoms and this absorption spectrum represents the collective absorptions of all individual hydrogen atoms.

CHEMISTRY IN ACTION

Discovering the chemical composition of the Sun and the planets

In 1814, Joseph von Fraunhofer, a German physicist and skilled glassmaker, noticed a series of dark regions in the visible spectrum of the Sun. He made a number of systematic measurements of these lines; in total, cataloguing more than 570 distinct lines.

About 45 years later, Gustav Kirchhoff, another German physicist, and Robert Bunsen, a German chemist, noticed that the lines Fraunhofer documented coincided with the characteristic emission lines of a number of elements. They correctly deduced that the dark regions are absorption lines of the chemical elements in the Sun. Thus they were able to determine the chemical composition of the Sun. These absorption lines are known as Fraunhofer lines (Figure 3.4.6) in his honour.

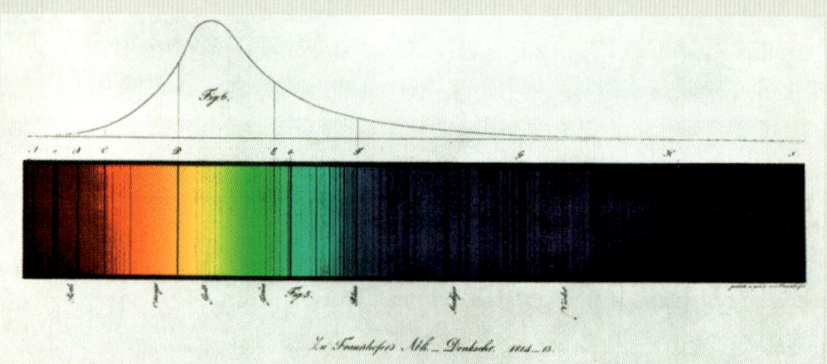

FIGURE 3.4.6 Fraunhofer's original documentation of the Sun's absorption lines

These principles are still employed today in astronomical spectroscopy where astronomers measure the electromagnetic spectra emitted by stars and other hot celestial bodies to determine their chemical composition.

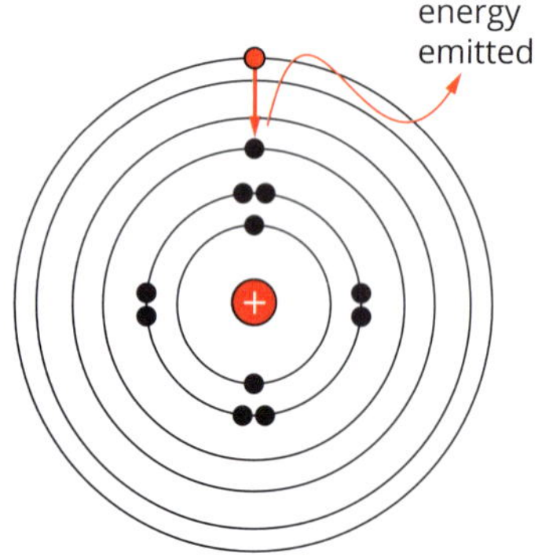

FIGURE 3.4.7 An excited electron quickly returns to a lower energy level, emitting electromagnetic energy such as light of a particular energy.

EMISSION SPECTRA

Heating an element vigorously can cause an electron to absorb energy and jump to a higher energy state. Shortly afterwards, the electron returns to a lower energy state. As the electron falls to a lower energy shell, it emits energy in the form of light that corresponds exactly to the difference in energy between the higher and the lower energy levels (Figure 3.4.7).

The electron can return to its original ground state in a number of different pathways. The electron may return directly to the ground state or may move to other energy levels before returning to the ground state (Figure 3.4.8). For example, an electron in the $n = 4$ shell may move to the $n = 2$ shell before returning to the $n = 1$ ground state.

As these energy levels represent distinct, quantised energies, each transition the electron makes is associated with the emission of a different quantity of energy in the form of different colours of light. These colours will be consistent for all hydrogen atoms.

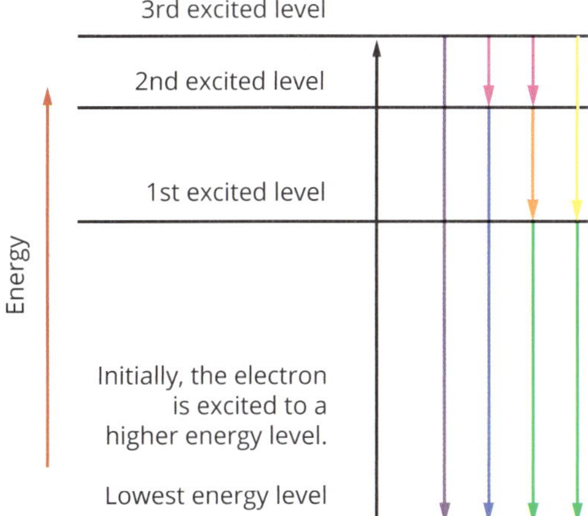

FIGURE 3.4.8 Emission of energy as electrons move from a higher to a lower energy state

So if a sample of hydrogen atoms is heated to an excited state and its valence electron jumps to some higher energy level, moments later as it returns to a lower energy level it will emit a colour of light with an energy that perfectly corresponds to the energy difference between the two levels. The individual colours of light that are emitted by the hydrogen atoms are known as **emission lines** and collectively these emission lines form an **emission spectrum**. The emission spectrum for hydrogen is shown in Figure 3.4.9.

FIGURE 3.4.9 The emission spectrum of hydrogen

Even though a hydrogen atom has only a single electron, there are several lines in the spectrum because electrons can jump between different energy levels. As is the case in absorption spectra, this emission spectrum represents the collective emissions of many individual hydrogen atoms.

The coloured lines in an emission spectrum represent electrons losing energy as they move back down to lower energy levels in the atom.

CHEMFILE

How helium got its name

Helium was discovered on the Sun before being discovered on Earth. The French astronomer Jules Janssen discovered helium in 1868 while studying the light from a solar eclipse in India similar to the solar eclipse shown in Figure 3.4.10.

FIGURE 3.4.10 The emission spectrum of helium was first detected in the sunlight of a solar eclipse like the one shown here.

Although the spectrum showed the full range of colours, the bright yellow line in the helium spectrum (Figure 3.4.11) stood out. The line could not be matched to the line spectra of any of the known elements. It was concluded that the line belonged to an unknown element. This element was named helium after the Greek sun god, Helios.

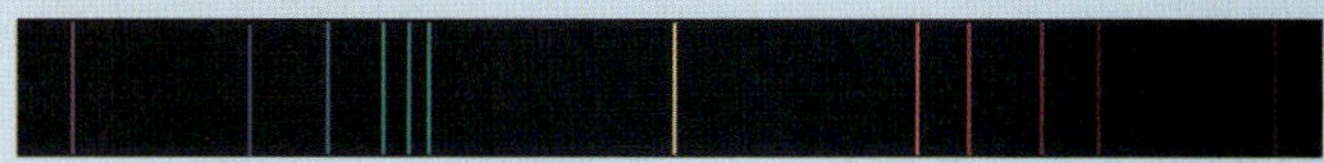

3.4.11 The emission spectrum of helium is made up of lines ranging from violet to red in colour.

3.4 Review

SUMMARY

- The Bohr model of the atom was the first atomic model to explain the origin of emission spectra.
- Bohr's model of the atom introduced the quantisation of energy, that is, the energy of the electrons only exist at certain discrete levels.
- The lowest or minimum energy state of an atom is called the ground state.
- An elevated or higher energy state an atom can find itself in is called an excited state.
- When an electron absorbs energy (e.g. heat or light), the electron can jump from one energy level (shell) to a higher energy level.
- Radiation from each portion of the electromagnetic spectrum can be described in terms of its frequency, wavelength and energy. Different colours of light have different frequencies, wavelengths and energies.
- The range of colours of light absorbed by electrons moving into higher energy levels are called absorption spectra.
- When an electron falls from a higher energy level to a lower energy level, it emits energy in the form of light.
- The range of colours of light emitted by electrons moving into lower energy levels is called emission spectra.
- Each element has a unique absorption/emission spectrum.
- Each line in the absorption/emission spectrum corresponds to a specific electron transition between two energy levels.

KEY QUESTIONS

1 Explain how emission spectra provide evidence for electron shells in the Bohr model.

2 Explain why the lines in an emission spectrum are different colours.

3 Which of the following represents the electron configuration of a carbon atom in its ground state?

A 2,4
B 2,5
C 2,3,1
D 2,4,2

4 Which of the following represents the electron configuration of a carbon atom in its excited state?

A 2,4
B 2,5
C 2,3,1
D 2,4,2

5 Do a sodium ion (2,8) and a neon atom (2,8) display the same emission spectra? Explain your answer.

6 Do all isotopes of an element display the same emission spectra? Explain your answer.

3.5 Spectroscopy

As each element has a different number of protons in the nucleus, the attraction of the nucleus for electrons will vary from element to element. No two elements have energy levels of exactly the same energy, so an emission or absorption spectrum is characteristic of a particular element. It may be used as a 'fingerprint' to identify the elements present in a substance.

The study of the interaction between matter and electromagnetic radiation is known as **spectroscopy** and is the basis for a number of qualitative and quantitative analytical techniques.

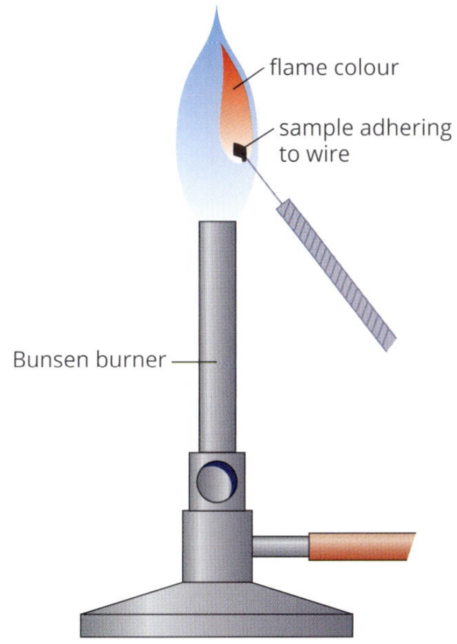

FIGURE 3.5.1 Performing a flame test. A moist wire has been dipped in the sample and then placed in the flame. A fine spray of solution from a spray bottle could be used instead.

FLAME TESTS

Chemists use the fact that some metals produce particular colours when they are heated as a convenient and simple method of analysis. These different colours are caused by the different electron transitions occurring in the different atoms. The metallic elements present in a compound can often be determined just by inserting a sample of the compound into a non-luminous Bunsen burner flame in a test known as a **flame test** (Figure 3.5.1).

The metal is identified by comparing the flame colour with the characteristic colours produced by metals. Some examples of the flame colours can be seen in Figure 3.5.2.

Table 3.5.1 below shows some of the flame colours that can be used to identify the presence of metals using a flame test. It is important to note that only a small number of metals produce flame colours as the emission lines of some elements are not within the visible spectrum of light.

FIGURE 3.5.2 The colour of a flame is different in the presence of different metal compounds and can be used to identify the presence of these metals. The flame colours shown here indicate the presence of (left to right) lithium (red), strontium (red), sodium (yellow), copper (green) and potassium (lilac).

TABLE 3.5.1 Common metal flame-test colours

Metal	Flame colour
sodium	yellow
strontium	scarlet
copper	green
barium	yellow-green
lithium	crimson
calcium	red
potassium	lilac

While the flame test provides a quick analysis of the chemical composition of a sample, there are a number of limitations with this technique.

- The flame test is a qualitative test only.
- Only a small range of metals are detectable by the flame test.
- Metals in low concentrations may be difficult to observe.
- Mixtures of metals will produce confusing results.

ATOMIC ABSORPTION SPECTROSCOPY

There is a direct mathematical relationship between the concentration of the sample being analysed and the intensity of light it absorbs. This relationship is used in the technique of **atomic absorption spectroscopy (AAS)**.

In AAS, the sample being analysed is vaporised in a flame to produce free atoms in their ground state. This cloud of ground-state atoms then passes through the optical path of a light source, which produces the emission spectrum of the particular chemical being examined.

As the light source produces light of the same energy required to raise ground-state atoms to an excited state, the cloud of gaseous atoms absorb some portion of the incident light.

The intensity of the absorbed light is measured, giving an indication of the concentration in the original sample.

Atomic absorption spectroscopy is employed by many modern-day chemists (Figure 3.5.3) and addresses some of the limitations of flame tests.

- AAS provides both a qualitative and quantitative test.
- More than 70 elements can be analysed by AAS.
- AAS can detect elements in concentrations as low as micrograms per litre.
- AAS is highly selective and is regularly used with mixtures.

Flame tests and atomic absorption spectroscopy can be used to identify the presence of certain atoms in a chemical sample.

FIGURE 3.5.3 An analytical chemist uses AAS to determine the concentration of metals in food.

CHEMFILE

Sir Alan Walsh

The modern form of AAS was largely developed in the early 1950s by a team of Australian researchers at the CSIRO lead by Sir Alan Walsh.

At the time, the main analysis technique focused on the emission of characteristic radiation, rather than the absorption of it. After a 'eureka' moment in his garden, Walsh constructed a working prototype the very next day. Over the next few years, Walsh and his researchers released a number of publications detailing their new technique.

Unfortunately, there was very little interest from other scientists. Walsh's reaction, as documented in a 1970 interview was, 'It was only because, really, of the cold shoulder treatment we were given that we got a bit mad and got involved in proving a point. So if the world had reacted differently, we might have come out of this with nothing'.

Alan Walsh was knighted in 1977 for 'services to science', the year of his retirement from the CSIRO in Melbourne.

FIGURE 3.5.4 Sir Alan Walsh, the inventor of atomic absorption spectroscopy

3.5 Review

SUMMARY

- Flame tests are based on the ability of electrons to release absorbed energy as they move from higher to lower energy levels.
- Flame tests are performed by inserting a sample in a non-luminous Bunsen burner flame.
- Flame tests can be used to detect the presence of a small number of metal elements.
- Atomic absorption spectroscopy (AAS) is based on the ability of electrons to absorb energy as they move from lower to higher energy levels.
- AAS can be used to detect a large range of elements, even in samples with low concentrations.

KEY QUESTIONS

1 What is spectroscopy?

2 Describe the advantages and disadvantages of a flame test.

3 In atomic absorption spectroscopy, what is the source of the energy that excites the atoms in the sample being analysed?

4 What are some of the advantages of atomic absorption spectroscopy over a flame test?

5 Flares are small pyrotechnic devices used for signalling or illumination. Red flares are traditionally used in maritime environments to indicate distress. Use Table 3.5.1 on page 67 to suggest which metal is likely present in the flare's chemical composition.

Chapter review

03

KEY TERMS

absorbance
absorption line
absorption spectrum
anion
atomic absorption spectroscopy (AAS)
atomic radius
Bohr diagram
Bohr model
cation
core charge
effective nuclear charge
electrical conductivity
electromagnetic radiation
electromagnetic spectrum
electron configuration
electron shell
electronegativity
emission line
emission spectrum
energy level
excited state
first ionisation energy
flame test
ground state
group
inert
ion
ionisation
ionisation energy
main group element
metallic character
metalloid
noble gas
octet rule
outermost shell
period
quantised
spectroscopy
successive ionisation energy
transition metal
valence electron
valence shell

Electronic structure of atoms

1 What is the fundamental difference between the planetary model of the atom put forward by Rutherford and Bohr's model of the atom?

2 Bohr's model of the atom describes an atom as consisting of rapidly moving electrons at a relatively large distance from a very small central nucleus. What is there *between* those electrons and the nucleus?

Electron arrangement in the periodic table

3 Use the periodic table in Figure 3.2.2 on page 47 to answer these questions.

a In which groups of the periodic table are the following found?

i B
ii Cl
iii Na
iv Ar
v Si
vi Pb

b In which periods of the periodic table are the following found?

i K
ii F
iii He
iv H
v U
vi P

c What is the name, symbol and electron configuration of the following?

i second element in group 14
ii second element in period 2
iii element that is in group 18 and period 3

4 Determine the elements with the following electron configurations.

a 2,2
b 2,7
c 2,8,4
d 2,8,8,2

5 Name an element with properties similar to those of:

a carbon
b magnesium
c iodine
d phosphorus.

6 Main group elements are elements that belong to specific groups in the periodic table. Which groups are these?

7 How many valence electrons are in atoms of elements found in the following groups?

a group 1
b group 15
c group 17
d group 2

Trends in the periodic table

8 Across a period, the number of subatomic particles in an atom increases but the size of an atom decreases. Why?

9 Account for the fact that it takes more energy to remove an electron from the outer shell of atoms of:

a phosphorus than magnesium

b fluorine than iodine.

10 Consider the elements in period 2 of the periodic table: lithium, beryllium, boron, carbon, nitrogen, oxygen, fluorine. Describe the changes in the following qualities that occur across the period.

a the sizes of atoms

b metallic character

c electronegativity

11 From each set of elements, select the element that has the largest first ionisation energy.

a phosphorus, arsenic, nitrogen

b silicon, chlorine, sulfur

c bromine, chlorine, sulfur

12 Explain which of oxygen, chlorine and fluorine will have the largest first ionisation energy.

13 Why is the ionisation energy of an atom always measured for gaseous atoms?

14 When a sodium atom forms a cation, is the radius of the resultant ion larger or smaller than the original atom? Explain the reasoning for your answer.

15 When a fluorine atom forms an anion, the radius of the ion is larger than the original atom. Suggest a reason why this is so.

Quantisation of energy

16 In a hydrogen atom, which electron shell will the electron be in if the atom is in the ground state?

17 For any given atom, which is likely to require the greater amount of energy and why: the excitation of the atom or the ionisation of the atom?

18 Suggest a reason for the slightly blue colour present in a non-luminous Bunsen burner flame.

19 When Kirchhoff and Bunsen reviewed Joseph von Fraunhofer's catalogue of absorption lines in sunlight, they found absorption lines characteristic of oxygen. Considering there is no oxygen in outer space and very little in the Sun, explain the likely cause of these absorption lines.

20 Why does an emission spectrum contain a number of lines of different colours?

Spectroscopy

21 Why is it important to conduct a flame test in a non-luminous flame?

22 What would be the easiest way to produce the characteristic radiation required for atomic absorption spectroscopy?

23 Why would a sample containing a number of different metals not be suitable for analysis by flame test but acceptable for analysis by atomic absorption spectroscopy?

Connecting the main ideas

24 New models for the atom evolve as scientists become aware of inconsistencies between current models and experimental data. Outline the problems with the existing model of the atom that led to the modifications suggested by the following scientists.

a Rutherford

b Bohr

25 What concepts or ideas are these two versions of the periodic table below trying to convey?

a

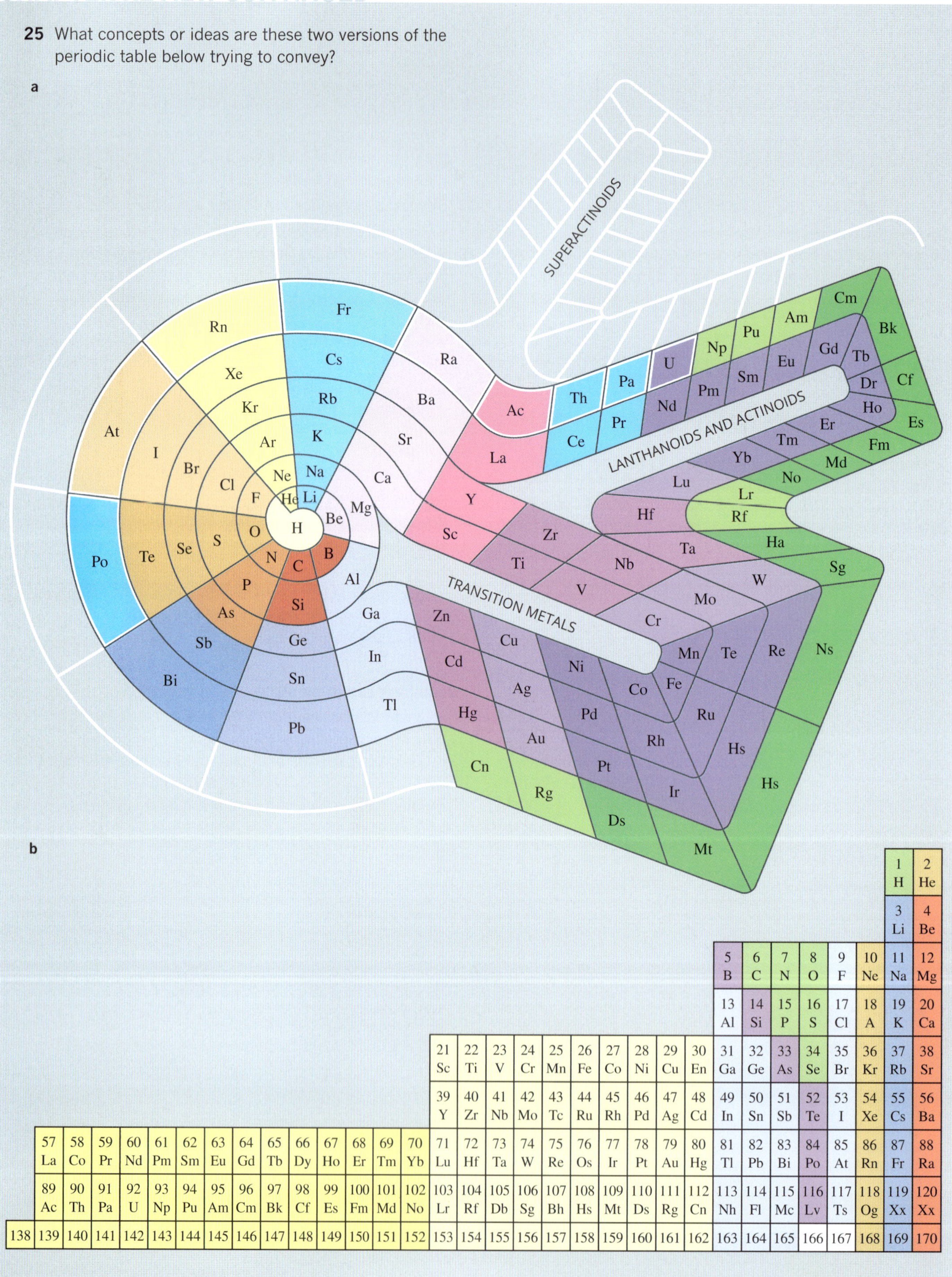

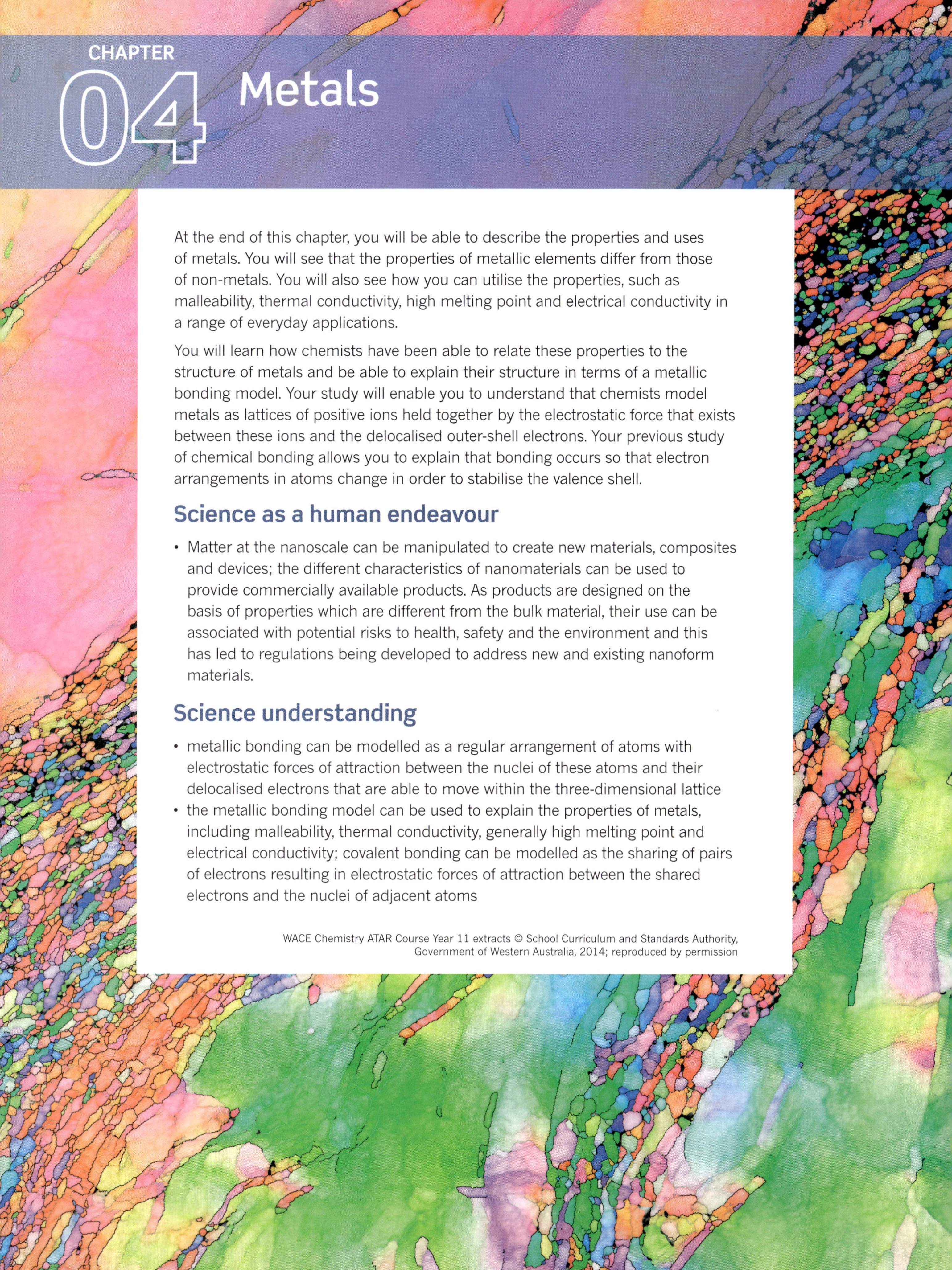

CHAPTER 04

Metals

At the end of this chapter, you will be able to describe the properties and uses of metals. You will see that the properties of metallic elements differ from those of non-metals. You will also see how you can utilise the properties, such as malleability, thermal conductivity, high melting point and electrical conductivity in a range of everyday applications.

You will learn how chemists have been able to relate these properties to the structure of metals and be able to explain their structure in terms of a metallic bonding model. Your study will enable you to understand that chemists model metals as lattices of positive ions held together by the electrostatic force that exists between these ions and the delocalised outer-shell electrons. Your previous study of chemical bonding allows you to explain that bonding occurs so that electron arrangements in atoms change in order to stabilise the valence shell.

Science as a human endeavour

- Matter at the nanoscale can be manipulated to create new materials, composites and devices; the different characteristics of nanomaterials can be used to provide commercially available products. As products are designed on the basis of properties which are different from the bulk material, their use can be associated with potential risks to health, safety and the environment and this has led to regulations being developed to address new and existing nanoform materials.

Science understanding

- metallic bonding can be modelled as a regular arrangement of atoms with electrostatic forces of attraction between the nuclei of these atoms and their delocalised electrons that are able to move within the three-dimensional lattice
- the metallic bonding model can be used to explain the properties of metals, including malleability, thermal conductivity, generally high melting point and electrical conductivity; covalent bonding can be modelled as the sharing of pairs of electrons resulting in electrostatic forces of attraction between the shared electrons and the nuclei of adjacent atoms

4.1 Properties of metals

More than 80% of the elements in the periodic table are metals.

Metals have been important to human beings since early times. The development of civilisation can be measured by the way we have used metals. The Copper Age (5000–3000 BCE) was followed by the Bronze Age (3000–1000 BCE) and the Iron Age (from 1000 BCE).

Gold, silver and copper can be found on Earth in an almost pure form. These metals were employed by prehistoric humans to make ornaments, tools and weapons. As humans' knowledge of metallurgy (the science of modifying metals) has developed, metals have played a central role in fields as diverse as construction, agriculture, art, medicine and transport.

The diverse properties of different metals make them suitable for many purposes. Table 4.1.1 shows the properties and uses of some metals. For example, titanium (Figure 4.1.1) is a very strong, relatively unreactive metal with a low density that is close to that of bone. Consequently, it is used in surgical implants that can last up to 20 years with little effect on the body. Titanium is also used in the aerospace industry, in art and architecture, and in sporting equipment such as golf clubs.

TABLE 4.1.1 Properties and uses of some metals

Metal	Properties	Uses
iron	soft, malleable, magnetic, good thermal and electrical conductor, fairly reactive, readily forms alloys	can corrode and is usually converted to more stable steel, which is used in buildings and bridges, automobiles, machinery and appliances
aluminium	low density, relatively soft when pure, excellent thermal and electrical conductor, malleable and ductile, good reflector of heat and light, readily forms alloys	saucepans, frying pans, drink cans, cooking foil, food packaging, roofing, window frames, appliance trim, decorative furniture, electrical cables, aircraft and boat construction
titanium	very strong, high melting point, low density, low reactivity, readily forms alloys	medical devices within the body, wheelchairs, computer cases; lightweight alloys are used in high-temperature environments such as spacecraft and aircraft
gold	shiny gold appearance, excellent thermal and electrical conductor, unreactive, readily forms alloys	electrical connections, jewellery, monetary standard, dentistry

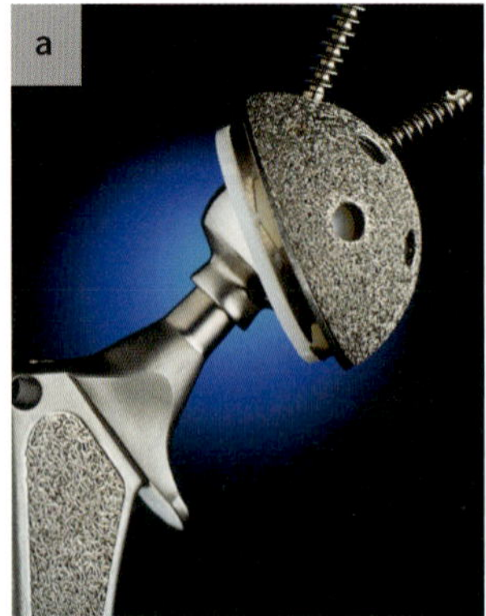

FIGURE 4.1.1 Titanium has many uses: (a) a replacement hip socket, (b) the spectacular curved space museum building in Moscow and (c) the SR-71 Blackbird reconnaissance aircraft. The SR-71 aircraft was the fastest piloted aircraft that used oxygen directly from the atmosphere.

In this section, you will examine the properties of metals. Then, in the next section, you will learn about the bonding model that chemists have developed to explain these properties. This model has helped chemists and materials engineers to understand why metals behave the way they do and how metals can be modified to create useful new materials.

PROPERTIES OF METALS

Table 4.1.2 gives the properties of some metals and non-metals. Despite the different properties of metals, most metals:

- exhibit a range of melting points and relatively high boiling points
- are good **conductors** of electricity
- are good conductors of heat
- generally have high densities.

TABLE 4.1.2 Properties of some metallic and non-metallic elements

Element	Melting point (°C)	Boiling point (°C)	Electrical conductivity ($MS\,m^{-1}$)*	Thermal conductivity ($J\,s^{-1}\,m^{-1}\,K^{-1}$)†	Density ($g\,mL^{-1}$)
Metals					
gold	1063	2970	45	310	19.3
iron	1540	3000	9.6	78	7.86
mercury	–39	357	1	8.4	13.5
potassium	64	760	14	100	0.86
silver	961	2210	60	418	10.5
sodium	98	892	21	135	0.97
Non-metals					
carbon (diamond)	3550	‡	10^{-17}	–	3.51
oxygen	–219	183	–	0.026	1.15 (liquid)

*$MS\,m^{-1}$ = megasiemens per metre.
†Thermal conductivity measures the conductance of heat.
‡Diamond sublimes (changes straight from a solid to a gas) when heated.

Not all metals have all of these properties. Mercury is a liquid at room temperature—it has an unusually low melting point. The group 1 elements (the **alkali metals**) have some properties that make them different from most other metals. They are all soft enough to be cut with a knife and they react vigorously with water to produce hydrogen gas. Both mercury and the group 1 elements exhibit most of the other properties listed above and are classified as metals.

Metals also generally have the following characteristics in common.

- They are **malleable**—they can be shaped by beating or rolling.
- They are **ductile**—they can be drawn into a wire.
- They are lustrous or reflective when freshly cut or polished.
- They are often hard, with high **tensile strength** (the force required to pull a material to the point where it breaks).
- They have low ionisation energies (a measure of the energy required to remove an electron from a gaseous atom) and electronegativities (a measure of the tendency of an atom to attract a bonding pair of electrons).

These properties can allow different metals to be used together in order to solve many engineering problems. The power transmission tower in Figure 4.1.2 is made of a few metals to take advantage of their different properties.

FIGURE 4.1.2 This power transmission tower relies on the strength of iron in steel for its structural integrity. The electricity cables are made from aluminium, utilising its ductility and electrical conductivity.

Most metals are good conductors of electricity and heat. They are malleable, ductile, and have high tensile strength and low ionisation energies.

Generally, metals are shaped for use in different applications by hammering, exploiting their malleability. Some metals, such as gold, copper and aluminium, are very malleable at room temperature. Other metals, such as iron, must be heated before they can be shaped.

Most metals are similar in appearance, being lustrous (reflective) and silvery-grey. Gold and copper are notable exceptions. Gold is a yellow coloured metal; copper is reddish.

CHEMFILE

Beryllium

Beryllium is the fourth element in the periodic table. It is one of the lightest metals, its density being two-thirds the density of aluminium. Beryllium is non-magnetic and has six times the stiffness of steel. The Space Shuttle and the Spitzer Space Telescope both use beryllium due to its strength and light weight. NASA's next-generation James Webb Space Telescope shown in Figure 4.1.3, scheduled for launch in 2018, will depend on a 6.5 metre mirror constructed using beryllium to see objects 200 times fainter than those previously visible.

FIGURE 4.1.3 Some of the 18 mirror segments of the James Webb Space Telescope. The mirrors are supported by beryllium ribs that maintain the mirror's shape under extreme conditions.

FORMING METAL IONS

In Chapter 3, you saw that there is a general tendency for atoms to combine so that they have eight electrons in the outer shell. This is called the octet rule.

Metallic elements are found on the left-hand side of the periodic table. The atoms of metals are generally larger than the atoms of non-metallic elements within a period and the core charge of their atoms is lower. The core charge of an atom is a measure of the attractive force felt by the outer-shell electrons towards the nucleus. It takes less energy to remove electrons from an outer shell when an atom is large, so the ionisation energy of metals is usually lower than for non-metals in the same period. As a consequence, metal atoms tend to lose their outer-shell electrons to form positive ions, called cations.

Metals tend to lose valence electrons and form cations.

Atoms of simple metals have one, two or three electrons in their outer shell. The cations that are formed when these metal atoms lose these valence electrons have a stable noble gas electron configuration, with eight electrons in their outer shell.

Worked example 4.1.1

DETERMINING CHARGES

Determine the charge of a calcium cation.	
Thinking	**Working**
Unreacted calcium atoms have the same number of protons and electrons.	Atomic number (Z) of calcium is 20: number of protons is 20, number of electrons is 20.
The electrons in an atom are in shells.	Shell configuration of calcium: 2,8,8,2
Only the outer-shell electrons will be lost.	Outer shell contains two electrons, $20 - 2 = 18$ electrons remaining.
Cation charge = number of protons – number of electrons	Cation charge $= 20 - 18 = +2$

Worked example: Try yourself 4.1.1

DETERMINING CHARGES

Determine the charge of an aluminium cation.

TRANSITION METALS

Between group 2 and group 13 in the periodic table is a block of elements known as the transition metals (Figure 4.1.4.). They include metals such as iron and nickel that are used to build bridges, cars and railway lines, and precious metals such as silver and gold that have ornamental and economic uses. Most transition metals are silver-coloured and are similar in appearance, as can be seen in Figure 4.1.5.

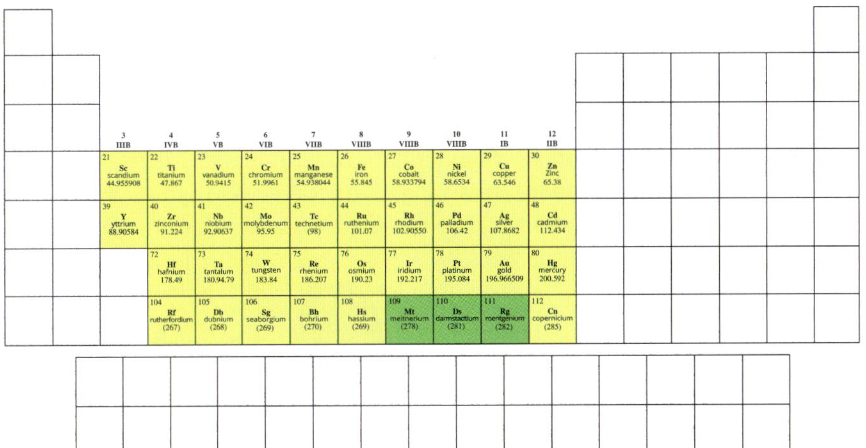

FIGURE 4.1.4 The transition metals are situated in the centre of the periodic table.

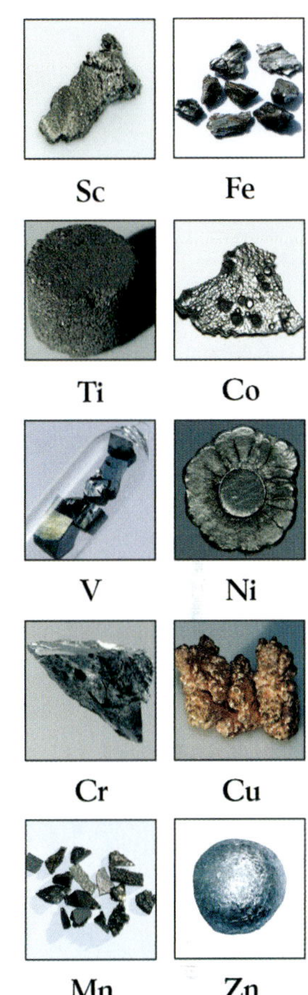

FIGURE 4.1.5 The first row of transition metals

The transition metals are very important to Australian industry. All of the metals in the first row are found in Australia and many are being mined today.

Iron is by far the most important metal to us. Nearly 10 times more iron is mined than all other metals combined. Iron obtained directly from a blast furnace is relatively **brittle** (meaning it snaps easily) and corrodes easily. Carbon and transition metals are combined with iron to produce mixtures or alloys, called **steel**. Steel has more desirable characteristics than pure iron. (Alloys are covered in more detail in Section 4.4.)

Copper is one of the few transition metals that is mainly used in its pure form. It is highly conductive and so it is used for most of the millions of kilometres of electrical wires that enable the transmission of electric energy for heating, lighting, telephones, radio and television.

Transition metals are not only important for industry, they are also important for life. All the transition metals in the first row, except scandium and titanium, are essential for animal life. Your body relies on the presence of trace elements to carry out certain biochemical reactions. For example, chromium, which you get from meat and bread, assists in the production of energy from glucose.

Properties of transition metals

Compared to the main group metals, transition metals have the following properties.

- They tend to be harder.
- They have higher densities.
- They have higher melting points.
- Some of them have strong magnetic properties.

The hardness, higher densities and higher melting points are due to the atoms of transition metals generally being a smaller size due to their greater core charge. This allows them to pack together more tightly with stronger bonds.

The high tensile strength of transition metals makes them suitable for use in the construction of buildings, cars, bridges and numerous other objects.

EXTENSION

Transition metal compounds

Transition metal compounds display a wide range of different colours. They are extensively used as pigments in paints, and to colour glass, ceramics and enamel. In Figure 4.1.6, the colours used by the artists are caused by the different transition metals present and the colours are still as vivid today as when they were painted.

FIGURE 4.1.6 (a) Arthur Streeton's *Golden summer, Eaglemont*, 1889. (b) An Australian Aboriginal painting

Ochre is a type of hard clay that contains iron oxides and hydroxides, which can be found naturally in many colours, including red, pink, white and yellow. Ground into a powder and mixed with liquids, ochre forms a paste that has been used for millennia by Aboriginal and Torres Strait Islander people for body decoration, cave painting, bark painting and other artwork.

The colours of many gemstones are also due to the presence of transition metals. For example, sapphires (Figure 4.1.7) contain traces of titanium and iron in a crystal lattice of aluminium oxide. The particles in a crystal lattice are arranged in ordered and symmetrical arrangements, which are repeated at regular intervals.

FIGURE 4.1.7 Blue sapphires get their colour from impurities of titanium and iron.

The colours arise when electrons within the metal ions in the compounds absorb light of particular wavelengths and move to higher energy levels. Absorbance of light with some wavelengths and transmission of light with other wavelengths results in the compounds appearing coloured. By contrast, compounds of group 1 and group 2 metals are usually colourless.

4.1 Review

SUMMARY

- Metals have the following characteristic properties:
  - high boiling points
  - good conductors of electricity in solid and liquid states
  - malleability and ductility
  - high densities
  - good conductors of heat
  - lustrous appearance
  - low electronegativities
  - low ionisation energies
  - react by losing electrons.
- The main differences between the properties of main group and transition metals are:
  - transition metals are harder
  - transition metals are more dense
  - transition metals have higher melting points
  - some transition metals have strong magnetic properties
  - transition metal compounds tend to be brightly coloured.

KEY QUESTIONS

1 Determine the charge of the cations formed from the following metals if they lost all of their outer-shell electrons.
  a Li
  b Mg
  c Al

2 **a** Potassium is classed as a metal. Which of its properties are similar to those of the metal gold? In what ways is it different?
  b Identify another element in Table 4.1.2 on page 75 that has similar properties to potassium.
  c Identify another metal in Table 4.1.2 on page 75 that has similar properties to gold.
  d Where are these four metals in the periodic table?

3 **a** Which metals would you select if you wanted a good electrical conductor?
  b What other factors might influence your choice?

4 Sodium and iron have very different physical properties. Explain why this is, based on where these metals are found in the periodic table.

5 Suggest some properties not included in Table 4.1.2 on page 75 that you would need to consider before choosing between aluminium and iron for building a bridge.

4.2 Metallic bonding

In this section, you will learn how the properties of metals can be explained in terms of the structure of the particles in metals. You will also learn about the bonding model that chemists have developed to explain these properties. The metallic bonding model has helped chemists and materials engineers to understand why metals behave as they do and how metals can be modified to create useful new materials.

CONNECTING PROPERTIES AND STRUCTURE

The properties of metals are listed in Table 4.2.1. Each of these properties gives some information about the structure and bonding of particles in metals.

TABLE 4.2.1 The physical properties of metals and resulting conclusions about metal structure and bonding

Property	What this tells us about structure
Metals are usually hard and tend to have high boiling points.	The forces between the particles must be strong.
Metals conduct electricity in the solid state and in the **molten** (melted) liquid state.	Metals have charged particles that are free to move.
Metals are malleable and ductile.	The attractive forces between the particles must be stronger than the repulsive forces between the particles when the layers of particles are moved.
Metals generally have high densities.	The particles are closely packed in a metal.
Metals are good conductors of heat.	There must be a way of quickly transferring energy throughout a metal object.
Metals are lustrous or reflective.	Free electrons are present, so metals can reflect light and appear shiny.
Metals tend to react by losing electrons.	Electrons must be relatively easily removed from metal atoms.

Chemists have developed a model for the structure of metals called the **metallic bonding model** that explains all the properties mentioned so far. You can deduce from the information in Table 4.2.1 that the metallic bonding model must include:

- charged particles that are free to move and conduct electricity
- strong forces of attraction between atoms throughout the metal structure
- some electrons that are relatively easily removed.

METALLIC BONDING MODEL

Electrons are the particles that enable metals to conduct electricity. They are able to move within the **lattice** (tightly packed arrangement) of metal atoms. Negatively charged electrons can be lost from the outer shell of metal atoms, forming positive ions (cations). As shown in Figure 4.2.1, the freed electrons **delocalise** (spread through a large area) to form a 'sea' of electrons throughout the entire metal structure and are strongly attracted to the metal cations.

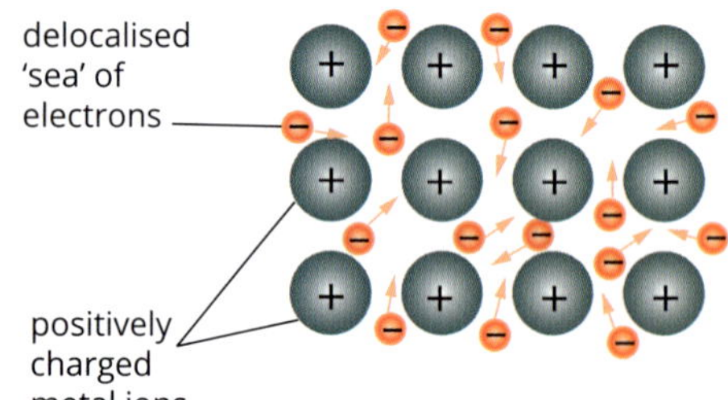

FIGURE 4.2.1 The metallic bonding model. Positive metal cations are surrounded by a mobile 'sea' of delocalised electrons. This diagram shows just one layer of metal ions.

Chemists believe that a solid sample of a metal can be explained by the following ideas.

- Positive ions are arranged in a closely packed structure. This structure is described as a regular, three-dimensional network of positive ions. The cations occupy fixed positions in the lattice.
- Negatively charged electrons move freely throughout the lattice. These electrons are called **delocalised electrons** because they belong to the lattice as a whole, rather than staying in the shell of a particular atom.

- The delocalised electrons come from the outer shells of the atoms. Inner-shell electrons are not free to move throughout the lattice and remain firmly bonded to individual cations.
- The positive cations are held in the lattice by the electrostatic force of attraction between these cations and the delocalised electrons. This attraction extends throughout the lattice and is called **metallic bonding**.

Together, these ideas make up the metallic bonding model. An example of how a metal, such as sodium, could be represented using this model is shown in Figure 4.2.2.

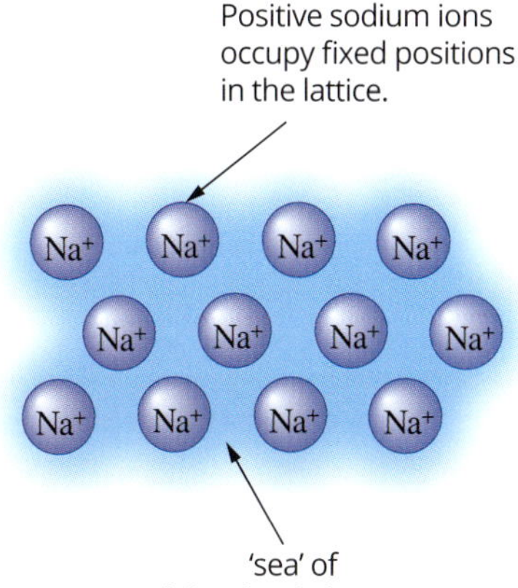

FIGURE 4.2.2 A representation of a sodium metal lattice. Each sodium atom loses its one valence electron. This electron is shared with all atoms in the lattice to form a 'sea' of delocalised electrons.

In the metallic bonding model, positive metal cations are surrounded by a sea of delocalised electrons.

EXPLAINING THE PROPERTIES OF METALS

Table 4.2.2 shows how the metallic bonding model is consistent with the relatively high boiling point, electrical **conductivity**, malleability and ductility of metals.

TABLE 4.2.2 Physical properties of metals and explanations from the metallic bonding model

Property	Explanation	
Metals are hard and have relatively high boiling points.	Strong electrostatic forces of attraction between positive metal ions and the sea of delocalised electrons hold the **metallic lattice** together.	
Metals are good conductors of electricity.	Free-moving delocalised electrons will move towards a positive electrode and away from a negative electrode in an electric circuit.	
Metals are malleable and ductile.	When a force causes metal ions to move past each other, layers of ions are still held together by the delocalised electrons between them.	

Other properties of metals

Metals generally have a high density. The cations in a metal lattice are closely packed. The density of a metal depends on the mass of the metal ions, their radius and the way in which they are packed in the lattice.

Metals are good conductors of heat. When the delocalised electrons bump into each other and into the metal ions, they transfer energy to their neighbour. Heating a metal gives the ions and electrons more energy and they vibrate more rapidly. The electrons, being free to move, transmit this energy rapidly throughout the lattice.

Metals are lustrous. Because of the presence of free electrons in the lattice, metals reflect light of all wavelengths and appear shiny.

Metals tend to react by losing electrons. The delocalised electrons in metals may participate in reactions anywhere on the metal's surface. The reactivity of a metal depends on how easily electrons can be removed from its atoms. This is covered in more detail in Section 4.3.

Limitations of the metallic bonding model

Although this model of metallic bonding explains many properties of metals, some cannot be explained so simply. These include the:

- range of melting points, hardness and densities of different metals
- differences in electrical conductivities of metals
- magnetic nature of metals such as cobalt, iron and nickel.

In order to deal with such questions, you would need a more complex model of metallic bonding, which is beyond the scope of this book.

Worked example 4.2.1

EXPLAINING THE CONDUCTIVITY OF ALUMINIUM

With reference to the electron configuration of aluminium, explain why solid aluminium can conduct electricity.	
Thinking	**Working**
Using the atomic number of the element, determine the electron configuration of its atoms. (You may need to refer to a periodic table.)	Al has an atomic number of 13. This means a neutral atom of aluminium has 13 electrons. The electron configuration is 2,8,3.
From the electron configuration, find how many outer-shell electrons are lost to form cations that have a stable, noble gas electron configuration. These electrons become delocalised.	Al has three electrons in its outer shell. Al atoms will tend to lose these three valence electrons to form a cation with a charge of +3. The outer-shell electrons become delocalised and form the sea of delocalised electrons within the metal lattice.
An electric current occurs when there are free-moving charged particles.	If the Al is part of an electric circuit, the delocalised electrons are able to move through the lattice towards a positively charged electrode.

Worked example: Try yourself 4.2.1

EXPLAINING THE CONDUCTIVITY OF MAGNESIUM

With reference to the electron configuration of magnesium, explain why solid magnesium can conduct electricity.

4.2 Review

SUMMARY

- Metallic bonding is the electrostatic force of attraction between a lattice of positive ions and delocalised valence electrons. The lattice of cations is surrounded by a 'sea' of delocalised electrons.
- The metallic bonding model can be used to explain the properties of metals, including their malleability, thermal conductivity, generally high melting point and electrical conductivity.

KEY QUESTIONS

1 The properties of calcium mean that it is classed as a metal.
   a Draw a diagram to represent a calcium metal lattice.
   b Describe the forces that hold this lattice together.

2 Barium is an element in group 2 of the periodic table. It has a melting point of 850°C and conducts electricity in the solid state. Describe how the properties of barium can be explained in terms of its bonding and structure.

3 Graphite is a non-metallic substance that can be lustrous and conducts electricity and heat. It is not malleable, but breaks if a force is applied.
   a What properties does graphite share with metals?
   b What inferences can you make about the structure of graphite given it shares these properties with metals?

4.3 Reactivity of metals

FIGURE 4.3.1 When water is dropped onto metallic potassium, hydrogen gas is produced.

In the previous section, you learnt that metals have many common properties. Metallic elements can also have very different properties. These include their **reactivity** with water, acids and oxygen. Some metals are extremely reactive and others are much less so.

This section will look at how the reactivity of different metals can be determined experimentally and explore some of the periodic patterns that exist.

DETERMINING THE REACTIVITY OF METALS

Reactivity with water

The way metals react with water can indicate their relative reactivity.

Figure 4.3.1 shows the reaction of potassium, a group 1 metal, with water. Enough heat is generated to instantly melt the potassium and ignite the hydrogen. The vigour of the reaction is an indication of the reactivity of the metal. Potassium has high reactivity with water, which is characteristic of the group 1 metals.

Table 4.3.1 describes the reaction of some group 1 and group 2 metals with water. In each case, a reaction results in the formation of hydrogen gas.

TABLE 4.3.1 Reaction of selected group 1 and 2 metals with water

Period	Group	Element	Reaction with water
3	1	sodium	reacts vigorously, producing enough energy to melt the sodium, which fizzes and skates on the water surface
4	1	potassium	reacts violently, making crackling sounds as the heat evolved ignites the hydrogen produced by the reaction
5	1	rubidium	explodes violently on contact with water
3	2	magnesium	will not react with water at room temperature but will react with steam
4	2	calcium	reacts slowly with water at room temperature

From these and other experimental observations, generalisations can be made.

- Metals in group 1 of the periodic table (i.e. Na, K and Rb) are more reactive in water than those in group 2 (i.e. Mg and Ca).
- Going down a group, the reactivity of the metal in water increases.

The reactivity of metals in water increases down a group and decreases across the period from left to right.

CHEMFILE

Reactivity of group 1 metals

Group 1 metals are so reactive that they must be handled with great care. They need to be stored under oil to prevent the metal coming into contact with moisture in the atmosphere.

FIGURE 4.3.2 Potassium metal is stored under oil to prevent contact with moisture.

Transition metals

Transition metals are generally less reactive with water than group 1 and group 2 metals are. For example, iron reacts fairly slowly with water. Gold and platinum are essentially unreactive.

Reactivity with acids

The reactivity of different metals with acids follows the same general patterns as the reactivity of metals with water. Metals are normally more reactive with acids than with water. More metals react with acids and the reactions tend to be more energetic.

FIGURE 4.3.3 Metals reacting with an equal amount of dilute hydrochloric acid. From left to right: magnesium ribbon, iron filings and copper turnings

Metals can be placed in an order of their relative reactivity. In Figure 4.3.3, the reactions of magnesium, iron and copper with hydrochloric acid are shown. The large amount of bubbling and the mist produced show that magnesium is the most reactive metal of the three, whereas copper is the least. The bubbles are due to the production of hydrogen gas.

magnesium + hydrochloric acid → magnesium chloride + hydrogen gas

This equation can also be written using chemical formulae:

$$Mg(s) + 2HCl(aq) \rightarrow MgCl_2(aq) + H_2(g)$$

Reactivity with oxygen

Many metals also react with oxygen. The group 1 metals all react rapidly with oxygen. Figure 4.3.4 shows sodium metal burning in a container of pure oxygen. The metal atoms and oxygen molecules rearrange to form a new compound, sodium oxide. The word equation for this reaction is:

sodium + oxygen → sodium oxide

This equation can be written using the chemical formula:

$$4Na(s) + O_2(g) \rightarrow 2Na_2O(s)$$

FIGURE 4.3.4 Sodium burning in pure oxygen

The oxygen in sodium oxide is present as negatively charged oxide ions, O^{2-}. Negatively charged ions are called anions.

The group 2 metals also react with oxygen to form oxides, although not as rapidly as group 1 metals. Heat is usually required to start the reaction.

While the transition metals are less reactive with oxygen than the metals in groups 1 and 2, their reactions are also important. Iron forms rust (iron oxide) when exposed to oxygen and water over a period of time. Many transition metals needed by society cannot be found in nature as pure elements but often exist as oxides.

Iron, copper, titanium and aluminium are all mined as the oxides and must be processed to obtain the finished metal. Figure 4.3.5 shows a cluster of crystals of rutile, the oxide from which titanium is extracted. The production of iron from iron oxide is covered in detail in Section 4.4.

FIGURE 4.3.5 Rutile is a mineral that contains titanium dioxide.

Gold and platinum, which are much less reactive than most other metals, are found in the Earth's crust in their pure form. Gold is often found in rock formations called seams alongside quartz, as shown in Figure 4.3.6.

FIGURE 4.3.6 Gold, an unreactive metal, exists in the Earth's crust in its metallic elemental form.

REACTIVITY SERIES OF METALS

Chemists have used experimental data from the reactions of metals to produce a **reactivity series of metals**, as shown in Figure 4.3.7. Group 1 metals are at the top of the series, while transition metals appear at the bottom.

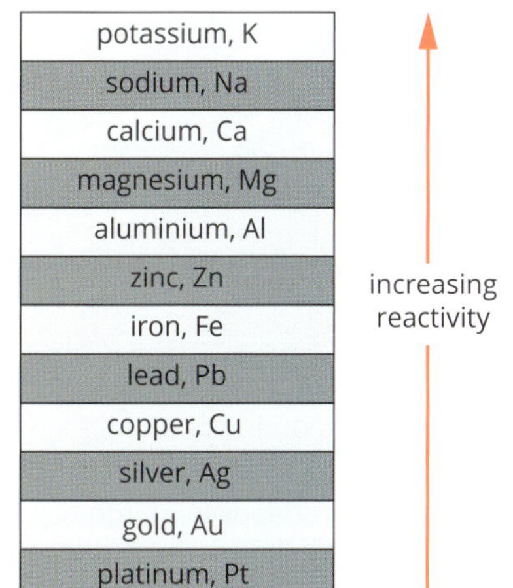

FIGURE 4.3.7 The reactivity series of some common metals

Reasons for different reactivities of metals

The larger the metal atoms, the more reactive the metals are. The further the valence electrons are from the nucleus, the more easily they are removed and the more easily they are involved in chemical reactions.

In general, the reactivity of main group metals increases going down a group in the periodic table and decreases across a period. This trend in reactivity can be explained in terms of the relative attractions of valence electrons to the nucleus of atoms.

When metals react, their atoms tend to form positive ions by donating one or more of their valence electrons to other atoms. The metal atoms that require less energy to remove electrons tend to be the most reactive. The most reactive metals tend to be those with the largest atomic radii and therefore the lowest ionisation energies, which are found in the bottom left-hand corner of the periodic table.

EXTENSION

Extracting iron from iron ore

Modern society is very dependent on iron. About 98% of world iron production is used to make steel. The steel in turn is used in bridges, buildings and all forms of transport. It also has many other uses.

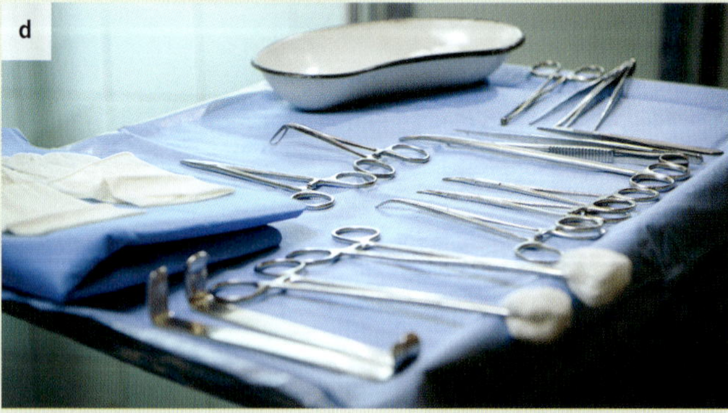

FIGURE 4.3.8 Steel is used in (a) the Sydney Harbour Bridge, (b) building frames in construction, (c) train tracks and (d) surgical instruments.

Australia is the world's largest exporter of iron ore (a natural compound containing a metal). Australia exported a record 767 million tonnes of iron ore in 2015. Most of the identified deposits of iron ore in Australia—almost 93% (totalling 64 billion tonnes)—are found in Western Australia. Massive deposits of iron ore in the Pilbara region of Western Australia are mined by open-cut methods (Figure 4.3.9).

Iron ore is composed mainly of iron(III) oxide combined with rocky material. The iron must be extracted from the ore before it can be used to make steel. In Australia, the iron oxides in iron ore are usually in the form of haematite (Fe_2O_3).

FIGURE 4.3.9 Iron ore is sourced from open-cut mines in Australia.

The raw materials used in the extraction of iron are:

- iron ore
- coke (a form of carbon)
- limestone
- air.

Extraction of iron from ore and the removal of unwanted materials is carried out in a tall, bottle-shaped tower called a blast furnace. A building containing a modern blast furnace and external conveyer belts is shown in Figure 4.3.10. These furnaces are heated to very high temperatures and are operated continuously for many years.

FIGURE 4.3.10 A blast furnace used to extract iron from iron ore. Iron ore is added continuously to the top of the furnace by the conveyor belts on the left-hand side.

Figure 4.3.11 shows a diagram of the inside of a blast furnace. Pre-heated air is blasted into the bottom part of the furnace, while solid 'charges' (scoops) of iron ore, coke and limestone are continuously added to the top. The construction of the furnace causes different temperature zones where different reactions can take place.

As the air rises through the furnace and meets the descending charge, oxygen reacts with coke to produce carbon dioxide, which then reacts further with the coke to produce carbon monoxide. These reactions are represented by the following equations:

carbon + oxygen → carbon dioxide

$$C + O_2 \rightarrow CO_2$$

carbon dioxide + carbon → carbon monoxide

$$CO_2 + C \rightarrow 2CO$$

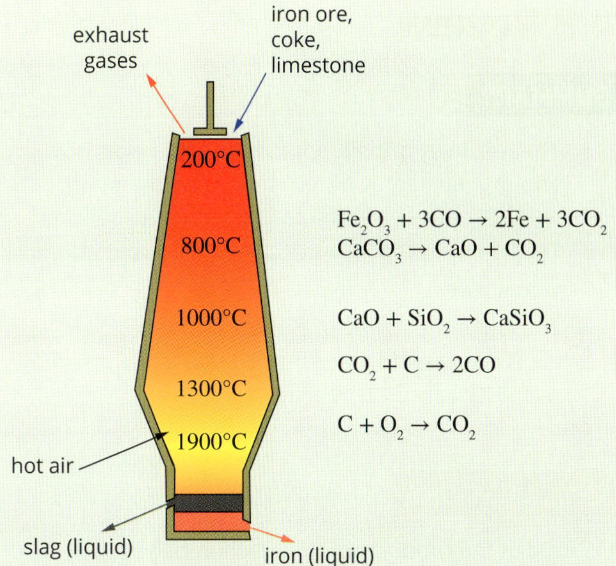

FIGURE 4.3.11 A diagram of a blast furnace. Hot air jets blast air in at the bottom. Iron ore, coke and limestone are added at the top.

The first reaction releases considerable heat energy and helps to maintain the high temperature that provides the fast rate of reaction required for metal production in the furnace. No external source of heat is needed to achieve temperatures up to 1800°C within the furnace.

Extraction of iron from iron oxide occurs in a series of steps in which carbon monoxide is the main reactant. The steps occur in different temperature zones within the blast furnace. The process is summarised by the single chemical equation:

carbon monoxide + iron oxide → iron + carbon dioxide

$$3CO + Fe_2O_3 \rightarrow 2Fe + 3CO_2$$

Iron ore is not pure iron oxide. It also contains rocky material such as silica (silicon dioxide), alumina and manganese oxides. These would not melt in the heat of the furnace and would eventually clog it up.

The limestone ($CaCO_3$) added to the top of the furnace breaks down in the furnace to form calcium oxide (CaO) and carbon dioxide. Calcium oxide can react with the unwanted materials to form a new compound called slag. Most of the slag is calcium silicate ($CaSiO_3$) due to the high amount of silica in iron ore.

Both the iron and slag are molten (melted) and sink to the bottom of the furnace. The slag is less dense and floats on top of the molten iron. This provides a cover that prevents the iron from reacting with the incoming air and re-forming iron oxide.

Holes at the base of the furnace are opened and the molten iron and slag are drained out and separated. In a steel works, the iron is usually transferred, while still molten, directly to a steel-making furnace. The slag may be used as a road-surfacing material or to manufacture cement.

4.3 Review

SUMMARY

- The reactivity of different metals with water, acid and oxygen can be determined experimentally.
- Group 1 metals are very reactive with water.
- Metals tend to be more reactive with acids than with water.
- The experimental results of metals reacting with acids are used to place metals in an order of reactivity called the reactivity series.
- The most reactive metals are found at the bottom left-hand side of the periodic table.
- Although the reactivities of transition metals vary, they are usually less reactive than the main group metals.

KEY QUESTIONS

1 Predict the products of the reaction between potassium and water, writing a chemical equation and a word equation for this reaction.

2 **a** Describe the trend in reactivity with water down the group 1 metals.

b What atomic property accounts for this trend in reactivity?

3 Pieces of iron, zinc and gold metal were each placed into test-tubes containing hydrochloric acid. The table below summarises the observations from each test. On the basis of the reactions, list the metals in order of reactivity.

Metal	Evidence of reaction
iron	Bubbles of gas are slowly produced.
	Iron metal disappears after a very long time.
zinc	Bubbles of gas are rapidly produced.
	Zinc metal disappears.
gold	No signs of a chemical reaction.
	Gold does not change appearance.

4 Using the reactivity series of some common metals (Figure 4.3.7 on page 85), determine whether calcium, platinum or aluminium would react most vigorously with oxygen in the air.

4.4 Modifying metals

The way in which a metal can be used is determined by the metal's physical and chemical properties. Although some metals are valuable in their pure state, most metals need to be modified to make them more useful.

Modified metals have a wide range of applications. In Figure 4.4.1, you can see a number of uses of modified metals. In each case, a metal with desirable properties has been chosen and improved by the processes covered in this section.

A metal can be modified in different ways, including:

- through alloy production
- by heat treatment
- by the formation of nano-sized structures.

In this section, you will look at the different ways metals are modified and the effect of these modifications on the metal's properties and uses.

FIGURE 4.4.1 (a) Modified metals are used in aeroplanes, which need to be made from materials that are strong, durable and light. (b) The Oscar statuettes are made of britannium (a mixture of tin, copper and antimony) and then coated successively in layers of copper, nickel, silver and gold. (c) Blacksmiths use heat and hammering tools to modify the properties of the metal used for horseshoes.

MAKING ALLOYS

Often, metals are mixed with small amounts of another substance, usually a metal or carbon. The substances are melted together, mixed and then allowed to cool. The resultant solid is an alloy.

By varying the composition of alloys, you can obtain materials with specific properties. Generally, an alloy is harder and melts at a lower temperature than the pure metal. This is because atoms of different sizes are now included in the metal lattice. As these atoms do not pack in the same way as the main metal, they will not allow the lattice to shift and bend in the same way. This disruption of the regular metallic lattice also accounts for the lowered melting point.

Steel: alloys of iron

Currently, almost all of the iron mined around the world is used to make the alloy steel. The simplest steel is made by adding a small amount of carbon to iron to make carbon steel.

Carbon steel is a type of **interstitial alloy**. In interstitial alloys, a small proportion of an element with significantly smaller atoms is added to a metal. The added atoms sit in interstices (very small spaces) between metal cations in the metallic lattice, as shown in Figure 4.4.2.

Carbon steel is generally harder and less malleable than pure iron. Varying the amount of carbon in the mixture produces steels with different properties, as shown in Table 4.4.1. This allows the steel with the most suitable properties to be used in specific applications.

TABLE 4.4.1 A summary of the properties and uses of some different carbon steels

Type of carbon steel	Percentage of carbon	Properties	Typical use
low-carbon steel	less than 0.3%	strong, easily shaped	bridges, buildings, ships and vehicles
medium-carbon steel	0.3–0.45%	increased hardness and tensile strength, decreased ductility	large machinery parts
high-carbon steel	0.45–0.75%	very strong, more brittle	springs and high-strength wires
very high-carbon steel	up to 2.5%	hard, more brittle	cutting tools

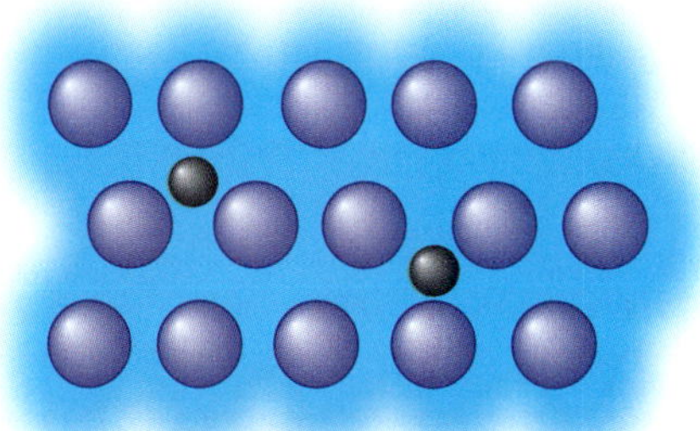

FIGURE 4.4.2 Steel is an interstitial alloy of iron and carbon. Note the relative sizes and position of the atoms in an interstitial alloy.

Other elements can also be added in addition to carbon to make steels with improvements to different properties. Examples of some of these are listed in Table 4.4.2.

TABLE 4.4.2 Some elements that are alloyed with iron to make steels

Alloying element	General effects on properties	Example
manganese	• increases strength and toughness	bicycle frames
chromium	• increases hardness and tensile strength • resists corrosion	stainless steel for cutlery, kitchen sinks
nickel	• increases toughness, tensile strength and hardness • resists corrosion	stainless steel
cobalt	• improves magnetic properties • resists high temperatures	alnico magnets, jet propulsion engines

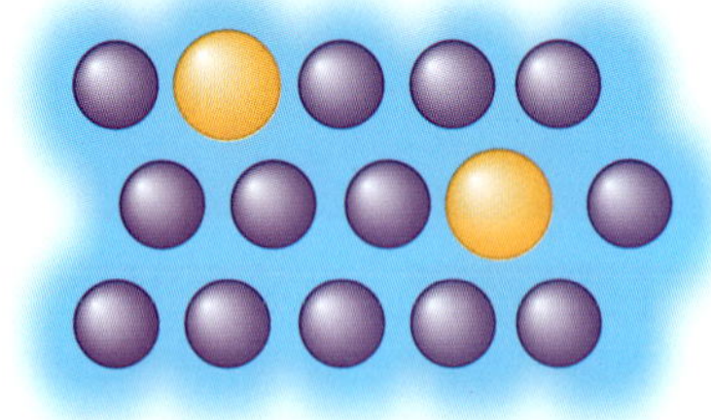

FIGURE 4.4.3 The iron, nickel and chromium atoms in stainless steel are relatively similar in size. The alloy is harder than any of these metals alone.

The steels in Table 4.4.2 are not considered to be interstitial alloys, as the added atoms are too big to fit into the spaces in the metallic lattice. The atoms of the elements added replace some of the iron cations and the mixture is called a **substitutional alloy**. In general, the metallic elements added to make substitutional alloys have fairly similar chemical properties and form cations of a similar size to the main metal.

Stainless steel is a substitutional alloy of nickel and chromium. The nickel and chromium atoms take the place of some of the iron atoms in the lattice (Figure 4.4.3). All the metal cations are attracted to the sea of electrons, so the lattice is still strongly bonded. However, because the different kinds of atoms are slightly different in size, the layers within the lattice cannot move as easily past each other. This makes the alloy harder and less malleable than pure iron.

When other metals or carbon are added to melted metals, the substance that is produced is called an alloy. Alloys tend to have improved properties compared to the original metal, making them more useful in society.

CHEMFILE

Gold

Pure gold is 24-carat gold. It is so soft that it can be moulded by hand. Gold in jewellery is a substitutional alloy (Figure 4.4.4). Nine-carat gold is 9/24 gold, which is 37.5% gold and 62.5% other metals. White gold is an alloy of gold and a metal such as nickel, manganese or palladium. The purity of white gold is also expressed in carats.

FIGURE 4.4.4 Gold jewellery is typically made from alloys. Pure gold (24-carat purity) is not often used for jewellery because it is too soft.

Common alloys

Many common materials are made of alloys in order to make the metal more suited to a particular purpose. Metals are blended together to combine properties, such as strength, colour, reflectivity and chemical stability. The resulting alloy will have different properties from the individual elements. Examples of substitutional alloys are shown in Table 4.4.3.

WORK HARDENING AND HEAT TREATMENT

The way a metal object is prepared also affects how it behaves. Many metals are prepared in the liquid state and then cooled. The rate at which a metal is cooled affects the properties of the solid.

The model that you have been developing for the structure of metals describes the arrangement of particles within a single metal **crystal**. A crystal is a region in a solid where the particles are arranged in a regular way. A sample of solid metal consists of many small crystals. Each crystal is a continuous regular arrangement of cations surrounded by a sea of delocalised electrons, but the arrangement of individual crystals with respect to one another is random, like those shown in Figure 4.4.5. As you can see in Figure 4.4.5, the regular lattice is disrupted at the point where one crystal meets another.

TABLE 4.4.3 Substitutional alloys found in everyday objects

Alloy	Metals used	
Australian 20-cent coin	copper (75%) and nickel (25%)	
Australian $2 coin	copper (92%), aluminium (6%) and nickel (2%)	
dental mercury amalgam	mercury and zinc	
brass	copper (65%) and zinc (35%)	
bronze	copper (90%) and tin (10%)	

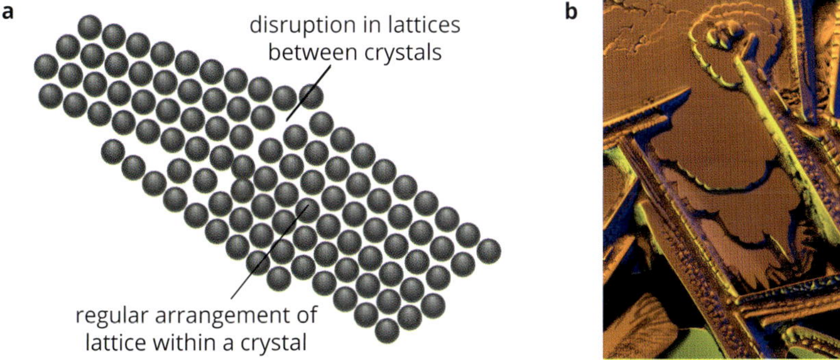

FIGURE 4.4.5 (a) Metals are made of a number of crystals, which have boundaries between them. (b) An electron micrograph shows the individual crystals that formed as a molten metal mixture cooled down and solidified.

The way a metal behaves—its malleability and brittleness—will depend on the size and arrangement of the crystals. Generally, smaller crystals result in harder metals as there is less free movement of layers of cations over each other. Smaller crystals also have more areas of disruption between them, and this usually means that these metals will be more brittle.

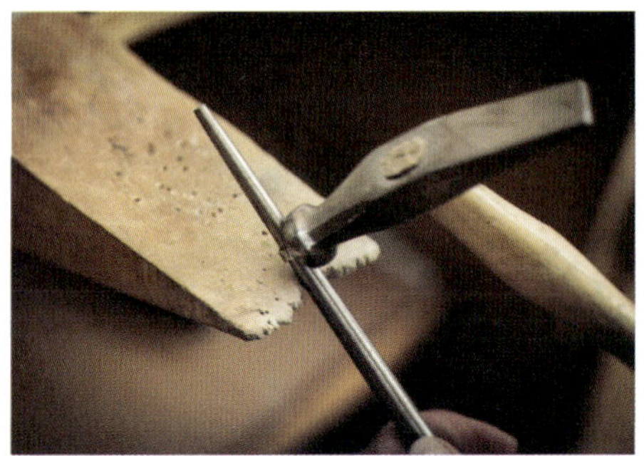

FIGURE 4.4.6 The crystal structure of metals like this gold ring can be altered by work hardening. Care is needed as the gold becomes more brittle through this process.

The crystal structure of metals can be altered in a number of ways. Two of these ways are **work hardening** and **heat treatment**.

Work hardening

Hammering or working cold metals causes the crystals to rearrange as they are pushed and deformed. This can result in the hardening of the metal as the crystals are flattened out and pushed closer together. Figure 4.4.6 shows a gold ring being hammered on a round tool, called a triblet, to harden and strengthen it.

Heat treatment

The physical properties of a metal can be altered by controlled heating and cooling. The three main methods of heat treatment (**annealing**, **quenching** and **tempering**) of metals and the effects on the metal's properties and structure are summarised in Table 4.4.4.

CHEMFILE

Work hardening with paperclips

You can see work hardening by bending a paperclip. If you bend it once, it remains fairly pliable, but if you bend it backwards and forwards several times, it snaps. Bending rearranges the crystal grains, making the metal harder but more brittle.

TABLE 4.4.4 Summary of heat treatment methods and their effect on metal properties

Treatment	Process	Effect on metal structure	Effect on metal properties
annealing	A metal is heated to a moderate temperature and allowed to cool slowly.	Larger metal crystals form.	The metal is softer with improved ductility.
quenching	A metal is heated to a moderate temperature and cooled quickly (sometimes by plunging into water).	Tiny metal crystals form.	The metal is harder and brittle.
tempering	A quenched metal is heated (to a lower temperature than is used for quenching) and allowed to cool.	Crystals of intermediate size form.	The metal is hard but less brittle.

When metals are heated above a critical temperature, the individual crystals merge. When the metal is allowed to cool, the crystals re-form. The rate of cooling determines how large the new crystals will be. Faster cooling leads to smaller crystals; slower cooling allows more time for crystals to grow larger.

Steels respond well to heat treatment. Annealing steels reduces their strength or hardness, increases uniformity of crystals and so reduces stresses and restores ductility. Hardening of steels by quenching and tempering increases the strength and wear properties (Figure 4.4.7).

CHEMISTRY IN ACTION

Swords

The art of producing a superior sword blade was a highly valued skill in ancient Japan. It required knowledge of the properties of alloys and the effects of heat treating and work hardening. The blade was heated and plunged into cold water to harden the edge.

To avoid the blade becoming too brittle, the sword was then tempered. The wavy line along the blade shows the interface between hard and flexible regions within the metal.

FIGURE 4.4.7 Japanese swordsmiths had to understand the properties of alloys and the effects of heat treating and work hardening.

METALLIC NANOMATERIALS

Metallic nanomaterials consist of metal atoms arranged to make structures that have a least one dimension of 1–100 nm (1 nm = 10^{-9} m). Metallic nanomaterials have very different properties when compared to a bulk sample of the same type of metal. Scientists have long been fascinated by the effects of nanomaterials. For example, gold–silver alloyed nanoparticles change the colour of glass (Figure 4.4.8).

As you learnt in Section 1.2 on page 8, emerging technologies have allowed scientists to manipulate metals at the nanoscale and create different forms of metallic nanomaterials.

FIGURE 4.4.8 As light passes through this stained-glass window containing gold–silver alloyed nanoparticles, it causes the glass to change colour.

Forms of metallic nanomaterials

Materials exist in many different forms at the nanoscale level, just as they do on the macroscale level. These forms include:

- particles
- rods
- wires
- tubes.

Metallic nanoparticles

Nanoparticles are usually spherical particles that range 1–100 nm in size. They have a very high surface area to volume ratio compared to the bulk material of similar volume.

The gold nugget shown in Figure 4.4.9 consists of many billions of cations and delocalised electrons. The properties of gold and other metals, such as electrical and thermal conductivity and metallic lustre, are explained in terms of the movement of delocalised electrons.

FIGURE 4.4.9 Bulk gold has a characteristic colour and typical metallic properties.

However, scientists have found that in a metal crystal with less than a certain number of atoms, the number of valence electrons released is too small to behave as a 'sea' of delocalised electrons. Metal nanoparticles that contain only 100 or so atoms have properties in-between those of metals and non-metals. In general, the nanoparticles have different optical properties and are more sensitive to heat.

For example, gold takes on a ruby colour when the gold particles are reduced to nano-size. You can see this in the stained-glass window in Figure 4.4.8. The deep yellow colour is caused by gold–silver alloyed nanoparticles.

Metallic nanorods

Nanorods are nanoscale rods in which each dimension ranges 1–100 nm. They have a length to width ratio of 3:1–5:1 (Figure 4.4.10). Nanorods of metals such as gold and silver have been synthesised in the laboratory. Their applications are diverse, including in display technologies and microelectronics, powering everything from solar cells to mobile phones.

FIGURE 4.4.10 A group of nanorods

Metallic nanowires

A **nanowire** is a nano-sized wire. The diameter of a nanowire is measured on the nanoscale, but its length is unrestricted. Nanowires differ from nanorods because they are much longer, as shown in Figure 4.4.11. Platinum nanowires have promising applications as catalysts and in electronics.

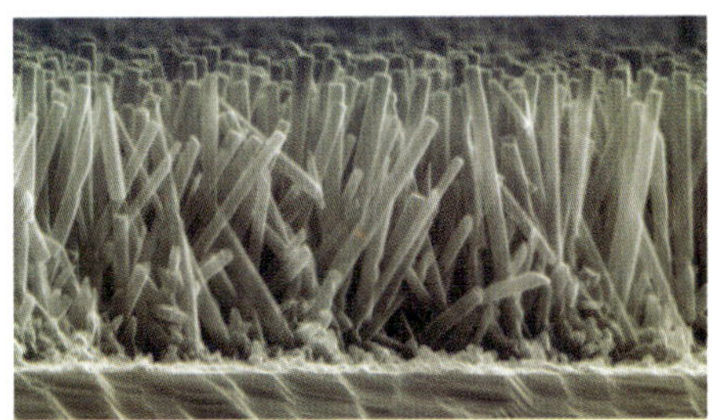

FIGURE 4.4.11 Zinc oxide nanowires

Uses of metallic nanomaterials

Nanomaterials have unique electrical, catalytic, magnetic, mechanical, thermal and imaging characteristics. This makes them attractive for use in medical, pharmaceutical, electronic and different engineering sectors.

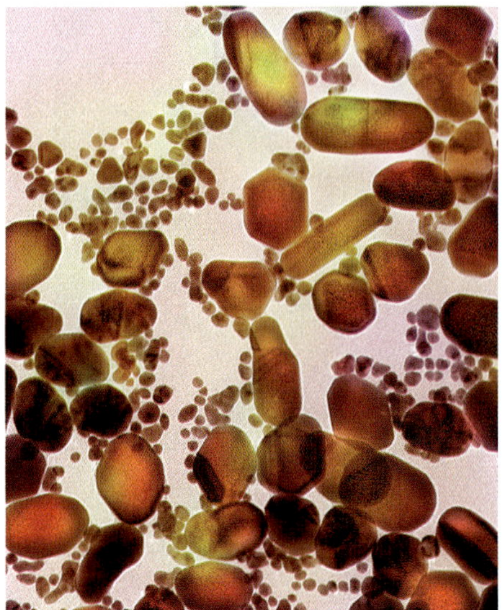

FIGURE 4.4.12 Applications of gold nanoparticles include cancer treatment research and detecting biological toxins.

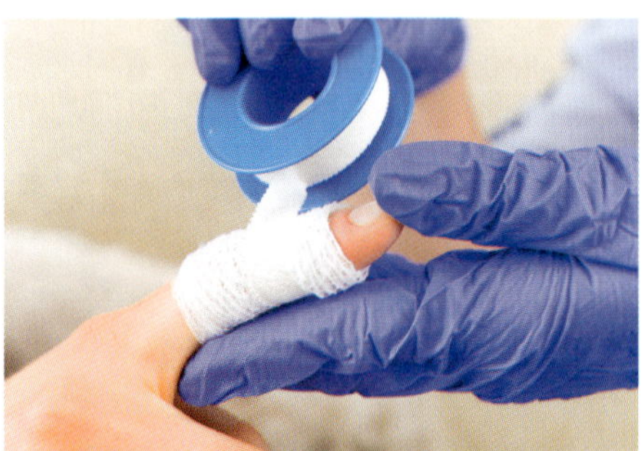

FIGURE 4.4.13 A wound dressing embedded with silver nanoparticles can be used to kill bacteria.

FIGURE 4.4.14 A nanotechnology copper-based solder has advantages in situations where long life and reliability are critical.

Gold nanoparticles in cancer research

Gold nanoparticles are the subject of substantial research with a wide range of applications (Figure 4.4.12). One area of development is in using gold nanoparticles as a targeted chemotherapy treatment method.

Gold nanoparticles can be attached to molecules of a tumour-killing agent known as tumour necrosis factor (TNF). The nanoparticles hide the molecule from the body's immune system.

The nanoparticles carrying the TNF tend to accumulate in cancer tumours, allowing TNF to destroy tumours. The nanoparticles do not appear to accumulate in other regions of the body, which means healthy cells are not affected.

Silver nanoparticles kill bacteria

Silver ions have long been known to kill bacteria. The ions can rapidly penetrate bacterial membranes and interact with proteins in the bacteria, destroying the cell structure of the bacteria and preventing them from reproducing.

Technology has enabled silver nanoparticles to be included in many different types of wound dressings. When the dressing (Figure 4.4.13) comes into contact with moisture from the wound, silver nanoparticles are slowly but continuously released from the wound pad. They then enter the wound and kill bacteria.

In similar antibacterial applications, Samsung has created and marketed a material called Silver Nano™, which adds silver nanoparticles to the surfaces of household appliances. Silver nanoparticles have been embedded in the surfaces of plastic storage bins, as well as in fabrics used by astronauts, babies and outdoor enthusiasts.

Copper nanoparticles go to space

Solder is a filler metal used to join two or more metals. Solders are essential in plumbing and metal constructions, including in satellites and spacecraft. For most of history, solders have contained a high amount of lead. Concerns about the toxicity of lead have driven the development of lead-free solder.

The complex electronics in satellites, such as the solar-powered satellite in Figure 4.4.14, must be reliable and efficient over a very long time. Space scientists have developed a nanotechnology copper-based solder that offers far superior performance over the materials currently in use. It is expected that the new solder material will produce joins with up to 10 times the electrical and thermal conductivity of current solders, with a wide range of space and defence applications.

Iron nanoparticles remove pollution in groundwater

Iron nanoparticles are being used to clean up the pollutant carbon tetrachloride (CCl_4) from groundwater (Figure 4.4.15). Carbon tetrachloride is a manufactured toxic chemical that has been shown to cause cancer in animals. Spills of carbon tetrachloride can spread through soil and create large areas of contamination.

Iron nanoparticles can quickly and effectively break down carbon tetrachloride to a mixture of relatively harmless products. It may be possible to inject the nano-sized iron deep into the ground where it can treat contaminated groundwater.

FIGURE 4.4.15 Nanoparticles of iron are being investigated as a way of eliminating a range of environmental pollutants.

4.4 Review

SUMMARY

- Most metals can be modified to alter their properties and make them more useful.
- An alloy is a mixture of a metallic element with other elements.
- Alloys are often harder and melt at a lower temperature than pure metals.
- Heat treatment changes the size and arrangement of crystals in metals.
- Metallic nanomaterials are nano-sized particles of metals that range 1–100 nm in size, where 1 nm = 10^{-9} m.
- Different forms of metallic nanomaterials include nanoparticles, nanorods and nanowires.
- Metallic nanomaterials have applications in medical, pharmaceutical, electronic, environmental and engineering sectors.

KEY QUESTIONS

1 Consider the alloys that make up a 20-cent coin and high-carbon steel.
  a List the elements present.
  b Draw a diagram to show how the atoms of these elements are arranged in the alloy.

2 Dentists fit partial dentures by means of small metal hooks. The hooks attach the denture to the remaining teeth. The hooks are easily bent at first to fit snugly in individual mouths. However, if the hooks are bent backwards and forwards too often, then they become brittle and snap. Explain why this happens.

3 Before shaping aluminium into objects by hammering, it is usually annealed to make it more malleable. The final product of this shaping process is then usually quenched to increase its strength. Use the metallic bonding model to explain why:
  a annealing makes aluminium more malleable
  b quenching can increase the strength of the final aluminium object.

4 During the manufacture of a metal object, many different modifications may be needed. Propose a series of modifications to produce a steel chisel that has a flexible and strong shaft with a hard blade.

5 Classify the metallic nanoparticles shown in the illustration.

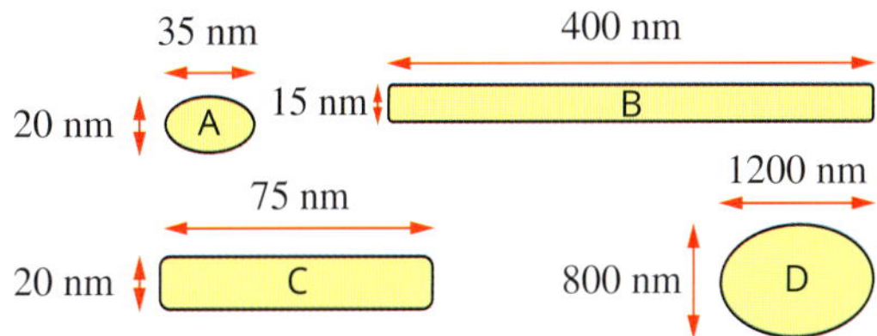

6 Iron nanoparticles must be stored away from oxygen as they will react quickly to produce iron oxide. Write a word equation for the reaction of iron nanoparticles with oxygen.

Chapter review

KEY TERMS

alkali metal
annealing
brittle
conductivity
conductor
crystal
delocalise
delocalised electron
ductile
heat treatment
interstitial alloy
lattice
malleable
metallic bonding
metallic bonding model
metallic nanomaterials
metallic lattice
molten
nanorod
nanowire
quenching
reactivity
reactivity series of metals
steel
substitutional alloy
tempering
tensile strength
work hardening

Properties of metals

1 Which of the following metals would have similar properties to beryllium?
Ca, Cs, Cu, Pb, Mg, Zn, Sr, K

2 Use the data in Table 4.1.2 on page 75 to answer the following questions.
- **a** Which metal is the best conductor of heat?
- **b** Why is this metal not used in saucepans?
- **c** What metals are used to make saucepans?

3 Which property most clearly distinguishes the metals from the non-metals listed in Table 4.1.2 on page 75?

4 What do you think is the most important property of each of the following metals that has led to its widespread use?
- **a** aluminium
- **b** copper
- **c** iron

5 The atomic number of calcium is 20. How many electrons are in an atom of calcium and in a Ca^{2+} cation?

6 Determine the electron configuration of an aluminium atom and the configuration of its most stable cation.

7 What is the meaning of the term 'ductile' when referring to metals?

Metallic bonding

8 Use the metallic bonding model to explain each of the following observations.
- **a** Copper wire conducts electricity.
- **b** A metal spoon used to stir a boiling mixture becomes too hot to hold.
- **c** Iron has a high melting point, 1540°C.
- **d** Lead has a density of $11.4\,g\,mL^{-1}$, which is much higher than for a non-metal such as sulfur.
- **e** Copper can be drawn out to form a wire.

9 Consider the metallic bonding model used to describe the structure and bonding of metals.
- **a** What is meant by the following terms?
  - **i** delocalised electrons
  - **ii** a lattice of cations
  - **iii** metallic bonding
- **b** Which electrons are delocalised in a metal?

10 Describe the arrangement of particles in a metal wire and how they allow the wire to conduct electricity.

11 Use a diagram to describe what is meant by the term 'metallic lattice'.

Reactivity of metals

12 Look at the periodic table at the end of the book.
- **a** Name a metal that would have similar properties to calcium.
- **b** In which part of the periodic table are magnetic metals found?

13 Which of the following metals would you expect to be the least reactive with water?
aluminium, sodium, rubidium

14 When a reactive metal is added to water, bubbles or fizzing can be observed. Explain the appearance of the bubbles.

15 The image to the right shows similar-sized pieces of iron and silver in test-tubes of sulfuric acid of the same concentration. Describe the reactivity of the two metals and identify which metal is on the left and which is on the right.

16 Observations of the reaction of metals A, B and C with dilute hydrochloric acid are summarised below. Identify which metal is aluminium, copper and sodium.

- Metal A does not react with dilute hydrochloric acid.
- Metal B is stored in oil because of its high reactivity, and it undergoes a violent reaction with dilute hydrochloric acid.
- Metal C reacts very slowly with dilute hydrochloric acid.

17 Identify the following statements about the reactions of metals as true or false.

a All metals react with acid to produce hydrogen gas.
b Hydrogen is a flammable gas.
c The most reactive metals are located at the top of a group in the periodic table.
d A metal that is very reactive with water is likely to be less reactive with a solution of a dilute acid.
e When metals react with oxygen in the air, they release energy.

18 Write the word equation for the reaction that occurs when magnesium is heated strongly in air.

Modifying metals

19 Alloys are modified metals. In each example, state the metals used and one property that is different from the original metals.

a Australian one dollar coin
b solder
c 9-carat gold
d stainless steel
e dental mercury amalgam

20 A student wishes to make an iron needle more malleable. She heats three needles strongly in a Bunsen burner flame. She then treats each needle differently.

- Needle 1 is allowed to cool slowly on the bench.
- Needle 2 is cooled quickly by dropping it into a beaker of cold water.
- Needle 3 is cooled quickly in the beaker of water then reheated in the flame before being allowed to cool more slowly on the bench.

Place the resulting needles in order of least malleable to most malleable.

21 Horseshoes are often made from steel that is worked into shape by a process of heating and hammering the metal. Explain how this process results in a better horseshoe than one simply made of iron.

22 A scientist synthesises a new material in the laboratory. It is a solid substance with a diameter of 15 nm and a length of 2300 nm. Which form of nanomaterial is this substance?

23 A scientist synthesised a tiny particle for use in solar cells, with a diameter of 8.34×10^{-7} m. Would this particle be classified as a nanomaterial?

24 Explain why a nanowire can be longer than 3000 nm but still be considered a metallic nanomaterial.

Connecting the main ideas

25 Metals have many uses in modern society.

a i Name one metal that is used in large quantities in the building industry.
ii What properties make it suitable for this use?
iii What properties limit its use in buildings?
b i Name one metal that is used in large quantities for making electrical wires.
ii What properties make it suitable for use in wires?
c i Name two metals that are used in large quantities in the jewellery trade.
ii What properties make them suitable for this use?

26 Draw a concept map using the following terms: metal, cation, delocalised electron, lattice, electrostatic attraction.

27 Some metals are found as elements in nature; others are found as compounds, combined with other elements such as oxygen and sulfur in ores. Australia has natural reserves of many metals and ores, with mining producing large quantities of metals, including aluminium, copper, gold, iron and silver.

a Give the chemical symbol for each of these metals.
b In which group and period of the periodic table are each of these metals found?
c Which of these metals are found in nature as elements rather than compounds?
d Which of these metals are transition elements?
e Which of these elements is the rarest?

28 It has been shown using X-rays that metals have a crystalline structure in the solid state.

a Explain, using the metallic bonding model, why metals form crystals.
b What problems can this crystalline structure of metals cause?

29 The boiling points of three metals—sodium, potassium and calcium—are given in the table below.

Metal	Boiling point (°C)
Na	892
K	760
Ca	1490

a In which group and period of the periodic table are these metals found?
b Write the electron configuration for each of the three metals.
c Use the metallic bonding model to suggest why:
i sodium has a higher boiling point than potassium
ii calcium has a much higher boiling point than potassium.

30 Aluminium is the most abundant metal in the Earth's crust. Why was aluminium not available before 1886?

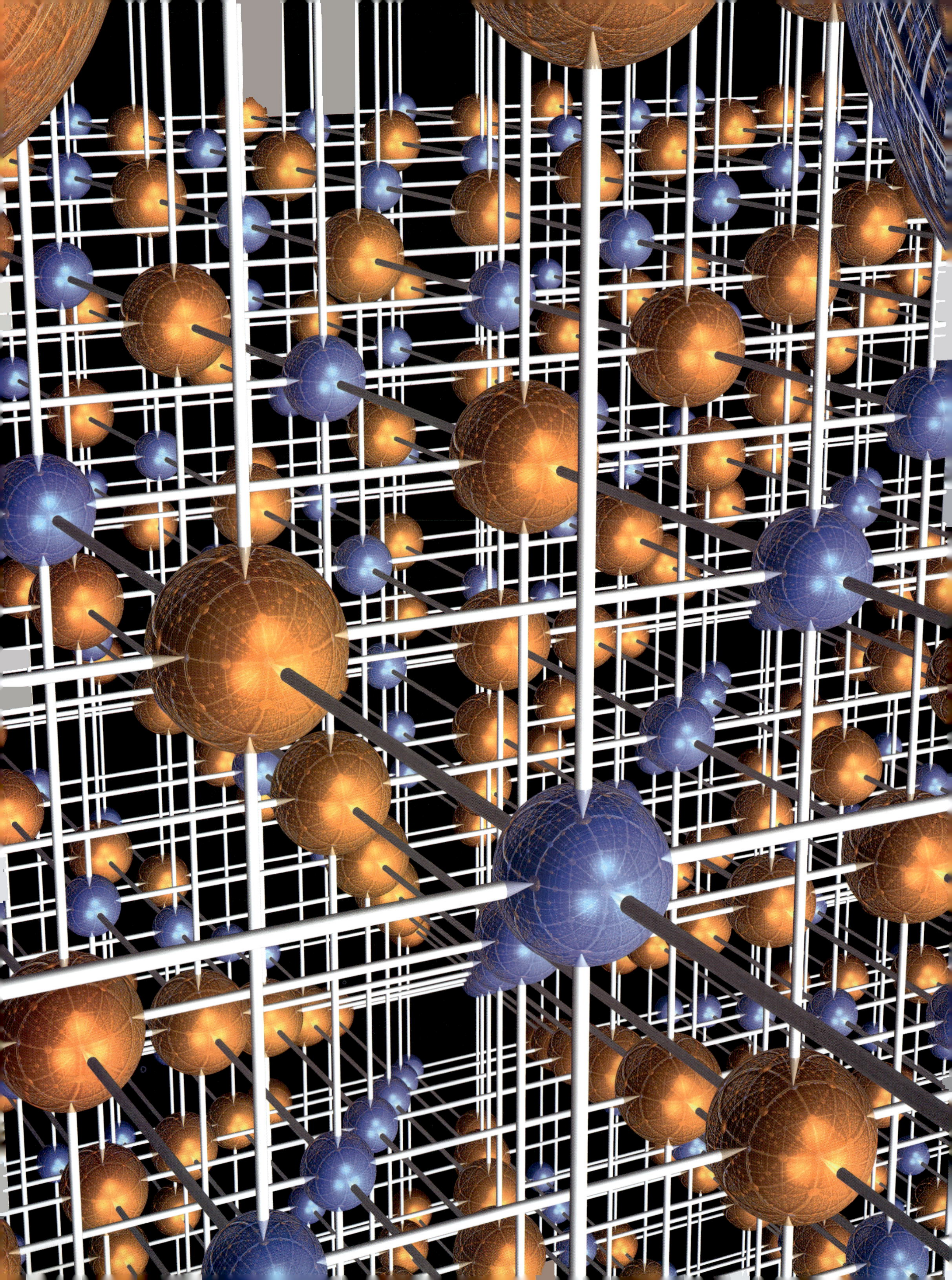

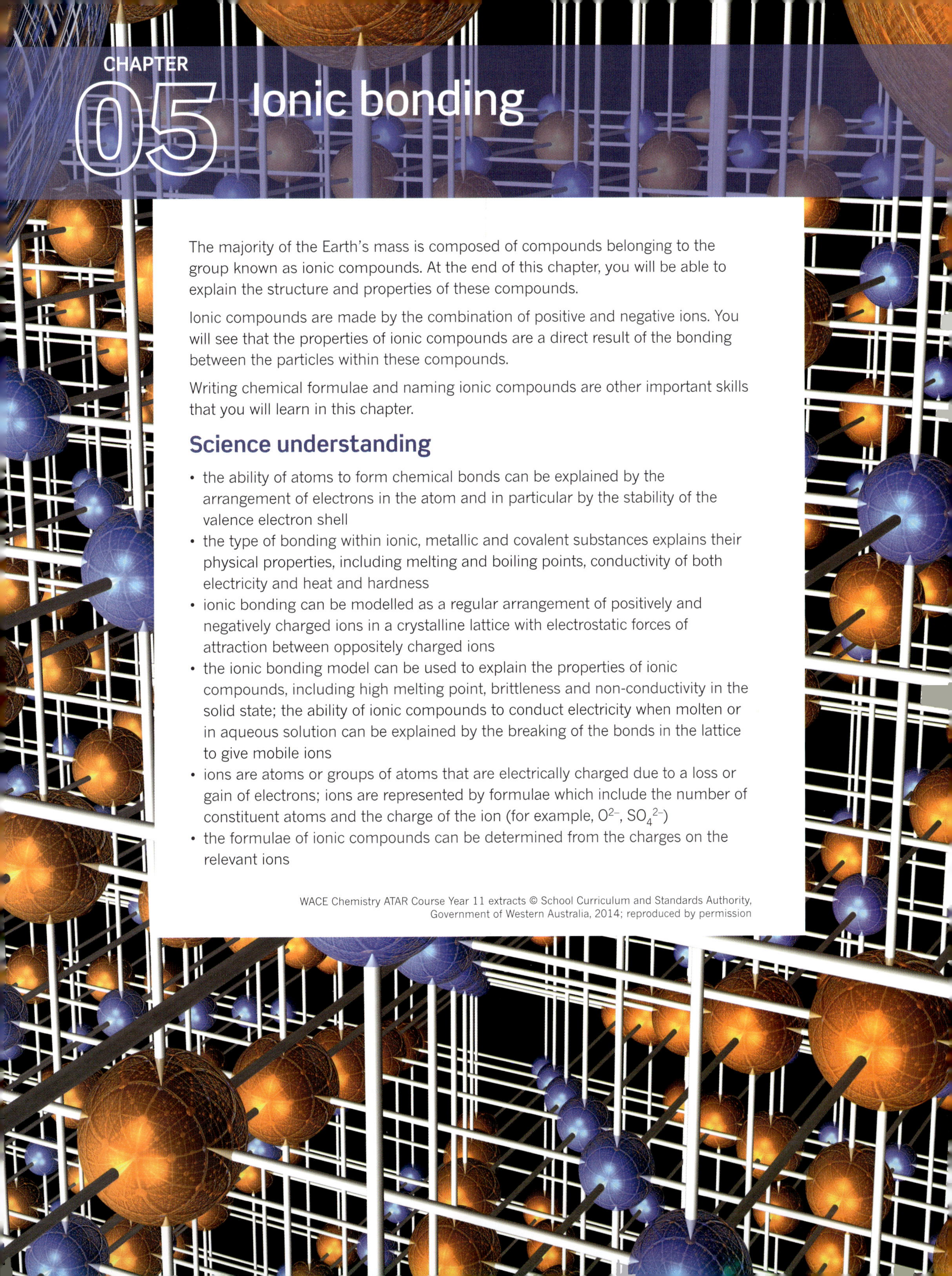

CHAPTER

05 Ionic bonding

The majority of the Earth's mass is composed of compounds belonging to the group known as ionic compounds. At the end of this chapter, you will be able to explain the structure and properties of these compounds.

Ionic compounds are made by the combination of positive and negative ions. You will see that the properties of ionic compounds are a direct result of the bonding between the particles within these compounds.

Writing chemical formulae and naming ionic compounds are other important skills that you will learn in this chapter.

Science understanding

- the ability of atoms to form chemical bonds can be explained by the arrangement of electrons in the atom and in particular by the stability of the valence electron shell
- the type of bonding within ionic, metallic and covalent substances explains their physical properties, including melting and boiling points, conductivity of both electricity and heat and hardness
- ionic bonding can be modelled as a regular arrangement of positively and negatively charged ions in a crystalline lattice with electrostatic forces of attraction between oppositely charged ions
- the ionic bonding model can be used to explain the properties of ionic compounds, including high melting point, brittleness and non-conductivity in the solid state; the ability of ionic compounds to conduct electricity when molten or in aqueous solution can be explained by the breaking of the bonds in the lattice to give mobile ions
- ions are atoms or groups of atoms that are electrically charged due to a loss or gain of electrons; ions are represented by formulae which include the number of constituent atoms and the charge of the ion (for example, O^{2-}, SO_4^{2-})
- the formulae of ionic compounds can be determined from the charges on the relevant ions

5.1 Properties and structures of ionic compounds

FIGURE 5.1.1 Many gemstones are made from ionic compounds.

In this chapter, you will study the structure and properties of a group of substances called **ionic compounds**. Ionic compounds are made by the combination of atoms, or groups of atoms, where electron(s) are transferred from one to another. In doing so, these particles become ions, hence ionic compounds are compounds formed from ions.

These materials are very common in the natural world because the Earth's crust is largely made up of complex ionic compounds. Most rocks, minerals and gemstones (Figure 5.1.1) are ionic compounds. Soil contains weathered rocks mixed with decomposed organic material, so soil contains large quantities of ionic compounds.

Ceramics, kitchen crockery and bricks are made from clays. Clays are formed by the weathering of rocks, so these materials also contain ionic compounds. Kitchen crockery and bricks contain mixtures of different ionic compounds. Table salt (sodium chloride) is a pure ionic compound.

PROPERTIES OF IONIC COMPOUNDS

If you think about the characteristics of rocks, kitchen crockery and table salt, you will recognise that these materials, and therefore ionic compounds, have some properties in common.

Table 5.1.1 lists some properties of typical ionic compounds. These compounds can be found in materials you might encounter in everyday life. Note that the compounds listed are simple ionic compounds, whereas rocks, ceramics and bricks contain more complex ionic compounds.

TABLE 5.1.1 Properties of typical ionic compounds

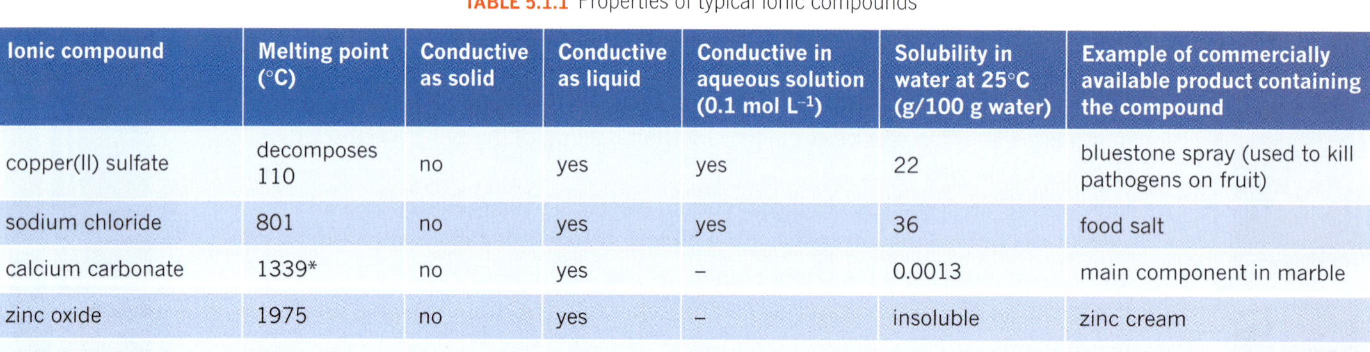

Ionic compound	Melting point (°C)	Conductive as solid	Conductive as liquid	Conductive in aqueous solution (0.1 mol L^{-1})	Solubility in water at 25°C (g/100 g water)	Example of commercially available product containing the compound
copper(II) sulfate	decomposes 110	no	yes	yes	22	bluestone spray (used to kill pathogens on fruit)
sodium chloride	801	no	yes	yes	36	food salt
calcium carbonate	1339*	no	yes	–	0.0013	main component in marble
zinc oxide	1975	no	yes	–	insoluble	zinc cream
sodium hydroxide	318	no	yes	yes	114	oven cleaner

*Melting point determined under pressure to prevent decomposition of compound.

Lists of data about ionic compounds, such as those in the Table 5.1.1, have allowed chemists to summarise their properties.

Generally, ionic compounds:

- have high melting and boiling points and they are all solids at room temperature
- are hard but brittle (fragile), unlike metals and therefore they are neither malleable nor ductile
- do not conduct electricity in the solid state
- are good conductors of electricity in the liquid state or when dissolved in water
- vary from very soluble to insoluble in water.

DEDUCING THE STRUCTURE OF A COMPOUND FROM ITS PROPERTIES

You will remember from Chapter 4 that the properties of metals indicate something about the structure of metals; that is, how particles are arranged in a metallic lattice. The properties of ionic compounds also provide evidence for how the particles are arranged in ionic compounds.

From Table 5.1.1, you can see that some of the physical properties of ionic compounds are different from those of metals. Chemists have concluded that the particles in ionic compounds, and how they are arranged in the solid state, are different from those present in metals.

Table 5.1.2 lists the observed properties of one particular ionic compound, sodium chloride (table salt). Beside each property is a description of the nature of the particles and therefore the types of forces between the particles. These descriptions can be inferred from the different properties of sodium chloride.

TABLE 5.1.2 Properties of sodium chloride and the information this provides about its structure

Property	What this tells us about structure
high melting point	Forces between the particles are strong.
hard, brittle crystals	Forces between the particles are strong.
does not conduct electricity in the solid state	No free-moving charged particles are present in solid sodium chloride.
conducts electricity in the molten state	Free-moving charged particles are present in molten sodium chloride.

In summary, these properties apply to ionic compounds generally.

- The forces between the particles are strong.
- There are no free-moving electrons present, as there are in metals.
- There are charged particles present, but in the solid state they are not free to move.
- When an ionic compound melts, the charged particles are free to move and then the compound will conduct electricity.

THE IONIC BONDING MODEL

Now that you understand some of the details of the structure of ionic compounds, the next step is to work out how the particles in these compounds are arranged in the solid state.

When two atoms or groups of atoms react to form ionic compounds, the following steps occur.

- One set of atoms, generally metal atoms, lose electrons and so become positively charged ions (called cations).
- These lost electrons are transferred to another set of atoms.
- The other set of atoms, generally non-metal atoms, become negatively charged ions (called anions).

You will remember from Chapter 3, page 44, how electrons arrange themselves into shells around the nucleus. Atoms are at their most stable when they have a filled valence shell. Therefore, an atom's ability to form a cation or anion depends on how many electrons it needs to gain or lose to achieve this stable arrangement.

Cations and anions then arrange themselves in the following way.

- Large numbers of cations and anions combine to form a three-dimensional lattice (or **crystal lattice**).
- The three-dimensional lattice is held together strongly by electrostatic forces of attraction between the oppositely charged ions. The electrostatic force of attraction holding the ions together is called **ionic bonding**.

- In the case of sodium chloride, sodium (electron configuration 2,8,1) loses one electron, so it has a full valence shell (electron configuration 2,8). Losing one electron makes it positively charged and it is given the symbol Na^+. Each chlorine atom normally has the electron configuration of 2,8,7. Therefore, it only needs to gain one electron to obtain a full valence shell (electron configuration 2,8,8). Gaining one electron makes it negatively charged and it becomes an anion, Cl^-.
- In sodium chloride, in order to maximise the forces of attraction, each sodium ion is surrounded by six chloride ions and each chloride ion is surrounded by six sodium ions, as shown in Figure 5.1.2.
- Even though each chloride ion is close to another chloride ion (Figure 5.1.2), the attractive force they have towards the sodium ions outweighs the repulsive force from the chloride ions so the lattice is held together quite strongly.

Ionic bonding is the strong, electrostatic attraction that exists between positively charged cations and negatively charged anions.

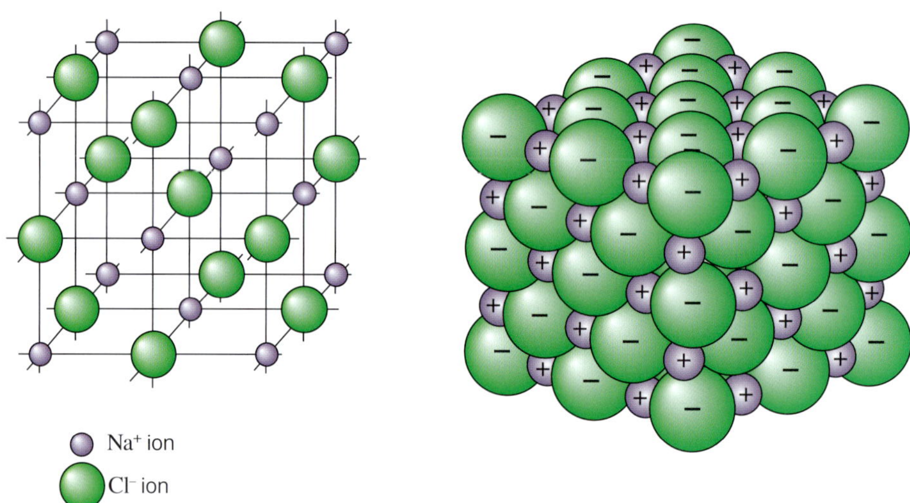

FIGURE 5.1.2 Two representations of part of the crystal lattice of the ionic compound sodium chloride. Forces of attraction between oppositely charged ions result in strong bonding.

The formula of sodium chloride

The **chemical formula** of sodium chloride is written as NaCl. However, it is important to note that in a solid sample of an ionic compound, such as sodium chloride, individual pairs of sodium and chloride ions do not exist. The solid is also not built up of discrete NaCl molecules.

Instead, the solid is made up of a continuous lattice of alternating Na^+ and Cl^- ions. NaCl is the formula unit for this compound; it represents the simplest whole number ratio of the ions that the substance is composed of. All sodium ions are an equal distance from six chloride ions and all chloride ions are an equal distance from six sodium ions. The overall ratio of sodium ions to chloride ions in the lattice is 1:1, so that is why the formula is written as NaCl.

5.1 Review

SUMMARY

- Ionic compounds form a crystal lattice made up of positive and negative ions.
- Cations are positive ions and are generally metals.
- Anions are negative ions and are generally non-metals.
- The three-dimensional lattice is held together strongly by electrostatic forces of attraction between the cations and anions. The electrostatic forces of attraction are called ionic bonding.
- In the case of sodium chloride, in order to maximise the forces of attraction, each sodium ion (Na^+) is surrounded by six chloride ions (Cl^-) and each chloride ion is surrounded by six sodium ions.
- Ionic compounds have high melting and boiling points. They are solids at room temperature.
- Ionic compounds are hard but brittle. Unlike metals, they are neither malleable nor ductile.
- Ionic compounds do not conduct electricity in the solid state.
- Ionic compounds are good conductors of electricity in the liquid state or when dissolved in water.
- In water, ionic compounds vary from very soluble to insoluble.

KEY QUESTIONS

1 Some properties of four different substances are described below. Which substance is most likely to be an ionic compound?

A Substance A has a melting point of 842°C and conducts electricity at 700°C.

B Substance B has a melting point of 308°C. It does not conduct electricity at 250°C but will conduct electricity at 350°C.

C Substance C has a melting point of 180°C and can be drawn into a wire.

D Substance D is a white solid that melts at 660°C and will not conduct electricity at 700°C.

2 Sodium chloride does not conduct electricity in the solid state but does conduct when molten (liquid).

a How could you use Figure 5.1.2 to explain why solid sodium chloride does not conduct electricity?

b Explain why molten sodium chloride conducts electricity.

3 Explain why aluminium would be more likely to form a cation than an anion.

4 Explain why the crystal lattice is held together so strongly in sodium chloride, even though the number of cations is equal to the number of anions.

5.2 Using the ionic bonding model to explain properties

In Chapter 4, you saw that the metallic bonding model represents the structure of metals as a lattice of positively charged metal ions held together by delocalised electrons. This model explains many of the properties of metals, such as why metals generally have high melting points and conduct electricity in the solid state.

In this section, you will see how the ionic bonding model explains the properties of ionic compounds.

FIGURE 5.2.1 Bricks made from the ionic compound magnesium oxide are used to line furnaces and kilns.

HIGH MELTING POINTS

To melt an ionic solid such as sodium chloride, energy is required to allow the ions to break free and move. Sodium chloride has a high melting point (801°C). This indicates that a large amount of energy is needed to overcome the electrostatic attraction between oppositely charged ions and allow them to move freely. Therefore, the ionic bonds between the positive sodium ions and negative chloride ions must be strong, which explains why a high temperature is required to melt solid sodium chloride.

The high melting point of ionic compounds is put to use in the bricks that line furnaces and kilns (Figure 5.2.1) and in the ceramic materials used to make brake discs for high-performance cars (Figure 5.2.2).

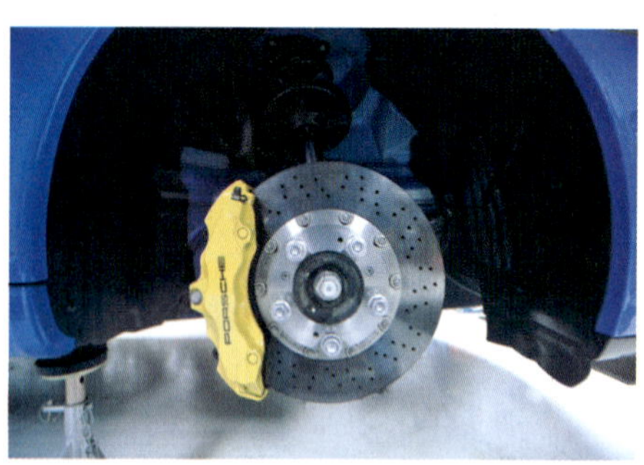

FIGURE 5.2.2 Ceramic brake discs work more effectively than steel ones at high temperatures. A ceramic brake disc contains ionic compounds that have very high melting temperatures and withstand the heat produced by braking better than metals.

HARDNESS AND BRITTLENESS

There are strong electrostatic forces of attraction between ions in an ionic compound, so a strong force is needed to disrupt the crystal lattice. Therefore, one of the properties of ionic compounds is that they are hard. This means that a sodium chloride crystal cannot be scratched easily.

The strength of house bricks, concrete bridges and cobbled streets can be attributed to the ionic bonding within their structures.

Although a salt crystal is hard, a strong impact such as a hammer blow will shatter the crystal. Therefore, it is said to be brittle. This is because the layers of ions will move relative to each other due to the force of the blow.

CHEMFILE

Early tools

Some of the earliest tools used by humans were axes, spearheads and coarse needles used for weaving. Each application required a material that was hard and could be shaped. Certain types of rocks that are composed of ionic compounds served this purpose well.

Figure 5.2.3 shows a primitive axe head used during the Stone Age. The Stone Age ended at different times in different parts of the world, as humans learnt to smelt (fuse or melt) metals such as copper from ore to create more refined and lighter tools. Their metallic nature meant that they were not as hard as the original tools (made of rock composed of ionic compounds).

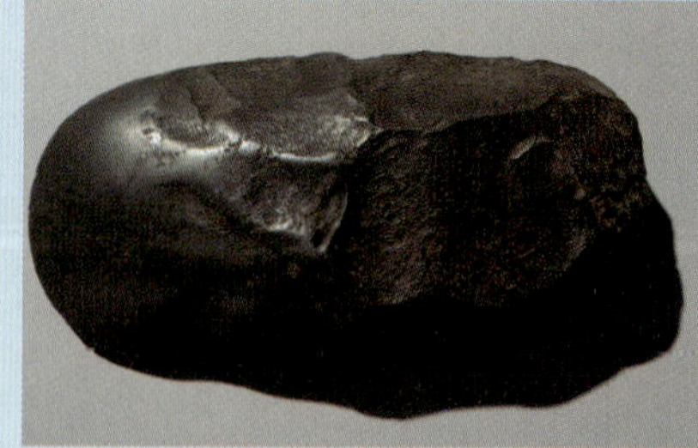

FIGURE 5.2.3 The hardness of ionic compounds is the reason why axes were once made from rocks.

During this movement, ions of like charge are shifted so they are next to each other, as seen in Figure 5.2.4. The resulting repulsion between the similarly charged ions causes the crystal to shatter.

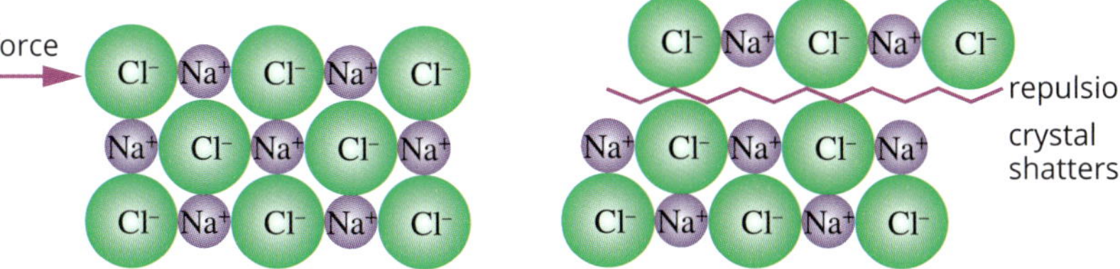

FIGURE 5.2.4 A lattice of an ionic compound shattering. Note that just before shattering, the Cl^- ions are adjacent to other Cl^- ions and the Na^+ ions are also next to each other.

FIGURE 5.2.5 A coffee cup being smashed

Materials made from clay, such as kitchen crockery (Figure 5.2.5), ceramic tiles and bricks, are hard, but they are also brittle.

CHEMFILE

Porcelain

Porcelain is a type of chinaware made from a clay called kaolin. Figure 5.2.6 shows a porcelain dinner setting. Clay is weathered rock and consists of a mixture of complex ionic compounds. The chemical formula for kaolin is $Al_2Si_2O_5(OH)_4$. The ions in kaolin are aluminium (Al^{3+}), silicate ($Si_2O_5^{2-}$) and hydroxide (OH^-).

Porcelain is made by moulding the object from kaolin and then heating it to about 1300°C in a kiln. Because it is made from ionic compounds, the resulting porcelain is hard but brittle. Be careful not to drop the family heirloom chinaware as it will most likely shatter!

FIGURE 5.2.6 Porcelain cups and plates are made from clay, which contains ionic compounds. Therefore, porcelain is hard and brittle.

ELECTRICAL CONDUCTIVITY

In the solid form, ions in sodium chloride are held in the crystal lattice and are not free to move, so solid sodium chloride does not conduct electricity. Remember that for a substance to conduct electricity, it must contain charged particles that are free to move. Figure 5.2.7 shows how the particles are arranged in an ionic compound in solid form.

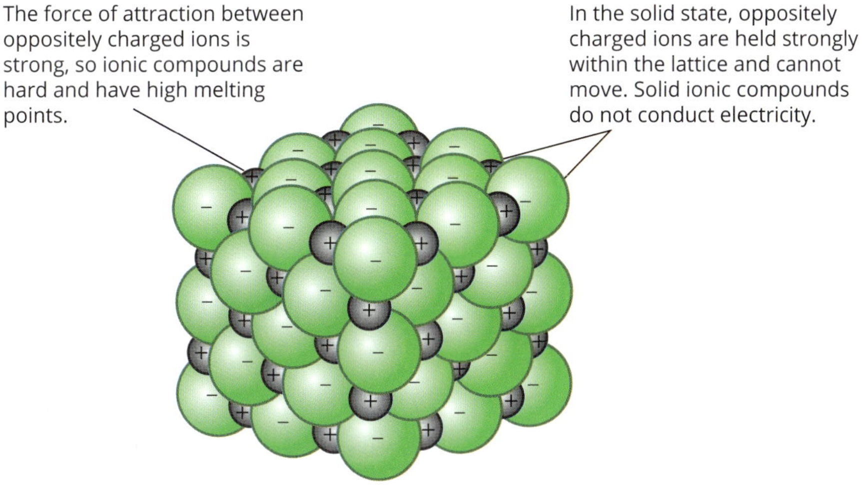

The force of attraction between oppositely charged ions is strong, so ionic compounds are hard and have high melting points.

In the solid state, oppositely charged ions are held strongly within the lattice and cannot move. Solid ionic compounds do not conduct electricity.

FIGURE 5.2.7 The arrangement of ions within a solid ionic compound form a crystal lattice.

FIGURE 5.2.8 A ceramic insulator on the post of an electric fence

The non-conducting property of ionic compounds is used in ceramic insulators, which are used to keep high-voltage power lines insulated from electricity poles and electric fence wires (Figure 5.2.8).

When solid ionic compounds melt, the ions become free to move, enabling the cations and anions in the molten compound to conduct electricity.

Similarly, when ionic compounds dissolve in water, ionic bonds in the lattice are broken and the ions are separated and move freely in solution.

When an electric current is applied to either a molten ionic compound or a solution of the compound in water, positive ions move towards the negatively charged electrode (cathode) and negative ions move towards the positively charged electrode (anode), resulting in an electric current as shown in Figure 5.2.9.

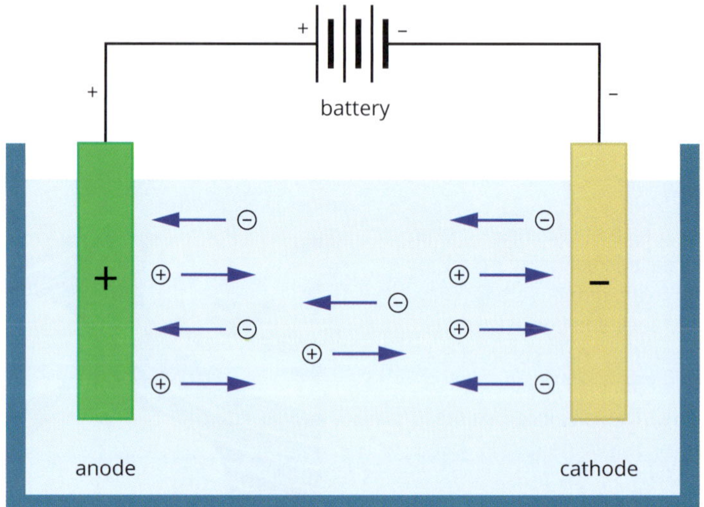

FIGURE 5.2.9 A molten ionic compound will conduct an electric current.

EXTENSION

Electrolysis

When an electric current passes through a molten ionic compound, or a solution of the compound, chemical reactions occur at the positive and negative electrodes. This process is known as electrolysis and is used to make a variety of chemicals. Chemicals formed by electrolysis are often difficult to obtain by other means.

For example, the final stage in the industrial production of aluminium involves passing an electric current through a molten liquid containing aluminium oxide (alumina). Aluminium metal is deposited at the negatively charged electrode.

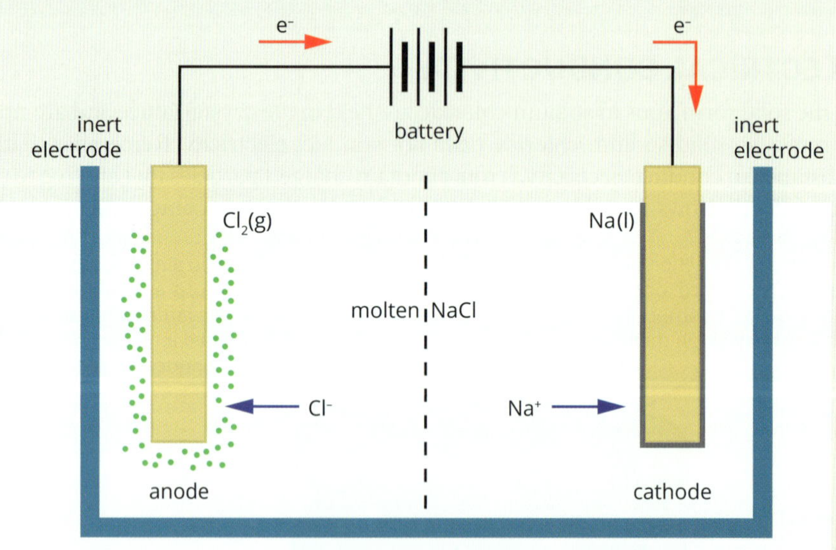

FIGURE 5.2.10 The electrolysis of molten sodium chloride produces sodium metal and chlorine gas.

Sodium metal is a starting material for the manufacture of a range of organic chemicals, including dyes. It is also produced on a large scale by electrolysis. In this case, electricity is passed through molten sodium chloride to make sodium. During this process, sodium ions move towards the negative electrode where they become sodium metal, and chloride ions move to the positive electrode and are converted into chlorine gas (Figure 5.2.10).

For an object to conduct electricity, it must contain charged particles that are free to move. Solid ionic compounds are made up of a crystal lattice, so the particles are not free to move and they cannot conduct electricity. When molten (melted) or dissolved in water, the charged particles are free to move and conduct.

A solution or molten substance that conducts electricity by means of the movement of ions is called an **electrolyte**.

SOLUBILITY

Some ionic compounds are very soluble in water, whereas others are insoluble. When a soluble ionic compound is added to water, the ions break away from the ionic lattice and mix with the water molecules. If an insoluble compound is added to water, the ions remain bonded together in the ionic lattice and do not form a solution.

Whether an ionic compound is soluble or insoluble depends on the relative strength of the forces of attraction between the:

- positive and negative ions in the lattice
- water molecules and the ions.

You will look at the solubility of ionic compounds in water in more detail in Chapter 15.

USES OF IONIC COMPOUNDS

Ionic compounds have a wide variety of uses. Some of these uses are related directly to the physical properties of the compounds, such as hardness, high melting points and their ability to conduct electricity in a solution. A number of ionic compounds and their uses in relation to their characteristic properties are described below.

Hardness

- Calcium phosphate is a constituent of bone tissue that gives it strength.
- Calcium sulfate in the form of gypsum is used to make plasterboard for lining the walls and ceilings of houses.
- Granite, limestone and sandstone are used as building stone.
- Bricks, tiles and crockery are made from clay, a material that contains particles held together by strong ionic bonding.

High melting point

- Magnesium oxide and other ionic compounds are used to line furnaces.
- Ceramics are materials that contain a mixture of strong ionic and covalent bonds. They are used in some engine parts.

Electrical conductivity

- Ammonium chloride is used as an electrolyte in dry cell batteries. Electrolytes contain ions in solution and this allows a current to flow in the battery.
- Potassium hydroxide is used as an electrolyte in the 'button' cells used in small electronic devices such as watches and calculators.

Other uses of ionic compounds

- Sodium hydrogencarbonate (bicarbonate of soda or baking soda) is used in baking to cause cakes to rise when placed in a hot oven. This occurs because the compound decomposes when heated to produce carbon dioxide gas.
- Sodium chloride has many uses, including as a flavouring agent and preservative in food.
- Sodium hypochlorite is used as a bleach and in swimming pools to kill microorganisms.

CHEMISTRY IN ACTION

Salt of the Earth

Salt is added to many foods as a preservative or to 'improve' the flavour. It is rarely essential to add salt—a diet that has meat and milk provides enough sodium for our needs (Tables 5.2.1 and 5.2.2). It is the sodium part of sodium chloride that we should limit in our diets.

TABLE 5.2.1 The sodium content of some natural foods

Food	Sodium content (g Na/100 g food)
beef	0.06
chicken	0.1
fish	0.1
crab meat	0.35
rice	–
cow's milk	0.06
celery	0.15
silverbeet	0.7
fresh fruit	–
most vegetables	0.1

TABLE 5.2.2 The sodium content of some processed foods

Food	Sodium content (g Na/100 g food)
sausage	1.2
ham	1.1
cheddar cheese	0.7
meat pie	0.5
tomato sauce	1.2
soy sauce	7.0
bread	0.5
potato chips	1.0
olives	2.3
salted nuts	4.0
doughnuts	0.5
ice-cream	0.08

A healthy adult requires about 2 grams of sodium per day to replace the sodium lost in urine and perspiration (sweat). Salt plays a role in the passage of nerve impulses and helps control the movement of water in and out of cells. A salt deficiency can lead to muscular cramps, loss of appetite and reduced brain function.

Doctors are increasingly critical of the amount of salt in our diets because of the connection between dietary salt and hypertension, or high blood pressure. The average Australian consumes 11 grams of salt daily. When you consider that a take-away meal can contain about 3.5 grams of salt, it is easy to see how you could reach this level. Food manufacturers have responded to public demand by decreasing the salt content in processed foods.

The body attempts to maintain the salt concentration within narrow limits. After a salty meal, you feel thirsty—your body's response to the increased salt concentration in the blood. You drink water to dilute the salt. This excess fluid remains in your body until the kidneys have removed the excess salt.

Ancient civilisations recognised that we need salt in our diets. Settlements were based in regions where salt occurred naturally or on salt trade routes. Wars were even fought over salt, with Roman soldiers being paid a salt allowance (the word 'salary' comes from the Latin word for salt). Salt is harvested from salt lakes, salt mines or the sea (Figure 5.2.11). Solar energy is commonly used to extract salt from seawater.

FIGURE 5.2.11 Harvesting salt from evaporation lakes

Salinity

Salinity is the presence of salt in groundwater and soil. Plants, like animals, can only tolerate a limited range of salt concentrations. Some plants growing on sand dunes tolerate higher amounts of salt than others. However, if the salt concentration in the soil is too high, even those plants die.

Salinity is caused by increasing amounts of groundwater (water found under the surface of the Earth). As groundwater increases, the water table (water level) rises, bringing dissolved salts, in this case sodium chloride, closer to the surface. The increased salinity of water near the soil surface leads to poor plant growth and reduces the productivity of agricultural land that is used for grazing animals and growing crops.

The rise in the water table has been caused by excess water from irrigation flowing down through the soil, and the clearing of trees, which would otherwise have absorbed this water through their roots.

Reducing salinity

A number of organisations are working to reduce soil and water salinity. Strategies include:

- using irrigation water more efficiently
- improving drainage on the surface of the land
- improving drainage under the surface
- growing trees to soak up excess water
- sealing irrigation channels to prevent leakage.

These are long-term projects, although reduced salinity is apparent in some areas already.

One property of salt that has the potential to cause environmental damage is its ability to dissolve in water (Figure 5.2.12).

FIGURE 5.2.12 Land that is affected by salinity

5.2 Review

SUMMARY

- Ionic compounds are hard and have high melting and boiling points. This is because of the strong forces of attraction between the positively and negatively charged ions in the ionic lattice.
- When an ionic compound is hit, the ions move within the lattice so that like-charged ions line up opposite each other and then repel, causing the lattice to be disrupted. This makes ionic compounds brittle.
- Ionic compounds do not conduct electricity in the solid state. Although the solid ionic lattice contains charged particles, the particles are not free to move.
- When ionic compounds are dissolved in water or are in molten form, the charged particles are free to move, which means they can conduct electricity.
- In water, ionic compounds vary from very soluble to insoluble. The solubility depends on whether the forces between the water molecules and the ions in the lattice are strong enough to pull the ions out of the lattice.
- Ionic compounds are useful because of their physical properties such as hardness and high melting points.

KEY QUESTIONS

1 Why do ionic compounds have such high melting and boiling points?

2 **a** Explain why the lattice breaks apart in an ionic compound when hit with a heavy blow.

b How does this give ionic compounds their brittle property?

3 Use the ionic bonding model to explain why ionic compounds that have been heated above their melting point conduct electricity but will not when in solid form.

4 Sodium chloride or common salt is an essential part of our diet. Throughout history, it has also been used to preserve food in the absence of refrigeration. So important was sodium chloride to daily life that many words or expressions in the English language are derived from the word 'salt'. 'Salary' and 'salinity' are two of them. What others can you think of?

5.3 Formation of ionic compounds

Some of the reactions that occur between metals and non-metals to form ionic compounds are very vigorous. The reaction between sodium and chlorine to form sodium chloride produces a lot of heat. You will remember from Chapter 4 that sodium is very reactive. At high temperatures, the production of sodium chloride from sodium metal and chlorine gas is very explosive, producing a flame and large amounts of energy.

In the previous sections, you saw that the positive and negative ions that are formed in this reaction are arranged to form a three-dimensional lattice. In this section, you will learn how these positive and negative ions are formed from the atoms. You will also learn how the ratio of each type of atom in the compound is determined.

FORMING IONS

When forming an ionic compound, two things occur.

- One group of atoms loses electrons to form positively charged ions (cations). These are generally metals.
- Another group of atoms gains electrons to form negatively charged ions (anions). These are generally non-metals.

From Chapter 3, you will remember that most metals have low ionisation energies and low electronegativities. This means that the valence electrons on a metal atom are weakly held and can be removed easily.

Non-metals, on the other hand, have a higher ionisation energy and higher electronegativity. This means that non-metal atoms have a high affinity for valence electrons and they are not easily removed.

This is why in the reaction that forms many ionic compounds, it is the non-metal atom that takes one or more valence electrons from the metal atoms.

Most importantly, the ions that are formed usually have a full valence shell—a stable electron configuration. Most of the main group elements can accommodate eight electrons in their valence shell. So in the process of forming ions, both positive and negative, the tendency is for ions to end up with eight electrons in their outer shell (valence shell). You will recall that this is known as the **octet rule**.

Noble gases (group 18) are elements that already have the most stable valence-shell configuration. Therefore, another way of thinking about the octet rule is that atoms tend to gain or lose electrons to gain a stable electron configuration identical to that of the noble gas nearest to them on the periodic table. The formation of stable ions is a powerful driving force in reactions between metals and non-metals when they produce ionic compounds.

For example, when sodium reacts with chlorine, each sodium atom loses one electron and each chlorine atom gains one electron.

After the reaction:

- the sodium ion has the stable electron shell configuration of 2,8 (the same as a neon atom)
- the chloride ion has the stable electron shell configuration of 2,8,8 (the same as an argon atom).

Figure 5.3.1 illustrates how, when sodium reacts with chlorine, an electron is lost by a sodium atom and gained by a chlorine atom. A diagram of this type is called an **electron transfer diagram**.

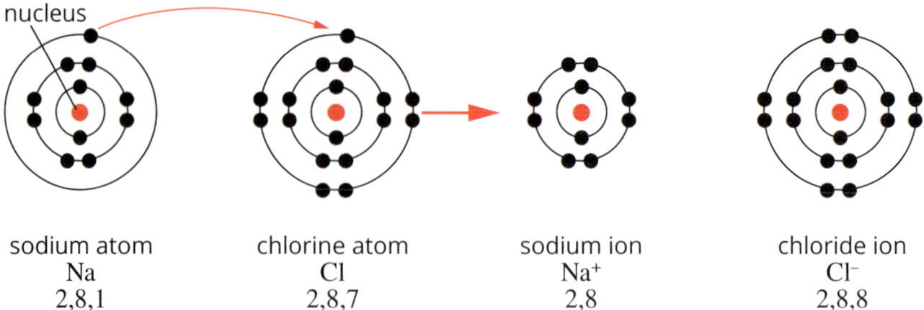

FIGURE 5.3.1 Electron transfer diagram showing the formation of sodium and chloride ions

The reaction between lithium and oxygen atoms is illustrated in Figure 5.3.2. In this reaction, an oxygen atom needs to gain two electrons to have eight electrons in its outer shell and form a stable ion. To allow this to happen, one oxygen atom will react with two lithium atoms, taking one electron from each atom.

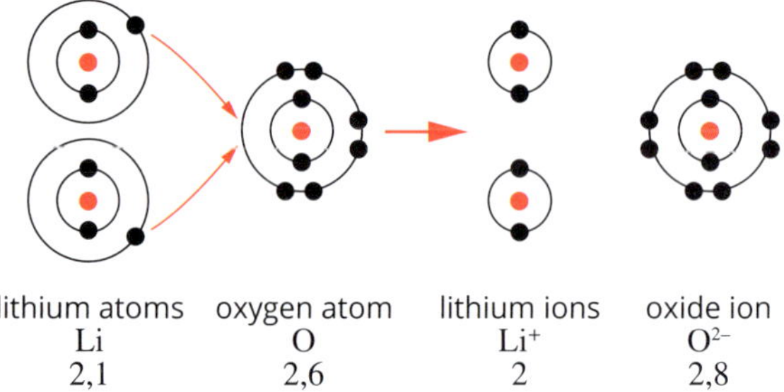

FIGURE 5.3.2 An electron transfer diagram showing the formation of lithium and oxide ions

After the reaction, there are just two electrons in what is now the outer shell of the Li^+ ion, which is the same as the electron configuration of a helium atom.

In Chapter 3, you learnt that atoms of elements in group 1 of the periodic table have electron configurations with one electron in the valence shell. Therefore, elements in this group often form cations with a charge of +1 as they readily lose this one outer-shell electron. For example, potassium has an electron configuration of 2,8,8,1. It will readily lose the one electron in the valence shell to have an electron configuration similar to that of argon (the nearest noble gas). The potassium ion now has an electron configuration of 2,8,8.

Metals in group 2 of the periodic table have electron configurations with two electrons in their valence shells. Therefore, they readily form ions with a charge of +2 as they lose these electrons. Magnesium is a metal in group 2 of the periodic table. The loss of the two valence electrons from an atom of magnesium is shown in Figure 5.3.3.

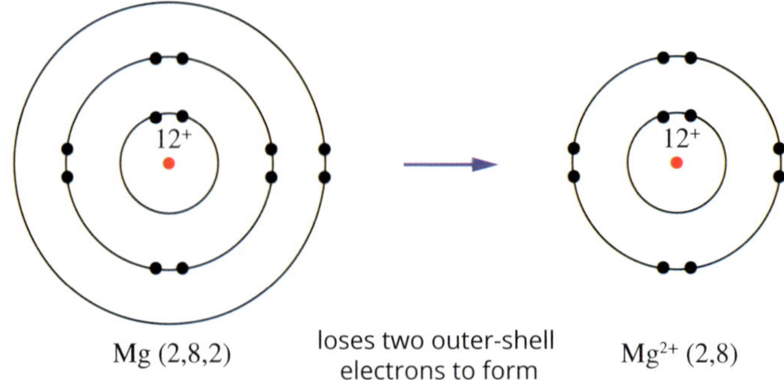

FIGURE 5.3.3 When two valence electrons are lost from a magnesium atom, the Mg^{2+} ion formed has a complete outer shell that contains eight electrons.

EXTENSION

Beryllium is an exception in group 2

Beryllium is found in group 2 of the periodic table. However, beryllium behaves very differently from the other alkaline earth metals in this group.

Beryllium has an electron configuration of 2,2. It could lose two electrons to form an ion similar to lithium with just one remaining valence shell (that is, full with two electrons). However, it does not do this because beryllium has a very high electronegativity compared with the rest of the group 2 metals. It tends to hold on to these two electrons. Beryllium mostly forms covalent compounds (which you will learn about in Chapter 6). For example, beryllium covalently bonds with two chlorine atoms to form beryllium chloride ($BeCl_2$). Beryllium chloride is used in the production of beryllium by electrolysis.

Non-metals in group 16 have six electrons in their valence shells, therefore they gain two electrons to form anions with a charge of −2.

Group 17 non-metals have seven electrons in their valence shell. They readily gain one electron to fill the valence shell according to the octet rule. This means they form anions with a charge of −1.

Atoms are at their most stable when they have a full valence shell. Most of the main group elements can accommodate eight valence electrons. This is known as the octet rule.

ELECTRON TRANSFER DIAGRAMS

Electron transfer diagrams can be used to show how electrons are transferred from metallic atoms to non-metallic atoms to form an ionic compound. Figure 5.3.4 shows such a diagram for the reaction of magnesium with oxygen.

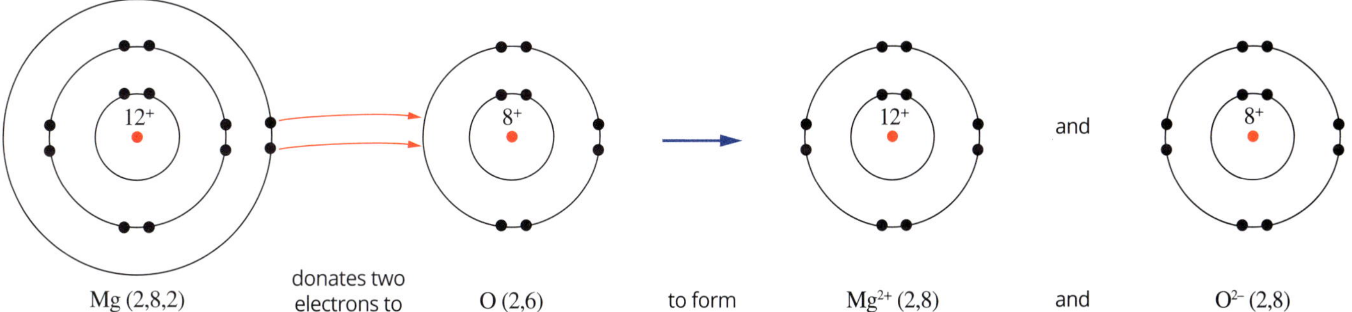

FIGURE 5.3.4 An electron transfer diagram showing the transfer of electrons when magnesium atoms react with oxygen atoms

When an atom loses electrons, it becomes more positively charged as the number of protons no longer equals the number of electrons. The ion formed is written with a superscript + sign indicating the charge. If there are two electrons lost it is written as 2+. When an atom gains electrons, in the case of anions, they become negatively charged. This is written with a superscript −. For example, if three electrons are gained, the charge is written as 3− in superscript.

Figure 5.3.5 shows the diagram for the reaction of magnesium with chlorine. A magnesium atom has an electron configuration of 2,8,2 so it will readily lose its two outer-shell electrons. A magnesium ion (Mg^{2+}) is formed so that it now has the electron configuration of the nearest noble gas, neon (2,8). Each chlorine atom has an electron configuration of 2,8,7, so each will gain one outer-shell electron. A chloride ion (Cl^-) has the electron configuration of the noble gas argon (2,8,8). Because magnesium readily donates two electrons, it is found with two chlorine atoms, each chlorine atom taking one electron from magnesium.

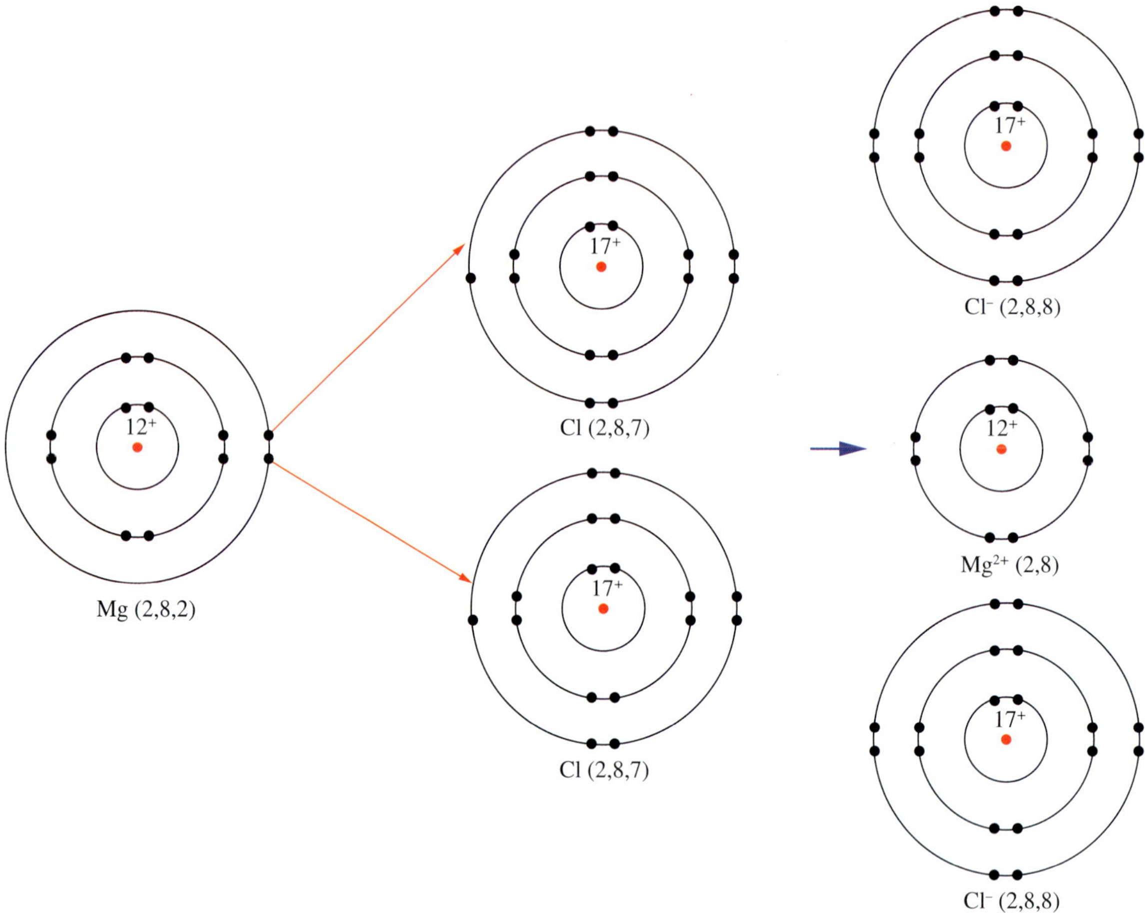

FIGURE 5.3.5 Electron transfer diagram showing the formation of the ionic compound magnesium chloride

To make sure you understand how ions are formed when metals and non-metals react, it can be useful to write equations for reactions that include the electron configurations of the reactants and products, as shown in Worked example 5.3.1.

Worked example 5.3.1

WRITING EQUATIONS FOR REACTIONS BETWEEN METALS AND NON-METALS ATOMS

Write an equation for the reaction between lithium and nitrogen atoms. Show the electron configurations for each element before and after the reaction.	
Thinking	**Working**
Write the symbol and electron configuration for the metal atom.	Li (2,1)
How many electrons will the metal atom lose from its outer shell when it reacts?	1
Write the symbol and electron configuration of the metal ion that will be formed.	Li^{+} (2)
Write the symbol and electron configuration for the non-metal atom.	N (2,5)
How many electrons will the non-metal atom gain in its outer shell when it reacts?	3
Write the symbol and electron configuration of the non-metal ion that will be formed.	N^{3-} (2,8)
The total number of electrons lost by metal atoms must equal the total number of electrons gained by non-metal atoms. What is the lowest number ratio of metal atoms to non-metal atoms that will allow this to happen?	metal atom : non-metal atom = 3:1
Using the ratio of metal ion : non-metal ion calculated above, write a balanced equation for the reaction. Show the electron configurations for both the reactant atoms and the product ions.	3Li (2,1) + N (2,5) → $3Li^{+}$ (2) + N^{3-} (2,8)

Worked example: Try yourself 5.3.1

WRITING EQUATIONS FOR REACTIONS BETWEEN METALS AND NON-METALS ATOMS

Write an equation for the reaction between calcium and phosphorus atoms. Show the electron configurations for each element before and after the reaction.

5.3 Review

SUMMARY

- During the formation of ionic compounds, metal atoms lose electrons to form positively charged ions (cations).
- During the formation of ionic compounds, non-metal atoms gain electrons to form negatively charged ions (anions).
- The ions present in an ionic compound have a stable electron configuration identical to that of the noble gas nearest to them on the periodic table.
- Electron transfer diagrams can be used to represent the formation of an ionic compound from its elements.
- When an ionic compound is formed from positively charged metal ions and negatively charged non-metal ions, the ions combine in proportions that produce an ionic compound with an overall zero charge.

KEY QUESTIONS

1 Indicate whether the following atoms will form cations or anions and explain why.
  - **a** calcium
  - **b** nitrogen
  - **c** fluorine
  - **d** aluminium
  - **e** phosphorus

2 Use diagrams similar to Figure 5.3.1 on page 112 to show the formation of ions in the reactions between:
  - **a** potassium and fluorine
  - **b** magnesium and sulfur
  - **c** aluminium and fluorine
  - **d** sodium and oxygen
  - **e** aluminium and oxygen.

3 Why are group 2 metals of the periodic table likely to form cations with a charge of +2?

4 Explain why potassium chloride has the formula KCl, whereas the formula of calcium chloride is $CaCl_2$.

5 Using the technique demonstrated in Worked example 5.3.1 on page 115, write an equation for the reaction between the following metal and non-metal atoms. Show the electron configurations for each element before and after the reaction.
  - **a** sodium and chlorine atoms
  - **b** magnesium and oxygen atoms
  - **c** aluminium and sulfur atoms

5.4 Chemical formulae of simple ionic compounds

You have seen that ionic compounds contain oppositely charged ions that are arranged in three-dimensional lattices. The ions can have different charges. For example, aluminium forms ions with a +3 charge, whereas oxygen forms ions with a −2 charge.

In this section, you will learn how to use your knowledge of the charges on ions to write an overall formula for an ionic compound.

WRITING THE FORMULA OF AN IONIC COMPOUND

Because ionic compounds are electrically neutral, the total number of positive charges on the metal ions must equal the total number of negative charges on the non-metal ions. This is the most important guiding principle when you are trying to work out the formula of an ionic compound.

This can be seen with the formula of the ionic compound sodium chloride.

- A sodium ion (Na^+) has a +1 charge.
- A chloride ion (Cl^-) has a −1 charge.
- Therefore, in a crystal of sodium chloride, the ratio of sodium ions to chloride ions is 1:1 and the formula of sodium chloride is NaCl.

Using the same steps, you can work out the formula of magnesium chloride.

- A magnesium ion (Mg^{2+}) has a +2 charge.
- A chloride ion (Cl^-) has a −1 charge.
- Therefore, in a crystal of magnesium chloride, two chloride ions are needed to provide two negative charges so that they balance the +2 charge on every magnesium ion. Therefore, the ratio of magnesium ions to chloride ions in the crystal is 1:2 and the formula of magnesium chloride is $MgCl_2$.

Figure 5.4.1 illustrates how formulae for some other ionic compounds can be determined.

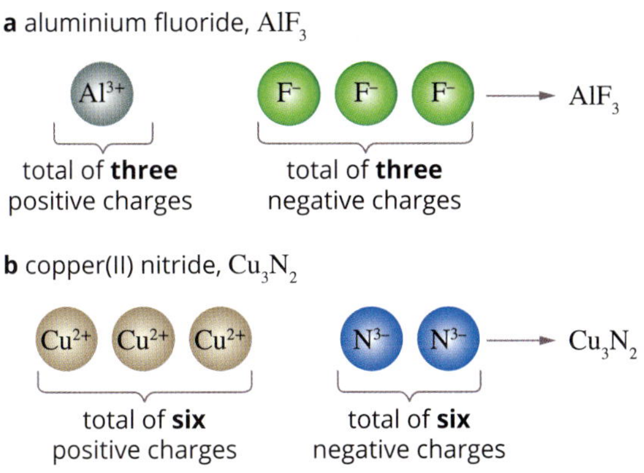

FIGURE 5.4.1 How to deduce chemical formulae from the charges on ions

Tables 5.4.1 and 5.4.2 list some of the more common positively and negatively charged ions. You may use these when you are writing formulae for ionic compounds.

TABLE 5.4.1 Names and formulae of some common cations

Charge		
+1	+2	+3
caesium, Cs^+ hydrogen, H^+ lithium, Li^+ potassium, K^+ rubidium, Rb^+ silver, Ag^+ sodium, Na^+	barium, Ba^{2+} calcium, Ca^{2+} cobalt(II), Co^{2+} copper(II), Cu^{2+} iron(II), Fe^{2+} lead(II), Pb^{2+} magnesium, Mg^{2+} manganese(II), Mn^{2+} nickel, Ni^{2+} strontium, Sr^{2+} zinc, Zn^{2+}	aluminium, Al^{3+} chromium(III), Cr^{3+} iron(III), Fe^{3+}

TABLE 5.4.2 Names and formulae of some common anions

Charge		
−1	−2	−3
bromide, Br^- chloride, Cl^- fluoride, F^- iodide, I^-	oxide, O^{2-} sulfide, S^{2-}	nitride, N^{3-}

Rules for writing chemical formulae

Here are some simple rules to follow when you are writing chemical formulae.

- Write the symbol for the positively charged ion first.
- Use subscripts to indicate the number of each ion in the formula. Write the subscripts after the ion they refer to.
- If there is just one ion present in the formula, omit the subscript '1'.
- Do not include the charges on the ions in the balanced formula.

These rules are illustrated in Figure 5.4.2.

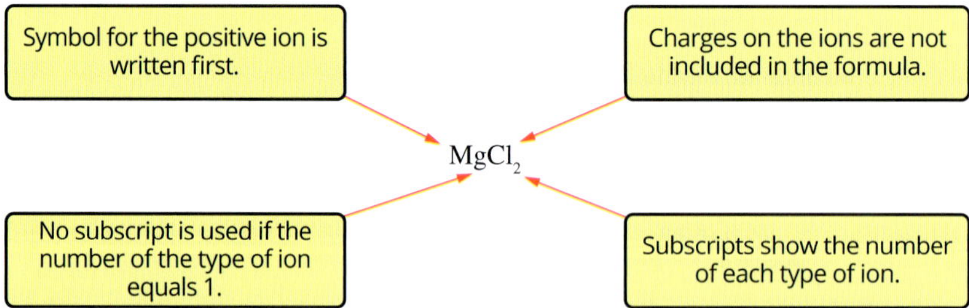

FIGURE 5.4.2 This diagram summarises the information provided by a chemical formula.

Worked example 5.4.1

STEPS IN WRITING A CHEMICAL FORMULA

Determine the chemical formula of the ionic compound formed between zinc and nitride ions. You may need to refer to Tables 5.4.1 and 5.4.2.	
Thinking	**Working**
Write the symbol and charge of the two ions forming the ionic compound.	Zn^{2+} and N^{3-}
Calculate the lowest common multiple of the two numbers in the charges of the ions.	$2 \times 3 = 6$
Calculate how many positive ions are needed to equal the lowest common multiple.	three Zn^{2+} ions
Calculate how many negative ions are needed to equal the lowest common multiple.	two N^{3-} ions
Use the answers from the previous two steps to write the formula for the ionic compound. Write the symbol of the positive ion first. (Note that 1 is not written as a subscript.)	Zn_3N_2

Worked example: Try yourself 5.4.1

STEPS IN WRITING A CHEMICAL FORMULA

Determine the chemical formula of the ionic compound formed between barium and fluoride ions. You may need to refer to Tables 5.4.1 and 5.4.2.

5.4 Review

SUMMARY

- When considering the formula of an ionic compound, the total number of positive charges on the metal ions must equal the total number of negative charges on the non-metal ions.
- Follow these rules when writing formulae of ionic compounds.
  - The symbol for the positively charged ion is written first.
  - Subscripts are used to indicate the number of each ion in the formula.
  - The charges on the ions are not included in the balanced formula as they cancel out.

KEY QUESTIONS

1 A_2B is an ionic compound. Both of the ions in A_2B have the same electron configuration as an argon atom. What is the identity of A_2B?

A calcium chloride
B potassium sulfide
C calcium sulfide
D sodium oxide

2 Listed below are pairs of metal and non-metal ions. When the ions in each pair combine to form ionic compounds, what will be the ratio of metal ion to non-metal ion?

a K^+ and S^{2-}
b Al^{3+} and F^-
c Ca^{2+} and N^{3-}
d Al^{3+} and P^{3-}
e Mg^{2+} and Cl^-

Use the information in Tables 5.4.1 and 5.4.2 on page 118 to answer Questions **3** and **4**.

3 Write formulae for the following ionic compounds.

a sodium chloride
b potassium bromide
c zinc chloride
d potassium oxide
e barium bromide
f aluminium iodide
g silver bromide
h zinc oxide
i barium oxide
j aluminium sulfide

4 Name the ionic compounds with the following formulae.

a KCl
b CaO
c MgS
d K_2O
e NaF

5.5 Writing formulae of more complex ionic compounds

The chemical formulae you have written for ionic compounds so far contain simple ions—ions that contain only one atom of an element. In this section, you will look at the formulae of some compounds containing more complex ions. You will also learn rules to help you name ionic compounds.

FORMULAE CONTAINING POLYATOMIC IONS

Simple ions contain only one atom, as you have seen in Section 5.4. However, other ions contain two or more atoms, which may be of different elements. These ions are called **polyatomic ions**.

The following rules apply to polyatomic ions.

- If different elements are present, then they are combined in a fixed ratio.
- The group of atoms behaves as a single unit with a specific charge.
- Subscripts are used to indicate the number of each kind of atom in the ion.

For example, a carbonate ion (CO_3^{2-}) contains one carbon atom and three oxygen atoms combined together to form an ion. The carbonate ion has a charge of −2. Other polyatomic ions are nitrate (NO_3^-), hydroxide (OH^-) and phosphate (PO_4^{3-}). The formulae of a number of polyatomic ions can be seen in Table 5.5.1.

TABLE 5.5.1 Common polyatomic cations and anions

Charge			
+1	−1	−2	−3
ammonium, NH_4^+	cyanide, CN^- dihydrogenphosphate, $H_2PO_4^-$ ethanoate (acetate), CH_3COO^- hydrogencarbonate, HCO_3^- hydrogensulfate, HSO_4^- hydroxide, OH^- nitrate, NO_3^- nitrite, NO_2^- permanganate, MnO_4^-	carbonate, CO_3^{2-} chromate, CrO_4^{2-} dichromate, $Cr_2O_7^{2-}$ hydrogenphosphate, HPO_4^{2-} oxalate, $C_2O_4^{2-}$ sulfate, SO_4^{2-} sulfite, SO_3^{2-}	phosphate, PO_4^{3-}

The common name for the hydrogencarbonate ion is bicarbonate.

If more than one polyatomic ion is required in a formula to balance the charge, then it is placed in brackets with the required number written as a subscript after the brackets. Some examples are:

- magnesium nitrate, $Mg(NO_3)_2$
- aluminium hydroxide, $Al(OH)_3$
- ammonium sulfate, $(NH_4)_2SO_4$.

Note that brackets are not required for the formula of sodium nitrate ($NaNO_3$), where there is only one NO_3^- ion present for each sodium ion.

Polyatomic ions are made up of two or more atoms charge. They need to be written within brackets if there is more than one present in an ionic formula. Subscripts are used to indicate the ratio of ions in the crystal lattice.

A formula that is expressed in terms of the simplest whole-number ratio of particles (in this case the particles are ions) is called an **empirical formula**.

The formulae of several other ionic compounds containing polyatomic ions are shown in Figure 5.5.1.

a magnesium nitrate, $Mg(NO_3)_2$

Mg^{2+} — total of **two** positive charges

NO_3^- NO_3^- — total of **two** negative charges

→ $Mg(NO_3)_2$

b iron(III) sulfate, $Fe_2(SO_4)_3$

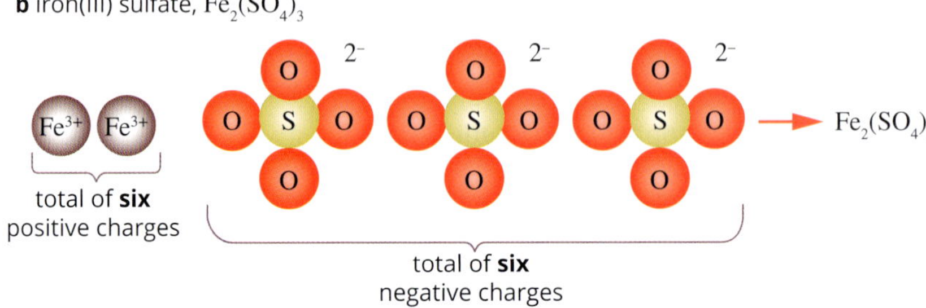

FIGURE 5.5.1 The chemical formulae of ionic compounds containing polyatomic ions

Formulae involving elements with multiple valencies

The **valency** of an atom is a measure of its combining ability with other atoms. For main group metals the valency is equal to the numeral of the ion's charge. Some transition metals, however, can form ions with different charges. For example, copper can form Cu^+ ions with a charge of +1 and Cu^{2+} ions with a charge of +2.

Some other metals with variable valency are:

- lead (Pb^{2+} and Pb^{4+})
- iron (Fe^{2+} and Fe^{3+})
- tin (Sn^{2+} and Sn^{4+}).

For compounds of these metals, you need to specify the charge on the ion when naming the compound. This is done by placing a Roman numeral immediately after the metal in the name of the compound. For example:

- iron(II) chloride contains the Fe^{2+} ion and so the formula is $FeCl_2$
- iron(III) chloride contains the Fe^{3+} ion and so the formula is $FeCl_3$
- copper(I) sulfide contains the Cu^+ ion and so the formula is Cu_2S.

Some transition metals can form ions with variable charges. To indicate the charge on the metal, write a Roman numeral in brackets immediately after the metal in the name of the compound.

NAMING IONIC COMPOUNDS

There are some basic conventions that are followed when naming ionic compounds.

- The name of a positively charged ion (cation) remains unchanged. For example, the cation of a sodium atom is called a sodium ion; the cation of an aluminium atom is an aluminium ion.
- For simple non-metal ions (anions), the name of the ion is similar to that of the atom, but ends in '-ide'. For example, the anion of the chlorine atom is chloride; the anion of the oxygen atom is oxide.
- For polyatomic anions containing oxygen, the name of the ion will usually end in '-ite' or '-ate'. For example, the NO_2^- ion is called a nitrite ion; the NO_3^- ion is called a nitrate ion. (For two different ions of the same element with oxygen, the name of the ion with the smaller number of oxygen atoms usually ends in '-ite' and the one with the larger number of oxygen atoms ends in '-ate'.)

5.5 Review

SUMMARY

- When writing formulae for ionic compounds, the following rules apply.
  - Ions that contain two or more atoms of different elements are called polyatomic ions.
  - If a chemical formula contains more than one polyatomic ion, the formula of the ion is placed in brackets with the number of ions written as a subscript after the brackets.
- When naming ionic compounds, the following rules apply.
  - The name of the metal ion is the same as the name of the metal.
  - Simple non-metal ions take the name of the atom, but end in '-ide'.
  - Polyatomic anions containing oxygen usually end in '-ite' or '-ate'.
  - For metals that form ions with different charges, the charge on the ion is shown by placing a Roman numeral after the name of the metal.
- The valency of a metal is equal to the charge of its ions.

KEY QUESTIONS

Use the information in Tables 5.4.1 and 5.4.2 on page 118 and Table 5.5.1 on page 121 to answer these questions.

1 Write formulae for the following ionic compounds.

a sodium carbonate
b barium nitrate
c aluminium nitrate
d calcium hydroxide
e zinc sulfate
f potassium hydroxide
g potassium nitrate
h zinc carbonate
i potassium sulfate
j barium hydroxide

2 Write formulae for the following ionic compounds.

a caesium chloride
b iron(III) oxide
c copper(II) oxide
d chromium(III) sulfate
e iron(II) oxide
f aluminium oxide
g calcium nitrate

3 Name the following ionic compounds.

a $Mg(OH)_2$
b Na_2CO_3
c $FeSO_4$
d $CuSO_4$
e $Ba(NO_3)_2$
f Cu_2SO_4
g $Fe_2(SO_4)_3$
h NH_4NO_3
i Na_2HPO_4

Chapter review

05

KEY TERMS

chemical formula
crystal lattice
electrolyte
electron transfer diagram
empirical formula
ionic bonding
ionic compound
polyatomic ion
valency

Properties and structures of ionic compounds

1 Describe an experiment you could carry out to demonstrate each of the following properties of the compounds given. In each case, describe what you would expect to observe.

a Solid magnesium chloride does not conduct electricity.

b Molten sodium chloride does conduct electricity.

c Solid sodium chloride is hard and brittle.

2 State whether the following properties relate to metallic lattices only, ionic lattices only or both metallic and ionic lattices.

a They contain both positively and negatively charged particles.

b The lattice is held together by forces of attraction between positively and negatively charged particles.

c They are hard.

d They are brittle.

e They conduct electricity in the molten state.

3 Copy and complete the following diagram to show what happens when ionic compounds are formed from metallic and non-metallic elements.

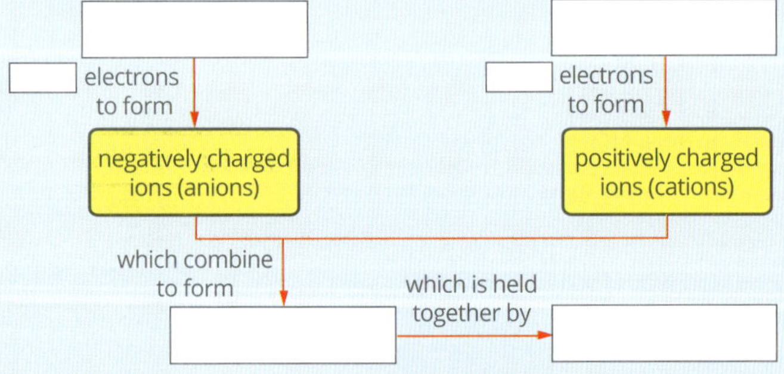

Using the ionic bonding model to explain properties

4 Use the ionic bonding model to explain the following properties of ionic compounds.

a They generally have high melting points.

b They are hard and brittle.

c They do not conduct electricity in the solid state but will conduct when molten or dissolved in water.

5 The melting point of sodium chloride is 801°C, whereas that of magnesium oxide is 2800°C.

a What particles are present in the two solids?

b Which solid has the stronger forces between its particles?

c Give some possible differences in the structure and bonding of the two solids that would explain the large difference in melting points.

6 Describe what happens to the forces between particles as sodium chloride is heated and melts.

7 The following is a list of some of the uses of ionic compounds. For each use, explain what property of ionic compounds enables them to be used in this way.

a insulators on electrified fences

b bricks for building the wall of a house

c one of the chemicals in a battery

Formation of ionic compounds

8 Use diagrams to show the electron transfer that occurs when:

a lithium reacts with chlorine

b magnesium reacts with fluorine

c potassium reacts with sulfur

d magnesium reacts with nitrogen.

9 Give the electron configurations of the following ions.

a Na^+

b O^{2-}

c Mg^{2+}

d N^{3-}

10 The electron configurations of some metallic and non-metallic elements are given. (The symbols shown for the elements are not their real ones.) Write formulae for the compounds they are most likely to form if they react together. The first example has been done for you.

a	A: 2,1	B: 2,6	A_2B
b	C: 2,8,3	D: 2,7	
c	E: 2,8,8,2	F: 2,8,6	
d	G: 2,8,8,1	H: 2,5	
e	K: 2,8,2	L: 2,6	

11 Refer to the periodic table at the end of the book and, for each general formula given, identify two elements that will react to form an ionic compound with that formula. (Remember the metal ion, as represented by X, is written first in each formula.)

  a XY_2

  b XY

  c X_2Y

  d X_3Y

  e XY_3

  f X_3Y_2

12 Explain why elements in group 17 of the periodic table are likely to forms ions with a –1 charge.

13 Write an equation for the reaction between the following metal and non-metal atoms. Show the electron configurations for each element before and after the reaction.

  a magnesium and chlorine atoms

  b aluminium and oxygen atoms

Chemical formulae of simple ionic compounds

14 Write the empirical formula for the ionic compound formed in the reaction between:

  a potassium and bromine

  b magnesium and iodine

  c calcium and oxygen

  d aluminium and fluorine

  e calcium and nitrogen.

15 Write the empirical formula for each of the following ionic compounds.

  a copper(I) chloride

  b silver oxide

  c lithium nitride

  d potassium iodide

16 What do subscripts in the formula of an ionic compound tell you about the arrangement of the metal and non-metal ions?

Writing formulae of more complex ionic compounds

17 The elements X, Y and Z form ionic compounds when they react with other elements. The following compounds are formed: Ca_3X_2, Y_2CO_3 and Al_2Z_3.

  a What is the charge of the ion formed by:

    i element X?

    ii element Y?

    iii element Z?

  b Use these charges on the ions to write correct chemical formulae for the:

    i sulfate salt of Y

    ii potassium salt of Z

    iii ionic compound formed between X and Y

    iv ionic compound formed between Y and Z.

18 Write the empirical formulae for the following ionic compounds.

  a copper(I) nitrate

  b chromium(II) fluoride

  c potassium carbonate

  d magnesium hydrogencarbonate

  e nickel(II) phosphate

19 Name the ionic compounds with the following chemical formulae.

  a $(NH_4)_2CO_3$

  b $Cu(NO_3)_2$

  c $CrBr_3$

Connecting the main ideas

20 A student compares the structure and bonding in metals and ionic compounds and makes the following statements. Comment on each of these statements, explaining clearly why you either agree or disagree.

  a Metals and ionic solids both contain positive ions in a regular arrangement.

  b In metals and ionic solids, there is attraction between one particle and all the neighbouring particles of opposite charge.

  c In metals and ionic solids, there will be forces of repulsion between particles with like charges.

  d In metals and ionic solids that contain metal cations, energy has been used to remove valence electrons from metal atoms.

  e A metal will conduct electricity, whereas an ionic solid will not because electrons are much smaller than negative ions.

21 Construct a concept map to show the connection between these terms: metal, non-metal, atom, valence electron, anion, cation, electrostatic attraction, ionic bonding.

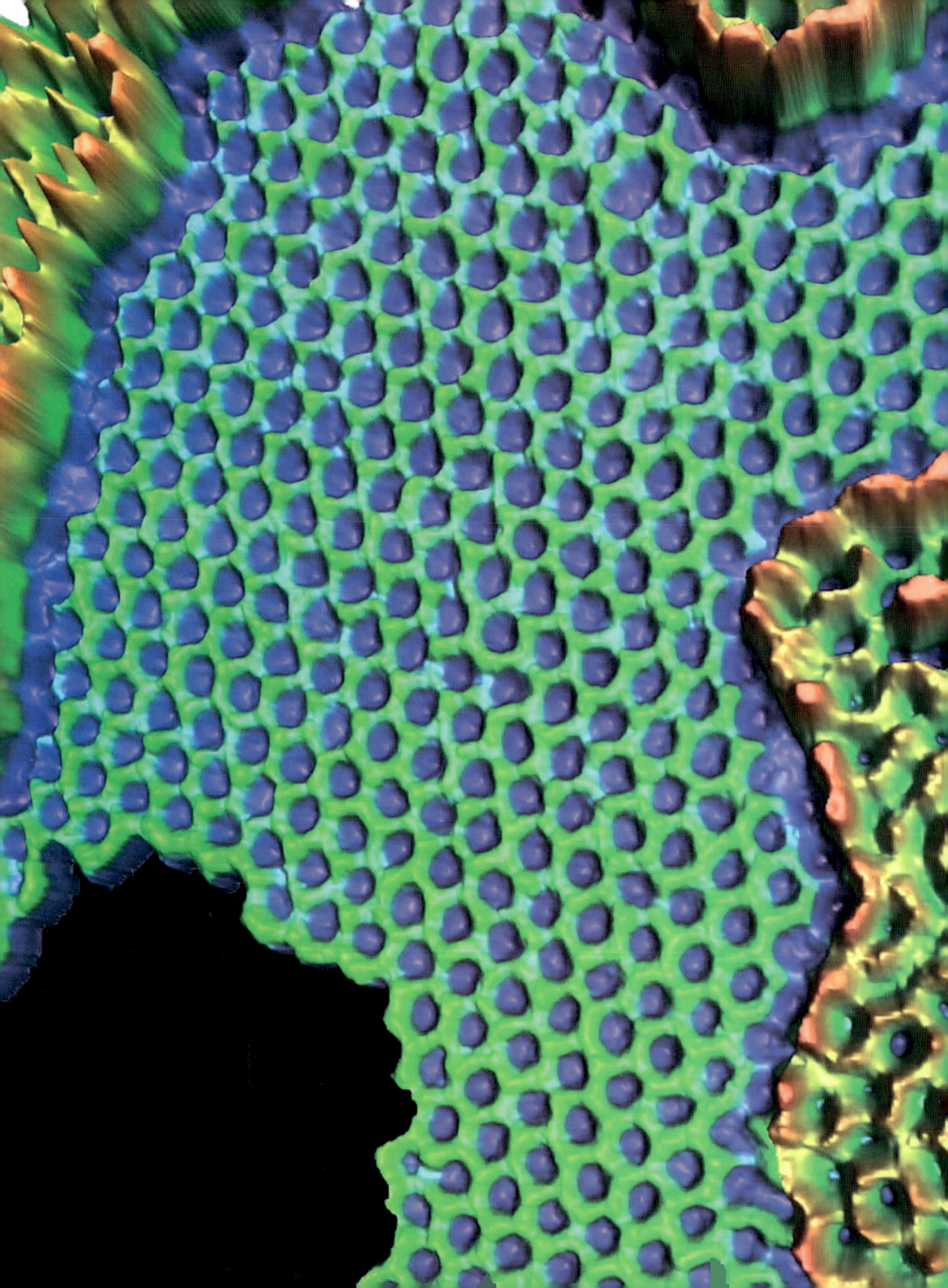

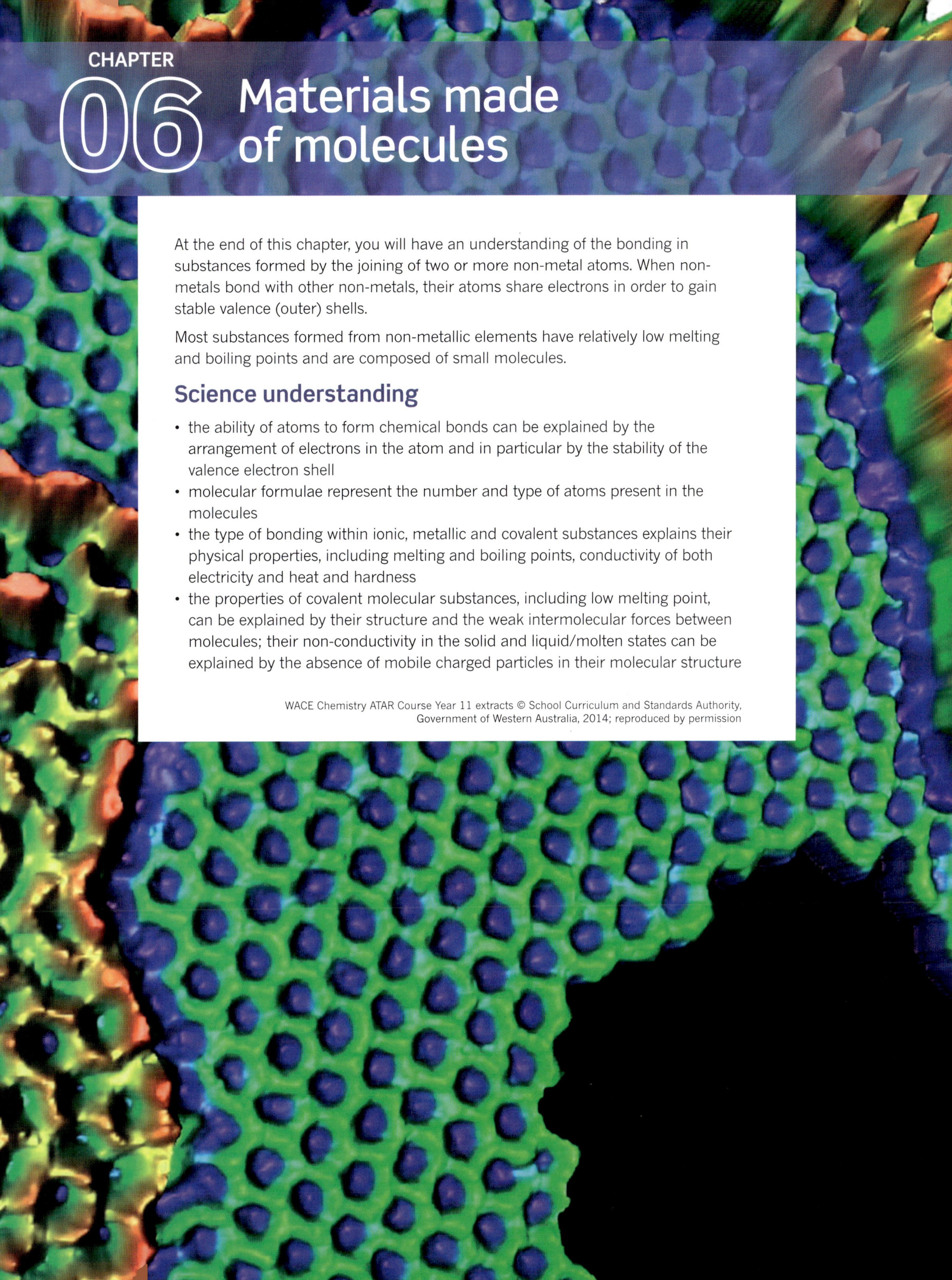

CHAPTER

06 Materials made of molecules

At the end of this chapter, you will have an understanding of the bonding in substances formed by the joining of two or more non-metal atoms. When non-metals bond with other non-metals, their atoms share electrons in order to gain stable valence (outer) shells.

Most substances formed from non-metallic elements have relatively low melting and boiling points and are composed of small molecules.

Science understanding

- the ability of atoms to form chemical bonds can be explained by the arrangement of electrons in the atom and in particular by the stability of the valence electron shell
- molecular formulae represent the number and type of atoms present in the molecules
- the type of bonding within ionic, metallic and covalent substances explains their physical properties, including melting and boiling points, conductivity of both electricity and heat and hardness
- the properties of covalent molecular substances, including low melting point, can be explained by their structure and the weak intermolecular forces between molecules; their non-conductivity in the solid and liquid/molten states can be explained by the absence of mobile charged particles in their molecular structure

6.1 Properties of non-metallic substances

In Chapter 4, you saw that the bonding between atoms in metallic elements is called metallic bonding. When metal atoms combine with atoms of non-metallic elements, the compounds contain another form of bonding; ionic bonding as described in Chapter 5. In this chapter, you will look at the chemical bonding that occurs when atoms of non-metals combine with each other.

EXAMPLES OF NON-METAL COMPOUNDS

Although there are fewer non-metals than metals in the periodic table, the atoms of non-metals form a much larger number of compounds than metals. Water, carbon dioxide, caffeine, sugar and cooking oil are just a few examples of common compounds formed from non-metals.

PROPERTIES OF NON-METALLIC ELEMENTS AND COMPOUNDS

The properties of compounds can tell you a lot about their chemical structures. Table 6.1.1 lists properties of a range of common substances. The substances chosen include a metal, ionic compounds and non-metal compounds. The information in Table 6.1.1 is useful because it allows you to:

- identify the characteristic properties of each category of substances; for example, metals conduct electricity as solids and liquids
- deduce information about the chemical structures of these substances. For example, if a substance has a high melting point, the bonding between particles must be strong. If a substance conducts electricity, it must contain charged particles that are free to move.

TABLE 6.1.1 Properties of some common metals, ionic compounds and non-metal compounds

Substance	Formula	Melting point (°C)	Boiling point (°C)	Conducts electricity as a solid	Conducts electricity as a liquid
copper	Cu	1084	2562	yes	yes
copper(II) chloride	$CuCl_2$	498	993	no	yes
table salt (sodium chloride)	NaCl	801	1413	no	yes
laughing gas (nitrous oxide)	N_2O	−90	−88	no	no
water	H_2O	0	100	no	no
ammonia	NH_3	−77	−33	no	no

From Table 6.1.1, you can see the following properties.

- Metals such as copper have high melting points and conduct electricity both as solids and as liquids. As you saw in Chapter 4, this indicates that strong bonds exist in metals and that they contain charged particles (delocalised electrons) that are free to move.
- The ionic compounds copper(II) chloride and sodium chloride have high melting points and conduct electricity as liquids but not as solids. In Chapter 5, you learnt that ionic solids have strong bonds and that they contain charged particles (ions) that are free to move in a liquid form but not in a solid form.

- The non-metal compounds ammonia, nitrous oxide (laughing gas) and water have low melting points and do not conduct electricity. Non-metal compounds can be soft in the solid state. Two conclusions about bonding in non-metal compounds that can be drawn from these properties are shown in Figure 6.1.1.

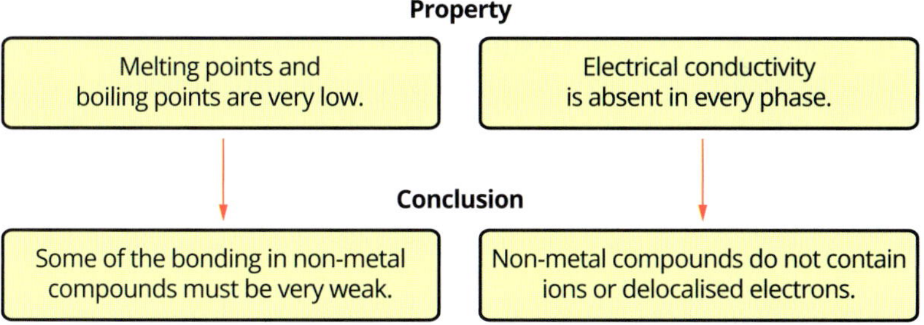

FIGURE 6.1.1 These conclusions can be drawn from the properties of non-metallic substances.

MOLECULES

Further insight into the chemical structure of non-metallic substances can be gained by considering what happens when you boil water in a beaker. Figure 6.1.2 shows that each water molecule contains two hydrogen atoms bonded to one oxygen atom. In-between the molecule are weak forces holding the particles to each other.

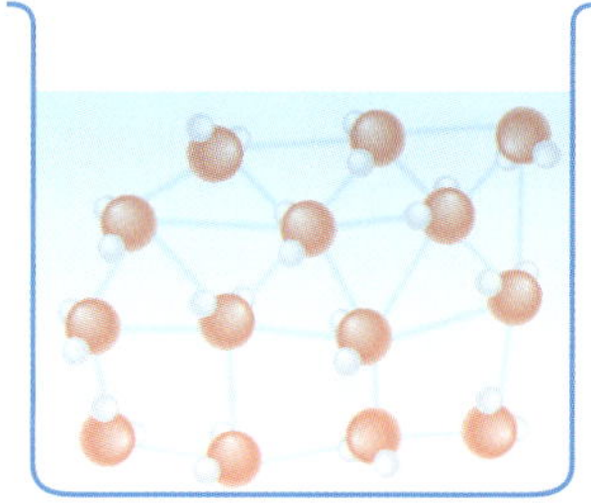

FIGURE 6.1.2 This representation shows the forces of attraction between molecules of water in a beaker.

When water boils (Figure 6.1.3), the weak forces of attraction between molecules must be overcome. However, the water molecules do not separate into hydrogen and oxygen. Rather, the water vapour that is formed still contains molecules in which two hydrogen atoms are bonded to one oxygen atom.

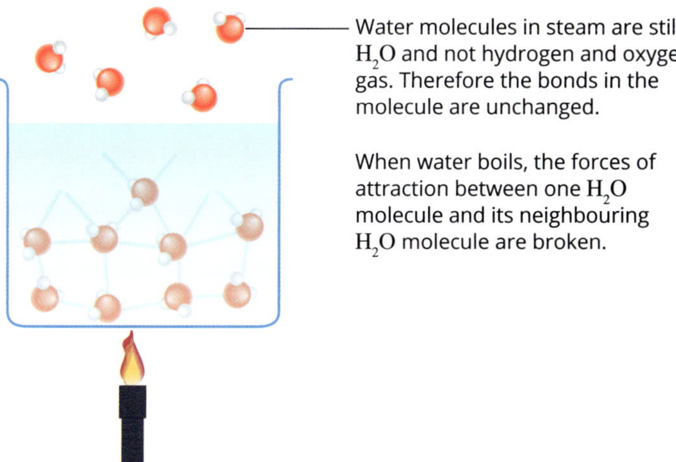

FIGURE 6.1.3 Changes occur to the forces of attraction between molecules in water when water starts to boil.

CHEMFILE

Big molecules with a spark

The covalent molecules we consider in this chapter are mostly small ones made of relatively few atoms such as water (H_2O), carbon dioxide (CO_2) and ammonia (NH_3), but molecules can have hundreds or even thousands of atoms. **Plastics** (polymers) such as polyvinyl chloride (PVC) are covalent molecular compounds with long chains of carbon atoms bonded together. These large molecular compounds generally have the properties expected of molecular substances, such as low melting points and no electrical or heat conductivity, but a new class of plastics is exciting much research interest. This class of plastics has a property we usually associate with metals—electrical conductivity.

Their electrical conductivity arises due to the particular way the atoms are bonded to each other. A sequence of alternating single and double bonds are formed (double bonds are considered in Section 6.2). This type of bonding creates a situation where electrons can move along the chain of carbon atoms. In effect, these electrons can be described as delocalised. These delocalised electrons can travel through the plastic and so allow the material to conduct electricity. There are many potential applications for these conducting plastics including solar cells, biosensors and printing electronic circuits.

The valence shell is the highest energy shell (outer shell) of an atom that contains electrons.

The particles of water shown in Figure 6.1.3 are examples of **molecules**. A molecule is a discrete (separate from others) group of atoms bonded together with a known formula.

Molecules can be found in some elements as well as in some compounds. The element oxygen, for example, forms O_2 molecules. Water can be referred to as a molecular compound with a **molecular formula** of H_2O. The forces of attraction between the hydrogen and oxygen atoms within water molecules are referred to as **intramolecular bonds**, or more simply bonds, and the forces of attraction between neighbouring water molecules are referred to as **intermolecular forces**.

The intramolecular bonds between the hydrogen and oxygen atoms in water molecules are strong compared to the intermolecular forces between the water molecules. It is the intermolecular forces that are broken when molecular substances such as water boil or melt. This allows the molecules to separate from each other while the atoms remain bound to one another within the molecules.

CHEMISTRY IN ACTION

Helium

The Sun is made up of about 24% helium and 75% hydrogen (Figure 6.1.4).

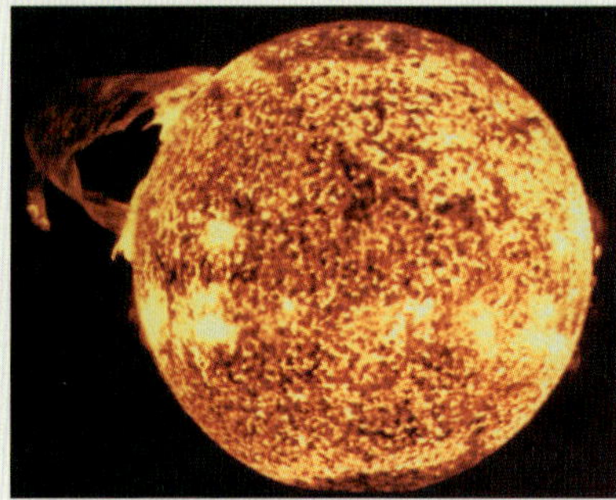

FIGURE 6.1.4 The Sun is made up of about 24% helium and 75% hydrogen. The first evidence for the existence of helium came from a spectral line observed in sunlight during a solar eclipse in 1868.

Helium is an interesting case study of bonding because there are almost no intermolecular forces between its atoms. Helium atoms are small and they have two electrons in their valence shell. The first shell can only hold two electrons, so this means the valence shell is full and the atoms are very stable. As a consequence, helium atoms do not form molecules.

The forces of attraction between helium atoms are almost non-existent. Helium has to be cooled to –269°C before it turns to a liquid. As a liquid, it can climb up the sides of its container and escape! Extraction of helium takes advantage of its extremely low boiling point. Most helium is extracted by fractional distillation from natural gas which can have as much as 7% helium, as opposed to the approximate 5.2 ppm (parts per million by volume) in the atmosphere. As natural gas is cooled, the other components liquefy and can be removed, leaving helium gas.

Although helium is the second most abundant element in the universe (second to hydrogen), it is very rare on Earth. It is used in liquid form (that is, at very low temperature) in Magnetic Resonance Imaging machines (MRIs) for medical purposes and particle accelerators such as the Large Hadron Collider, as well as in party balloons. Concerns about dwindling supplies of helium sparked debates about restricting its use to medical and scientific purposes—so no more party balloons. Present rates of use suggest supplies will last to about 2040, however, in 2016 a large reserve of helium was discovered in the Great Rift Valley in Tanzania which may have enough to extend reserves for one or two decades. As well, the methodology used to discover this reservoir has the potential to uncover other reserves in regions geologically similar to the Rift Valley.

6.1 Review

SUMMARY

- Many substances contain only non-metal atoms. Some of these substances are elements, while others are compounds.
- Non-metallic elements and compounds usually have low boiling points and do not conduct electricity.
- Many non-metallic elements and compounds are composed of molecules. Molecules are discrete groups of atoms of known formula, bonded together.
- There are two important types of forces of attraction in molecular compounds—the bonds within the molecules (intramolecular bonds) and the attractions between molecules (intermolecular forces)
- The melting and boiling points of non-metallic elements and compounds depend on the intermolecular forces between molecules. Generally, these molecular substances have low melting and boiling points and, when solid, the solids tend to be soft. This indicates they have weak forces of attraction between molecules.
- In general, non-metallic elements and compounds do not conduct electricity because they do not contain free-moving charged particles (neither delocalised electrons nor ions).

KEY QUESTIONS

1. Identify the non-metallic elements and compounds from this list: $MgSO_4$, NO_2, H_2, Ni, $CaCl_2$, Br_2.
2. Write the definition of a molecule.
3. Explain the following general properties of non-metals.
   a. Non-metals do not conduct electricity.
   b. Non-metals have low melting and boiling points.
4. When sugar is gently heated, it turns into a clear liquid. If the liquid is heated strongly, it turns black and a gas is produced. Explain what is happening to the two types of forces of attraction in sugar when it is heated. Use the terms 'intermolecular forces' and 'bonds' in your answer.

6.2 Covalent bonding

This section examines a series of simple molecules to help you to understand the concept of a **covalent bond**, which is formed when non-metallic atoms share electrons. Using your knowledge of the valence shell electron arrangements of non-metallic atoms, you will be able to predict the molecules that different elements can form.

COVALENT BONDS

When atoms bond, they become more stable if they gain a valence (outer) shell of eight electrons by combining with other atoms (the octet rule).

Commonly, when atoms of non-metals combine, electrons are shared so that each atom has eight electrons in its valence shell. Molecules formed in this way are more stable than the separate atoms.

Non-metallic atoms have a relatively high number of electrons in their valence shells and they tend to share rather than to transfer electrons. Covalent bonding occurs when electrons are shared between atoms. The positively charged nuclei of the neighbouring atoms each have an attraction for the shared electrons so this keeps the atoms held together in the molecule.

Single covalent bonds

When atoms share two electrons, one from each atom, the covalent bond formed is called a **single covalent bond**. Two examples of substances that contain single bonds are hydrogen and chlorine.

Hydrogen

Hydrogen atoms have one electron. The valence shell for a hydrogen atom can hold a maximum of two electrons. A hydrogen atom can bond to another hydrogen atom to form a molecule of H_2, as shown in Figure 6.2.1. It should be noted that because the valence shell for a hydrogen atom can hold a maximum of two electrons, it does not follow the octet rule when forming bonds. It only needs two electrons to fill the valence shell.

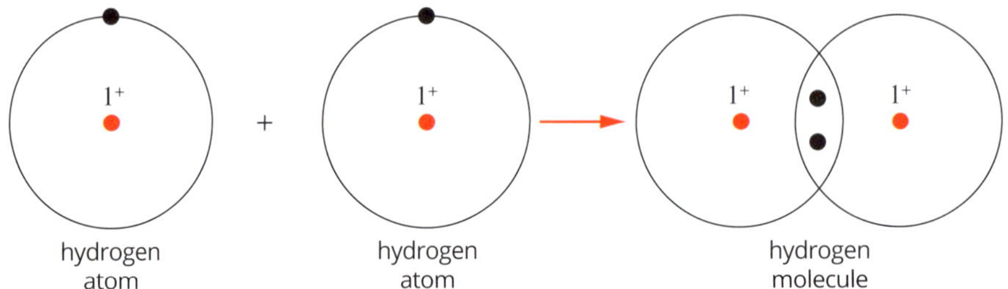

FIGURE 6.2.1 A covalent bond is formed when two hydrogen atoms share two electrons, one from each atom.

In the molecule that is formed:

- two hydrogen atoms share two electrons, one from each atom, to form a single covalent bond
- the atoms of hydrogen are now strongly bonded together by two electrons (an electron pair) in their valence shells.

The hydrogen molecule can be represented as H_2. Molecules that contain two atoms like this are called **diatomic molecules**.

Two alternative ways of representing a hydrogen molecule are shown in Figure 6.2.2.

Remember that the nucleus of a hydrogen atom only contains a single proton. In a hydrogen molecule, the shared pair of electrons is attracted to nuclei of both hydrogen atoms. This means the two electrons will spend most of their time between the two nuclei instead of orbiting their own proton within its nuclei. This electrostatic attraction to the electrons holds the molecule together.

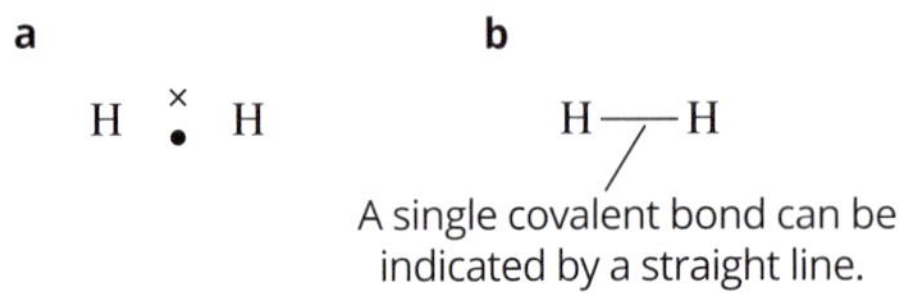

FIGURE 6.2.2 A hydrogen molecule can be represented by (a) showing the shared electrons as dots or crosses, or (b) showing the shared electron pair (the bond) as a line.

This model of a hydrogen molecule is consistent with the observed properties of hydrogen gas. The covalent bonds in the molecules are strong, but the intermolecular forces in hydrogen (the attractions between one molecule and the surrounding molecules) are weak. This explains the low melting temperature of −259°C for hydrogen. Hydrogen does not conduct electricity as it does not contain ions or delocalised electrons.

Hydrogen is an example of a covalent molecular substance.

Chlorine

A chlorine atom has an electron configuration of 2,8,7. It requires one more electron to achieve eight electrons in its valence shell.

One chlorine atom can share an electron with another chlorine atom to form a molecule of chlorine with a single covalent bond. As a result, both atoms gain valence shells of eight electrons as shown in Figure 6.2.3. This is an example of the application of the octet rule.

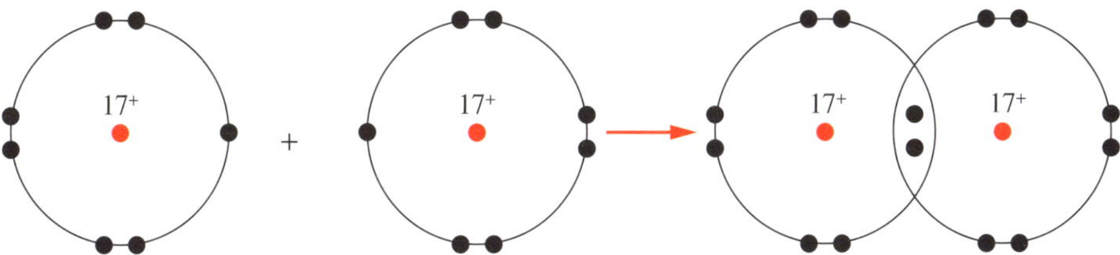

FIGURE 6.2.3 Two chlorine atoms share one electron each to form a chlorine molecule. Note that usually only valence-shell electrons are shown in these diagrams.

Lewis structure diagrams

Chemists often use **Lewis structure diagrams** (also known as **electron dot diagrams**). Lewis structure diagrams only show the valence-shell electrons of an atom because only electrons in the valence shells are involved in bonding. In a Lewis structure diagram, lines can be used to represent the two electrons in a covalent bond. One line represents one pair of electrons.

The Lewis structure diagram also allows you to distinguish between bonding electrons and non-bonding electrons. A chlorine molecule has one pair of bonding electrons. The valence-shell electrons that are not involved in bonding are called the **non-bonding electrons**. Each chlorine atom has six non-bonding electrons, present as three pairs of electrons. Pairs of non-bonding electrons are known as **lone electron pairs** (or **lone pairs**).

Figure 6.2.5 shows two versions of a Lewis structure for a molecule of chlorine. It is easy to see that each chlorine atom in the molecule has eight electrons in its valence shell. In dot diagrams, you can use crosses in place of the dots (as in the case of diagram **a**) if you want to show which atoms have 'provided' the electrons.

Double covalent bonds

In a **double covalent bond**, two pairs of electrons (four electrons in total) are shared between the atoms, rather than just one pair.

For example, the oxygen molecule contains a double covalent bond. The electron configuration of an oxygen atom is 2,6. Each oxygen atom requires two electrons to gain a stable valence shell containing eight electrons. Therefore, when one oxygen atom bonds to another one, each atom shares two electrons.

CHEMFILE

Hydrogen airships

Hydrogen has a low density. This was once thought to make it suitable for use in airships. Zeppelins were a type of rigid airship that was used as a mode of transport during the early 1900s. However, their popularity as a way of travel decreased after the hydrogen gas in the zeppelin *Hindenburg* caught fire in 1937, killing many onboard.

FIGURE 6.2.4 The German passenger zeppelin *Hindenburg* exploded during its attempt to dock at the Lakehurst Naval Air Station in the United States.

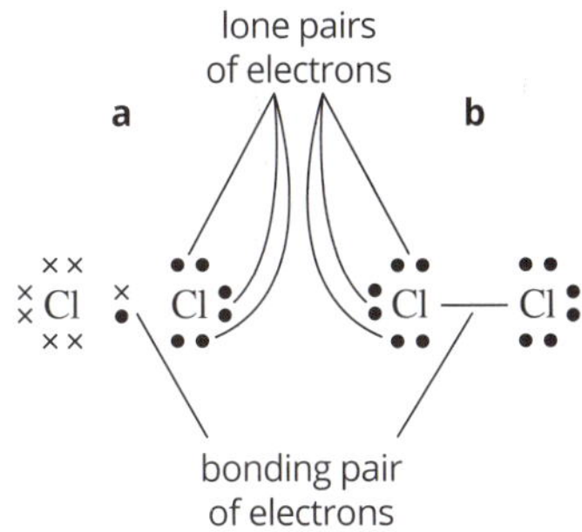

FIGURE 6.2.5 Two different ways of representing the chlorine molecule. They show the valence shell electrons only. (a) Lewis structure diagrams: the electrons can be presented by either dots or crosses (b) valence structures: lines are used to represent pairs of electrons.

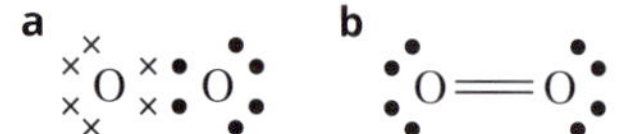

FIGURE 6.2.7 (a) The Lewis structure diagram shows that O_2 has a double covalent bond. Four electrons are shared and each oxygen has two non-bonding electron pairs. (b) Alternatively, the bonding electrons are represented by two lines, one line per bond (pair).

CHEMFILE

The strong triple covalent bond in N_2

The triple covalent bond in nitrogen gas (N_2) is relatively strong and not easily broken. This means that nitrogen gas is relatively unreactive. Nitrogen is an essential element in living organisms because it is a major component of proteins and other biological molecules. Even though 78% of air is nitrogen gas, very few organisms can make use of the nitrogen because it is so unreactive. Only nitrogen-fixing microorganisms are able to convert nitrogen gas into soluble nitrogen-containing compounds. These compounds are then absorbed by plants, allowing nitrogen to be passed up the food chain.

In the early 20th century, German chemist Fritz Haber invented a process for converting nitrogen gas and hydrogen gas into ammonia, which is used to make synthetic fertilisers. This enabled more food to be grown to feed a growing world population.

As you can see in Figure 6.2.6, each oxygen atom in the molecule now has eight valence-shell electrons, four of these are bonding electrons and four are non-bonding electrons.

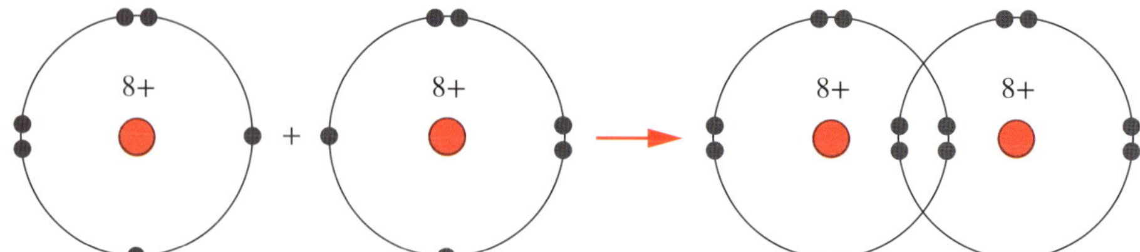

FIGURE 6.2.6 In oxygen molecules, each oxygen atom contributes two electrons to the bond between the atoms.

In Figure 6.2.7, you can see two different versions of Lewis structures that represent the oxygen molecule.

Triple covalent bonds

A **triple covalent bond** occurs when three electron pairs are shared between two atoms. For example, the nitrogen molecule contains a triple bond. The electron configuration of nitrogen is 2,5. A nitrogen atom requires three electrons to achieve eight electrons in its valence shell. When it bonds to another nitrogen atom, each atom contributes three electrons to the bond that forms, as shown in Figure 6.2.8. The Lewis structure of the nitrogen molecule is shown in Figure 6.2.9.

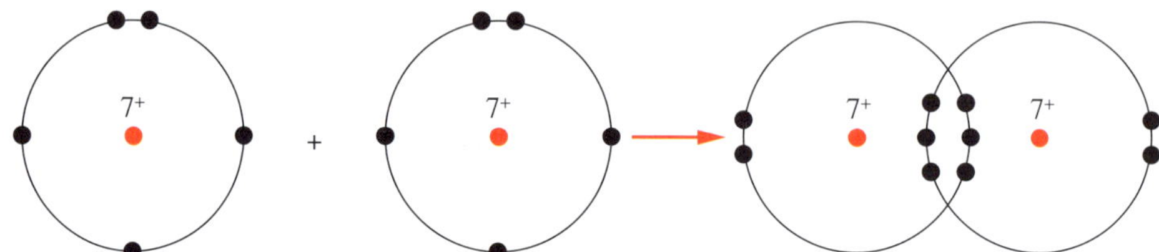

FIGURE 6.2.8 Nitrogen atoms contribute three electrons each to form a triple covalent bond in a molecule of N_2.

:N≡N:

FIGURE 6.2.9 The triple covalent bond in a nitrogen molecule can be shown as three parallel lines.

MOLECULAR COMPOUNDS

A diatomic molecule contains two atoms. The molecules discussed so far have been diatomic molecules that contain atoms of the same elements.

Covalent bonds can also form between atoms of different elements. Hydrogen chloride (HCl) is a simple example (Figure 6.2.10). A hydrogen atom requires one electron to gain a stable valence shell, as does a chlorine atom. They can share an electron each and form a single covalent bond.

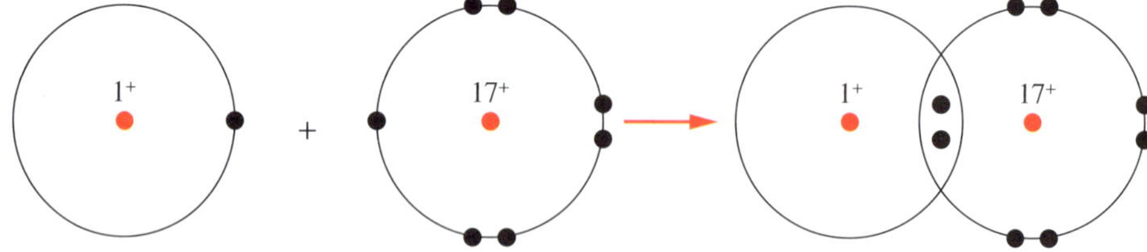

FIGURE 6.2.10 A pair of electrons is shared in the formation of a molecule of HCl.

Polyatomic molecules

If the atoms of one element require a different number of valence electrons from the atoms of the element it is bonding with, the molecule that is formed may not be a simple diatomic molecule. Molecules made up of more than two atoms are called **polyatomic molecules**. Two examples of polyatomic molecules are water and methane.

Water

When a compound forms between hydrogen and oxygen, an oxygen atom shares two electrons and a hydrogen atom shares one. To resolve this imbalance, two hydrogen atoms each share one electron with an oxygen atom.

As you can see in the Lewis structure shown in Figure 6.2.11, a water molecule contains:

- two single covalent bonds, each containing a shared electron pair
- four non-bonding electrons on the oxygen atom.

FIGURE 6.2.11 A water molecule has two single covalent bonds. Electrons can be represented as dots and crosses, or simply, as here, by dots.

Methane

When a compound forms between carbon and hydrogen, four hydrogen atoms are needed to provide the four electrons required in order to have eight electrons in the valence shell of a carbon atom (Figure 6.2.12). The molecule formed has a chemical formula of CH_4 and is called methane.

FIGURE 6.2.12 In a methane molecule, a carbon atom shares one electron with each of four hydrogen atoms to gain eight electrons in its valence shell.

EXTENSION

Polyatomic ions

You have considered Lewis structures for polyatomic molecules but there are also polyatomic ions, for example hydroxide ion (OH^-), ammonium ion (NH_4^+), nitrate ion (NO_3^-) and sulfate ion (SO_4^{2-}). Polyatomic ions are made up of two or more atoms. The atoms in a polyatomic ion bond to each other covalently in the same way as atoms in a polyatomic molecule and can be represented by Lewis structures.

When drawing the Lewis structure for a polyatomic ion, the number of electrons available for sharing is determined by the number of valence electrons from each atom in the ion as well as the electrons either gained or lost as indicated by the charge on the ion. For example, in the hydroxide ion, the oxygen atom has six valence electrons and the hydrogen has one. Hydroxide has a charge of negative one so it has gained an electron. This means there are a total of eight electrons to share between the oxygen and hydrogen atoms. The hydrogen atom needs one more electron to complete its valence shell (remember it only needs two electrons) and the oxygen atom needs two electrons to complete its valence shell (oxygen follows the octet rule). Thus, the oxygen atom and hydrogen atom share a pair of electrons to form a single bond. Its Lewis structure is shown in Figure 6.2.13. Note the eight electrons in the valence shells of the atoms.

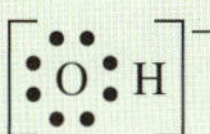

FIGURE 6.2.13 Lewis structure for the hydroxide ion. A single covalent bond forms between the atoms.

Consider the ammonium ion. Nitrogen has five valence electrons and each hydrogen has one valence electron (thus four electrons from the hydrogen atoms). The +1 charge tells us that when the atoms bonded to form the ion one electron was lost, so there are eight valence electrons available for bonding. Figure 16.2.14 shows the central nitrogen atom sharing a pair of electrons with each of the four hydrogen atoms.

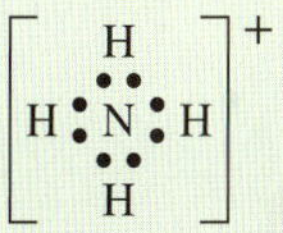

FIGURE 6.2.14 Lewis structure for the ammonium ion. The central N atom has an octet and each H atom has a pair of electrons.

Similarly, NO_3^- has 24 valence electrons available for sharing and oxygen atoms form a double bond and two single bonds around the central N atom (Figure 6.2.15).

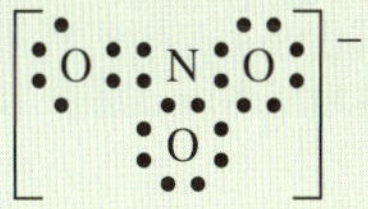

FIGURE 6.2.15 Lewis structure for the nitrate ion. The central N atom forms a single bond with two oxygen atoms and a double bond with one oxygen atom. All atoms have an octet.

SHAPES OF MOLECULES

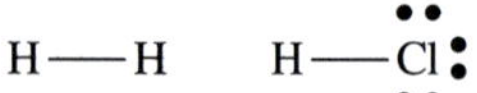

FIGURE 6.2.16 H_2 and HCl both contain only one bond, so they are linear in shape.

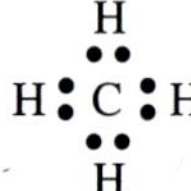

FIGURE 6.2.17 The carbon atom in methane has four pairs of electrons in its valence shell. These electron pairs repel each other.

The shapes of diatomic molecules such as H_2 and HCl are simple. With two atoms only, a diatomic molecule is linear, as you can see in Figure 6.2.16.

Lewis structures can be used as a starting point in determining the shape of molecules. These diagrams show the valence-shell electron pairs in the molecule. Electron pairs have a negative charge and they repel each other. The electron pairs arrange themselves as far away from each other as possible.

In the case of methane, the Lewis structure diagram (Figure 6.2.17) shows four valence-shell electron pairs around the carbon atom. Repulsion between the electron pairs means that the hydrogen atoms form a **tetrahedral shape** around the carbon atom (Figure 6.2.18). In this molecule, the bonds between the carbon and hydrogen atoms are at an angle of 109.5° to each other.

You will have a more detailed look at the shapes of molecules in Chapter 12.

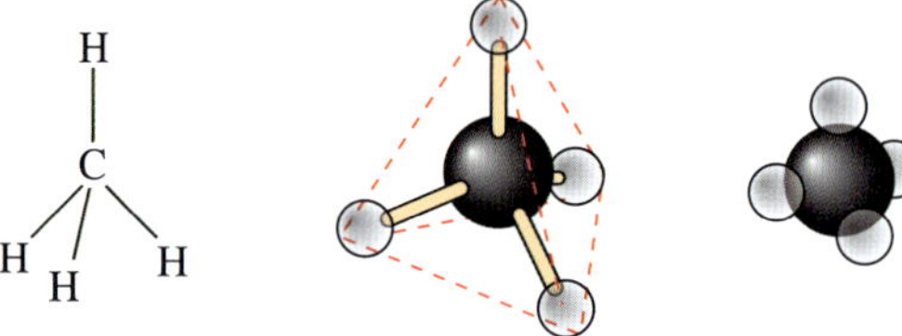

FIGURE 6.2.18 Three types of three-dimensional representations of a methane molecule show that hydrogen atoms form a tetrahedral arrangement around the central carbon atom.

Representations of molecules

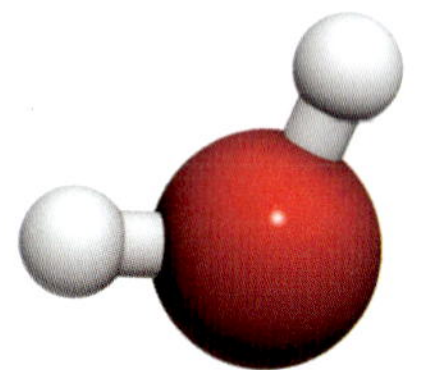

FIGURE 6.2.19 Ball-and-stick model of a water molecule

FIGURE 6.2.20 Space-filling model of a water molecule

A Lewis structure diagram does not show the shape of a molecule, but there are several other accepted styles for representing molecules that do. A **ball-and-stick model**, as shown in Figure 6.2.19, displays the shape and the type of bonds (single, double or triple) in a molecule. The atoms are represented by spheres that are connected by rods, representing the bonds. The spheres represent the centre of the atom, not the whole space the atom occupies.

A **space-filling model** (Figure 6.2.20) also uses spheres but this time the spheres represent the whole atom, including its electron charge cloud. Different coloured spheres are used in both types of models to identify the different elements.

Each different representation of a molecule has its advantages. A Lewis structure diagram is often drawn to determine the number and type of bonds present, while the ball-and-stick and space-filling models show the shape of the molecule and the position of atoms. Table 6.2.1 compares different representations of a PF_3 molecule.

TABLE 6.2.1 Comparison of different representations of phosphorus trifluoride, PF_3

	Lewis structure with electrons as dots	Lewis structure with lines for bonds	Ball-and-stick	Space-filling
diagram	F P F F	F P F F		
model best used for	determining the formula of the molecule and the type of bonds present	simplifying Lewis structure diagram	displaying the molecule shape	showing the relative size and position of the atoms in the molecule
limitation of model	does not show the relative size of atoms or shape of the molecule	does not show the relative size of atoms or shape of the molecule	shows the shape but not the relative sizes of the atoms	shows the relative size and position of the atoms but does not show bond angles or types of bonds

Worked example 6.2.1

LEWIS STRUCTURE DIAGRAMS

Draw a Lewis structure diagram of methane (CH_4).	
Thinking	**Working**
Write the electron configuration of the atoms in the molecule.	C electron configuration: 2,4 H electron configuration: 1
Determine how many electrons each atom requires for a stable valence shell.	C requires 4 electrons. H requires 1 electron.
Draw a Lewis structure diagram of the likely molecule, ensuring that each atom has a stable valence shell. Valence electrons not involved in bonding will be in non-bonding (lone) pairs.	Draw a Lewis structure diagram of the molecule. H •• H : C : H •• H

Worked example: Try yourself 6.2.1

LEWIS STRUCTURE DIAGRAMS

Draw a Lewis structure diagram of ammonia (NH_3).

NON-OCTET MOLECULES

As you have seen, when hydrogen atoms form covalent bonds, they only need two electrons to complete their valence shell. The atoms beryllium (Be) and boron (B) only have two or three electrons respectively in their valence shells. Therefore, when these electrons 'pair up' to form bonds, beryllium is limited to a total of four electrons and boron six electrons in the outer shell. However, due mainly to the small size of these atoms, this arrangement is stable enough for the molecules to form, even though they do not follow the octet rule.

For example, for boron trifluoride (BF_3), each fluorine atom (electron configuration 2,7) forms one covalent bond. Boron has an electron configuration of 2,3. There are only three electrons in the valence shell of boron. Each of these is shared with one of three fluorine atoms to form covalent bonds. This results in only three pairs of electrons (six electrons) around the boron atom in BF_3 (Figure 6.2.21). Boron is said to have an incomplete octet but the molecule still forms.

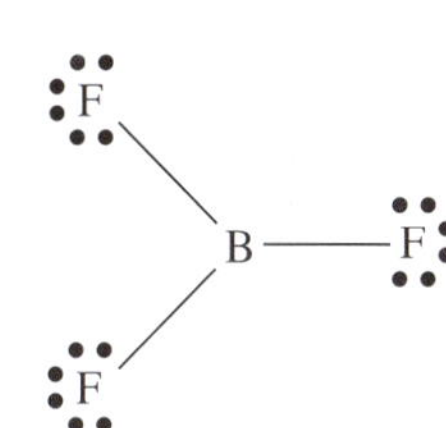

FIGURE 6.2.21 Lewis structure of boron trifluoride (BF_3)

6.2 Review

SUMMARY

- The atoms in molecules are held together by covalent bonds.
- A covalent bond involves the sharing of electrons.
- Covalent bonds form between non-metallic atoms, often enabling the atoms to gain valence shells containing eight electrons (except hydrogen which gains a valence shell containing two electrons).
- A single covalent bond forms when two atoms share a pair of electrons.
- A double covalent bond forms when two atoms share two pairs of electrons.
- A triple covalent bond forms when two atoms share three pairs of electrons.
- Valence-shell electrons that are not involved in bonding are called lone pairs.
- Lewis structure (electron dot) diagrams show the valence electron arrangements of atoms in a molecule.

KEY QUESTIONS

1 How many covalent bonds are formed between atoms in these diatomic molecules?
- **a** H_2
- **b** N_2
- **c** O_2
- **d** F_2

2 Draw Lewis structure diagrams for each of the following molecules.
- **a** fluorine (F_2)
- **b** hydrogen fluoride (HF)
- **c** water (H_2O)
- **d** tetrachloromethane (CCl_4)
- **e** phosphine (PH_3)
- **f** butane (C_4H_{10})
- **g** carbon dioxide (CO_2)

You can use lines to show bonding pairs of electrons if you wish.

3 What is the maximum number of covalent bonds an atom of each of the following elements can form?
- **a** F
- **b** O
- **c** N
- **d** C
- **e** H
- **f** Ne

4 When oxygen forms covalent molecular compounds with other non-metals, the Lewis structures that represent the molecules of these compounds all show each oxygen atom with two lone pairs of electrons. Why are there always two lone pairs?

5 Suggest the most likely formula of the compound formed between the following pairs of elements.
- **a** C, Cl
- **b** N, Br
- **c** Si, O
- **d** H, F
- **e** P, F

6 Identify whether each of the following statements about methane (CH_4) is true or false.
- **a** Methane is a molecular compound.
- **b** A molecule of methane contains four atoms.
- **c** There are 20 atoms in four molecules of methane.
- **d** The forces of attraction between the carbon and hydrogen atoms in methane are intermolecular forces.

7 Identify the type of representation of a molecule that best fits the following descriptions.
- **a** shows the shape of the molecule and type of atoms present
- **b** shows the arrangement of valence electrons between atoms in a molecule as dots
- **c** shows the relative size and position of atoms and the shape of molecule

Chapter review

KEY TERMS

ball-and-stick model
covalent bond
diatomic molecule
double covalent bond
electron dot diagram
intermolecular force
intramolecular bond
Lewis structure diagram
lone electron pair
molecular formula
molecule
non-bonding electron
plastic
polyatomic molecule
single covalent bond
space-filling model
tetrahedral shape
triple covalent bond

Properties of non-metallic substances

1 Hydrogen chloride (HCl) exists as a gas at room temperature. What can you conclude about the strength of the intermolecular forces in pure hydrogen chloride?

2 Non-metallic substances generally do not contain free-moving charged particles. What does this indicate about the general physical properties of non-metallic substances?

3 Solid carbon dioxide sublimes when it is heated; that is, it goes straight from a solid to a gas. Explain what happens to the forces of attraction in carbon dioxide when it sublimes. Use the terms 'intermolecular forces' and 'bonds' in your answer.

4 Explain why most substances that have an odour are covalent molecular substances.

5 Use the properties of the four substances given in the table below to determine which of A, B, C and D are magnesium chloride, copper, naphthalene (C_8H_{10}) and phosphorus trihydride. Give a brief reason for your decision.

Substance	Electrical conductivity in aqueous solution	Melting point (°C)	Hardness when solid
A	no	80	soft
B	yes	714	brittle
C	not soluble in water	1084	hard
D	no	–133	soft

Covalent bonding

6 Select the statement that best describes the way hydrogen atoms bond to each other.
A One hydrogen atom donates an electron to another hydrogen atom to form a molecule.
B Hydrogen atoms form a lattice with delocalised electrons.
C Hydrogen atoms share electrons to obtain a complete valence shell of eight electrons.
D Two hydrogen atoms share an electron each to form a hydrogen molecule.

7 In a molecule of nitrogen (N_2), how many of the following are there?
a bonding electrons
b covalent bonds
c non-bonding electrons

8 The formula of a molecule is XY_4. Select the alternative that could match this formula.
A OH_4
B CH_4
C HBr_4
D CO_4

9 Atom X has an electron configuration of 2,6. Atom Y has a configuration of 2,7. What is the likely molecular formula of the compound that will form between X and Y?

10 Silicon forms a compound with element X with formula SiX_4. Give the most likely number of valence electrons in atoms of X so that the bonding follows the octet rule. Explain your answer.

11 Explain why neon atoms do not form covalent bonds.

12 How many lone pairs are in each of the following molecules?

a H_2
b NH_3
c HCl
d O_2

13 Draw Lewis structures (electron dot diagrams) for each of the following molecules and identify the number of bonding and non-bonding electrons in each molecule.

a HBr
b H_2O
c CF_4
d C_2H_6
e PF_3
f Cl_2O
g CH_4
h H_2S

14 The atoms in molecules of nitrogen, oxygen and fluorine are held together by covalent bonds.

a How are the bonds in these molecules similar?
b How are they different?

15 Experimental evidence shows that the double bond between the two oxygen atoms in O_2 is much stronger than a single bond between two oxygen atoms in a compound such as hydrogen peroxide (H_2O_2).

a Draw Lewis structure diagrams and structural formulae for O_2 and H_2O_2.
b Explain why the oxygen double bond is stronger than the oxygen single bond.
c Why does oxygen not form a triple bond or three single covalent bonds?

Connecting the main ideas

16 Classify the following substances as metallic, ionic or molecular: $CuCl_2$, Ag, HCl, H_2O, Cu, CaS, NH_3.

17 The labels have fallen off the jars of two white powders in a laboratory. One jar contains the ionic compound calcium chloride and the other the covalent molecular compound glucose. Describe a simple laboratory test that can be done to identify the powders. Both are soluble in water.

18 Describe how the atoms in a covalent molecule are held together.

19 Identify whether each of the following statements about carbon dioxide (CO_2) is true or false.

a Carbon dioxide is a molecular compound.
b A molecule of carbon dioxide contains three atoms.
c There are 99 oxygen atoms in 33 molecules of carbon dioxide.
d The forces of attraction between the carbon and oxygen atoms in carbon dioxide are covalent bonds.
e There are two single covalent bonds in a molecule of carbon dioxide.
f There are four lone pairs of electrons in a molecule of carbon dioxide.

CHAPTER

07 Carbon

Carbon is in group 14 of the periodic table and has four valence electrons. It is in period 2 of the periodic table and is therefore a relatively small atom. The combination of these two facts makes carbon a unique element. Life on Earth is based on molecules that contain carbon. The hardest substance known to humans is a natural form of carbon—diamond.

Diamond is a pure substance made entirely from carbon atoms. This chapter will explore the model of how carbon atoms are arranged in diamond that has been developed to explain diamond's unique properties such as hardness and stability at high temperatures. Carbon exists in several other forms that are very different from diamond. This chapter will also describe the models of the arrangements of atoms in these other forms of carbon, including graphite and relatively newly discovered substances called fullerenes.

Science understanding

- the properties of covalent network substances, including high melting point, hardness and electrical conductivity, are explained by modelling covalent networks as three-dimensional structures that comprise covalently bonded atoms
- elemental carbon exists as a range of allotropes, including graphite, diamond and fullerenes, with significantly different structures and physical properties

7.1 Carbon lattices

Diamonds are highly valued in our society. Diamonds are a form of pure carbon, but they are not the only form that carbon can take. Graphite, charcoal, graphene and fullerenes are also made entirely from carbon but their properties are very different from those of diamond. Scientists have found many exciting uses for these other forms of carbon. As the properties of these forms of carbon are quite different, it is not surprising that their chemical structures are also different.

ABUNDANCE AND PROPERTIES OF CARBON

Carbon is a fascinating element for many reasons.

- It is a vital component of all living systems.
- It is the 11th most abundant element in the universe.
- It has three isotopes: ^{12}C (98.9% abundant), ^{13}C (1.1% abundant) and ^{14}C (traces).
- It has one of the highest melting points of any element. It undergoes **sublimation**, changing from a solid state directly to a gas state at temperatures above 3500°C.
- It is a non-metal, but a number of forms can conduct electricity.
- It can form single, double and triple covalent bonds with several other elements.
- It can form large molecules and lattice structures by bonding to itself.

ALLOTROPES

Some elements can exist with their atoms in several different structural arrangements called **allotropes**, which give them different physical forms. In different allotropes, the atoms are bonded to each other in different, specific ways. This gives them significantly different properties from other allotropes of the same element.

An example of an element that exists as allotropes is oxygen. Oxygen gas consists of diatomic molecules with the formula O_2. Each oxygen atom in this arrangement is bound to one other oxygen atom. Ozone is another molecule containing only oxygen. Ozone molecules have the formula O_3 and consist of a central oxygen atom bound to two other oxygen atoms. Figure 7.1.1 shows the structure of these two molecules. As both contain only oxygen atoms, they are both allotropes of oxygen.

FIGURE 7.1.1 Oxygen and ozone are two molecules that contain only oxygen atoms.

 Allotropes are different forms of the same element.

ALLOTROPES OF CARBON

Diamonds (Figure 7.1.2) and graphite (Figure 7.1.3) are both minerals made of the same single element—carbon. Graphene and fullerenes are new materials that are also allotropes of carbon.

FIGURE 7.1.2 Diamond is the hardest naturally occurring substance.

FIGURE 7.1.3 Natural graphite is soft and black.

Table 7.1.1 summarises some information about the structure, properties and uses of the three most common allotropes of carbon: diamond, graphite and **amorphous** carbon.

TABLE 7.1.1 Comparison of properties of some of the allotropes of carbon

Allotrope	Structure	Properties	Uses
diamond	covalent network lattice, each carbon surrounded by four other carbon atoms in a tetrahedral arrangement	• very hard • sublimes • non-conductive • brittle	• jewellery • cutting tools • drills
graphite	covalent layer lattice, each carbon bonded to three other carbons, one delocalised electron per carbon atom	• conductive • slippery • soft • greasy material	• lubricant • pencils • electrodes • reinforcing fibres
amorphous carbon	irregular structure of carbon atoms; many varieties exist with many different, non-continuous packing arrangements	• conductive • non-crystalline • cheap	• printing ink • carbon black filler • activated charcoal • photocopying

Diamond

Diamond is the hardest naturally occurring substance known.

Diamond does not contain small, discrete (individual) molecules. Instead, the carbon atoms bond to each other to form a continuous three-dimensional structure called a **covalent network lattice**. There are no weak intermolecular forces present, only strong covalent bonds. This is what gives the diamond its hardness and high sublimation point.

In general, substances that have a network lattice structure have very high melting points or decomposition temperatures. They are also very hard because the atoms are held firmly in fixed positions in the lattice.

Diamond is made up of carbon atoms that bond with four neighbouring carbon atoms forming a covalent network lattice. This structure makes diamond extremely hard.

In Chapter 12 you will learn in more detail that molecules have particular shapes that arise because the electron pairs in the outer shell position themselves as far away from each other as possible. When an atom has four electron pairs in its outer shell, the electron pairs position themselves in a tetrahedral shape. In the covalent network lattice for diamond shown in Figure 7.1.4, you can also see that individual atoms within diamond form single covalent bonds to four other carbon atoms in a tetrahedral arrangement.

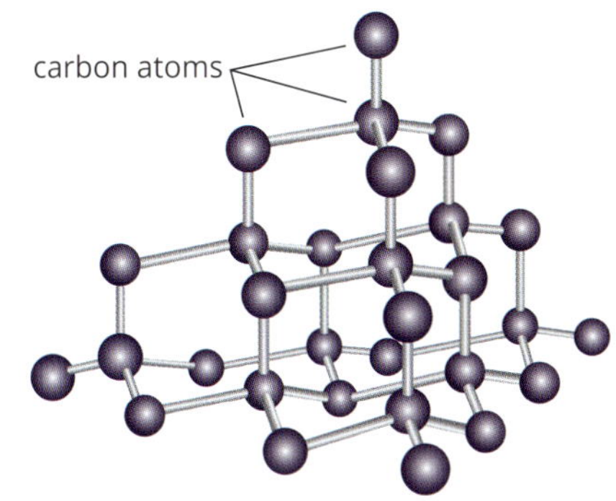

FIGURE 7.1.4 The structure of diamond showing each carbon atom with four single covalent bonds to neighbouring atoms

FIGURE 7.1.5 Diamond-tipped drills used to drill through rock in the fracking industry

The model we have for the structure of diamond can be used to explain its properties.

- Single covalent bonds between carbon atoms are strong bonds. The entire structure of a diamond consists of a continuous network of these bonds, making diamond very hard and rigid.
- There are no small molecules in diamond, so there are no weak forces between the atoms. There are only strong covalent bonds between carbon atoms and this makes the sublimation point very high (about 3500°C).
- The rigidity means that diamonds are brittle and break rather than bend.
- Diamond does not conduct electricity because it does not contain any charged particles that are free to move.
- Because the atoms in diamonds are held together very strongly, the thermal conductivity is extremely high. It is five times greater than that of copper, leading to some specialty electronic uses where diamond is used to transfer heat away from some important electrical components.

The crystalline appearance of diamonds and their high refractive index make them sparkle and has made them extremely popular as jewellery, but the hardness of diamond also lends itself to industrial uses. Many industrial cutting and drilling tools for working with tough materials are diamond tipped. The drill tips in Figure 7.1.5 are used to drill through rock in the fracking industry. They contain small pieces of diamond that improve the hardness and durability of the tool.

CHEMFILE

Impact diamonds

'We are speaking about trillions of carats', trumpeted the 2012 headline from the British *Daily Mail*. It was in reference to a 100 kilometre meteorite crater, the Popigai Crater, in Russia that could supply world markets with diamonds for 3000 years. The now closed Mirny mine, shown in Figure 7.1.6, is also in Russia. This open-cut mine is more than 500 metres deep and has yielded diamonds worth more than $20 billion since 1951.

FIGURE 7.1.6 The Mirny diamond mine is more than 500 metres deep.

It is thought that the impact of a large meteorite created enough heat and pressure in the Popigai Crater to make diamonds. Russian scientists are reported to have known of this deposit since 1971, but kept details hidden until supplies from other sources began to run out. Diamonds formed from a meteorite strike, like those in Figure 7.1.7, are referred to as 'impact diamonds'. They can be almost as large as a 20-cent coin, and are prized for their extreme hardness.

FIGURE 7.1.7 High-quality 'impact diamonds' can be almost the size of a 20-cent coin.

Silica

Diamond is not the only covalent network lattice substance. Silicon, silicon carbide and silicon dioxide (silica) are examples of other covalent network lattices. In the same way as carbon, elemental silicon can also form this type of lattice as they both have four valence electrons, making it possible for the atoms to join continuously to themselves, or other atoms such as oxygen and carbon.

Silicon dioxide is one of the most common substances on our planet. It is present in nature as quartz and sand, and is the major component of glass. Silicon dioxide forms a covalent network lattice in which each silicon atom is bonded tetrahedrally to four oxygen atoms and each oxygen atom is bonded to two silicon atoms (Figure 7.1.8).

FIGURE 7.1.8 This model of silica shows the highly regular arrangement of atoms. Silicon is represented by the black spheres and oxygen by the red spheres.

CHEMISTRY IN ACTION

Not just sand but a valued mineral

Silica is mined in Western Australia as part of its mineral sands industry. This industry, based mainly in the south-west region of Western Australia, produces a range of products, mostly for export, including titanium-based mineral sands, zirconium-based mineral sands and silicon-based products.

High-grade silica (silicon dioxide) is mined for uses including the manufacture of a range of glass products, computer screens, plasma display screens, and the abrasive silicon carbide (SiC). Fumed silica (microscopic droplets of amorphous silica fused into three-dimensional structures) is another product of the mineral sands industry. It can be produced during the production of fused zirconia (a crystalline form of zirconium dioxide, ZrO_2) from the mineral sand zircon (zirconium silicate, $ZrSiO_4$). The reaction that occurs in the furnace to produce zirconia is:

$$ZrSiO_4 \rightarrow ZrO_2 + SiO_2$$

Silica fume is used for the production of low cement refractories (refractory materials keep their strength at high temperatures) that are used for construction purposes. It also has applications as a thickening agent, free-flow agent in powders, viscosity adjustment agent in paints, printing inks and adhesives, and light abrasive in products such as toothpaste.

Another important product from the mining of silica sands is silicon itself. Silicon is used for a range of products including silicones (silicon-based polymers), silicon chips for the electronics industry, optical glass for optical fibres and liquid crystal displays and photovoltaic cells. The silicon is produced by reacting the silica with carbon as summarised by:

$$SiO_2 + 2C \rightarrow Si + 2CO$$

This reaction does not occur directly but rather by three intermediate steps as follows:

$$SiO_2 + C \rightarrow SiO + CO$$
$$SiO + 2C \rightarrow SiC + CO$$
$$SiC + SiO_2 \rightarrow Si + SiO + CO$$

The silicon monoxide (SiO) for the final step reacts with oxygen from the air to produce fumed silica (SiO_2) which is sold as a by-product.

CHEMFILE

In the pink

Diamonds are mined in Western Australia in the far north in the east Kimberley region at Argyle Diamond Mine. It has been operating since 1983 and is one of the world's largest diamond suppliers and the largest supplier of naturally coloured diamonds (Figure 7.1.9). Argyle diamonds are particularly known for their pink and red colours.

Diamonds form in the Earth's mantle at depths of about 140 km to 190 km under pressures ranging from 4.5 GPa to 6 GPa (gigapascals) and temperatures from 900°C to 1500°C. Pure diamonds, where the network lattice of carbon atoms is perfect, don't absorb light and appear colourless. While most other naturally coloured diamonds get their colour from the presence of impurities in the network structure, the pink colour arises from distortion in the crystal lattice. These distortions cause the diamond to absorb light in the green region of the visible spectrum, reflecting light from the red part of the spectrum. Yellow diamonds arise due to the presence of nitrogen atoms in the crystal lattice, and blue occurs from the presence of boron atoms.

FIGURE 7.1.9 Rough and uncut diamonds including a colourless diamond and a range of coloured diamonds

Graphite

Graphite is a very different form of carbon. As you can see in Figure 7.1.10, the carbon atoms in graphite are in layers. There are strong covalent bonds between the carbon atoms in each layer. However, there are weak forces, called dispersion forces, between the layers (You will learn more about dispersion forces in Chapter 12). As a consequence, it is hard in one direction but quite slippery and soft in another direction. The structure of graphite is referred to as a **covalent layer lattice**.

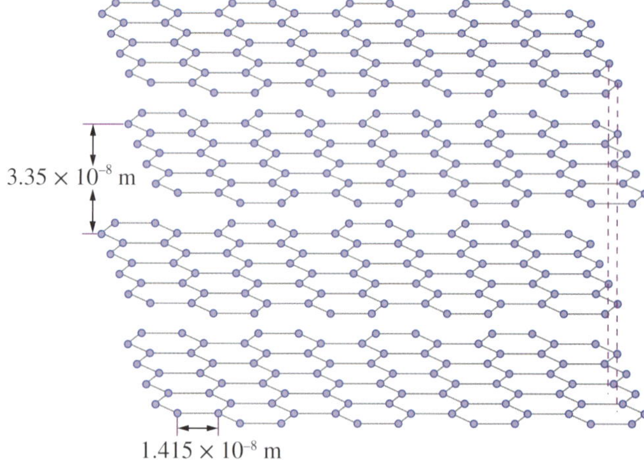

FIGURE 7.1.10 Graphite has a covalent layer lattice structure. The carbon atoms within each layer are covalently bonded to each other. Weak dispersion forces exist between the layers.

Other properties of graphite also support the model of a covalent layer lattice structure.

- The strong covalent bonds between the atoms in each layer explain graphite's resistance to heat. Graphite sublimes at a temperature of about 3600°C.
- Each carbon atom is bonded to three other carbon atoms. The fourth valence electron from each atom is able to move within the layer. The electrical conductivity of graphite is due to these delocalised electrons.

The conductivity of graphite makes it suitable for applications such as battery electrodes where conductivity is required but a metal is not suitable.

In graphite, each carbon atom is covalently bonded to three other carbon atoms. The layered network structure contains delocalised electrons. Bonds within the layers are strong but bonds between layers are weak dispersion forces.

Graphite can also be used as a lubricant. The weak dispersion forces between layers allow these layers to slide over each other and to reduce the friction between moving parts, such as in locks or machinery.

Graphite is also used as an additive to improve the properties of rubber products and it can be woven into a fibre. This helps to reinforce plastics. Figure 7.1.11 shows spun graphite fibre, which can be used to make strong composite materials such as those used in tennis racquets, fishing rods and racing-car shells.

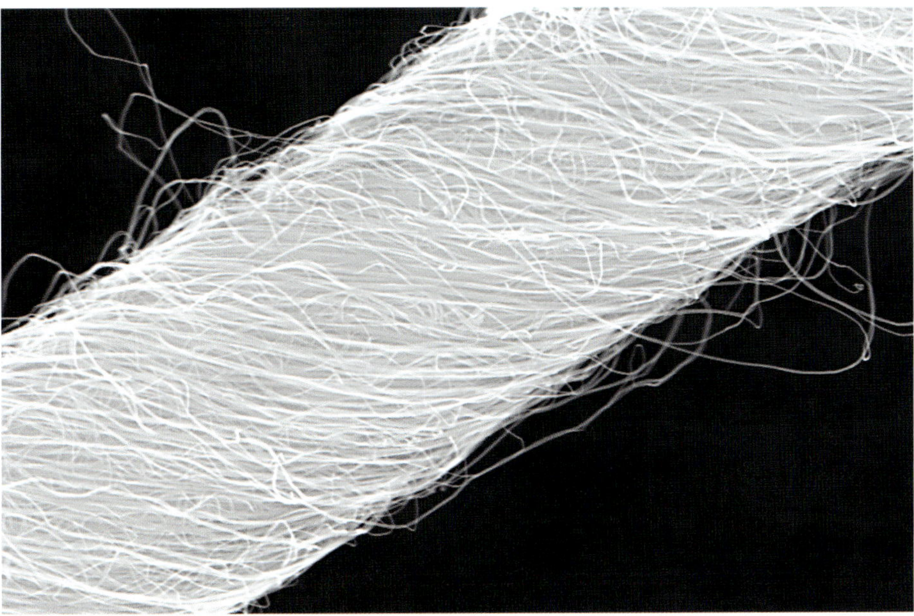

FIGURE 7.1.11 Graphite fibre can be used to reinforce plastics.

CHEMFILE

Black-lead pencils

In 1564, a very pure deposit of graphite was discovered in England. The graphite was so stable that it could be cut into thin, square sticks that could be used for writing. String was wrapped around the graphite to make the first pencils. Later the string was replaced with wood (Figure 7.1.12).

FIGURE 7.1.12 An early pencil that consisted of a strip of graphite placed between pieces of wood

The pencils were so effective that during the Napoleonic Wars, the English were considered to have a technological advantage because their pencil-written communications were far more effective than the French equivalents. Napoleon commissioned a French inventor, Nicholas-Jacques Conte, to develop an alternative to pure graphite. The mixtures of clay and powdered graphite that he designed are the basis for the 'lead' in modern pencils.

EXTENSION

Amorphous forms of carbon

Charcoal (Figure 7.1.13) and carbon black (Figure 7.1.14) are examples of amorphous carbon that has no consistent structure. Amorphous carbon contains irregularly packed, tiny crystals of graphite and other non-uniform arrangements. Lumps of charcoal are produced for use as a fuel, while carbon black is used to make printer ink.

FIGURE 7.1.13 Lumps of charcoal are produced for use as a fuel.

FIGURE 7.1.14 Carbon black is used in printer ink.

Amorphous carbon can be formed from the combustion of wood and other plant matter when there is a limited supply of air. There are several other types of amorphous carbon, including soot, which can be seen in Figure 7.1.15, being emitted from an industrial chimney.

FIGURE 7.1.15 Soot being emitted from an industrial chimney

Each form of amorphous carbon has its uses and some have been used by society for centuries. Since the Middle Ages it has been common to produce charcoal in ovens. Figure 7.1.16 shows a number of beehive-shaped ovens that were used to produce charcoal from timber. These ovens were built between 1876 and 1879.

FIGURE 7.1.16 These ovens in Nevada, USA, were built between 1876 to 1879 to make charcoal.

Uses of carbon black

Carbon black is a refined type of amorphous carbon in which the particle size is more uniform. Most carbon black is used to reinforce rubber products such as tyres and hoses, causing their black appearance. The surface interaction between the fine carbon particles and the rubber molecules increases the strength and toughness of the product.

Many printer and photocopier toners contain carbon black particles mixed with a binder polymer and other additives. More than 9 million tonnes of carbon black is used annually worldwide.

Uses of activated charcoal

Charcoal can be 'activated' by heating it to high temperatures in the presence of an inert gas. Activated charcoal particles are so porous that it is estimated that a 1 gram sample has a surface area similar to that of an Australian Rules football oval. Figure 7.1.17 shows a microscope image of activated charcoal.

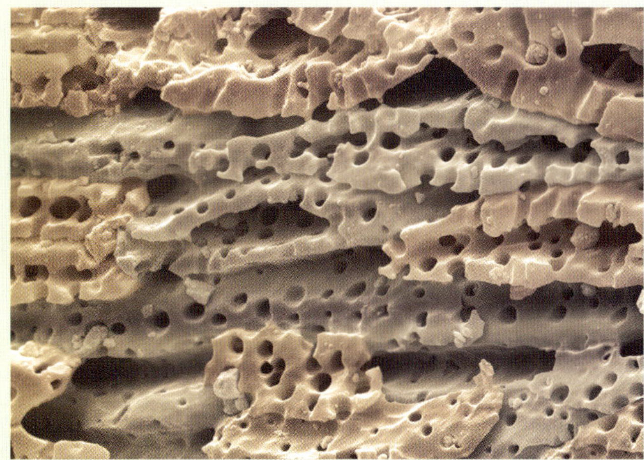

FIGURE 7.1.17 A microscope image of activated charcoal shows that it contains many pores and hollows.

Activated charcoal can adsorb impurities onto its porous surface. The impurities are trapped in the pores of the activated charcoal particles by weak attractive forces, such as dispersion forces. This makes activated charcoal useful as:

- water filters
- odour absorbent inserts in shoes
- a treatment for a drug overdose. Activated charcoal is pumped into the victim's stomach to adsorb the harmful drug molecules.

CHEMFILE

Production of biochar

Most scientists agree that rising levels of carbon dioxide in the atmosphere, which are a result of human activities, are causing climate change. One area of research at the CSIRO to help combat rising carbon dioxide levels is the production of biochar.

Biochar is a high-carbon, porous and fine-grained residue produced by placing biomass (plant and forest waste) in a trench and covering it with soil. The biomass is allowed to smoulder, burning very slowly in the absence of oxygen. The product of this process is carbon and not carbon dioxide.

Biochar can increase the fertility of soils and agricultural production, using waste that would otherwise have become carbon dioxide. Figure 7.1.18 shows farmers adding biochar to the soil in a field.

FIGURE 7.1.18 Biochar added to soil improves the soil by supplying carbon and trapping nutrients.

7.1 Review

SUMMARY

- Carbon can be found in the Earth's crust in the form of diamond, graphite and charcoal. The structures and properties of these allotropes are very different.
- In diamond, each carbon atom is covalently bonded to another four carbon atoms in a tetrahedral shape, forming a covalent network lattice structure. Diamond sublimes at a high temperature, is extremely hard and has a sparkling, crystalline appearance.
- In graphite, each carbon atom is covalently bonded to three other carbon atoms. The layered network structure contains delocalised electrons. Bonds within the layers are strong but bonds between layers are weak dispersion forces. Graphite is slippery, conducts electricity and sublimes at a high temperature.

KEY QUESTIONS

1 Why can carbon form so many different compounds?

2 **a** What is meant by the word 'sublime'?
b Explain why diamond and graphite only sublime at temperatures over 3500°C.

3 Explain the following properties of diamond and graphite in terms of the models for their respective structures.
a hardness or softness
b ability or inability to conduct electricity

4 Explain the following in terms of the models for structures of graphite and diamond.
a Graphite is used as a lubricant.
b Diamond is often used as an edge on saws and a tip on drills.

7.2 Carbon nanomaterials

Rockets are costly and dangerous, so imagine taking an elevator ride into space. While this concept might sound like science fiction, scientists are considering this idea very seriously. Japanese company Obayashi has announced that they will have a space elevator up and running by the year 2050. The elevator would reach 96 000 km into space and use robotic cars powered by magnetic linear motors. The concept is only possible because of the properties of a recently discovered allotrope of carbon—carbon **nanotubes**.

FIGURE 7.2.1 Artist's impression of a space elevator made of a carbon nanotube

Diamond, graphite and amorphous carbon have long been recognised as allotropes of carbon. However, since the 1970s, scientists have discovered how to make a new range of carbon allotropes that are examples of nanomaterials.

You will remember from Chapter 1 that nanomaterials are particularly interesting because they have a very high surface area to volume ratio, leading to some unique or enhanced properties.

FULLERENES

In the late 1970s, while working at the Australian National University in Canberra, Dr Bill Burch discovered a new allotrope of carbon. He manufactured the substance, but didn't investigate its structure. In 1985 a major scientific breakthrough occurred when analysis of this allotrope by scientists Harry Kroto, Richard Smalley and Robert Curl (who were awarded a Nobel Prize for their work) found that it contained molecules with a mass equivalent to 60 carbon atoms. They proposed that the allotrope was made up of molecules containing a roughly spherical group of carbon atoms arranged in a series of pentagons and hexagons, similar to the shape of a soccer ball, as you can see in Figure 7.2.2. This structure was later confirmed using crystallographic analysis techniques.

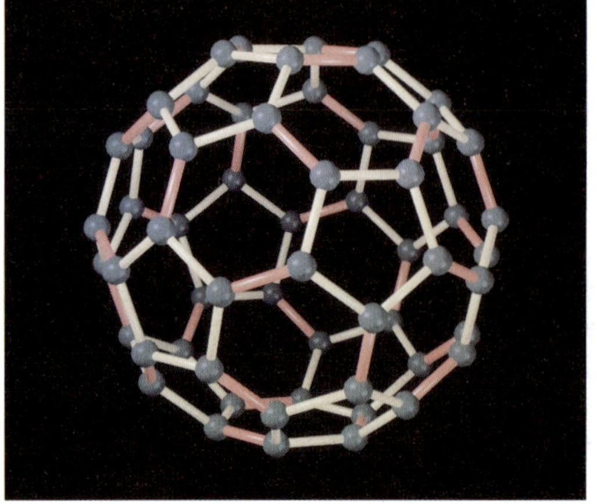

FIGURE 7.2.2 The structure of a fullerene has a similar pattern to the surface of some soccer balls.

Fullerenes are an allotrope of carbon where the atoms are arranged in a series of pentagons and hexagons.

FIGURE 7.2.3 The Biosphere in Montreal, Canada, is a museum designed by Buckminster Fuller.

Scientists have since found further variations of this molecule. These molecules have similar structures to the geodesic designs of architect Richard Buckminster 'Bucky' Fuller. They are called fullerenes, although they are more commonly referred to as **buckyballs**. Figure 7.2.3 shows the Biosphere in Montreal, Canada—a museum designed by Buckminster Fuller.

Fullerenes have three covalent bonds to each carbon atom and in some ways appear to be similar to graphite. This leaves delocalised electrons in the structure and the possibility of electrical conductivity. Although fullerenes were initially just a curiosity, scientists predict that they have significant potential in a number of fields such as composite materials and **photovoltaic cells** (solar panels). The most stable fullerene molecule involves 60 carbon atoms bonded into an approximately spherical shape that is known as buckminsterfullerene or C_{60}.

CHEMFILE

Fullerenes in flexible photovoltaic cells

Traditional photovoltaic cells (the solar panels you see on many rooftops) are made from highly refined and purified silicon crystals. Manufacturing these crystals is complex and the cells produced are rigid and brittle.

Research into fullerenes has led to alternative types of photovoltaic cells, known as polymer solar cells. Polymer solar cells (Figure 7.2.4) use alternate layers of fullerene molecules instead of silicon. The cell produced is lighter than conventional cells and offers the advantage of being flexible. A flexible cell could match the curved shape of a caravan roof or the cabin of a boat. The initial problem with these cells was their low efficiency. Research into fullerenes and other aspects of the cells continues to improve their efficiency.

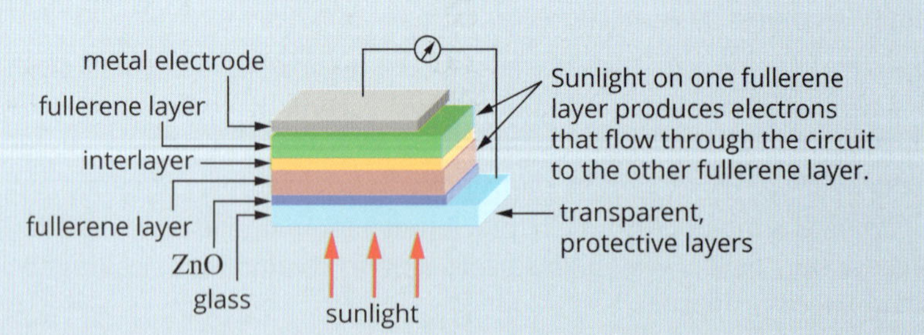

FIGURE 7.2.4 Polymer solar cells are constructed from transparent covering layers and alternate layers of fullerenes.

GRAPHENE AND NANOTUBES

Two other allotropes of carbon that are being heavily researched are nanotubes, which are regarded as part of the fullerene family, and graphene.

Graphene

You have seen that graphite has a layered structure. Graphene is best described as a single layer of graphite (Figure 7.2.5). Graphene is a single layer sheet with the same arrangement as those stacked in graphite. It is a very new material and was first isolated in 2004.

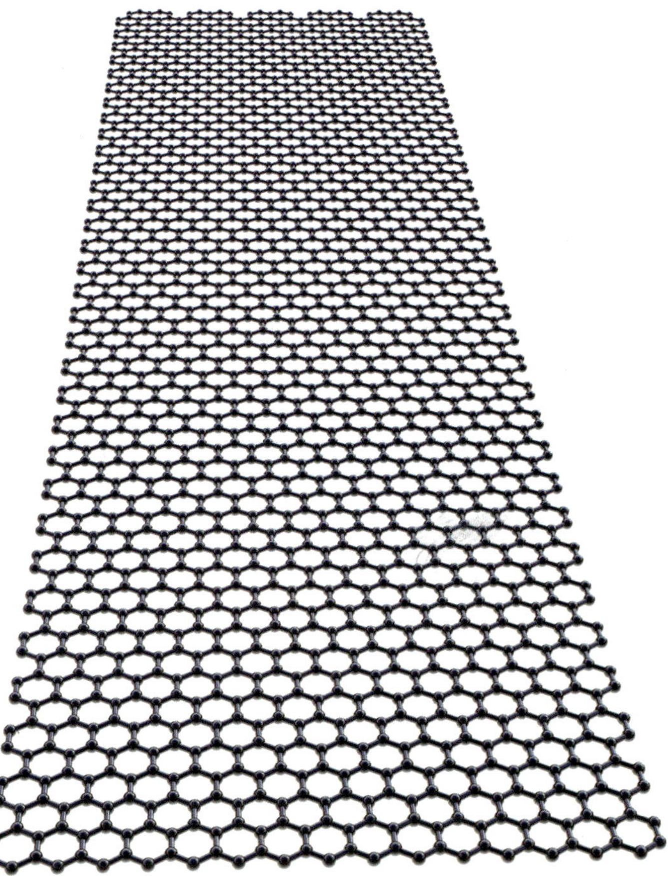

FIGURE 7.2.5 Graphene is a single layer sheet with the same arrangement as those stacked in graphite.

Graphite is soft, due to the weak dispersion forces between its layers. Graphene is only a single layer and retains the electrical conductivity of graphite but it is an extremely strong and tough material.

Graphene has many potential uses.

- It could replace silicon as the basis for computer chips and circuits due to its high electrical conductivity.
- It could be used in desalination plants. Water under pressure can pass through the thin layer but dissolved impurities cannot.
- It could be used to construct electrodes where it is an advantage for an electrode to be a non-metal.
- It could be used in organic photovoltaic cells.
- It could be used to reinforce composite materials because of its strength.

An interesting feature of graphene is that, because it is a single layer, every carbon atom is available for reaction from two sides at any instant during a chemical reaction.

Nanotubes

Nanotubes are closely related to graphene. They are called nanotubes because they have a long, hollow structure with walls formed from graphene. The diameter of these cylinders is very small, around 1 nanometre (10^{-9} metre) wide, while they can be millions of times longer. They can be capped on the end of each cylinder by half a fullerene molecule as shown in Figure 7.2.6.

Nanotubes can be single-walled or multi-walled. A multi-walled nanotube has smaller tubes sitting inside larger tubes.

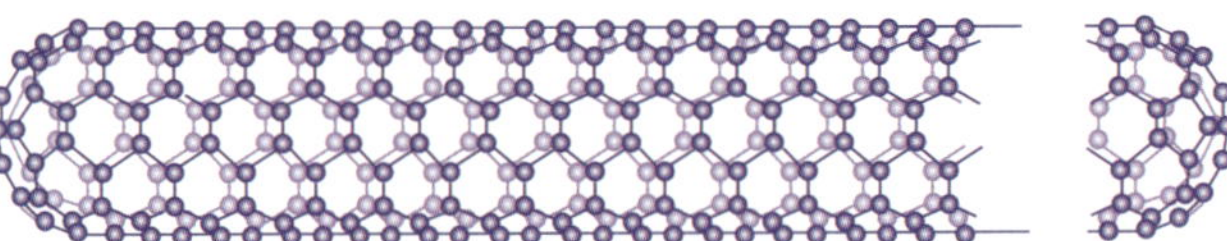

FIGURE 7.2.6 A carbon nanotube can be regarded as a sheet of graphene rolled into a cylinder and capped on the ends by half a fullerene molecule.

Scientists are interested in nanotubes because of their:

- unique strength
- electrical conductivity (the conductivity of nanotubes depends on their shape; some are conductors and others are semi-conductors)
- thermal conductivity
- strong forces of attraction to each other.

Nanotubes hold great promise in fields such as optics, nanotechnology and electronics. Their extraordinary strength and thermal and electrical conductivity suggest they may be useful as additives in various structural materials.

POTENTIAL OF CARBON NANOMATERIALS

Carbon nanomaterials offer huge gains in performance and properties over some other materials in current use and have a broad range of potential applications. The carbon–carbon bonds in these structures are very strong and there are no weak points in a single layer of graphene or a nanotube.

Carbon nanotubes are:

- up to 300 times stronger than steel. Rope made from nanotubes with a diameter of 1 cm could support a weight of more than 1000 tonnes. Nanotubes are already being used in high-performance sporting equipment
- better conductors of electricity than silver. Since nanotubes are essentially 'wires' that are much narrower in diameter than metal wire, they offer the possibility for extreme miniaturisation of electrical circuits
- better thermal conductors than diamond. Nanotubes could be used to transfer heat away from electrical components
- stronger than Kevlar fibres. Stain-resistant nanofabrics that never require washing are already available. You could even carry water in the pockets of a vest made from this material
- capable of adsorbing more gas or impurities than activated charcoal.

Perhaps the best way to highlight the potential of nanomaterials is with two exciting examples.

Volvo concept car

Volvo has embarked on a radical new design for an electric car that aims to harness the properties of nanomaterials. As shown in Figure 7.2.7, most of the steel in the car has been replaced with carbon nanotube sheets. These sheets have the advantage of lightweight strength but they can also serve as a giant battery for the car. Fullerenes are incorporated in the carbon sheets, allowing them to act as photovoltaic cells, supplying the energy needed to recharge and power the car.

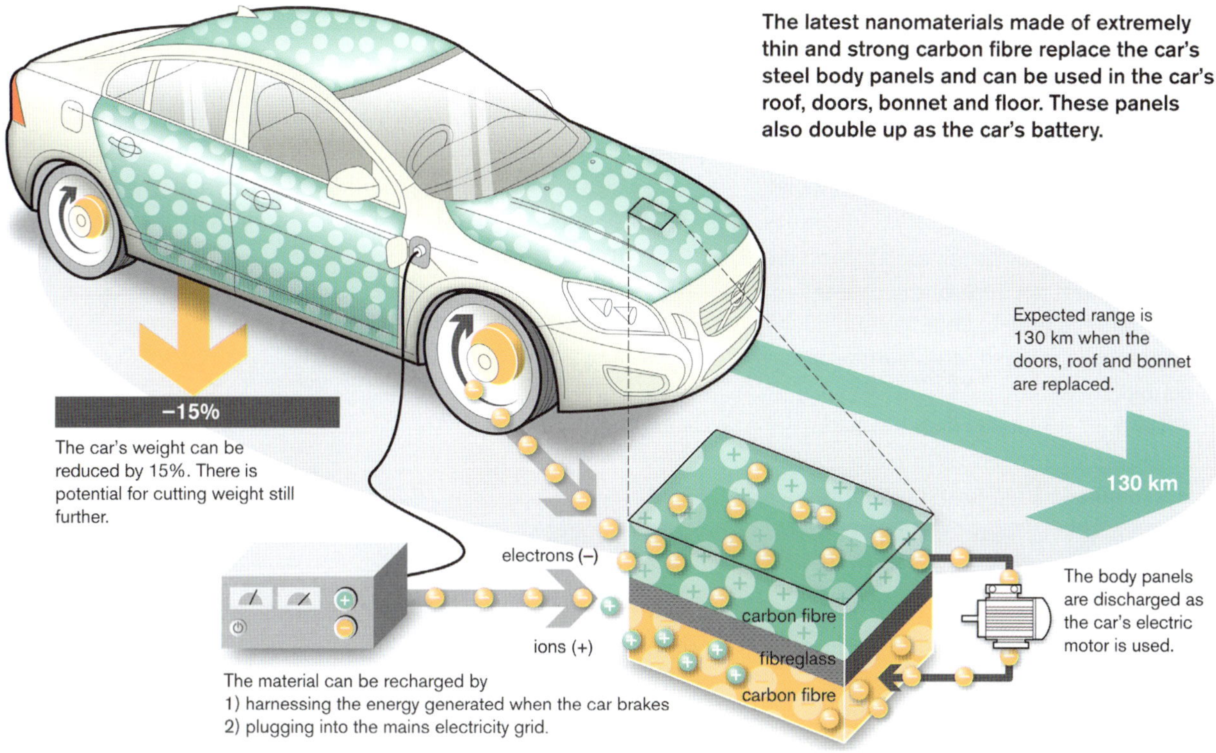

FIGURE 7.2.7 The Volvo concept car

Solar aircraft

In 2015, the Swiss-designed plane *Solar Impulse-2* (Figure 7.2.8) set off to become the first solar-powered aeroplane to circumnavigate the globe. The plane is powered by solar cells and the lightweight strength for the structure comes from an assortment of carbon nanomaterials and composites. *Solar Impulse-2* completed its 42 000 km journey in July 2016, more than 16 months after it began.

FIGURE 7.2.8 The *Solar Impulse-2*

7.2 Review

SUMMARY

- Over the past 40 years, scientists have developed a range of new carbon allotropes called fullerenes.
- Fullerenes are examples of nanomaterials and are of interest to scientists because of the properties that their high surface area to volume ratio offer, including high tensile strength and high electrical and heat conductivity.
- Spherical fullerenes are known as buckyballs. Fullerenes can also be tubular, as in nanotubes. Graphene is a single-layered form of graphite.
- In all these allotropic forms, the carbon atoms are bonded to three other carbon atoms as shown in in the table below.
- Potential applications of fullerenes include fibres and fabrics, electrical circuits, photovoltaic cells and filtration systems.

Buckyball	Graphene	Nanotube

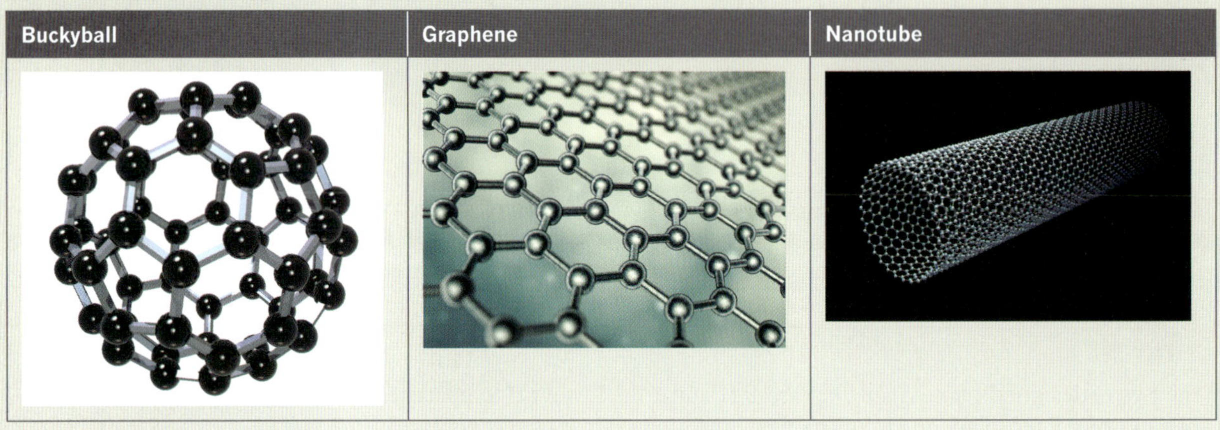

KEY QUESTIONS

1 Describe how the structures and properties of fullerenes are both similar to and different from those of graphite.

2 Describe the bonding within a C_{60} buckyball.

3 Describe the bonding within a graphene sheet.

4 Explain why carbon nanotubes are being considered for use in the construction of a space elevator.

Chapter review

07

KEY TERMS

allotrope
amorphous
buckyball
covalent layer lattice
covalent network lattice
diamond
graphite
nanotube
photovoltaic cell
sublimation

Carbon lattices

1. 'Carbon forms several allotropes.' Explain the meaning of this statement.
2. Graphite has unusual properties. It sublimes at a high temperature, it conducts electricity and it can be used as a lubricant. Use a diagram to explain why graphite has these properties.
3. Charcoal is often used by chemists to decolourise solutions. Explain how the structure of charcoal removes compounds from solutions.
4. Why can graphite be used as a lubricant under conditions of very low temperature and conditions of very high temperature?
5. Why does diamond have such a high sublimation point?
6. Silica (silicon dioxide), like diamond, has a high melting point, is very hard and is a non-conductor of electricity. Explain, in terms of its bonding, why it shares these properties with diamond.
7. The structures of methane and diamond are shown in the image below. Each carbon atom in methane (CH_4) has a tetrahedral arrangement of atoms around it. A carbon atom in diamond also has a tetrahedral arrangement. However, the two substances have very different properties.

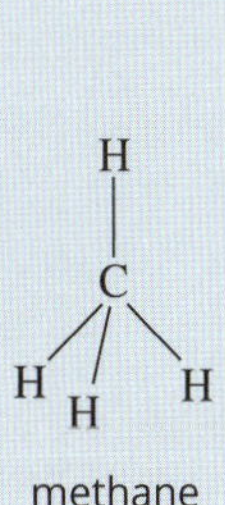

methane

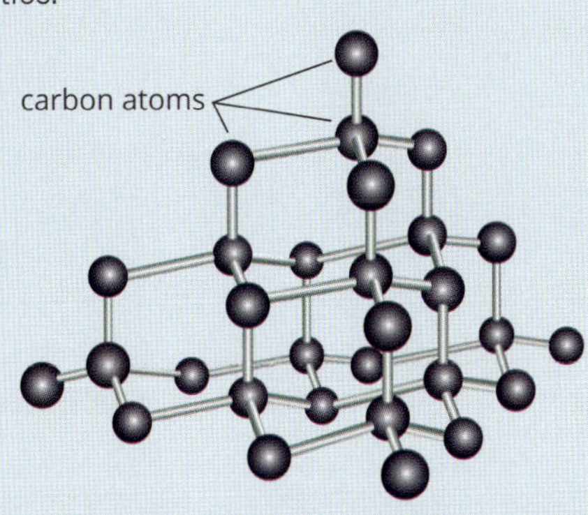

diamond

   a. Describe all of the types of bonding that would be present in each substance.
   b. Use the types of bonding present in each substance to explain the different properties you would expect each to have.

Carbon nanomaterials

8. Describe the geometry of the bonds around carbon atoms in diamond and graphene.
9. In terms of the bonding between the carbon atoms, explain why graphene is so strong.
10. Explain why fullerenes can have high electrical conductivity.
11. Describe the structure of a carbon nanotube.
12. Why are carbon nanotubes very strong?
13. A possible use for carbon nanotubes is as supports for catalysts. A catalyst may be spread, atom by atom, on the surface of a carbon nanotube. Explain why this has the potential to be a very effective form of catalyst.

Connecting the main ideas

14. Identify the pairs of allotropes from the following substances:

    CO_2, O_2, H_2O, S_6, CO, H_2O_2, CO, O_3, S_8

15 Copy and complete the table below.

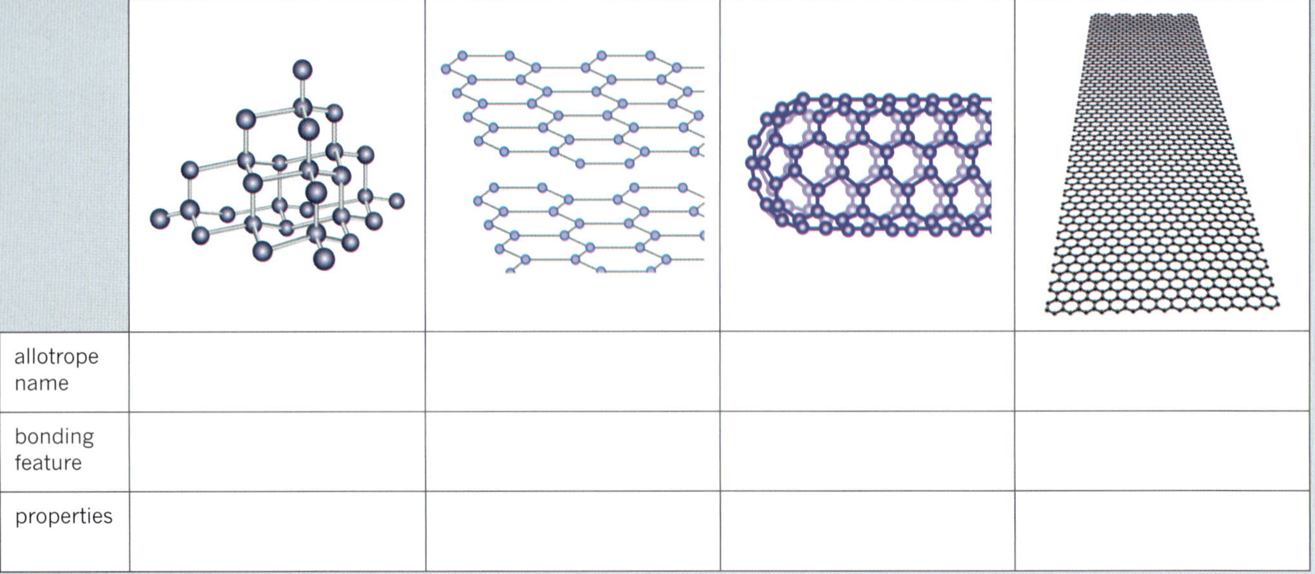

allotrope name				
bonding feature				
properties				

16 Compare and contrast the structure and properties of graphite and graphene.

17 The image below shows an experiment to determine the electrical conductivity of diamond.

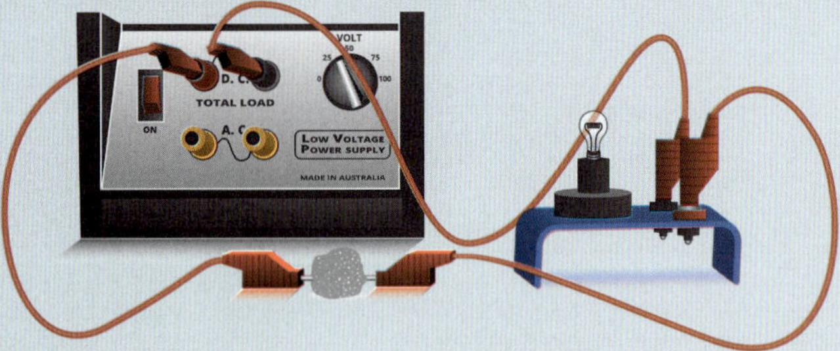

a Explain briefly how this test works and what is required for a substance to conduct electricity.

b Classify each allotrope of carbon as either conductive or non-conductive of electricity.

18 Germanium can form a number of allotropes including one called germanene, which is similar in structure to graphene. Explain why it is possible for germanium to form a structure similar to that of graphene.

19 **a** Explain why silicon carbide has a similar structure to diamond.

b Predict the physical properties of silicon carbide.

20 Explain the electrical conductivity of graphite, solid copper and an aqueous solution of sodium chloride by reference to similarities and differences in the models used to describe their structures.

CHAPTER 08

Organic compounds

At the end of this chapter, you will appreciate the significance of the compounds that are formed between carbon and hydrogen. Carbon is only the 11th most abundant element in the universe, yet it forms more compounds than all other elements except hydrogen. Carbon is present in living things and many non-living things.

The prevalence of carbon in living things is why the study of the compounds of carbon and hydrogen is known as organic chemistry. You make use of these compounds in so many aspects of your life: from the polymer case of your telephone to the pasta that you may eat for dinner tonight.

In this chapter, you will start to explore some of the many families of carbon compounds. However, you will only scratch the surface of a varied and fascinating area of chemistry.

Science understanding

- hydrocarbons, including alkanes, alkenes and benzene, have different chemical properties that are determined by the nature of the bonding within the molecules
- IUPAC nomenclature is used to name straight and simple branched alkanes and alkenes from C_1–C_8
- molecular structural formulae (condensed or showing bonds) can be used to show the arrangement of atoms and bonding in covalent molecular substances
- alkanes, alkenes and benzene undergo characteristic reactions such as combustion, addition reactions for alkenes and substitution reactions for alkanes and benzene

8.1 Alkanes

Carbon forms more compounds than all other elements combined because they have certain properties.

- Carbon has four valence electrons so carbon can potentially form covalent bonds with four different atoms.
- Carbon atoms can form strong covalent bonds with other carbon atoms.
- The covalent bonds formed can be a combination of single, double or triple bonds.

Because of this, carbon can bond to itself to form molecules of varied length and shape. The different structures have different properties and applications.

Carbon-based molecules are all around you. Caffeine, petrol, pesticides, plastics and artificial flavours are all carbon-based compounds. Many of these compounds are produced from **crude oil**. Crude oil is produced by the effects of heat and pressure on dead animals, plants and microorganisms trapped in the Earth's crust, buried beneath sediment formed over millions of years. The branch of chemistry that studies the chemistry of carbon compounds is known as organic chemistry. This section focuses on crude oil as a source of carbon compounds and on compounds found in crude oil that are formed between carbon and hydrogen: the **hydrocarbons**.

CRUDE OIL

The amount of carbon on Earth is essentially fixed, with most of the carbon atoms on Earth having been here for billions of years. It is the location of these carbon atoms that changes over time. Whether they are currently part of a sunflower seed, a molecule of carbon dioxide, a human being or a plastic chair, you can be certain that the carbon atoms have been in a different compound previously.

Origin of crude oil

Fossil fuels, such as coal, oil and natural gas, come from the remains of plants and animals. When prehistoric marine microorganisms, such as bacteria and plankton, died and were buried by sands millions of years ago, these organisms accumulated as organic sediment and gradually became part of the Earth's crust. Over millions of years, this organic material was affected by high temperatures and pressures, causing the oils and fats to be converted into hydrocarbons. This mixture of hydrocarbons is crude oil.

Crude oil has a low density, which means it could migrate upwards through the crust, where it often became trapped beneath impervious (unable to be passed through) rock (Figure 8.1.1). Accumulation of oil and gas under the rock creates an oil field.

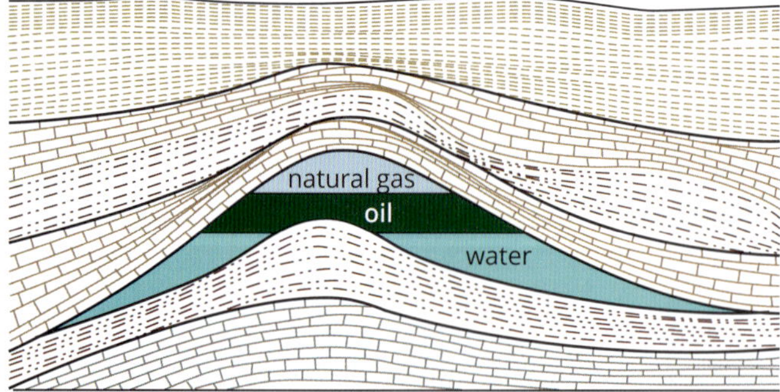

FIGURE 8.1.1 Typical structure of the impervious rock that traps oil and natural gas underground and creates an oil field

Crude oil is not used in its raw state. It is transported from oil fields to oil refineries where it undergoes fractional distillation in a fractionating tower as shown in Figure 8.1.2. In this process, the crude oil is separated into its various components, or fractions. Each fraction is made up of a range of hydrocarbons with similar boiling points and hence molecular masses. The components of crude oil that are obtained by fractional distillation are used for a wide range of purposes, but presently more than 90% of them are used for fuels.

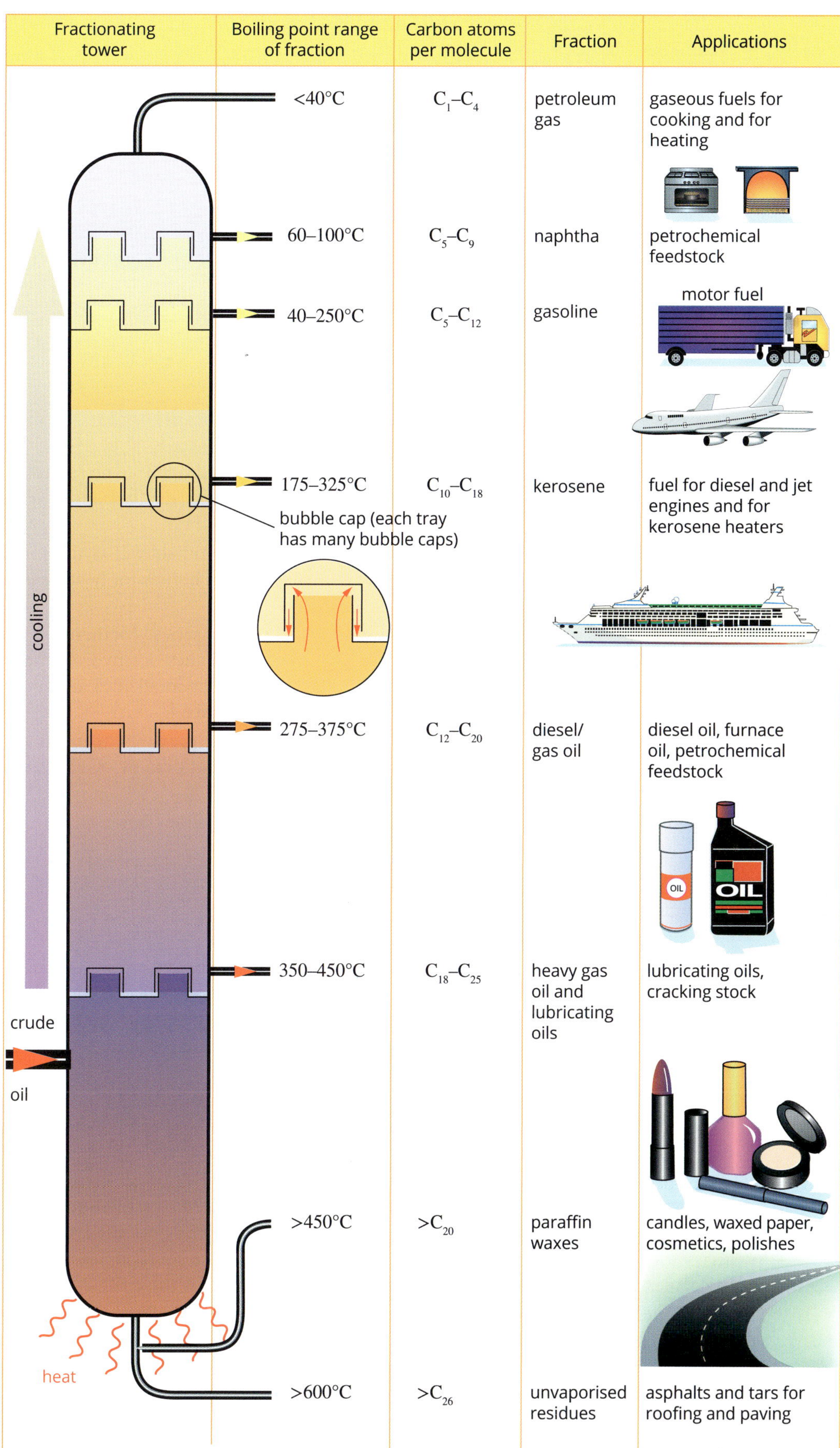

FIGURE 8.1.2 Fractions found in crude oil

Further processing of the crude oil fractions is needed to make these components even more useful.

Heavier fractions undergo a process called **cracking**, which is shown in Figure 8.1.3. This process breaks the larger hydrocarbon molecules into smaller molecules using heat and a catalyst. The smaller molecules and especially those with carbon–carbon double bonds (called alkenes) are needed by the petrochemical industry.

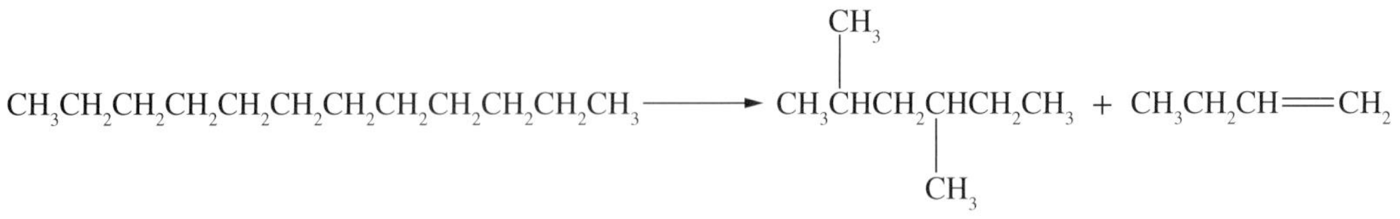

$$CH_3CH_2CH_2CH_2CH_2CH_2CH_2CH_2CH_2CH_2CH_2CH_3 \longrightarrow CH_3CH(CH_3)CH_2CH(CH_3)CH_2CH_3 + CH_3CH_2CH{=}CH_2$$

FIGURE 8.1.3 Cracking of a larger hydrocarbon molecule results in two smaller molecules, one of which has a carbon–carbon double bond.

ALKANES

Hydrocarbons are composed only of carbon and hydrogen. The simplest hydrocarbon is methane. In methane, one carbon atom is covalently bonded with four hydrogen atoms to form a molecule with the formula CH_4.

Methane is a colourless, odourless gas that is very flammable. It is the main component of natural gas, which is used in many Australian homes for heating and cooking.

Hydrocarbons can be classified into several groups or series. Methane is the first of a series of compounds known as the **alkanes**. Alkanes are hydrocarbons that contain only single bonds. All the carbon–carbon bonds in alkanes are single covalent bonds. Molecules such as this are said to be **saturated**.

Each member of the alkane series differs from the previous member by a $-CH_2-$ unit. A series of molecules in which each member differs by $-CH_2-$ from the previous member is known as a **homologous series**.

Compounds that are members of the same homologous series have:

- a similar structure
- a pattern to their physical properties
- similar chemical properties
- the same general formula.

Alkanes have only weak forces between the molecules because they are 'non-polar'. You will learn more about polar and non-polar molecules in Chapter 12. As the number of carbons and the size of the molecules increase within this homologous series, the strength of these forces increases, so the melting and boiling points of the alkanes increase.

Table 8.1.1 shows the first three members of the alkane series. Note that the bonds around each carbon atom adopt a tetrahedral shape and the boiling points increase as the molecules become larger.

The naming of the members of each homologous series follows set conventions. The **stem name**, or parent name, in propane (prop-) indicates there are three carbon atoms in the molecule. This same stem name (prop-) is used in other series for the member that contains three carbon atoms. The stem names used for molecules with between one and 10 carbon atoms are listed in Table 8.1.2.

Alkanes are named by adding -ane after the stem name. For example, an alkane that contains eight carbon atoms is called octane. The molecular formula of octane is C_8H_{18}. The alkanes have the general formula C_nH_{2n+2}, where n stands for the number of carbon atoms. If an alkane molecule has 12 carbon atoms, the number of hydrogen atoms is $2n + 2 = 2 \times 12 + 2 = 26$. The molecular formula is therefore $C_{12}H_{26}$.

TABLE 8.1.1 Structure, properties and some uses of the first three alkanes

Name and molecular formula	Structural formula	Properties	Uses
methane, CH_4	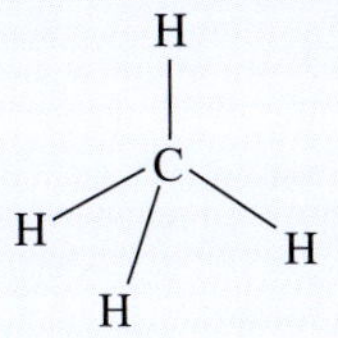	non-polar, gas, boiling point (BP) −164°C	cooking, Bunsen burners, gas heating
ethane, C_2H_6	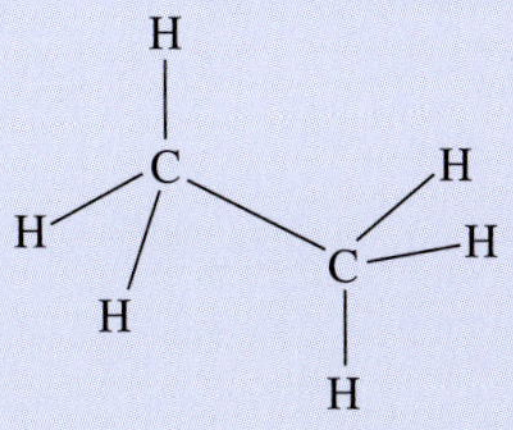	non-polar, gas, BP −87°C	conversion to ethene
propane, C_3H_8	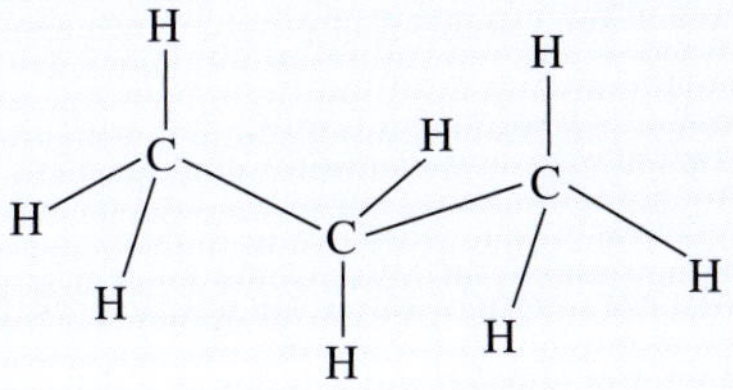	non-polar, gas, BP −42°C	liquid petroleum gas (LPG)

TABLE 8.1.2 Stem names used for molecules with between one and 10 carbon atoms

Stem (parent) name	Number of carbon atoms
meth-	1
eth-	2
prop-	3
but-	4
pent-	5
hex-	6
hept-	7
oct-	8
non-	9
dec-	10

The name of an alkane ends in -ane and alkanes have the general formula C_nH_{2n+2}.

Writing formulae of alkanes

Sometimes you want to show more information about a molecule than just its overall composition. You can use a variety of ways to write the formulae of carbon compounds. Table 8.1.3 shows different ways of representing ethane and butane.

TABLE 8.1.3 Different ways of representing alkanes

Alkane	Molecular formula	Lewis structure diagram (electron diagram)	Condensed structural formula (semi-structural formula)	Structural formula
ethane	C_2H_6	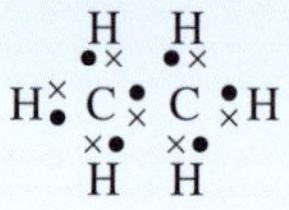	CH_3CH_3	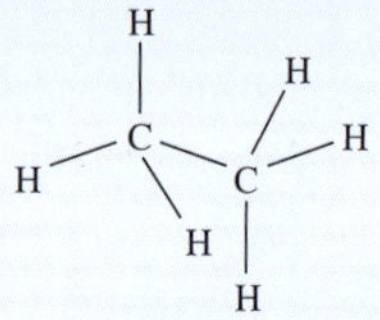
butane	C_4H_{10}	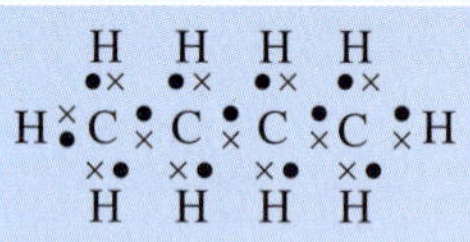	$CH_3CH_2CH_2CH_3$ or $CH_3(CH_2)_2CH_3$	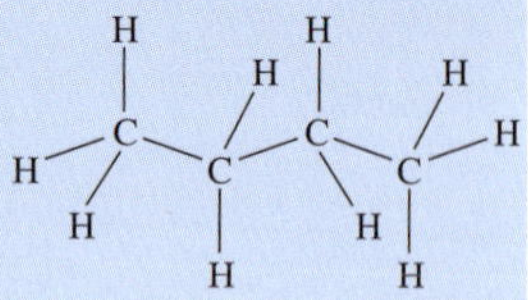

Structural formulae show all the bonds in a molecule, but lone pairs can be omitted. **Condensed structural formulae** are also known as semi-structural formulae and show the atoms that are connected to each carbon atom, but do not show the bonds.

Representing three-dimensional molecules on a two-dimensional computer screen or page is a challenge for chemists. For example, each carbon in an alkane has four bonds in a tetrahedral arrangement. When the shape of the molecule is not important, it is often drawn in a simplified two-dimensional format as shown in Figure 8.1.4.

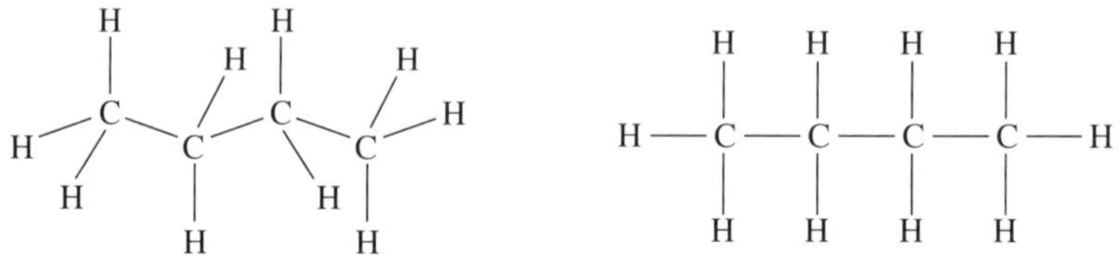

FIGURE 8.1.4 Molecules such as butane are often drawn in two dimensions for simplicity.

Structural isomers of alkanes

There is only one molecule that can be formed with the molecular formula of methane (CH_4). This is also the case for ethane (C_2H_6) and propane (C_3H_8). However, alkanes that have four or more carbon atoms have more than one possible structure.

Figure 8.1.7 shows two different molecules that have the molecular formula C_4H_{10}. Molecules that have the same molecular formula but have different arrangements of atoms are said to be **structural isomers** of each other. The more atoms in the molecule, the more possible isomers there are.

```
a
      H   H   H   H
      |   |   |   |
  H — C — C — C — C — H
      |   |   |   |
      H   H   H   H

b
      H   H   H
      |   |   |
  H — C — C — C — H
      |   |   |
      H   |   H
      H — C — H
          |
          H
```

a: $CH_3(CH_2)_2CH_3$ b: $(CH_3)_3CH$ or $CH_3CH(CH_3)_2$

FIGURE 8.1.7 There are two structural isomers that have the molecular formula C_4H_{10}.

Figure 8.1.7a shows the straight-chain isomer of butane, so named because the four carbon atoms are bonded in a continuous chain. Figure 8.1.7b is the branched isomer because you could consider it as a straight-chain alkane, with three carbon atoms in the chain, with a $-CH_3$ **side group** attached. A $-CH_3$ side group is called a methyl group, as its structure is similar to that of methane with one less hydrogen. A $-CH_2CH_3$ side group is called an ethyl group.

CHEMFILE

Have you got the patience for slow science?

The longest running laboratory experiment in the world was set up in 1927 at the University of Queensland in Brisbane. The long-running 'pitch drop' experiment is designed to show that pitch, the name given to a mixture of highly viscous hydrocarbons, is actually a liquid. Pitch, otherwise known as bitumen (Figure 8.1.5), appears solid on first inspection and shatters if you strike it with a hammer.

FIGURE 8.1.5 A sample of pitch at room temperature appears solid and shatters when hit with a hammer.

Professor Thomas Purnell set up the pitch experiment as shown in Figure 8.1.6 by placing a sample of pitch into a funnel encased in a sealed container. He wanted to demonstrate to students that pitch was indeed a liquid—just a very, very viscous one.

Over time, the pitch has settled and slowly flowed through the funnel producing a drop, on average, once every decade. The ninth drop was recorded in April of 2014.

Today bitumen is used extensively, forming the main component of road surfaces. Under normal conditions, the road surface appears solid. However, heating reduces the viscosity and road workers are able to work the pliable bitumen.

FIGURE 8.1.6 Professor Thomas Purnell set up the famous University of Queensland 'pitch drop' experiment in 1927. Professor John Mainstone, shown here, oversaw the experiment for many years.

Therefore, Figure 8.1.7b is named methylpropane, as it can be regarded as a propane molecule with a methyl group attached. Methyl and ethyl groups are examples of **alkyl groups**. Alkyl groups have one less hydrogen atom than the corresponding alkane of the same name, so the general formula of an alkyl group is $-C_nH_{2n+1}$.

Table 8.1.4 lists the names of the alkyl groups with one to five carbon atoms.

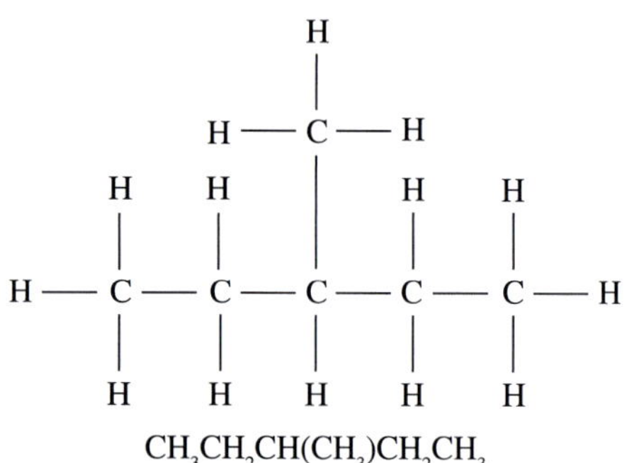

FIGURE 8.1.8 The structural and condensed structural formula of a branched alkane

TABLE 8.1.4 Names of the alkyl groups with one to five carbons

Alkyl group	Corresponding alkane	Name of alkyl group
$-CH_3$	methane	methyl
$-CH_2CH_3$	ethane	ethyl
$-CH_2CH_2CH_3$	propane	propyl
$-CH_2CH_2CH_2CH_3$	butane	butyl
$-CH_2CH_2CH_2CH_2CH_3$	pentane	pentyl

When writing the condensed structural formula of a branched alkane, alkyl groups are written in brackets (Figure 8.1.8).

Structural isomers are compounds with different physical and chemical properties. Table 8.1.5 shows the isomers of hexane and their respective melting points and boiling points. The boiling and melting points are different because the forces of attractions between the different versions of hexane molecules are affected by the structure of the different molecules (isomers).

TABLE 8.1.5 Melting and boiling points of the isomers of hexane

Isomer	Melting point (°C)	Boiling point (°C)
$CH_3—CH_2—CH_2—CH_2—CH_2—CH_3$ hexane	–95.3	68.7
$CH_3—CH_2—CH(CH_3)—CH_2—CH_3$ 3-methylpentane	–118.0	63.3
$CH_3—CH_2—CH_2—CH(CH_3)—CH_3$ 2-methylpentane	–153.7	60.3
$CH_3—CH(CH_3)—CH(CH_3)—CH_3$ 2,3-dimethylbutane	–128.6	58
$CH_3—CH_2—C(CH_3)_2—CH_3$ 2,2-dimethylbutane	–99.8	49.7

CHEMFILE

Burning ice

Large amounts of methane gas are stored around the world trapped as solid methane hydrate. These methane hydrates look very similar to ice with one main difference—they burn.

The burning ice is a result of the methane gas being released from the crystal structure of the ice as it melts (Figure 8.1.9).

Large deposits of methane hydrate are trapped in the Arctic regions of the Earth and deep in the oceans where low temperatures and high pressures trap methane. This source of methane is produced by the anaerobic decomposition of organic material by bacteria. The lattice structure of methane hydrate is shown in Figure 8.1.10.

Oil companies have known of the existence of methane hydrate as far back as the 1930s. However it is only now, with growing concerns over dwindling fossil fuel supplies and global warming, that scientists are endeavouring to find a way to safely use these large fuel sources.

One fear is that with rising ocean temperatures the deposits of methane hydrate throughout the oceans will melt, releasing large amounts of methane into the atmosphere. Methane is a major contributor to the enhanced greenhouse effect, so a release of this size is likely to accelerate the effects of global climate change.

FIGURE 8.1.9 The combustion of blocks of methane hydrate gives the appearance of burning ice. As the ice melts, methane gas is released from the lattice structure, providing the fuel for the combustion reaction.

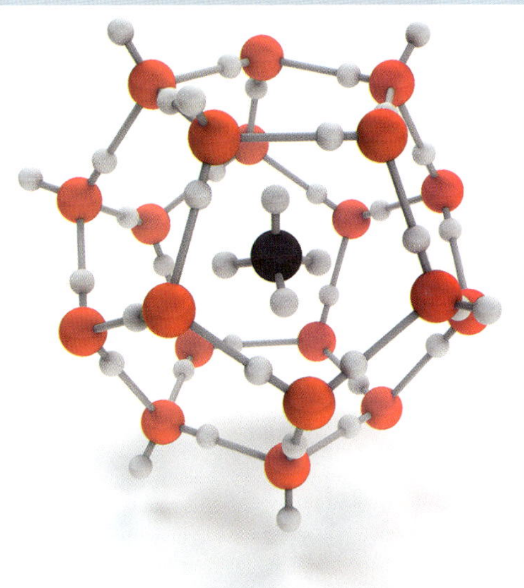

FIGURE 8.1.10 Structure of methane hydrate. Methane molecules are trapped inside the hexagonal structure of ice.

NAMING ALKANES

The systematic naming of carbon compounds is controlled by the International Union of Pure and Applied Chemists (IUPAC). Under the IUPAC system, the name of the compound provides details of its structure. The following rules apply when naming alkanes.

1. Identify the longest unbranched carbon chain and count the number of carbon atoms in that chain. This gives the molecule its stem name.
2. If there are branches to be added to the name, number the carbon atoms in the chain from the end of the chain that will give the smallest numbers to branching groups.
3. Identify the type and the number of the alkyl groups attached to the chain.
4. Place the number and position of each of the alkyl groups at the beginning of the compound's name.
5. If two identical side chains are present, use 'di-' as a prefix; for three use 'tri-'.
6. If there are different alkyl side chains, list them in alphabetical order at the start of the name, with their numbers to indicate their respective positions.

Carefully check that you have identified the longest unbranched carbon chain. Sometimes the longest carbon chain is not drawn in a straight line.

The following steps show the process of naming the isomer of hexane shown in Figure 8.1.11.

FIGURE 8.1.11 The IUPAC system will be used to name this isomer of hexane.

1 Identify the longest unbranched carbon chain.

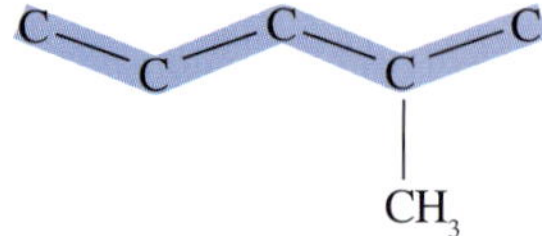

2 Number the carbons, starting from the end closest to the branch.

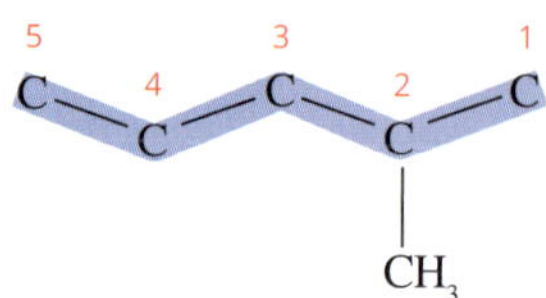

3 Name the side branches and main chain.

five carbon atoms = pentane

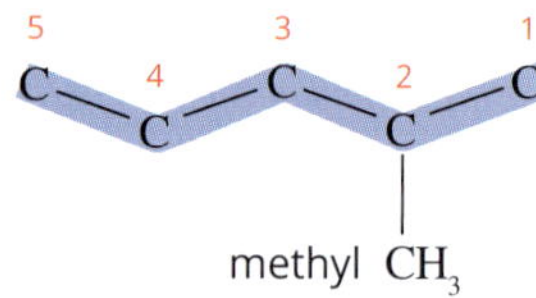

4 Combine all components to write the full name: 2-methylpentane.

Figure 8.1.12 shows examples of applications of these rules. Note the use of the prefix 'di-' to indicate the presence of two methyl side branches and the numbering to indicate their position along the longest continuous carbon chain (for example, 3,3- indicates that both side branches come off the third carbon atom).

Note that in the names, numbers are separated by commas, numbers are separated from letters with a hyphen. You also only need to use numbers when there is more than one possible position on the chain for the groups. For example, you need to include the '2' in 2-methylpentane because the methyl group could be on the second or the third carbon atom. But for methylpropane, no number is required as the only possible place for the methyl group to be attached to the chain is the second carbon atom.

$CH_3-CH_2-CH(CH_3)-CH_2-CH_3$

3-methylpentane

$CH_3-CH_2-C(CH_3)_2-CH_2-CH_3$

3,3-dimethylpentane

$CH_3-CH(CH_3)-CH(CH_3)-CH_3$

2,3-dimethylbutane

FIGURE 8.1.12 IUPAC systematic names for three alkanes. Note that there are no spaces in the names of these compounds.

Worked example 8.1.1

IUPAC NAMING SYSTEM FOR ALKANES

Write the systematic name for the following molecule.

$$CH_3-CH_2-CH_2-\underset{\displaystyle |}{\overset{\displaystyle CH_3}{CH}}-CH_2-CH_3$$

Thinking	Working
Identify the longest carbon chain in the molecule. The stem name of the molecule is based on this longest chain.	There are six carbons in the longest chain. The stem name is based on hexane.
Number the carbon atoms starting from the end closest to the branch.	$CH_3-CH_2-CH_2-CH(CH_3)-CH_2-CH_3$ numbered from right to left: 6 5 4 3 2 1
Identify the branch.	The side chain is a methyl group.
Combine all components.	The name of the molecule is 3-methylhexane.

Worked example: Try yourself 8.1.1

IUPAC NAMING SYSTEM FOR ALKANES

Write the systematic name for the following molecule.

$$CH_3-CH_2-\overset{\displaystyle CH_3}{\overset{|}{CH}}-CH_2-CH_3$$

8.1 Review

SUMMARY

- Hydrocarbons are compounds containing carbon and hydrogen only.
- Crude oil is the source of many hydrocarbons. These hydrocarbons are separated from one another by fractional distillation.
- In a homologous series, each member has one more $-CH_2-$ unit than the previous member. Members of a homologous series have similar structures and chemical properties and the same general formula. Alkanes are an example of a homologous series.
- Alkanes are saturated hydrocarbon molecules that contain only single bonds and have a general formula C_nH_{2n+2}.
- Structural isomers are molecules with the same molecular formula but different arrangements of atoms.
- Hydrocarbon molecules can be drawn using structural formulae or condensed structural formulae.
- The IUPAC naming system is used to provide systematic names for hydrocarbon molecules. Names are based on the longest unbranched carbon chain.

KEY QUESTIONS

1 From which one or more of the following sources does crude oil originate?
 A plant material
 B forest animals that have recently died
 C dinosaurs only
 D marine microorganisms

2 Methane is the smallest hydrocarbon molecule.
 a What is the molecular formula of methane?
 b Why is methane a hydrocarbon?
 c Why does the carbon atom in a methane molecule bond to four, rather than two, three, five or any other number of hydrogen atoms?
 d Draw the structural formula of methane. Why does methane have this particular arrangement of hydrogen atoms around each carbon atom?

3 Answer the following questions relating to the hydrocarbon with the molecular formula, C_3H_8.
 a What is the name of this hydrocarbon?
 b Draw the structural formula of this hydrocarbon.
 c Write the condensed structural formula of this hydrocarbon.

4 Write the systematic names of these alkanes based on their condensed structural formulae.
 a $CH_3CH_2CH_2CH_3$
 b $CH_3CH_2CH_2CH_2CH_2CH_2CH_3$
 c $CH_3CH_2CH_2CH(CH_3)_2$
 d $CH_3CH(CH_3)CH_2CH_3$
 e $CH_3CH(CH_3)CH_2CH(CH_3)_2$

5 Write the systematic names of the alkanes shown here based on their structural formulae.

a

```
    H   H   H   H
    |   |   |   |
H — C — C — C — C — H
    |   |   |   |
    H   H   H   H
```

b

```
        H
        |
    H — C — H
    H   |   H
    |   |   |
H — C — C — C — H
    |   |   |
    H   H   H
```

c

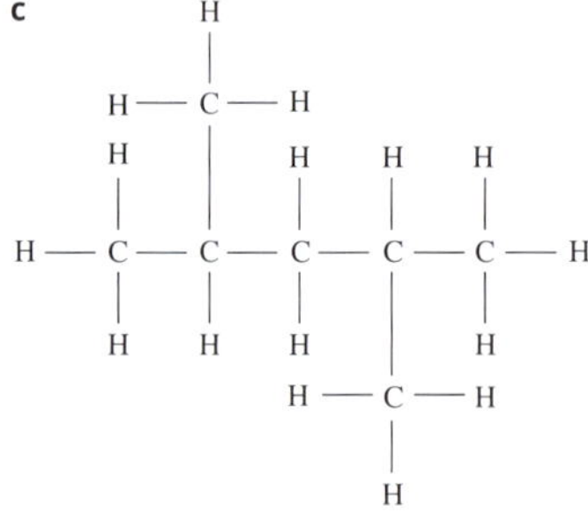

d

```
    H
    |
H — C — H
    |   H   H   H
    |   |   |   |
H — C — C — C — C — H
    |   |   |   |
    H   H   |   H
        H — C — H
            |
        H — C — H
            |
            H
```

6 Draw the structural formulae of these alkanes based on their systematic names.
 a hexane
 b 3-methylhexane
 c 3,3-dimethylpentane
 d 3-ethyl-2-methylpentane

8.2 Alkenes

Carbon forms many compounds with hydrogen in which there are double or even triple bonds between the carbon atoms. These compounds are called **unsaturated** hydrocarbons. A double bond is formed when two pairs of electrons are shared and a triple bond is formed when three pairs of electrons are shared. Some of the most useful carbon compounds are unsaturated. This section focuses on the structures and naming of hydrocarbon compounds with carbon–carbon double bonds.

ALKENES

Alkenes are a homologous series of hydrocarbons with at least one carbon–carbon double bond. The general formula for the alkenes with one double bond is C_nH_{2n}. The presence of a carbon–carbon double bond means the alkenes are described as unsaturated molecules. Alkenes are more reactive than alkanes, which do not have any carbon–carbon double bonds.

The name of an alkene ends in -ene and alkenes have the general formula C_nH_{2n}.

The simplest alkene is ethene (C_2H_4). The next member in the series is propene (C_3H_6), which has an additional $–CH_2–$ unit. The first three members of the alkene homologous series are shown in Table 8.2.1. Note that the angle of the bonds around a carbon atom with a double bond is 120°. All alkenes are non-polar and hence do not dissolve in water.

TABLE 8.2.1 Structure, properties and some uses of the first three alkenes

Name and molecular formula	Structural formula	Properties	Uses
ethene, C_2H_4		non-polar, gas, boiling point (BP) –78.4°C	in the manufacture of a wide range of chemicals
propene, C_3H_6		non-polar, gas, BP –47.7°C	in the manufacture of propene oxide and polymers
butene, C_4H_8		non-polar, gas, BP –6.3°C	in the manufacture of butanol and polymers

Writing formulae of alkenes

Like alkanes, there is a variety of ways of writing the formulae of alkenes. Table 8.2.2 shows different ways of representing ethene and propene.

TABLE 8.2.2 Different ways of representing alkenes

Alkene	Molecular formula	Electron dot diagram (Lewis structure)	Condensed structural formula (semi-structural formula)	Structural formula
ethene	C_2H_4	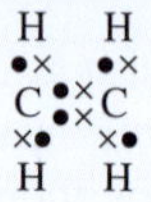	CH_2CH_2	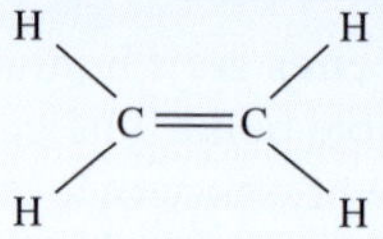
propene	C_3H_6	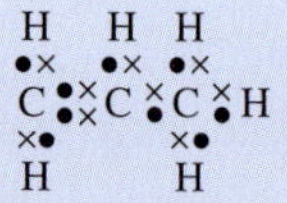	CH_2CHCH_3	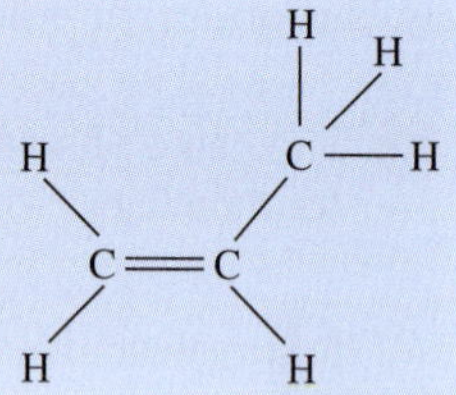

Structural isomers of alkenes

For alkene molecules that contain more than three carbon atoms, structural isomers will exist. As you can see in Figure 8.2.1, isomers may result from branches in the carbon chain, or if the carbon–carbon double bond is in different positions.

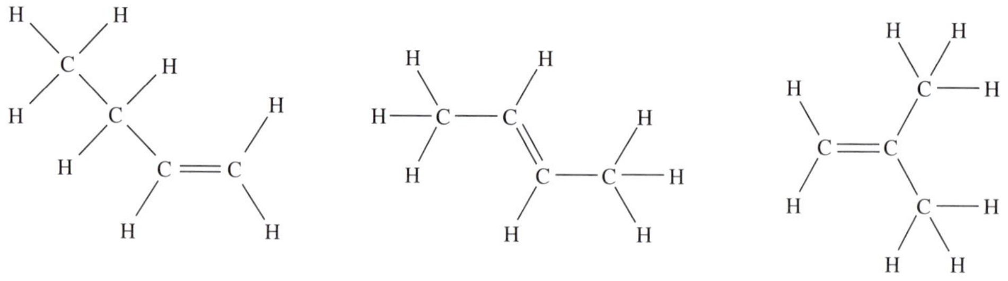

but-1-ene

but-2-ene

methylpropene

FIGURE 8.2.1 Isomers of butene. Isomers of alkenes can have branches in the carbon chain or the double bond at different points in the carbon chain.

When naming alkene isomers, the carbons are numbered from the end of the carbon chain that gives the lowest number to the first carbon in the double bond. The location of the double bond in the molecule is indicated by this number. In Figure 8.2.1, you can see that the double bond in but-1-ene is between carbons one and two. The double bond in but-2-ene is between carbons two and three.

Cis-trans (geometric) isomers of alkenes

In ***cis-trans*** **(geometric) isomers** of alkenes, the atoms in the isomers are joined in the same order but have a different arrangement in space. This occurs because the atoms associated with a carbon-carbon double bond are fixed in position.

Two atoms joined by a single bond can rotate freely around the single bond. At first glance, it may look as if the atoms in the two molecules shown in Figure 8.2.2 have a different spatial arrangement. However, they are the same molecule because free rotation occurs around the single bond.

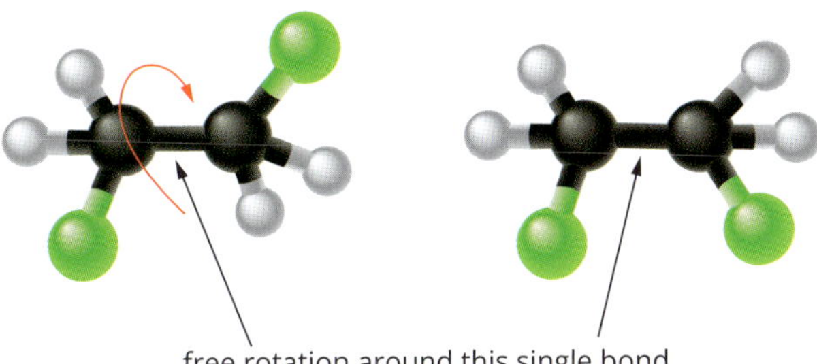

FIGURE 8.2.2 Atoms can freely rotate around a carbon–carbon single bond. These two models are representations of the same molecule.

On the other hand, two atoms that are joined by a double bond do not rotate freely because the double bond locks the atoms firmly in position. The models in Figure 8.2.3 represent two molecules that cannot be rotated to form the same structure.

FIGURE 8.2.3 Atoms cannot freely rotate around a carbon–carbon double bond. These two models do not represent the same molecule.

Cis-trans, or geometric, isomerism can occur when there are two different groups attached to each carbon atom involved in the double bond (Figure 8.2.4).

- In *cis* isomers, the same group on each carbon is located on the same side of the double bond.
- In *trans* isomers, the same group on each carbon is located on opposite sides of the double bond.

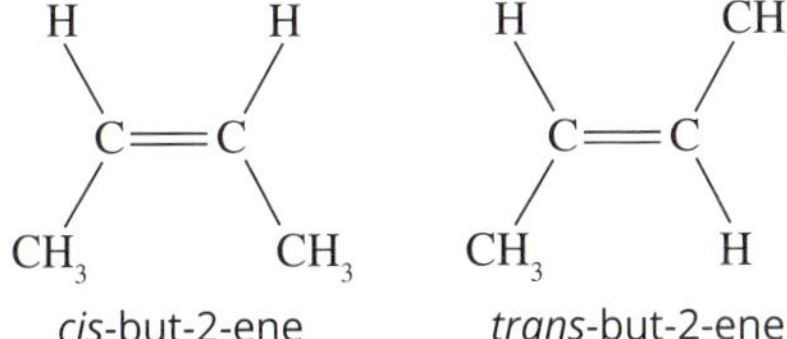

FIGURE 8.2.4 *Cis* and *trans* isomers of but-2-ene. The two molecules have the same condensed structural formula, but they have different structures due to a different spatial arrangement of atoms.

NAMING ALKENES

The following rules apply when naming alkenes.

1. Identify the longest unbranched carbon chain. The chain must include the double bond.
2. Number the carbon atoms in the chain from the end of the chain that will give the smallest numbers to the double-bonded carbon atoms.
3. Name the alkyl groups after the alkane from which they are derived.
4. Identify the position of the double bond by the number of the first carbon atom involved in the bond. Use the suffix '-ene' to indicate the presence of a double bond, e.g. hex-2-ene.
5. List the number and position of each of the alkyl groups at the beginning of the compound's name in alphabetical order.
6. If two identical side chains are present, use 'di-' as a prefix; for three use 'tri-'.
7. Identify whether the structure exhibits geometric isomerism and if so, whether it is a *cis* or *trans* isomer.

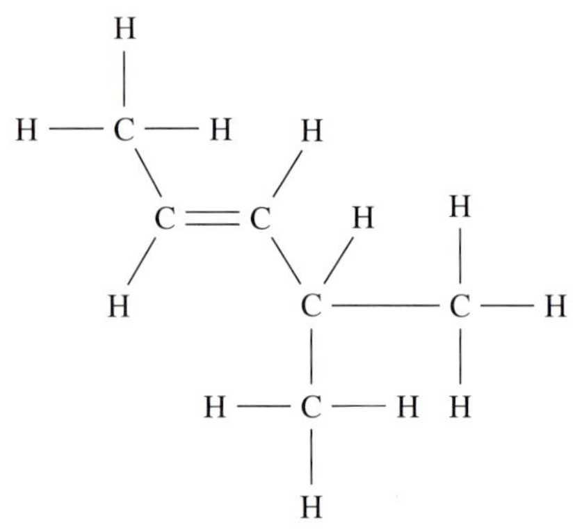

FIGURE 8.2.5 The IUPAC system will be used to name this isomer of hexene.

The following steps show the process of naming the isomer of hexene shown in Figure 8.2.5.

1 Identify the longest carbon chain that contains the double bond. The name of the molecule is based on this chain.

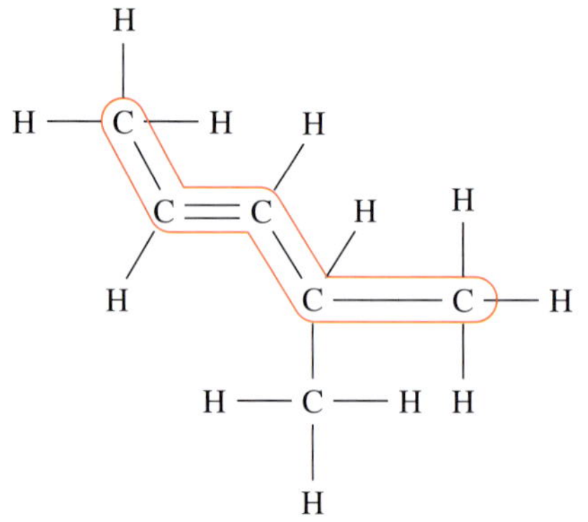

The longest chain contains five carbons—pentene.

2 Number the carbons, starting from the end closest to the double bond. Note the position of the double bond.

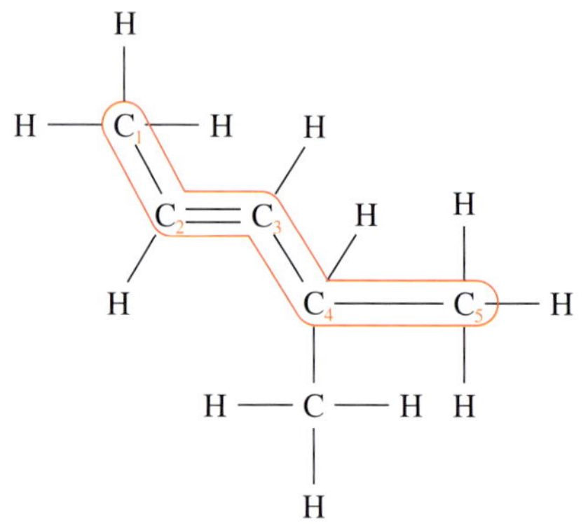

It is a five-carbon chain with the double bond starting at C number 2.

3 Name each side branch and find the number of the carbon that it is on.

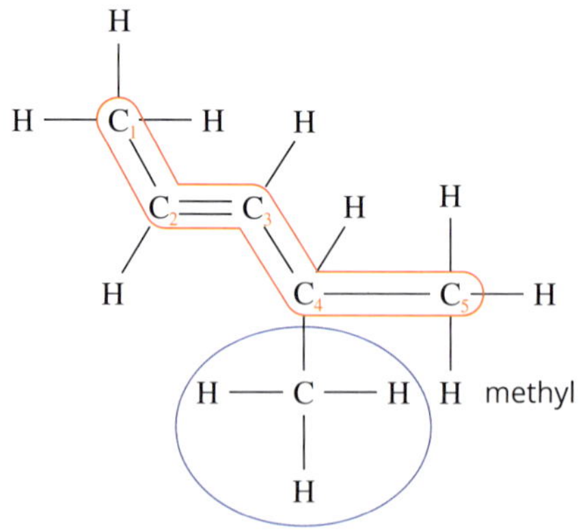

methyl

There is a methyl group on C number 4.

4 Combine all components to write the full name.

4-methylpent-2-ene

5 Identify whether the structure exhibits *cis-trans* isomerism and, if so, whether it is a *cis* or *trans* isomer.

- It is a *trans* isomer.
- *trans*-4-methylpent-2-ene

CHEMFILE

How do we see? *Cis-trans* isomerism of retinal

In 1967, Georges Wald was awarded a share of the Nobel Prize for Physiology and Medicine for his work with pigments in the retina. He discovered that vitamin A (retinal) was critical to our vision and investigated the way in which it absorbs light.

Our vision begins when molecules of retinal absorb a small packet of light called a photon. This absorption causes the *cis* form of retinal to be converted to the *trans* form (Figure 8.2.6), changing the shape of the retinal molecule.

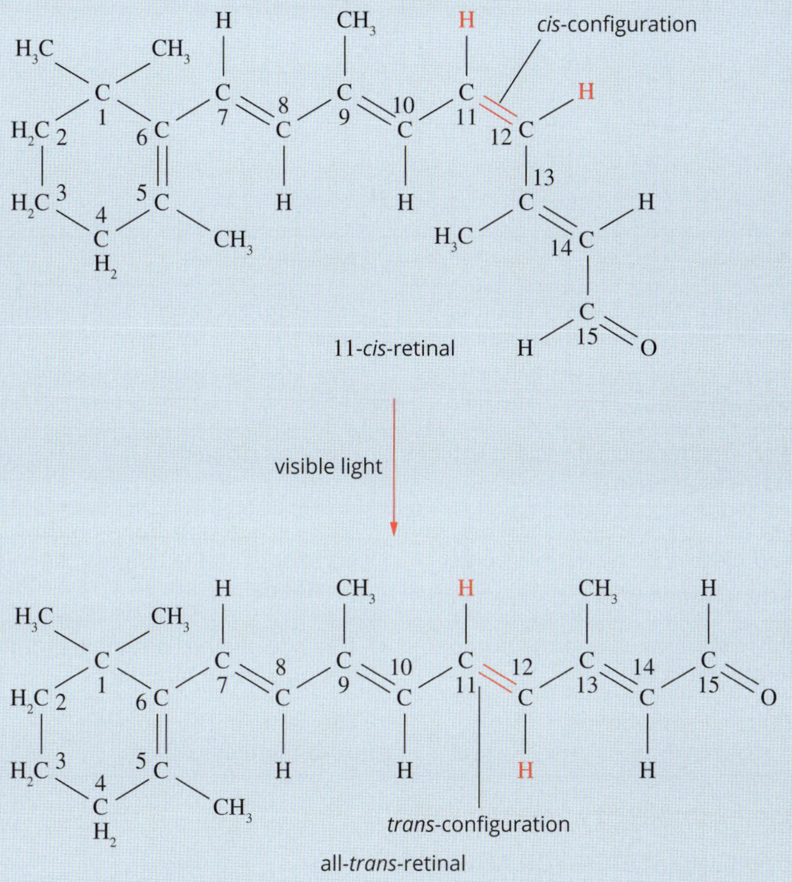

cis-configuration

11-*cis*-retinal

visible light

trans-configuration

all-*trans*-retinal

FIGURE 8.2.6 When *cis*-retinal absorbs a photon of light, it is converted to the *trans* isomer.

The *cis* form of retinal is able to bond tightly to the protein opsin in our eyes. After it has reacted with light and been converted to the *trans* form, it no longer fits closely within the protein, as its shape has changed. This causes the opsin protein to change its shape, and begins a signalling pathway that sends an impulse along the optic nerve to the brain. This is how the light that enters our eyes is converted to vision in our brains.

Worked example 8.2.1

IUPAC NAMING SYSTEM FOR ALKENES

Write the systematic name for the following molecule.

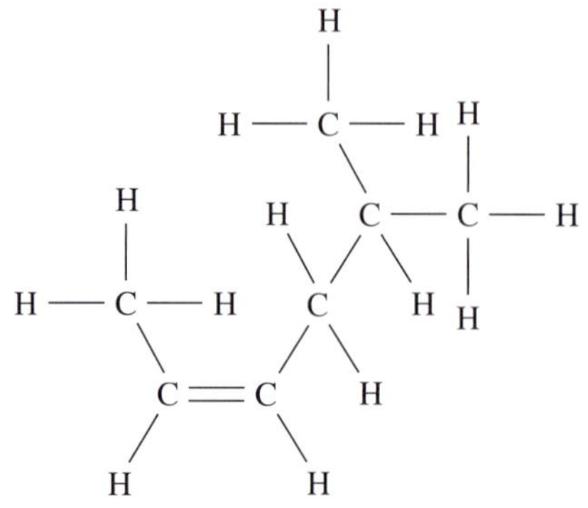

Thinking	Working
Identify the longest carbon chain in the molecule that contains the double bond. The name of the molecule is based on this longest chain.	There are six carbons in the longest chain. The name is based on hexene.
Identify whether the compound exhibits geometric isomerism.	It is a *cis* isomer.
Number the carbon atoms starting from the end closest to the double bond. Note the position of any double bond.	There is a double bond on carbon number 2, so the longest chain is hex-2-ene.
Identify each branch and the number carbon that it is on.	The side chain is a methyl group and it is on carbon number 5.
Combine all components.	The name of the molecule is *cis*-5-methylhex-2-ene.

Worked example: Try yourself 8.2.1

IUPAC NAMING SYSTEM FOR ALKENES

Write the systematic name for the following molecule.

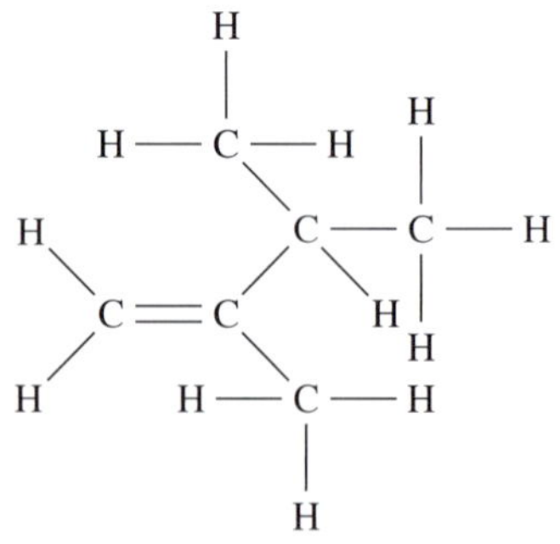

8.2 Review

SUMMARY

- Unsaturated hydrocarbons are compounds containing carbon and hydrogen only, with a carbon–carbon double bond.
- Alkenes have the general formula C_nH_{2n}.
- Names are based on the longest unbranched carbon chain. The position of the double bond determines which end to start numbering from.
- Alkenes can demonstrate structural isomerism and *cis-trans* (geometric) isomerism.

KEY QUESTIONS

1 List the following alkenes in order from lowest to highest number of atoms: pentene, ethene, methylpropene, octene, propene.

2 **a** How many carbon atoms are in methylpropene?
  b How many hydrogen atoms are in methylpropene?
  c Of which straight chain alkene is methylpropene an isomer?

3 Which of the following is the correct condensed structural formula of pent-1-ene?
  A C_5H_{12}
  B $CH_3CH_2CH_2CHCH_2$
  C $CH_3CH_2CH_2CCH$

4 Write the systematic names of these unsaturated hydrocarbons, based on their formulae.
  a 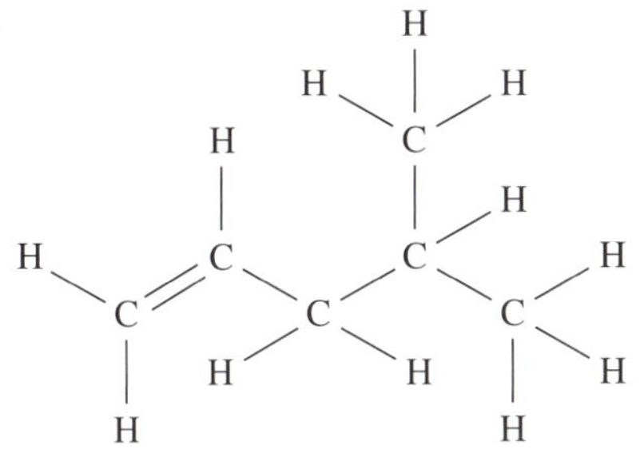
  b $CH_3CH_2CHCHCH_3$
  c

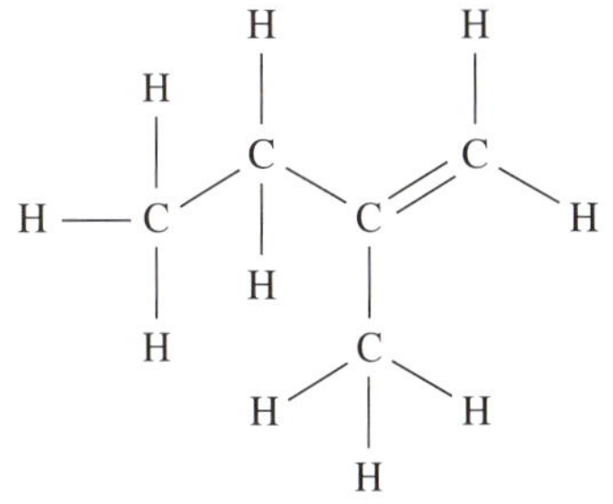

5 Draw the structural formulae of these unsaturated hydrocarbons based on their systematic names.
  a *cis*-but-2-ene
  b hex-1-ene
  c 3-methylpent-1-ene
  d 4,4-dimethylpent-1-ene

8.3 Benzene

In 1825, Michael Faraday isolated a new hydrocarbon, named benzene (C_6H_6). The structure and properties of benzene puzzled scientists for 40 years. Although its molecular formula was found to be C_6H_6, scientists were unable to draw a structure that matched the observed properties of benzene (Figure 8.3.1).

$CH_2{=}CH{-}C{\equiv}C{-}CH{=}CH_2$ $\qquad$ $CH{\equiv}C{-}CH_2{-}C{\equiv}C{-}CH_3$

FIGURE 8.3.1 These are two molecular structures with the formula C_6H_6. Neither of these structures accounts for the chemical properties of benzene.

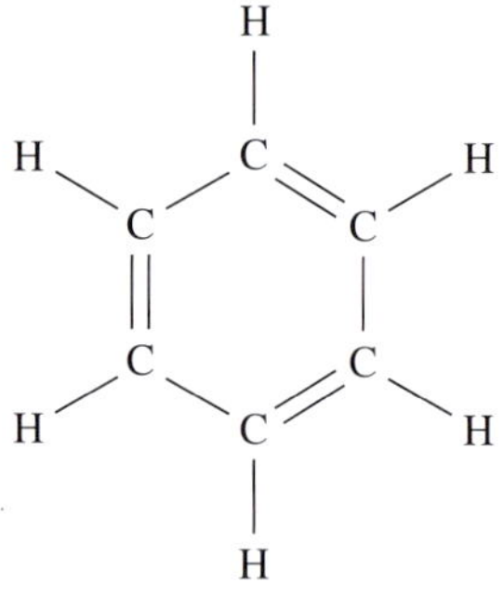

FIGURE 8.3.2 The Kekulé structure for benzene is a ring of six carbon atoms, joined by alternating single and double bonds. This structure was later found to be inconsistent with its chemical reactivity.

Benzene was found to be unexpectedly stable. August Kekulé made a major contribution to the puzzle in 1865. He proposed a ring structure in which alternate carbon atoms were joined by double bonds (Figure 8.3.2).

In his own words, Kekulé described a dream that led him to the discovery of benzene's structure:

> 'Again the atoms were gambolling before my eyes. This time the smaller groups kept modestly in the background. My mental eye, rendered more acute by repeated visions of this kind, could now distinguish larger structures of manifold conformation, long rows, sometimes more closely fitted together, all twining and twisting in snakelike motion. But look. What was that? One of the snakes had seized hold of its own tail and the form whirled mockingly before my eyes. As if by a flash of lighting I awoke.'

It is now known that benzene has a ring structure, but it does not have alternating double and single bonds. The electrons that make up the double bonds are not in fixed positions: they are delocalised and can move through the ring. In fact, the carbon-to-carbon bonds are all the same length and the molecule is more stable than expected of an unsaturated molecule (Figure 8.3.3).

a

b delocalised electrons

c

FIGURE 8.3.3 Three representations of a benzene molecule: (a) as an alternating combination (resonance) of two 'Kekulé structures', (b) visualising the delocalised electrons and (c) the convenient way the structural formula can be written.

Benzene is one of the components of petrol. It was once widely used as a solvent but, because it is carcinogenic (increases the risk of cancer), it has been phased out of this role. Polystyrene, phenol and benzoic acid are examples of the many commercial products that are manufactured from reactions involving benzene.

Molecules such as benzene, which can be represented as containing a ring of carbon atoms with alternating single and double bonds, are referred to as aromatic hydrocarbons, a name that comes from the fact that many of them have sweet or pleasant odours. The term 'arenes' is often used for aromatic hydrocarbons.

CHEMISTRY IN ACTION

Mothballs in space

Mothballs traditionally contained the active ingredient naphthalene. Naphthalene consists of a fused pair of benzene rings and has the formula $C_{10}H_8$ (Figure 8.3.4). It was first discovered in the early 1820s, when a white solid with a pungent odour was distilled from coal tar. It has been used as a household fumigant, as it is toxic to the moths that attack fabric. Indeed, although mothballs now contain the safer compound 1,4-dichlorobenzene, naphthalene was used for many years for this purpose. Its effectiveness as a repellent is such that it is produced in trace amounts by termites, which use it to repel ants and poisonous fungi. Unfortunately, naphthalene is now known to carry health risks. It is carcinogenic to humans and animals and exposure to large amounts may damage or destroy red blood cells.

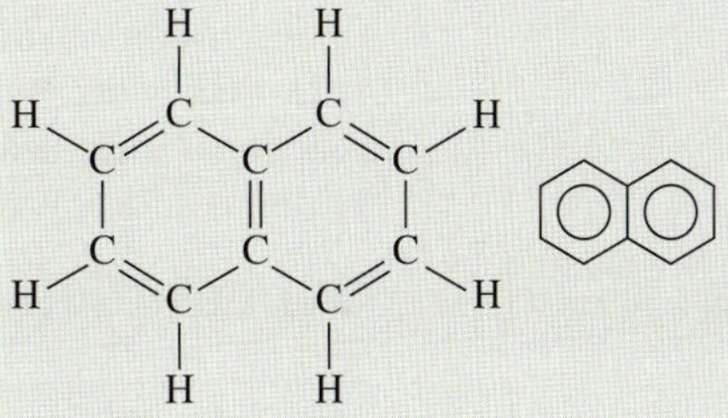

FIGURE 8.3.4 Two different representations of the compound naphthalene. Naphthalene consists of two fused benzene rings.

Interestingly, naphthalene has also been detected in space, having been observed by astronomers as part of the spectrum of Unidentified Infrared Emissions. In space, it is protonated naphthalene (containing an extra proton) that is responsible for a large part of the infrared emissions. It is estimated that more than 20% of all the carbon in the universe might be associated with hydrocarbons such as naphthalene that contain multiple benzene rings.

8.3 Review

SUMMARY

- Benzene has the formula C_6H_6.
- Benzene consists of a six-sided carbon ring, with six delocalised electrons contained within the ring. All C–C bonds are of the same length, intermediate between a single and double C–C bond.

KEY QUESTIONS

1 Which one of the following best describes the bonds between carbon atoms in benzene?
  A All are single bonds.
  B All are double bonds.
  C They are alternating single and double bonds.
  D They are intermediate between single and double bonds.

2 In 1865, August Kekulé proposed a planar structure for benzene in which the carbon atoms were bonded in a hexagonal ring, with alternating double and single bonds. Which of the following best explains why his description of the structure is incorrect?
  A The carbon atoms adopt a circular arrangement, rather than a hexagon.
  B A benzene molecule contains only double bonds between the carbon atoms.
  C One electron from each carbon is free to move around the ring.
  D There are three adjacent double bonds in a benzene molecule.

8.4 Reactions of hydrocarbons

Hydrocarbons take part in a number of different types of reactions. All hydrocarbons undergo combustion, reacting with oxygen gas in exothermic reactions. For this reason, they are very useful as fuels. Alkanes and benzene also undergo substitution reactions; unsaturated alkenes, being more reactive than saturated hydrocarbons, undergo addition reactions.

Substitution reactions

Alkanes can react with halogens, such as chlorine and bromine, in the presence of ultraviolet (UV) light. In these reactions, a carbon–hydrogen bond is broken and replaced by a carbon–halogen bond. This type of reaction is called a substitution reaction. An example of a substitution reaction is shown below in Figure 8.4.1.

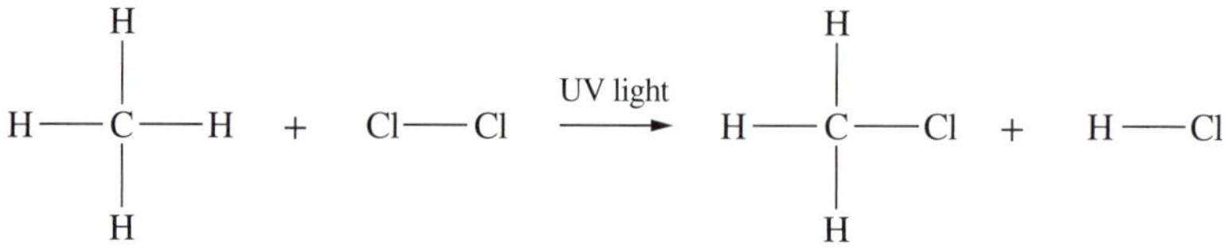

$$CH_4 + Cl{-}Cl \xrightarrow{\text{UV light}} CH_3Cl + H{-}Cl$$

FIGURE 8.4.1 Methane reacts with chlorine gas. A chlorine atom is substituted for a hydrogen atom on the methane molecule.

If an excess of the halogen is present (i.e. **excess reactant**) and the reaction is allowed to proceed to completion, further substitution reactions can take place. For example, when there is a large excess of chlorine present relative to methane, the favoured product is tetrachloromethane, with each of the four hydrogen atoms having been replaced by chlorine atoms (Figure 8.4.2).

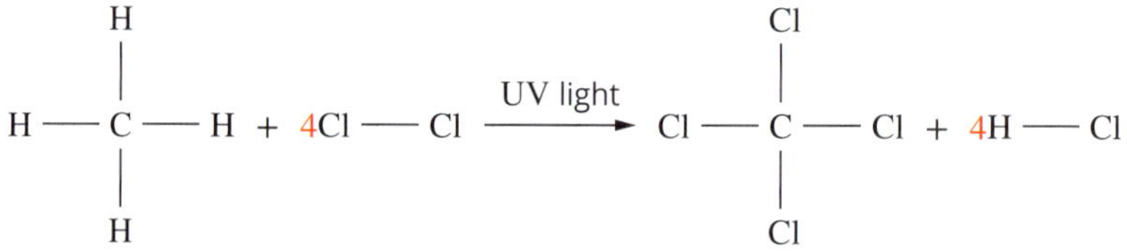

$$CH_4 + 4Cl{-}Cl \xrightarrow{\text{UV light}} CCl_4 + 4H{-}Cl$$

FIGURE 8.4.2 When methane reacts with an excess of chlorine gas, four chlorine atoms are substituted for hydrogen atoms on the methane molecule.

Benzene also undergoes substitution reactions in which one of the hydrogen atoms is replaced by another atom or a group of atoms. Alkane substitution reactions require UV light, however, substitution reactions of benzene also use catalysts, such as aluminium chloride or aluminium bromide, as indicated in Figure 8.4.3 below.

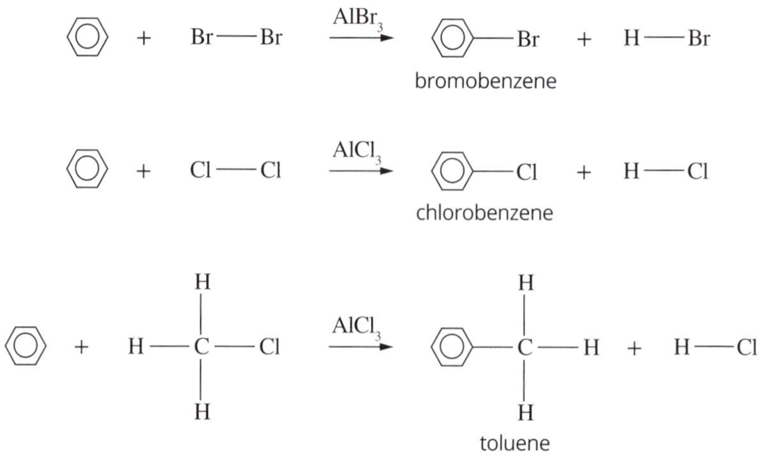

$$C_6H_6 + Br{-}Br \xrightarrow{AlBr_3} C_6H_5{-}Br + H{-}Br$$

bromobenzene

$$C_6H_6 + Cl{-}Cl \xrightarrow{AlCl_3} C_6H_5{-}Cl + H{-}Cl$$

chlorobenzene

$$C_6H_6 + CH_3Cl \xrightarrow{AlCl_3} C_6H_5{-}CH_3 + H{-}Cl$$

toluene

FIGURE 8.4.3 Benzene undergoes catalysed substitution reactions, which involve hydrogen atoms being replaced by Br or Cl atoms or a methyl ($-CH_3$) group.

Alkanes and benzene undergo substitution reactions, normally in the presence of UV light. In these reactions, a hydrogen atom is replaced by another atom or group of atoms.

Addition reactions

The double carbon–carbon covalent bond in alkenes has a significant effect on the chemical properties of this homologous series. Alkenes are generally more reactive compounds, reacting more readily and with more chemicals than alkanes, which contain only single bonds. The reactions of alkenes usually involve the addition of a small molecule to the double bond of the alkene. These reactions are called **addition reactions**.

This is what happens during addition reactions.

- Two reactant molecules combine to form one product molecule.
- The carbon–carbon double bond 'opens up' to become a single bond.
- An unsaturated compound becomes saturated.
- The atoms of the small molecule adding to the alkene are 'added across the double bond', so that one atom or group from the molecule forms a bond to the carbon atoms on each end of the double bond.

Unlike substitution reactions, there is no inorganic product formed. All the atoms in the reactants end up in the final product.

Four different types of addition reactions of alkenes will be explained in detail.

Reaction of alkenes with hydrogen

Alkenes react with hydrogen gas in the presence of a metal catalyst, such as nickel, to form alkanes. This reaction is known as a hydrogenation reaction and forms a saturated alkane. The reaction shown in Figure 8.4.4 is hydrogenation of ethene with hydrogen gas to produce ethane.

> Alkenes undergo addition reactions. In these reactions, the double carbon–carbon bond is broken and a new atom or group of atoms is added to each carbon atom.

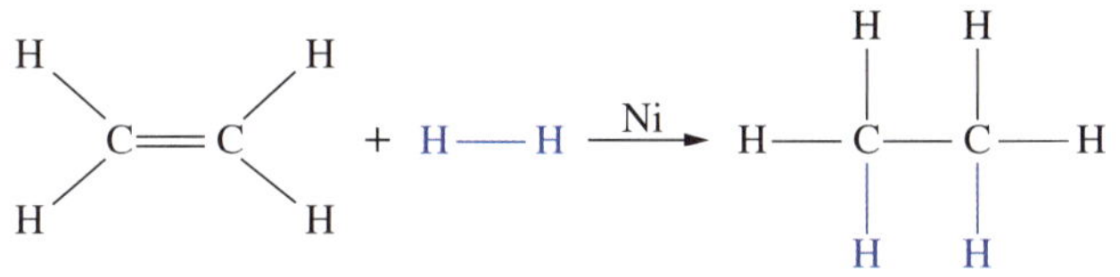

$$H_2C{=}CH_2 + H{-}H \xrightarrow{Ni} H{-}CH_2{-}CH_2{-}H$$

FIGURE 8.4.4 Addition reaction of ethene with hydrogen to form the corresponding saturated alkane, ethane

The reaction is too slow for the reaction to proceed at room temperature without a catalyst.

Reaction of alkenes with halogens

Figure 8.4.5 shows the reaction of ethene with bromine. The halogen adds across the double bond of the molecule, so in the product there is one bromine atom attached to each carbon atom.

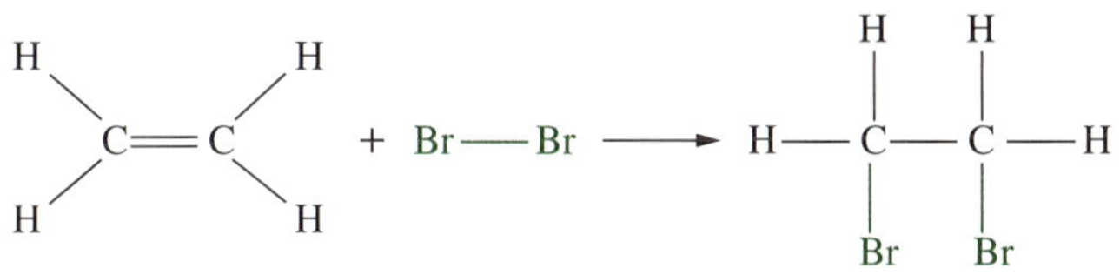

$$H_2C{=}CH_2 + Br{-}Br \longrightarrow H{-}CHBr{-}CHBr{-}H$$

FIGURE 8.4.5 Addition reaction of ethene with bromine

This reaction proceeds at room temperature without a catalyst. Other halogens such as Cl_2 and I_2 also undergo addition reactions with alkenes.

Reactions of alkenes with hydrogen halides

Alkenes also react with hydrogen halides (HCl, HBr, HF, HI) by addition reactions. Figure 8.4.6 shows the reaction of but-2-ene with hydrogen chloride to produce a single product. In this reaction, a hydrogen atom adds to one of the carbon atoms in the carbon–carbon double bond and the chlorine atom adds to the other carbon atom.

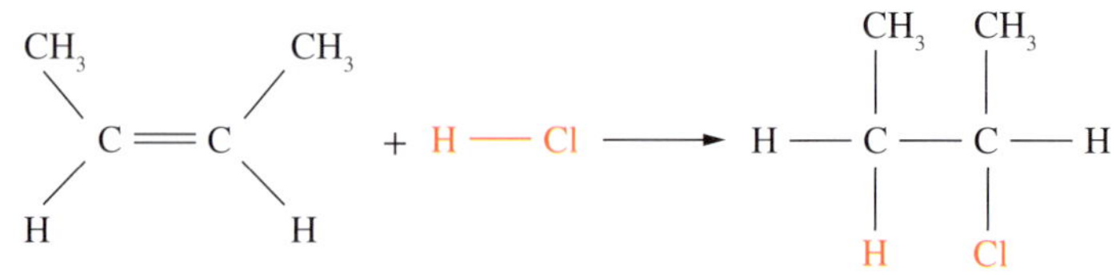

$$CH_3CH{=}CHCH_3 + H{-}Cl \rightarrow CH_3CH_2CHClCH_3$$

FIGURE 8.4.6 The addition reaction of but-2-ene with hydrogen chloride

However, as Figure 8.4.7 shows, when hydrogen chloride is reacted with but-1-ene, two products are possible.

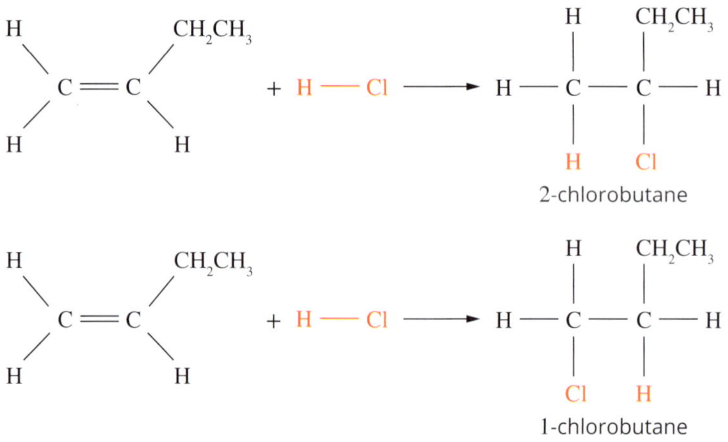

$$CH_2{=}CHCH_2CH_3 + H{-}Cl \rightarrow CH_3CHClCH_2CH_3$$

2-chlorobutane

$$CH_2{=}CHCH_2CH_3 + H{-}Cl \rightarrow CH_2ClCH_2CH_2CH_3$$

1-chlorobutane

FIGURE 8.4.7 Addition reaction of but-1-ene with hydrogen chloride. Two isomers are possible as products.

The addition reaction can produce two isomers. In one product, the hydrogen atom from the hydrogen chloride molecule has been added to the carbon atom in the carbon–carbon double bond at the end of the but-1-ene molecule (C1). In the other product, the hydrogen atom has been added to a carbon atom at the other end of the carbon–carbon double bond (C2).

When you react an asymmetrical alkene such as but-1-ene with an asymmetrical reactant, isomers are produced. More of one isomer is usually produced than the other. The reasons for this are beyond the scope of this course (see the Extension on page 183).

Reactions of alkenes with water

Alkenes react with water under specific conditions to form the corresponding alcohol. For example, ethanol can be produced by an addition reaction of ethene and water, using a catalyst to increase the rate of the reaction. Figure 8.4.8 shows the addition reaction of steam and ethene, using a phosphoric acid catalyst. This reaction is used extensively in industry for the production of ethanol.

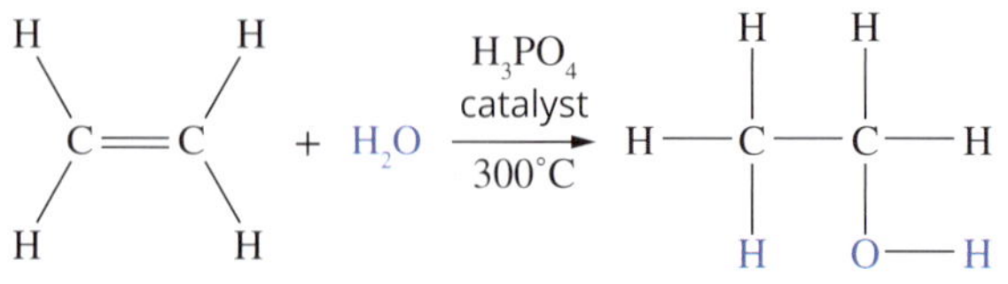

$$CH_2{=}CH_2 + H_2O \xrightarrow[300^\circ C]{H_3PO_4 \text{ catalyst}} CH_3CH_2{-}O{-}H$$

FIGURE 8.4.8 Addition reaction of ethene with water in the presence of a phosphoric acid catalyst

The reaction is carried out at 300°C. The gaseous reactants are passed over a solid bed of the catalyst and gaseous ethanol is formed.

The reaction of ethene with steam is often described as a hydration reaction. Hydration reactions are reactions that involve water as a reactant. In the addition reaction, water is 'added' across the double bond. The reaction is used for the commercial manufacture of ethanol because it is a one-step process that uses little energy, apart from initial heating. The catalyst is a solid, whereas the reactants and products are gases, which means it is easy for manufacturers to remove the product from the reaction mixture, leaving the catalyst intact.

EXTENSION

Markovnikov's rule

Markovnikov's rule describes the way that asymmetrical reactants, such as hydrogen halides (HCl, HBr and HI), add to asymmetrical alkenes. This rule states that the electron-deficient atom (the hydrogen atom due to its lower electronegativity) will add to the carbon atom with more hydrogen atoms already bonded to it, and the electron-rich atom (the halogen atom due to its higher electronegativity) will add to the carbon atom with fewer hydrogen atoms already bonded to it.

For example, when HCl is added to prop-1-ene, there are two possible products. The product that is more likely to be formed involves the hydrogen atom being added to the first carbon atom, and the chlorine atom being added to the second carbon atom, as shown in Figure 8.4.9.

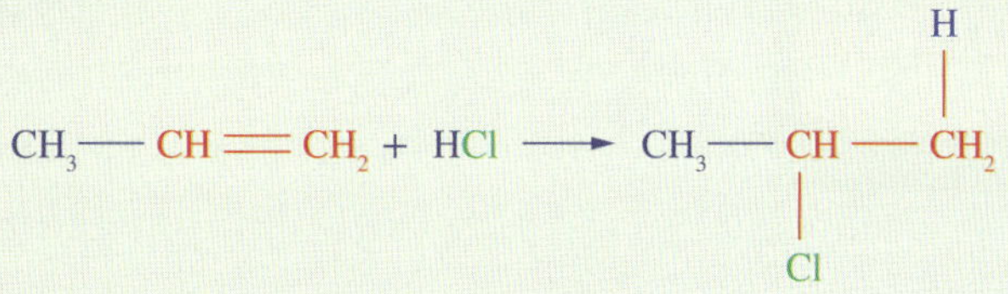

$$CH_3-CH=CH_2 + HCl \longrightarrow CH_3-\underset{\displaystyle Cl}{\underset{|}{CH}}-\overset{\displaystyle H}{\overset{|}{CH_2}}$$

FIGURE 8.4.9 When HCl is added to propene, the H and Cl atoms are added to each side of the double bond. The hydrogen atom is added to the carbon atom with the greater number of hydrogen atoms already attached, according to Markovnikov's rule.

Test for unsaturation

Bromine is used in a standard test to determine whether a substance contains molecules that are unsaturated (contain carbon–carbon double bonds). An aqueous bromine solution is orange in colour. When a few drops are added to an alkane, which is a saturated compound, there is no immediate reaction and the colour does not change. This is because substitution reactions are slow and require a catalyst. However, when a few drops are added to an unsaturated compound, such as an alkene, an addition reaction occurs and the orange colour disappears almost instantly as the bromine is consumed (Figure 8.4.10).

FIGURE 8.4.10 Test for unsaturation. Adding a few drops of orange-coloured bromine solution to hexane (right) produces no immediate reaction. In sunflower oil (left), the colour of the bromine solution disappears almost immediately because molecules in the sunflower oil contain carbon–carbon double bonds, which undergo addition reactions with bromine (Br_2).

CHEMFILE

Addition polymers

Under some conditions, alkenes undergo an addition reaction with themselves to produce long chains, known as addition polymers. Polymers are long molecules that contain hundreds or thousands of smaller repeating units called monomers. The reaction of the monomer ethene with itself to form polyethene, shown in Figure 8.4.11, is the simplest example of the addition polymerisation process. Several thousand ethene monomers usually react to make one molecule of polyethene.

Large square brackets, and the subscript *n*, are used to simplify the drawing of long polymer molecules. The value of *n* may vary within each polymer molecule, but the average molecular chain formed might contain as many as 20 000 carbon atoms. Polymers really are very large molecules!

Polypropene, polyvinyl chloride (PVC) and polystyrene are all addition polymers that are commonly found in a range of different applications (Figure 8.4.12).

ethene monomers

polyethene segment

OR

usually represented like this

FIGURE 8.4.11 Thousands of ethene monomers join together to make one chain of polyethene.

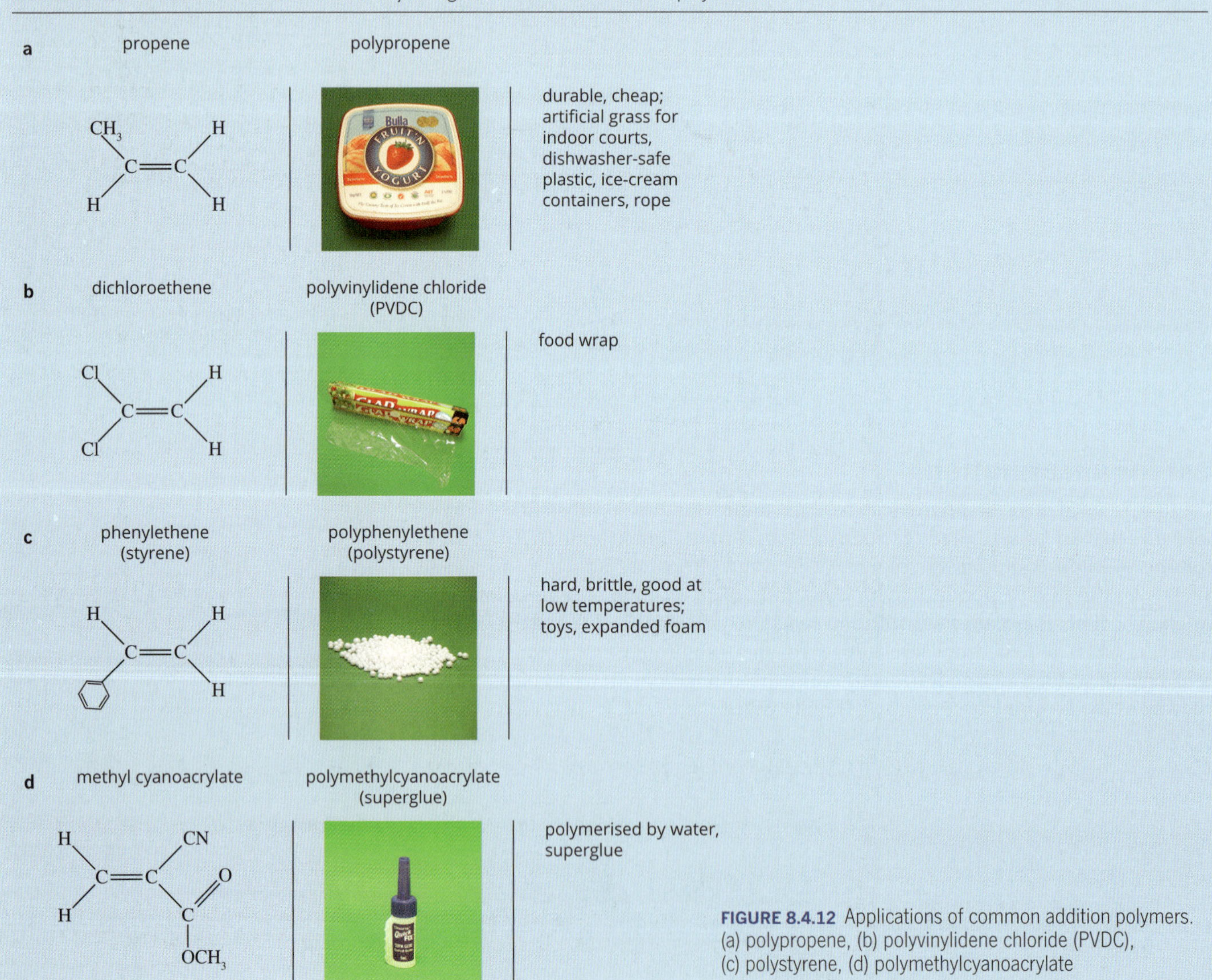

FIGURE 8.4.12 Applications of common addition polymers. (a) polypropene, (b) polyvinylidene chloride (PVDC), (c) polystyrene, (d) polymethylcyanoacrylate

CHEMISTRY IN ACTION

Making margarine

The presence of double bonds (unsaturation) in long hydrocarbon molecules decreases their melting and boiling points. Vegetable oils such as olive oil and canola oil contain high proportions of polyunsaturated and mono-unsaturated fats and as a result are liquids at room temperature.

To prepare margarine, the melting point of vegetable oils can be increased, thereby hardening them, through using an addition reaction. By reacting the oil with hydrogen gas in the presence of a nickel catalyst, the double bonds are broken and replaced with single carbon–carbon bonds. The reaction conditions are carefully controlled so that some, not all, of the carbon–carbon double bonds are broken. In this way, the consistency and texture of the margarine at room temperature can be controlled.

FIGURE 8.4.13 Vegetable oils contain a high degree of unsaturation (many double bonds) and therefore have low melting points and are liquids at room temperature. By using an addition reaction with hydrogen, the double bonds can be broken, increasing the melting point and creating a product with the texture and consistency of margarine.

COMBUSTION

Hydrocarbons are used widely as fuels. The reaction between a fuel and oxygen is a combustion reaction. The burning of petrol in a car engine and the use of natural gas for cooking are examples of combustion.

Complete combustion

If the supply of oxygen is plentiful, the products of combustion will be carbon dioxide and water. This is called **complete combustion**. Combustion reactions release significant amounts of energy, which is why they are popular as fuels. Their use as fuels is discussed further in Chapter 11.

The equations for the complete combustion of two alkanes, methane and propane, are:

methane: $CH_4(g) + 2O_2(g) \rightarrow CO_2(g) + 2H_2O(g)$

propane: $C_3H_8(g) + 5O_2(g) \rightarrow 3CO_2(g) + 4H_2O(g)$

Alkenes and benzene can also undergo combustion reactions. The equations for the combustion of ethene and benzene, for example, are:

ethene: $C_2H_4(g) + 3O_2(g) \rightarrow 2CO_2(g) + 2H_2O(g)$

benzene: $2C_6H_6(g) + 15O_2(g) \rightarrow 12CO_2(g) + 6H_2O(g)$

Incomplete combustion

When the supply of oxygen is limited, carbon monoxide, which is poisonous, can be formed instead of carbon dioxide. This can be dangerous when running engines in confined spaces. Carbon monoxide is hard to detect because it is colourless and odourless. Workers cleaning out industrial tanks, for example, can sometimes collapse, not realising that a dangerous level of carbon monoxide is present.

The following equation shows the **incomplete combustion** of propane:

$$2C_3H_8(g) + 7O_2(g) \rightarrow 6CO(g) + 8H_2O(g)$$

8.4 Review

SUMMARY

- Alkanes undergo substitution reactions with halogens in the presence of UV light.
- Benzene undergoes substitution reactions with halogens and alkyl halides in the presence of a catalyst.
- Alkenes undergo addition reactions with halogens, hydrogen, hydrogen halides and water. The double bond is broken resulting in the formation of a single C–C bond.
- Alkanes and alkenes undergo combustion reactions. In excess oxygen, these combustion reactions produce carbon dioxide and water vapour, whereas in limited oxygen, carbon monoxide is produced instead of carbon dioxide.

KEY QUESTIONS

1 Which one of the following describes the reaction that occurs between an alkane and chlorine in the presence of ultraviolet light?

A combustion reaction
B substitution reaction
C elimination reaction
D addition reaction

2 Write a balanced equation for a reaction between methane and Br_2 in the presence of UV light.

3 Write a balanced equation for the reaction between but-1-ene and Cl_2.

4 Write a balanced equation for the reaction between benzene and Br_2 in the presence of a suitable catalyst.

5 If bromine is readily decolourised by a particular compound, which one of the following could be a possible identity of that compound?

A propane
B 4-methyl-pent-2-ene
C methyl-benzene
D 3-methyl-pentane

Chapter review

KEY TERMS

addition reaction
alkane
alkene
alkyl group
complete combustion
condensed structural formula
cracking
crude oil
excess reactant
fossil fuel
cis-trans (geometric) isomer
homologous series
hydrocarbon
incomplete combustion
saturated
side group
stem name
structural formula
structural isomer
unsaturated

Alkanes

1 Why can carbon form so many compounds?

2 Which one of the following best describes a homologous series?

A compounds in which successive members differ by one carbon and two hydrogen atoms

B a series of compounds that exist in different physical forms

C compounds in which each member differs from the previous one by a $–CH_3$ group

D compounds with the same molecular formula but different arrangements of atoms

3 The formula of a hydrocarbon is $C_{16}H_{34}$.

a To which homologous series does it belong?

b What is the formula of the next hydrocarbon in the homologous series?

c What is the formula of the previous hydrocarbon in the same homologous series?

4 Draw the structural formulae and give the systematic names for each of the following.

a CH_3CH_3

b $CH_3CH(CH_3)_2$

c $CH_3CH_2CH(CH_3)CH_2CH_3$

d $CH_3(CH_2)_3CH_3$

5 Draw the structural formulae of the following alkanes.

a 3-methylpentane

b 2,2-dimethylbutane

c 2,3-dimethylpentane

6 Hexane is a liquid at room temperature but propane is a gas. Explain why.

Alkenes

7 Classify each of the following hydrocarbons as alkanes or alkenes.

a C_2H_6

b C_3H_6

c $C_{20}H_{42}$

d C_5H_{10}

e C_5H_{12}

8 Draw the structural formulae and give the systematic names of all the isomers of butene.

9 Write the systematic names of the following unsaturated hydrocarbons.

a

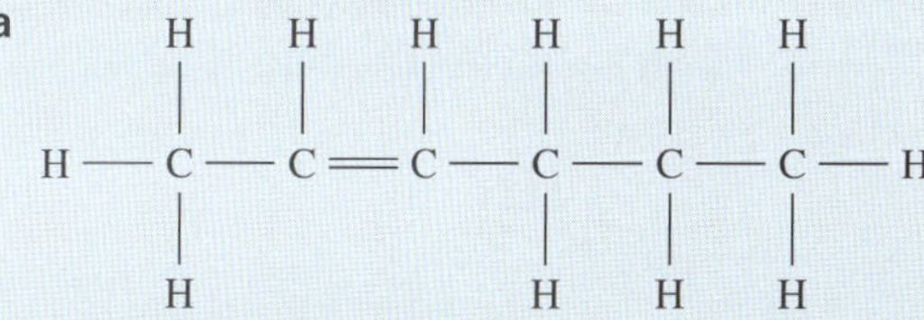

b $CH_3CHCHCH(CH_3)CH_3$

c C_3H_6

Benzene

10 Draw the Kekulé structure of benzene and describe why it is an inaccurate representation.

11 Which of the following is most likely to contain an aromatic (benzene) ring?

A $C_6H_{11}Br$

B C_6H_5Br

C C_6H_{14}

D $C_6H_6Br_6$

Reactions of hydrocarbons

12 Write a balanced chemical equation for the complete combustion of the following alkanes. Remember to include the states of the reactants and products.

- **a** gaseous methane
- **b** liquid hexane
- **c** gaseous methylpropane
- **d** solid $C_{31}H_{64}$

13 Write a balanced chemical equation for each of the following reactions.

- **a** complete combustion of ethene gas
- **b** gaseous propene bubbled into aqueous bromine solution
- **c** gaseous but-2-ene reacted with steam in the presence of a phosphoric acid catalyst
- **d** gaseous ethene reacted with gaseous hydrogen chloride

14 You are given two hydrocarbons—hexane and hex-2-ene. Describe a chemical test by which you could distinguish between them.

15 The compound below can be synthesised (made) by using either an addition or a substitution reaction. Write equations to represent both methods.

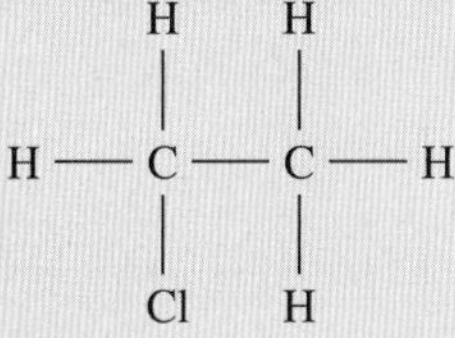

16 Butane is reacted to completion (all the hydrogen atoms are substituted with Br atoms) with an excess of bromine water in the presence of UV light. How many molecules of bromine are needed for each molecule of butane used?

17 Ethanol (CH_3CH_2OH) can be produced from ethene using an addition reaction. Write an equation to represent this reaction.

Connecting the main ideas

18 Explain the following.

- **a** The first member of the alkene homologous series is ethene, not methene.
- **b** Carbon compounds usually have four covalent bonds around each carbon atom.

19 Labels on margarine and oil often include one of the following terms: polyunsaturated, mono-unsaturated, saturated. What does each term mean?

20 A compound is found to have the formula $C_{10}H_{20}$. What would you expect to observe if a few drops of bromine water were added? (Assume that the compound is not a cyclic compound.)

CHAPTER 09

The mole

Chemists in fields as diverse as environmental monitoring, pharmaceuticals and fuel production routinely carry out chemical reactions in their work. It is important for them to be able to measure specific quantities of chemicals quickly and easily, in part because the amount of products formed depends on the amount of reactants.

At the end of this chapter, you will have a greater understanding of the way in which chemists measure quantities of chemicals, in particular, the way they can accurately determine the number of particles in samples of elements and compounds simply by weighing them. This is essential for designing and producing materials, including cosmetics, fuels, fertilisers, pharmaceuticals and building materials.

Science understanding

- the mole is a precisely defined quantity of matter equal to Avogadro's number of particles
- the mole concept relates mass, moles and molar mass and, with the Law of Conservation of Mass, can be used to calculate the masses of reactants and products in a chemical reaction
- percentage composition of a compound can be calculated from the relative atomic masses of the elements in the compound and the formula of the compound
- molecular formulae represent the number and type of atoms present in the molecules

9.1 Masses of particles

Atoms are particles so small that they cannot be counted or weighed individually or even in groups of thousands or millions. For example, the number of carbon atoms in this dot, · , printed on this page, is approximately 1×10^{20} atoms. The mass of one atom is incredibly small. For example, one atom of carbon has a mass of approximately 2×10^{-23} g.

Figure 9.1.1 shows a model of a glucose molecule. Glucose is a type of sugar used as an energy source by almost all living organisms on Earth. One glucose molecule contains 6 carbon atoms, 12 hydrogen atoms and 6 oxygen atoms bonded together. One glucose molecule has an actual mass of 3×10^{-22} g. This means that the teaspoon of glucose crystals shown in Figure 9.1.1 contains approximately 1.4×10^{22} glucose molecules.

Such small masses are not easily measured and can be inconvenient to use in calculations. This section will introduce you to the ways scientists determine and use the masses of different particles.

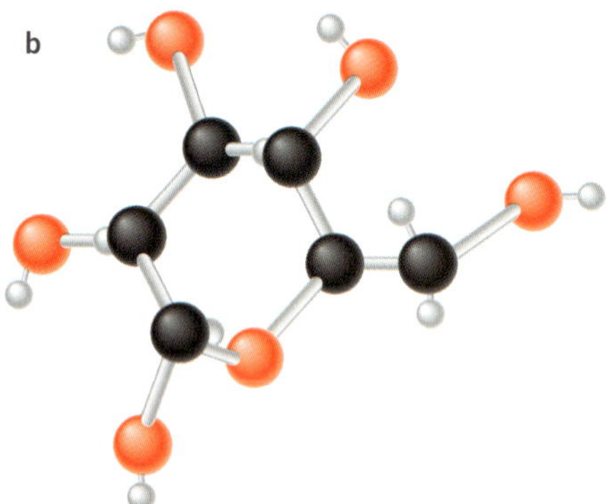

FIGURE 9.1.1 A teaspoon of glucose crystals contains an incredibly large number of extremely small glucose molecules. The single glucose molecule pictured contains 6 carbon atoms (black), 12 hydrogen atoms (white) and 6 oxygen atoms (red) bonded together.

RELATIVE MASSES

The masses used most frequently in chemistry are relative masses, rather than actual masses. The standard to which all masses are compared is the mass of an atom of the common isotope of carbon, **carbon-12** or ^{12}C, which is given a mass of exactly 12.

> Recall that isotopes are atoms of the same element that have different numbers of neutrons in their nucleus. So isotopes have the same atomic number but different mass number.

RELATIVE MOLECULAR MASS

Some elements and compounds exist as molecules, for example oxygen (O_2), nitrogen (N_2) and carbon dioxide (CO_2). For these substances, a **relative molecular mass** can be determined.

The relative molecular mass, symbol M_r, is equal to the sum of the relative atomic masses of the atoms in the molecule. Remember, you can obtain relative atomic masses (or standard atomic weights) of elements from the periodic table.

> The relative molecular mass is the mass of one molecule of that substance relative to the mass of a ^{12}C atom taken as 12 units exactly.

Worked example 9.1.1

CALCULATING THE RELATIVE MOLECULAR MASS OF MOLECULES

Calculate the relative molecular mass of carbon dioxide (CO_2).

Thinking	Working
Use the periodic table to find the relative atomic mass for the elements represented in the formula.	$A_r(C) = 12.01$ $A_r(O) = 16.00$
Determine the number of atoms of each element present, taking into consideration any brackets in the formula.	1 × C atom 2 × O atoms
Determine the relative molecular mass by adding the appropriate relative atomic masses.	$M_r = 1 \times A_r(C) + 2 \times A_r(O)$ $= 1 \times 12.01 + 2 \times 16.00$ $= 44.01$

Worked example: Try yourself 9.1.1

CALCULATING THE RELATIVE MOLECULAR MASS OF MOLECULES

Calculate the relative molecular mass of nitric acid (HNO_3).

RELATIVE FORMULA MASS

Some compounds, such as sodium chloride (NaCl) and magnesium oxide (MgO), do not exist as molecules but rather as ionic lattices. For ionic compounds, the term **relative formula mass** is used. Relative formula mass, like relative molecular mass, is calculated by taking the sum of the relative atomic masses of the elements in the formula.

The relative formula mass is the mass of a formula unit relative to the mass of an atom of ^{12}C taken as 12 units exactly. It is numerically equal to the sum of the relative atomic masses of the elements in the formula.

Worked example 9.1.2

CALCULATING THE RELATIVE FORMULA MASS OF IONIC COMPOUNDS

Calculate the relative formula mass of magnesium hydroxide ($Mg(OH)_2$).	
Thinking	**Working**
Use the periodic table to find the relative atomic mass for the elements represented in the formula.	$A_r(Mg) = 24.31$ $A_r(O) = 16.00$ $A_r(H) = 1.008$
Determine the number of atoms of each element present, taking into consideration any brackets in the formula.	1 × Mg atom 1 × 2 = 2 O atoms 1 × 2 = 2 H atoms
Determine the relative formula mass by adding the appropriate relative atomic masses.	Relative formula mass $= 1 \times A_r(Mg) + 2 \times A_r(O) + 2 \times A_r(H)$ $= 24.31 + 2 \times 16.00 + 2 \times 1.008$ $= 58.33$

Worked example: Try yourself 9.1.2

CALCULATING THE RELATIVE FORMULA MASS OF IONIC COMPOUNDS

Calculate the relative formula mass of copper(II) nitrate ($Cu(NO_3)_2$).

Relative molecular mass and relative formula mass, like relative atomic mass, have no units, because they are based on the masses of atoms of elements compared with the mass of the carbon-12 isotope.

9.1 Review

SUMMARY

- The carbon-12 isotope is assigned a mass of exactly 12 units, and the relative atomic masses of atoms of elements are compared to this.
- The relative molecular mass of molecules, or relative formula mass of ionic compounds, is calculated from the sum of the relative atomic masses of their constituent elements.

KEY QUESTIONS

1 Calculate the relative molecular mass of:

a sulfuric acid (H_2SO_4)

b ammonia (NH_3)

c ethane (C_2H_6).

2 Calculate the relative formula mass of:

a potassium chloride (KCl)

b sodium carbonate (Na_2CO_3)

c aluminium sulfate ($Al_2(SO_4)_3$).

9.2 Introducing the mole

FIGURE 9.2.1 Each of these ice cubes contains more than 10^{23} water (H_2O) molecules. Molecules and atoms are so small, and the numbers of them in everyday samples are so large, that it would be very inconvenient to count them individually.

It is often essential for chemists to be able to measure an exact number of particles of an element or compound. However, the particles in elements and compounds are so small that it would be difficult to count atoms, ions or molecules individually. If it were possible to count individual particles, the numbers in even very small samples would be huge and very inconvenient to work with.

The ice cubes shown in Figure 9.2.1 each contain more than 10^{23} water molecules (H_2O). As each water molecule is composed of two hydrogen atoms and one oxygen atom, the number of individual atoms in each ice cube is greater than 10^{23}. A quantity that allows chemists to measure accurate amounts of extremely small particles is required.

In this section, you will learn about the quantity used by chemists: the mole.

THE CHEMIST'S COUNTING UNIT

FIGURE 9.2.2 Twelve eggs is one dozen, 24 eggs is two dozen and six eggs is half a dozen.

A dozen is a convenient quantity for buying the eggs shown in Figure 9.2.2. For atoms, ions and molecules that are much smaller than eggs, a quantity that describes a much larger number is needed.

The quantity used by scientists is the **mole**. The mole is often referred to as the '**amount** of substance' and is given the symbol n, and the unit mol.

So, n(glucose) = 2 mol is read as 'the amount of glucose molecules is 2 moles'.

Chemists use the mole as a counting measure. Figure 9.2.3 shows some quantities that you would be very familiar with such as pair, dozen and ream. One dozen is equal to 12 and two dozen equals 24, 20 dozen equals 240 and half a dozen equals six. In the same way, chemists know that 1 mole is equivalent to a certain number and that 2 moles, 20 moles and half a mole are all multiples of that number.

FIGURE 9.2.3 Convenient quantities that you would be very familiar with. A pair of shoes equals two, a dozen roses equals 12, a ream of paper equals 500 sheets.

Information provided by molecular formulae

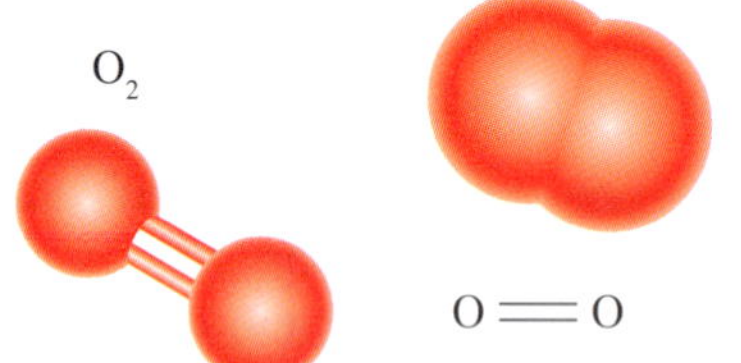

FIGURE 9.2.4 Four different ways chemists represent the oxygen molecule: a formula (O_2), a structural formula (O=O) and two coloured molecular models. Each formula or model shows that one oxygen molecule contains two oxygen atoms.

When referring to a mole of a substance, it is important to indicate which particle is being specified. Figure 9.2.4 shows four different ways to represent the oxygen molecule. The oxygen molecule is most commonly described by the **formula** O_2.

The expression, '1 mole of oxygen' is ambiguous because it could describe 1 mole of oxygen atoms (O) or 1 mole of oxygen molecules (O_2). As there are two atoms in each oxygen molecule, 1 mole of oxygen molecules will contain 2 moles of oxygen atoms. As you learnt in Chapter 6, the molecular formula of a substance indicates the number of atoms of each element in one molecule of a substance. For example, in 1 mole of oxygen gas, there are 2 moles of oxygen atoms.

So if you know the number of moles of a certain substance, the formula of the substance allows you to also know how many moles of the atoms or ions that make up that substance. Some examples of this is are given in Table 9.2.1.

TABLE 9.2.1 Examples of the use of the mole as a measure of the number of particles

Number of moles of substance	Information that can be obtained from the moles of the substance
1 mole of hydrogen gas (H_2)	1 mole of hydrogen molecules (H_2) 2 moles of hydrogen atoms (H)
2 moles of water (H_2O)	2 moles of water molecules (H_2O) 4 moles of hydrogen atoms (H) 2 moles of oxygen atoms (O)
2 moles of calcium chloride ($CaCl_2$)	2 moles of Ca^{2+} ions 4 moles of Cl^- ions
10 moles of glucose ($C_6H_{12}O_6$)	10 moles of glucose ($C_6H_{12}O_6$) molecules 60 moles of carbon atoms (C) 120 moles of hydrogen atoms (H) 60 moles of oxygen atoms (O)

Worked example 9.2.1

CALCULATING THE NUMBER OF MOLES OF ATOMS GIVEN THE NUMBER OF MOLES OF MOLECULES

Calculate the amount, in mol, of hydrogen atoms in 3.6 mol of sulfuric acid (H_2SO_4).	
Thinking	**Working**
List the data given in the question next to the appropriate symbol. Include units.	The number of mol of hydrogen atoms is the unknown so: $n(H) = ?$ $n(H_2SO_4) = 3.6$ mol
Calculate the amount, in mol, of hydrogen atoms from the amount of sulfuric acid molecules and the molecular formula.	$n(H) = n(H_2SO_4) \times 2$ $= 3.6 \times 2$ $= 7.2$ mol

Worked example: Try yourself 9.2.1

CALCULATING THE NUMBER OF MOLES OF ATOMS GIVEN THE NUMBER OF MOLES OF MOLECULES

Calculate the amount, in mol, of hydrogen atoms in 0.75 mol of water (H_2O).

AVOGADRO'S NUMBER

One mole of any substance is defined as the same number of particles as there are atoms in exactly 12 grams of carbon-12. This number of atoms has been experimentally determined to be 6.022141×10^{23} atoms. That's 602 214 100 000 000 000 000 000 atoms!

This number is commonly rounded to 6.022×10^{23} and is referred to as **Avogadro's number** or Avogadro constant. It is given the symbol N_A.

 $N_A = 6.022 \times 10^{23}$ mol^{-1}

Therefore, 1 mole of particles contains 6.022×10^{23} particles.

CHEMFILE

Avogadro's number

It is very difficult to imagine just how big Avogadro's number really is, especially when atoms, ions and molecules are so small. Here are some examples to help:

- 6.022×10^{23} grains of sand, placed side by side, would stretch from Earth to the Sun and back about seven million times.
- A computer counting 10 billion times every second would take two million years to reach 6.022×10^{23}.
- 6.022×10^{23} of the marshmallows shown in Figure 9.2.5 would cover Australia to a depth of 900 km!

FIGURE 9.2.5 One mole of any substance contains 6.022×10^{23} particles.

Avogadro's number is an enormous number, but the extremely small size of atoms, ions and molecules means that 1 mole of most elements and compounds does not have a large mass or take up a large volume.

For example, in Figure 9.2.6 you can see that 1 mole of water, that is 6.022×10^{23} water molecules, has a volume of only 18 mL, whereas, 1 mole of table salt (NaCl), has a mass of 58.5 grams.

FIGURE 9.2.6 One mole of sodium chloride and 1 mole of water contain the same number of particles, 6.022×10^{23}.

Given that 1 mol of a substance contains 6.022×10^{23} particles, then:

- 2 mol of a substance contains $2 \times (6.022 \times 10^{23}) = 1.204 \times 10^{24}$ particles
- 0.3 mol of a substance contains $0.3 \times (6.022 \times 10^{23}) = 1.81 \times 10^{23}$ particles
- 4.70×10^{23} particles $= \dfrac{4.70 \times 10^{23}}{6.022 \times 10^{23}} = 0.781$ mol
- 7.35×10^{24} particles $= \dfrac{7.35 \times 10^{24}}{6.022 \times 10^{23}} = 12.2$ mol.

As you can see from Figure 9.2.7, a mathematical relationship exists between the number of particles, N, and the amount of substance in moles, n. This relationship can be written as:

$$n = \frac{N}{N_A}$$

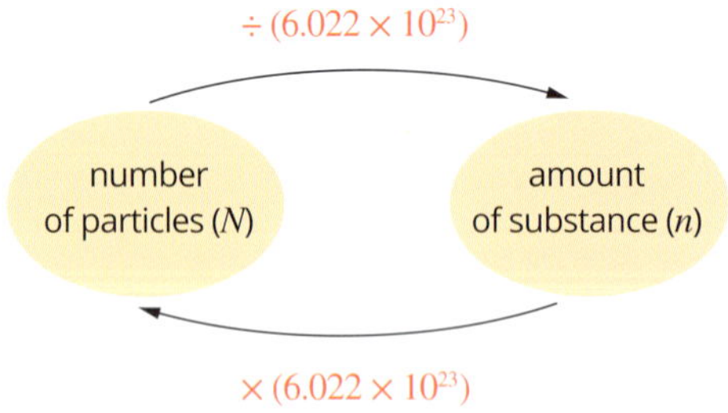

$\div (6.022 \times 10^{23})$

number of particles (N) → amount of substance (n)

amount of substance (n) → number of particles (N)

$\times (6.022 \times 10^{23})$

FIGURE 9.2.7 Relationship between number of particles and amount of substance in moles

Calculations using the mole and Avogadro's number

Three quantities have been introduced so far:

- the mole, which is given the symbol n and the unit mol
- Avogadro's number, which is given the symbol N_A and has the value 6.022×10^{23}
- the actual number of particles (atoms, ions or molecules), which is given the symbol N.

The mathematical relationship that links the three quantities is $n = \frac{N}{N_A}$.

The following worked examples show you how to use this relationship in calculations.

Worked example 9.2.2

CALCULATING THE NUMBER OF MOLECULES

Calculate the number of molecules in 3.5 moles of water (H_2O).	
Thinking	**Working**
List the data given in the question next to the appropriate symbol. Include units.	The number of water molecules is the unknown, so: $N(H_2O) = ?$ $n(H_2O) = 3.5\,mol$ $N_A = 6.022 \times 10^{23}$
Rearrange the formula to make the unknown the subject.	$n = \frac{N}{N_A}$ so $N(H_2O) = n \times N_A$
Substitute in data and solve for the answer.	$N(H_2O) = n \times N_A$ $= 3.5 \times 6.022 \times 10^{23}$ $= 2.1 \times 10^{24}$ molecules

Worked example: Try yourself 9.2.2

CALCULATING THE NUMBER OF MOLECULES

Calculate the number of molecules in 1.6 moles of carbon dioxide (CO_2).

Worked example 9.2.3

CALCULATING THE NUMBER OF ATOMS

Calculate the number of oxygen atoms in 2.5 mol of oxygen gas (O_2).	
Thinking	**Working**
List the data given in the question next to the appropriate symbol. Include units.	The number of oxygen atoms is the unknown so: $N(O) = ?$ $n(O_2) = 2.5\,mol$ $N_A = 6.022 \times 10^{23}$
Calculate the amount, in mol, of oxygen atoms from the number of oxygen molecules and the molecular formula.	$n(O) = n(O_2) \times 2$ $= 2.5 \times 2$ $= 5.0\,mol$
Rearrange the formula to make the unknown the subject.	$n = \frac{N}{N_A}$ so $N(O) = n \times N_A$
Substitute in data and solve for the answer.	$N(O) = n \times N_A$ $= 5.0 \times 6.022 \times 10^{23}$ $= 3.0 \times 10^{24}$ atoms

Worked example: Try yourself 9.2.3

CALCULATING THE NUMBER OF ATOMS

Calculate the number of hydrogen atoms in 0.35 mol of methane (CH_4).

Worked example 9.2.4

CALCULATING THE NUMBER OF MOLES OF PARTICLES GIVEN THE NUMBER OF PARTICLES

Calculate how many moles of ammonia are in 2.5×10^{22} of ammonia molecules.	
Thinking	**Working**
List the data given in the question next to the appropriate symbol. Include units.	The number of mol of ammonia molecules is the unknown so: $n(NH_3) = ?$ $N(NH_3) = 2.5 \times 10^{22}$ molecules $N_A = 6.022 \times 10^{23}$
Rearrange the formula to make the unknown the subject.	$n = \frac{N}{N_A}$ n is the unknown so rearrangement is not required.
Substitute in data and solve for the answer.	$n(NH_3) = \frac{N}{N_A} = \frac{2.5 \times 10^{22}}{6.022 \times 10^{23}}$ $= 0.042$ mol

Worked example: Try yourself 9.2.4

CALCULATING THE NUMBER OF MOLES OF PARTICLES GIVEN THE NUMBER OF PARTICLES

Calculate how many moles of magnesium are in 8.1×10^{20} of magnesium atoms.

9.2 Review

SUMMARY

- A mole is a convenient quantity for counting particles. The mole is given the symbol *n* and the unit mol.
- One mole is defined as the amount of substance that contains the same number of 'specified' particles as there are atoms in 12 g of carbon-12.
- The number of particles in 1 mol is given the symbol N_A. This is known as Avogadro's number and has the numerical value of 6.022×10^{23}.
- The formula $n = \frac{N}{N_A}$ can be used or rearranged to calculate the amount or number of specified particles in a sample.

KEY QUESTIONS

1 Calculate the number of:
   a atoms in 2.0 mol of sodium atoms (Na)
   b molecules in 0.10 mol of nitrogen molecules (N_2)
   c atoms in 20.0 mol of carbon atoms (C)
   d molecules in 4.2 mol of water molecules (H_2O)
   e atoms in 1.0×10^{-2} mol of iron atoms (Fe)
   f molecules in 4.62×10^{-5} mol of CO_2 molecules.

2 Calculate the number of moles in:
   a 3.0×10^{23} molecules of water (H_2O)
   b 1.5×10^{23} atoms of neon (Ne)
   c 4.2×10^{25} atoms of iron (Fe)
   d 4.2×10^{25} molecules of ethanol (C_2H_5OH).

3 Calculate the amount, in mol, of:
   a sodium atoms in 1.0×10^{20} of sodium atoms
   b aluminium atoms in 1.0×10^{20} of aluminium atoms
   c chlorine molecules in 1.0×10^{20} of chlorine molecules.

4 Calculate the amount, in mol, of:
   a chlorine atoms in 0.4 mol of chlorine (Cl_2)
   b hydrogen atoms in 1.2 mol of methane (CH_4)
   c hydrogen atoms in 0.12 mol of ethane (C_2H_6)
   d oxygen atoms in 1.5 mol of sodium sulfate (Na_2SO_4).

9.3 Molar mass

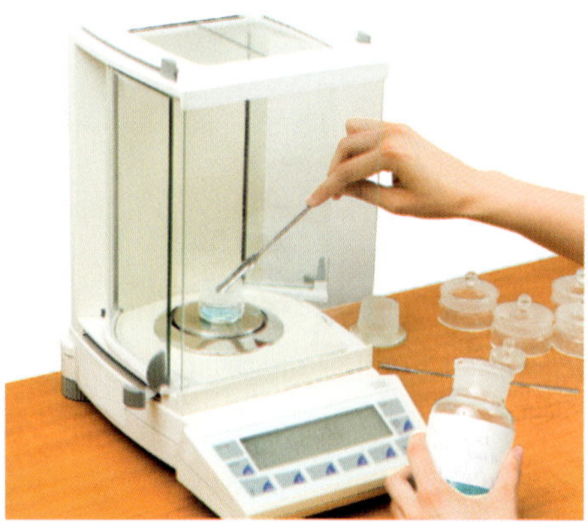

FIGURE 9.3.1 A digital balance is a piece of laboratory equipment used for weighing.

Chemical laboratories often contain a balance like the one in Figure 9.3.1, which is used for weighing. If a chemist knows that a specific mass of a substance always contains a specific number of particles, it is possible to easily weigh a sample of the substance and calculate the exact number of particles present in the sample.

In this section you will learn about how the amount of a substance, measured in moles, is related to the mass of the substance.

MOLAR MASS

Chemists have defined the mole so that the number of moles of a substance can be determined by simply measuring its mass.

The particles of different elements and compounds have different masses. Therefore, the masses of 1 mole of different elements or compounds will also be different. This is like saying that the mass of one dozen oranges will be greater than the mass of one dozen mandarins because one orange is heavier than one mandarin. The mass, in grams, of 1 mole of a particular element or compound is known as its **molar mass**. It is given the symbol, M, and the unit, $g\,mol^{-1}$.

Remember that a mole is defined as the amount of substance that contains the same number of specified particles as there are atoms in 12.00 g of carbon-12.

- 1 atom of ^{12}C has a relative atomic mass of 12 exactly.
- 1 mole of atoms of ^{12}C has a mass of 12 *grams* exactly.

Naturally occurring carbon is mainly composed of the ^{12}C isotope, but because there is a small amount of the ^{13}C isotope also present, the molar mass of carbon is $12.01\,g\,mol^{-1}$.

Consider an atom of ^{12}C and an atom of ^{24}Mg. Carbon-12 has been assigned a relative isotopic mass of 12 exactly. On that scale, the relative isotopic mass of ^{24}Mg is 24. Since 1 mole of ^{12}C atoms weighs exactly 12 g, then 1 mole of ^{24}Mg must weigh twice as much, 24 g.

Table 9.3.1 shows you how to calculate the molar masses of some common substances.

- The molar mass of an element is the mass of 1 mole of the element. It is equal to the relative atomic mass of the element expressed in grams.
- The molar mass of a compound is the mass of 1 mole of the compound. It is equal to the relative molecular mass or relative formula mass of the compound expressed in grams.
- The molar mass is given the symbol, M, and the unit, $g\,mol^{-1}$.

TABLE 9.3.1 Calculating the molar mass of a substance by adding the relative atomic masses for each atom present in the substance based on the molecular or ionic formula

Substance	Relative atomic masses	Molar mass of substance
Na	Na: 22.99	$= 22.99\,g\,mol^{-1}$
O_2	O: 16.00	$= 2 \times 16.00$ $= 32.00\,g\,mol^{-1}$
H_2O	H: 1.008 O: 16.00	$= (2 \times 1.008) + 16.00$ $= 18.016\,g\,mol^{-1}$
CO_2	C: 12.01 O: 16.00	$= 12.01 + (2 \times 16.00)$ $= 44.01\,g\,mol^{-1}$
$NaNO_3$	Na: 22.99 N: 14.01 O: 16.00	$= 22.99 + 14.01 + (3 \times 16.00)$ $= 85.00\,g\,mol^{-1}$

From the calculations in Table 9.3.1 and the photograph of 1 mole of some common substances in Figure 9.3.2, you can see that 1 mole of each substance has a different mass.

FIGURE 9.3.2 One mole of each substance has a different mass.

Counting by weighing

A useful relationship links the amount of a substance (n), in mol, its molar mass (M), in g mol^{-1}, and the given mass of the substance (m), in grams.

Mass of a given amount of substance (g) = amount of substance (mol) × molar mass (g mol^{-1}).

This can be written as $m = n \times M$ and rearranged to the following relationship.

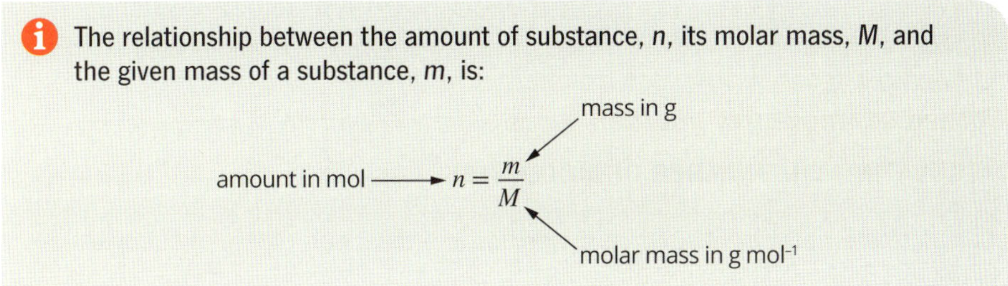

$$n = \frac{m}{M}$$

Worked example 9.3.1

CALCULATING THE MASS OF A SUBSTANCE

Calculate the mass of 0.35 mol of magnesium nitrate ($Mg(NO_3)_2$).	
Thinking	**Working**
List the data given to you in the question. Remember that whenever you are given a chemical formula, you can calculate the molar mass.	$m(Mg(NO_3)_2) = ?$ g $n(Mg(NO_3)_2) = 0.35$ mol $M(Mg(NO_3)_2) = 24.31 + (2 \times 14.01) + (6 \times 16.00)$ $= 148.33$ g mol^{-1}
Calculate the mass of magnesium nitrate using: $n = \frac{m}{M}$	$n = \frac{m}{M}$ so $m = n \times M$ $m(Mg(NO_3)_2) = 0.35 \times 148.33$ $= 52$ g

Worked example: Try yourself 9.3.1

CALCULATING THE MASS OF A SUBSTANCE

Calculate the mass of 4.68 mol of sodium carbonate (Na_2CO_3).

You are now in a position to count atoms by weighing. When you use the mole, you are effectively counting the number of particles in a substance. The total number of particles present in a substance is equal to the number of moles of the substance multiplied by 6.022×10^{23}.

Some calculations require you to use both of the formulae $n = \frac{m}{M}$ and $n = \frac{N}{N_A}$. Worked example 9.3.2 is such a calculation.

Worked example 9.3.2

CALCULATING THE NUMBER OF MOLECULES

Calculate the number of CO_2 molecules present in 22.0 g of carbon dioxide.	
Thinking	**Working**
List the data given to you in the question. Convert mass to grams if required. Remember that whenever you are given a formula you can calculate the molar mass.	$N(CO_2) = ?$ $M(CO_2) = 12.0 + (2 \times 16.0) = 44.01\ g\,mol^{-1}$ $m(CO_2) = 22.0\ g$
Calculate the amount, in mol, of CO_2 using: $n = \frac{m}{M}$	$n(CO_2) = \frac{m}{M}$ $= \frac{22.0}{44.01}$ $= 0.4999\ mol$
Calculate the number of CO_2 molecules using: $n = \frac{N}{N_A}$	$n = \frac{N}{N_A}$ so $N = n \times N_A$ $N(CO_2) = 0.4999 \times 6.022 \times 10^{23}$ $= 3.01 \times 10^{23}$ molecules

Worked example: Try yourself 9.3.2

CALCULATING THE NUMBER OF MOLECULES

Calculate the number of sucrose molecules in a teaspoon (4.2 g) of sucrose ($C_{12}H_{22}O_{11}$).

CHEMISTRY IN ACTION

The sting of a bee

The formula for pentyl ethanoate, $C_7H_{14}O_2$, is represented by the structure shown in Figure 9.3.3.

Pentyl ethanoate ($C_7H_{14}O_2$) is the compound that gives bananas their characteristic odour. It is also a pheromone released by bees when they sting. A pheromone is a chemical produced by an animal or insect that changes the behaviour of another animal or insect of the same species. Bees release pentyl ethanoate as an alarm pheromone to alert other bees to danger. The compound is released near the sting shaft and attracts other bees to the area, where the group behaves defensively, charging and stinging. Smoke masks the alarm pheromone, interrupting the defensive response and calming the bees, allowing beekeepers an opportunity to work with the beehive.

Each time a bee stings, one-thousandth of a milligram (1.0×10^{-6} g) of pentyl ethanoate is released. Knowing the mass of pentyl ethanoate released in a bee sting enables chemists to calculate the number of pentyl ethanoate molecules in each sting using the relationship between mass and mole.

As the molar mass of pentyl ethanoate ($C_7H_{14}O_2$) is equal to 130.2 g mol^{-1}, using the formula $n = \frac{m}{M}$, 1.0×10^{-6} g of pentyl ethanoate is equal to 7.7×10^{-9} mol.

From Section 9.2, you know that the number of particles can be calculated using the formula $N = n \times N_A$. The number of pentyl ethanoate molecules in each bee sting is therefore equal to 4.6×10^{15}.

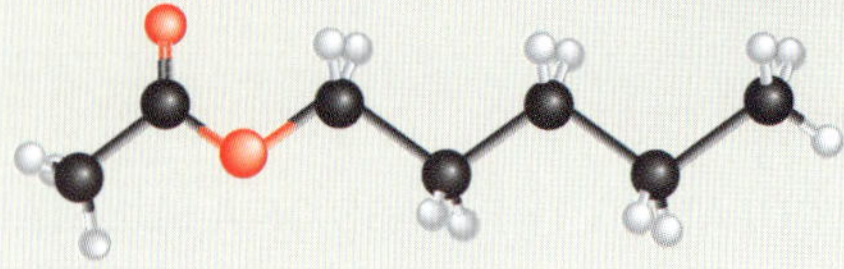

FIGURE 9.3.3 Pentyl ethanoate is a compound released when bees sting.

9.3 Review

SUMMARY

- The molar mass of an element or compound is the mass, in grams, of 1 mole of that element or compound. Molar mass is given the symbol *M* and the unit g mol^{-1}.
- The molar mass of an element or compound has the same numerical value as the relative mass of the element or compound.
- The formula $n = \frac{m}{M}$ can be used or rearranged to calculate the mass, amount or molar mass of an element or compound.

KEY QUESTIONS

1 Calculate the molar mass of the following, to 1 decimal place.
 a nitrogen (N_2)
 b ammonia (NH_3)
 c sulfuric acid (H_2SO_4)
 d iron(III) nitrate ($Fe(NO_3)_3$)
 e ethanoic acid (CH_3COOH)
 f sulfur atoms (S)
 g vitamin C (ascorbic acid, $C_6H_8O_6$)
 h hydrated copper(II) sulfate ($CuSO_4{\cdot}5H_2O$)

2 Calculate the mass of the following, to 3 significant figures.
 a 1.00 mol of sodium atoms (Na)
 b 2.00 mol of oxygen molecules (O_2)
 c 0.100 mol of methane molecules (CH_4)
 d 0.250 mol of aluminium oxide (Al_2O_3)

3 Calculate the amount of the following, in mol, to the appropriate number of significant figures.
 a H atoms in 5.0 g of hydrogen (H)
 b H molecules in 5.0 g of hydrogen (H_2)
 c Al atoms in 2.7 g of aluminium (Al)
 d CH_4 molecules in 0.4 g of methane (CH_4)
 e O_2 molecules in 0.10 g of oxygen (O_2)
 f O atoms in 0.10 g of oxygen (O_2)
 g P_4 molecules in 1.2×10^{-3} g of phosphorus (P_4)
 h P atoms in 1.2×10^{-3} g of phosphorus (P_4)

4 Calculate the number of atoms in the following, to 2 significant figures.
 a 23 g of sodium (Na)
 b 4.0 g of argon (Ar)
 c 0.243 g of magnesium (Mg)
 d 10.0 g of gold (Au)

5 **a** Calculate the number of molecules in:
  i 16 g of oxygen (O_2)
  ii 2.8 g of nitrogen (N_2).
 b Calculate the number of oxygen atoms in 3.2 g of sulfur dioxide (SO_2).
 c Calculate the total number of atoms in 288 g of ammonia (NH_3).

9.4 Percentage composition

FIGURE 9.4.1 Chemical formulae are part of the language of chemistry.

Compounds are substances that contain two or more different elements. The relative proportions of each element in a compound can be expressed as a formula, showing either the ratio of atoms of each element in a compound or the actual numbers of atoms in a molecule.

The formulae of compounds are an important part of the language of chemistry. You would be familiar with some of the formulae shown in Figure 9.4.1. Formulae are used to represent chemicals in equations and on numerous other occasions.

In this section you will learn how to use the formula of a compound to calculate percentage composition, that is, the percentage of each element in terms of mass.

PERCENTAGE COMPOSITION

The **percentage composition** of a given compound tells you the proportion by mass of the different elements in that compound. The proportion of each element is expressed as a percentage of the total mass of the compound (Figure 9.4.2).

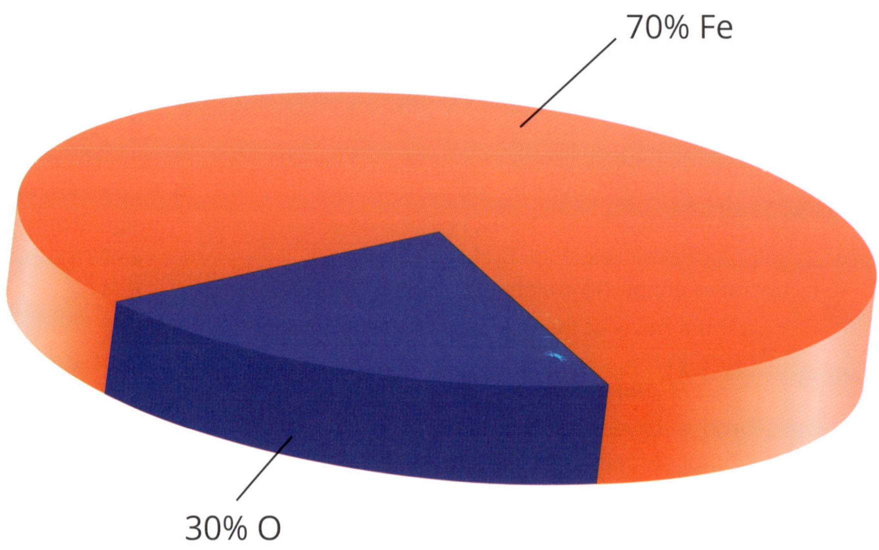

FIGURE 9.4.2 The pie chart shows the percentage composition, by mass, of iron(III) oxide (Fe_2O_3).

Being able to determine percentages by mass is important in chemistry. For example, the iron ore mined in Western Australia is iron(III) oxide, and has the formula Fe_2O_3. A company that is producing iron from this iron ore will want to know the mass of iron that can be extracted from a given quantity of iron ore. The pie chart in Figure 9.4.2 shows that the percentage of iron in iron(III) oxide is 70%. This means that the company can extract a maximum of 70 g iron per 100 g of iron ore.

If the chemical formula of a compound is known, the percentage composition can be determined from the molar masses of the elements and compound using the following formula.

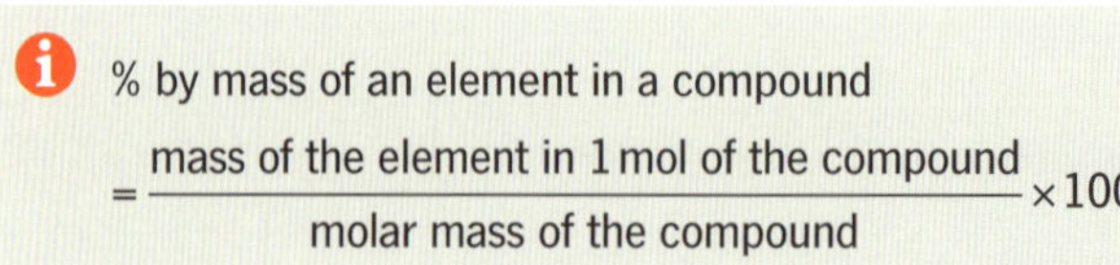

% by mass of an element in a compound

$$= \frac{\text{mass of the element in 1 mol of the compound}}{\text{molar mass of the compound}} \times 100$$

CHEMFILE

Elemental analysis of magnetite

If the chemical formula of a compound is unknown, the percentage composition by mass may be found using **elemental analysis** in a laboratory. An example is the analysis of the mineral magnetite, one of the three most commonly occurring oxides of iron. The most magnetic of all naturally occurring minerals on Earth, magnetite was used by the Chinese to make the first magnetic compass.

Found in many different types of igneous and metamorphic rocks, large deposits are located in the Pilbara region of Western Australia. Elemental analysis allowed chemists to determine the percentage composition of magnetite as 72.4% iron and 27.6% oxygen (Figure 9.4.3).

FIGURE 9.4.3 Elemental analysis of the mineral magnetite has allowed chemists to determine the percentage composition to be 72.4 % iron and 27.6% oxygen.

Worked example 9.4.1

CALCULATING PERCENTAGE COMPOSITION

Calculate the percentage by mass of aluminium in alumina (Al_2O_3).	
Thinking	**Working**
Find the molar mass of the compound.	$M(Al_2O_3) = (2 \times 26.98) + (3 \times 16.00)$ $= 101.96\,g\,mol^{-1}$
Find the total mass of the element in 1 mole of the compound.	mass of Al in 1 mol $= 2 \times M(Al)$ $= 2 \times 26.98$ $= 53.96\,g$
Find the percentage by mass of the element in the compound.	% by mass of Al in Al_2O_3 $= \frac{\text{mass of Al in 1 mol of } Al_2O_3}{\text{molar mass of } Al_2O_3} \times 100$ $= \frac{53.96}{101.96} \times 100$ $= 52.92\%$

Worked example: Try yourself 9.4.1

CALCULATING PERCENTAGE COMPOSITION

Calculate the percentage by mass of nitrogen in ammonium nitrate (NH_4NO_3).

9.4 Review

SUMMARY

- The percentage, by mass, of an element in a compound can be calculated from the mass of the element in 1 mole of the compound and the molar mass of the compound.

KEY QUESTIONS

1 Calculate the percentage by mass of:
   - **a** iron in iron(III) oxide (Fe_2O_3)
   - **b** uranium in uranium oxide (U_3O_8)
   - **c** nitrogen in ammonium chloride (NH_4Cl)
   - **d** oxygen in copper(II) nitrate ($Cu(NO_3)_2$).

2 Nitrogen is an essential element required for healthy plant growth. Each of the following compounds is a fertiliser used by farmers and market gardeners. By calculating the percentage by mass of nitrogen in each, determine the fertiliser that has the highest nitrogen content.
   - **a** urea (NH_2CONH_2)
   - **b** ammonium nitrate (NH_4NO_3)
   - **c** ammonium sulfate ($(NH_4)_2SO_4$)

3 Calculate the percentage by mass of each element in the following compounds.
   - **a** silicon carbide (SiC)
   - **b** water (H_2O)
   - **c** potassium phosphate (K_3PO_4)
   - **d** sucrose ($C_{12}H_{22}O_{11}$)
   - **e** gold sulfate ($Au_2(SO_4)_3$)

Chapter review

09

KEY TERMS

amount
Avogadro's number
carbon-12
elemental analysis
formula
molar mass
mole
percentage composition
relative formula mass
relative molecular mass

Masses of particles

1 Determine the relative molecular mass (M_r) of:
  - **a** water (H_2O)
  - **b** white phosphorus (P_4)
  - **c** carbon monoxide (CO).

2 Determine the relative formula mass of:
  - **a** zinc bromide ($ZnBr_2$)
  - **b** barium hydroxide ($Ba(OH)_2$)
  - **c** iron(III) carbonate ($Fe_2(CO_3)_3$).

Introducing the mole

3 For each of the following numbers of molecules, calculate the amount of substance, in mol.
  - **a** 4.50×10^{23} molecules of water (H_2O)
  - **b** 9.00×10^{24} molecules of methane (CH_4)
  - **c** 2.3×10^{28} molecules of chlorine (Cl_2)
  - **d** one molecule of sucrose ($C_{12}H_{22}O_{11}$)

4 For each of the following amounts of molecular substances, calculate the:
  - **i** number of molecules
  - **ii** total number of atoms.
  - **a** 1.45 mol of ammonia (NH_3)
  - **b** 0.576 mol of hydrogen sulfide (H_2S)
  - **c** 0.0153 mol of hydrogen nitrate (HNO_3)
  - **d** 2.5 mol of sucrose ($C_{12}H_{22}O_{11}$)

Molar mass

5 How would the molar mass (M) of a compound differ from its relative molecular mass (M_r)?

6 What is the molar mass (M) of each of the following?
  - **a** iron (Fe)
  - **b** sulfuric acid (H_2SO_4)
  - **c** sodium oxide (Na_2O)
  - **d** zinc nitrate ($Zn(NO_3)_2$)
  - **e** glycine (H_2NCH_2COOH)
  - **f** aluminium sulfate ($Al_2(SO_4)_3$)
  - **g** hydrated iron(III) chloride ($FeCl_3 \cdot 6H_2O$)

7 What is the mass of each of the following?
  - **a** 0.060 mol of ethane (C_2H_6)
  - **b** 0.32 mol of glucose ($C_6H_{12}O_6$)
  - **c** 6.8×10^{-3} mol of urea ($(NH_2)_2CO$)
  - **d** 6.12 mol of copper atoms (Cu)

8 What is the amount, in mol, of each of the following?
  - **a** carbon atoms in 1.201 g carbon
  - **b** sulfur molecules (S_8) in 10.0 g sulfur
  - **c** methane molecules (CH_4) in 20.0 g methane
  - **d** aspirin molecules ($C_6H_4(OCOCH_3)COOH$) in 300 mg aspirin
  - **e** aluminium oxide (Al_2O_3) in 3.5 tonnes of aluminium oxide (1 tonne = 1000 kg)

9
  - **a** If 6.022×10^{23} atoms of calcium have a mass of 40.1 g, what is the mass of one calcium atom?
  - **b** If 1.0 mol of water molecules has a mass of 18.0 g, what is the mass of one water molecule?
  - **c** What is the mass of one molecule of carbon dioxide?

10 For each of the following molecular substances, calculate the:
  - **i** amount of substance in moles
  - **ii** number of molecules
  - **iii** total number of atoms.
  - **a** 4.2 g of phosphorus (P_4)
  - **b** 75.0 g of sulfur (S_8)
  - **c** 0.32 g of hydrogen chloride (HCl)
  - **d** 2.2×10^{-2} g of glucose ($C_6H_{12}O_6$)

11 What mass of iron (Fe) would contain as many iron atoms as there are molecules in 20.0 g water (H_2O)?

12 For each of the following ionic substances calculate the amount of:
  - **i** substance, in moles
  - **ii** each ion, in moles.
  - **a** 5.85 g of NaCl
  - **b** 45.0 g of $CaCl_2$
  - **c** 1.68 g of $Fe_2(SO_4)_3$

13
  - **a** If 0.50 mol of a substance has a mass of 72.0 g, what is the mass of 1.0 mol of the substance?
  - **b** If 6.0×10^{22} molecules of a substance have a mass of 10.0 g, what is the molar mass of the substance?

14 Calculate the molar mass of a substance if:
  - **a** 2.0 mol of the substance has a mass of 80.0 g
  - **b** 0.10 mol of the substance has a mass of 9.8 g
  - **c** 1.7 mol of the substance has a mass of 74.8 g
  - **d** 3.50 mol of the substance has a mass of 371 g.

15 Which of the following metal samples has the greatest mass?

A 100 g copper

B 4.0 mol of iron atoms

C 1.2×10^{24} atoms of silver

16 A new antibiotic has been isolated and only 2.0 mg is available. The molar mass is found to be $12.5\ kg\,mol^{-1}$.

a Express the molar mass in $g\,mol^{-1}$.

b Calculate the amount of antibiotic, in mol.

c How many molecules of antibiotic have been isolated?

Percentage composition

17 Calculate the percentage by mass of each element in:

a Al_2O_3

b $Cu(OH)_2$

c $MgCl_2{\cdot}6H_2O$

d $Fe_2(SO_4)_3$

e perchloric acid ($HClO_4$).

18 Determine the percentage by mass of carbon in:

a naphthalene ($C_{10}H_8$)

b ethanoic acid (CH_3COOH)

c urea (NH_2CONH_2)

d aspirin ($C_6H_4(OCOCH_3)COOH$).

Connecting the main ideas

19 a How many molecules of water are there in 5.0 moles of ice?

b How many moles of Cu atoms in 2.4×10^{24} atoms of copper?

c What is the mass of a magnesium atom?

d What is the mass of a water molecule? (Molar mass of H_2O is 18.02 g.)

20 a The formula for gold sulfate is $Au_2(SO_4)_3$. For every 1 mole of $Au_2(SO_4)_3$, how many moles are there of:

i gold atoms?

ii sulfur atoms?

iii oxygen atoms?

b What is the molar mass, *M*, of gold sulfate?

Use your answer from part **b** to answer the following. Show your working out.

c Calculate the number of moles of gold sulfate in 27.5 g of $Au_2(SO_4)_3$.

d What is the mass of 0.660 mole of $Au_2(SO_4)_3$?

e What is the percentage of gold in gold sulfate?

21 The following questions are based on the hydrated salt, copper(II) nitrate hexahydrate, $Cu(NO_3)_2{\cdot}6H_2O$.

a What is the molar mass, *M*, of $Cu(NO_3)_2{\cdot}6H_2O$?

b 30.0 g of the crystals are weighed out. How many moles of the hydrated salt is this?

c What is the percentage of water in copper(II) nitrate hexahydrate?

22 An analytical chemist studying the noble gases, He, Ne, Ar, Kr, Xe and Rn (group 18 in the periodic table), is sent a sealed glass cylinder with the following label:

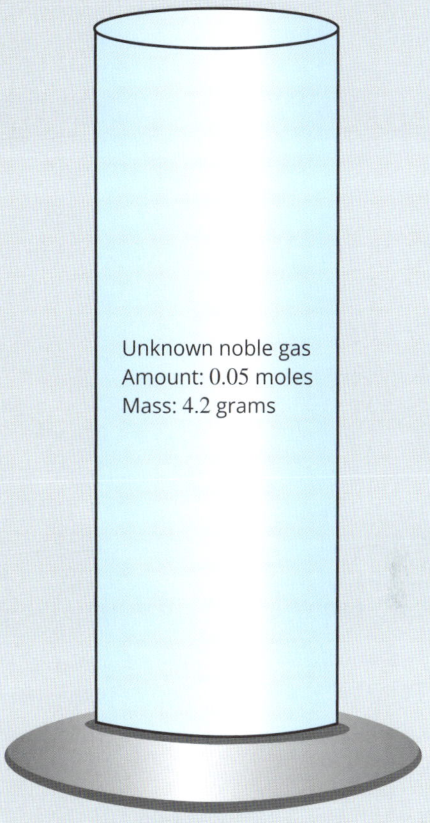

Which noble gas was the unknown gas? (Show your reasoning and calculation.)

23 As the science of chemistry progressed, chemists such as Avogadro, Lavoisier and Dalton, used a fairly simple (compared to modern electronic methods) way of measuring atomic masses. They would take equal volumes of gases, weigh them, and compare their masses to a standard atomic mass such as oxygen. The formula was:

$$\frac{\text{mass of oxygen}}{\text{atomic mass of oxygen}} = \frac{\text{mass of unknown gas X}}{\text{atomic mass of unknown gas X}}$$

During an experiment carried out in 1809 to determine the atomic mass of X, the following data was obtained by Dalton in his laboratory:

mass of oxygen = 1.0 g

atomic mass of oxygen = 16.0 g

mass of unknown gas X = 2.2 g

a Calculate the atomic mass of X.

b Suggest one reason why this technique may give incorrect results. (Hint: think about the formulae of some gases that you have encountered.)

CHAPTER

10 Energy changes in chemical reactions

When a piece of wood burns, we observe the release of energy mainly as heat and light. The energy is released to the surroundings and the burnt wood and ash will contain less energy than before it was burned. The energy that was once stored in the wood has been released to the environment. In a similar way, stored energy in the food that we eat can be released to provide us with energy that the organs in our bodies require to be able to function.

In this chapter, you will learn about energy transformation processes like these: how seemingly 'invisible' energy from chemicals can be released to produce energy, and also how some chemical reactions absorb heat energy from the surroundings, actually increasing the amount of this stored 'chemical' energy.

By the end of this chapter, you will have a greater understanding of how energy changes occur as a result of chemical reactions, and be able to represent these changes in a range of ways, including chemical equations and energy diagrams.

Science understanding

- chemical reactions can be represented by chemical equations; balanced chemical equations indicate the relative numbers of particles (atoms, molecules or ions) that are involved in the reaction
- chemical reactions and phase changes involve enthalpy changes, commonly observable as changes in the temperature of the surroundings and/or the emission of light
- endothermic and exothermic reactions can be explained in terms of the Law of Conservation of Energy and the breaking of existing bonds and forming of new bonds; heat energy released or absorbed by the system to or from the surroundings, can be represented in thermochemical equations
- the mole is a precisely defined quantity of matter equal to Avogadro's number of particles

10.1 Exothermic and endothermic reactions

FIGURE 10.1.1 The energy released by the combustion of wood in a fire is easily seen and felt.

Chemical reactions occur when particles (atoms, molecules or ions) collide and are rearranged to form new particles. Chemical reactions involve energy changes. As the reactant particles are rearranged, the chemical energy of the reactants is also changed.

In some chemical reactions, including the combustion of fuels, the rearrangement of atoms causes energy to be released to the surroundings (Figure 10.1.1). In other chemical reactions, energy is absorbed from the surroundings as the chemical reaction takes place. The energy change in some reactions is very small and can only be determined with specialised equipment.

In this section, you will learn about the energy changes that occur during chemical reactions. You will also learn how to classify chemical reactions based on their energy changes.

CHEMICAL ENERGY

There are many different forms of energy. For example, you are in contact with forms of heat (thermal energy), light, sound energy and electrical energy every day. You can probably see, hear or feel some of them as you are reading these pages.

All substances have a form of energy called **chemical energy**. It is stored in the chemical bonds between atoms and molecules. This energy results from:

- attractions between electrons and protons in atoms
- repulsions between nuclei
- repulsions between electrons
- movement of electrons
- vibrations and rotations around bonds.

However, unlike heat, light or sound energy, you cannot directly observe this energy; you can only experience the effect of changes of this stored chemical energy. For example, when you eat a meal, the bonds between the food molecules have stored energy that you can access to provide energy for other chemical and physical activities that take place in your body (Figure 10.1.2).

FIGURE 10.1.2 When you eat food, you access the chemical energy stored in the food. This energy powers all of the chemical reactions and physical activities that take place in your body.

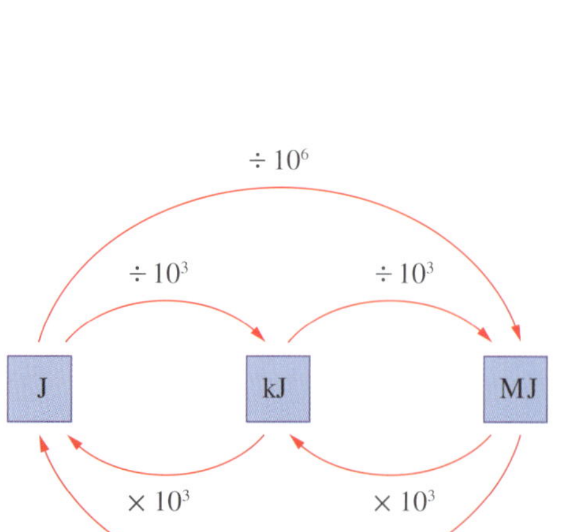

FIGURE 10.1.3 Converting between different energy units

The SI unit for energy is the joule, J. Other units also used for quantifying energy are kilojoules, kJ, and megajoules, MJ.

The relationship between joules, kilojoules and megajoules is:

$$1\,\text{MJ} = 1000\,\text{kJ} = 1\,000\,000\,\text{J}$$

Figure 10.1.3 shows how you can convert between different units of energy.

ENERGY CONSERVATION

The **law of conservation of energy** states energy cannot be created or destroyed. However, it can change forms. When energy is transformed from one form to another, the total amount of energy remains the same. So, if the chemical energy stored in a substance reduces, that energy must go somewhere. Likewise, a substance can't gain in chemical energy without absorbing that energy from another source.

Systems and surroundings

When we talk about energy changes in chemical reactions, we often refer to a system and its surroundings.

In chemistry, the **system** is usually the chemical reaction. When we say that energy is released or absorbed by a system, we are referring to energy changes that occur as bonds are broken and formed between the atoms of the elements involved in the reaction.

The **surroundings** are usually regarded as everything else. For example, the walls of a container in which a reaction takes place in the gas phase, or the water in a solution in which a reaction takes place, can be regarded as the surroundings for the reaction. Energy leaves the system (the reaction) and enters the surroundings, or leaves the surroundings and enters the system.

Energy changes during chemical reactions

The reactants in a chemical reaction have a certain amount of chemical energy stored in their bonds. The products that form as a result of the rearrangement of particles during the chemical reaction have different bonds and so have a different amount of chemical energy. Energy will be released or absorbed during the reaction depending on the relative energies of the bonds within the reactants and products.

In general, when particles are separated from each other, energy is required for this process to happen. The separated particles will therefore have more chemical energy than when they were together. This process can be seen in Figure 10.1.4. Conversely, if you consider particles coming together, the final arrangement with the particles joined, or bonded together, will have less energy than the separate particles. According to the law of conservation of energy, this 'lost' energy must go somewhere. In reality it is transferred to the surroundings. In Figure 10.1.4 this energy is represented by the wavy arrow, leaving the particles of the substance.

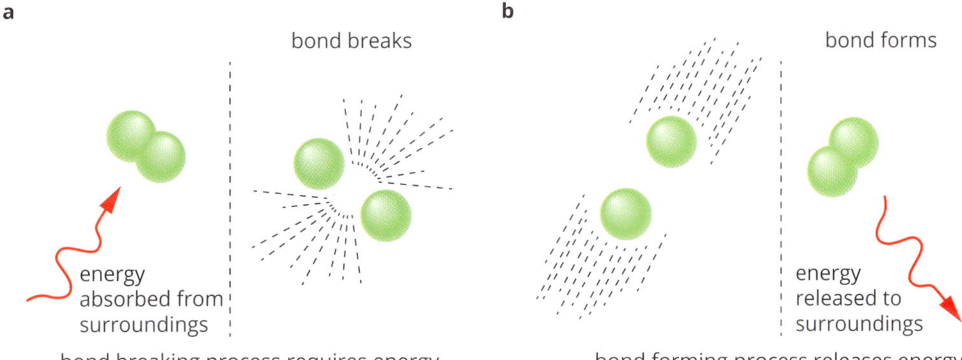

FIGURE 10.1.4 Energy changes when chemical bonds are broken and made

Often the energy released to or absorbed from the surroundings is in the form of heat (thermal energy). However, other forms of energy can also be released or absorbed including light, electricity and movement (kinetic energy).

CHEMISTRY IN ACTION

Glow-in-the-dark light sticks

You might have seen glow-in-the-dark hoops, necklaces and bracelets, similar to those shown in Figure 10.1.5, at festivals or concerts.

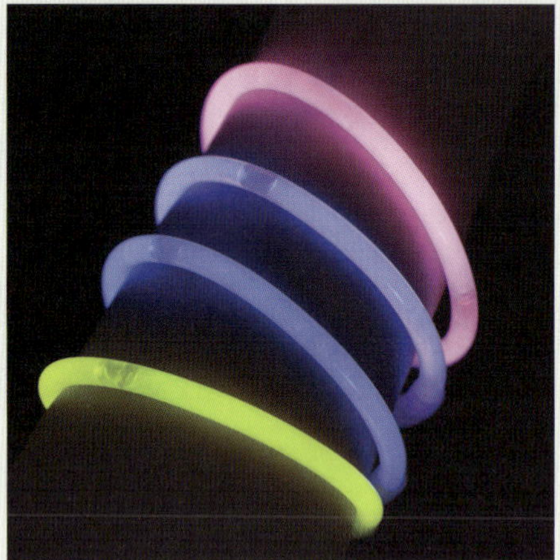

FIGURE 10.1.5 Glow-in-the-dark bracelets give off light that is the result of chemiluminescence.

Glow-in-the-dark bracelets contain chemicals held in separate containers. When these bracelets are bent, the containers break and the chemicals combine. Light is produced through a process called **chemiluminescence**. When the aqueous reactants (hydrogen peroxide in one container and diphenyl oxalate in another container) mix, energy is released from the reaction that occurs. This reaction is shown in Figure 10.1.6.

Instead of the energy from this reaction being released to the surroundings solely as heat, a carrier molecule transfers the energy to a **chemiluminescent** dye in the glow stick. The electrons in the dye are excited to higher energy levels. Light is emitted as these electrons return to their original lower energy levels. The light from the glow stick is the result of the emission spectrum (as discussed in Chapter 3) of the dye molecule.

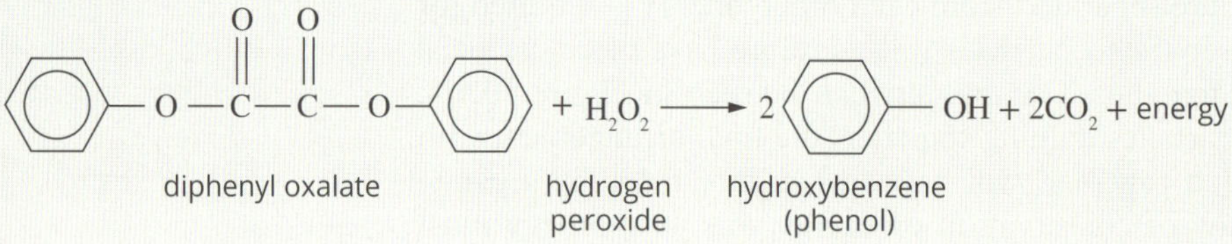

$$C_6H_5{-}O{-}\overset{\overset{O}{\|}}{C}{-}\overset{\overset{O}{\|}}{C}{-}O{-}C_6H_5 + H_2O_2 \longrightarrow 2\,C_6H_5{-}OH + 2CO_2 + \text{energy}$$

diphenyl oxalate | hydrogen peroxide | hydroxybenzene (phenol)

FIGURE 10.1.6 This reaction occurs in a glow stick.

CHEMFILE

Glow-worms

Glow-worms (Figure 10.1.7) apply similar chemical principles to chemiluminescence for their glow-in-the-dark bioluminescence. Three chemicals within the worm combine. However, they require oxygen to produce light. When the worm breathes, oxygen acts as the oxidising agent in the chemical reaction between the three reactants producing the bioluminescence. Worms are able to control the amount of 'glow' by breathing in more or less oxygen.

FIGURE 10.1.7 A female glow-worm. The luminescent abdominal organs are visible.

EXOTHERMIC AND ENDOTHERMIC SYSTEMS

Exothermic reactions

A chemical reaction in which energy is released is called an **exothermic** reaction. In this situation the total chemical energy of the products of the chemical reaction is less than the total chemical energy of the reactants. This 'lost' energy is released to the surroundings. Energy 'exits' the reaction system and the chemical reaction is called an exothermic reaction.

The released energy can be shown in a chemical equation by writing 'energy' on the product side of the arrow.

For example, the production of water from the reaction between hydrogen and oxygen gas is an exothermic reaction. This can be represented by the equation:

$$2H_2(g) + O_2(g) \rightarrow 2H_2O(l) + \text{energy}$$

Another example of an exothermic reaction is the combustion of methane gas:

$$CH_4(g) + 2O_2(g) \rightarrow CO_2(g) + 2H_2O(g) + \text{energy}$$

Heat is given off to the surroundings. All combustion reactions give off heat energy to the surroundings and are therefore exothermic reactions.

Endothermic reactions

When the total chemical energy of the products of a chemical reaction is greater than the total chemical energy of the reactants, energy is absorbed from the surrounding environment. Energy 'enters' the reaction system and the chemical reaction is called an **endothermic** reaction.

If an endothermic reaction takes place in a container, the container may feel cold to the touch. This is because the reaction system is absorbing heat from the surroundings, leaving the environment cooler.

The idea that a process that absorbs energy gets cooler can be confusing, but to avoid this confusion you have to look at the system and the surroundings separately. The energy that is gained by the system is the 'unobservable' stored chemical energy that cannot be experienced or directly measured. This energy is absorbed from the surroundings, which lose heat (i.e. get colder).

In a chemical equation of an endothermic reaction, the energy that is required can be written on the reactant side of the equation arrow.

For example, the **decomposition** of water by electrolysis (passing an electric current through the water) into hydrogen and oxygen is an endothermic process. This can be represented by the equation:

$$2H_2O(l) + \text{energy} \rightarrow 2H_2(g) + O_2(g)$$

> If the total chemical energy of the products is *less* than the total energy of the reactants, energy will be released from the system into the surroundings. This is called an exothermic reaction.
>
> If the total chemical energy of the products is *greater* than the total energy of the reactants, energy will be absorbed from the surroundings. This is called an endothermic reaction.

Changes of state

Changes of state (phase changes), such as a solid melting to form a liquid, are a type of physical change, rather than a chemical change. But these changes also involve energy being absorbed or released.

For example, the melting of ice into water requires the absorption of energy, making it an endothermic process. The boiling of water to produce water vapour is also endothermic.

Conversely, condensing a gas to a liquid and freezing a liquid to form a solid both release heat to the surroundings and so are exothermic processes. These processes are summarised in Figure 10.1.8.

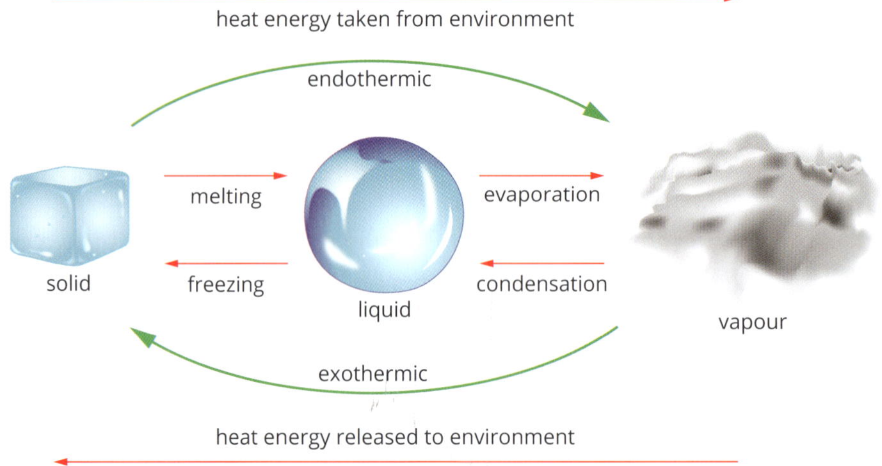

FIGURE 10.1.8 Energy changes associated with phase changes

CHEMFILE

Instant cold packs

Instant cold packs are often carried in first aid kits at sporting events (Figure 10.1.9). One type of cold pack contains a sealed bag of water surrounded by solid ammonium nitrate. When the cold pack is squeezed, the water bag is broken and ammonium nitrate dissolves in the water. The process is endothermic, absorbing heat from the surroundings and quickly lowering the pack's temperature:

$$NH_4NO_3(s) + \text{energy} \rightarrow NH_4NO_3(aq)$$

Notice that bonds (in this case ionic bonds) are being broken, hence energy is required to be absorbed from the surroundings. Writing this process as an ionic equation emphasises this point:

$$NH_4NO_3(s) + \text{energy} \rightarrow NH_4^+(aq) + NO_3^-(aq)$$

FIGURE 10.1.9 An endothermic reaction produces an instant cold pack.

10.1 Review

SUMMARY

- Energy is measured in J, kJ or MJ:
  1 MJ = 10^3 kJ = 10^6 J
- Chemical energy is stored in the bonds between atoms.
- Energy is conserved during a chemical reaction; energy cannot be created or destroyed.
- Chemical reactions and changes of state involve energy changes.
- A chemical reaction that releases energy to the surroundings is called an exothermic reaction.
- A chemical reaction in which energy is absorbed from the surroundings is called an endothermic reaction.
- All combustion reactions are exothermic reactions.

KEY QUESTIONS

1 Which of the following statements about combustion reactions is correct?

A Combustion reactions are usually exothermic, meaning energy is absorbed during the reaction.

B Combustion reactions are always exothermic, meaning energy is absorbed during the reaction.

C Combustion reactions are usually endothermic, meaning energy is absorbed by the system.

D Combustion reactions are always exothermic, meaning energy is released during the reaction.

2 Convert the following energy values to kJ.

a 0.180 MJ

b 1.5×10^6 J

c 10.0 J

d 2.0×10^{-3} J

3 Explain the difference between the terms 'system' and 'surroundings' in relation to a chemical reaction.

4 Explain the term 'endothermic' in relation to the total amount of chemical energy of the reactants and products.

10.2 Thermochemical equations, energy profile diagrams and enthalpy

As you saw in the previous section, the word 'energy' can be included in a combustion equation to show that energy is released. However, it is generally more useful for a chemist to be precise about the magnitude (size) of the energy change that takes place.

Thermochemical equations achieve this by including a sign and numerical value for the energy change that occurs in the reaction.

It is also useful for chemists to be able to show the energy changes that occur as a reaction proceeds. All reactions absorb some energy before they can proceed, even if energy is released overall. An **energy profile diagram** represents the energy changes that occur during the course of a reaction.

In this section, you will learn how to write and interpret thermochemical equations and draw energy profile diagrams. These representations are important for the rates of reactions topic later in Year 11, and when considering chemical equilibrium in Year 12.

However, before we consider this, we need to give a name to this stored chemical energy that we have been discussing in Section 10.1.

ENTHALPY

Enthalpy change is a measure of the amount of energy absorbed or released during chemical reactions. It is given the symbol ΔH and is determined by subtracting the enthalpy of the reactants from the enthalpy of the products.

The stored chemical energy of a substance is called its **enthalpy**. It is given the symbol H. All chemicals have a certain amount of stored chemical energy, or enthalpy. However, it is the change in enthalpy that occurs in a chemical reaction that is most important to consider.

The exchange of heat energy between the system and its surroundings is referred to as the **enthalpy change**, or **heat of reaction**, and is given the symbol ΔH.

For a chemical reaction:

$$\text{reactants} \rightarrow \text{products}$$

the enthalpy change (ΔH) is calculated by:

$$\Delta H = H_{\text{products}} - H_{\text{reactants}}$$

Enthalpy change in exothermic reactions

When the enthalpy of the products is less than the enthalpy of the reactants, energy is released from the system into the surroundings, so the reaction is exothermic. The system has lost energy, so ΔH has a negative value.

Therefore, for combustion reactions (which are exothermic reactions), $\Delta H < 0$ (Figure 10.2.1).

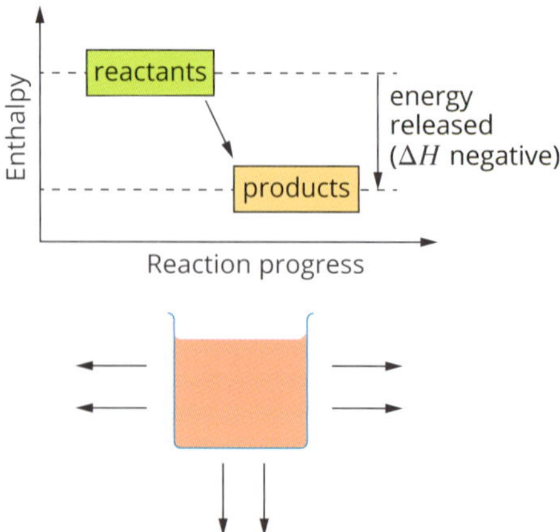

FIGURE 10.2.1 In the combustion of a fuel, the enthalpy of the reactants is greater than the enthalpy of the products, so energy is released into the surroundings during the reaction.

CHEMISTRY IN ACTION

Explosives—extreme exothermic reactions

Humans have been using chemicals to make explosions since 919 BCE. Chinese people first mixed saltpetre (potassium nitrate), sulfur and charcoal with explosive results. They quickly realised that there were many uses for this mixture, which later became known as gunpowder. It was put to military use and eventually led to the development of bombs, cannons and guns.

Today, explosives are an essential tool for mining and other engineering works, such as road construction, tunnelling, building and demolition (Figure 10.2.2).

Explosives transform chemical energy into large quantities of thermal energy very quickly in highly exothermic reactions with a large drop in the enthalpy of the chemicals involved. Although thermal energy is also released when fuels such as petrol and natural gas burn, the rate of combustion in these reactions is limited by the availability of oxygen gas to the fuel. In contrast, the compounds making up an explosive contain sufficient oxygen for a complete (or almost complete) reaction to occur very quickly.

FIGURE 10.2.2 An old bridge is demolished with the help of explosives.

Enthalpy change in endothermic reactions

When the total enthalpy of the products is greater than the total energy of the reactants, energy must be absorbed from the surroundings, so the reaction is endothermic. The system has gained energy, so ΔH has a positive value, i.e. $\Delta H > 0$ (Figure 10.2.3).

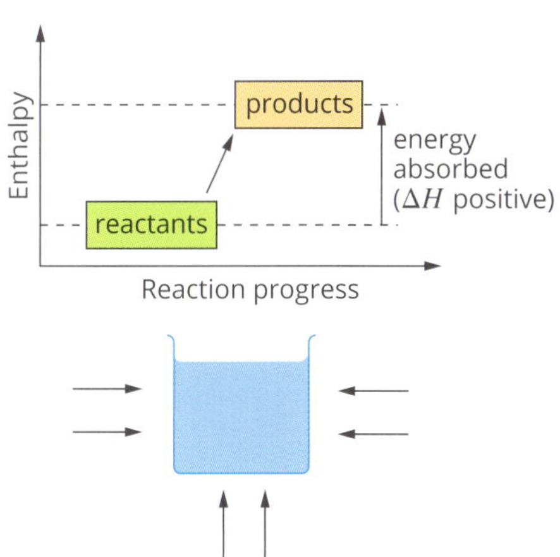

FIGURE 10.2.3 For an endothermic reaction, the enthalpy of the reactants is less than the enthalpy of the products so energy is absorbed during the reaction.

Thermochemical equations

The enthalpy change can be shown by writing the ΔH value to the right of the chemical equation. Such an equation is called a thermochemical equation. The ΔH value in a thermochemical equation usually has the units $kJ\,mol^{-1}$ (kilojoules per mole). This means that the amount of energy (in kJ) signified by the ΔH value corresponds to the mole amounts specified by the coefficients in the equation.

For example, respiration in most living things can be considered as a type of combustion reaction of glucose. Therefore, respiration is an exothermic reaction. Previously you have seen such situations written with the energy included in the equation:

$$C_6H_{12}O_6(aq) + 6O_2(g) \rightarrow 6CO_2(g) + 6H_2O(l) + \text{energy}$$

This can be written as the thermochemical equation shown below:

$$C_6H_{12}O_6(aq) + 6O_2(g) \rightarrow 6CO_2(g) + 6H_2O(l) \qquad \Delta H = -2803\,kJ\,mol^{-1}$$

This thermochemical equation tells you that when 1 mole of glucose reacts with 6 moles of oxygen to produce 6 moles of carbon dioxide and 6 moles of water, 2803 kJ of energy is released to the surroundings.

The equation may also be written to include the enthalpy in the equation.

$$C_6H_{12}O_6(aq) + 6O_2(g) \rightarrow 6CO_2(g) + 6H_2O(l) + 2803\,kJ\,mol^{-1}$$

The ΔH *is still* equal to $-2803\,kJ\,mol^{-1}$.

Note that the energy is written on the right-hand side of the equation as a 'product' in this exothermic reaction.

Enthalpy changes per mole of reactant

The combustion of hydrogen gas in oxygen is shown in the equation:

$$2H_2(g) + O_2(g) \rightarrow 2H_2O(l) \qquad \Delta H = -572\,kJ\,mol^{-1}$$

However, in this reaction there are 2 moles of hydrogen, and to compare enthalpy changes, you have to ensure that you are comparing equal quantities of the substance, in this case 1 mole of the chemical being combusted, i.e. kilojoules per 1 mole (kJ mol^{-1}).

In this case, the coefficients in the equation are divided by 2 so that there is only 1 mole of H_2 and the equation becomes:

$$H_2(g) + \frac{1}{2}O_2(g) \rightarrow H_2O(l) \qquad \Delta H = -286\,kJ\,mol^{-1}$$

Therefore the heat of combustion of hydrogen is given as $-286\,kJ\,mol^{-1}$.

Physical changes

Enthalpy changes also occur during physical changes, so thermochemical equations can be written for physical changes. Melting ice is an example of a physical change. It is an endothermic process, because heat must be applied to solid ice in order to break the bonds between the water molecules when it is converted it into liquid. The thermochemical equation for this reaction is:

$$H_2O(s) \rightarrow H_2O(l) \qquad \Delta H = +6\,kJ\,mol^{-1}$$

The ΔH value is positive because this is an endothermic reaction.

The equation may also be written to include the enthalpy in the equation.

$$H_2O(s) + 6\ kJ\ mol^{-1} \rightarrow H_2O(l)$$

The ΔH *is still* equal to $+6\,kJ\,mol^{-1}$.

Note that it is very important to always include state symbols in all thermochemical equations. Physical changes involve an enthalpy change, so the state of the species in a chemical reaction will affect the enthalpy change of the reaction.

CHEMISTRY IN ACTION

The chemistry behind an explosion

When chemical explosives such as ammonium nitrate, trinitrotoluene (TNT) and nitroglycerine decompose, they release large amounts of energy very quickly.

This is the thermochemical equation for the decomposition of nitroglycerine:

$$4C_3H_5N_3O_9(l) \rightarrow 12CO_2(g) + 10H_2O(g) + 6N_2(g) + O_2(g) \qquad \Delta H = -1456\,kJ\,mol^{-1}$$

The negative ΔH value indicates that this is an exothermic reaction.

Notice that 29 moles of gas (the total number of moles of all products) are produced from 4 moles of nitroglycerine. At atmospheric pressure, the reactant products would expand to fill a volume more than 10000 times larger than the volume of the nitroglycerine! During a blast, this gas is usually produced within a small hole into which the explosive has been placed, creating huge pressure that shatters the surrounding rock or structure.

Effect on ΔH of reversing a chemical reaction

Reversing a chemical equation changes the sign but not the magnitude of ΔH.

For example, methane (CH_4) reacts with oxygen gas to produce carbon dioxide gas and water in an exothermic reaction:

$$CH_4(g) + 2O_2(g) \rightarrow CO_2(g) + 2H_2O(l) \quad \Delta H = -890\,kJ\,mol^{-1}$$

If this reaction is reversed, the magnitude of ΔH remains the same because the enthalpies of the individual chemicals have not changed, but the sign changes to indicate that energy must be absorbed, i.e. it is an endothermic reaction:

$$CO_2(g) + 2H_2O(l) \rightarrow CH_4(g) + 2O_2(g) \quad \Delta H = +890\,kJ\,mol^{-1}$$

This demonstrates the law of conservation of energy and is true for all reversible reactions.

Worked example 10.2.1

INTERPRETING A THERMOCHEMICAL REACTION

The ΔH of the following reaction is $-2220\,kJ\,mol^{-1}$.

$C_3H_8(g) + 5O_2(g) \rightarrow 3CO_2(g) + 4H_2O(g)$

Rewrite the reaction, including the energy term.

Thinking	Working
The reaction is exothermic so show energy is a product in the reaction.	Energy term goes on right-hand side.
$\Delta H = -2220\,kJ\,mol^{-1}$ so add this value.	$C_3H_8(g) + 5O_2(g) \rightarrow 3CO_2(g) + 4H_2O(g) + 2220\,kJ\,mol^{-1}$

Worked example: Try yourself 10.2.1

INTERPRETING A THERMOCHEMICAL REACTION

The ΔH of the photosynthesis reaction shown below is $2803\,kJ\,mol^{-1}$.

$6CO_2(g) + 6H_2O(g) \rightarrow C_6H_{12}O_6(aq) + 6O_2(g)$

Rewrite the reaction, including the energy term.

Worked example 10.2.2

IDENTIFYING THE ENTHALPY CHANGE IN A THERMOCHEMICAL REACTION

Rewrite the reaction below, showing the enthalpy change separately.

$N_2(g) + 2O_2(g) + 33\,kJ\,mol^{-1} \rightarrow 2NO_2(g)$

Thinking	Working
The reaction is endothermic—energy is gained from the surroundings.	Enthalpy change is positive.
Energy absorbed = $33\,kJ\,mol^{-1}$	$N_2(g) + 2O_2(g) \rightarrow 2NO_2(g) \quad \Delta H = +33\,kJ\,mol^{-1}$

Worked example: Try yourself 10.2.2

IDENTIFYING THE ENTHALPY CHANGE IN A THERMOCHEMICAL REACTION

Rewrite the reaction, showing the enthalpy change separately.

$4C_3H_5N_3O_9(l) \rightarrow 12CO_2(g) + 10H_2O(g) + 6N_2(g) + O_2(g) + 1456\,kJ\,mol^{-1}$

A reaction cannot proceed unless the bonds in the reactants are broken.

FIGURE 10.2.4 When zinc comes into contact with hydrochloric acid, it reacts almost immediately. The reactants have sufficient energy to 'overcome' the activation energy barrier.

Activation energy

The energy required to break the bonds of reactants so that a reaction can proceed is called the **activation energy**. The activation energy is an energy barrier that must be overcome before a reaction can get started. (The concept of activation energy is discussed in more detail in Chapter 18.)

An activation energy barrier exists for both exothermic and endothermic reactions. If the activation energy for a reaction is very low, the chemical reaction can be initiated as soon as the reactants come into contact because the reactants already have sufficient energy for a reaction to take place. An example of this can be seen in the reaction between zinc and hydrochloric acid in Figure 10.2.4.

The reaction between zinc and hydrochloric acid produces hydrogen gas:

$$Zn(s) + 2HCl(aq) \rightarrow ZnCl_2(aq) + H_2(g)$$

As you can see in Figure 10.2.4, bubbles of hydrogen gas are vigorously produced as soon as zinc is added to the acid.

ENERGY PROFILE DIAGRAMS

The energy changes that occur during the course of a chemical reaction can be shown on an energy profile diagram.

The energy profile diagram for an exothermic combustion reaction like the one shown in Figure 10.2.5 indicates that the enthalpy of the products is always less than the enthalpy of the reactants. Overall, energy is released and so the ΔH valuc is negative. 'I'he energy profile also shows that, even in exothermic reactions, the activation energy must first be absorbed to start the reaction.

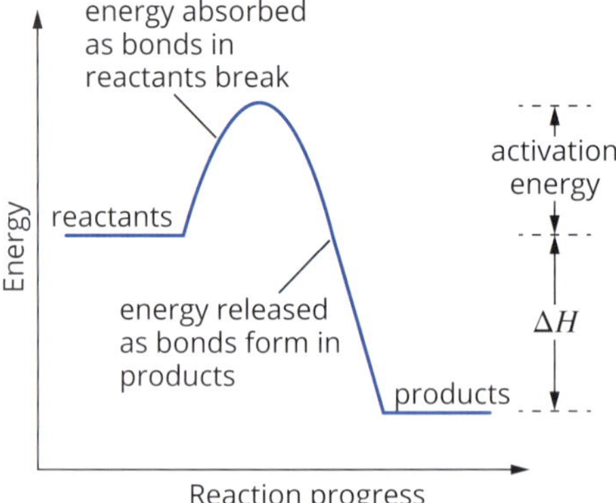

FIGURE 10.2.5 The characteristic shape of an energy profile diagram for an exothermic reaction

The energy profile diagram for an endothermic reaction (Figure 10.2.6) shows that the enthalpy of the products is greater than the enthalpy of the reactants. Overall, energy is absorbed and so the ΔH value is positive. The energy profile also shows the absorption of the activation energy before the release of energy as bonds form in the products.

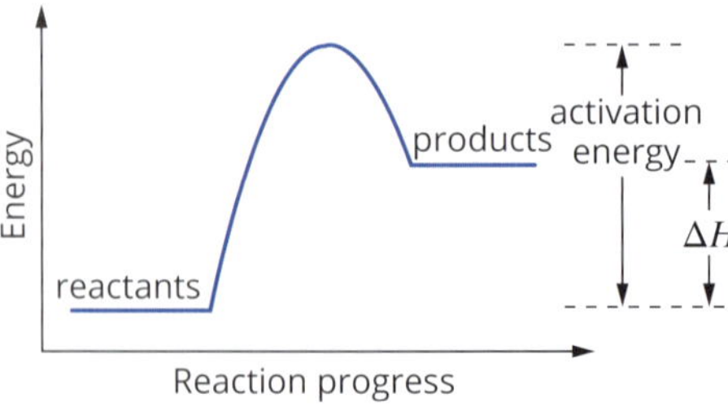

FIGURE 10.2.6 The characteristic shape of an energy profile diagram for an endothermic reaction

10.2 Review

SUMMARY

- Thermochemical equations include a ΔH value for a chemical reaction. The unit of ΔH is $kJ\,mol^{-1}$.
- The value of ΔH indicates the magnitude of the energy change and whether the energy is absorbed (a positive value) or released (a negative value).
- Reversing an equation causes the sign, but not the magnitude, of ΔH to change, as the reaction changes from exothermic to endothermic, or vice versa.
- States of matter must be included in thermochemical equations because changes of state involve enthalpy changes.
- Activation energy is the energy that must be absorbed to break the bonds in the reactants so that a chemical reaction can proceed. Both endothermic and exothermic reactions require activation energy.
- Energy profile diagrams show energy changes over the course of a reaction. They show the relative enthalpies of reactants and products and the activation energy.

KEY QUESTIONS

1 Explain what a negative ΔH value indicates about a chemical reaction, in terms of the relative enthalpies of the reactants and products.

2 When 1 mole of methane gas undergoes combustion in oxygen to produce carbon dioxide and water, 890 kJ of energy is released. Write a balanced thermochemical equation for this reaction.

3 The combustion of octane to form carbon dioxide and liquid water can be written as:

$$C_8H_{18}(g) + 12\frac{1}{2}O_2 \rightarrow 8CO_2(g) + 9H_2O(l) \quad \Delta H = -5450\,kJ\,mol^{-1}$$

The combustion of octane to form carbon dioxide and steam can be written as:

$$C_8H_{18}(g) + 12\frac{1}{2}O_2(g) \rightarrow 8CO_2(g) + 9H_2O(g)$$

How would the energy released by the combustion of 1 mole of octane to form steam compare with the energy released by 1 mole of octane to form liquid water?

4 The energy profile diagram below shows the energy changes during the reaction of hydrogen and iodine to form hydrogen iodide:

$$H_2(g) + I_2(g) \rightarrow 2HI(g)$$

a Is the reaction exothermic or endothermic?

b Describe the relative enthalpies of the reactants and products.

c Comment on the size of the activation energy compared with ΔH.

5 The reaction for photosynthesis is the opposite of the reaction for respiration shown on page 215. Write a thermochemical equation for photosynthesis.

Chapter review

10

KEY TERMS

activation energy
chemical energy
chemiluminescence
chemiluminescent
decomposition
endothermic
energy profile diagram
enthalpy
enthalpy change
exothermic
heat of reaction
law of conservation of energy
surroundings
system
thermochemical equation

Exothermic and endothermic reactions

1 Convert the following units of energy to the unit given (to 3 significant figures).
   a 2205 J to kJ
   b 0.152 kJ to J
   c 1 890 000 J to MJ
   d 0.0125 MJ to kJ

2 Decide whether the following processes are exothermic or endothermic. Give reasons for your answers.
   a burning of wood
   b melting of ice
   c recharging of a car battery
   d decomposition of plants in a compost heap

3 Explain why chemiluminescence is an example of an exothermic reaction.

Thermochemical equations, energy profile diagrams and enthalpy

4 Which one of the following statements is correct about the energy profile diagrams of both endothermic and exothermic reactions?
   A There is always less energy absorbed than released.
   B The enthalpy of the products is always less than the enthalpy of the reactants.
   C Some energy is always absorbed to break bonds in the reactants.
   D The ΔH value is the difference between the enthalpy of the reactants and the highest energy point reached on the energy profile.

5 Identify whether each of the following statements related to activation energy is true or false.
   a Activation energy is the energy required to break bonds in the reactants.
   b Reactions that start immediately do not have an activation energy.
   c Reactions that release energy overall do not need to absorb activation energy.
   d The match used to light a fire is providing activation energy.

6 Explain why reversing a chemical reaction reverses the sign of ΔH.

Connecting the main ideas

7 The combustion of butane gas in portable stoves can be represented by the thermochemical equation:

$$2C_4H_{10}(g) + 13O_2(g) \rightarrow 8CO_2(g) + 10H_2O(l) \quad \Delta H = -5772\ kJ\ mol^{-1}$$

   a How does the overall energy of the bonds in the reactants compare with the overall energy of the bonds in the products?
   b Draw an energy profile diagram for the reaction, labelling ΔH and activation energy.

8 The enthalpy change for the combustion of 1 mole of hydrogen is $-286\ kJ\ mol^{-1}$. Write a thermochemical equation for the complete combustion of hydrogen.

9 In a steelworks, carbon monoxide present in the exhaust gases of the blast furnace can be used as a fuel elsewhere in the plant. It reacts according to the equation:

$$CO(g) + \frac{1}{2}O_2(g) \rightarrow CO_2(g) \quad \Delta H = -283\ kJ\ mol^{-1}$$

   a Which has the greater total enthalpy: 1 mol of CO(g) and 0.5 mol of $O_2(g)$, or 1 mol of $CO_2(g)$?
   b Write the value of ΔH for the following equations.
      i $2CO(g) + O_2(g) \rightarrow 2CO_2(g)$
      ii $2CO_2(g) \rightarrow 2CO(g) + O_2(g)$

CHAPTER 11

Fuels and introduction to stoichiometry

In this chapter, you will learn how fuels are used to meet global energy needs and you will gain an appreciation of the chemistry that underpins decisions about the use of fuels. Combustion reactions are used to release useful heat energy from the chemical energy stored in fuels. You will explore how fuels vary in terms of the energy that they release when they are burnt.

You will also consider the environmental impact of using different types of fuels, including their carbon emissions as well as the other pollutants they release into the atmosphere. Current research being conducted into the production of renewable fuels and the potential for reducing the harmful impact of fossil fuels will also be discussed.

Science as a human endeavour

- There are differences in the energy output and carbon emissions of fossil fuels (including coal, oil, petroleum and natural gas) and biofuels (including biogas, biodiesel and bioethanol). These differences, together with social, economic, cultural and political values, determine how widely these fuels are used.

Science understanding

- fossil fuels (including coal, oil, petroleum and natural gas) and biofuels (including biogas, biodiesel and bioethanol) can be compared in terms of their energy output, suitability for purpose, and the nature of products of combustion
- the mole concept relates mass, moles and molar mass and, with the Law of Conservation of Mass, can be used to calculate the masses of reactants and products in a chemical reaction

11.1 Types of fuels

Fuels provide us with energy. They are substances that have chemical energy stored within them. As you learnt in Chapter 10, all chemicals contain stored energy, which is called enthalpy. What makes a **fuel** special is that this stored chemical energy can be released relatively easily.

Sugar is an example of a common fuel (Figure 11.1.1). A cube of table sugar (sucrose) can provide your body with 82 kilojoules of energy. This is about 1% of your daily energy needs. If sucrose is burnt, this energy is released as heat. The combustion of 1 kilogram of sucrose releases sufficient energy to melt more than 5 kilograms of ice and then boil all the liquid water produced.

FIGURE 11.1.1 Sugars, such as sucrose, are fuels for your body.

Although sugars provide energy for your body, you do not heat your home, power cars, or produce electricity by burning sugar! A range of other fuels such as wood, coal, oil, natural gas, LPG, ethanol and petrol (Figure 11.1.2) are used for these energy needs.

In this section, you will explore a range of different fuels and consider the chemical, economic and environmental factors that make them suitable for their purposes.

FIGURE 11.1.2 Petrol is just one type of fuel that is used each day to meet our energy needs.

THE NEED FOR FUELS

A fuel is a substance with stored energy that can be released relatively easily for use as heat or power. Although this chapter will focus on fuels with stored chemical energy, the term 'fuel' is also applied to sources of nuclear energy, such as uranium.

The use of fuels by society can be considered from a number of points of view, including at a:

- local level, e.g. the type of fuel used to heat your house
- national level, e.g. whether Australia's use of energy resources is sustainable
- global level, e.g. whether the use of fossil fuels (coal, oil and natural gas) is contributing to the enhanced greenhouse effect.

These are not separate issues. Choices made locally have regional and global effects. The decisions of global and national governments affect how and which fuels are used.

Units of energy

The international system of units (**SI units**) is a widely used system of measurement that specifies units for a range of quantities. The SI unit for energy is the joule, symbol J. As 1 J of energy is a relatively small amount, it is common to see the following units in use:

- kilojoules, $1\ kJ = 10^3\ J$
- megajoules, $1\ MJ = 10^6\ J$
- gigajoules, $1\ GJ = 10^9\ J$
- terajoules, $1\ TJ = 10^{12}\ J$.

Use of energy in Australia and the world

World energy consumption is around 4×10^{20} joules per year. The United States consumes a quarter of the world's energy. Australia consumes about one-hundredth of the world's energy. But energy consumption per person in Australia is only just below that of the United States. Figure 11.1.3 shows the ways in which Australians use energy. As you can see, heating and transportation account for 87% of Australia's total energy consumption.

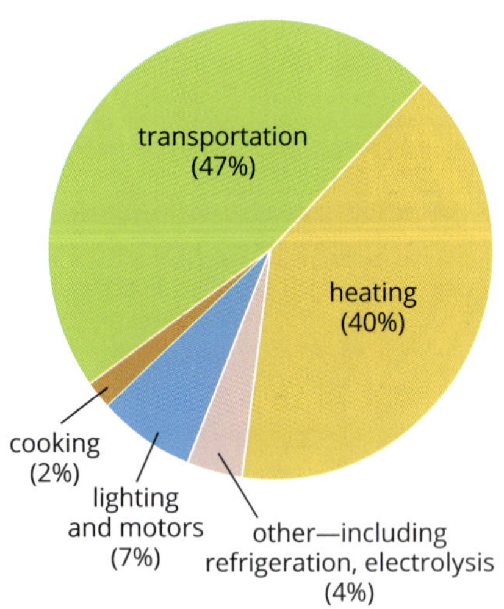

FIGURE 11.1.3 This pie chart shows how energy is used in Australia.

In Australia and around the world, most of the energy used for heating, electricity generation and powering vehicles comes from fossil fuels. About 86% of Australian electricity is generated from these fuels, with 73% from coal and 13% from natural gas.

Coal-fired power stations are the dominant source of the world's electricity because they are often the cheapest form of generation. Electricity from coal-fired power stations is reliable and coal is very abundant.

About 14% of Australia's electricity comes from renewable energy sources. Hydroelectricity contributes 7% of total electricity, and wind, biofuels and solar energy make up the other 7%.

Future energy needs

Burning wood was the dominant method of obtaining energy up to the middle of the 19th century. Wood supplies once seemed unlimited and, like fossil fuels today, satisfied most of the demands of the time.

Fossil fuels (coal, oil and gas) now provide nearly 90% of the world's energy needs. As members of a society that is heavily dependent on fossil fuels as a source of energy, we can at times find it hard to imagine obtaining energy from elsewhere.

The world first became aware that fossil fuels are a finite energy reserve during the 'oil crisis' of the early 1970s. Several Middle Eastern oil exporters restricted production for political reasons. This dramatically increased the cost of crude oil and caused huge increases in the price of petrol around the world.

Given the limited reserves and concerns about the link between fossil fuels and climate change, there is considerable interest in identifying and developing new energy sources. Many countries are already considering alternative sources of energy. The development of alternative sources for large-scale energy production is not a simple task as these replacement sources need to be reliable and cost effective. There is also a growing pressure for new sources to be sustainable and renewable, unlike the currently used fossil fuels. These factors are due to the growing worries about the effect of burning fossil fuels on climate change and the limited supplies of fossil fuels. Figure 11.1.4 shows the increase in world energy production from different sources.

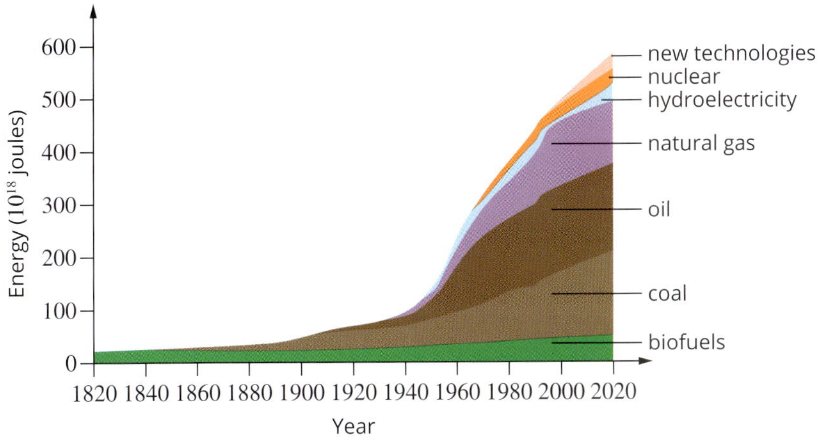

FIGURE 11.1.4 This graph shows the increase in world energy production.

FOSSIL FUELS

Non-renewable resources are those that are used faster than they can be replaced. Coal, oil and natural gas are non-renewable fuels. Reserves of fossil fuels are limited and they could eventually be exhausted.

There is an ongoing conflict between using non-renewable resources and **sustainability**. Sustainability is the combination of meeting the long-term ongoing needs of society, while still meeting the immediate and short-term demands. This concept incorporates the care of the physical environment as well as the economic and social wellbeing of future generations.

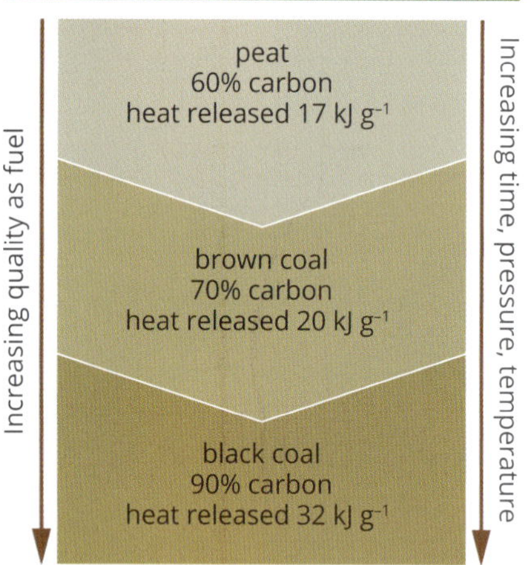

FIGURE 11.1.5 Steps in the formation of coal. Values of the carbon content and heat released upon combustion are for dried coal.

FIGURE 11.1.6 In Australia, brown coal is mined in the Latrobe Valley in Victoria.

Formation of fossil fuels

Coal, oil and natural gas were formed from ancient plants, animals and microorganisms. Buried under tonnes of mud, sand and rock, this once biological material has undergone complex changes to become the fossil fuels used by societies today. The organic matter still retains some of the chemical energy the plants originally accumulated by carrying out **photosynthesis**. Chemical energy in fossil fuels can be considered trapped solar energy.

Fossil fuels form over millions of years. This is why these fuels are considered non-renewable. Once reserves of the fossil fuels have been used, they will not be replaced in the foreseeable future.

Coal

As wood and other plant material turn into coal, gradual chemical changes occur. Wood is about 50% carbon and as it is converted into coal, the carbon content increases and the proportion of hydrogen and oxygen decreases. The wood progressively becomes peat, brown coal and then black coal (Figure 11.1.5). Coal is a mixture of large molecules made from carbon, hydrogen, nitrogen, sulfur and other elements.

The amount of water in coal decreases as these changes occur. When coal is burned, the energy released causes the water to vaporise, reducing the net amount of heat released. Black coal, which contains the least water and therefore the highest percentage of carbon, is a better fuel than brown coal or peat.

Although black coal is usually buried further underground than brown coal, its higher heat value often makes it economical to mine. Black coal is mined in the Collie region in Western Australia. The coal produced there is highly competitive when compared with natural gas and other coal deposits, as it has a low sulfur content and generates low levels of noxious gases during combustion. This black coal is used for domestic power generation as well as exported overseas. More than three-quarters of Australia's electricity is sourced from coal. The reaction that occurs when coal burns can be written as:

$$C(s) + O_2(g) \rightarrow CO_2(g)$$

This equation is a simplification of the electricity-creating process, as coal is not pure carbon. Significant quantities of ash are formed, as well as water vapour and sulfur dioxide.

Large brown coal deposits are located in the Latrobe Valley in Victoria (Figure 11.1.6). The power stations located next to these open-cut mines burn brown coal to generate electricity. Australia is the fifth largest producer of coal and the second largest exporter of coal in the world.

CHEMFILE

Forming crude oil

The main deposits of crude oil were formed from small marine animals (zooplankton) and plants (phytoplankton) that lived up to one billion years ago. Some crude oil deposits are estimated to be even older, as much as three to four billion years old. If a deposit of crude oil were trapped beneath a layer of impermeable rock, then a layer of natural gas would also form.

The first deposits of crude oil were discovered at the end of the 19th century in the United States. Today, the largest crude oil deposits are in Russia, Iran, Iraq and Saudi Arabia. New crude oil deposits are still being found throughout the world. The oldest deposits found so far are in Venezuela, where the oil is estimated to be almost four billion years old. However, only about 10% of the oil discovered is profitable to extract.

Permeable rocks contain tiny spaces through which liquid substances can move. Crude oil has a lower density than water, so oil migrates upwards through permeable rocks over time. Large deposits of oil are formed when portions of this migrating oil become trapped under impermeable rocks. To extract the crude oil, drilling into the impermeable rock has to take place (Figure 11.1.7). In most cases, the oil flows up by itself under high pressure that has gradually built up from when the oil was formed. As the extraction continues, the overall pressure drops and pumps are needed to extract the remaining deposit.

FIGURE 11.1.7 Operating drill during oil and gas exploration

Fuels from crude oil

Crude oil (also called petroleum) is a mixture of hydrocarbon molecules that are mostly alkanes, which were introduced in Chapter 8. Crude oil itself is of no use as a fuel, but its components can be made into many important fuels such as petrol, kerosene, diesel and liquefied petroleum gas (LPG).

The relative amounts of different alkanes in crude oil vary with the deposit. For example, oil from Bass Strait and the Carnarvon Basin, Western Australia, contains relatively few of the larger molecules needed to form lubricants and bitumen.

Crude oil is separated into a range of fractions by fractional distillation. Fractional distillation does not produce pure substances. Each fraction is still a mixture of hydrocarbon compounds. These fractions can be used as fuels, or treated further to produce more specific products through chemical processes.

Alkanes are hydrocarbons with the general formula C_nH_{2n+2}. Alkanes are commonly found in crude oil.

CHEMISTRY IN ACTION

Fractional distillation of crude oil

Fractional distillation uses heat to separate a mixture into a number of different parts or fractions. A number of different temperatures are produced because of the nature of the column in which the crude oil is placed. The temperature of the tower decreases gradually with increasing height. Within the tower are horizontal trays, each containing hundreds of bubble caps. Bubble caps impede (stop) the upwards movement of gases (Figure 11.1.8). As the vapour rises, it forces the caps up and it bubbles through condensed liquid in the trays. Those substances in the vapour that have boiling points almost equal to the temperature of the liquid in the trays condense and are collected. Consequently, fractions collected from trays higher in the tower will have lower boiling points.

The boiling point of a molecular compound depends on the strength of its intermolecular forces, which are stronger with increasing molecular mass. As a result, each fraction consists of alkanes within a specific mass range. Lighter alkanes (lower boiling point) condense near the top of the tower, whereas heavier alkanes (higher boiling point) condense near the bottom. The composition and boiling range of each fraction are summarised in Figure 11.1.8. For example, the petrol fraction that boils (and condenses) between 100°C and 250°C consists of alkanes containing eight to 12 carbon atoms; that is, C_8H_{18} to $C_{12}H_{26}$.

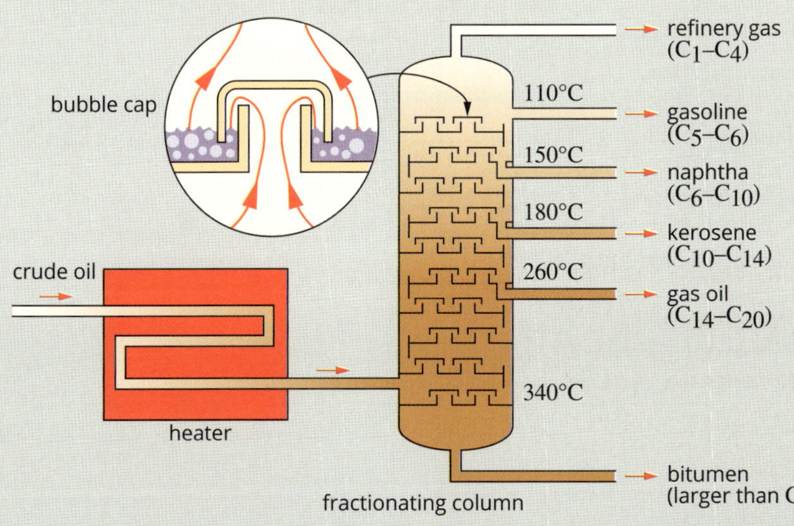

FIGURE 11.1.8 Fractional distillation of oil, showing the bubble caps that stop the upward movement of gases

Oil reserves

Australia contains relatively small amounts of oil reserves compared to the reserves in the Middle East. Figure 11.1.9 shows the current locations of Australian oil, gas and coal fields. These reserves are associated with giant offshore gas fields near the northern Western Australian coast, and in reserves in outback South Australia and in Bass Strait. Western Australia currently produces 71% of Australia's crude oil and this value is anticipated to rise due to new supply sources that have been found. However, Australia's production is set to decline gradually and will likely be exhausted later this century. Australia already imports more than 90% of the crude oil it uses. Importation of large amounts of oil has a significant impact on Australia's economy.

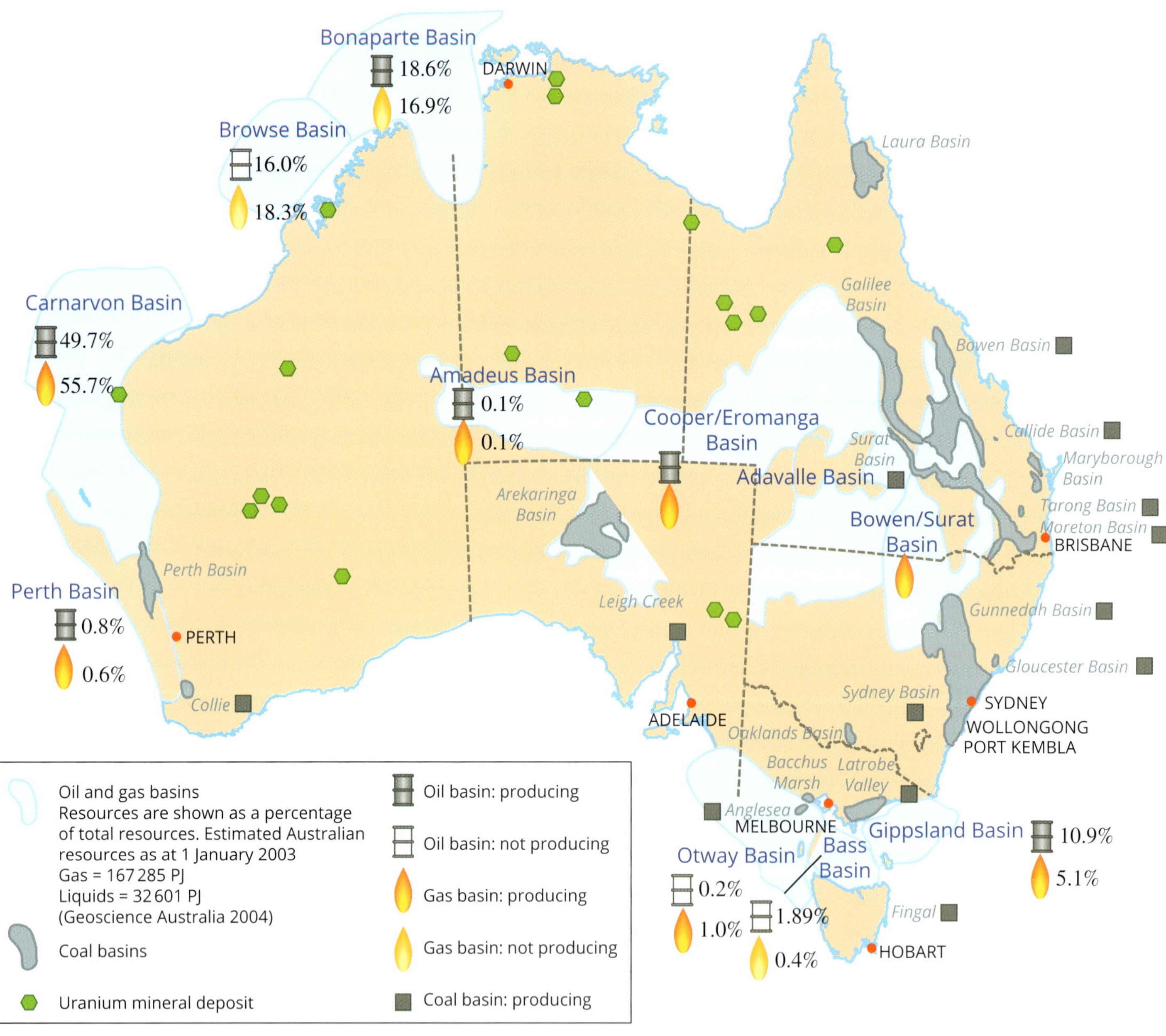

FIGURE 11.1.9 Location of Australian oil, gas and coal fields

Natural gas

Natural gas is another fossil fuel found in deposits in the Earth's crust. It is a popular fuel for home heating and cooking. Natural gas is mainly composed of methane (CH_4) together with small amounts of other hydrocarbons such as ethane (C_2H_6) and propane (C_3H_8). Water, sulfur, carbon dioxide and nitrogen may also be present in natural gas.

Natural gas can be found:

- in gas reservoirs trapped between layers of rocks
- as a component of petroleum deposits
- in coal deposits where it is bonded to the surface of the coal. Coal seams usually contain water and the pressure of the water can keep the gas adsorbed to the coal surface. Natural gas found this way is known as coal seam gas or CSG. It is a major component of the energy supplies of Queensland
- trapped in shale rock, where it is referred to as shale gas. Shale gas is mined in many parts of the United States.

Natural gas is accessed by drilling as with crude oil; drilling causes the natural gas to flow to the surface (Figure 11.1.10).

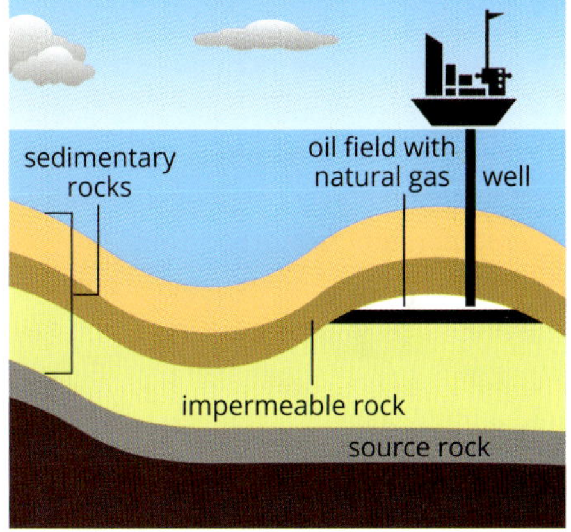

FIGURE 11.1.10 Natural gas deposits are often found trapped above crude oil. Once a well is sunk into the deposit, the natural gas flows to the surface.

Propane and butane gases can be separated from natural gas by fractional distillation. Propane and butane become liquids under pressure and are sold as **liquefied petroleum gas** (LPG). The natural gas remaining after the removal of propane and butane is used widely as a fuel for home heating and cooking.

LPG has applications as a motor fuel and in home gas bottles. As world oil prices rise, natural gas could increasingly replace petrol in cars and other vehicles. LPG is already used to power many of Transperth's buses.

Fracking

The extraction of natural gas from coal or shale deposits usually involves a process called **fracking**. Under pressure, the natural gas is adsorbed on the surface of the coal or shale. Fracking is used to fracture the rock or coal to release the natural gas.

Fracking begins with drilling a well into the deposit to access the trapped gas. The well is encased in steel and concrete to prevent leakage into local water supplies. Fracking fluid is then pumped down the well at extremely high pressures. This high-pressure fluid fractures the surrounding rock or coal, creating fissures through which gas can flow. This process is shown in Figure 11.1.11.

Materials used for fracking include sand, water and other chemicals. There are concerns about the potential impact of this process on the local environment and underground water supplies.

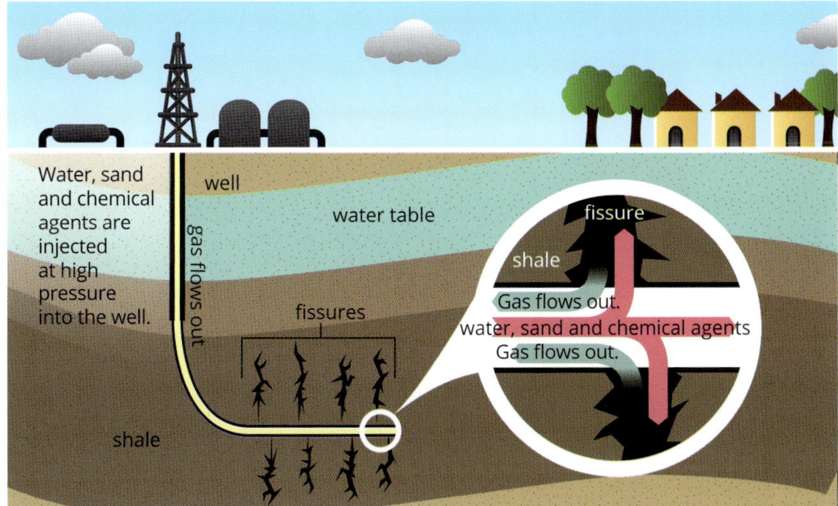

FIGURE 11.1.11 The fracking process: sand, water and other chemicals are injected into the deposit at high pressure to free the gas from the coal or shale.

CHEMISTRY IN ACTION

Debate surrounding coal seam gas

Coal seam gas (CSG) is a natural gas that is extracted from underground coal seams, where it is trapped in pores in the coal. Almost 30% of Australia's natural gas reserves come from coal seam gas. Reserves of CSG are found in multiple areas of Western Australia, as seen in Figure 11.1.12.

Various chemicals have been used as fracking fluids. Fracking fluids increase the permeability of the rock and therefore flow of gas to the surface. Fluids such as benzene, toluene, xylene and ethylbenzene were once commonly used. The use of these fluids has been banned in both New South Wales and Queensland because of concern over their effect on the environment. For example, these potentially carcinogenic compounds may escape and contaminate groundwater.

Water is now commonly used as a fracking fluid. Generally, large amounts of water are not available at the fracking site, so water needs to be transported in, which can have significant economic and environmental costs.

Since 2005, 15 exploration wells have been drilled in Western Australia to search for coal seam gas. The Environmental Protection Agency has not placed a ban on fracking, and instead assesses every possible project on a case-by-case basis. Currently there are still large offshore reserves of oil and gas that Western Australia relies on, however, it is predicted that coal seam gas will become industrialised in the near future, despite the concerns of many regional community members.

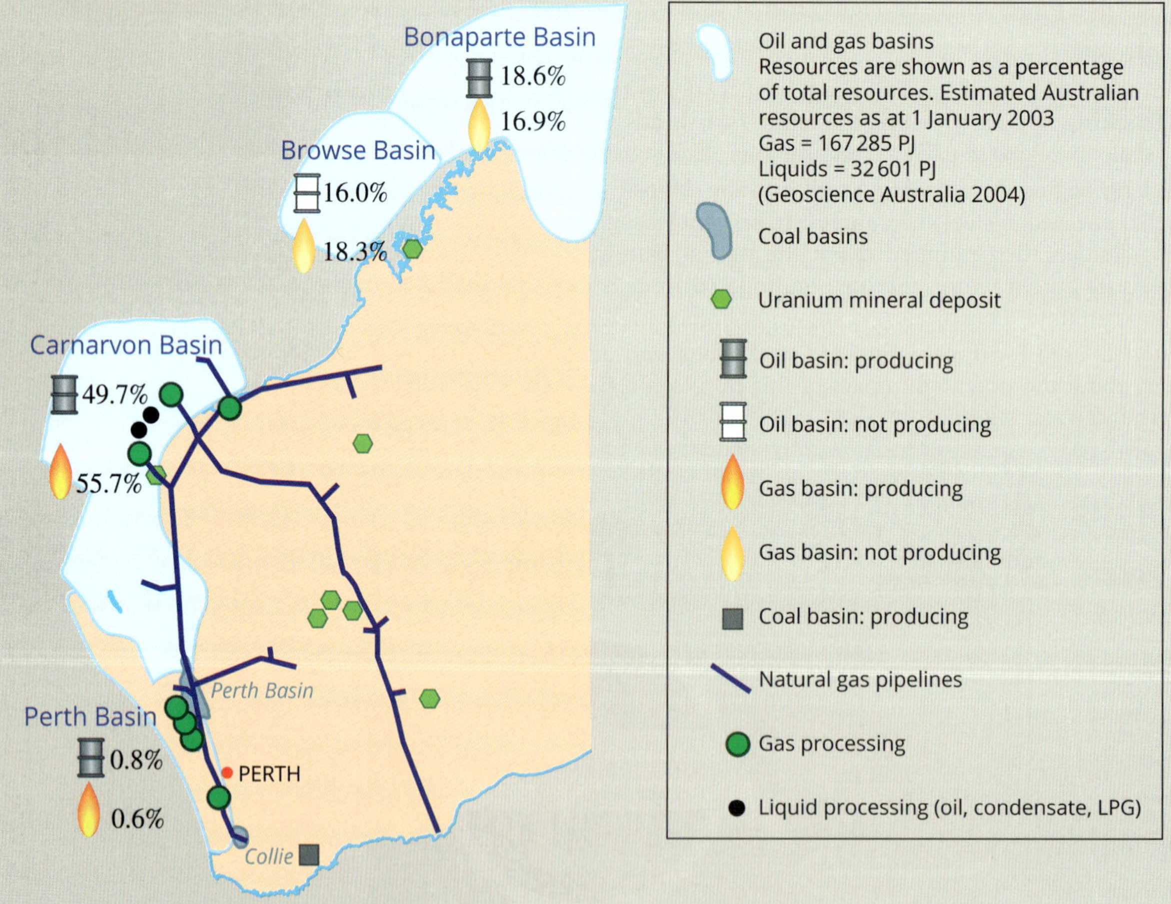

FIGURE 11.1.12 Location of gas basins in Western Australia. The basins on the mainland can be accessed by fracking.

BIOFUELS

Ideally, new sources of energy will be clean and **renewable**. Renewable energy is energy that can be obtained from natural resources that can be constantly replenished.

Biochemical fuels (or **biofuels**) are fuels derived from plant materials such as grains (maize, wheat, barley or sorghum), sugar cane (Figure 11.1.13) and vegetable waste, and vegetable oils. The three common biofuels are **biogas**, **bioethanol** and **biodiesel**. They are used alone or blended with fossil fuels such as petrol and diesel.

FIGURE 11.1.13 Harvesting sugar cane in Queensland. Sugar cane can be a source of the raw materials for the production of bioethanol.

As well as being renewable, biofuels are not considered to contribute significantly to an increase in the atmospheric carbon dioxide levels. The plant materials used in the generation of biofuels are produced by photosynthesis, which removes carbon dioxide from the atmosphere and produces glucose ($C_6H_{12}O_6$) in the following reaction:

$$6CO_2(g) + 6H_2O(l) \rightarrow C_6H_{12}O_6(aq) + 6O_2(g)$$

The plants convert the glucose into cellulose and starch. Although carbon dioxide is released back into the atmosphere when the biofuel is burnt, the net impact should be less than for fossil fuels. The use of biochemical fuels can therefore be described as **carbon neutral**, although there will be some carbon emissions from the manufacture and distribution of biochemical fuels. Therefore, there will always be some net emissions associated with their use.

Bioethanol

For thousands of years, humans have employed biological catalysts (**enzymes**) from yeasts to convert starches and sugars to ethanol. Enzymes catalyse the breakdown of the starch in grain crops (such as barley and wheat) to glucose. **Fermentation** is a process which uses other enzymes from yeasts to convert glucose and other small sugar molecules to ethanol (CH_3CH_2OH) and carbon dioxide:

$$C_6H_{12}O_6(aq) \rightarrow 2CH_3CH_2OH(aq) + 2CO_2(g)$$

Fermentation is an anaerobic process as it takes place in the absence of oxygen.

Bioethanol is used extensively in Australia, including E10 petrol which contains 10% ethanol. This mix can be used by most modern car engines and its use reduces the consumption of petrol derived from crude oil while also reducing the total amount of carbon dioxide that is released into the atmosphere. In addition, the use of E10 fuel reduces the quantity of particles such as unburnt carbon and gases like the oxides of nitrogen that are emitted by exhausts and contribute to air pollution.

Biogas

Biogas is released in the breakdown of organic waste by anaerobic bacteria. These bacteria decompose the complex molecules contained in substances such as carbohydrates and proteins into the simple molecular compounds carbon dioxide and methane. A digester (Figure 11.1.14) is a large tank filled with the anaerobic bacteria that digest the complex molecules to form biogas.

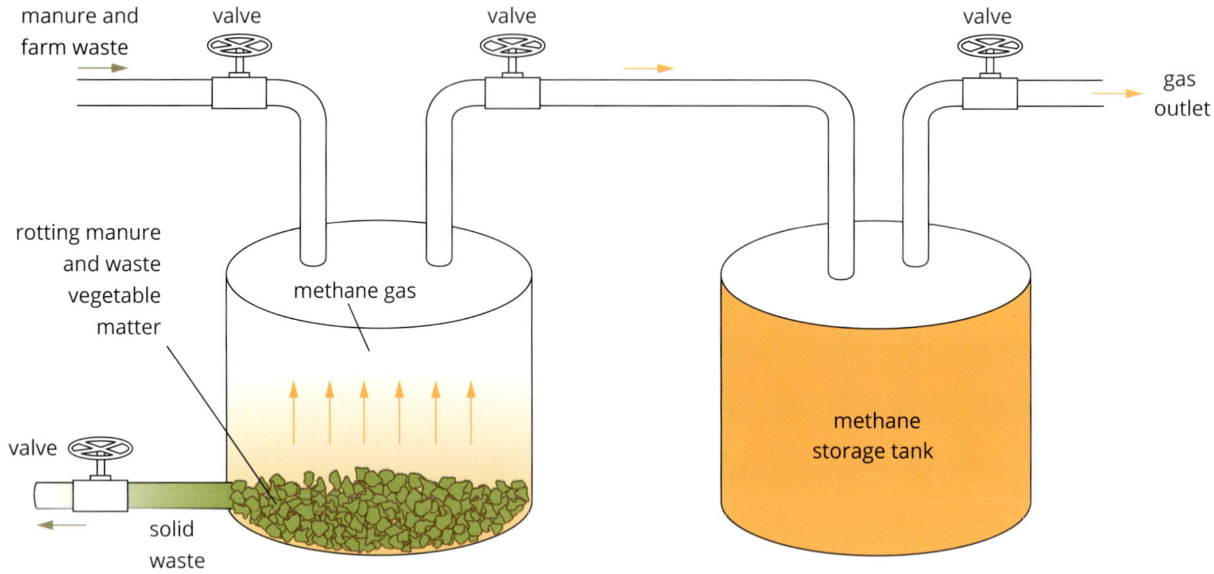

FIGURE 11.1.14 A digester is used in the production of biogas.

FIGURE 11.1.15 Pipes buried in this rubbish tip collect biogas.

A range of materials, including rotting rubbish (such as that seen in Figure 11.1.15) and decomposing plant material, can be used to produce biogas.

The composition of biogas depends on the original material from which it is obtained and the method of decomposition. The typical composition of a sample of biogas is shown in Table 11.1.1.

TABLE 11.1.1 Typical percentage composition of different molecules found in biogas

Gas	Formula	Percentage composition (by volume)
methane	CH_4	60
carbon dioxide	CO_2	32
nitrogen	N_2	4.5
hydrogen sulfide	H_2S	2
oxygen	O_2	1
hydrogen	H_2	0.5

As you can see from Table 11.1.1, biogas consists mainly of methane and carbon dioxide. Biogas can be used for heating and to power homes and farms. There are more than seven million biogas generators in China. Biogas generators are particularly suited to farms, as the waste from a biogas generator makes a rich fertiliser.

In the future, it is likely more energy will be obtained from biogas generated at sewage works, chicken farms, piggeries and food-processing plants. Your local rubbish tip also has the potential to supply biogas. The gas can be used directly for small-scale heating or to generate electricity.

Biodiesel

Biodiesel is a mixture of organic compounds called esters, which are produced by a chemical reaction between vegetable oils or animal fats and an alcohol, most commonly methanol (CH_3OH).

The usual raw material for the production of biodiesel is vegetable oil from sources such as soyabean, canola or palm oil. Recycled vegetable oil or animal fats can also be used. The structure of a typical biodiesel molecule is shown in Figure 11.1.16.

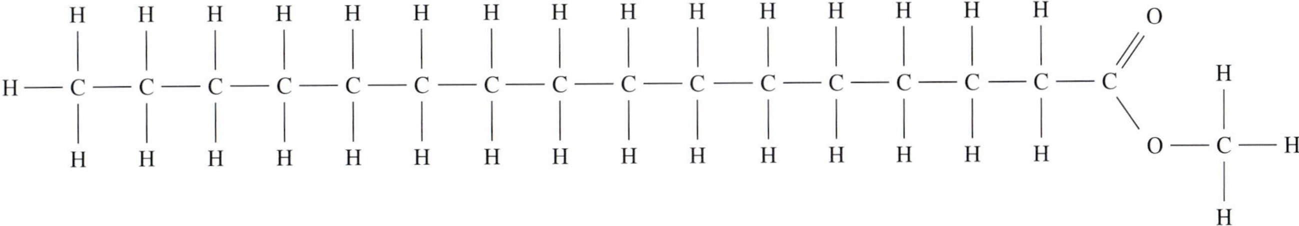

FIGURE 11.1.16 Structural formula of a typical biodiesel molecule

Petrodiesel is a hydrocarbon fraction obtained from crude oil. The chemical and physical properties of biodiesel are very similar to those of petrodiesel, so it is possible to run diesel vehicles on 100% biodiesel fuel.

A diesel engine is a form of internal combustion engine that does not require a spark to ignite the fuel in the engine cylinder. The high energy content of diesel makes it an excellent fuel for transportation and equipment designed to do heavy work. Therefore diesel engines are favoured in situations requiring high fuel efficiencies and for heavy haulage vehicles.

Diesel engines have a 20–40% better fuel economy than petrol engines because:

- diesel engines are typically more energy efficient than petrol engines
- diesel fuel has a higher density than petrol, so although the energy content of diesel and petrol, measured in $kJ\,g^{-1}$, is similar, diesel fuel yields more energy per litre.

Biodiesel is **biodegradable** and non-toxic and produces fewer pollutants in vehicle emissions. Like ethanol, the use of biodiesel does not add to the overall amount of carbon dioxide already present in the atmosphere as the creation of the raw materials that make biodiesel take carbon dioxide from the atmosphere.

CHEMFILE

Rudolf Diesel

Rudolf Diesel (1858–1913) was a German engineer who invented the 'oil engine' that was named after him. His prototype engine first operated in 1893 and was powered by peanut oil. Diesel was well aware of the potential value of an engine that could run on renewable fuel. He demonstrated his engine at the World Exhibition in Paris in 1900 and was awarded the Grand Prix—the highest prize.

Just before his death in 1913, Diesel stated: 'The diesel engine can be fed with vegetable oils and would help considerably in the development of agriculture of the countries which use it.'

Diesel also predicted: 'The use of vegetable oils for engine fuels may seem insignificant today. But such oils may become, in course of time, as important as petroleum and the coal tar products of the present time.'

FIGURE 11.1.17 Rudolf Diesel, inventor of the diesel engine

11.1 Review

SUMMARY

- A fuel is a substance with stored energy that can be released relatively easily.
- A fuel is considered to be non-renewable if it cannot be replenished at the rate at which it is consumed. Fossil fuels such as coal, oil and natural gas are non-renewable.
- Fossil fuels are produced over millions of years by the breakdown of biomass at high temperatures and pressures underground.
- Australia has large reserves of coal and natural gas.
- A fuel is considered to be renewable if it can be replenished at the rate at which it is consumed. Biofuels such as biogas, bioethanol and biodiesel are renewable.
- Biogas is formed by the anaerobic breakdown of organic waste.
- Bioethanol can be produced by fermentation of starches and sugars.
- Biodiesel is produced in a reaction between a vegetable oil or an animal fat and a small alcohol molecule such as methanol.
- Some of the non-renewable and renewable fuels in use in Australia are listed in Table 11.1.2.
- Diesel engines are used in many forms of transport and in heavy-duty equipment.

TABLE 11.1.2 Types of renewable and non-renewable fuels in use in Australia

Non-renewable fuels	Renewable fuels
coal oil liquefied petroleum gas (LPG) natural gas coal seam gas (CSG)	bioethanol biogas biodiesel

KEY QUESTIONS

1. What is the difference between a renewable and non-renewable fuel?
2. Give an example of a renewable fuel source and a non-renewable fuel source used in Australia.
3. In Australia, which resource is likely to last longest before it is depleted: coal, oil or natural gas? Explain your answer.
4. Wood from forests is a renewable resource that supplied global energy needs for thousands of years.
   a. Why is wood no longer sustainable as the major energy source for today's society?
   b. Is it possible to have a non-renewable and sustainable energy source? Explain.
5. Why is it necessary to treat crude oil by fractional distillation?
6. Why are CO_2 emissions from the use of biodiesel not considered as problematic as those produced from the use of petrodiesel?

11.2 Combustion reactions

Fuels contain stored chemical energy that can be harnessed to perform useful functions. The release of this energy occurs when the fuel is burnt in the presence of oxygen, a process known as **combustion**. The heat energy released during combustion provides heat for warmth and cooking, as well as acting as the source of electrical energy and mechanical energy for transport.

In this section, you will look at different fuels used to produce electricity or power vehicles and compare the environmental impact of these fuels.

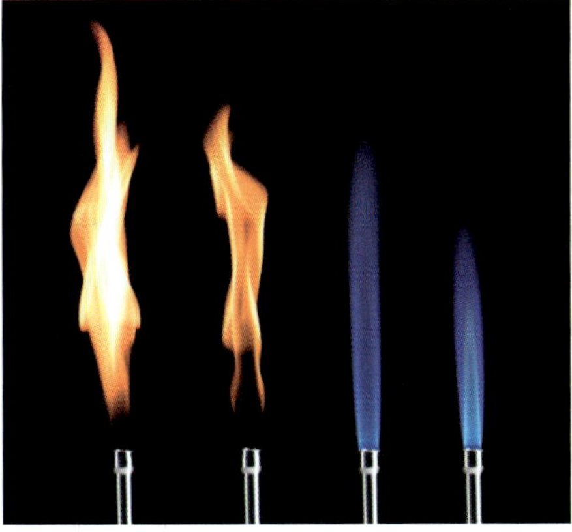

FIGURE 11.2.1 The yellow flame of a Bunsen burner is due to incomplete combustion and produces carbon as a product. The blue flame is a hotter flame that occurs when the collar hole is open and more oxygen is allowed into the reaction. Complete combustion can then occur.

COMBUSTION AS A CHEMICAL PROCESS

Combustion reactions are exothermic reactions in which the reactant combines with oxygen to produce oxides. The combustion of a hydrocarbon produces carbon dioxide and water, provided there is enough oxygen present. An example is the combustion of propane (C_3H_8), a major component of LPG, shown in the following equation:

$$C_3H_8(g) + 5O_2(g) \rightarrow 3CO_2(g) + 4H_2O(g) \qquad \Delta H = -2220\,kJ\,mol^{-1}$$

Note the following important features of this thermochemical equation.

- No atoms are created or destroyed in the reaction, so the numbers of atoms of each element are balanced on the two sides of the equation.
- The balanced equation tells you the number of molecules of oxygen that are required for the combustion of each molecule of propane.
- The enthalpy change for the reaction is negative, indicating that this is an exothermic reaction.
- The enthalpy change is given in kJ mol^{-1}. It tells you the energy released, in kJ, according to the coefficients given in the equation. (In this case, 1 mole of propane is combusted.)
- State symbols show the state of each reactant and product. (In combustion reactions, water is released as water vapour—a gas.)

Combustion of hydrocarbons

Combustion reactions can be described as complete or incomplete. The difference between the two is due to the amount of oxygen available to react with the fuel.

Complete combustion occurs when oxygen is plentiful. The only products are carbon dioxide and water.

An example is the complete combustion of methane:

$$CH_4(g) + 2O_2(g) \rightarrow CO_2(g) + 2H_2O(g)$$

When the oxygen supply is limited, incomplete combustion occurs. As less oxygen is available, not all of the carbon can be converted into carbon dioxide. Carbon monoxide and/or carbon are produced instead. The hydrocarbon burns with a yellow, smoky or sooty flame, due to the presence of glowing carbon particles. Figure 11.2.1 shows the appearance of the different flames of a Bunsen burner due to incomplete and complete combustion.

The equation for the incomplete combustion of methane to form carbon monoxide is:

$$2CH_4(g) + 3O_2(g) \rightarrow 2CO(g) + 4H_2O(g)$$

The complete combustion of hydrocarbons occurs when there is sufficient oxygen for the fuel to burn. The products of complete combustion are carbon dioxide and water. When oxygen is not plentiful, incomplete combustion occurs. The products of incomplete combustion are carbon monoxide and/or carbon and water.

CHEMFILE

Carbon monoxide poisoning

Carbon monoxide is a highly poisonous gas. It combines readily with haemoglobin, the oxygen carrier in blood. When attached to carbon monoxide, haemoglobin cannot transport oxygen around the body, which leads to oxygen starvation of tissues.

Even at concentrations as low as 10 parts per million (ppm), carbon monoxide can cause drowsiness, dizziness and headaches. At about 200 ppm, carbon monoxide can lead to death. The average carbon monoxide concentration in large cities, mostly due to incomplete combustion of fuels in cars, is now 7 ppm, but it can be as high as 120 ppm at busy intersections in heavy traffic.

FIGURE 11.2.2 Car exhaust gases can contain high levels of carbon monoxide as a result of incomplete combustion of fuels.

WRITING EQUATIONS FOR THE COMPLETE COMBUSTION OF FUELS

Chemical equations tell us a lot about reactions and therefore it is important that we write them correctly. Writing equations for the complete combustion of fuels containing carbon and hydrogen is relatively straightforward, because the products are always carbon dioxide and water.

Perhaps the most important of all combustion reactions involving fuels are those that occur when petrol is burnt. Petrol is a mixture of hydrocarbons, including octane.

The combustion reactions of octane (C_8H_{18}) and the other hydrocarbons in petrol power the internal combustion engines in most of Australia's 17.6 million motor vehicles.

Worked example 11.2.1

WRITING EQUATIONS FOR COMPLETE COMBUSTION OF HYDROCARBON FUELS

Write the equation, including state symbols, for the complete combustion of butane (C_4H_{10}).	
Thinking	**Working**
Add oxygen as a reactant and carbon dioxide and water as the products.	$C_4H_{10} + O_2 \rightarrow CO_2 + H_2O$
Balance carbon and hydrogen atoms, based on the formula of the hydrocarbon.	$C_4H_{10} + O_2 \rightarrow 4CO_2 + 5H_2O$
Find the total number of oxygen atoms on the product side.	Total O $= (4 \times 2) + 5$ $= 13$
If this is an odd number, multiply all of the coefficients in the equation by two, except for the coefficient of oxygen.	$2C_4H_{10} + O_2 \rightarrow 8CO_2 + 10H_2O$
Balance oxygen by adding the appropriate coefficient to the reactant side of the equation.	$2C_4H_{10} + 13O_2 \rightarrow 8CO_2 + 10H_2O$
Add state symbols.	$2C_4H_{10}(g) + 13O_2(g) \rightarrow 8CO_2(g) + 10H_2O(g)$

Worked example: Try yourself 11.2.1

WRITING EQUATIONS FOR COMPLETE COMBUSTION OF HYDROCARBON FUELS

Write the equation, including state symbols, for the complete combustion of hexane (C_6H_{14}).

A similar series of steps can also be used to write the combustion equations for other carbon-based fuels that contain oxygen; for example, alcohols.

Worked example 11.2.2

WRITING EQUATIONS FOR COMBUSTION REACTIONS OF ALCOHOLS

Write the equation, including state symbols, for the complete combustion of liquid ethanol (C_2H_5OH).	
Thinking	**Working**
Add oxygen as a reactant and carbon dioxide and water as the products.	$C_2H_5OH + O_2 \rightarrow CO_2 + H_2O$
Balance carbon and hydrogen atoms, based on the formula of the alcohol.	$C_2H_5OH + O_2 \rightarrow 2CO_2 + 3H_2O$
Find the total number of oxygen atoms on the product side. Then subtract the one oxygen atom in the alcohol molecule from the total number of oxygen atoms on the product side.	Total O on product side $= (2 \times 2) + 3$ $= 7$ Total O on product side − 1 in alcohol $= 7 - 1 = 6$
If this is an odd number, multiply all the coefficients in the equation by two, except for the coefficient of oxygen.	$C_2H_5OH + O_2 \rightarrow 2CO_2 + 3H_2O$
Balance oxygen by adding the appropriate coefficient to the reactant side of the equation.	$C_2H_5OH + 3O_2 \rightarrow 2CO_2 + 3H_2O$
Add state symbols.	$C_2H_5OH(l) + 3O_2(g) \rightarrow 2CO_2(g) + 3H_2O(g)$

Worked example: Try yourself 11.2.2

WRITING EQUATIONS FOR COMBUSTION REACTIONS OF ALCOHOLS

Write the equation, including state symbols, for the complete combustion of liquid methanol (CH_3OH).

WRITING EQUATIONS FOR INCOMPLETE COMBUSTION OF FUELS

When the supply of oxygen is insufficient, the combustion of fuels might not be complete. These incomplete combustion reactions can also be represented by equations. In general, for the incomplete combustion of hydrocarbons, as well as carbon-based fuels that contain oxygen, the products are carbon monoxide and/or carbon and water.

Worked example 11.2.3

WRITING EQUATIONS FOR INCOMPLETE COMBUSTION OF FUELS

Write an equation, including state symbols, for the incomplete combustion of ethane gas (C_2H_6) to form carbon monoxide and water vapour.	
Thinking	**Working**
Add oxygen as a reactant and carbon monoxide and water as the products.	$C_2H_6 + O_2 \rightarrow CO + H_2O$
Balance the carbon and hydrogen atoms, based on the formula of the hydrocarbon.	$C_2H_6 + O_2 \rightarrow 2CO + 3H_2O$
Balance oxygen by adding the appropriate coefficient to the reactant side of the equation.	$C_2H_6 + \frac{5}{2}O_2 \rightarrow 2CO + 3H_2O$
If oxygen gas has a coefficient that is half of a whole number, multiply all of the coefficients in the equation by two.	$2C_2H_6 + 5O_2 \rightarrow 4CO + 6H_2O$
Add state symbols.	$2C_2H_6(g) + 5O_2(g) \rightarrow 4CO(g) + 6H_2O(g)$

Worked example: Try yourself 11.2.3

WRITING EQUATIONS FOR INCOMPLETE COMBUSTION OF FUELS

Write an equation, including state symbols, for the incomplete combustion of liquid methanol (CH_3OH) to form carbon monoxide and water vapour.

ENVIRONMENTAL IMPACT

A discussion of the environmental impact of fuels needs to consider both the impact of emissions from the combustion of the fuel, and the impact on the environment of obtaining the fuel in the first place.

Emissions from fuel combustion

Carbon dioxide

As you have seen, carbon dioxide is produced when carbon-based fuels, including wood, coal, gas and ethanol, undergo combustion. Carbon dioxide is vital to life on Earth. Through the process of photosynthesis, in the presence of chlorophyll (a plant pigment) and sunlight, green plants convert carbon dioxide and water to glucose and oxygen. This glucose produced is the source of energy that all aerobic (oxygen-using) life forms utilise for energy.

Because large quantities of fuel are burnt every day to meet society's energy needs, the level of carbon dioxide production is high. Carbon dioxide is not a toxic gas at normal concentrations in our atmosphere, however, it can have many negative impacts on our environment as a whole.

When carbon dioxide dissolves in water, the majority of it will remain as unreacted $CO_2(aq)$. However, a small amount will react with water to form carbonic acid, H_2CO_3:

$$CO_2(aq) + H_2O(l) \rightarrow H_2CO_3(aq)$$

This will cause a slight acidity to large water bodies, and as such is the reason that rainwater is slightly acidic. An increase in atmospheric carbon dioxide concentration will cause an increase in the acidity of the oceans. This process is called **acidification**.

Due to its molecular structure, carbon dioxide can absorb and re-emit infrared radiation, also known as heat. This feature makes it a **greenhouse gas**.

Energy from the Sun heats the surface of the Earth. The Earth in turn radiates energy back towards space but greenhouse gases in the atmosphere absorb and re-radiate the energy in a process known as the **greenhouse effect**. The higher the concentration of a greenhouse gas, the more energy is trapped.

The greenhouse effect occurs naturally due to the gases present in the atmosphere. It has enabled the Earth to maintain its average temperature over time. Without the natural greenhouse effect moderating global temperatures, life would not be possible on Earth. However, scientists are concerned that increasing levels of greenhouse gases produced by our use of fossil fuels are causing global warming and triggering consequential shifts in weather patterns and climate. This is referred to as the **enhanced greenhouse effect**.

> The greenhouse effect is caused by heat being trapped in the Earth's atmosphere by greenhouse gases, which causes an increase in temperatures at the Earth's surface. As the amount of greenhouse gases in the Earth's atmosphere increases due to human activities, more heat is trapped, which is predicted to cause global changes in climate.

Methane, water vapour, nitrogen oxides and ozone are also greenhouse gases. Methane is 21 times more effective at trapping heat than carbon dioxide. The way in which greenhouse gases restrict heat radiation leaving the Earth is shown in Figure 11.2.3.

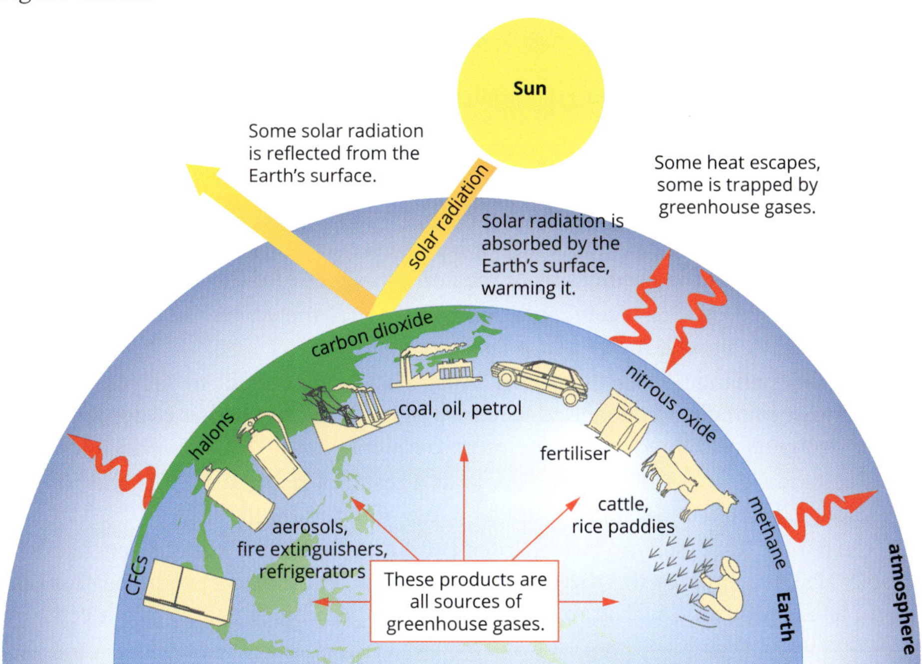

FIGURE 11.2.3 The greenhouse effect. Greenhouse gases help to maintain the temperature at the Earth's surface. Increased quantities of these gases as a result of human activities create an enhanced greenhouse effect.

The graph shown in Figure 11.2.4 on page 238 shows that the Earth is warming. This dramatic rise coincided with the Industrial Revolution in the late 18th and early 19th centuries. Presently, the concentration of carbon dioxide in the Earth's atmosphere is rising by 0.5% per year. There is overwhelming scientific evidence to conclude that the use of fossil fuels by industrialised countries is the main contributing factor to this increase. Many countries are choosing alternatives to fossil fuels to address these fears.

A further impact on the atmospheric carbon dioxide levels is the continued clearing of forests, particularly in tropical areas. Carbon dioxide is released when trees are burnt or left to decay. In addition, the clearing of the forests will result in less carbon dioxide being removed from the atmosphere by photosynthesis.

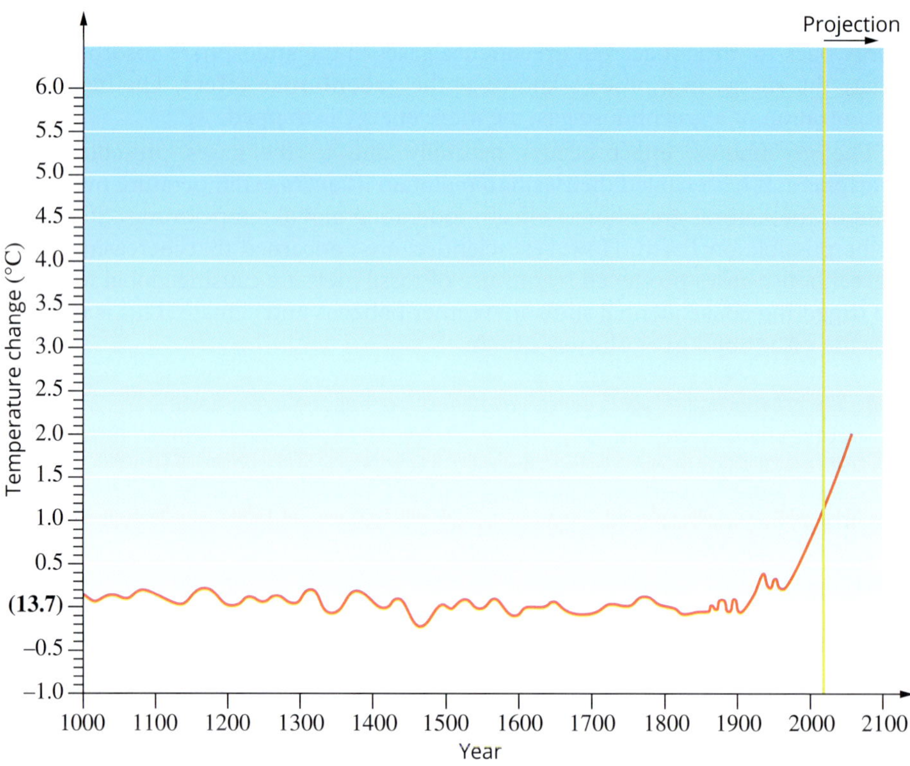

FIGURE 11.2.4 Change in the average surface temperature of the Earth from the year 1000 projected to 2100. Estimates of future temperature increases vary, depending on what assumptions are made.

Global temperatures are on course to rise between 2 and 4 degrees by the end of the century due to global warming. The agriculture in warm regions of the world, including Australia, could feel catastrophic consequences of this temperature rise. These are some of the predictions of the consequences of rapid global warming.

- Rising sea levels. Sea levels could rise by up to 1.2 metres by the year 2100. This is in part due to the melting of polar ice caps.
- Climatic changes. South-eastern Australia is predicted to have increased summer rainfall and decreased winter rainfall. Extreme weather conditions, such as droughts and floods, could be more frequent.
- Ecological changes. These may occur as a result of rising sea levels and climatic change. For example, there could be a loss of natural wetlands.

Each fuel discussed in this chapter produces carbon dioxide when it burns. Table 11.2.1 compares the theoretical mass of carbon dioxide produced from the complete combustion of 1 gram of each fuel and per unit of energy produced.

TABLE 11.2.1 Mass of CO_2 produced from the combustion of 1 gram of fuel and per megajoule of energy produced

Fuel	Mass of CO_2 (g) emitted per gram of fuel	Mass of CO_2 (g) per megajoule of energy produced (approx.)
coal	3.7	93
natural gas	2.8	56
LPG	3.0	65
petrol	3.1	73
ethanol	1.9	72
petrodiesel	3.4	71
biodiesel	2.8	75

CHEMFILE

Cleaner coal

Research is being conducted into ways of reducing carbon dioxide emissions from coal-fired power stations. These include:

- storing carbon dioxide deep underground (geosequestration)
- removing water from brown coal by heating; the volatile hydrocarbons that are driven off are also burnt to produce electricity
- absorbing carbon dioxide into solvents (chemical capture).

There are several methods for converting coal to liquid and gaseous fuels. Flash pyrolysis has been developed by CSIRO for converting coal to oil. In this process, crushed coal is heated to 600°C. The tar that forms is reacted with hydrogen to give a type of crude oil.

Other emissions

Carbon dioxide and water vapour are not the only products formed when fuels are burnt. Fuels may contain elements such as sulfur, which burns to form sulfur dioxide, or the high temperatures of combustion can lead to reactions with nitrogen in air. Table 11.2.2 lists the more common pollutants formed.

TABLE 11.2.2 Pollutants formed from fuel combustion

Emission	Formation	Comments
sulfur dioxide (SO_2)	sulfur in fuel reacts with oxygen: $S(s) + O_2(g) \rightarrow SO_2(g)$	• causes respiratory problems • leads to the formation of **acid rain**
nitrogen oxides (NO and NO_2)	nitrogen in fuel reacts with oxygen, or nitrogen in air reacts due to high temperatures: $N_2(g) + O_2(g) \rightarrow 2NO(g)$ $2NO(g) + O_2(g) \rightarrow 2NO_2(g)$	• causes respiratory problems • leads to the formation of other pollutants and acid rain
ozone (O_3)	nitrogen oxides react with oxygen at high temperatures	• causes respiratory problems
particulates, such as ash	combustion of impurities in fuel	• irritant • adheres to houses and plants
carbon monoxide	incomplete combustion of fuel	• poisonous gas
organic chemicals, such as methanal and ethanal	reactions of other organic chemicals in fuel	• toxic and carcinogenic compounds

The level of each pollutant mentioned in Table 11.2.2 varies with the composition of the fuel and the efficiency of the particular combustion process. However, it is possible to generalise.

- Ash from burning is usually more of a problem with coal than other fuels. Ash is produced when coal is burnt without any purification or removal of impurities.
- Sulfur levels are lower in natural gas and LPG than in liquid or solid fossil fuels.
- The molecules in petrol are larger than the molecules in natural gas, LPG and ethanol. As a consequence, the combustion of petrol tends to be less complete. This produces more carbon monoxide and particulates.
- Although biofuels are renewable, they can still produce the same pollutants as fossil fuels when burnt.

Environmental impact of sourcing the fuel

All fuels have to be mined or produced and the environmental impact of sourcing a fuel needs to be considered. Table 11.2.3 shows some of the sources of the fuels described in this chapter.

TABLE 11.2.3 Sources of fuels

Fuel source	Notes
 Biogas collected under domes from a sewage plant	Biogas is often collected from sewage farms and rubbish tips. The gas collection minimises emissions associated with these sites. Because methane is much more effective as a greenhouse gas than carbon dioxide, it is better to collect the methane in biogas and combust it to produce carbon dioxide than to release it directly to the atmosphere.
 Oil rig	Oil rigs, once in place, can operate with low impact on the environment, but the issues created when a spill or explosion occurs can be significant.
  Open-cut coal mine	Coal mines can be open cut or underground. Open-cut mines, such as in the Latrobe Valley in Victoria, are damaging to the local environment.
 Harvesting a wheat crop for bioethanol production	Bioethanol is produced from crops. Growing crops requires energy expenditure and the use of resources such as water and fertiliser. Intensive farming can lead to land degradation and erosion. These are larger issues if crops are grown solely to produce ethanol, but of less concern if waste from food crops is used as a raw material. Diverting crops from fuel production could also drive up the cost of food produced from those crops.

CHEMFILE

The Gorgon Gas Project

The $74.7 billion Gorgon Gas Project (Figure 11.2.5) is located on Barrow Island, a class A nature reserve 85 km off the Pilbara coast of Western Australia. At this site, liquid natural gas (LNG) is processed from nearby subsea fields for exportation and some domestic use. The project was approved by the Environmental Protection Agency (EPA) in 2007, with many stipulations placed on the project's construction and operation phases. These protections placed limits on the amount of pollutants, like CO_2, mercury, nitrous oxides and hydrogen sulfides, that would be produced at the plant.

FIGURE 11.2.5 The Gorgon Gas Project, Barrow Island, Western Australia

Additionally, Barrow Island is home for 24 animal species that are native to the island (including the Barrow Island Euro, shown in Figure 11.2.6) and others which are classified as vulnerable or endangered, as well as a large variety of flora. This large array of biodiversity is in part due to the isolation of the island, which was separated from mainland Australia about 8000 years ago. This isolation has meant that no introduced predators to Australia have reached the island. As a result, companies that are involved in the project have designed the plant and other amenities to reduce the impact on the environment, as well as keeping a tally on the local animal populations and promoting a sustainable approach to biodiversity in the wider community.

FIGURE 11.2.6 The Gorgon Gas Project has been designed to reduce the environmental impact on Barrow Island's flora and fauna, such as this Barrow Island Euro.

11.2 Review

SUMMARY

- Fuels such as petrol, natural gas, biogas and bioethanol undergo combustion reactions in excess oxygen to form carbon dioxide and water.
- The combustion of fuels can produce a range of other pollutants such as carbon monoxide, sulfur dioxide, nitrogen oxides and particulates.
- Petrol produces more energy per gram than LPG or bioethanol. However, bioethanol and LPG produce less carbon dioxide and particulates in emissions.
- Biofuels offer several environmental advantages: CO_2 is absorbed during the growth of crops used in their production, they can be replenished and they can be produced from material that would have otherwise been waste.
- A shift to large-scale production of biofuels could place a strain on resources and available farmland.

KEY QUESTIONS

1. Write a balanced equation for the complete combustion of liquid benzene (C_6H_6).
2. Write a balanced equation for the incomplete combustion of ethanol (C_2H_5OH) when carbon monoxide is formed.
3. Write a balanced equation for the combustion of hydrogen sulfide (H_2S) in excess oxygen.
4. How many molecules of oxygen are required to react with each molecule of heptane when it burns completely?
5. Describe how the following human activities affect the concentration of carbon dioxide in the atmosphere.
   - **a** burning biofuels
   - **b** burning petrol
   - **c** clearing native forest
   - **d** growing crops to produce biofuels
   - **e** using windpower rather than fossil fuels
   - **f** increased industrialisation
6. Explain the effects that the following gases have in the atmosphere.
   - **a** carbon dioxide
   - **b** carbon monoxide
   - **c** methane
   - **d** water vapour
   - **e** sulfur dioxide

11.3 Calculations involving fuels

COMPOSITION OF FUELS

The amount of carbon present in a fuel will influence the amount of energy released by the fuel, the amount of oxygen required for the fuel to combust completely and the amount of carbon dioxide released during the combustion of the fuel. Therefore, we need to know the composition of fuels, or more specifically, the amount of carbon in the fuel, to be able to calculate the energy output of the fuel as well as the carbon dioxide emissions.

Worked example 11.3.1

CALCULATING THE CARBON CONTENT IN COAL

Calculate the mass of carbon present in 1600 tonnes of brown coal that has a carbon content of 75.0%.	
Thinking	**Working**
Change the percentage to a fraction and multiply by the mass of the substance.	$m(\text{C}) = \frac{75.0}{100} \times 1600 = 1120$ tonnes

Worked example: Try yourself 11.3.1

CALCULATING THE CARBON CONTENT IN COAL

Calculate the mass of carbon present in 1300 tonnes of black coal that has a carbon content of 90.0%.

The mass of carbon in compounds used as fuels can be calculated using the formula of the compound and the relative atomic masses (standard atomic weights) of the elements contained in the fuels. Knowing the IUPAC system for naming alkanes will help you solve these problems.

Worked example 11.3.2

CALCULATING THE CARBON CONTENT IN PURE COMPOUNDS

Calculate the mass of carbon in 20.0 kg of octane.	
Thinking	**Working**
Write the formula of the compound and calculate the relative molecular mass.	$M(\text{C}_8\text{H}_{18}) = (8 \times 12.01) + (18 \times 1.008)$ $= 114.2$
Calculate the percentage of carbon in this total.	$\%(\text{C}) = \frac{8 \times 12.01}{114.2} \times 100$ $= 84.13\%$
Calculate the unknown mass of carbon using this percentage.	$m(\text{C}) = \frac{84.13}{100} \times 20.0\text{ kg}$ $= 16.8\text{ kg}$

Worked example: Try yourself 11.3.2

CALCULATING THE CARBON CONTENT IN PURE COMPOUNDS

Calculate the mass of carbon in 25.0 kg of butanol, C_4H_9OH.

> ℹ The heat of combustion is usually measured at conditions of 298 K and 100 kPa.

> ℹ Only fuels that exist as pure substances can have their heat of combustion measured in $kJ\,mol^{-1}$.

ENERGY CONTENT OF FUELS

All substances contain chemical energy. The **heat of combustion** of a fuel is defined as the enthalpy change that occurs when a specified amount (e.g. 1 g, 1 L, 1 mol) of the fuel burns completely in oxygen. Therefore the energy content of a fuel used in combustion can be represented by the fuel's heat of combustion. Heat of combustion is usually measured at conditions of 298 K (25°C) and 100 kPa, (which means that the water produced should be shown in the liquid state). The heat of combustion can be given the symbol ΔH_c. The chemical energy available from a substance is referred to as its **energy content**. Fuels are examples of substances with high energy contents.

Many fuels, including wood, coal and kerosene, are mixtures of chemicals and do not have a specific chemical formula or molar mass. This means their heat of combustion cannot be expressed in $kJ\,mol^{-1}$. Therefore, it is measured only as $kJ\,g^{-1}$, $kJ\,L^{-1}$ or MJ/tonne.

The heats of combustion for some common elements and compounds present in fuels are listed in Table 11.3.1. Heat energy is released during combustion, so ΔH_c always has a negative value.

Note that these are theoretical values and, when these fuels are combusted, not all the energy is released as heat energy, and not all of the heat energy liberated can be harnessed for its intended use. However, even if the energy transformation is not 100% efficient, we can still use these values to compare the energy released from the combustion of different compounds.

TABLE 11.3.1 Heats of combustion for some common elements and compounds

Substance	Heat of combustion, ΔH_c ($kJ\,mol^{-1}$)
methane	−890
ethane	−1560
propane	−2220
butane	−2886
octane	−5450
methanol	−725
ethanol	−1367
hydrogen	−286
carbon (graphite)	−394

As Worked example 11.3.3 shows, you can use the data in Table 2.3.1 to calculate the energy released on combustion of a specified mass of one of the fuels. The energy released when n mol of a fuel burns is given by the equation:

$$\text{energy} = n \times \Delta H_c$$

Worked example 11.3.3

CALCULATING ENERGY RELEASED BY A SPECIFIED MASS OF A PURE FUEL

Calculate the amount of energy released when 3.60 kg of butane (C_4H_{10}) is burnt in an unlimited supply of oxygen.

Thinking	Working
Calculate the number of moles of the compound using $n = \frac{m}{M}$.	$n(C_4H_{10}) = \frac{m}{M}$ $= \frac{3.60 \times 10^3}{58.12}$ $= 61.94$
Multiply the number of moles by the heat of combustion. (Give answer to 3 significant figures.)	energy $= n \times \Delta H_c$ $= 61.94 \times 2886$ $= 1.79 \times 10^5$ kJ

Worked example: Try yourself 11.3.3

CALCULATING ENERGY RELEASED BY A SPECIFIED MASS OF A PURE FUEL

Calculate the amount of energy released when 5.40 kg of propane (C_3H_8) is burnt in an unlimited supply of oxygen.

Energy content per gram

The energy content of a pure substance can be represented in units of kilojoules per gram, as well as in kilojoules per mole.

For a pure substance, the heat of combustion per gram can be calculated by simply dividing the heat of combustion per mole ($kJ\,mol^{-1}$) by the molar mass of the substance.

For example, for ethanol (which is being burnt in a spirit burner in Figure 11.3.1):

heat of combustion per mole = $-1367\,kJ\,mol^{-1}$

molar mass = $46.068\,g\,mol^{-1}$

heat of combustion per gram = $\frac{-1367}{46.0} = -29.7\,kJ\,g^{-1}$

For fuels that are mixtures, approximate values for the heat of combustion per gram are shown in Table 11.3.2.

Energy content per tonne

The quantities of fuel consumed and the amount of energy produced are often so large that the unit megajoules per tonne, MJ/tonne, is a more useful way of expressing the heat of combustion. For example, the Hazelwood power station in Victoria consumed about 13 million tonnes of coal in one year prior to being shut down.

$1\,kJ = 10^{-3}\,MJ$

$1\,g = 10^{-6}\,tonne$

Therefore, $1\,kJ\,g^{-1} = 10^{-3}\,MJ/10^{-6}\,t$

This can be simplified to $1\,kJ\,g^{-1} = 10^{3}\,MJ/tonne$.

FIGURE 11.3.1 Ethanol burning in a spirit burner. Combustion of 1 gram of ethanol releases almost 30 kJ of energy.

TABLE 11.3.2 Approximate heat of combustion values for some fuel mixtures

Substance	Heat of combustion ($kJ\,g^{-1}$) (approx.)
wood, dried	–18
peat, dried	–25
brown coal, dried	–30
black coal, dried	–35

EXTENSION

Energy density

Liquid fuels, such as petrol, are normally sold by volume rather than by mass. For these fuels, it is often convenient to refer to the heat of combustion per litre. Table 11.3.3 shows heat of combustion per litre for some common liquid fuels. The energy released when one litre of fuel undergoes complete combustion is often called the fuel's energy density.

TABLE 11.3.3 Energy densities for some common liquid fuels

Substance	Energy density ($kJ\,L^{-1}$)
petrol (unleaded 91)	–34 200
kerosene	–36 500
diesel fuel	–38 000
heating oil	–38 500
ethanol	–23 400

STOICHIOMETRY AND REACTIONS OF FUELS

Due to the growing concerns of fuel combustion on the Earth's environment, it is important to be able to compare the amount of carbon dioxide and other toxic products that will be released from the combustion of different fuel options. Scientists use calculations to predict the amount of carbon dioxide released into the atmosphere when carbon-based fuels are used, or the amount of toxic products that would be produced from fuels that contain impurities such as sulfur.

These calculations, which involve the use of the mole concept (Chapter 9), combined with an understanding of chemical equations, are called **stoichiometric calculations**. In this chapter, these calculations will be used for reactions involving fuels, but stoichiometry is important in all areas of chemistry.

Stoichiometry is the study of ratios of moles of substances. Stoichiometric calculations are based on the law of conservation of mass: *In a chemical reaction, the total mass of all products is equal to the total mass of all reactants.*

Another way of expressing this law is that, in a chemical reaction, atoms are neither created nor destroyed.

Equations and reacting masses

Consider the equation for the reaction that occurs when methane burns in oxygen:

$$CH_4(g) + 2O_2(g) \rightarrow CO_2(g) + 2H_2O(g)$$

The coefficients used to balance the equations show the ratios between the reactants and products involved in the reaction—the **mole ratio**. The equation indicates that 1 mole of $CH_4(g)$ reacts with 2 moles of $O_2(g)$ to form 1 mole of $CO_2(g)$ and 2 moles of $H_2O(g)$. In more general terms, the amount of oxygen used will always be double the amount of methane used, double the amount of carbon dioxide produced and the same as the amount of water vapour produced.

$$\frac{n(O_2)}{n(CH_4)} = \frac{2}{1} \quad \frac{n(CO_2)}{n(O_2)} = \frac{1}{2} \text{ and } \frac{n(H_2O)}{n(O_2)} = \frac{2}{2} = 1$$

In general for calculations, you will be given, or you will be able to work out from data, the number of moles of one chemical in the reaction. This is called the 'known'—you know the amount (in moles) of this substance. You will then need to calculate the number of moles of another chemical, the 'unknown'. The ratio(s) of the moles, as shown by the coefficients in the equation, allow you to carry out this calculation.

The relationship used can be written as:

$$n(\text{unknown chemical}) = \frac{\text{coefficient of unknown chemical}}{\text{coefficient of known chemical}} \times n(\text{known chemical})$$

Worked example 11.3.4

USING MOLE RATIOS

How many moles of carbon dioxide are generated when 24 moles of propane are burnt completely in oxygen?	
Thinking	**Working**
Write a balanced equation for the reaction.	$C_3H_8(g) + 5O_2(g) \rightarrow 3CO_2(g) + 4H_2O(g)$
Note the number of moles of the known substance (in this case propane).	$n(C_3H_8) = 24$ mol
Write a mole ratio for the chemicals being investigated. $\frac{\text{coefficient of unknown}}{\text{coefficient of known}}$	$\frac{n(CO_2)}{n(C_3H_8)} = \frac{3}{1}$
Calculate the number of moles of the unknown substance using: $n(\text{unknown}) = \text{mole ratio} \times n(\text{known})$	$n(CO_2) = \frac{3}{1} \times 24$ $= 72$ mol

Worked example: Try yourself 11.3.4

USING MOLE RATIOS

How many moles of carbon dioxide are generated when 0.50 moles of butane are burnt completely in oxygen?

MASS-MASS STOICHIOMETRY

Calculations usually require you to start and finish with masses rather than moles, as this is how quantities of chemicals are often measured. To calculate the number of moles from the mass of a substance, we can use this relationship:

$$\text{moles} = \frac{\text{mass in grams}}{\text{molar mass}}$$

which is written as

$$n = \frac{m}{M}$$

and to calculate a mass, we can rearrange this relationship to become:

$$m = n \times M$$

Worked example 11.3.5

SOLVING MASS–MASS STOICHIOMETRIC PROBLEMS

Calculate the mass of carbon dioxide produced when 5400 g of propane (C_3H_8) is burnt completely in oxygen.	
Thinking	**Working**
Write a balanced equation for the reaction.	$C_3H_8(g) + 5O_2(g) \rightarrow 3CO_2(g) + 4H_2O(g)$
Calculate the number of moles of the known substance (in this case propane) using: $n = \frac{m}{M}$	$n(C_3H_8) = \frac{5400}{44.094}$ $= 122.5$ mol
Find the mole ratio: $\frac{\text{coefficient of unknown}}{\text{coefficient of known}}$	$\frac{n(CO_2)}{n(C_3H_8)} = \frac{3}{1}$
Calculate the number of moles of the unknown (in this case carbon dioxide) substance using: $n(\text{unknown}) = \text{mole ratio} \times n(\text{known})$	$n(CO_2) = \frac{3}{1} \times 122.5$ $= 367.4$ mol
Calculate the mass of the unknown substance using: $m = n \times M$	$m(CO_2) = 367.4 \times 44.01$ $= 16\,169$ g ≈ 16.2 kg

Worked example: Try yourself 11.3.5

SOLVING MASS–MASS STOICHIOMETRIC PROBLEMS

Calculate the mass of carbon dioxide produced when 3600 g of butane (C_4H_{10}) is burnt completely in oxygen. (The molar mass of butane is 58.12 g mol^{-1}.)

CHOOSING A FUEL

A wide range of fuels is used for many different purposes. As we have seen in this chapter, the factors that need to be considered when selecting a fuel for a particular purpose include:

- energy released per unit mass or unit volume
- availability and cost of fuel
- ease of transport
- hazards to people and the environment associated with its use and its waste products
- social, economic, cultural and political values that can affect the choice of a fuel.

We can find many examples of these factors on fuel choice. Liquid hydrogen was chosen for the *Saturn* rockets used in the US space program because it provides large quantities of energy per gram. As space rockets need to carry their own fuel, this was a crucial factor.

One of the key concerns for fuel choice for electricity production is efficiency. The term **energy efficiency** is used to describe the percentage of energy from a source that is converted to useful energy. For example, if the efficiency of a solar panel on a roof (Figure 11.3.2) is listed as 17%, it means that 17% of the energy

FIGURE 11.3.2 Commercial solar cells convert solar energy to electrical energy with an efficiency of 12–18%.

arriving on the panel from the Sun is transformed to electrical energy. The other 83% is converted to other forms of energy. The largest proportion of the Sun's energy reaching the solar cells is converted into heat energy that simply increases the temperature of the cells.

In Australia, electrical energy is produced from several different fuels.

Electricity from coal

The combustion of coal generates more than three-quarters of Australia's electricity. Rather than transport coal to every factory, office and household, the chemical energy is converted to electrical energy at a power station. Electricity is transmitted easily from the power station by metal cables and wires to other regions. The reaction occurring when coal burns can be written as:

$$C(s) + O_2(g) \rightarrow CO_2(g)$$

The energy released from the combustion of coal is about $32\,kJ\,g^{-1}$.

A number of energy transformations occur in a coal-fired power station.

- Coal is burnt—chemical energy in coal is converted to thermal energy.
- Heat from the burning coal is used to boil water—thermal energy from the burning coal becomes thermal energy in steam.
- Steam is passed through a turbine—thermal energy in the steam becomes mechanical energy as the turbine spins. (This is the least efficient energy transformation in the sequence.)
- Electricity is produced from a generator that is driven by the turbine—mechanical energy is converted to electrical energy.

Figure 11.3.3 illustrates how the thermal energy released by the coal is converted to electrical energy.

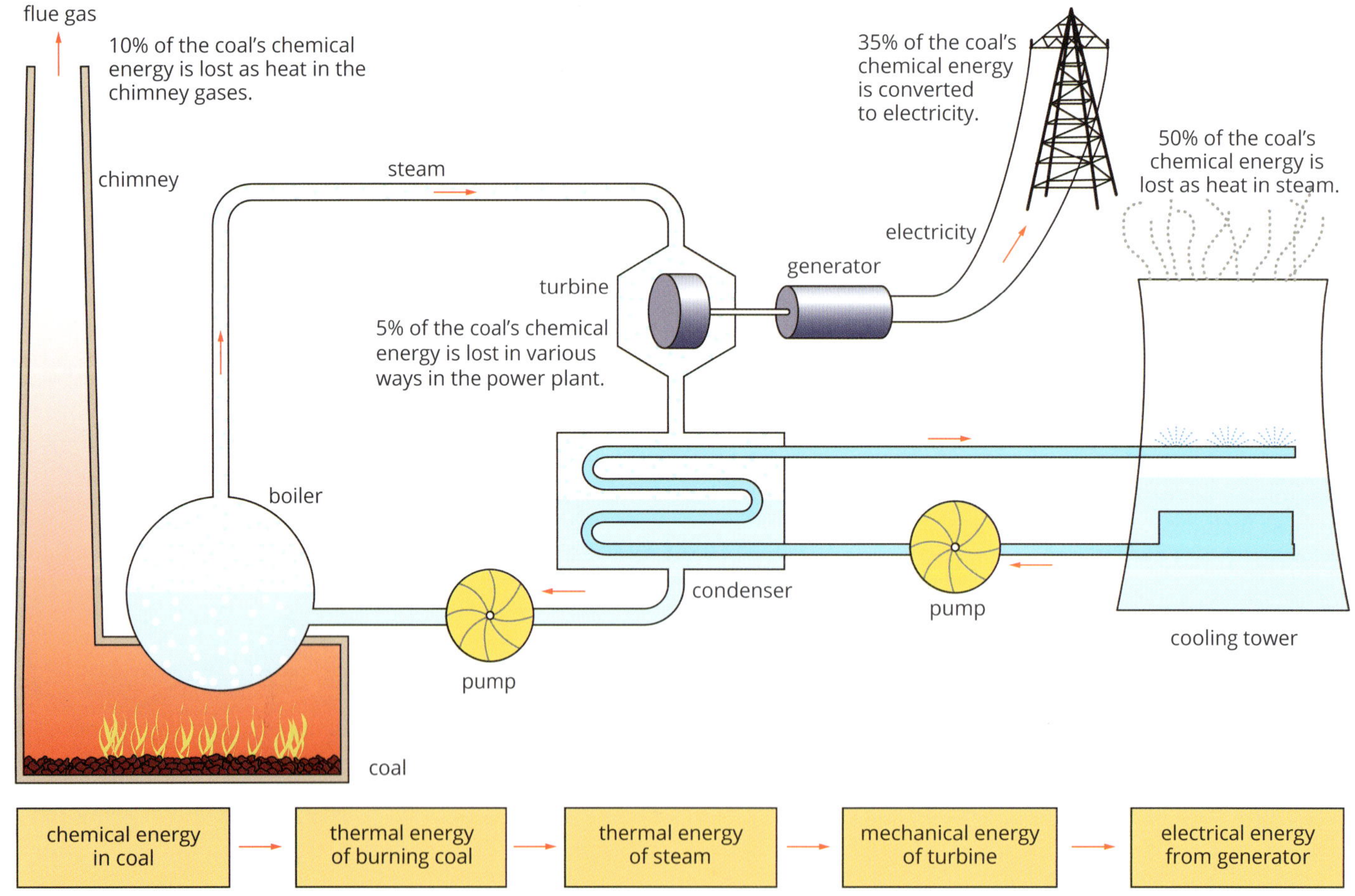

FIGURE 11.3.3 A representation of the process of burning coal to produce electricity. The chemical energy in the coal undergoes several transformations before electricity is produced.

The overall efficiency of a coal-fired power station is 30–40%. The combustion of brown coal is usually at the lower end of this efficiency range. Energy is lost during each step of the process, mainly as heat.

Electricity from natural gas

Natural gas is also used in a number of locations in Western Australia, including Pinjar and Kwinana, to generate electricity for both residential and industrial purposes. In a gas-fired power plant, methane and other small alkanes are burnt to release energy. As shown in Figure 11.3.4, the hot gases produced by combustion cause air to expand in a combustion turbine to generate electrical energy. This is a simpler process than in a coal-fired plant where the thermal energy is used to produce steam.

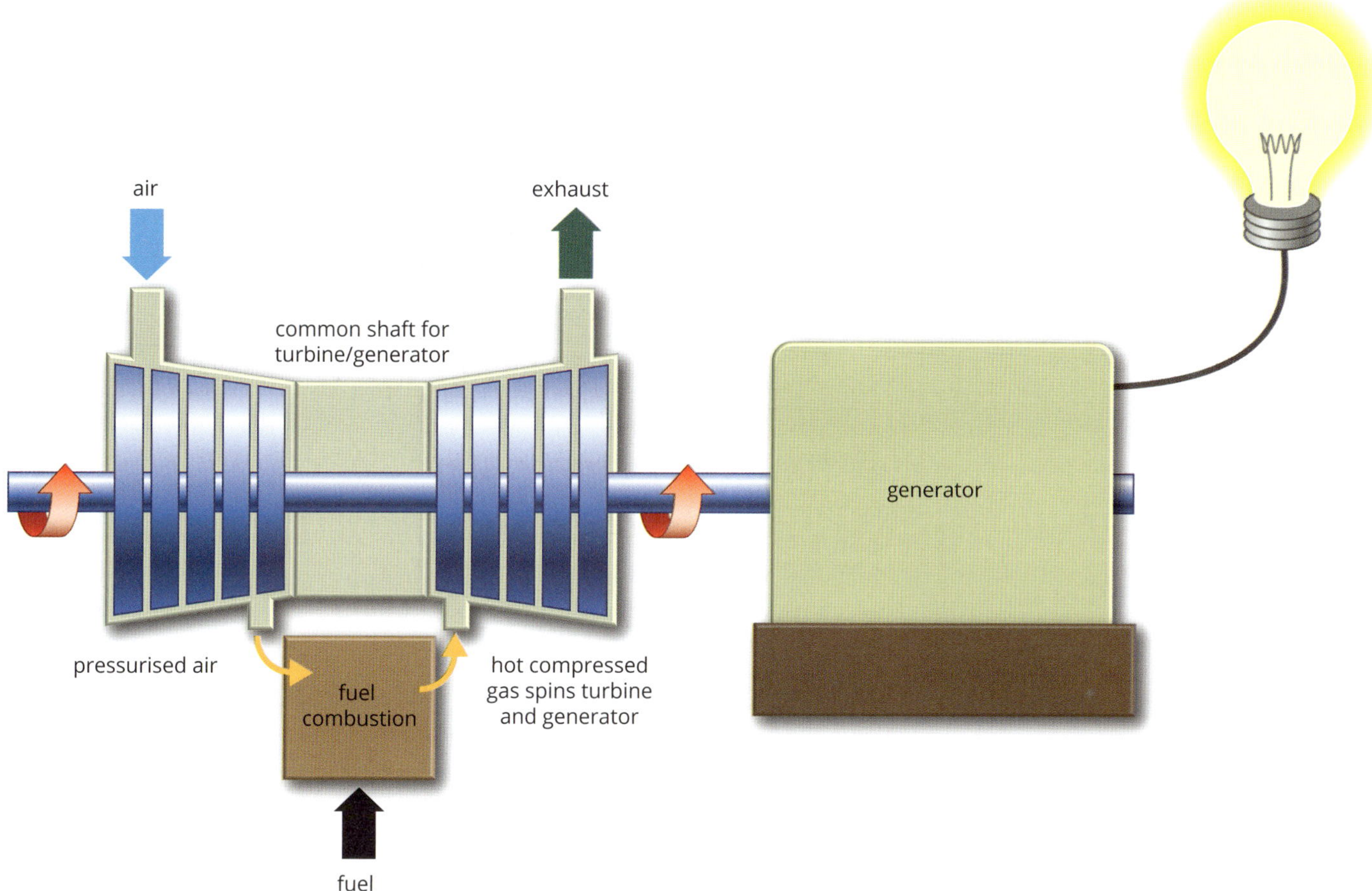

FIGURE 11.3.4 In a gas-fired power plant, the hot gases produced expand air in a combustion turbine to generate electricity.

The composition of natural gas varies but the main combustion reaction involves methane. The equation is:

$$CH_4(g) + 2O_2(g) \rightarrow CO_2(g) + 2H_2O(l)$$

The combustion of 1.00 mole of methane releases 890 kJ of energy, equivalent to 55.6 $kJ\,g^{-1}$. This is a significantly higher value than that of coal.

A gas-fired plant is more efficient than a coal-fired power station, reaching efficiencies just over 40%. Gas-fired plants also emit less carbon dioxide and particulate matter (small solid particles of solid combustion products) per unit of energy released. An added advantage of gas-fired plants is that the output can be varied at short notice. This allows the operators to adjust to the fluctuating power usage of consumers. Most Australian states have gas-fired plants but many of the plants are small-scale ones. Coal seam gas is the source of methane used in some states.

Electricity from biogas

Biogas is a renewable fuel that can be used to generate electricity, usually in small-scale electricity generators rather than large power plants. These smaller generators are often located at the site where the biogas is produced. For example, sewage works commonly burn biogas produced in a generator to supply some of their power needs.

The main reaction occurring in the combustion of biogas is the same reaction as methane burning in a gas-fired power station. The energy released per gram of biogas is less than that of natural gas because the methane content in biogas is significantly lower.

Berrybank Farm near Ballarat, Victoria, is an example of the innovative use of biogas. More than \$2 million has been spent on building infrastructure to collect the manure from 20 000 pigs. The manure is fed into a digester that produces two useful products: biogas and fertiliser. The biogas is used to fire generators, like the one shown in Figure 11.3.5b, that produce an estimated \$180 000 of electricity annually.

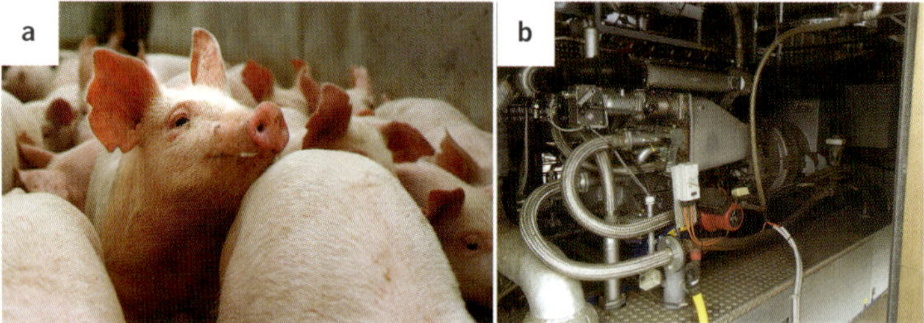

FIGURE 11.3.5 (a) Some of the pigs on the Berrybank Farm near Ballarat; (b) one of the generators that uses the biogas fuel

FUEL FOR TRANSPORT

Crude oil is the source of most of the fuel we use for transport. Crude oil is a mix of alkanes. The alkanes in crude oil are separated into a series of fractions (parts) by fractional distillation. Some of these fractions are important fuels, such as liquid petroleum gas (LPG), petrol, kerosene and petrodiesel. LPG can also be separated from natural gas.

Petrol

Perhaps the most important of all hydrocarbon combustion reactions are those that occur when petrol is burnt. Petrol is a mixture of hydrocarbons, including octane, and the combustion reactions of these chemicals power most of Australia's 17.6 million motor vehicles. The equation for the combustion of octane is:

$$2C_8H_{18}(l) + 25O_2(g) \rightarrow 16CO_2(g) + 18H_2O(l)$$

The combustion of 1.00 mole of octane releases 5450 kJ of energy, equivalent to 47.8 kJ g^{-1}. Combustion occurs in the cylinder of a car engine. The hot gases formed push the piston in the engine, enabling the car to move. A typical piston is shown in Figure 11.3.6.

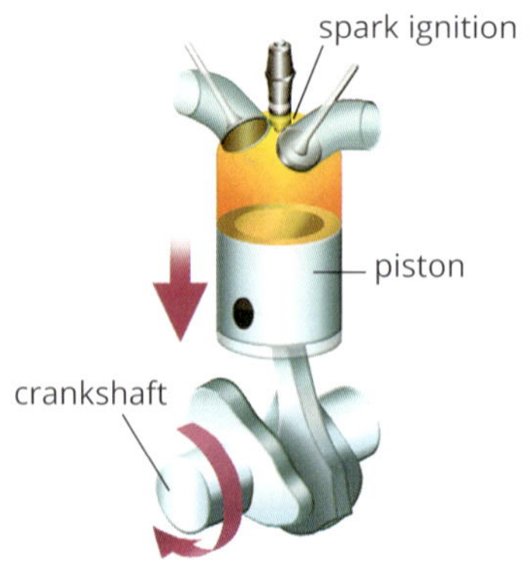

FIGURE 11.3.6 The combustion of octane (C_8H_{18}) and the other hydrocarbons in petrol pushes the pistons in internal combustion engines.

The efficiency of a petrol engine in a new car can be as high as 25%. The operation of a piston in a car engine can be seen in Figure 11.3.7.

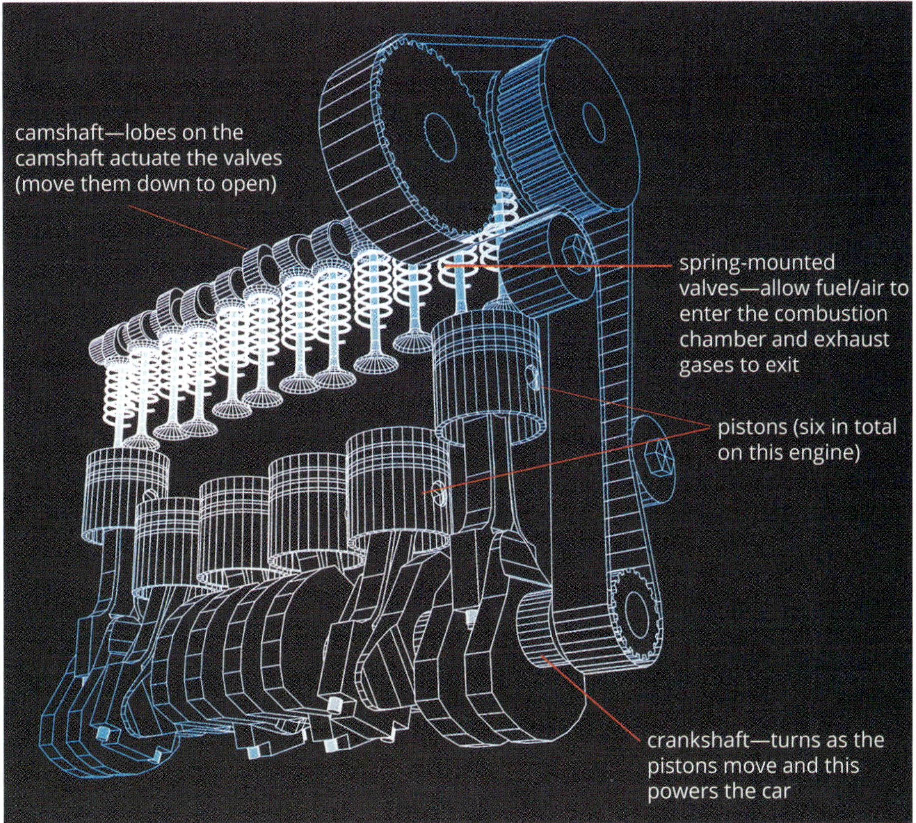

FIGURE 11.3.7 In a standard engine, fuel and air are mixed in cylinders and then compressed, and the mixture is ignited. The explosion drives the piston down, which turns the crankshaft and powers the car. The flow of fuel and exhaust gases is controlled by spring-mounted valves shown in the centre.

Liquefied petroleum gas

Liquefied petroleum gas (LPG) can also be used in cars. Most of the vehicles that use LPG as a fuel have a standard petrol engine with a fuel tank and fuel injection system modified to suit a gaseous fuel. The equation for the combustion of propane, a major component of LPG, is:

$$C_3H_8(g) + 5O_2(g) \rightarrow 3CO_2(g) + 4H_2O(l)$$

The combustion of 1.00 mole of propane releases 2220 kJ of energy, equivalent to 50.5 kJ g^{-1}.

In Australia, LPG is a significantly cheaper fuel than petrol, yet its popularity is still limited. There are a few reasons for this.

- Most new vehicles are designed to run on petrol; therefore, the owner has to pay around $2000 for a conversion.
- The LPG fuel tank takes up boot space.
- There are fears that LPG cylinders might explode if the vehicle crashes.
- The prices of fuels fluctuate, so often it is difficult to do meaningful price comparisons.

Bioethanol

Australia imports more than 90% of its fuel requirements. This reliance on other countries, combined with concerns over the greenhouse emissions of fossil fuels, has sparked interest in the production of the renewable biofuels such as bioethanol.

Bioethanol can be produced from crops such as sugar cane. However, sugar cane is also needed for sugar production so there are limits to the amounts of bioethanol that can be produced in this way. Instead, researchers are trialling less valuable sources of sugar and starch for bioethanol production.

The Manildra Group refinery plant at Bomaderry in New South Wales, shown in Figure 11.3.8, is one of Australia's largest ethanol refineries. At this plant, flour and starch are produced from wheat and sold for use in food manufacture. The waste that remains still contains high levels of starch, which is converted to ethanol.

FIGURE 11.3.8 Ethanol refinery at Bomaderry (near Nowra), New South Wales

The equation for the combustion of ethanol is:

$$C_2H_5OH(l) + 3O_2(g) \rightarrow 2CO_2(g) + 3H_2O(l)$$

The combustion of 1.00 mole of ethanol releases 1367 kJ of energy, equivalent to 29.7 kJ g^{-1}. As Table 11.3.4 shows, the energy content of ethanol is about 62% that of petrol, so a larger mass (and volume) of ethanol is required to provide the same amount of energy. At a simple level, the lower energy content of ethanol can be regarded as the result of the carbon atoms in an ethanol molecule being partly oxidised ('partly burnt'). This is due to the presence of oxygen in the ethanol molecule.

Energy density is the energy released when 1 litre of fuel undergoes complete combustion.

TABLE 11.3.4 Energy content and energy density of vehicle fuels

Fuel	Energy content (kJ g^{-1})	Energy density (kJ L^{-1})
methane	55.6	23 500 (liquefied)
propane (LPG component)	50.5	29 400 (liquefied)
butane (LPG component)	49.6	29 800 (liquefied)
octane (petrol fraction)	47.8	33 400
ethanol	29.7	23 400
petrodiesel	48.1	40 000
biodiesel	37.8	36 000

11.3 Review

SUMMARY

- Heat of combustion, ΔH_c, indicates the maximum amount of energy that can be released when a specified amount of fuel undergoes complete combustion. Common units are kJ mol^{-1}, kJ g^{-1} and MJ/tonne.
- The amount of energy released by different fuels can be compared by referring to heats of combustion.
- For a pure fuel with a heat of combustion, ΔH_c, measured in kJ mol^{-1}, the energy released when n mol of the fuel burns is given by the equation:

$$\text{energy} = n \times \Delta H_c$$

- For a fuel that is not pure with a heat of combustion, ΔH_c, measured in kJ g^{-1}, the energy released when m grams of the fuel burns is given by the equation:

$$\text{energy} = m \times \Delta H_c$$

- The amount of carbon present in a fuel will influence the amount of oxygen needed for combustion and thus the amount of carbon dioxide produced.
- Stoichiometry is the study of ratios of moles of substances, with stoichiometric calculations being based on the law of conservation of mass.
- Using stoichiometric ratios, you can predict the amount of CO_2 and other harmful emissions created from the combustion of different fuels.
- The combustion reactions of fuels are used to produce electricity and to power vehicles.
- In a coal-fired power station, thermal energy from coal creates steam that is used to turn a turbine and generate electricity. The efficiency of this process is between 30 and 40%.
- Electricity can also be produced from the combustion of natural gas or biogas.
- Petrol engines can use petrol or petrol blended with ethanol. Engines can also be modified to run on LPG.
- Biodiesel, which is sourced from biomass, provides an alternative to petrodiesel.
- Table 11.3.5 compares the advantages and disadvantages of some fuels.

TABLE 11.3.5 Advantages and disadvantages of fuels described in this chapter

Fuel	Advantages	Disadvantages
coal	• large reserves • relatively high energy content	• non-renewable • high level of emissions • less easily transported than liquid or gaseous fuels
natural gas	• more efficient than coal for electricity production • easy to transport through pipes • relatively high energy content	• non-renewable • limited reserves • polluting (but less than coal and petrol)
biogas	• renewable • made from waste • reduces waste disposal • low running costs • CO_2 absorbed during photosynthesis	• low energy content • supply of waste raw materials limited
petrol	• high energy content • ease of transport	• non-renewable • polluting (but less than coal) • limited reserves
LPG	• low cost • easily separated from natural gas • relatively high energy content; fewer particulates produced than petrol	• non-renewable • polluting (but less than petrol)
bioethanol	• renewable • can be made from waste • CO_2 absorbed during photosynthesis • burns smoothly • fewer particulates produced than petrol	• limited supply of raw materials from which to produce it • lower energy content than petrol • may require use of farmland otherwise used for food production
biodiesel	• renewable • can be made from biomass • no modifications to vehicles required	• limited supply of raw materials for production • lower energy content than petrodiesel

Review (*Continued*)

KEY QUESTIONS

1 How many moles of water are produced when 50 moles of butanol (C_4H_9OH) are burned in an unlimited supply of oxygen?

2 Information about two hydrocarbon fuels, propane and octane, is given in the table below.

Characteristic	Propane (C_3H_8)	Octane (C_8H_{18})
heat of combustion ($kJ\,mol^{-1}$)	–2220	–5450
molar mass ($g\,mol^{-1}$)	44.0	114.0

Calculate the heat of combustion of each fuel in $kJ\,g^{-1}$ and use your answer to state which fuel produces more energy per kilogram.

3 Calculate the mass of carbon in 1.00 kg of butane (C_4H_{10}).

4 Using the information in Tables 11.3.1 on page 244 and 11.3.2 on page 245, calculate the amount of energy released when the following amounts of each fuel undergo complete combustion.

a 250 g of methane

b 9.64 kg of propane

c 403 kg of ethanol

d 573 t of dried brown coal

5 Octane (C_8H_{18}) is a component of petrol. It burns in oxygen to produce carbon dioxide and water. Energy is released during this reaction. The equation for this reaction is:

$$2C_8H_{18}(g) + 25O_2(g) \rightarrow 16CO_2(g) + 18H_2O(g)$$

a Calculate the mass of oxygen required to react with 200 g of octane.

b Calculate the mass of carbon dioxide produced in part **a**.

6 A combustion reaction occurred in which 1.00 kg of butane (C_4H_{10}) reacted with excess oxygen gas.

a Calculate the mass of oxygen, in kg, required for the complete combustion of this amount of butane.

b Calculate the mass of carbon dioxide, in kg, produced in this reaction.

7 Classify each of the following as an advantage or a disadvantage of the use of bioethanol compared to petrol as an energy source.

a less CO_2 impact overall

b lower energy content ($kJ\,g^{-1}$)

c can be produced from waste products

d renewable resource

e greater amount of CO_2 emitted to travel a set distance

8 **a** The water content of brown coal is 60–70%. What implications does this high water content have for the energy released from burning the coal?

b What pre-treatment could raise the energy content per gram of brown coal consumed in a power station?

c If some electricity produced from burning coal was used to reduce its water content, then how could this energy cost be minimised?

d What impact do impurities such as sulfur have on generating electricity from brown coal?

Chapter review

11

KEY TERMS

acid rain
acidification
biodegradable
biodiesel
bioethanol
biofuel
biogas
carbon neutral
combustion
energy content
energy density
energy efficiency
enhanced greenhouse effect
enzyme
fermentation
fracking
fuel
greenhouse effect
greenhouse gas
heat of combustion
liquefied petroleum gas (LPG)
mole ratio
natural gas
non-renewable
petrodiesel
photosynthesis
renewable
SI units
stoichiometric calculation
stoichiometry
sustainability

Types of fuels

1 The world has become very dependent on the products of the petrochemical industry, but the raw materials of crude oil and natural gas are likely to be virtually exhausted by 2100. Assuming that current production remains unchanged and no alternative sources are available, suggest the impact of the lack of raw materials on our lifestyle.

2 Refer to Figure 11.1.4 on page 223 to answer the following questions.

- **a** In the year 2000, what type of fuel provided the most energy on a global scale?
- **b** Rank the different fuels in order from most to least amount of energy expected to be produced in 2020.

3 Why are fossil fuels considered to be non-renewable?

4 Refer to Figure 11.1.5 on page 224 to answer the following questions.

- **a** What type of coal takes the longest amount of time to form?
- **b** Of peat, brown coal and black coal, which releases the least amount of heat energy?
- **c** Which type of coal is a better quality fuel?

5 The following energy sources are used across the world: coal, bioethanol, biodiesel, natural gas. Which one is likely to run out the fastest?

6 Explain why bioethanol is sometimes described as a 'carbon neutral' fuel. Use chemical equations for photosynthesis, fermentation and combustion to support your answer.

7 Use the terms 'methane', 'oxygen', 'bacteria' and 'carbon dioxide' to explain the formation and composition of biogas.

Combustion reactions

8 Write a balanced equation for the complete combustion of butanol (C_4H_9OH), a commonly used biofuel.

9 Write a balanced equation for the incomplete combustion of butane (C_4H_{10}) where carbon monoxide is formed.

10 The fact that global warming is taking place is now generally accepted. Conduct some research using the internet to discover some of the consequences of global warming. Give one example each of the effect on:

- **a** the polar ice caps
- **b** changing weather patterns
- **c** crop production
- **d** extinction of plant or animal species.

Calculations involving fuels

11 Calculate the energy released when 5.00 kg of methane (CH_4) is burnt in an unlimited supply of oxygen. The heat of combustion of methane is $-890\ kJ\ mol^{-1}$. (Give your answer in megajoules, MJ.)

12 Propane burns in oxygen according to the equation:

$$C_3H_8(g) + 5O_2(g) \rightarrow 3CO_2(g) + 4H_2O(g)$$

6.70 g of propane was burned in excess oxygen.

- **a** What mass of carbon dioxide would be produced?
- **b** What mass of oxygen would be consumed in the reaction?
- **c** What mass of water would be produced?

13 Ethanol can be used as a substitute for petrol. The heat of combustion for ethanol (CH_3CH_2OH) is $-1367\ kJ\ mol^{-1}$.

- **a** Write a balanced equation to represent the complete combustion of ethanol in air.
- **b** Calculate the energy released when 5.00 g of ethanol burns completely in air.
- **c** Calculate the mass of carbon dioxide released in this process.

14 Methane and propane are used as alternative transport fuels. The heats of combustion of methane and propane are $-890\ kJ\ mol^{-1}$ and $-2220\ kJ\ mol^{-1}$ respectively. Calculate the mass, in kg, of each gas required to release 100 000 kJ of heat energy and calculate the mass of carbon dioxide gas produced in the process. Compare the differences between the two fuels.

15 **a** Explain what 'E10 petrol' is.

b How does the introduction of E10 help with the potential shortage of crude oil?

16 What are some of the ways in which energy is lost, leading to a reduced efficiency, in a coal-fired power station?

Connecting the main ideas

17 In a steelworks, carbon monoxide present in the exhaust gases of the blast furnace can be used as a fuel elsewhere in the plant. It reacts according to the equation:

$$CO(g) + \frac{1}{2}O_2(g) \rightarrow CO_2(g) \quad \Delta H = -283\,kJ\,mol^{-1}$$

a Which has the greater total enthalpy: 1 mol of CO(g) and 0.5 mol of O_2(g) or 1 mol of CO(g) and 1 mol of CO_2(g)?

b Write the value of ΔH for the following equations:

i $2CO(g) + O_2(g) \rightarrow 2CO_2(g)$

ii $2CO_2(g) \rightarrow 2CO(g) + O_2(g)$

18 In a car engine, small amounts of the petrol are vaporised, enter the cylinder and are ignited by a spark. If 12.0 mg of octane is released into the cylinder, what mass of oxygen would be required to be present for complete combustion?

19 An indoor gas heater burns propane, (C_3H_8) at a rate of 12.7 grams per minute.

a Calculate the minimum amount of oxygen per minute that needs to be available for the complete combustion of the propane.

b What will happen if the flow of oxygen per minute is insufficient for the complete combustion of the propane?

20 Ethanol is produced industrially by reacting ethene with water using a phosphoric acid catalyst at 300°C:

$$C_2H_4(g) + H_2O(g) \rightarrow CH_3CH_2OH(g)$$

a Explain whether ethanol produced by this method is a biochemical fuel.

b Describe how ethanol, which is classified as a biochemical fuel, could be produced.

21 Methane, ethanol, octane and biodiesel are used to power vehicles.

a List the fuels in order of their energy content per gram (from highest to lowest). You may need to refer to Table 11.3.4 on page 252.

b The emissions of carbon dioxide per gram from the combustion of bioethanol are less than that from octane (petrol). However, a car using bioethanol produces more carbon dioxide when driving the same distance as a car using octane. Which one of the follow could be the best explanation for this difference?

A Bioethanol is more efficient than octane.

B The energy content of bioethanol is less than that of octane.

C The temperature of the engine favours the combustion of octane.

D More energy transformations are required in the combustion of bioethanol.

22 Conduct some research on the internet to find out which nations are the top 10 consumers of energy.

23 The 2015 Paris Agreement was an international response to global warming in which all nations were asked to commit to keeping the global average temperature rise to below 2°C, through reductions in greenhouse gas emissions.

a What impact would adoption of the Paris Agreement have on Australia?

b Discuss the role that biofuels could play in helping Australia meet its target for reducing greenhouse gas emissions.

UNIT 1 • CHEMICAL FUNDAMENTALS: STRUCTURE, PROPERTIES AND REACTIONS

REVIEW QUESTIONS

Section 1: Multiple choice

1 Which one of the following statements best describes an element in group 17 of the periodic table?
- **A** an ion with a charge of negative one
- **B** an element that gains one electron to achieve a full valence shell
- **C** an element that can form covalent network solids
- **D** a noble gas

2 Which one of the following is the correct electron configuration of oxygen?
- **A** 2,8,8
- **B** 2,2,6
- **C** 2,6
- **D** 2,8

3 An element X forms a chloride with the formula XCl_3. Which one cannot be element X?
- **A** Al
- **B** Fe
- **C** N
- **D** Sr

4 Which one of the following solids is classified as a molecular solid?
- **A** silicon (Si)
- **B** alumina (Al_2O_3)
- **C** bronze (a mixture of Cu and Sn)
- **D** dry ice (solid CO_2)

5 In which one of the following solids are both ionic and covalent bonds present?
- **A** iodine
- **B** lead iodide
- **C** ammonium chloride
- **D** hydrogen iodide

6 What is the correct IUPAC name of the substance represented by the structure below?

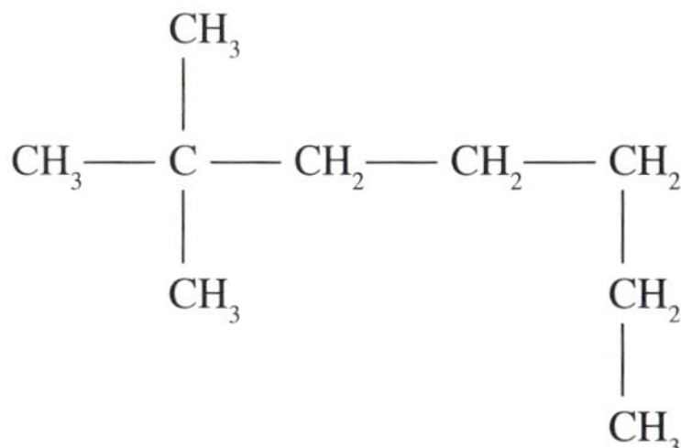

- **A** 2,2-dimethylheptane
- **B** 2,2,4-trimethylhexane
- **C** 1-ethyl-1,3,3-trimethylbutane
- **D** octane

7 Which sequence of steps is most likely to be used to separate a pure sample of salt from a mixture of salt and charcoal?
- **A** distillation, evaporation, filtration, crystallisation
- **B** filtration, crystallisation, dissolution, evaporation
- **C** dissolution, filtration, evaporation, crystallisation
- **D** dissolution, evaporation, filtration, distillation

8 Which expression shows the mass of a nitrogen molecule?
- **A** $\frac{28.0}{6.022 \times 10^{23}}$
- **B** 14.0 g
- **C** $\frac{6.022 \times 10^{23}}{28.0}$
- **D** 28.0 g

9 Which one of the following lists only non-renewable sources of energy?
- **A** natural gas, fuel oil, hydroelectric power
- **B** coal, biomass to produce ethanol, oil
- **C** crude oil, wood, natural gas
- **D** natural gas, coal, bottled gas (LPG)

10 From the table below, identify two elements that are isotopes. (Select A, B, C or D.)

Element	Number of protons	Number of electrons	Number of neutrons
W	20	21	21
X	19	18	19
Y	19	21	19
Z	20	19	20

- **A** elements X and W
- **B** elements X and Y
- **C** elements W and Z
- **D** elements Y and W

11 $C_6H_{14} + Br_2 + \text{UV light} \rightarrow C_6H_{13}Br + HBr$

The above reaction is an example of:
- **A** addition.
- **B** substitution.
- **C** combustion.
- **D** sublimation.

12 A solution may be best described as:
- **A** a pure substance of constant composition.
- **B** a homogeneous mixture of uniform composition.
- **C** a substance that can be purified by filtration.
- **D** a heterogeneous mixture of variable composition.

Section 2: Short answer

1 a Write the formula for the following compounds.
   i gold(III) sulfide
   ii strontium hydroxide
   iii ammonium phosphate
  b Write the name for the following substances.
   i $Fe_2(SO_4)_3$
   ii P_2O_5

2 One type of sugar found in fruit is fructose.
The molecular formula for fructose is $C_6H_{12}O_6$.
The structural formula is:

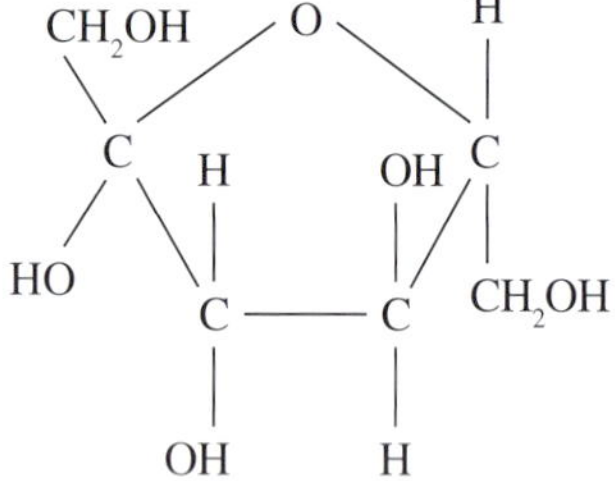

  a What is the mass of **1 mole** of fructose? Provide your answer to 4 significant figures.
  b What is the mass of **1 molecule** of fructose? Provide your answer to 2 significant figures.
  c Why do fructose molecules attract each other?
  d What type of chemical bonds exist between the fructose molecules in the solid sugar?

3 a Draw the **electron dot structure** for the following molecules including all bonding and non-bonding electrons.
   i water (H_2O)
   ii dichlorodifluoromethane (CCl_2F_2)
  b Draw the **structural formula** for the following molecules.
   i *cis*–dibromoethene ($C_2H_2Br_2$)
   ii carbon disulfide (CS_2)

4 a Using IUPAC nomenclature, name the following organic compounds.

   i
```
       H              H
       |             /
   H — C — C ═ C
       |    |        \
       H    |         H
       H — C — H
            |
            H
```

   ii
```
   F        H
    \      /
     C ═ C
    /      \
   H        F
```

   iii
```
                 CH3                  CH3
                  |                    |
   CH3 — CH2 — C — CH2 — CH — CH — CH3
                  |              |
                 CH3            CH2
                                 |
                                CH3
```

  b State which of the compounds from part **a** (i, ii and/or iii) would undergo an addition reaction with chlorine gas.
  c Write the **molecular formula** for compound (iii) in part **a**.

5 a Calculate the percentage of water in aluminium nitrate hexahydrate, $Al(NO_3)_3 \cdot 6H_2O$. Give your answer to 3 significant figures.
  b Calculate how many moles are in 1.00 kg of aluminium nitrate hexahydrate.

6 Copy and complete the table.

Reaction	$\Delta H_{reaction}$	Exothermic or endothermic
$2H_2(g) + O_2(g) \rightarrow 2H_2O(g) + 287\ kJ\ mol\ L^{-1}$		
$H_2S(g) + 90\ kJ\ mol\ L^{-1} \rightarrow H_2(g) + S(s)$		

Section 3: Extended answer

1 Materials are pure substances with distinct measurable and observable properties. These properties can be related to the types of chemical bonds between the atoms, ions and molecules from which matter is composed.
Electrical conductivity depends on whether these bonds are metallic, ionic, covalent network or covalent molecular.
In terms of chemical bonding, explain how the following solids are electrical conductors or insulators (non-conductors). Labelled diagrams can be used in your answers.
  a copper (Cu)
  b sodium chloride (NaCl)
  c water (ice, H_2O)
  d graphite (C)
  e diamond (C)

2 Butane (C_4H_{10}) is a gaseous hydrocarbon fuel that is often used as an energy source for portable stoves, among many other things.

a Draw the structural formula of butane.

b There is another hydrocarbon that has the formula C_4H_{10}.
Draw the structural formula for that isomer and name the compound.

c Using the molecular formula (C_4H_{10}), write the balanced equation for the complete combustion of butane in air.
The enthalpy of reaction for the above equation is $-5750\ kJ\ mol^{-1}$.

d Is the combustion of butane exothermic or endothermic?

e Rewrite your equation including the heat (enthalpy) of reaction as a reactant or product.

f Calculate the mass of carbon dioxide produced for every 1.00 kg of butane burned.

Butane undergoes a slow substitution reaction with gaseous chlorine.

g Using molecular or structural formulae, write the equation for the reaction between 1 mole of butane and 1 mole of fluorine.

h Write the IUPAC name of the organic product.

An alkene with four carbon atoms is but–1–ene.

i Draw the structural formula for but–1–ene.

j Using structural formulae, write the equation for the reaction between but–1–ene and bromine.

k Name the organic product formed in the reaction in part **j**.

UNIT

2 Molecular interactions and reactions

Students develop their understanding of the physical and chemical properties of materials, including gases, water and aqueous solutions, acids and bases. Students explore the characteristic properties of water that make it essential for physical, chemical and biological processes on Earth, including the properties of aqueous solutions. They investigate and explain the solubility of substances in water, and compare and analyse a range of solutions. They learn how rates of reaction can be measured and altered to meet particular needs, and use models of energy transfer and the structure of matter to explain and predict changes to rates of reaction. Students gain an understanding of how to control the rates of chemical reactions, including through the use of a range of catalysts.

Learning outcomes

By the end of this unit, students:

- understand how models of the shape and structure of molecules and intermolecular forces can be used to explain the properties of substances, including the solubility of substances in water
- understand how kinetic theory can be used to explain the behaviour of gaseous systems, and how collision theory can be used to explain and predict the effect of varying conditions on the rate of reaction
- understand how models and theories have developed based on evidence from a range of sources, and the uses and limitations of chemical knowledge in a range of contexts
- use science inquiry skills to design, conduct, evaluate and communicate investigations into the properties and behaviour of gases, water, aqueous solutions and acids and bases, and into the factors that affect the rate of chemical reactions
- evaluate, with reference to empirical evidence, claims about chemical properties, structures and reactions
- communicate, predict and explain chemical phenomena using qualitative and quantitative representations in appropriate modes and genres.

CHAPTER 12

Intermolecular forces

Covalent molecular substances are the only substances that can be solids, liquids or gases at room temperature. They can be hard or soft, flexible or brittle, sticky or oily—almost any consistency. This broad range of properties makes molecular substances very useful materials. The properties of covalent molecular substances can generally be explained by the strong covalent bonds inside the molecules and much weaker bonds between molecules.

During this chapter, you will become familiar with the valence shell electron pair repulsion (VSEPR) theory and use of this theory to predict the shape and polarity of molecules. This is because knowing the shapes of molecules is vital when it comes to predicting and explaining the type and strength of forces acting between the molecules. You will learn how intermolecular forces determine many of the physical properties of covalent molecular substances and use these predictions to infer information about the nature and strength of bonding between molecules.

In addition, you will learn to identify intermolecular forces, including dipole–dipole forces, hydrogen bonding and dispersion forces.

Science understanding

- the valence shell electron pair repulsion (VSEPR) theory and Lewis structure diagrams can be used to explain, predict and draw the shapes of molecules
- observable properties, including vapour pressure, melting point, boiling point and solubility, can be explained by considering the nature and strength of intermolecular forces within a covalent molecular substance
- the polarity of molecules can be explained and predicted using knowledge of molecular shape, understanding of symmetry and comparison of the electronegativity of atoms involved in the bond formation
- the shape and polarity of molecules can be used to explain and predict the nature and strength of intermolecular forces including dispersion forces, dipole–dipole forces and hydrogen bonding

12.1 Shapes of molecules

The shapes of molecules are critical in determining many physical properties of covalent molecular substances. In particular, molecular shape affects vapour pressure, melting point, boiling point and solubility. This is because the shape of a molecule determines how it interacts with other molecules.

For very large molecules, such as DNA, proteins and enzymes, shape plays a key role in how they behave chemically and biologically. For example, the twisted double helix of DNA allows the molecule to coil up tightly so that it fits inside the nucleus of a cell.

To understand the shape of large molecules like DNA, scientists use complex techniques, such as X-ray crystallography or powerful computer simulations. However, the shape of small molecules can be predicted using a relatively simple model known as the **valence shell electron pair repulsion (VSEPR) theory**. In this section, you will see how VSEPR theory can be used to predict the shape of molecules such as the water molecule shown in Figure 12.1.1.

FIGURE 12.1.1 Water molecules (H_2O) have a distinctive shape that is responsible for many of its properties. The valence shell electron pair repulsion theory accurately predicts the shape of water molecules.

VALENCE SHELL ELECTRON PAIR REPULSION THEORY

As the name suggests, the VSEPR theory uses our knowledge of the valence electrons in the atoms of a molecule to predict the shape of the molecule. The VSEPR theory is based on the principle that negatively charged electron pairs in the outer shell of an atom repel each other. As a consequence, these electron pairs are arranged as far away from each other as possible.

Electron pair repulsion

In general, atoms in covalent molecules are most stable when they have eight electrons in their valence shell (the octet rule). These eight electrons are arranged into four pairs of electrons.

For example, the carbon atom in methane is covalently bonded to four hydrogen atoms. In this arrangement, the carbon atom shares a pair of electrons with each hydrogen atom. The four pairs of electrons give the carbon atom a stable octet as shown in Figure 12.1.2. In this diagram, the dots represent electrons donated to the single covalent bond by the carbon atom. The crosses represent electrons donated to the single covalent bond by the hydrogen atoms. (Note that dots and crosses are used here to distinguish the electrons of the hydrogen atoms from the electrons in the carbon atom valence shell. It is also correct to use dots for all electrons.)

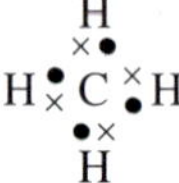

FIGURE 12.1.2 This electron dot diagram (Lewis structure) of a methane (CH_4) molecule shows the four electron pairs surrounding the central carbon.

Hydrogen atoms are an exception to the octet rule. Hydrogen atoms are stable with just two electrons because this gives them a full valence shell, or in other words, the stable electron configuration of a helium atom.

The VSEPR theory states that the electron pairs in methane repel each other so that they are as far apart as possible. In effect, this means that the angle between the bonds is as large as possible. This is a situation where it might help to imagine the electrons as a 'cloud' of negative charge—remembering that negative charges will always repel other negative charges. This repulsion between the electron pairs results in a tetrahedral shape as shown in the ball-and-stick model in Figure 12.1.3. The tetrahedral shape ensures that the electron pairs are as far from each other as possible with angles of 109.5° between all of the single bonds.

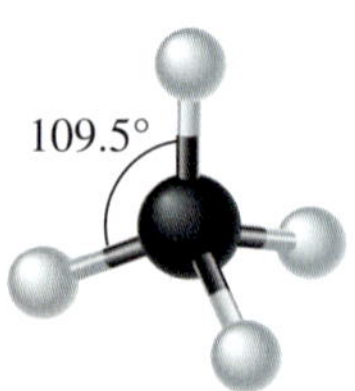

FIGURE 12.1.3 In a methane molecule (CH_4), the electron pairs in the single covalent bonds repel each other. The repulsion forces the bonds as far apart as possible, arranging the hydrogen atoms into a tetrahedral shape at an angle of 109.5° to each other.

Lone pairs of electrons

Not all electron pairs in molecules exist as covalent bonds. Some electrons form a non-bonding pair of electrons known as a lone pair of electrons. In the VSEPR theory, lone pairs of electrons are treated in the same way as electron pairs in covalent bonds in order to determine the shape of a molecule because they still take up space around the atom.

In the ammonia molecule shown in Figure 12.1.4, the nitrogen atom has a stable octet made up of one lone pair of electrons and three single bonds. The four electron pairs repel each other to form a tetrahedral arrangement. However, the lone pair is

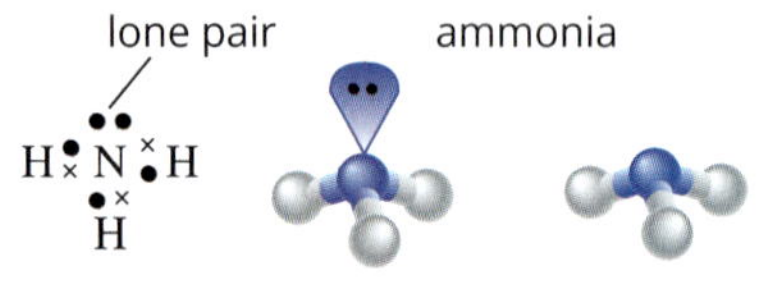

FIGURE 12.1.4 The electron dot diagram of an ammonia molecule (NH_3) shows that the nitrogen atom has one lone pair of electrons and three covalent bonds. These four electron pairs form a tetrahedral arrangement around the atom due to the repulsion of the electron pairs. The result is a pyramidal molecule.

not included when describing the shape of the molecule. Instead, the three hydrogen atoms are described as forming a pyramidal arrangement with the nitrogen atom. The lone pair occupies slightly more space than the bonding electrons (because it is not shared with another atom and is therefore closer to the carbon atom) so the three single covalent bonds are pushed closer together. The bond angle is therefore now slightly less than 109.5°.

In water molecules, the oxygen atom has a stable octet made up of two lone pairs and two single bonds. The four electron pairs repel each other to form a tetrahedral arrangement. This causes the two hydrogen atoms to form a V-shape or bent arrangement with the oxygen atom as shown in Figure 12.1.5. With two lone pairs in the molecule, the two single covalent bonds are pushed closer together. In this case, the bond angle around the central atom is also less than 109.5°.

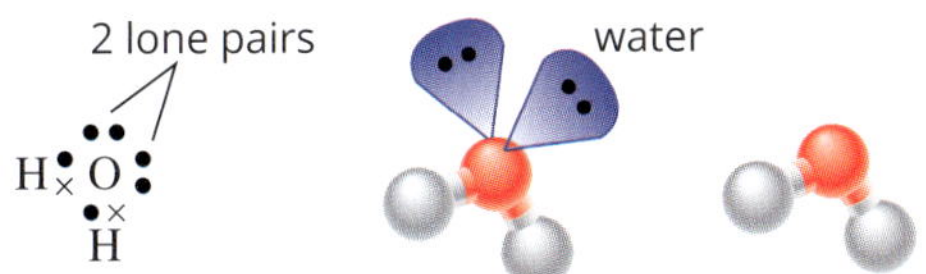

FIGURE 12.1.5 The electron dot diagram of a water molecule (H_2O) shows that the oxygen atom has two lone pairs of electrons and two covalent bonds. These four electron pairs form a tetrahedral arrangement around the atom due to the repulsion of the electron pairs. The result is a V-shaped or bent molecule.

In a hydrogen fluoride molecule, the fluorine atom has a stable octet made up of three lone pairs and one single bond. The four electron pairs repel each other to form a tetrahedral arrangement. Therefore, the hydrogen and fluorine atoms form a linear molecule, as you can see in Figure 12.1.6.

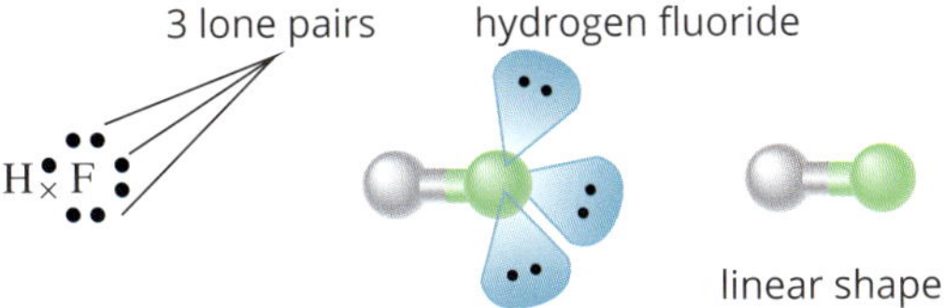

FIGURE 12.1.6 The electron dot diagram of a hydrogen fluoride molecule (HF) shows that the fluorine atom has three lone pairs of electrons and one covalent bond. These four electron pairs form a tetrahedral arrangement around the atom due to the mutual repulsion of the electron pairs. The result is a linear molecule.

Worked example 12.1.1

PREDICTING THE SHAPE OF MOLECULES

Predict the shape of a molecule of phosphine (PH_3).	
Thinking	**Working**
Draw the electron dot diagram for the molecule.	H:P̈:H with H below P (electron dot diagram)
Count the number of bonds and lone pairs on the central atom.	There are three bonds and one lone pair.
Determine how the groups of electrons will be arranged to get maximum separation.	Because there are four electron pairs, the groups will be arranged in a tetrahedral arrangement.
Deduce the shape of the molecule by considering the arrangement of just the atoms.	The phosphorus and hydrogen atoms are arranged in a pyramidal shape.

CHEMFILE

That's a-maser-ing!

Ammonia is not just a useful household cleaning product; it can also be used to create microwave lasers—'masers'. A pyramidal ammonia molecule with its hydrogen atoms pointing down can suddenly flip so that the hydrogen atoms are pointing up. This is a bit like an umbrella flipping inside-out on a windy day. When the ammonia molecule flips, it can release microwave radiation.

Charles Hard Townes and James Power Gordon used this property of ammonia to create the first maser, as shown in Figure 12.1.7. Their work laid the foundation for other types of masers, which are now used in radio telescopes to study the far reaches of the universe. Charles Hard Townes was awarded the Nobel Prize in Physics in 1964 for this work.

FIGURE 12.1.7 Charles Hard Townes and James Power Gordon invented the first maser using ammonia gas.

Lone pairs of electrons influence a molecule's shape but are not considered a part of the shape.

Worked example: Try yourself 12.1.1

PREDICTING THE SHAPE OF MOLECULES

Predict the shape of a molecule of hydrogen sulfide (H_2S).

Double bonds, triple bonds and valence shell electron pair repulsion theory

A double bond contains two pairs of electrons. A triple bond contains three pairs of electrons. In the VSEPR theory, double and triple bonds are considered to be just one 'cloud' of electrons, so the double or triple bond is treated in the same way as a single bond.

For example, if a central atom has two single bonds and one double bond, then the three sets of bonds will repel each other to get maximum separation. This results in a molecular shape known as trigonal planar because the atoms form a triangle in one plane. An example of this structure is the methanal molecule (CH_2O) shown in Figure 12.1.8. Trigonal planar molecules have bond angles of 120°.

methanal, CH_2O
(trigonal planar)

FIGURE 12.1.8 Methanal has a central atom that forms a double bond with an oxygen atom and single bonds with two hydrogen atoms. The bonds repel each other to form a trigonal planar arrangement.

If the central atom has two double bonds, then the two double bonds repel each other. This results in a linear molecule like carbon dioxide (CO_2), shown in Figure 12.1.9. A linear molecule has a bond angle of 180°.

FIGURE 12.1.9 In a carbon dioxide molecule, the carbon atom forms double bonds with two oxygen atoms. The two double bonds repel each other. This results in a linear molecule.

Finally, if the central atom has a single bond and a triple bond, as in hydrogen cyanide (HCN), then the molecule is also linear (Figure 12.1.10).

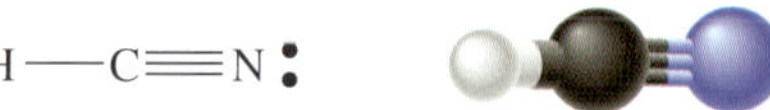

hydrogen cyanide, HCN
(linear)

FIGURE 12.1.10 A hydrogen cyanide molecule is linear. In this case, the central carbon atom forms a triple bond with the nitrogen atom and a single bond with the hydrogen atom.

Molecules that do not obey the octet rule

Not all atoms obey the octet rule. For example, boron forms stable molecules with just three electron pairs in its valence shell. In borane (BH_3), the central boron atom is bonded to three hydrogen atoms, but has no lone pairs. There are only three pairs of electrons in this molecule. These three pairs repel each other to form a trigonal planar molecule (Figure 12.1.11).

borane, BH_3 trigonal planar shape

FIGURE 12.1.11 Borane (BH_3) has a central boron atom attached to three hydrogen atoms.

The VSEPR theory can be used to predict the shape of these molecules. The VSEPR theory predicts that the three electron pairs will move as far away from each other as possible into a trigonal planar shape.

Boron trichloride has a similar arrangement. The three chlorine atoms are located 120° from each other around the central boron atom, as can be seen in Figure 12.1.12.

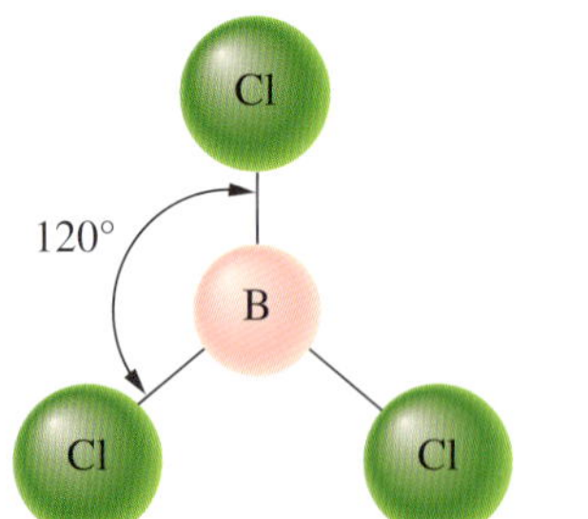

FIGURE 12.1.12 Boron trichloride (BCl_3) has a trigonal planar shape.

Beryllium forms stable molecules with just two electron pairs in its valence shell. In beryllium hydride (BeH_2), the central beryllium atom forms single bonds with two hydrogen atoms, but has no lone pairs of electrons. The two single bonds repel each other to form a linear molecule, as shown in Figure 12.1.13.

H Be H

beryllium hydride, BeH_2 linear shape

FIGURE 12.1.13 Beryllium hydride has a central beryllium atom attached to two hydrogen atoms. There are only two pairs of electrons in this molecule. These two pairs repel each other to form a linear molecule.

12.1 Review

SUMMARY

- The shapes of molecules play an important role in determining the intermolecular forces between molecules and are related to the physical properties of substances.
- The shapes of simple molecules can be predicted by the valence shell electron pair repulsion (VSEPR) theory as shown in Table 12.1.1.
- The VSEPR theory is based on the principle that electron pairs in the outer shell of an atom repel each other. As a consequence, these electron pairs are arranged as far away from each other as possible.
- Electron pairs can exist in covalent bonds or as lone pairs of electrons. Lone pairs and covalent bonds repel each other.
- Lone pairs influence a molecule's shape but are not considered a part of the shape.
- Double and triple bonds are treated in the same way as single bonds and lone pair electrons in VSEPR theory.

TABLE 12.1.1 Summary of shapes of molecules

Example	Methane, CH_4	Ammonia, NH_3	Water, H_2O	Hydrogen fluoride, HF
electron dot diagram	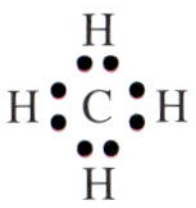	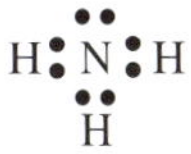	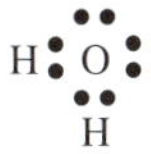	
number of single bonds	4	3	2	1
number of lone pairs	0	1	2	3
ball-and-stick model				
shape	tetrahedral	pyramidal	V-shaped or bent	linear

KEY QUESTIONS

1 Explain VSEPR theory and how it is used to determine the shape of molecules.

2 How many electron pairs are there in the valence shell of the fluorine atom in a hydrogen fluoride molecule? State how many single bonds and lone pairs are present.

3 Draw the Lewis structure for each of the following molecules.

a H_2S
b HI
c CCl_4
d PH_3
e CS_2

4 Describe the shape of each of the molecules in Question **3**.

5 Copy and complete the table below, determining the shape of each of the molecules with the specific number of bonds listed.

Molecule	Number of single bonds on the central atom	Number of lone pairs of electrons on the central atom	Shape of the molecule
a	4	0	
b	3	1	
c	2	2	

12.2 Properties of covalent molecular substances

The physical properties of covalent molecular substances are primarily determined by the strength of the forces between the molecules. These forces are known as intermolecular forces. The shape of a molecule is one factor that determines the strength of intermolecular forces. However, the covalent bonds inside the molecules also play an important role.

Both the shape of a molecule and the nature of its covalent bonds must be taken into account when predicting the strength of intermolecular forces. This section examines how these factors determine the strength of intermolecular forces and therefore determine important physical properties such as vapour pressure, melting point, boiling point and solubility.

INTERMOLECULAR FORCES

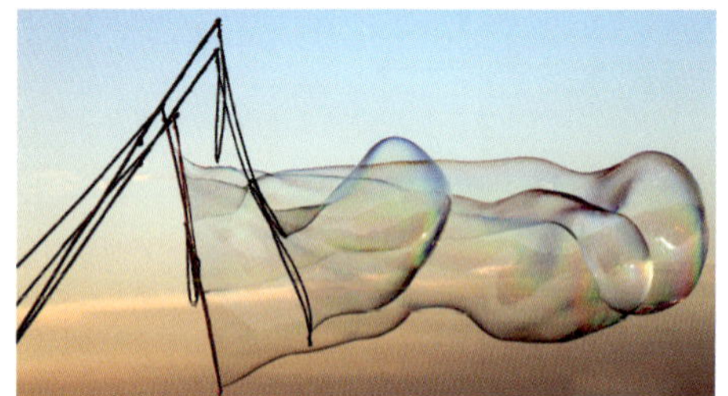

FIGURE 12.2.1 These bubbles are held together by intermolecular forces.

Intermolecular forces exist between molecules and their effects are seen everywhere. If you have ever struggled to flatten out a sheet of cling wrap, then you've experienced the effects of intermolecular forces. Glue is an example of how intermolecular forces can be used to stick things together. The bubbles shown in Figure 12.2.1 are also held together by intermolecular forces.

If you take a closer look at liquids, you will notice further evidence of intermolecular bonding. Intermolecular bonds create the **surface tension** that allows insects to walk on water, as shown in Figure 12.2.2.

FIGURE 12.2.2 The surface tension created by intermolecular forces allows insects to walk on water.

Strength of intermolecular forces

Intermolecular forces are 10–100 times weaker than the strong intramolecular forces such as ionic, metallic and covalent bonds. For example, diamond is made up of a lattice of carbon atoms that are bonded with strong covalent bonds. This extremely strong lattice is the reason why diamond has a very high melting point and is the hardest natural substance on Earth.

The atoms in covalent molecules such as water (H_2O) and carbon dioxide (CO_2) are also held together by strong covalent bonds. However, covalent molecular substances tend to have much lower melting and boiling points than ionic, metallic and covalent network substances. This is because the forces *between* the molecules are much weaker. It is the weak intermolecular forces that are broken when a covalent molecular substance is converted from a solid to a liquid or a liquid to a gas.

The vapour pressure of a substance is also related to its intermolecular forces. When a liquid is put into a closed container, some of the particles escape from the surface liquid. These particles form a gas or a vapour that exerts a pressure on the container due to their collisions with the container wall. This pressure is known as the **vapour pressure** (Figure 12.2.3).

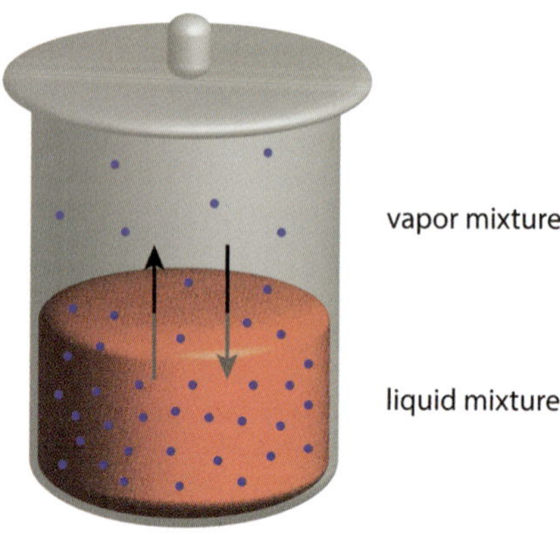

FIGURE 12.2.3 Vapour pressure is the pressure that the gaseous molecules exert on the closed container walls when the rates of evaporation and condensation become equal.

Liquids with stronger intermolecular forces have lower vapour pressures. For example, water has a vapour pressure at room temperature of 3.17 kPa. On the other hand, the liquid used as a lighter fluid (butane) has a much higher vapour pressure, of approximately 200 kPa. This reflects the fact that the intermolecular forces in water are much stronger than the intermolecular forces in butane. (kPa stands for kilopascals, one of many units used for pressure—normal atmospheric pressure is 101 kPa.)

In a liquid with strong intermolecular forces, the molecules are held together tightly, making it harder for them to escape from the surface of the liquid. Therefore, vapour pressure is also lower.

When a liquid is heated, it shows a greater tendency to evaporate as the molecules have a higher average kinetic energy that enables them to overcome the intermolecular forces holding the molecules together. As the temperature increases, a greater proportion of the particles will have sufficient kinetic energy to escape from the liquid to produce the gaseous phase, and the vapour pressure of the liquid increases (Figure 12.2.4).

The boiling point of a liquid is defined as the temperature at which the liquid's vapour pressure reaches the atmospheric pressure of the surroundings.

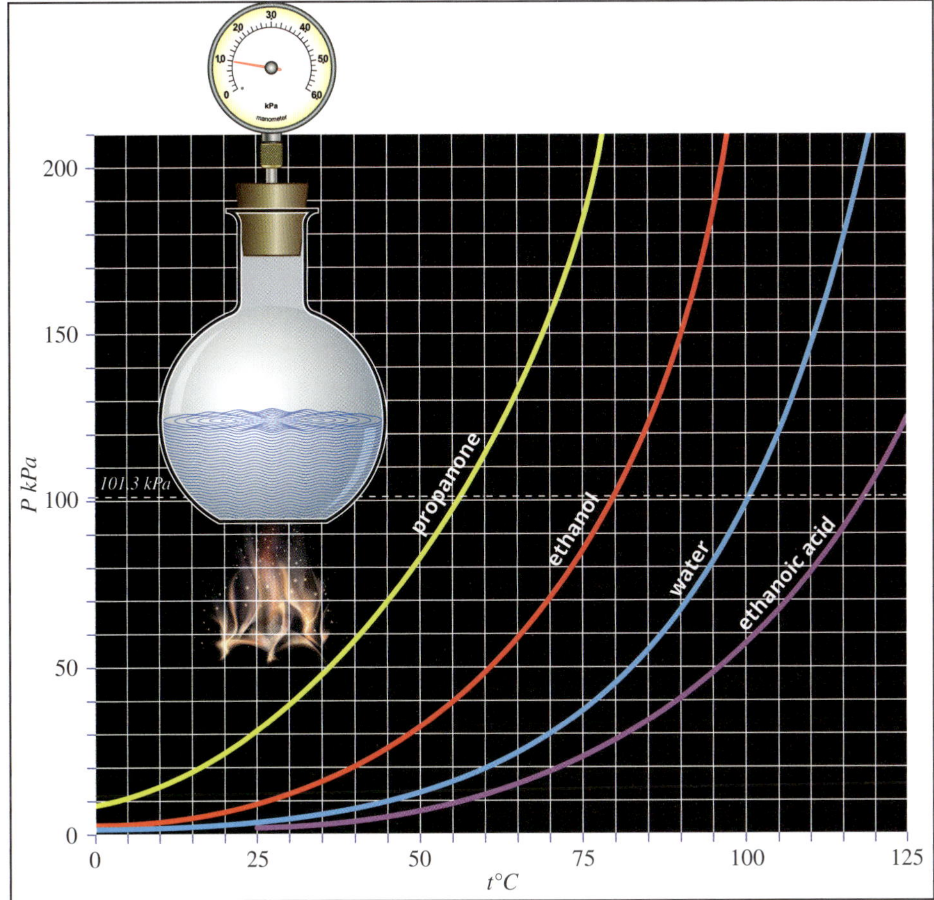

FIGURE 12.2.4 As temperature increases, the vapour pressure increases. The boiling points of liquids can be predicted by identifying the temperature at which the vapour pressure is equal to atmospheric pressure. Here, at an atmospheric pressure of 101.3 kPa, ethanol has a boiling point of about 80°C and water about 100°C.

> When liquids change to gases, the intermolecular bonds are broken. For example, when liquid water changes to gaseous water, the intermolecular bonds holding the water molecules to other water molecules are broken, not the covalent bonds between the hydrogen and oxygen atoms.

ELECTRONEGATIVITY AND POLARITY

Intermolecular forces are an example of electrostatic forces. You have already seen many other examples of how electrostatic forces form bonds at the atomic level. The attractions between positive and negative ions in an ionic crystal lattice (Chapter 5) and the attraction of shared electrons and nuclei in a covalent bond (Chapter 6) are all examples of electrostatic forces.

The electrostatic attraction between molecules works in a similar way. In intermolecular forces, the electrostatic attraction is between positive and negative charges in the molecules. These charges appear as a result of uneven electron distributions within the molecules. The following sections examine how the shape of the molecule and the electronegativity of its atoms can cause these uneven electron distributions.

Polarity of diatomic molecules

Electronegativity is the key factor that determines the electron distribution in molecules. Electronegativity is the tendency of an atom in a covalent bond to attract electrons. Electronegativity increases from left to right across the periods of the periodic table and decreases down the groups of the table, as shown in Figure 12.2.5. You will remember seeing these patterns in Chapter 3.

Electronegativity increases →

Electronegativity decreases ↓

Key: 2.20 H — electronegativity, symbol

Period \ Group	1	2	3	4	5	6	7	8	9	10	11	12	13	14	15	16	17	18
1	2.20 H																	He
2	0.98 Li	1.57 Be											2.04 B	2.55 C	3.04 N	3.44 O	3.98 F	Ne
3	0.93 Na	1.31 Mg											1.61 Al	1.90 Si	2.19 P	2.58 S	3.16 Cl	Ar
4	0.82 K	1.00 Ca	1.36 Sc	1.54 Ti	1.63 V	1.66 Cr	1.55 Mn	1.83 Fe	1.88 Co	1.91 Ni	1.90 Cu	1.65 Zn	1.81 Ga	2.01 Ge	2.18 As	2.55 Se	2.96 Br	Kr
5	0.82 Rb	0.95 Sr	1.22 Y	1.33 Zr	1.6 Nb	2.16 Mo	1.9 Tc	2.2 Ru	2.28 Rh	2.20 Pd	1.93 Ag	1.69 Cd	1.78 In	1.96 Sn	2.05 Sb	2.1 Te	2.66 I	Xe
6	0.79 Cs	0.89 Ba	La*	1.3 Hf	1.5 Ta	2.36 W	1.9 Re	2.2 Os	2.20 Ir	2.28 Pt	2.54 Au	2.00 Hg	1.62 Tl	2.33 Pb	2.02 Bi	2.0 Po	2.2 At	Rn
7	0.7 Fr	0.9 Ra	Ac**	Rf	Db	Sg	Bh	Hs	Mt	Ds	Rg	Cn	Nh	Fl	Mc	Lv	Ts	Og

Lanthanoids*	1.12 Ce	1.13 Pr	1.14 Nd	1.13 Pm	1.17 Sm	1.2 Eu	1.2 Gd	1.1 Tb	1.22 Dy	1.23 Ho	1.24 Er	1.25 Tm	1.1 Yb	1.27 Lu
Actinoids**	1.3 Th	1.5 Pa	1.38 U	1.36 Np	1.28 Pu	1.13 Am	1.28 Cm	1.3 Bk	1.3 Cf	1.3 Es	1.3 Fm	1.3 Md	1.3 No	Lr

FIGURE 12.2.5 Periodic table with electronegativity values. Electronegativity increases from left to right across the periods and decreases down the groups.

Non-polar diatomic molecules

When two atoms form a covalent bond, you can regard the atoms as competing for the electrons being shared between them. If the two atoms in a covalent bond are the same (i.e. have identical electronegativities), then the electrons are shared equally between the two atoms. This is the case for diatomic molecules such as chlorine (Cl_2), oxygen (O_2), hydrogen (H_2) and nitrogen (N_2).

Bonds with an equal distribution of valence (outer shell) electrons are said to be **non-polar** because there is no charge on either end of the molecule.

Figure 12.2.6 shows the electron distribution in the non-polar fluorine molecule (F_2). The molecule has a high **electron density** between the two fluorine atoms, forming the covalent bond. However, the valence electrons are distributed evenly between the two atoms, making the bond non-polar.

Electron density is the measure of the probability of an electron being present at a particular location within an atom. In molecules, areas of electron density are commonly found around the atom, and its bonds.

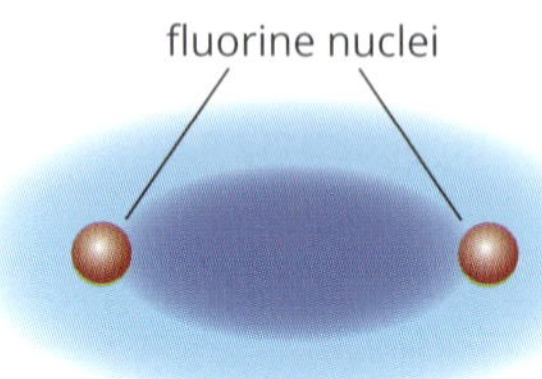

FIGURE 12.2.6 Fluorine molecules have a symmetric distribution of electrons and are therefore non-polar.

Polar diatomic molecules

If the covalent bond is between atoms of two different elements, then the electrons will stay closer to the most electronegative atom as it has a stronger pull on the electrons in the bond. An example is the hydrogen fluoride molecule (HF), shown in Figure 12.2.7. Molecules with an imbalanced electron distribution are said to be **polar**.

A fluorine atom (electronegativity = 3.98) is more electronegative than a hydrogen atom (electronegativity = 2.20). Therefore, in a hydrogen fluoride molecule the electrons tend to stay closer to the fluorine atom. This imbalance in the electron distribution means the fluorine atom is negatively charged and the hydrogen atom is positively charged. A molecule with two oppositely charged poles is said to be polar. The fluorine atom is described as having a partial negative charge, which is represented with the Greek letter delta as $\delta-$. The hydrogen atom is described as having a partial positive charge, $\delta+$. The separation of the positive and negative charges is known as an **electric dipole**, or simply as a **dipole** as they have two oppositely charged poles at each end of the molecule.

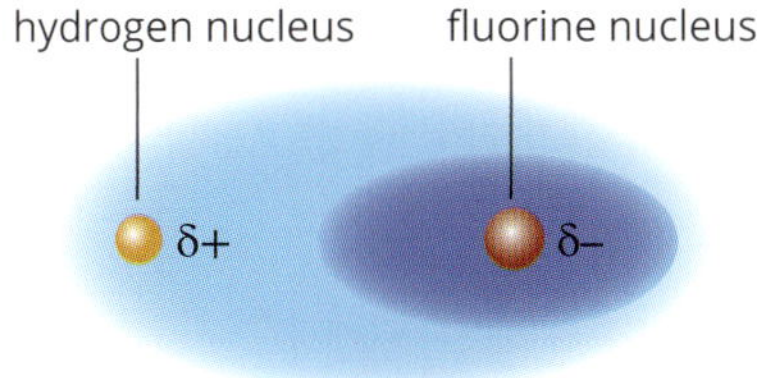

HF contains a polar bond; fluorine is more electronegative than hydrogen.

FIGURE 12.2.7 The electron distribution in hydrogen fluoride is asymmetric because of the different electronegativities of the hydrogen and fluorine atoms. Hydrogen fluoride is an example of a polar molecule.

Hydrogen fluoride is polar because it has a permanent electric dipole created by the different electronegativities of the two atoms. All diatomic molecules that are made up of atoms with different electronegativities are polar to some extent. The level of **polarity** will depend on the difference between the electronegativities of the two atoms. The greater the difference between the electronegativities, the greater the polarity of the molecule.

The partial charges on polar molecules are different from the charges on ions. They can be considered in terms of the slightly negative end of the bond having a larger share of the electrons, but as the atom still has its normal number of protons in the nucleus, overall this end of the bond can be considered as having an excess of electrons. Conversely, the atom at the slightly positive end of the bond has lost some of its share of the electrons that would have balanced the positive charge of the protons. One way to think about this is that this end is slightly 'naked' in terms of electrons—exposing the positive charge beneath.

> As the difference in electronegativities of two atoms increases, the covalent bond formed between them increases in polarity. In other words, it becomes more like an ionic bond.

The partial charges on a polar molecule will always add to give a total charge of zero, since the charges on polar molecules are due to the unequal sharing of electrons between atoms, so the molecules are still neutral overall. There are no electrons lost or gained as happens in the formation of ions. The partial charges on a polar molecule are also a lot smaller than the charges on ions. That is why intermolecular bonds are much weaker than ionic bonds.

It is not just the covalent bonds in diatomic molecules that can be polar. Covalent bonds in larger molecules can have some degree of polarity. The polarity of any covalent bond can be compared by examining the difference in the electronegativities of the atoms involved in the bond.

Note that if the electronegativity of the atoms bonded together is very similar, the bond can be considered as non-polar, even if there is a small difference in electronegativity. For example, the difference between the electronegativities of carbon and hydrogen is 2.55 – 2.20 = 0.35. Because this is a relatively small difference, C–H bonds can be considered to be non-polar. This is important as C–H bonds are very common in hydrocarbons and other organic molecules that you will encounter later in the course.

Worked example 12.2.1

COMPARING THE POLARITY OF COVALENT BONDS

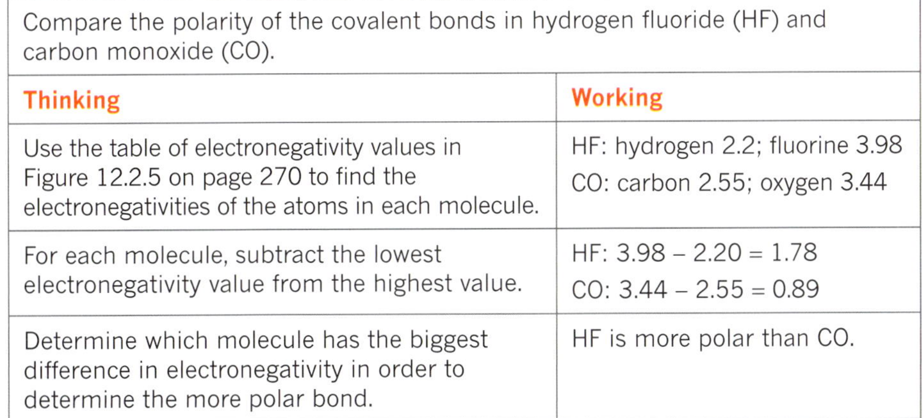

Compare the polarity of the covalent bonds in hydrogen fluoride (HF) and carbon monoxide (CO).	
Thinking	**Working**
Use the table of electronegativity values in Figure 12.2.5 on page 270 to find the electronegativities of the atoms in each molecule.	HF: hydrogen 2.2; fluorine 3.98 CO: carbon 2.55; oxygen 3.44
For each molecule, subtract the lowest electronegativity value from the highest value.	HF: 3.98 – 2.20 = 1.78 CO: 3.44 – 2.55 = 0.89
Determine which molecule has the biggest difference in electronegativity in order to determine the more polar bond.	HF is more polar than CO.

Worked example: Try yourself 12.2.1

COMPARING THE POLARITY OF COVALENT BONDS

Compare the polarity of the covalent bonds in nitrogen monoxide (NO) and hydrogen chloride (HCl).

Polarity of polyatomic molecules

Determining the polarity of polyatomic molecules (those with more than two atoms) is a little more complicated. This is because the polarity of polyatomic molecules depends on the shape of the molecule as well as the polarity of the covalent bonds. As a general rule:

- **symmetrical molecules** are non-polar
- **asymmetrical molecules** are polar.

In this context, it is the symmetry of the polar bonds (dipoles) that is important and symmetry may need to be considered in more than one dimension. One way to consider this is, in the diagrams that follow, imagine that the electrons are being pulled towards the slightly negative end of the dipoles; then consider whether the direction and strength of the individual dipoles (pull of electrons) cancel each other out. If they do, this molecule would be classified as symmetrical and would be non-polar (an analogy for this would be a dead heat in a 'tug-of-war' for electrons). If the dipoles are not balanced out (remembering there might be up to four to consider) then the molecule is considered asymmetrical and the molecule would have a net dipole and be polar.

Non-polar molecules

Even molecules with polar covalent bonds can be non-polar if the molecule is symmetrical.

In methane, the carbon atom is slightly more electronegative than the hydrogen atoms. Therefore, the carbon atom attracts the electrons more strongly than each hydrogen atom and has a partial negative charge, leaving hydrogen with a partial positive charge. However, the methane molecule is a perfect tetrahedron and is therefore symmetrical. The symmetry of the molecule means that the individual dipoles of the covalent bonds cancel each other perfectly. The result is a molecule with no overall dipole. It is non-polar.

The methane molecule shown in Figure 12.2.8 is an example of a non-polar molecule with polar covalent bonds.

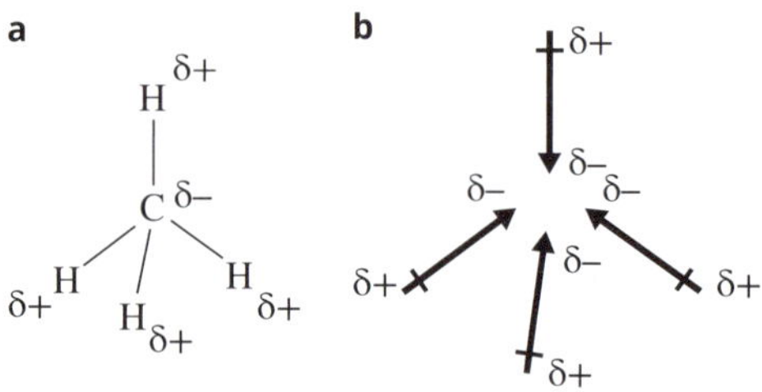

FIGURE 12.2.8 (a) Structure of a methane molecule showing the partial charges on the atoms. (b) The individual dipoles are distributed symmetrically around the molecule.

Polar molecules

In asymmetrical molecules, the individual dipoles of the covalent bonds do not cancel each other. This results in a net dipole, making the overall molecule polar.

The chloromethane molecule shown in Figure 12.2.9 is an example of an asymmetric molecule, shown by its **valence structure**. The chlorine atom is more electronegative than the carbon atom. Therefore, the chlorine atom attracts electrons from the carbon atom while the carbon atom attracts electrons from the hydrogen atoms. The individual dipoles of the covalent bonds are shown in the central image in Figure 12.2.9. These add to give the molecule a net dipole.

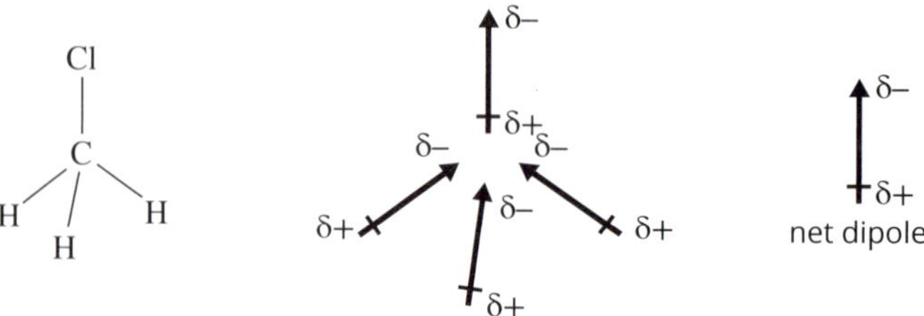

FIGURE 12.2.9 Chloromethane is a polar molecule because the individual dipoles of the covalent bonds add to give a net dipole.

Table 12.2.1 shows some more examples of how symmetry determines the polarity of covalent molecules.

TABLE 12.2.1 Examples of polar and non-polar covalent molecules

Molecule	Valence structure	Symmetrical/ asymmetrical	Polar/non-polar
methanal (formaldehyde)	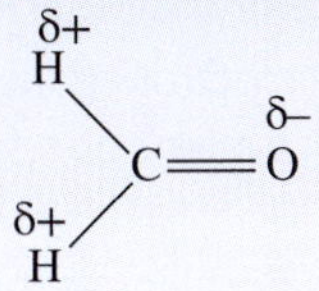 δ+ H, δ+ H bonded to C; C=O δ−	asymmetrical	polar
carbon dioxide	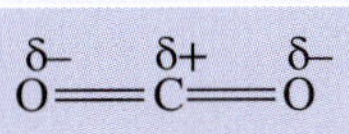 δ− O=C=O δ− (δ+ on C)	symmetrical	non-polar
tetrafluoromethane	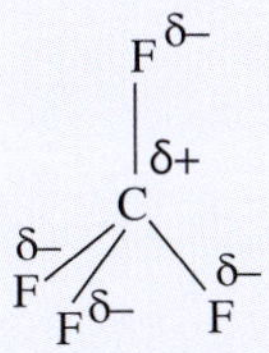 C δ+ bonded to four F δ−	symmetrical	non-polar
water	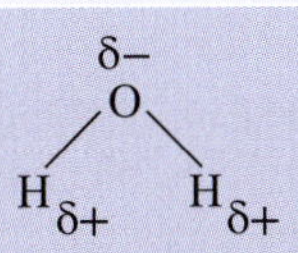 O δ− bonded to two H δ+	asymmetrical	polar
ammonia	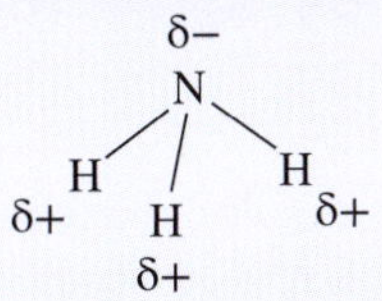 N δ− bonded to three H δ+	asymmetrical	polar

The polarity of a molecular substance has a significant effect on its solubility. In general, substances that are polar tend to dissolve in polar solvents. Non-polar substances tend to dissolve in non-polar solvents. You can remember this with the saying 'like dissolves like'. You will learn more about the effect of intermolecular forces on the solubility of substances in Chapter 15 and Chapter 16.

CHEMFILE

How a microwave oven works

Microwave ovens use the polarity of molecules (especially water molecules) to heat food. The microwave oven irradiates the food with microwaves. The microwaves produce an electric field that interacts with the polar molecules. The electric field causes the molecules to vibrate several million times per second. This gives the molecules extra kinetic energy. As the kinetic energy of the molecules increases, the temperature increases. Therefore, the molecules heat up and cook the food.

12.2 Review

SUMMARY

- Many of the physical properties of covalent molecular substances are determined by intermolecular forces. These properties include solubility, melting point, vapour pressure and boiling point.
- Intermolecular forces are broken when liquids change to gas.
- Intermolecular forces are 10–100 times weaker than strong intramolecular forces such as ionic, covalent and metallic bonds.
- Shape and polarity of molecules are factors in determining the strength of intermolecular forces.
- As the difference in electronegativities of two atoms increases, a covalent bond increases in polarity.
- Diatomic molecules containing the same type of atom are non-polar.
- In general, symmetrical molecules are non-polar and asymmetrical molecules with polar bonds are polar.

KEY QUESTIONS

1 Covalent bonds can form between the following pairs of elements in a variety of compounds. Use the electronegativity values given in Figure 12.2.5 on page 270 to identify the atom in each pair that would have the largest share of bonding electrons.

a S and O
b C and H
c C and N
d N and H
e F and O
f P and F

2 The greater the differences in electronegativity between two atoms, the more polar is the bond formed between them.

a Which of the examples in Question **1** would be the most polar bond?
b Which of the examples in Question **1** would be the least polar bond?

3 Use the electronegativity values in Figure 12.2.5 on page 270 to order the polarity of the covalent bonds between the following atoms from highest to lowest: carbon–nitrogen, nitrogen–nitrogen, hydrogen–nitrogen, sulfur–nitrogen, oxygen–nitrogen.

4 Label the following diagram of a tetrachloromethane molecule to show the dipoles between the atoms.

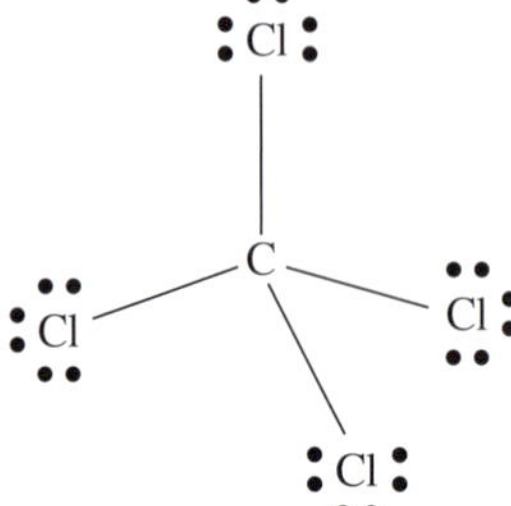

5 Examine the symmetry of each of these general diagrams of molecular structures, and determine if the molecules are likely to be polar or non-polar. The white, black and red circles represent atoms with different electronegativities.

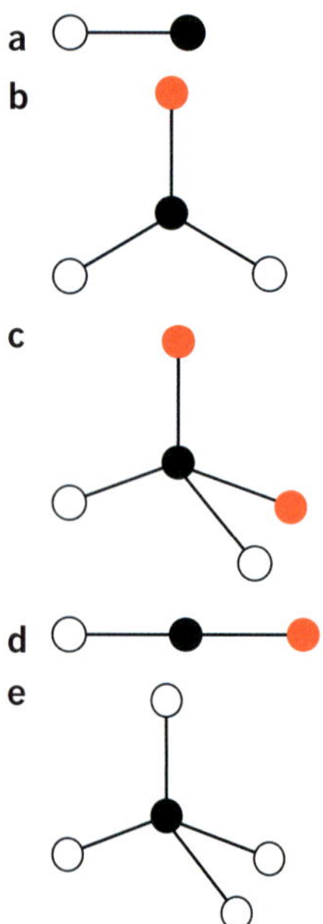

6 Select the correct description of how the electrons are distributed in a polar, diatomic molecule.

A The electrons are located halfway between the two atoms.
B On average, the electrons are distributed evenly across the entire molecule.
C On average, the electrons spend more time near the most electronegative atom.
D On average, the electrons spend more time near the least electronegative atom.

12.3 Types of intermolecular forces

Many factors determine the strength of intermolecular forces, including the size, shape and polarity of molecules. These factors not only determine the strength of the intermolecular forces in a substance, they also determine the types of intermolecular forces.

There are three main types of intermolecular forces:

- dipole–dipole forces
- hydrogen bonding
- dispersion forces.

In this section, you will examine the nature of these three types of intermolecular forces and their role in determining the physical properties of covalent molecular substances.

DIPOLE-DIPOLE FORCES

Dipole–dipole forces only occur in polar molecules. These forces result from the attraction between the positive and negative ends of the polar molecules, as shown in Figure 12.3.1.

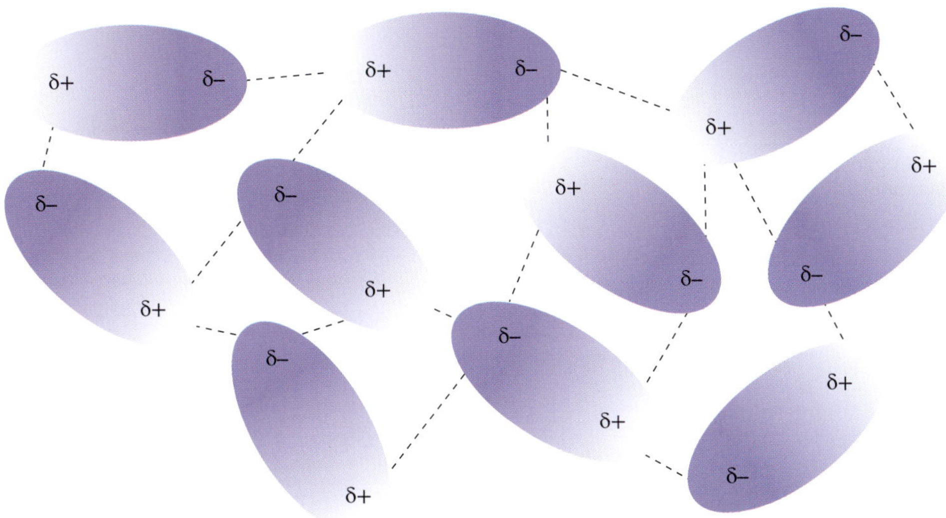

FIGURE 12.3.1 The positive and negative ends of polar molecules attract each other.

Dipole–dipole forces are relatively weak since the partial charges (the δ+ and δ–) on the molecules are small. However, the more polar a molecule is, the stronger the dipole–dipole forces are. The polarity will be larger when there is a large difference in the electronegativities of the atoms or a high degree of asymmetry in the shape of the molecule causing a large imbalance in the bond dipoles.

Molecules are more polar when there is a greater difference in electronegativity of the atoms or a large asymmetry in the shape of the molecule. Molecules that are more polar have stronger dipole–dipole forces between them.

The strength of the dipole–dipole forces in molecules is directly related to the melting and boiling points of the substance. The stronger the dipole–dipole forces, the higher the melting and boiling points. This is because dipole–dipole forces bond the molecules together in the solid or liquid. Stronger dipole–dipole forces require more energy (i.e. higher temperatures) to break these bonds and change the solid to a liquid or the liquid to a gas.

For example, compare methanal (CH_2O) and ethane (CH_3CH_3), shown in Figure 12.3.2. Methanal is asymmetrical and therefore a polar molecule. This results in dipole–dipole forces between the molecules. Methanal's boiling point is −19°C. On the other hand, the ethane molecule is symmetrical and therefore non-polar. This means there are no dipole–dipole forces and its boiling point is much lower at −88.5°C.

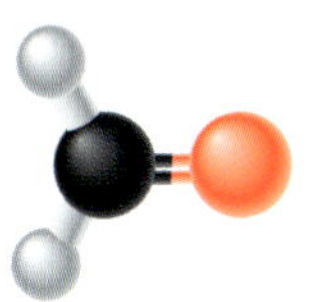
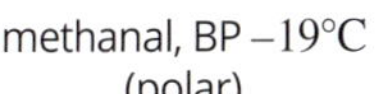
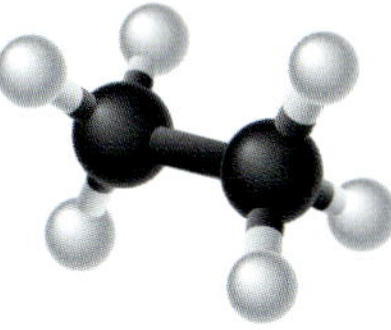

methanal, BP −19°C (polar) ethane, BP −88.5°C (non-polar)

FIGURE 12.3.2 Molecular structures of methanal and ethane

Hydrogen bonding

Hydrogen bonding is a special form of dipole–dipole force. Hydrogen bonding only occurs between molecules in which a hydrogen atom is covalently bonded to an oxygen, a nitrogen or a fluorine atom.

Oxygen, nitrogen and fluorine atoms are small and highly electronegative. Therefore, they strongly attract the electrons in a covalent bond. This creates a significant partial positive charge on a hydrogen atom bonded to these atoms. The partial positively charged hydrogen atom is then attracted to lone pairs of electrons in the nitrogen, oxygen and fluorine atoms of neighbouring molecules, as shown in Figure 12.3.3. Remember that hydrogen only has one electron and this electron is strongly attracted to the nitrogen, oxygen or fluorine atom in the covalent bond within the molecule that contains the hydrogen atom. This leaves this partially positive hydrogen atom 'exposed' due to its lack of electrons and it is attracted to a lone pair of electrons on a nitrogen, oxygen or fluorine atom in another molecule.

The result is a relatively strong intermolecular bond that is approximately 10 times stronger than a dipole–dipole bond, but about one-tenth the strength of an ionic or a covalent bond.

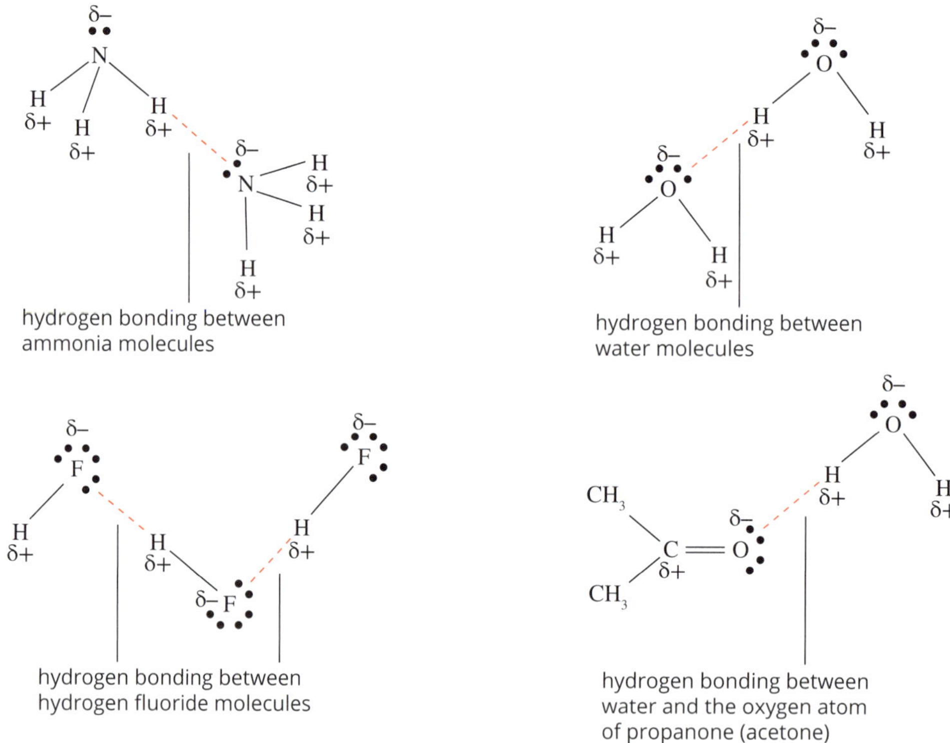

FIGURE 12.3.3 Examples of polar hydrogen bonding between molecules

The presence of hydrogen bonds results in higher melting and boiling points. Figure 12.3.4 demonstrates the effect of hydrogen bonding on boiling point by comparing methanol to ethane and methanal. Recall that ethane is non-polar so it has no dipole–dipole forces between molecules. Methanal has dipole–dipole forces but does not have hydrogen bonding. Methanol has hydrogen bonding between molecules because of the hydrogen atom attached to the oxygen atom. As a result, the boiling point of methanol (64.7°C) is significantly higher than the boiling points of methanal (–19°C) and ethane (–88.5°C).

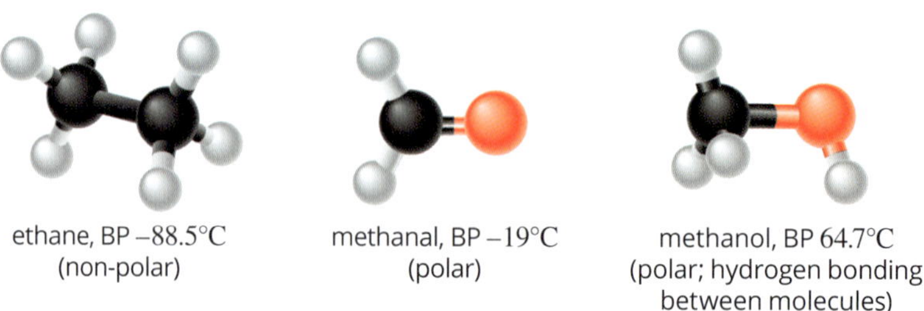

FIGURE 12.3.4 Molecular structures of ethane, methanal and methanol

The two key requirements for hydrogen bonding are:

1 a hydrogen atom covalently bonded to an oxygen, a nitrogen or a fluorine atom
2 a lone pair of electrons on the nitrogen, oxygen or fluorine atoms of neighbouring molecules.

It is both the very high electronegativities and small atomic radii of oxygen, nitrogen and fluorine atoms, combined with the single electron of the hydrogen atom, that cause the formation of the hydrogen bonds. Chlorine atoms also have a very high electronegativity but do not form hydrogen bonds because of their larger atomic radius, which reduces the concentration of negative charge around the atom.

Hydrogen bonding occurs in molecules where a hydrogen atom is bonded to an oxygen, a nitrogen or a fluorine atom.

Hydrogen bonding in water

Water is a very common substance but it has some unusual properties. One unusual property of water is that ice floats. Ice floats because ice is less dense than liquid water. For most other substances, the solid form is denser than the liquid.

The fact that ice is less dense than liquid water can be explained by hydrogen bonding. Water has two hydrogen atoms attached to an oxygen atom in a V-shape. Therefore, the hydrogen atoms can form hydrogen bonds with the lone pairs of electrons on the oxygen atoms of neighbouring molecules as shown in Figure 12.3.5. In this way, a water molecule can form hydrogen bonds with four other water molecules.

The hydrogen bonding holds the water molecules in ice in a regular crystal lattice. In this lattice, the molecules are held further apart than in liquid water, as shown in Figure 12.3.6. As a result, ice is less dense and therefore floats in liquid water.

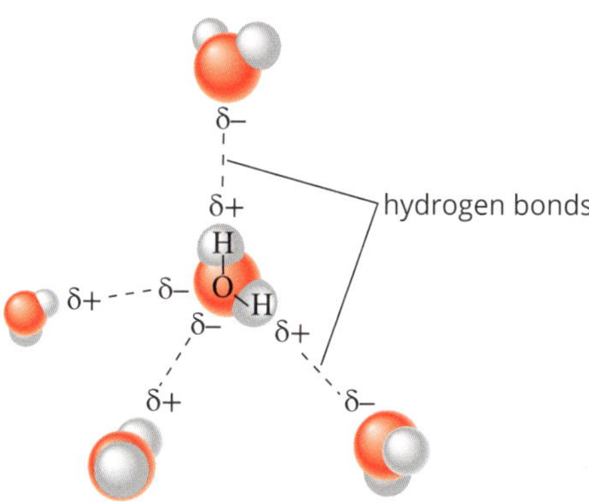

FIGURE 12.3.5 Each water molecule can form hydrogen bonds with four other water molecules. The two lone pairs of electrons on the molecule bond to hydrogen atoms on two neighbouring molecules. In addition, the two hydrogen atoms on the molecule form bonds with lone pairs of electrons in two neighbouring molecules.

FIGURE 12.3.6 Ice floats because the water molecules form a crystal lattice in which the molecules are spaced more widely apart than in liquid water. This arrangement means ice is less dense than liquid water.

When ice melts and forms liquid water, the density of the water increases rapidly. This is because the open crystal lattice collapses and the water molecules pack together more tightly.

However, as the temperature of the water increases further, water molecules in the liquid move and vibrate more rapidly. The movement causes the molecules to spread further apart. As the molecules move further apart, the liquid becomes less dense.

CHEMFILE

Snowflake symmetry

No two snowflakes are exactly alike but all snowflakes have something in common. Every snowflake has six symmetrical arms. This phenomenon is the result of hydrogen bonding. The hydrogen bonding between the water molecules creates a crystal lattice with hexagonal symmetry. As a result, snowflakes show the same hexagonal symmetry as the crystal lattice.

FIGURE 12.3.7 A snowflake with hexagonal symmetry

Dispersion forces

Dipole–dipole forces and hydrogen bonding explain the intermolecular forces in polar substances. However, they do not explain the existence of intermolecular forces in non-polar substances.

We know that intermolecular forces are present in non-polar substances because non-polar substances form liquids and solids. Without intermolecular forces, there would be nothing to hold the molecules together and non-polar substances would only exist as gases. However, there are many non-polar compounds that are liquids at room temperature, such as vegetable oil, and even non-polar solids, such as candle wax. In fact, all non-polar substances form liquids or solids if cooled to a low enough temperature. Even hydrogen liquefies below −259°C.

The forces of attraction between non-polar molecules are known as **dispersion forces**. Dispersion forces are caused by **temporary dipoles** in the molecules that are the result of random movement of the electrons surrounding the molecule. These temporary dipoles are also known as **instantaneous dipoles**.

Dispersion forces are always present between molecules, no matter whether they are polar or non-polar, as electrons are constantly in motion within atoms (Figure 12.3.8).

(a) In a molecule, the electrons are constantly moving. In the case of a non-polar molecule, such as fluorine, the electrons spend an equal amount of time around each atom.

F F

(b) Occasionally, the electrons gather more closely together at one end of the molecule, causing one end of the molecule to become negative and the other end to become positive. This is known as a temporary dipole.

(c) These temporary dipoles can then induce (create) dipoles in the neighbouring molecules.

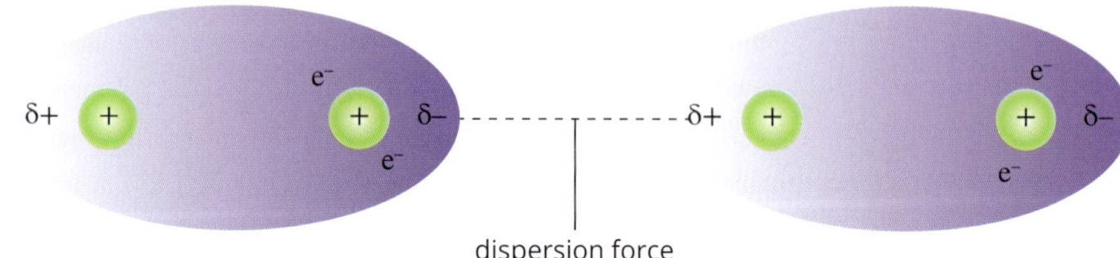

(d) The neighbouring molecules then induce dipoles in their neighbours and so on. The temporary dipoles attract each other to create the intermolecular forces known as dispersion forces.

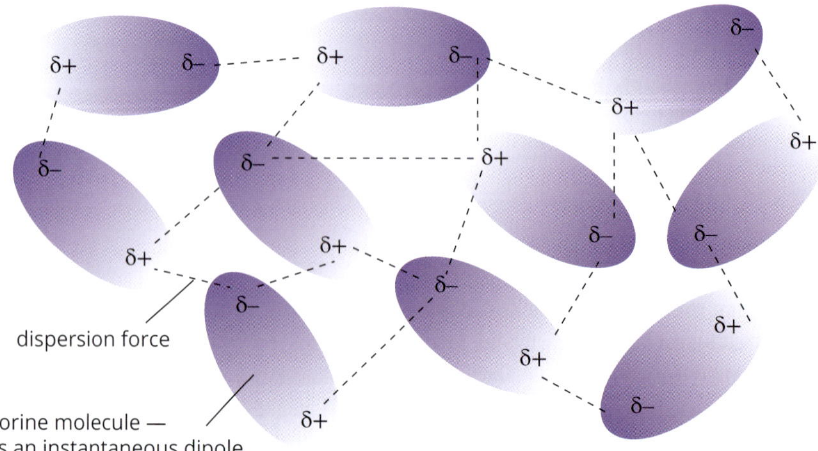

FIGURE 12.3.8 How dispersion forces form within non-polar substances. These diagrams show the forces forming between fluorine molecules.

Strength of dispersion forces

The strength of dispersion forces increases as the size of the molecule increases. Larger molecules have a larger number of electrons. It is easier to produce temporary dipoles in molecules with large numbers of electrons. Since larger molecules have stronger dispersion forces, they have higher melting and boiling points.

Table 12.3.1 shows the boiling points of the halogens (group 17), which all form non-polar, diatomic molecules. The only forces between their molecules are dispersion forces. You can see that as the molecules increase in size and the dispersion forces become stronger, the boiling points of the substances increase.

TABLE 12.3.1 The effect of dispersion forces and the size of atoms on the boiling points of some molecules

Molecule	Molecular mass	Number of electrons	Boiling point (°C)
fluorine (F_2)	38.0	18	–188
chlorine (Cl_2)	70.9	34	–35
bromine (Br_2)	159.8	70	59
iodine (I_2)	253.8	106	184

a

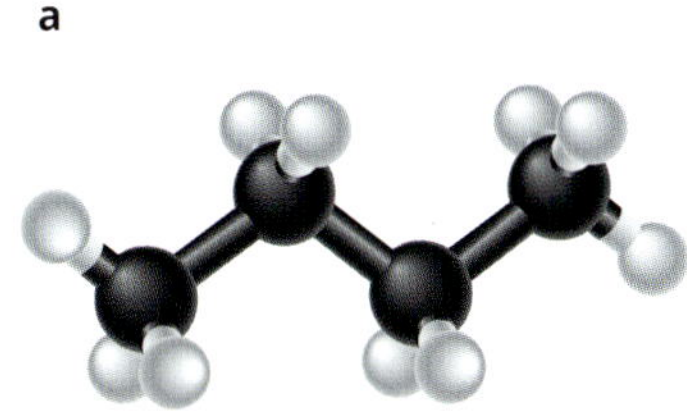

butane, C_4H_{10} (BP –0.5°C)

b

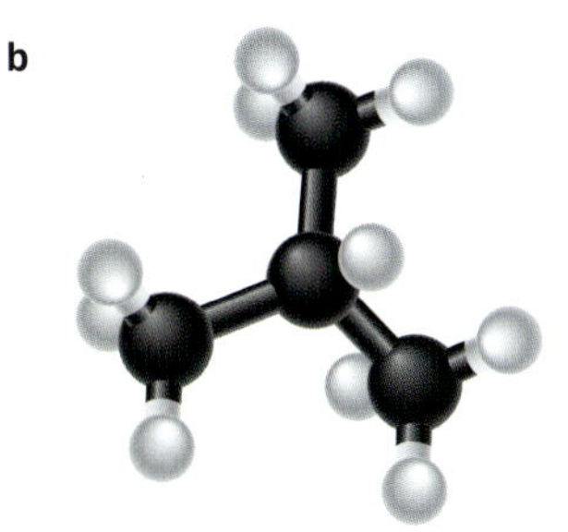

methylpropane, C_4H_{10} (BP –11°C)

FIGURE 12.3.9 Butane and methylpropane have different boiling points because their molecules are different shapes.

The shape of a molecule also influences the strength of the dispersion forces. Molecules that form long chains will tend to have stronger dispersion forces than more compact molecules with similar numbers of electrons.

For example, butane and methylpropane (Figure 12.3.9) both contain four carbon atoms and 10 hydrogen atoms. The boiling point of butane is –0.5°C, while the boiling point of methylpropane is –11°C. The higher boiling point of butane is because of the different shapes of the two molecules; butane is a long molecule while methylpropane is compact. Being less compact and long means butane has more contact area to interact with its neighbouring molecules to form stronger dispersion forces.

It is important to remember that dispersion forces occur between polar molecules as well as non-polar molecules. However, because dispersion forces are generally weaker than dipole-dipole forces, and much weaker than hydrogen bonds, in substances with small polar molecules, dispersion forces do not have a significant influence on the properties of these substances. However, in large molecules, the dispersion forces can dominate over the dipole–dipole forces and hydrogen bonding.

Table 12.3.2 shows the boiling points of three polar molecules: hydrogen chloride (HCl), hydrogen bromide (HBr) and hydrogen iodide (HI). Hydrogen chloride is the most polar of these molecules and therefore has the strongest dipole–dipole forces. However, hydrogen iodide has the highest boiling point because it is a larger molecule and therefore has stronger dispersion forces.

TABLE 12.3.2 Comparison of the boiling points of three polar molecules

Substance	Molecular mass	Number of electrons	Boiling point (°C)
hydrogen chloride (HCl)	36.5	18	–85.1
hydrogen bromide (HBr)	81.0	36	–66.8
hydrogen iodide (HI)	127.9	54	–35.4

The boiling point of hydrogen fluoride (19.5°C) is significantly higher than any of these other compounds. This is because the hydrogen bonding between hydrogen fluoride molecules is much stronger than both the dispersion forces and the dipole–dipole forces between the other molecules listed in Table 12.3.2.

12.3 Review

SUMMARY

- The melting and boiling points of covalent molecular substances increase as the strength of the intermolecular bonding increases.
- There are three main types of intermolecular bonds: dipole–dipole forces, hydrogen bonds and dispersion forces.
- Dipole–dipole forces are only present between polar molecules and are the result of the attraction between the positive and negative ends of polar molecules.
- The polarity of polyatomic molecules depends on the electronegativity of the atoms in the molecule and the symmetry of the molecule.
- The greater the polarity of a molecule, the stronger the dipole–dipole forces.
- Hydrogen bonding occurs between highly polar molecules in which hydrogen atoms are covalently bonded to an oxygen, a nitrogen or a fluorine atom.
- A hydrogen bond is formed between the hydrogen atom in one molecule and lone pairs of electrons in an oxygen, a nitrogen or a fluorine atom in another molecule.
- Hydrogen bonding is the reason why ice is less dense than liquid water.
- Hydrogen bonds are the strongest of the three main types of intermolecular forces.
- Dispersion forces are the result of attraction between temporary dipoles that form in molecules.
- Temporary dipoles are due to random fluctuations in the distribution of electrons in molecules.
- Dispersion forces occur between polar and non-polar molecules.
- Dispersion forces are stronger between larger molecules because it is easier to create temporary dipoles in molecules with a larger number of electrons.

KEY QUESTIONS

1 Identify which of the following substances would contain dipole–dipole forces between the molecules: bromine (Br_2), hydrogen chloride (HCl), methane (CH_4), tetrachloromethane (CCl_4), chloromethane (CH_3Cl).

2 Refer to Figure 12.2.5 on page 270 for electronegativity values and select which one of the following substances would have the strongest dipole–dipole forces between the molecules in the liquid state.

A fluorine (F_2)
B carbon monoxide (CO)
C hydrogen chloride (HCl)
D methane (CH_4)

3 Consider the following substances. Identify whether the molecules are attracted to one another by dispersion forces, dipole–dipole forces or hydrogen bonds.

a NH_3
b $CHCl_3$
c CH_3Cl
d OF_2
e HBr
f H_2S
g HF
h H_2O

4 In ice, each water molecule is surrounded, at equal distances, by four other water molecules. In each case, there is an attraction between the positive hydrogen atom on one water molecule and a lone pair associated with the oxygen atom of another water molecule. Draw a diagram to show the arrangement of four water molecules around another water molecule.

5 'Cloudy ammonia' is often used as a cleaning solution in bathrooms. This solution contains ammonia (NH_3) dissolved in water. Draw a diagram to represent hydrogen bonding between a water molecule and an ammonia molecule.

6 Identify the types of intermolecular bonding in the following substances.

a methane (CH_4)
b methanol (CH_3OH)
c chloromethane (CH_3Cl)
d methylamine (CH_3NH_2)
e propane ($CH_3CH_2CH_3$)

Chapter review

12

KEY TERMS

asymmetrical molecule
dipole
dipole–dipole force
dispersion force
electric dipole
electron density
hydrogen bond
instantaneous dipole
non-polar
polar
polarity
surface tension
symmetrical molecule
temporary dipole
valence shell electron pair repulsion (VSEPR) theory
vapour pressure

Shapes of molecules

1 Match the molecular formula to the correct molecular shape.

Molecular formula	Molecular shape
phosphorus trichloride (PCl_3)	tetrahedral
hypochlorous acid (HOCl)	linear
trichloromethane ($CHCl_3$)	V-shaped
hydrogen fluoride (HF)	pyramidal

2 Examine the following valence structure and use the VSEPR theory to predict the shape of the molecule.

$H-\ddot{N}=\ddot{O}:$

3 How many electrons are involved in the bonding in a molecule of beryllium fluoride?

4 All of the following molecules have four pairs of electrons around the central atom. Classify the molecular shapes as tetrahedral, pyramidal or V-shaped.

a 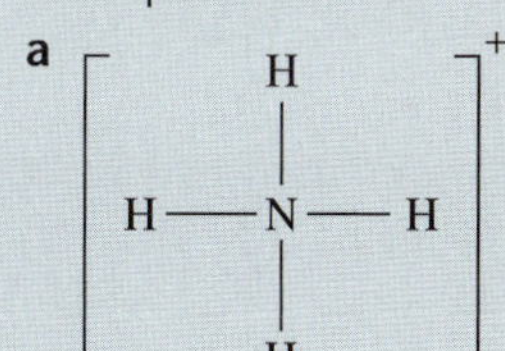

b 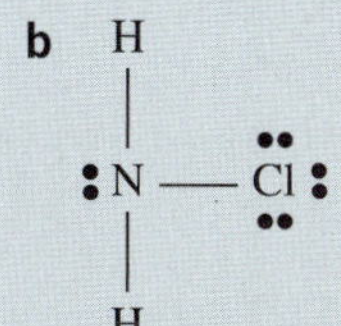

c 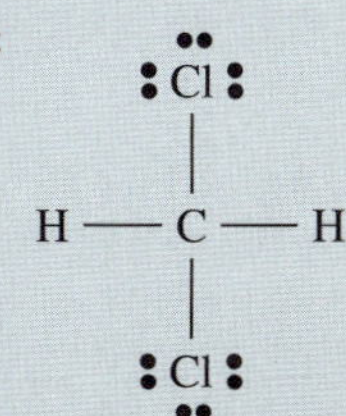

d :F — O — F:

e :Cl — P — Cl:
with :Cl: bonded below P

Properties of covalent molecular substances

5 Are the following molecules polar or non-polar? Draw structural formulae to help you decide.

a CS_2
b Cl_2O
c SiH_4
d CH_3Cl
e CH_3CH_3
f CCl_4

6 Use the electronegativities from Figure 12.2.5 on page 270 to determine which of the following molecules contains the most polar bonds.

A CO_2
B H_2O
C H_2
D H_2S
E NH_3

7 For each of the molecules listed in Question **6**, state whether they are polar or non-polar.

8 Order the following covalent bonds from most polar to least polar.
Si–O, N–O, F–F, H–Br, O–Cl

Types of intermolecular forces

9 For each of the following structures, state:
- **i** whether the molecule is polar or non-polar
- **ii** whether the strongest intermolecular forces of attraction between molecules of each type would be dispersion forces, hydrogen bonding or dipole–dipole forces.

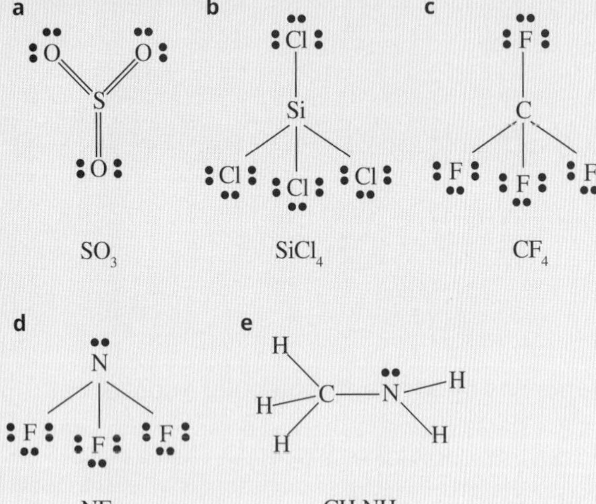

10 Consider solid samples of the following compounds. In which cases will the only forces between molecules in the samples be dispersion forces?
- **A** tetrachloromethane (CCl_4)
- **B** sulfur dioxide (SO_2)
- **C** carbon dioxide (CO_2)
- **D** hydrogen sulfide (H_2S)

11 The melting points of four halogens are given in the table below. Describe and explain the trend in melting points of these elements.

Halogen	Melting point (°C)
fluorine (F_2)	−220
chlorine (Cl_2)	−101
bromine (Br_2)	−7
iodine (I_2)	114

12 Consider the two compounds OF_2 and CF_4. OF_2 has a boiling point of −145°C and CF_4 has a boiling point of −128°C. Between molecules of which compound would the intermolecular forces of attraction be greater? Explain your answer.

13 The mass of a hydrogen fluoride molecule is similar to the mass of a neon atom. However, the boiling points of these substances are very different. The boiling point of hydrogen fluoride is 19.5°C whereas that of neon is −246°C. Explain the difference in this property of the two substances.

14 At room temperature, CCl_4 is a liquid whereas CH_4 is a gas.
- **a** Which substance has the stronger intermolecular attractions?
- **b** Explain the difference in the strengths of the intermolecular attractions.

15 Fluorine (F_2) is a gas at room temperature whereas iodine (I_2) is a solid. With respect to intermolecular forces, what factor best explains this difference?

16 Explain the difference between a permanent dipole and a temporary dipole. Your explanation should describe how the dipoles are formed and the type of intermolecular bonding that results.

Connecting the main ideas

17 Consider the following list of molecules:
N_2, Cl_2, O_2, NH_3, HCl, CH_4, H_2O, CO_2, CCl_4, $CHCl_3$
- **a** Select one substance from the list above that matches each of the descriptions below. Draw a Lewis diagram for each example, showing bonding and non-bonding electron pairs.
  - **i** a molecule that contains one triple bond
  - **ii** a molecule that contains one double bond
  - **iii** a molecule that contains two double bonds
- **b** From the list, which substances contain:
  - **i** polar molecules?
  - **ii** symmetrical molecules?
  - **iii** molecules with hydrogen bonding between them?

18 Water is a polar molecule. Explain how this fact shows that water is not a linear molecule.

19 Explain why covalent molecular substances generally have lower melting and boiling points than ionic and metallic substances.

20 Select the substance that will have the highest boiling point.
- **A** hydrogen fluoride (HF)
- **B** hydrogen chloride (HCl)
- **C** hydrogen bromide (HBr)
- **D** hydrogen iodide (HI)

21 Select the physical property of a substance that is not determined by intermolecular forces.
- **A** boiling point
- **B** vapour pressure
- **C** electrical conductivity
- **D** solubility

22 Select the compound below that has the highest melting point.
- **A** propane (C_3H_8)
- **B** octane (C_8H_{18})
- **C** decane ($C_{10}H_{22}$)
- **D** icosane ($C_{20}H_{42}$)

CHAPTER

13 Chromatography

The term ‘chromatography’ refers to a set of techniques that can be used to separate and analyse the components in a mixture. Separation is performed by passing the mixture through a medium that allows the components of the mixture to move through at different rates.

In this chapter, you will learn about the basic principles of chromatography. You will also learn how these basic principles are applied in the instrumental techniques of thin-layer chromatography (TLC), high-performance liquid chromatography (HPLC) and gas chromatography (GC), which allow extremely sensitive and rapid analysis of the components in a mixture.

Science as a human endeavour

- Chromatographic techniques, including thin-layer chromatography (TLC), gas chromatography (GC), and high-performance liquid chromatography (HPLC), are used to determine the components of a wide range of mixtures in various settings. The decision to use a particular chromatographic technique depends on a number of factors, including the properties of the substances being separated, the amount of substance available for analysis and the sensitivity of the equipment. Chromatographic techniques have a wide range of analytical and forensic applications, including monitoring air and water pollutants, drug testing of urine and blood samples, and testing for food additives and quality.

Science understanding

- data from chromatography techniques, including thin-layer chromatography (TLC), gas chromatography (GC), and high-performance liquid chromatography (HPLC), can be used to determine the composition and purity of substances; the separation of the components is caused by the variation in strength of the interactions between atoms, molecules or ions in the mobile and stationary phases

13.1 Principles of chromatography

Chromatography is a widely-used technique used to separate and analyse the substances present in a mixture. It can be used to analyse numerous inorganic and organic substances such as contaminants in water, toxic gases in air, additives and impurities in food, and drugs present in blood.

In this section, you will learn about the underlying principles of chromatography and also about three simple forms of the technique called **paper chromatography**, **thin-layer chromatography** and **column chromatography**.

CHEMISTRY IN ACTION

Drug testing in sport

Since competitive sports began, some people have sought to gain an advantage by whatever means available. For example, athletes in ancient Greece and Rome consumed specially prepared lizard meat in the hope that it would improve their performance and give them a competitive edge.

Today, drug problems in sport usually result from the inappropriate use of therapeutic substances and manipulation of substances such as hormones and electrolytes that occur naturally in an animal's or athlete's system. In horse racing, the use of illegal 'go-fast' or 'stopper' drugs is now relatively rare because of their ease of detection by an analytical technique called gas chromatography.

Gas chromatography is an extremely sensitive type of chromatography. Urine samples are routinely taken from athletes competing in major events to ensure that the athletes are not benefiting from the use of illegal, performance-enhancing drugs. These samples are analysed by gas chromatography in combination with a mass spectrometer as the detector. You will learn more about gas chromatography later in this chapter.

FIGURE 13.1.1 A chemist prepares samples from athletes for drug testing.

HOW CHROMATOGRAPHY WORKS

All methods of chromatography have a:

- **stationary phase**
- **mobile** (moving) **phase**.

You can perform a simple chromatography experiment by dipping the end of a stick of chalk into water-soluble black ink and then standing the chalk in a beaker containing a small amount of water, as seen in Figure 13.1.2.

As the water carries the ink up the chalk, you will see that the ink separates into bands of different colours. Each band contains one of the substances, or **components**, present in the ink mixture. The pattern of bands or spots is called a **chromatogram**.

In this simple chalk-and-ink exercise, the stationary phase is the chalk and the mobile phase is the water.

FIGURE 13.1.2 The pattern of bands produced when this stick of chalk is dipped in black ink and then placed in water is called a chromatogram.

As the components in the ink are swept upwards over the stationary phase by the solvent, they undergo a continual process of **adsorption** onto the solid stationary phase, followed by **desorption** and dissolving into the mobile phase. The ability of the components to stick to the stationary phase will depend upon the polarity of the stationary phase and the component molecules. Similarly, the attraction of the components to the solvent molecules is determined by their polarity.

The rate of movement of each component depends mainly upon:

- how strongly the component adsorbs onto the stationary phase
- how readily the component dissolves in the mobile phase.

The components separate because they undergo these two processes to different degrees. In the example involving the chalk and ink shown in Figure 13.1.2, the blue dye in the ink is more soluble in the mobile phase than the red dye, and bonds less strongly with the stationary phase than the red dye. Therefore the blue dye in the ink has moved faster up the piece of chalk than the red dye, resulting in their separation.

The term 'adsorption' describes the adhesion of molecules or substances to the surface of a solid or liquid. The stationary phase in chromatography is either a solid or a liquid. The mobile phase is in either a liquid or a gas state.

Water, the mobile phase, is a polar solvent. The blue dye moves more quickly with the water up the stationary phase than the red dye, indicating that the blue dye contains compounds of molecules the more polar than those in the red dye.

PAPER AND THIN-LAYER CHROMATOGRAPHY

In the laboratory, paper chromatography is performed on high-quality absorbent paper, similar to filter paper, as the stationary phase. Figure 13.1.5 shows paper chromatography in use for the separation of the components of different inks. Thin-layer chromatography (TLC) is very similar to paper chromatography. In this case the stationary phase is a thin layer of a fine powder, such as alumina (aluminium oxide), spread on a glass or plastic plate. Both techniques are useful for **qualitative analysis**, in which the chemicals that are present in a mixture are identified.

FIGURE 13.1.5 Paper chromatography can be used to separate mixtures of inks into their separate components.

CHEMFILE

Colour writing

While studying the coloured materials (pigments) in plants, Russian botanist Mikhail Tsvet (Figure 13.1.3) developed the separation technique known as chromatography. The word chromatography means 'colour writing' and comes from the Greek *khromatos*, meaning 'of colour', and *graphos*, meaning 'written'.

FIGURE 13.1.3 Mikhail Tsvet was the inventor of chromatography.

Passionate about botany, Tsvet discovered that different pigments appeared as different coloured bands. In 1903, chlorophyll and xanthophyll were the only known plant pigments. Tsvet produced chromatograms similar to the one shown in Figure 13.1.4 and discovered two forms of chlorophyll and eight other pigments.

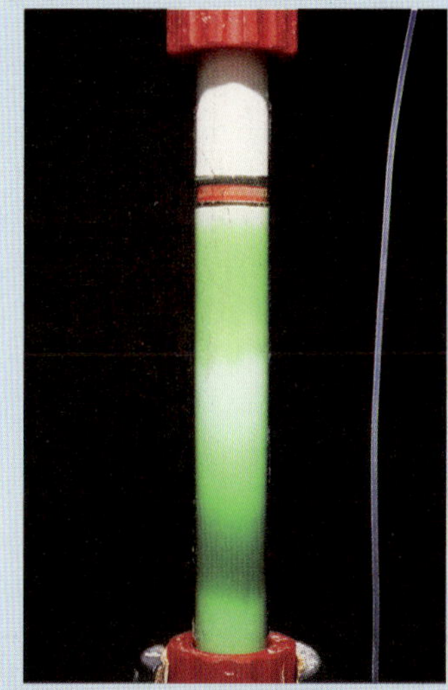

FIGURE 13.1.4 Chromatogram of a plant sample similar to the chromatograms obtained by Mikhail Tsvet

Chromatography is such an important technique that two Nobel prizes have been awarded for research based mainly on this analytical method.

Paper and thin-layer chromatography in practice

In both paper and thin-layer chromatography, a small spot of the solution of the sample to be analysed is placed on one end of the chromatography paper or plate.

The position of this spot is called the **origin**. The paper or plate is then placed in a container with solvent. The origin must be a little above the level of the solvent so that the components can be transported up the paper or plate and not dissolve into the liquid in the container. As the solvent rises up the paper or plate, the components of each sample separate depending on their attraction to the stationary phase and their solubility in the solvent, as shown in Figure 13.1.6.

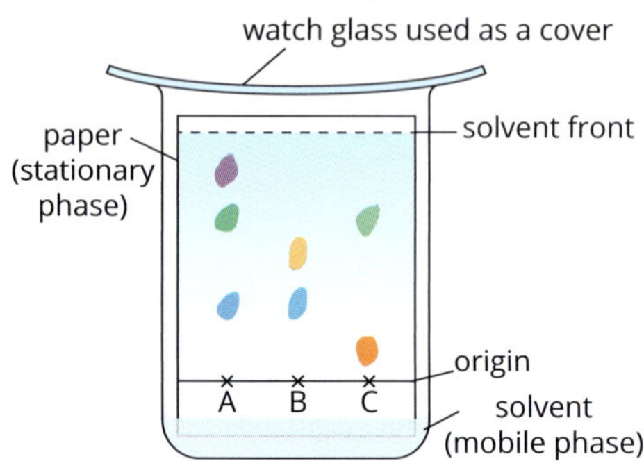

FIGURE 13.1.6 Paper chromatography of three different food colours (A, B and C)

Identifying the components of a mixture

The components in a mixture can be identified by chromatography in one of two ways:

1 Include standards of known chemicals on the same chromatogram as the unknown sample and comparing the resulting positions of the unknown components with those of the known samples.

2 Calculate the **retardation factor** (R_f) of the sample and comparing these with the R_f values of known samples.

Method 1: Using standards

In this method, you need to know what chemicals might be present in the sample. For example, if you wish to find out whether a vitamin tablet contains vitamins A and D, a sample of the vitamin tablet can be placed alongside samples of each of vitamins A and D on the same chromatogram. If spots from the tablet sample move the same distance from the origin as the spots from the pure samples, then the tablet is likely to contain the vitamins.

The sample and standards are 'run' on the same chromatogram because the distances moved from the origin will depend on the distance moved by the solvent front. The further the solvent front is allowed to travel, the further the spots travel.

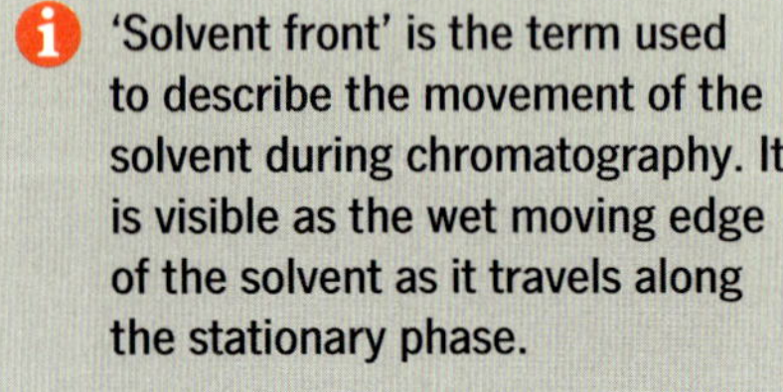

'Solvent front' is the term used to describe the movement of the solvent during chromatography. It is visible as the wet moving edge of the solvent as it travels along the stationary phase.

Method 2: Calculating R_f values

Another way of identifying the components of a mixture is by comparing the distance they travel up the stationary phase to the distance travelled by the solvent front. This is expressed as a retardation factor, R_f, for a component:

$$R_f = \frac{\text{distance the component travelled from the origin}}{\text{distance the solvent front travelled from the origin}}$$

You can see from the chromatogram in Figure 13.1.7 that:

- R_f values will always be less than one
- the component most strongly adsorbed onto the stationary phase moves the shortest distance and has the lowest R_f value.

Each component has a characteristic R_f value for the conditions under which the chromatogram was obtained. By comparing the R_f values of components of a particular mixture with the R_f values of known substances determined under identical conditions, you can identify the components present in a mixture.

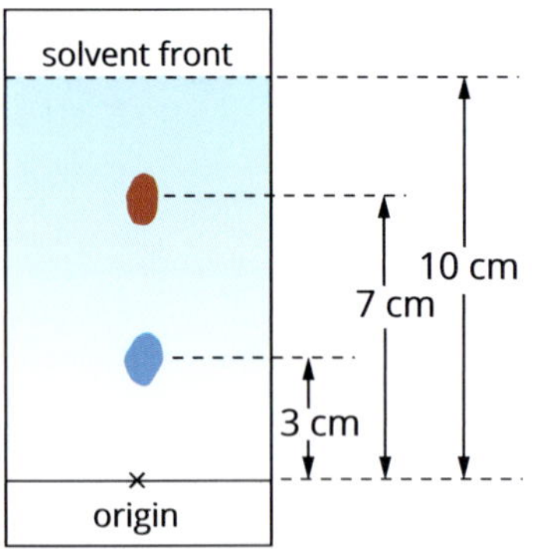

FIGURE 13.1.7 A chromatogram of a sample that consists of two components. The distances from the origin enable R_f calculations.

In this method, the actual distance moved by the solvent front is no longer critical as the proportion of the distance moved from the origin (the R_f value) stays the same provided the conditions under which the chromatogram is obtained are the same.

This means the R_f values of unknown spots can be compared against a table of R_f values of common materials. However, changes in the temperature, the type of stationary phase, the amount of solvent vapour around the paper or plate and the type of solvent will all change the R_f value for a particular chemical.

For the R_f value of a component to be comparable to a set of standard R_f values, they must be determined under identical conditions.

Worked example 13.1.1

CALCULATING R_f VALUES

Calculate the R_f value of the red component in Figure 13.1.7.	
Thinking	**Working**
Record the distance the component has moved from the origin and the distance the solvent front has moved from the origin.	distance from origin of red component = 7 cm distance from origin of solvent front = 10 cm
$R_f = \frac{\text{distance of component from origin}}{\text{distance of solvent front from origin}}$	$R_f(\text{red component}) = \frac{7}{10}$ $= 0.7$

Worked example: Try yourself 13.1.1

CALCULATING R_f VALUES

Calculate the R_f value of the blue component in Figure 13.1.7.

In chromatograms of inks, plant pigments and food dyes, the components can be seen easily; however, most compounds are colourless and must be made visible. Many organic compounds fluoresce and appear blue when viewed under ultraviolet light. For other compounds, the chromatogram can be sprayed with a chemical that reacts to form coloured or fluorescent compounds. For example, a chemical called ninhydrin reacts with amino acids to give blue- and brown-coloured compounds that are easily detected on the chromatograph.

The choice between paper and thin-layer chromatography depends upon the sample being analysed. Table 13.1.1 lists the advantages of each method.

TABLE 13.1.1 A comparison of the advantages of paper and thin-layer chromatography

Paper chromatography	Thin-layer chromatography
cheap	detects smaller amounts
little preparation	better separation of less polar compounds
more efficient for polar and water-soluble compounds	corrosive materials can be used
easy to handle and store	a wide range of stationary phases is available

For a particular combination of stationary phase and mobile phase, many different chemicals may have similar R_f values. Paper and thin-layer chromatography only provide a guide to the identity of a chemical. Further testing using high-performance liquid chromatography (HPLC) or gas chromatography (GC) may be required to confirm the identity of a chemical. Section 13.2 will introduce you to HPLC and GC, and how they can be used to determine the presence *and* concentration of chemicals in a mixture.

CHEMISTRY IN ACTION

Thin-layer chromatography in forensic analysis

Many forensic investigations use thin-layer chromatography (TLC) to analyse materials recovered from the scene of an investigation. TLC is useful in detecting chemicals of forensic concern, including chemical weapons, explosives and illicit drugs.

Scientists can use a computerised and portable TLC machine that can be taken anywhere and used to analyse multiple samples at a time. These analyses can be completed within 30 minutes, allowing analyses to be performed at the scene rather than in a dedicated laboratory.

The technique is inexpensive, reliable, fast, and easy to perform. The method was refined in the late 1960s so that it could reliably measure the amounts of compounds.

High-performance thin-layer chromatography

High-performance TLC (HPTLC) is an enhanced form of TLC that automates the different steps, increases the resolution achieved and allows more accurate measurements of the quantities of components present in a sample. It provides better resolution of chemically similar compounds than conventional TLC, and requires smaller samples. However, HPTLC requires specialised analysis equipment, and so is still not as popular or widespread as conventional TLC.

When a sample undergoes analysis, there may be uncertainty in the size of the drop of sample and its position at the origin when deposited by hand. This uncertainty has been overcome in HPTLC by replacing the manual method with piezoelectric devices and inkjet printers.

The plate can also be developed using two different solvents. Initially, the sample is applied to the plate and developed with the first solvent. Then the plate is rotated by 90° and developed with a second solvent. This procedure increases the resolution of the separation and produces more accurate R_f values.

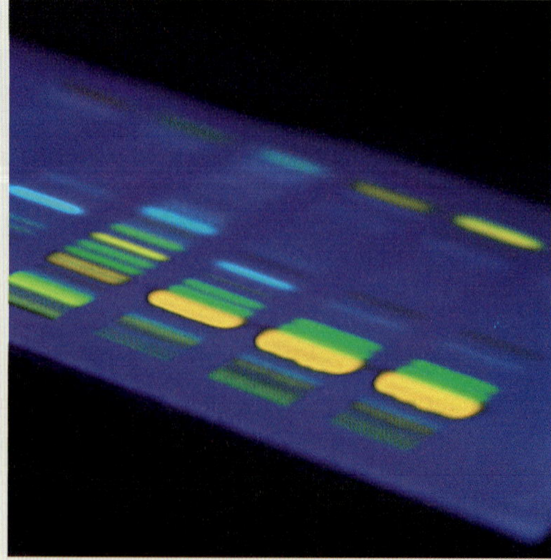

FIGURE 13.1.8 These glowing bands, seen under fluorescent lighting, are due to the components of mixtures separated on a high-performance thin-layer chromatography plate.

COLUMN CHROMATOGRAPHY

Figure 13.1.9 shows another form of chromatography—column chromatography—which can also be used to separate the components of a mixture.

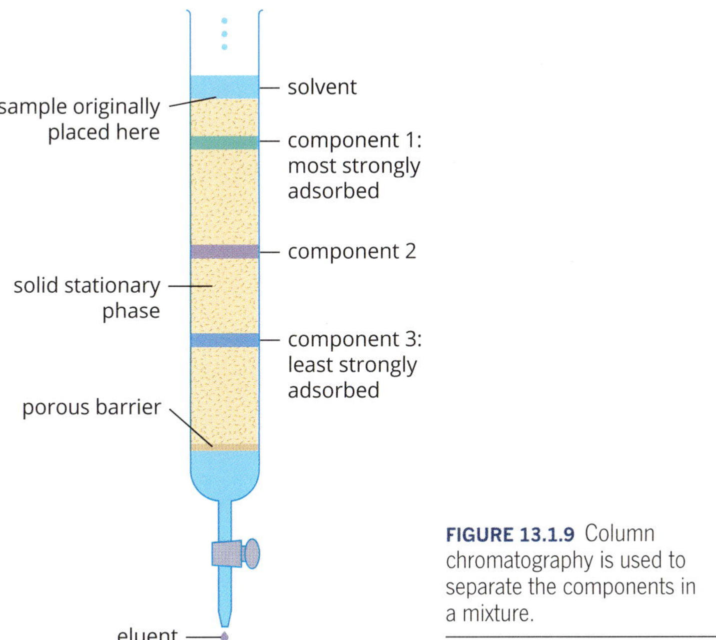

FIGURE 13.1.9 Column chromatography is used to separate the components in a mixture.

The solid stationary phase is packed into a glass column. The sample mixture is applied carefully to the top of the packed solid, and a solvent, which acts as the mobile phase, is dripped slowly onto the column from a reservoir above. A tap at the bottom of the column allows the solvent, which is called the **eluent**, to leave the column at the same rate as it enters it at the other end. Figure 13.1.10 shows how column chromatography can be used to separate and collect different components in a black ink.

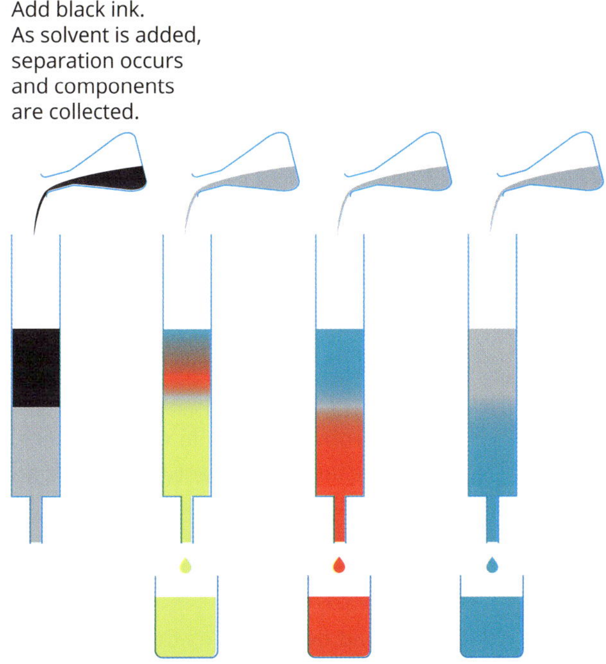

FIGURE 13.1.10 Column chromatography can be used to separate and collect the components in black ink. The yellow dye in the ink is more soluble in the mobile phase and adsorbs less strongly to the stationary phase than the red and blue dyes. The yellow dye in the ink moves faster through the column than the red and blue dyes, resulting in separation of the dyes.

13.1 Review

SUMMARY

- Chromatography is a technique commonly used to separate and identify the components in a mixture.
- Paper and thin-layer chromatography are simple forms of this technique.
- All chromatographic techniques involve a mobile phase and a stationary phase.
- Components separate during chromatography as a consequence of how strongly they adsorb to the stationary phase and desorb back into the mobile phase.
- Components in a mixture have differing affinities for the mobile and stationary phases.
- In paper and thin-layer chromatography, the components in a mixture are identified by comparison with known standards and determination of R_f values.

KEY QUESTIONS

1 Paper chromatography is used to separate the pigments in a plant leaf. A spot of each pigment is placed on a sheet of chromatography paper and the chromatogram is run inside a closed jar that is partly filled with ethanol. Which one of the following options is the mobile phase in this instance?

A pigments
B ethanol
C paper
D ethanol vapour in the closed jar

2 Match each term with its correct description.

Term	Description
adsorption	the breaking of the attraction between a substance and the surface to which the substance is adsorbed
desorption	the components of a mixture undergo adsorption to this phase
components	the different compounds in the mixture, which can be separated by chromatography
polar molecule	the attraction of one substance to the surface of another
mobile phase	a molecule that acts as a dipole; it has one or more polar covalent bonds, with the charge being distributed asymmetrically
stationary phase	the solvent that moves over the stationary phase in chromatography

3 An extract from a plant was analysed by thin-layer chromatography with a non-polar solvent. The chromatogram obtained is shown in the figure below. The table below gives the R_f values of some chemicals commonly found in plants measured under the same conditions.

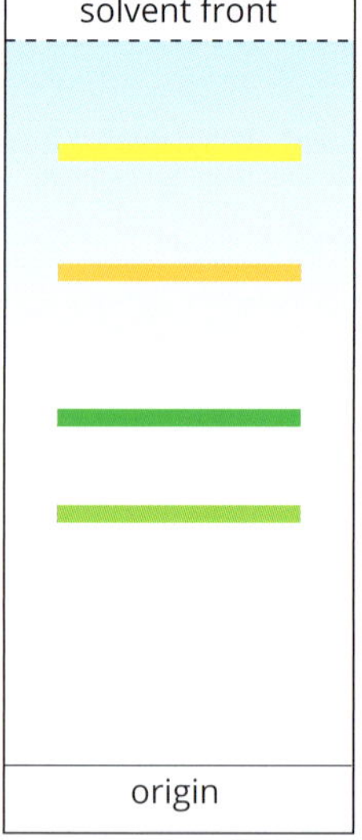

Chemical	R_f
xanthophyll	0.67
β-carotene	0.82
chlorophyll a	0.48
chlorophyll b	0.35
leutin	0.39
neoxanthin	0.27

a Using the figure above, measure and record the distance from the origin to the centre of each band, and the distance of the solvent front from the origin.
b Calculate the R_f value of each band.
c Compare R_f values for the bands with the R_f values in the table above and name the chemicals present in the extract.
d If water had been used as the solvent, would the chromatogram be likely to have a similar appearance? Explain your reasoning.

4 Phenacetin was once an ingredient in analgesic drugs, but it is not used now because it causes liver damage. It is soluble in chloroform and is colourless. A chemist wishes to analyse a brand of analgesic by thin-layer chromatography to determine whether it contains phenacetin. Outline the steps in the analysis. (Assume that a sample of pure phenacetin is available to the chemist.)

5 Use the following terms to complete the summary of column chromatography. Not all terms will be used.

adsorb
desorb
longest
mobile
shortest
stationary

During column chromatography, the components of the sample ________________ onto the stationary phase and ________________ into the liquid mobile phase. A component that adsorbs most strongly to the ________________ phase and is least soluble in the ________________ phase would be expected to take the ________________ time to pass through the column.

13.2 Advanced applications of chromatography

In Section 13.1, you learnt the principles of chromatography and how the techniques of paper, thin-layer and column chromatography can be used to separate the components in a mixture.

In these techniques, a liquid mobile phase is used to carry the components of the mixture past a solid stationary phase. The components undergo a continual process of adsorption to the stationary phase and desorption back into the mobile phase. The components of a mixture undergo the processes of adsorption and desorption to different degrees, depending on the strength of their attraction to the stationary phase and the mobile phase. As a result, the components separate as they move past the stationary phase.

In this section, you will learn how these principles are used as the basis for two modern analytical techniques, high-performance liquid chromatography and gas chromatography, which are highly sensitive and capable of detecting minuscule amounts of a compound.

HIGH-PERFORMANCE LIQUID CHROMATOGRAPHY

High-performance liquid chromatography (HPLC) is a chromatographic technique based on column chromatography that allows scientists to perform extremely sensitive analyses of a wide range of mixtures. It is commonly used for the separation and identification of complex mixtures of similar compounds, such as contaminants that are soluble in water, drugs in blood and hydrocarbons in oil samples.

High-performance liquid chromatography (HPLC) is sometimes referred to as high-pressure liquid chromatography.

HPLC is now used routinely for environmental, pharmaceutical and industrial analyses. For example, it can be used to analyse for the presence and concentration of dioxins, insecticides, pesticides and oil spills in water. It is also used to determine the presence of pesticides in food or to detect the presence of drugs in blood.

Figure 13.2.1 shows the apparatus used for HPLC. The basic principles of HPLC are the same as for column chromatography described earlier. There are two main differences between simple column chromatography and HPLC.

- The particles in the solid used in the HPLC column are often 10–20 times smaller than those used in column chromatography, allowing more frequent adsorption and desorption of the components, giving much better separation of similar compounds.
- The small particle size used in HPLC creates a considerable resistance to the flow of the mobile phase and so the solvent is pumped through the column under high pressure.

A range of solids is available for use in HPLC columns, some with chemicals specially bonded to their surfaces to improve the separation of particular classes of compounds.

In HPLC, the components are usually detected by passing the eluent stream through a beam of ultraviolet (UV) light. Many organic compounds absorb UV light, so when an organic compound passes in front of the beam of light a reduced signal is picked up by a detector. The amount of light received by the detector is recorded on a chart that moves slowly at a constant speed or sent to a computer. The resulting trace is called a chromatogram. Each component forms one peak in the chromatogram.

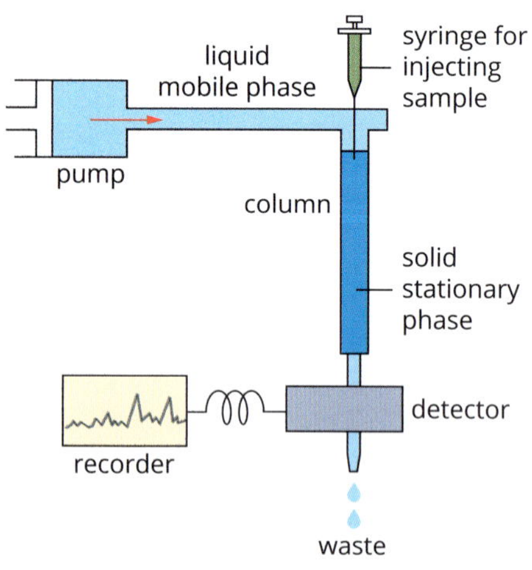

FIGURE 13.2.1 The construction of a high-performance liquid chromatograph

Figure 13.2.2 indicates how the chromatogram is produced as the components pass through the instrument. Figure 13.2.3 shows the HPLC instrument in use.

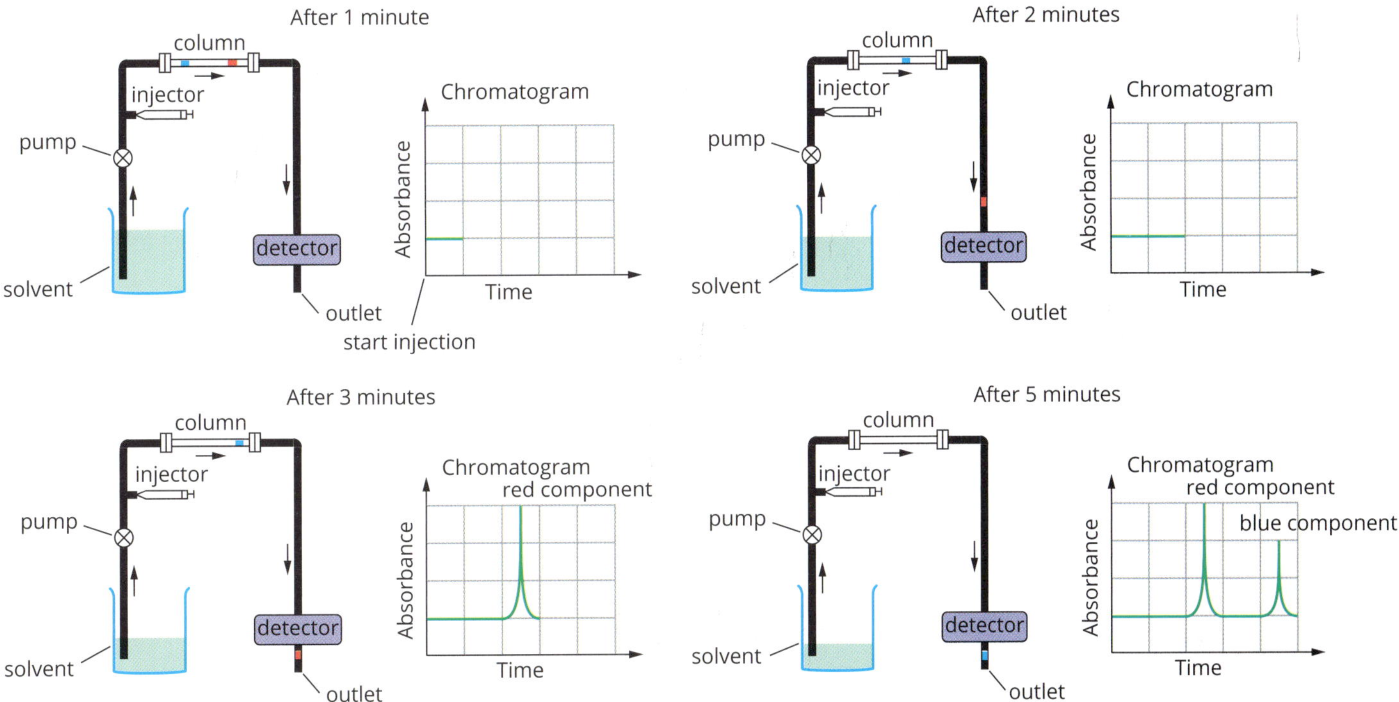

FIGURE 13.2.2 Separation of components of a sample using a high-performance liquid chromatograph

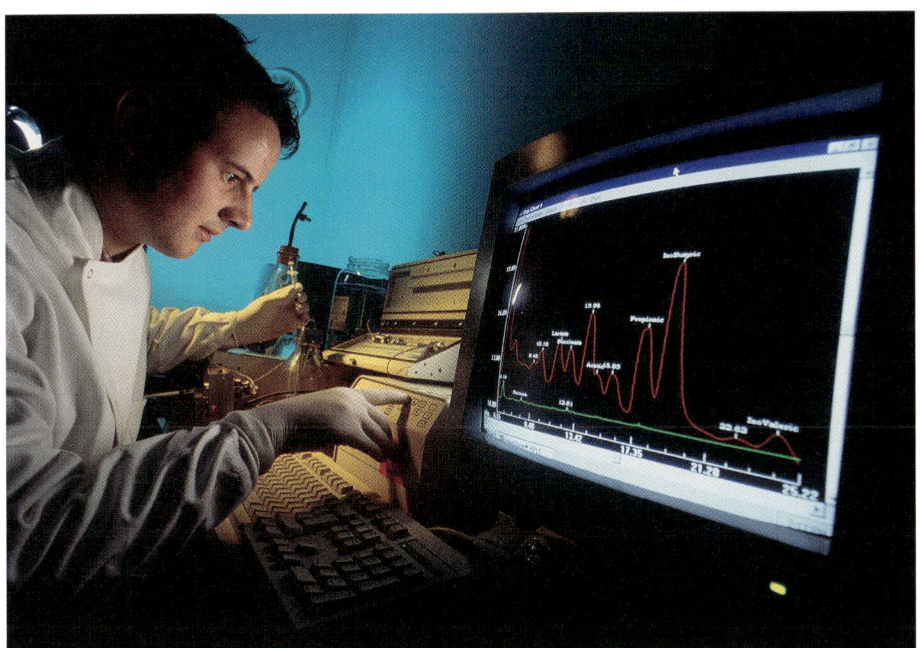

FIGURE 13.2.3 A research chemist using high-performance liquid chromatography to analyse the components in a mixture for medical research

GAS CHROMATOGRAPHY

The most sensitive of the chromatographic techniques is **gas chromatography (GC)**. It is capable of detecting as little as 10^{-12} g of a component. However, it is limited to components that can be readily vapourised without decomposing. Such compounds usually have relative molecular masses of less than 300. HPLC, on the other hand, can separate compounds with relative molecular masses that are as high as 1000 or even more.

Gas chromatography is an extremely sensitive analytical technique. However, its use is limited to compounds that vaporise without decomposing (usually with relative molecular masses of less than 300).

The extreme sensitivity of GC makes it ideal for the analysis of trace contaminants in samples or for the detection of tiny amounts of very potent components. For example, the urine samples routinely taken from athletes competing in major events are analysed by GC to ensure that the athletes are not benefiting from the use of illegal, performance-enhancing drugs.

A generalised diagram of the equipment used in GC is shown in Figure 13.2.4.

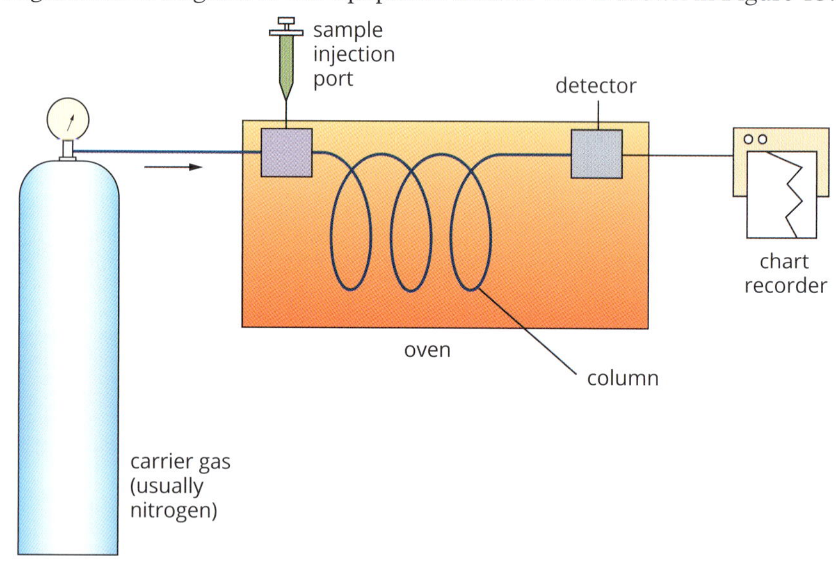

FIGURE 13.2.4 A gas chromatograph

Gas chromatography has the following features:

- The mobile phase is an unreactive gas, often nitrogen, called the **carrier gas**.
- A small amount of sample is injected into the top of the column through an **injection port**. The injection port is heated to vaporise the sample, which is then swept into the column by the carrier gas.
- The column is a series of loops of glass or metal that has an internal diameter of about 4 mm and is about 2–3 m long in total. It is heated. The column can be packed with a porous solid coated with an ester or liquid hydrocarbon with a high boiling point, or it can be packed with an adsorbent solid such as silica gel or alumina. The solid acts as a stationary phase.

As the sample passes through the instrument, the components of the sample repeatedly interact with the stationary phase and are swept forward by the carrier gas. Components that adsorb least strongly to the stationary phase are swept out first by the gas. As the components emerge from the end of the column, they are sensed by the detector.

CHEMFILE

The flame ionisation detector—an Australian invention

When the solid particles used in the column of a gas chromatograph are coated with a liquid hydrocarbon or ester with a high boiling point, the instrument is known as a gas–liquid chromatograph (GLC) and the liquid acts as the stationary phase.

Figure 13.2.5 shows one of the most useful detectors used in gas–liquid chromatography, the flame ionisation detector, invented in 1957 by an Australian, Ian McWilliam.

In this detector, organic compounds leaving the column are burnt in a hydrogen–oxygen flame. Ions produced in the flame are attracted to electrodes and cause a current to flow. This current is used to indicate the presence of the component and, after calibration, the component's concentration in the original sample.

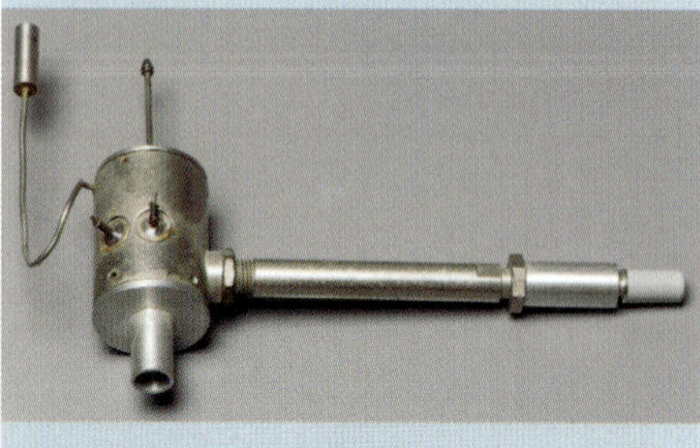

FIGURE 13.2.5 A flame ionisation detector

APPLICATIONS OF HPLC AND GC

Chemists use HPLC and GC in chemical analysis to answer two questions.

- What chemical is present in the sample? (qualitative analysis)
- How much of each chemical is present? (**quantitative analysis**)

HPLC and GC provide qualitative and quantitative analyses of components in a mixture. The size of the peaks is due to the amount of light absorbed by each component, and this can be used to calculate the relative amount of each component within the mixture.

Qualitative analysis

Qualitative analysis can be used to identify the components in a sample and determine its purity. A solution of a pure compound that is thought to be one of the components is injected into the HPLC or GC under the same conditions. The chromatogram of the standard is then compared with the chromatogram of the sample.

The time taken for a component to pass through the HPLC or GC column is called its **retention time (R_t)**, and is characteristic of the component for the conditions of the experiment. It is used in the same way as the retardation factor (R_f) in paper and thin-layer chromatography.

The same compound will have the same retention time if the conditions (temperature, mobile phase, stationary phase, flow rate, pressure etc.) remain the same. Retention times can therefore be used to identify the components causing the peaks on a gas or liquid chromatogram.

Example: Detecting pesticides

Pesticides can be used to prevent mould in food such as the lettuce used in salads. HPLC can be used to test waste-water samples from a farm for the presence of two pesticides: vinclozolin and carbendazin.

The high-performance liquid chromatogram of a water sample is shown in Figure 13.2.6. The two component pesticide peaks, marked 1 and 2, are identified by obtaining chromatograms of known pure compounds under exactly the same conditions as the sample.

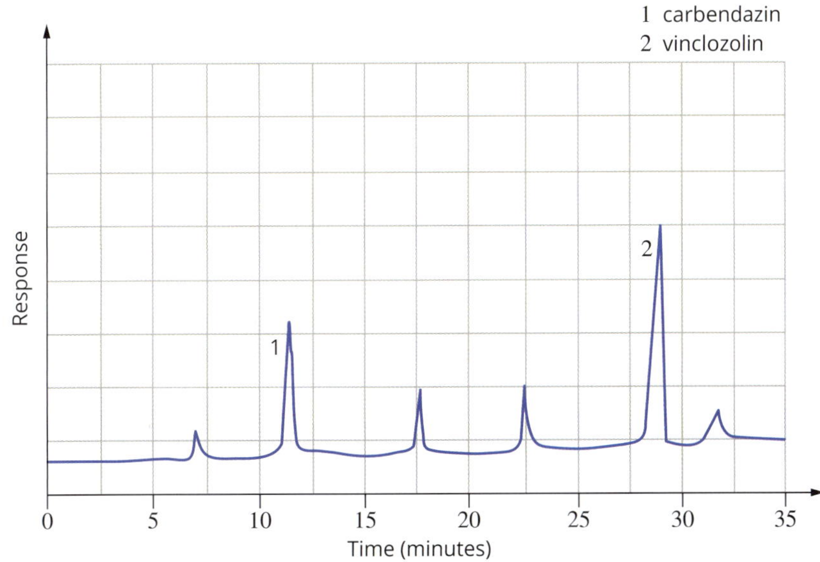

FIGURE 13.2.6 High-performance liquid chromatogram of a water sample from a farm

A compound can also be identified by adding a known compound to the sample (a procedure called 'spiking'). The chromatogram in Figure 13.2.7 was produced when the water sample was spiked with carbendazin. Note that peak 1 is much bigger in the spiked sample. There are no extra peaks, indicating that carbendazin must have been component 1 in the sample.

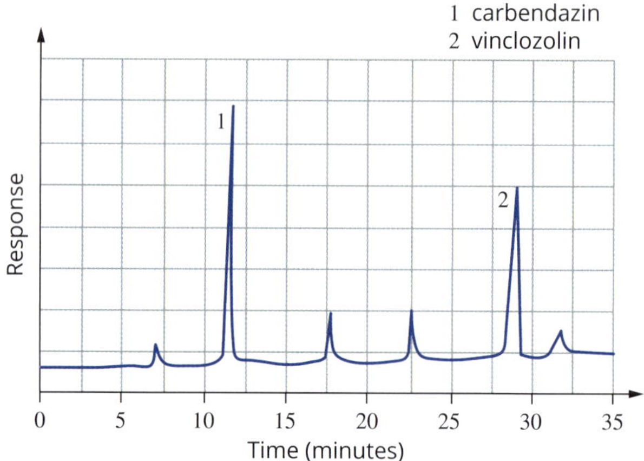

FIGURE 13.2.7 High-performance liquid chromatogram for a spiked water sample

Quantitative analysis

To determine the concentration of an individual component in a mixture, its peak area is compared with the peak areas of samples of the same chemical at known concentrations. A solution with an accurately known concentration is called a **standard solution**.

By plotting the peak areas against the concentrations of the standard solutions, a **calibration curve** can be drawn and used to determine unknown concentrations.

Worked example 13.2.1 shows how HPLC or GC can be used to find the concentration of a component in a mixture.

Worked example 13.2.1

CONCENTRATION OF A PESTICIDE

The concentration of the pesticide carbendazin in a sample of waste water from a farm was determined by HPLC. The chromatograms of a series of standards with accurately known concentrations of carbendazin were obtained under the same conditions as the sample.

The results from the chromatograms of the sample and the standards are shown in the following table. Determine the concentration of carbendazin in the sample of waste water. (Note that units of $ng\,g^{-1}$ are used because the concentrations are so low: $200\,ng\,g^{-1}$ is equivalent to just 0.2 ppm (parts per million) or 0.0000002% by mass.)

	Peak area	Concentration ($ng\,g^{-1}$); $1\,ng = 10^{-9}\,g$
Standard 1	0.95	20
Standard 2	2.20	60
Standard 3	3.60	100
Standard 4	5.00	140
Standard 5	7.00	200
Sample	3.00	?

The information that is obtained from a high-performance liquid chromatogram or a gas chromatogram is:

- retention time, R_t – which can be used to identify components in a mixture
- peak area – which allows the concentration of an individual component in a mixture to be determined by comparing its peak area with the peak areas of samples with known concentrations of the same chemical.

Thinking	Working
Construct a calibration graph by plotting peak area versus concentration.	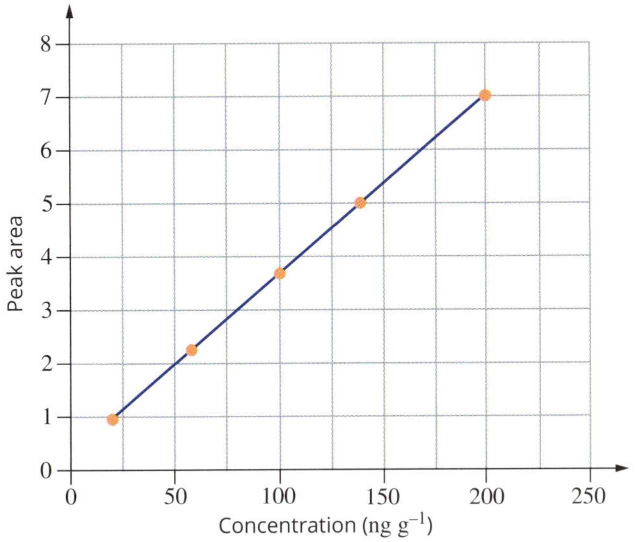
Mark where the peak area of the sample lies on the calibration curve by tracing a horizontal line to the curve and then a vertical line down to the *x*-axis.	Calibration graph for carbendazin in salad lettuce 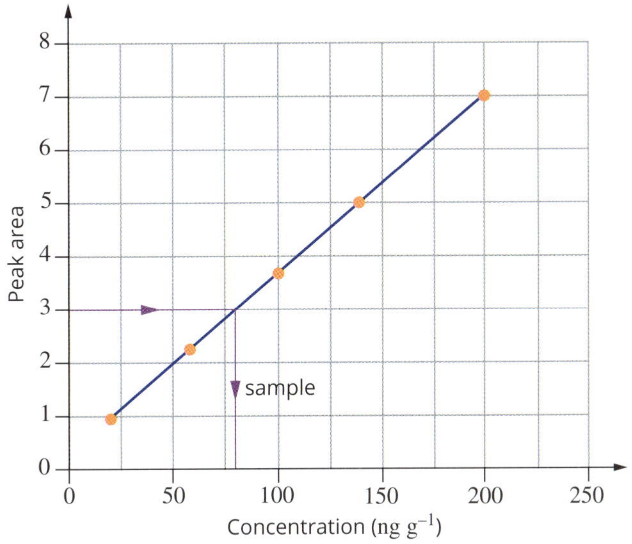
Determine the concentration of the sample by reading the concentration value for the unknown solution from the *x*-axis.	The concentration of carbendazin in the water sample is 80 ng g^{-1}.

Worked example: Try yourself 13.2.1

CONCENTRATION OF A PESTICIDE

Procymidone is a pesticide that can be used to prevent disease in oranges. Tests have been performed by HPLC to determine the concentration of this pesticide in a sample of orange juice.

The chromatograms of a series of standards with accurately known concentrations of procymidone were obtained under the same conditions as the sample.

The results from the chromatograms of the sample and the standards are shown in the following table.

Determine the concentration of procymidone in the sample of orange juice.

	Peak area	Concentration (mg kg^{-1})
Standard 1	10	0.5
Standard 2	20	1.0
Standard 3	30	1.5
Standard 4	40	2.0
Standard 5	50	2.5
Sample	15	?

CHEMISTRY IN ACTION

What's in your olive oil?

Gourmet cooks value olive oil (Figure 13.2.8) for its distinctive taste and aroma. Nutritionists favour it because it is rich in monounsaturated triglycerides, which are believed to lower blood cholesterol levels and reduce the risk of heart disease. But there is continuing frustration among Australian olive oil producers at the lack of regulation of olive oil sold in Australia. Imported olive oil has shown evidence of rancidity and, in some cases, substitution of less expensive oils such as corn, peanut and soybean for olive oil.

FIGURE 13.2.8 Pure olive oil is highly regarded for its aroma and taste, and because it is thought to reduce blood cholesterol levels.

The HPLC technique can be used by food scientists as the basis for an analytical test based on olive oil's unique composition. Chromatography separates the oils into their component triglycerides. An example of the chromatograms obtained for pure olive oil and an impure oil can be seen in Figure 13.2.9.

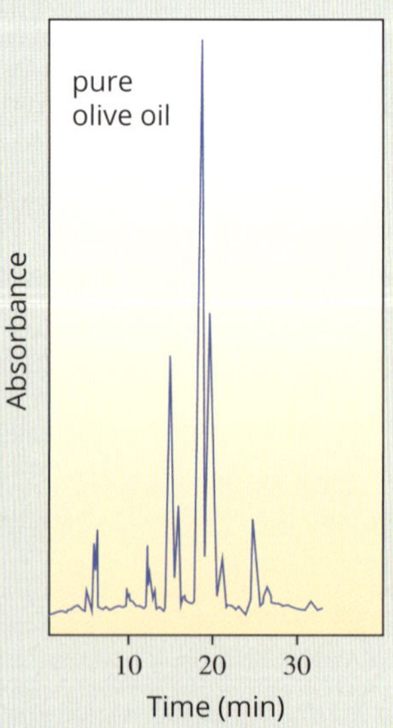

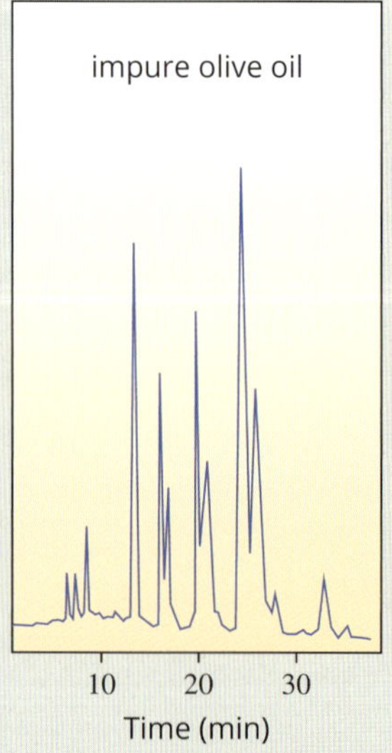

FIGURE 13.2.9 Peaks in the chromatogram of an impure oil indicate the presence of compounds found in corn oil, rather than olive oil.

COMPARING CHROMATOGRAPHIC TECHNIQUES

In Sections 13.1 and 13.2, you learnt about different chromatographic techniques that can be used for analysis: paper and thin-layer chromatography, HPLC and GC. What factors do chemists consider when choosing which technique to use in a particular situation?

The decision to use a particular technique depends on a number of factors, including:

- properties of the components being separated
- amount of sample available for analysis
- concentration of the component in the sample
- sensitivity of the technique
- time taken for analysis
- cost of equipment.

Some of the important features of the different chromatographic techniques are listed in Table 13.2.1. Chromatography is widely used for the drug testing of blood and urine samples, monitoring water and air pollution, and testing food quality.

TABLE 13.2.1 Features of different chromatographic techniques

Technique	Typical substances tested	Typical samples	Advantages	Disadvantages	Comments
paper and thin-layer chromatography	polar, water-soluble substances	drug detection, dyes in foodstuffs	very cheap, only basic laboratory equipment needed, easy to perform	poor precision and accuracy	samples need to be coloured or visible under UV light; otherwise, samples can be made visible using stains.
high-performance liquid chromatography	medium to high molecular mass organic compounds, e.g. pesticides, enzymes	foods, drugs, biological samples	high sensitivity and precision, small sample size, readily automated	moderately expensive instrument, trained technician needed to operate	samples must be soluble in a suitable solvent.
gas chromatography	low molecular mass organic compounds, e.g. acetone, aspirin	water, gases, foods, drugs, biological samples	high sensitivity and precision, small sample size, readily automated	moderately expensive instrument, trained technician needed to operate	samples must be vaporised without breaking down.

EXTENSION

Combining analytical techniques

Both GC and HPLC can be combined with another analytical technique, mass spectrometry (MS), which you learnt about in Chapter 2. Together, the techniques allow chemists to determine very small quantities and identify a wider range of materials, as well as gain valuable information about the structures of the compounds.

The advantage of the techniques of gas chromatography–mass spectrometry (GC–MS) and high-performance liquid chromatography–mass spectrometry (HPLC–MS) is that, by first passing the sample through a chromatograph, a complex sample can be separated into its many components. Each component then passes through the mass spectrometer, which can compare unknown mass spectra with an online library containing hundreds of thousands of different spectra.

GC–MS and HPLC–MS (Figure 13.2.10) have become essential techniques for many types of analyses.

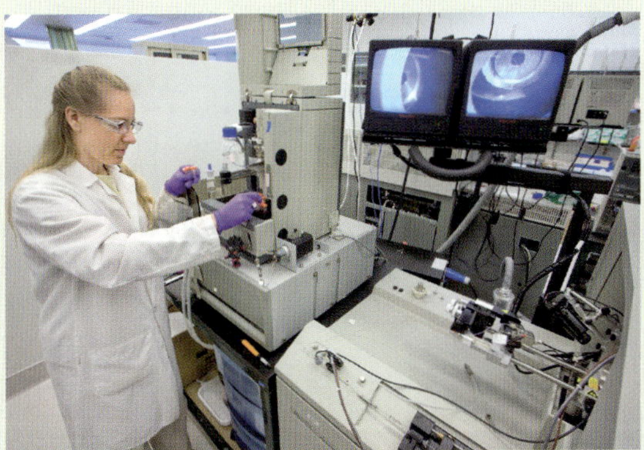

FIGURE 13.2.10 An HPLC–MS instrument in a laboratory

13.2 Review

SUMMARY

- High-performance liquid chromatography (HPLC) and gas chromatography (GC) are very sensitive techniques used for qualitative and quantitative analysis.
- In HPLC, the mobile phase is liquid under pressure; in GC, the mobile phase is a gas.
- In HPLC and GC, retention time is used to identify components in a mixture.
- The concentration of an individual component in a mixture can be determined by comparing its peak area on a chromatogram with the peak areas of samples with known concentrations of the same chemical analysed under identical conditions as the mixture.

KEY QUESTIONS

1 The retention time can be used in GC to determine:
  A the identity of the chemical.
  B the concentration of a chemical.
  C the amount of a chemical in a sample.
  D all of the above.

2 The concentration of a substance is most accurately determined in HPLC by measuring:
  A peak area.
  B peak height.
  C retention time.
  D R_f value.

3 Compounds A and B are equally soluble in water, but A is slightly more strongly adsorbed on silica powder. Compound C has a lower solubility in water than A and B. A sample containing A, B and C is mixed with water and injected into an HPLC instrument, which uses water as the solvent and silica as the stationary phase. List the order of the components as they are likely to emerge from the column.

4 Nitroglycerine patches are used to treat forms of heart disease. A sample from a patch was injected onto a gas–liquid chromatograph column. The peak corresponding to nitroglycerine had an area of 4.4 mm^2. The peak areas for three standard solutions were also measured, as shown in the following table.
  a Construct a calibration curve of peak area versus concentration of nitroglycerine.
  b Determine the concentration of nitroglycerine in the sample in $mg\,mL^{-1}$.

Nitroglycerine standards ($\mu g\,mL^{-1}$)	Peak area (mm^2)
2	3.0
4	5.8
6	8.8

5 The ethanol content of waste water from a brewery can be analysed by HPLC. The peak areas produced from HPLC analysis of a sample of water and a number of standard solutions of ethanol are shown in the table below.

	Ethanol (%)	Relative peak area
Water sample	?	5.6
Standard 1	0.2	1.6
Standard 2	0.4	3.2
Standard 3	0.6	4.8
Standard 4	0.8	6.4

  a Construct a calibration curve of relative peak area against concentration of ethanol (%).
  b Determine the percentage of ethanol in the water sample, correct to 1 decimal place.

Chapter review

13

KEY TERMS

adsorption
calibration curve
carrier gas
chromatogram
chromatography
column chromatography
components
desorption
eluent
gas chromatography (GC)
high-performance (or high-pressure) liquid chromatography (HPLC)
injection port
mobile phase
origin
paper chromatography
qualitative analysis
quantitative analysis
retardation factor (R_f)
retention time (R_t)
standard solution
stationary phase
thin-layer chromatography

Principles of chromatography

1 Use the following terms to complete the sentences about paper chromatography. Words can be used more than once.

above	mobile	phases
below	origin	sample
components	phase	stationary

In paper chromatography, the paper acts as the ______________ phase. A small spot of a solution is placed at one end of the paper, called the ______________. The sample solution contains a number of different coloured compounds, the ______________. The paper is suspended so that the end with the spot is ______________ the surface of the solvent. The solvent or ______________ phase moves up the ______________ phase. Different coloured spots are observed at various places on the paper, due to the separation of different ______________.

2 **a** Use the terms 'adsorbed' and 'absorbed' correctly in each of the sentences below.

- **i** Water was ______________ by the towel as the wet swimmer dried himself.
- **ii** A thin layer of grease ______________ onto the cup when it was washed in the dirty water.

b Explain the difference between the terms 'adsorbed' and 'absorbed'.

3 Describe how the rate of adsorption and desorption of a component that appears at the top of a paper chromatogram compares to that of a component that appears at the bottom of the chromatogram.

4 When a solution containing components X and Y is placed in a chromatography column, the rates at which X and Y move down the column are determined by several factors. Which one of the following factors does *not* directly affect the rate at which X and Y move down the column?

- **A** the polarity of the molecules of X and Y
- **B** the attractions of X and Y to the stationary phase
- **C** the solubility of X and Y in the solvent
- **D** the length of the chromatography column

5 The chromatogram of a dye is shown below. Calculate the R_f values for each of the blue, purple and yellow components. Give answers to 1 decimal place.

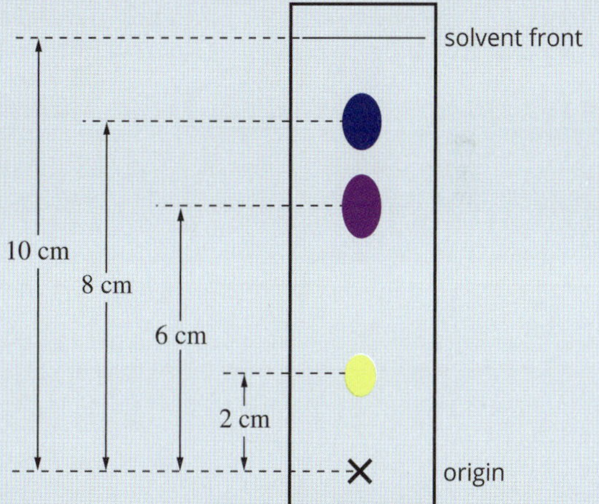

6 A sample of brown dye from a lolly is placed at the origin on a chromatography plate. The solvent front moves 9.0 cm from the origin. A blue component of the dye moves 7.5 cm and a red component 5.2 cm in the same time. Calculate the R_f values of the two components.

7 Consider the paper chromatogram of three food colours in Figure 13.1.6 on page 286.

- **a** Why must the level of the solvent be lower than the origin where spots of the mixture are originally placed?
- **b** Why are R_f values always less than one?
- **c** How many different components have been used to make colour A?
- **d** Which components present in colours B and C are also in colour A? Explain.
- **e** Which component of colour A is most strongly adsorbed on the stationary phase?
- **f** Calculate the R_f values of each component of colour C.

8 Thin-layer chromatography showed that the black dye used in a brand of writing ink contained blue, red, orange and yellow components. The R_f values of these substances in ethanol as solvent are 0.59, 0.32, 0.80 and 0.19, respectively.

a How far apart would the blue and yellow components be after the solvent front had moved 8.0 cm from the origin?

b When the red component had travelled 6.0 cm from the origin, how far would the orange component have travelled?

c Draw the chromatogram of the ink to scale after the solvent front had moved 15 cm from the origin.

9 The diagram below shows a thin-layer chromatogram of amino acids in a medicine.

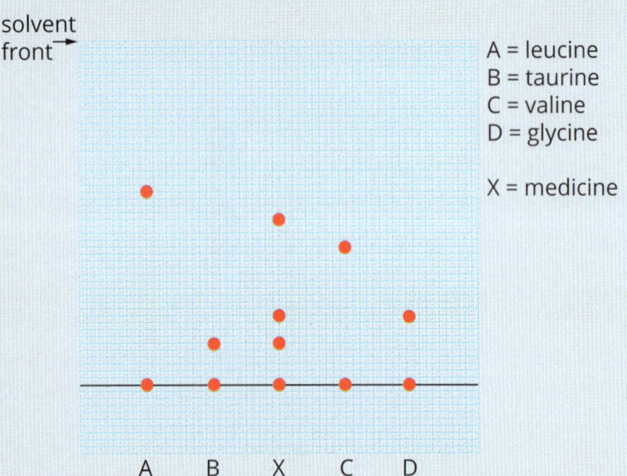

a Which amino acids are in the medicine?

b Amino acids are colourless. What technique could be used to make the amino acid spots visible?

c Calculate the R_f value of taurine.

d Which amino acid is bonded least strongly to the stationary phase?

10 The components in a plant extract were separated by simple column chromatography. The mobile phase used was water and the column was packed with alumina powder. Select from the following terms to complete the sentences that relate to the diagram of the column chromatography experiment.

A	smaller	used coarser alumina powder
B	low	increased the column length
C	high	decreased the column length
	larger	

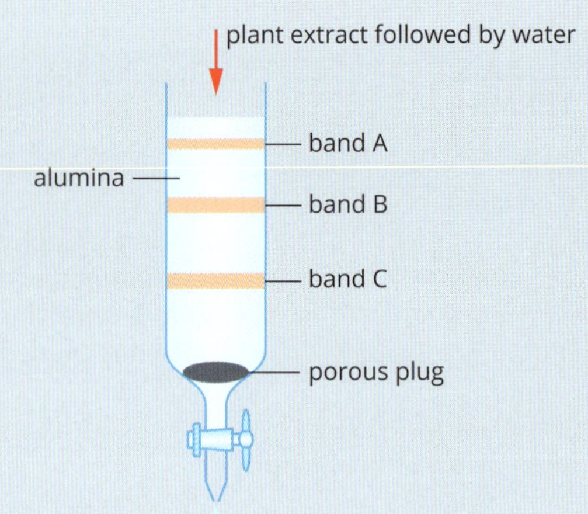

The component most strongly adsorbed to the stationary phase is band ____.

As band B begins to emerge from the column, it appears to separate into two bands. These two components would be more successfully separated if you ____________________.

HPLC differs from column chromatography because the particles in the stationary phase are ____________ and ____________ pressure is applied to the mobile phase.

Advanced applications of chromatography

11 Why is it necessary to measure the peak areas produced by a number of standard solutions when performing quantitative analysis using an HPLC instrument?

A HPLC does not directly produce measures of concentration.

B A graph of the actual number of standard solutions against the peak area can then be drawn.

C The peak areas are measured to ensure they remain constant throughout the analysis.

D The fewer standard solutions used, the more accurate the graph will be.

12 Which one of the following is the basis for the operation of HPLC?

A interaction of chemicals with both mobile and stationary phases

B movement of chemicals through a column using a carrier gas

C passage of chemicals through a heated column under pressure

D small molecules becoming trapped in gaps between solid particles in a packed column

13 On the basis of their chemical properties, which of the following gases would be least suitable as the carrier gas for gas chromatography?

A oxygen
B argon
C nitrogen
D helium

14 A chemist working for an environmental agency wishes to analyse the levels of a herbicide in a country stream using HPLC. Sort the following steps in the analytical process into the correct order.

A Measure the herbicide peak areas from the chromatograms of the standard solutions and the stream water sample.
B Obtain chromatograms of the stream water and standard solutions of the herbicide.
C Construct a calibration curve and mark the herbicide peak area from the stream water sample on it.
D Use R_t values to identify the herbicide peak on the chromatogram of stream water.
E Determine the herbicide concentration in the stream water sample.

15 A herbal tea extract was analysed by HPLC. The chromatogram obtained is shown in the graph below.

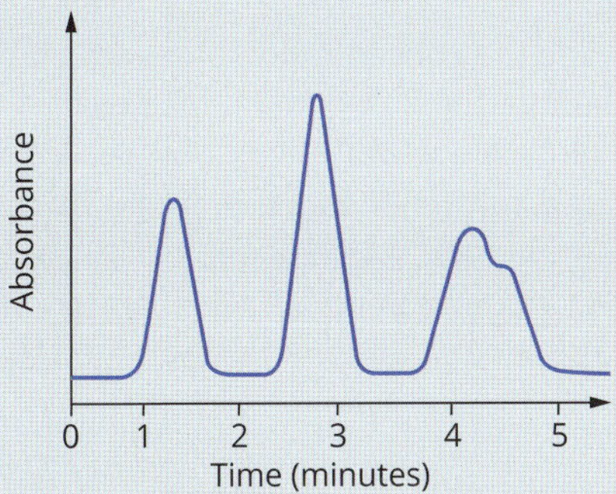

a Explain what information chemists can obtain from this chromatogram.
b How many components are evident?
c Briefly explain how the components are separated by the HPLC technique.

16 Australian wines are routinely tested for ethanol content. A quick and reliable method is by gas chromatography (GC). The peak areas produced from GC analysis of a sample of wine and a number of standard solutions of ethanol are shown in the following table.

	% ethanol	Relative ethanol peak area
Wine sample	?	5.6
Standard 1	2.0	1.6
Standard 2	4.0	3.2
Standard 3	6.0	4.8
Standard 4	8.0	6.4

a Plot a calibration curve of relative peak area against concentration of ethanol (%).
b Determine the percentage of ethanol in the wine sample, correct to 1 decimal place.

17 The concentration of benzene in a sample of petrol was determined by gas chromatography. A series of standards with accurately known concentrations of benzene were run under the same conditions as the sample. Data from the chromatograms of the sample and the standards are shown in the following table. Calculate the concentration of benzene in the sample.

% benzene	Peak area (cm^2)
0.2	1.0
0.4	2.4
0.8	4.6
1.2	7.0
Sample	3.6

Connecting the main ideas

18 A chemist working for a local council wishes to analyse the composition of a river where children swim. It is thought the water might contain high levels of the toxic organic chemical dioxin.

Which one of the following statements concerning the analysis of this water is correct?

A Paper chromatography would allow determination of the concentration of dioxin present but not confirm its identity.
B The results obtained from HPLC would provide identification and accurate concentrations of the dioxin present.
C GC can identify if dioxin is present but not its concentration.
D Paper chromatography would be suitable for identifying the dioxin and measuring its concentration.

19 Each chromatography technique has disadvantages associated with it. Complete the table below using the options provided to indicate the problem or problems associated with each technique. Options may be used more than once.

A It requires relatively large amounts of solvents to operate.

B It is difficult or impossible to obtain quantitative data.

C Expensive equipment is needed.

D Samples must be able to be dissolved in solvent.

Technique	Problem
paper or thin-layer chromatography	
high-performance chromatography (HPLC)	
gas chromatography (GC)	

20 The organophosphorus insecticide parathion has been widely used in mosquito-prone areas. An empty drum of the insecticide was found close to a major reservoir. The EPA was asked to analyse the water to determine whether it was a threat to human health. Levels above 0.01 ppm (parts per million) in water are a threat to human health. Parathion has an LD_{50} value (lethal dose to 50% of test animals exposed to this concentration per kilogram of their body mass) of 8 mg kg^{-1}. Parathion standards in water of 0 ppm, 10 ppm, 20 ppm and 30 ppm parathion were prepared and analysed by HPLC, as shown in the following graph. An undiluted sample of the reservoir water was run on the column under the same conditions.

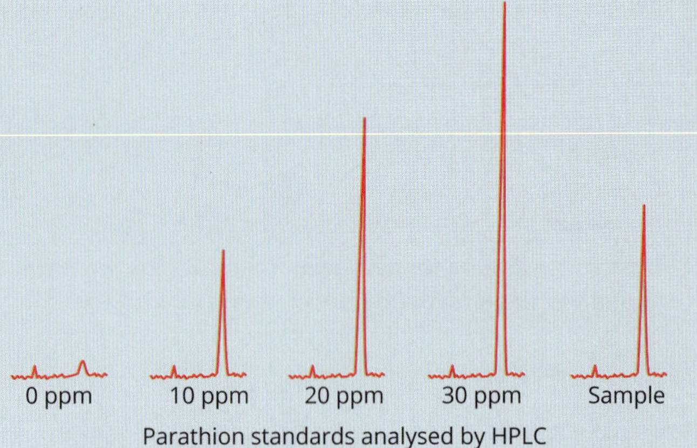

Parathion standards analysed by HPLC

a Complete the table below.

Standard	Peak height (cm)
Standard 0 ppm parathion	
Standard 10 ppm parathion	
Standard 20 ppm parathion	
Standard 30 ppm parathion	
Reservoir water	

b Construct a calibration curve for the analysis of parathion.

c Determine the concentration of parathion in the water sample. (Assume that peak height is a measure of the concentration of the insecticide.)

d Is the reservoir water within the legal limits for safe drinking? Support your answer with a calculation.

e What volume of water would a laboratory mouse, mass 150 g, need to drink to reach the LD_{50} dose?

CHAPTER 14

Gases

At the end of this chapter, you will be able to describe the properties of gases and relate these to the kinetic molecular theory. You will understand how the pressure and volume of a gas sample are measured and explain the relationship between volume, pressure and temperature. You will learn how the volume of a gas can be calculated from its amount, measured in moles.

Science understanding

- the behaviour of an ideal gas, including the qualitative relationships between pressure, temperature and volume, can be explained using the Kinetic Theory
- the mole concept can be used to calculate the mass of substances and volume of gases (at standard temperature and pressure) involved in a chemical reaction

14.1 Introducing gases

FIGURE 14.1.1 (a) Air is used to inflate vehicle tyres. Air is a mixture of gases and is easily compressed. When the car goes over a bump in the road, the air compresses slightly and absorbs the impact of the bump. (b) The gases that cause the smell of a freshly brewed cup of coffee rapidly fill an entire room. Gases mix readily and, unlike solids and liquids, occupy all available space.

There are countless everyday examples you observe of the behaviour of gases (Figure 14.1.1). Such examples tell us a great deal about the physical properties of gases—those properties that can be observed and measured without changing the nature of the gas itself.

In this section, you will learn about the properties and behaviour of gases.

PROPERTIES OF GASES

Table 14.1.1 summarises some of the properties of gases and compares them with the properties of solids and liquids. These observations can be used to develop a particle model of gas behaviour.

TABLE 14.1.1 Some properties of the three states of matter

	Gases	Liquids	Solids
Density	low density	high density	high density
Volume and shape	fill the space available, because particles move independently of one another	fixed volume, adopt shape of container, because particles are affected by attractive forces	fixed volume and shape, because particles are affected by attractive forces
Compressibility	compress easily	almost incompressible	almost incompressible
Ability to mix	rapidly mix together	slowly mix together unless stirred	do not mix unless finely divided

The low density of gases, relative to that of liquids and solids, suggests that the particles in a gas are much more widely spaced. The mass of any gas in a given **volume** will be less than the mass of a liquid or solid in the same volume. This is consistent with the observation that gases are easily compressed—a property fundamental to scuba diving, as seen in Figure 14.1.2. The air that the scuba diver breathes must be compressed to fit into a scuba tank carried on the diver's back.

FIGURE 14.1.2 The underwater activity of a scuba diver relies on the physical properties of gases, including their ability to be compressed.

Gases spread to fill the space available, as shown in Figure 14.1.3. This suggests that the particles of a gas move independently of each other.

The wide spacing and independent movement of particles explains why different gases mix rapidly.

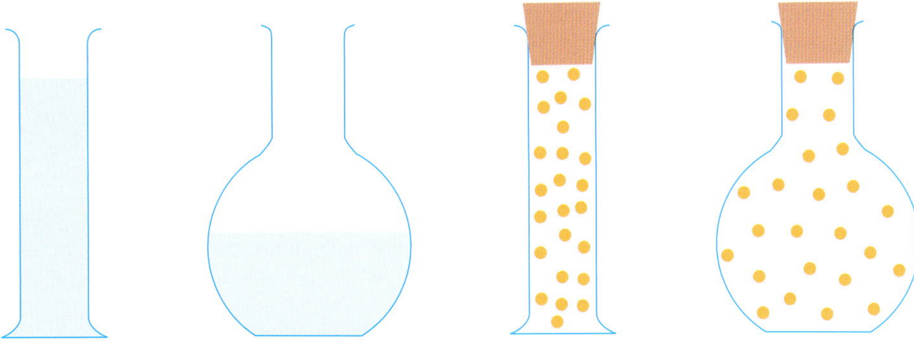

FIGURE 14.1.3 Both liquids and gases take the shape of the container they are in. However, a liquid has a fixed volume, whereas the volume of a gas expands to fill all available space in a container.

KINETIC MOLECULAR THEORY

Scientists have developed a model to explain gas behaviour based on the behaviour of the particles of a gas. This model is known as the **kinetic molecular theory** of gases. The theory is based on the following assumptions.

- Gases are composed of small particles, either atoms or molecules.
- The total volume occupied by the particles in a gas is negligible compared to the average distance between them. Consequently, most of the volume occupied by a gas is empty space.
- Gas particles move rapidly in random, straight-line motion.
- Particles collide with each other and with the walls of the container.
- The forces between particles are extremely weak.
- **Kinetic energy**, the energy of motion, can be transferred from one particle to another, but the total kinetic energy remains constant. Therefore, collisions between gas particles are **elastic collisions**—kinetic energy is conserved.
- The average kinetic energy of the particles increases as the temperature of the gas increases.

Ideal gas behaviour

A gas that conforms to the criteria of the kinetic molecular theory of gases is described as behaving as an **ideal gas**. In particular, in an ideal gas the total volume occupied by the particles in a gas is negligible compared to the total volume of the gas, and collisions between gas particles are elastic. Ideal gas behaviour allows us to predict the behaviour of gases in terms of how they respond to changes in pressure, volume or temperature.

Most gases that you encounter, such as oxygen, hydrogen and carbon dioxide, behave as ideal gases, especially at high temperatures and low pressures, when the space between the atoms or molecules of gas is maximised. As gases cool down, or are placed under higher pressures, the particles come closer together and the increased interactions between the particles cause the gases to show non-ideal behaviour. Likewise, gases that contain molecules with higher relative molecular masses will not behave ideally due to the larger size of the particles in the gas.

CHEMFILE

Airbags in cars

Front and side airbags are now standard equipment in new cars. Airbags contain a mixture of crystalline solids—sodium azide (NaN_3), potassium nitrate (KNO_3) and silica (SiO_2)—stored in a canister. When there is an accident, a sensor 'ignites' the sodium azide, initiating the first step in this reaction. Sodium metal and hot nitrogen gas are products of this energy-releasing reaction. The nitrogen gas rapidly expands, which inflates the nylon airbag. In the second part of the reaction, the molten sodium metal immediately reacts with the potassium nitrate, generating more nitrogen gas, as well as sodium oxide and potassium oxide, which are white powdery solids.

FIGURE 14.1.4 An airbag deploys in a test car within 30 milliseconds of impact, with a crash dummy taking the impact.

THE RELATIONSHIP BETWEEN MOLECULAR KINETIC ENERGY AND TEMPERATURE

According to the kinetic molecular theory, the average kinetic energy of gas particles in a sample is directly proportional to the temperature of the gas. This means that as the temperature of a gas sample increases, the average kinetic energy of the particles in the gas also increases. However, it is important to understand that at any given temperature, the following are true.

- The average kinetic energy of a gas does not depend on the chemical identity of the gas. For example, molecules of hydrogen gas and molecules of oxygen gas will have the same average kinetic energy at the same temperature.
- Not all particles within a sample will have the same kinetic energy. Some particles will have a lower kinetic energy and some will have a higher kinetic energy. The temperature reflects the average kinetic energy of all particles in the sample.

The distribution of the kinetic energies of particles in a gas at a given temperature is illustrated by the graph in Figure 14.1.5.

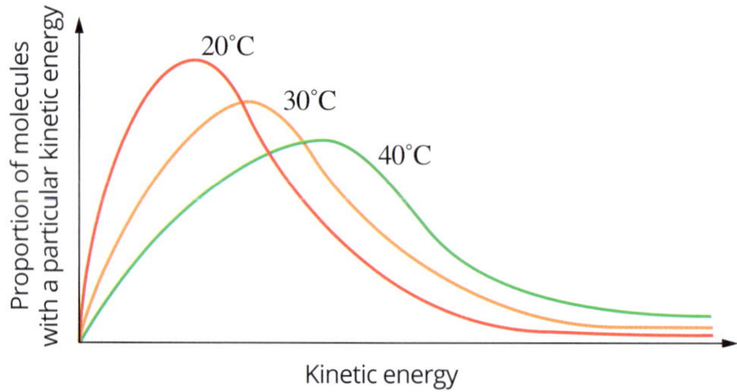

FIGURE 14.1.5 The distribution of molecular kinetic energies for three different temperatures

This graph is called a **Maxwell–Boltzmann distribution graph** and is also known as a **kinetic energy distribution diagram**. The total area under each curve represents the total number of molecules in a given sample. The areas under the three curves are equal, because the same number of molecules is being compared at three different temperatures.

Notice that, at 20°C, a greater proportion of molecules have a lower kinetic energy. The reverse is true at 40°C. At this higher temperature, a greater proportion of molecules have a higher kinetic energy.

From the figure above, you can see the following.

- The proportion of molecules with high kinetic energy increases with temperature.
- The average kinetic energy of a sample increases with temperature.
- Only a small proportion of molecules have very low or very high kinetic energy at any temperature.

The average kinetic energy of particles in gases is related to their average speed. Recall that different gases at the same temperature have the same average kinetic energy and that kinetic energy (KE) is given by the formula $KE = \frac{1}{2}mv^2$, where m = mass and v = velocity. Therefore, at a given temperature, particles with a higher mass on average travel at a lower velocity than particles with a lower mass.

Diffusion

Diffusion is the process whereby gases in a mixture spread out to uniformly fill the total volume available (Figure 14.1.6). The rate at which gases diffuse depends on the temperature of the gases, their molar masses, and the average velocity of their particles.

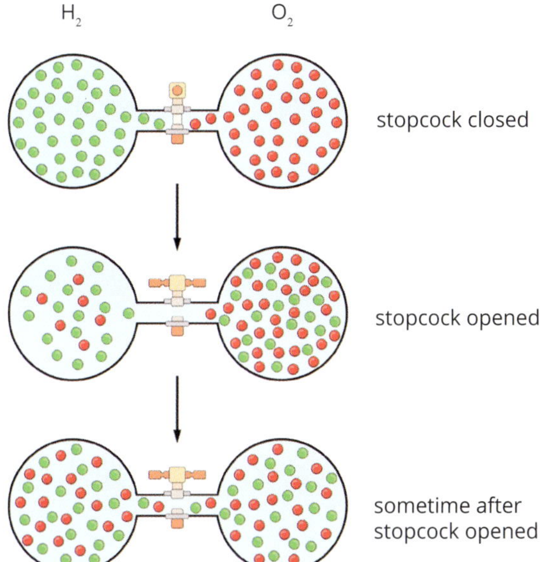

FIGURE 14.1.6 The diffusion of H_2 and O_2 particles in gaseous state when allowed to uniformly spread throughout a container. The gaseous particles will keep moving until a homogeneous mixture is achieved. Note that the hydrogen gas diffuses more quickly into the oxygen than the oxygen gas diffuses into the hydrogen.

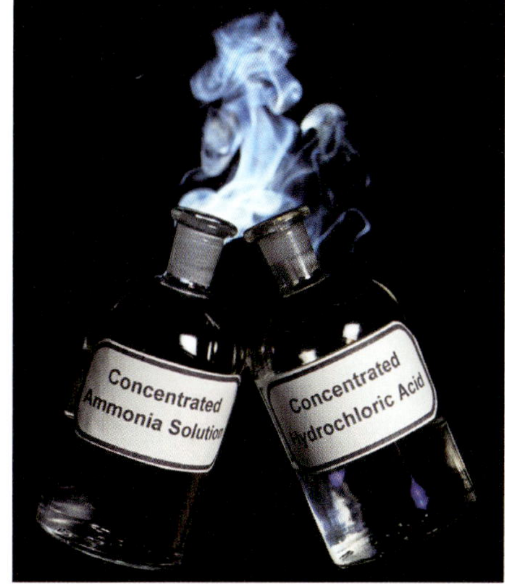

FIGURE 14.1.7 White ammonium chloride powder is formed as gases from concentrated ammonia solution and hydrochloric acid solution diffuse from their containers and react. Ammonia and hydrogen chloride are colourless gases.

You learnt earlier that as temperature increases in a specific sample of gas, the average kinetic energy of the particles also increases. In this case, the mass of the particles is held constant. Therefore, the velocity and speed of the particles increase with temperature.

Conversely, if the temperature of a sample of gases is held constant, the average kinetic energy of particles is constant. In this case, because $KE = \frac{1}{2}mv^2$, the lighter the particles of gas, the greater their velocity. At a given temperature, gases of lower molecular mass move at a greater velocity and, therefore, will diffuse more rapidly than gases of higher molecular mass. This is why the hydrogen gas shown in Figure 14.1.6 will diffuse more rapidly than the oxygen gas. Figure 14.1.7 shows an example of diffusion.

THE NATURE OF VOLUME AND PRESSURE

When you blow up a balloon, the balloon expands because you put more air into it and the rubber of the balloon stretches. If you then plunge that balloon into liquid nitrogen, as shown in Figure 14.1.8, the balloon shrinks dramatically. The temperature of the gas decreases, causing the volume to decrease.

FIGURE 14.1.8 A balloon filled with air at room temperature is dipped into liquid nitrogen at −196°C. The volume of the air inside the balloon decreases dramatically.

FIGURE 14.1.9 Pumping more air into a tyre increases the pressure in the tyre because more particles are being pumped into a nearly fixed volume.

Pumping more air into a tyre, as shown in Figure 14.1.9, increases the pressure inside the tyre because more particles are being forced into approximately the same volume inside the tyre.

Volume

Volume is the quantity that is used to describe the space that a substance occupies. Because a gas occupies the whole container that it is in, the volume of a gas is equal to the volume of its container.

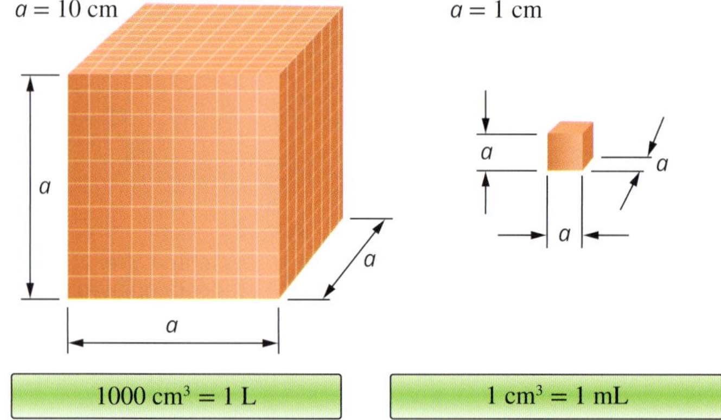

FIGURE 14.1.10 Cubic centimetres, millilitres and litres are the most commonly used units of volume.

There are several different units used for volume, some of which are represented in Figure 14.1.10. The common units are litre (L), millilitre (mL), cubic metre (m^3) and cubic centimetre (cm^3).

- 1 mL = 1 cm^3
- 1 L = 1×10^3 mL or 1000 mL
- 1 m^3 = 1000 L = 1×10^6 mL

Small volumes of gas are usually measured in millilitres (mL) or litres (L). Very large samples are measured in cubic metres (m^3).

Different units are used to describe volume. The relationship between these units is:

1000 mL = 1000 cm^3 = 1 L

1×10^6 mL = 1×10^6 cm^3 = 1 m^3

Worked example 14.1.1

CONVERTING VOLUME UNITS

A gas has a volume of 255 mL.
What is its volume in:

a cubic centimetres (cm^3)?
b litres (L)?
c cubic metres (m^3)?

Thinking	Working
Recall the conversion factors for each of the units of volume. Apply the correct conversion to each situation.	**a** The units of mL and cm^3 are equivalent. 1 mL = 1 cm^3 255 mL = 255 cm^3 **b** 1000 mL = 1 L Divide volume in mL by 1000 to convert to L. $255\text{ mL} = \frac{255}{1000}$ $= 0.255\text{ L}$ **c** 1×10^6 mL = 1 m^3 Divide volume in mL by 1×10^6 to convert to m^3. $255\text{ mL} = \frac{255}{1 \times 10^6}$ $= 2.55 \times 10^{-4}\text{ m}^3$

Worked example: Try yourself 14.1.1

CONVERTING VOLUME UNITS

A gas has a volume of 700 mL.
What is its volume in:

a cubic centimetres (cm^3)?

b litres (L)?

c cubic metres (m^3)?

Pressure

People often talk about exerting pressure on something as if it is some kind of force. In terms of the kinetic theory of gases, **pressure** is defined as the force exerted on a unit area of a surface by the particles of a gas as they collide with the surface.

The more a gas is compressed, the greater the number of collisions the gas particles will have with each other and the walls of their container. These collisions produce a force on the walls of the container, such as the inside of a tyre, which can be measured. The force per unit area is described as pressure.

The pressure of a gas can be measured in different ways. Figure 14.1.11 shows a pressure gauge like those attached to cylinders of compressed air used for scuba diving.

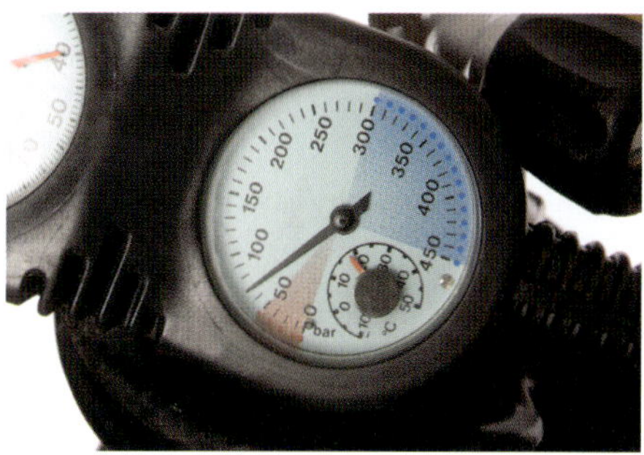

FIGURE 14.1.11 A pressure gauge like this one would be found on a cylinder of compressed air used in scuba diving.

Air is a mixture of gases including nitrogen, oxygen, carbon dioxide and argon. In a gaseous mixture, such as air, the measured air pressure is the sum of pressures of the individual gases in the mixture. In air, the nitrogen molecules collide with the walls of a container, exerting a pressure. In a similar way, the oxygen molecules exert a pressure, as do molecules of each gas present in the mixture. In the gaseous mixture of nitrogen and oxygen shown in Figure 14.1.12, the measured pressure is the sum of the **partial pressure** of oxygen and the partial pressure of nitrogen.

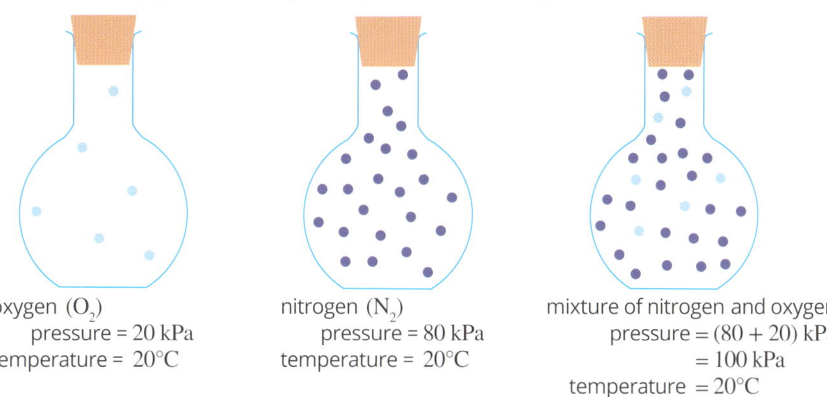

FIGURE 14.1.12 The total pressure of a mixture of gases is the sum of the partial pressures (individual pressures) of each of the gases in the mixture.

Units of pressure

Since pressure is the force exerted on a unit area of a surface, the relationship can be written as:

$$\text{pressure} = \frac{\text{force}}{\text{area}}$$

$$P = \frac{F}{A}$$

The units of pressure depend on the units used to measure force and area. Over the years, scientists in different countries have used different units to measure force and area, so there are a number of different units of pressure.

CHEMFILE

Torricelli's barometer

In the 17th century, the Italian physicist Evangelista Torricelli invented the earliest barometer, an instrument used to measure atmospheric pressure. It was a straight glass tube, closed at one end, containing mercury. The tube was inverted so that the open end was below the surface of mercury in a bowl as seen in Figure 14.1.13.

The column of mercury in Torricelli's barometer was supported by the pressure of the gas particles in the atmosphere colliding with the surface of the mercury in the open bowl.

At sea level, the top of the column of mercury was about 760 mm above the surface of the mercury in the bowl. Torricelli found that the height of the mercury column decreased when he took his barometer to higher altitudes in the mountains.

At higher altitudes, there are fewer air particles and therefore less frequent collisions on the surface area of mercury. The reduced pressure supports a shorter column of mercury.

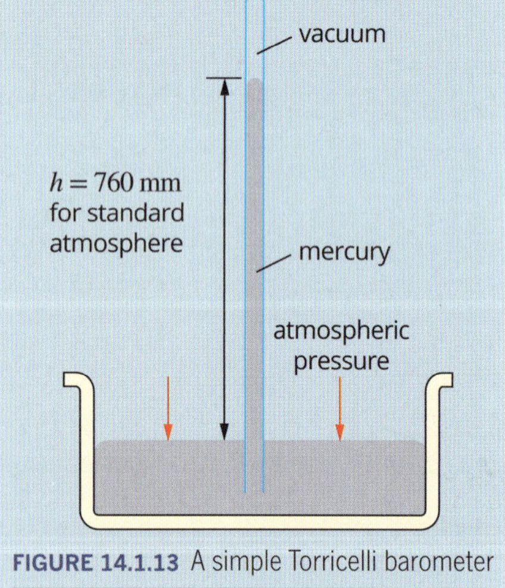

FIGURE 14.1.13 A simple Torricelli barometer

The SI unit for force is the newton (N) and for area is the square metre (m^2). Pressure in SI units is therefore newton per square metre ($N\,m^{-2}$). One newton per square metre is equivalent to a pressure of one **pascal** (Pa).

In 1982, the International Union of Pure and Applied Chemistry (IUPAC), the organisation responsible for naming chemicals and setting standards, adopted a standard for pressure equivalent to 100 000 Pa or 100 kPa. This gave rise to a new pressure unit, the **bar**, where 1 bar equals 100 kPa.

The use of mercury barometers in past years resulted in pressure often being measured in millimetres of mercury, or mmHg.

Another unit for pressure is the standard atmosphere (atm). One standard atmosphere (1 atm) is the pressure required to support 760 millimetres of mercury (760 mmHg) in a mercury barometer at 25°C. This is the average atmospheric pressure at sea level. One atmosphere equals 101.3 kPa.

Summary of pressure units

There are four common units of gas pressure: pascal, bar, millimetres of mercury and atmosphere (Table 14.1.2). These relationships can be used to convert pressure from one unit to another.

TABLE 14.1.2 Common units of gas pressure

Name of unit	Symbol for unit	Conversion to $N\,m^{-2}$
newton per square metre	$N\,m^{-2}$	
pascal	Pa	$1\,Pa = 1\,N\,m^{-2}$
kilopascal	kPa	$1\,kPa = 1 \times 10^3\,Pa = 1 \times 10^3\,N\,m^{-2}$
atmosphere	atm	$1\,atm = 101.3\,kPa = 1.013 \times 10^5\,N\,m^{-2}$
bar	bar	$1\,bar = 100\,kPa = 1.00 \times 10^5\,N\,m^{-2}$
millimetres of mercury	mmHg	$760\,mmHg = 1\,atm = 1.013 \times 10^5\,N\,m^{-2}$

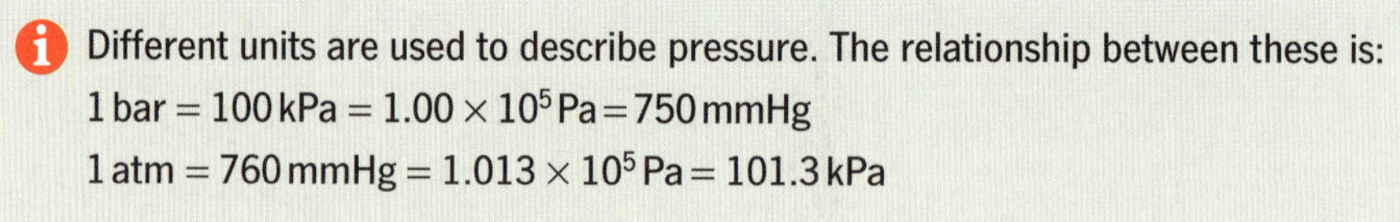

Different units are used to describe pressure. The relationship between these is:

$1\,bar = 100\,kPa = 1.00 \times 10^5\,Pa = 750\,mmHg$

$1\,atm = 760\,mmHg = 1.013 \times 10^5\,Pa = 101.3\,kPa$

Worked example 14.1.2

CONVERTING PRESSURE UNITS

Mount Everest is the highest mountain on Earth.

a The atmospheric pressure at the top of Mount Everest is 0.337 bar. What is the pressure in kilopascals (kPa)?

b The atmospheric pressure at the top of Mount Everest is 253 mmHg. What is the pressure in atmospheres (atm)?

c The atmospheric pressure at the top of Mount Everest is 0.333 atm. What is the pressure in kilopascals (kPa)?

d The atmospheric pressure at the top of Mount Everest is 253 mmHg. What is the pressure in bars?

Thinking	Working
a To convert bars to kilopascals, use the conversion relationship: 1 bar = 100 kPa To change bar to kPa, multiply the value by 100.	$0.337\,bar = 0.337 \times 100$ $= 33.7\,kPa$

b To convert millimetres of mercury to atmospheres, use the relationship: 1 atm = 760 mmHg To change mmHg to atm, divide the value by 760.	$253\,\text{mmHg} = \frac{253}{760}$ $= 0.333\ \text{atm}$
c To convert atmospheres to kilopascals, use the conversion relationship: 1 atm = 101.3 kPa To change atm to kPa, multiply the value by 101.3.	$0.333\,\text{atm} = 0.333 \times 101.3$ $= 33.7\,\text{kPa}$
d This can be done in two steps. First, convert millimetres of mercury to atmospheres. Use the conversion relationship: 760 mmHg = 1 atm To change mmHg to atm, divide the value by 760. Keep the answer in your calculator and proceed to the next step. Next, convert atmospheres to bar. Use the conversion relationship: 1 atm = 1.013 bar To change atm to bar, multiply the quotient from the previous step by 1.013.	$253\,\text{mmHg} = \frac{253}{760}$ $253\,\text{mmHg} = \frac{253}{760} \times 1.013$ $= 0.337\ \text{bar}$

Worked example: Try yourself 14.1.2

CONVERTING PRESSURE UNITS

Cyclone Yasi was one of the biggest cyclones in Australian history.

a The atmospheric pressure in the eye of Cyclone Yasi was measured as 0.902 bar. What was the pressure in kilopascals (kPa)?

b What was the pressure in the eye of Cyclone Yasi in atmospheres (atm) if it was known to be 677 mmHg?

c If the atmospheric pressure in the eye of Cyclone Yasi was 0.891 atm, what was the pressure in kilopascals (kPa)?

d The atmospheric pressure in the eye of Cyclone Yasi was 677 mmHg. What was the pressure in bars?

CHEMISTRY IN ACTION

Decompression chambers

Scuba diving and water pressure

If you swim at the water's surface, your body experiences a pressure of about 1 atm, due to the surrounding air. Below the surface, your body experiences an additional pressure due to the water. This additional pressure amounts to about 1 atm for every 10 m of depth. Therefore, at 20 m the pressure on your body is about 3 atm.

As the pressure on your body increases, the volume of your body cavities such as your lungs and inner ears decrease. This squeezing effect makes diving well below the water's surface without scuba equipment very uncomfortable. Scuba equipment overcomes this problem by supplying air from tanks to the mouth at the same pressure as that produced by the underwater environment.

Scuba diving and gas solubility

As the pressure in a diver's lungs increases during a dive, more gas dissolves in the blood. Nitrogen (N_2) is one of these gases. When a diver ascends, the pressure drops, the nitrogen becomes less soluble in the blood and so comes out of solution. If a diver ascends too quickly, the rapid pressure drop causes the nitrogen to come out of the blood as tiny bubbles (Figure 14.1.14a). This is similar to the bubbles of carbon dioxide you observe when you open a bottle of soft drink.

These bubbles cause pain in joints and muscles. If they form in the spinal cord, brain or lungs, they can cause paralysis or death. Divers suffering from this effect ('the bends') are put in a decompression chamber, like the one shown in Figure 14.1.14b. The chamber increases the pressure surrounding the diver's body, forcing any nitrogen bubbles to dissolve in the blood, and then slowly reduces the pressure back to 1 atm over an extended period of time.

Hyperbaric oxygen therapy (HBOT) is a medical treatment in which patients breathe pure oxygen while inside a decompression chamber at a pressure higher than 1 atm. Its medical uses include improved wound healing by reduction of swelling, infection control and the stimulation of new blood vessel growth. The hyperbaric unit at The Alfred Hospital in Melbourne is equipped with three pressurisable hyperbaric rooms used for treating the bends, ulcers, soft tissue infections, carbon monoxide poisoning and wounds that won't heal such as those resulting from diabetes or radiotherapy.

HBOT is also used by athletes for faster recovery from soft tissue injuries, to produce sharper performance and to maintain more intense training schedules. Tennis star Novak Djokovic uses a hyperbaric chamber to assist with recovery from long, exhausting matches. AFL clubs are interested in the use of hyperbaric oxygen therapy for treating soft tissue injuries. One of the earliest cases of hyperbaric oxygen treatment in football was for an ankle injury sustained by Carlton midfielder Fraser Brown in the 1995 preliminary final against North Melbourne. Every morning, in the week leading up to the Grand Final, he spent an hour in a chamber in the hyperbaric unit at The Alfred Hospital. Eventually selected to play in the Grand Final, Brown made it through the game with the help of strapping and pain-killing injections, helping Carlton win the premiership.

FIGURE 14.1.14 (a) Pressure differences during scuba diving can cause gases like nitrogen to become less soluble in the blood of scuba divers and lead to the bends. (b) A decompression chamber is used to treat divers with the bends.

EXTENSION

The relationship between volume, temperature and pressure

By performing experiments on gases, it is possible to establish some rules or laws that quantify the relationship between volume, pressure, temperature and the number of particles of gas. These relationships have become known as gas laws. The gas laws are beyond the scope of this course but are summarised below.

The relationship between gas volume and pressure

In 1662, Robert Boyle, an Irish chemist, showed by experiment that if the volume of a fixed amount of gas at constant temperature is halved, the pressure doubles. If the volume is tripled, the pressure drops to one third of its original value. Boyle concluded that, for a given amount of gas at constant temperature, the volume of the gas is inversely proportional to its pressure. The relationship between gas volume and pressure is known as Boyle's law.

Changing the volume of a fixed amount of gas at constant temperature causes a change in the pressure of the gas. The pressure of the gas in the syringe shown in Figure 14.1.15 increases as the plunger is pushed in and decreases as the plunger is pulled out.

This relationship is seen in the changing volume of a weather balloon as it rises to altitudes with much lower pressure than at ground level. A weather balloon filled with helium gas to a volume of 40 L at a pressure of 1 atm increases in volume to 200 L by the time it reaches an altitude with a pressure of 0.2 atm.

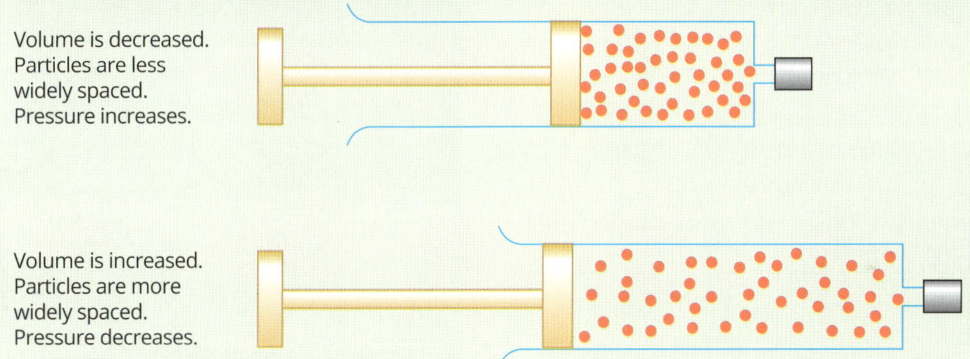

FIGURE 14.1.15 The pressure of the gas in the syringe is decreased when the plunger is pulled out. In this case, the syringe needle is capped, and no gas is allowed to enter or escape.

The relationship between gas volume and temperature

In 1787, Jacques Charles, experimentally determined that the volume of a fixed amount of gas is directly proportional to the absolute temperature, provided the pressure remains constant. The relationship between gas volume and temperature is known as Charles' law.

Table 14.1.3 shows the results of an experiment that investigates the relationship between temperature and the volume of a given amount of gas at constant pressure. In this experiment, the gas in a syringe is heated slowly in an oven. The pressure on the plunger of the syringe is held constant.

TABLE 14.1.3 Variation of volume with temperature

Temperature (°C)	20	40	60	80	100	120	140
Volume (mL)	60.0	64.1	68.2	72.3	76.4	80.5	84.6

EXTENSION *continued*

The graphs of these results, and the results of similar experiments with other gas samples, are linear, as shown in Figure 14.1.16. When the graph is extrapolated to a volume of 0 L, it crosses the temperature axis at −273.15°C. This has led scientists to develop a new temperature scale, known as the Kelvin scale or absolute temperature scale. On the Kelvin scale, each temperature increment is equal to one temperature increment on the Celsius scale, and 0°C is equal to 273.15 K.

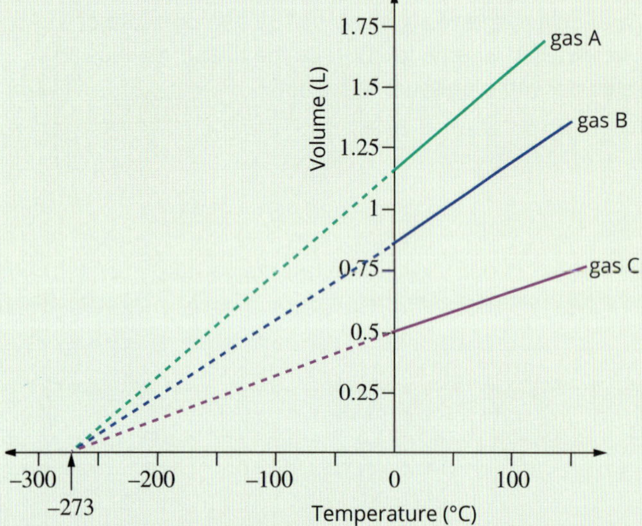

FIGURE 14.1.16 The variation of volume with temperature for fixed amount of three different gases at constant pressure

The relationship between temperature on the Celsius scale and temperature on the Kelvin scale is given by the equation:

$$T\text{ (in K)} = T\text{ (in °C)} + 273.15$$

The temperature 0 K (–273.15°C) is the lowest temperature theoretically possible. For this reason, 0 K is known as absolute zero. At this temperature, all molecules would have minimum kinetic energy and molecular motion will theoretically have stopped. The accepted convention for the value of absolute zero is –273.15°C.

When a weather balloon rises in the atmosphere, the pressure and temperature of the gas inside decrease. These changes have opposing effects on the volume of the balloon. The decrease in pressure causes the volume of the balloon to increase. To a lesser extent, the decrease in temperature causes the volume of the balloon to decrease. The effect is an overall increase in the volume of the balloon.

FIGURE 14.1.17 When a weather balloon rises in the atmosphere, its volume increases as a result of the decrease in atmospheric pressure.

14.1 Review

SUMMARY

- Gases have relatively low density, they are easily compressed and different gases mix rapidly together.
- Gas properties can be explained by the kinetic molecular theory.
- According to the kinetic molecular theory:
  - the volume of the particles in a gas is very small compared with the distance between the particles
  - the average kinetic energy of the particles in a gas is proportional to its temperature
  - gas particles are in rapid random motion, colliding with each other and the container wall
  - the forces between particles in a sample of gas are negligible.
- Gases that behave according to the kinetic molecular theory are called ideal gases.
- Diffusion is the process whereby gases in a mixture spread out to uniformly fill the total volume available.
- Pressure is defined as the force per unit area.
- The pressure exerted by a gas is caused by the collisions of gas particles against the wall of the container.
- To convert between different volume units, use these relationships:

  $1\,L = 1000\,mL = 1000\,cm^3$

  $1\,m^3 = 1 \times 10^6\,cm^3 = 1 \times 10^6\,mL = 1000\,L$
- To convert between different pressure units, use these relationships:

  $1\,bar = 100\,kPa = 1.00 \times 10^5\,Pa = 750\,mmHg$

  $1\,atm = 760\,mmHg = 1.013 \times 10^5\,Pa = 101.3\,kPa = 1.013\,bar$

KEY QUESTIONS

1 Use the kinetic molecular theory to explain the following observed properties of gases.
   - **a** Gases occupy all the available space in a container.
   - **b** Gases can be easily compressed compared with their corresponding liquid forms.
   - **c** A given volume of a gaseous substance weighs less than the same volume of the substance in the liquid state.
   - **d** Gases readily mix together.
   - **e** The total pressure of a mixture of gases is equal to the sum of the pressures exerted by each of the gases in the mixture.

2 Use the ideas of the kinetic molecular theory of gases to explain the following observations.
   - **a** Tyre manufacturers recommend a maximum pressure for tyres.
   - **b** The pressure in a car's tyres will increase if a long distance is travelled on a hot day.
   - **c** You can smell dinner cooking as you enter your house.
   - **d** A balloon will burst if you blow it up too much.

3 Four gases and their molar masses are listed below. If the gases are at the same temperature, sort them to show their rate of diffusion from fastest to slowest.
   - argon ($39.95\,g\,mol^{-1}$)
   - helium ($4.003\,g\,mol^{-1}$)
   - krypton ($83.80\,g\,mol^{-1}$)
   - neon ($20.18\,g\,mol^{-1}$)

4 In the kinetic molecular theory, pressure is described as the force per unit area of surface. Explain what happens to the pressure in each of the following situations.
   - **a** The temperature of a filled aerosol can is increased.
   - **b** A gas in a syringe is compressed.

5 Convert each of the following pressures to the units specified.
   - **a** 140 kPa to Pa
   - **b** 92 000 Pa to kPa
   - **c** 4.24 atm to mmHg and Pa
   - **d** 120 kPa to mmHg, atm and bar

6 Convert the following volumes to the unit specified.
   - **a** 2 L to mL
   - **b** 4.5 L to m^3
   - **c** 2250 mL to L
   - **d** 120 mL to L

14.2 Molar volume of a gas

In 1811, the Italian scientist Amedeo Avogadro proposed that, at the same temperature and pressure, equal volumes of all gases contain equal numbers of particles. This became known as **Avogadro's law**. The law can be summarised by the relationship:

$$V \propto n$$

The volume, V, occupied by a gas depends directly on the amount of gas, n, in moles, at a constant pressure and temperature.

This relationship is shown in Figure 14.2.1. Both syringes show a gas at a constant temperature and pressure. The volume doubles with twice the number of molecules of gas in the syringe.

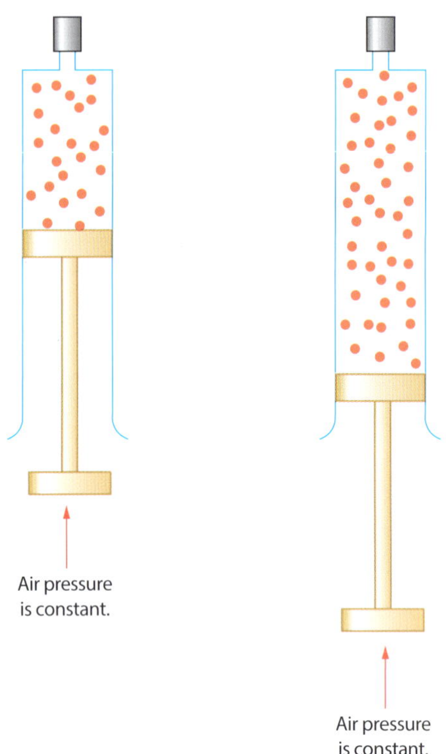

FIGURE 14.2.1 When the amount of gas in the syringe is doubled, the volume doubles, provided the pressure on the plunger and the temperature of the gas remain constant.

> At the same temperature and pressure, equal volumes of all gases contain equal numbers of particles. The law can be summarised by the relationship:
> $V \propto n$

CHEMFILE

Amedeo Avogadro

Avogadro's full name was Lorenzo Romano Amedeo Carlo Avogadro and his title was Count of Quaregna and Cerrato. Born in Italy in 1776, he worked for three years as a lawyer before leaving the legal profession to study physics and mathematics. This path eventually led him to formulate his hypothesis, which is now known as Avogadro's law: under identical conditions of temperature and pressure, equal volumes of gases contain an equal number of particles. It was not until five years after Avogadro's death that the Italian chemist Stanislao Cannizzaro presented Avogadro's work and he finally received credit for it.

FIGURE 14.2.2 Amedeo Avogadro (1776–1856)

If the amount of gas is fixed at 1 mole, the volume the gas occupies will depend almost entirely on its temperature and pressure. This volume is defined as the **molar volume**, V_m, of a gas. Molar volume is the amount of space, or volume, occupied by 1 mole of any gas.

The volume of 1 mol of gas, V_m, is equal to its total volume, V, divided by the number of moles, n, of gas present. This can be represented by the relationship:

$$V_m = \frac{V}{n} \text{ (at a given temperature and pressure)}$$

Or, by rearranging this expression:

$$n = \frac{V}{V_m} \text{ (at a given temperature and pressure)}$$

The molar volume of a gas varies with temperature and pressure. However, molar volume does not vary with the identity of the gas. For example, 1 mole of neon gas will occupy the same volume as 1 mole of nitrogen and 1 mole of ozone gas at the same temperature and pressure (Figure 14.2.3).

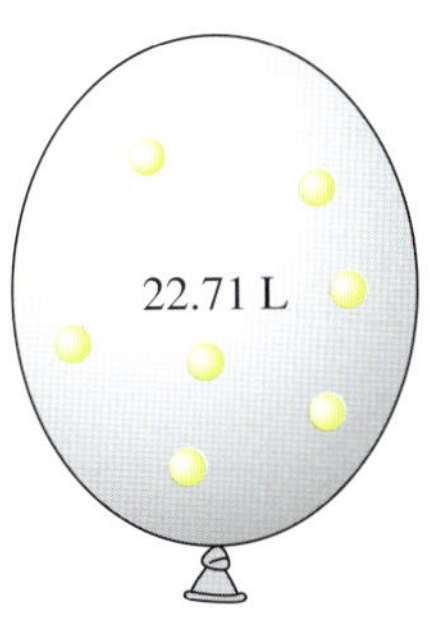

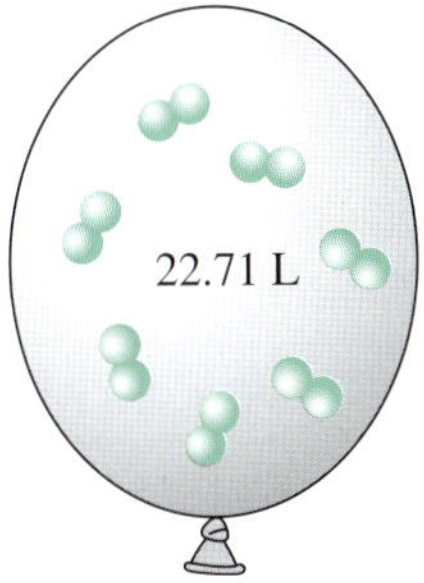

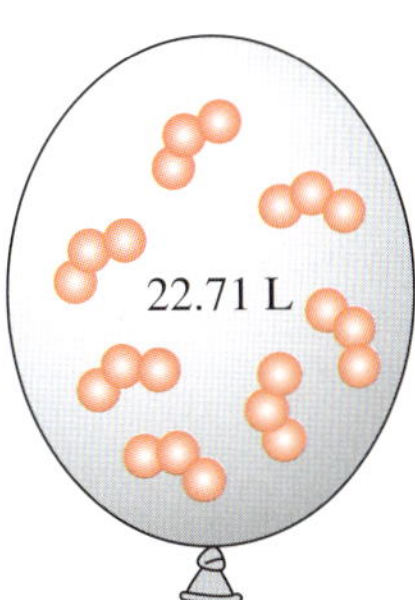

1 mole of neon, Ne
$P = 100$ kPa
$T = 0°C$
$V = 22.71$ L
number of atoms
$= 6.022 \times 10^{23}$
mass = 20.2 g

1 mole of nitrogen, N_2
$P = 100$ kPa
$T = 0°C$
$V = 22.71$ L
number of molecules
$= 6.022 \times 10^{23}$
mass = 28.0 g

1 mole of ozone, O_3
$P = 100$ kPa
$T = 0°C$
$V = 22.71$ L
number of molecules
$= 6.022 \times 10^{23}$
mass = 48.0 g

FIGURE 14.2.3 1 mole of any gas occupies a volume of 22.71 L at STP.

Standard conditions

Standard temperature and pressure (STP) refers to a temperature of 0°C (273.15 K) and a pressure of 100 kPa. Table 14.2.1 shows the molar volume of an ideal gas and of helium at STP.

TABLE 14.2.1 Molar volume at STP

Gas	Formula	Molar volume at STP (L mol^{-1})
Ideal gas	–	22.71
Helium	He	22.69

An ideal gas is a theoretical gas composed of particles that do not interact except during elastic collisions. At STP, most gases behave very like an ideal gas and, therefore, have a molar volume very like that of an ideal gas.

It is usual to assume that the molar volume of a gas is 22.71 L mol^{-1} at STP. From this value, the amount, in moles, of a gas given its volume at STP can be calculated.

From the table above, it can be seen that the volumes of real gases vary slightly at STP. The stronger the intermolecular forces between the gas molecules, the greater their volume varies from that of an ideal gas at STP.

The observation that gases only approximate ideal behaviour suggests that they do not behave exactly according to the kinetic molecular theory. Generally, a gas behaves more like an ideal gas at higher temperature and lower pressure. At STP, most gases behave like an ideal gas.

Although the molar volume of a gas at STP is commonly used, it is important to remember that 1 mole of any gas at the same temperature and pressure will occupy a fixed volume, regardless of the particular temperature and pressure. This is Avogadro's law.

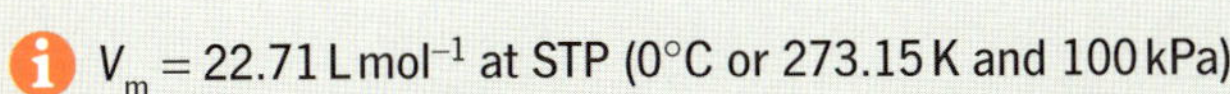
$V_m = 22.71$ L mol^{-1} at STP (0°C or 273.15 K and 100 kPa)

Worked example 14.2.1

CALCULATING THE VOLUME OF A GAS FROM ITS AMOUNT (IN MOL)

Calculate the volume, in L, occupied by 0.24 mol of nitrogen gas at STP. Assume that nitrogen behaves like an ideal gas.	
Thinking	**Working**
Rearrange $n = \frac{V}{V_m}$ to make volume the subject.	$n = \frac{V}{V_m}$ $V = n \times V_m$
Substitute in the known values where $V_m = 22.71\ \text{L mol}^{-1}$ (at STP) and solve.	$V_{(STP)} = n \times V_m$ $= 0.24 \times 22.71$ $= 5.450\ \text{L}$
Consider the units and significant figures. The answer should be given to the smallest number of significant figures in the measurement.	$V_{(STP)} = 5.5\ \text{L}$

Worked example: Try yourself 14.2.1

CALCULATING THE VOLUME OF A GAS FROM ITS AMOUNT (IN MOL)

Calculate the volume, in L, occupied by 3.50 mol of oxygen gas at STP. Assume that oxygen behaves like an ideal gas.

Worked example 14.2.2

CALCULATING THE AMOUNT OF A GAS FROM ITS VOLUME

Calculate the amount (in mol) of nitrogen gas in a volume of 6.1 L at STP. Assume that nitrogen behaves like an ideal gas.	
Thinking	**Working**
Use $n = \frac{V}{V_m}$ where $V_m = 22.71\ \text{L mol}^{-1}$ (at STP)	$n = \frac{V}{V_m}$ $= \frac{6.1}{22.71}$ $= 0.2686\ \text{mol}$
Consider the units and significant figures. The answer should be given to the smallest number of significant figures in the measurement.	$n = 0.27\ \text{mol}$

Worked example: Try yourself 14.2.2

CALCULATING THE AMOUNT OF A GAS FROM ITS VOLUME

Calculate the amount (in mol) of oxygen gas in a volume of 3.5 L measured at STP. Assume that oxygen behaves like an ideal gas.

14.2 Review

SUMMARY

- Standard temperature and pressure (STP) refers to a temperature of 0°C (273.15 K) and a pressure of 100 kPa.
- The molar volume, V_m, of a gas is the volume occupied by 1 mol of gas at a given temperature and pressure:

$$n = \frac{V}{V_m}$$

- The value of V_m is 22.71 L mol^{-1} at STP.
- Avogadro's law states that equal volumes of gas at the same temperature and pressure contain the same number of particles.

KEY QUESTIONS

1 Calculate the volume of the following gases at STP.
  - **a** 1.4 mol of chlorine (Cl_2)
  - **b** 1.0×10^{-3} mol of hydrogen (H_2)
  - **c** 1.4 g of nitrogen (N_2)

2 Calculate the number of moles of the following gas samples. All volumes are measured at STP.
  - **a** 2.80 L of neon (Ne)
  - **b** 50.0 L of oxygen (O_2)
  - **c** 140 mL of carbon dioxide (CO_2)

3 Which one of the following is the approximate volume of 1.4 mol of chlorine gas, Cl_2, at STP?
  - **A** 63.59 L
  - **B** 16.22 L
  - **C** 0.0616 L
  - **D** 32 L

14.3 Calculations involving reactions with gases

In this section, you will learn how you can combine the mole concept with your understanding of molar volume to calculate the mass or volume of a gas in a chemical reaction. This is useful in many situations, including determining the volume of carbon dioxide gas produced by fuels or the mass of water vapour produced in a combustion reaction.

Calculations that involve the use of the mole concept, combined with an understanding of chemical equations, are called stoichiometric calculations, which were introduced in Chapter 11. Stoichiometry is the study of ratios of moles of substances. Stoichiometric calculations are based on the law of conservation of mass.

Another way of expressing this is that, in a chemical reaction, atoms are neither created nor destroyed. Consequently, given the amount of one substance involved in a chemical reaction and a balanced equation for the reaction, you can calculate the amounts of all other substances involved.

 In a chemical reaction, the total mass of all products is equal to the total mass of all reactants.

EQUATIONS AND REACTING AMOUNTS

Consider the equation for the reaction that occurs when methane (CH_4) burns in oxygen:

$$CH_4(g) + 2O_2(g) \rightarrow CO_2(g) + 2H_2O(g)$$

The coefficients used to balance the equations show the ratios between the reactants and products involved in the reaction. The equation indicates that 1 mole of $CH_4(g)$ reacts with 2 moles of $O_2(g)$ to form 1 mole of $CO_2(g)$ and 2 moles of $H_2O(g)$. In more general terms, the amount of oxygen used will always be double the amount of methane used, double the amount of carbon dioxide produced and the same as the amount of water vapour produced.

$$\frac{n(O_2)}{n(CH_4)} = \frac{2}{1}, \frac{n(CO_2)}{n(O_2)} = \frac{1}{2} \text{ and } \frac{n(H_2O)}{n(O_2)} = \frac{2}{2} = 1$$

In general, for stoichiometric calculations you will be told, or you will be able to work out, the number of moles of one chemical in the reaction (the 'known'). You will then need to calculate the number of moles of another chemical (the 'unknown').

You can write the relationship between the known and unknown chemicals using ratios:

 $$\frac{n(\text{unknown chemical})}{n(\text{known chemical})} = \frac{\text{coefficient of unknown chemical}}{\text{coefficient of known chemical}}$$

Worked example 14.3.1

USING MOLE RATIOS

How many moles of carbon dioxide are generated when 5.5 moles of propane (C_3H_8) are burned completely in oxygen?	
Thinking	**Working**
Write a balanced equation for the reaction.	$C_3H_8(g) + 5O_2(g) \rightarrow 3CO_2(g) + 4H_2O(g)$
Note the number of moles of the known substance.	$n(C_3H_8) = 5.5\,mol$
Write a mole ratio for: $\frac{\text{coefficient of unknown}}{\text{coefficient of known}}$	$\frac{n(CO_2)}{n(C_3H_8)} = \frac{3}{1}$
Calculate the number of moles of the unknown substance using: $n(\text{unknown}) = \text{mole ratio} \times n(\text{known})$	$n(CO_2) = \frac{3}{1} \times 5.5$ $= 16.5\,mol$

Worked example: Try yourself 14.3.1

USING MOLE RATIOS

How many moles of carbon dioxide are generated when 2.55×10^3 moles of butane (C_4H_{10}) are burned completely in oxygen?

When carrying out any stoichiometric calculations, you must always clearly state the mole ratio for the reaction you are working with.

MASS–MASS STOICHIOMETRY

Calculations can require you to start and finish with masses rather than moles, as this is how quantities of chemicals are often measured.

Stoichiometric calculations generally follow the same pattern. The number of moles of a 'known' substance is calculated from data that is given to you, the **mole ratios** in the equation are used to find the number of moles of the 'unknown' substance, and the desired quantity of the unknown substance is then calculated.

Figure 14.3.1 shows the equations used to determine the volume, moles or mass of reactants or products in a chemical reaction involving gases. To determine these quantities for a given reaction, you must start with a balanced chemical equation.

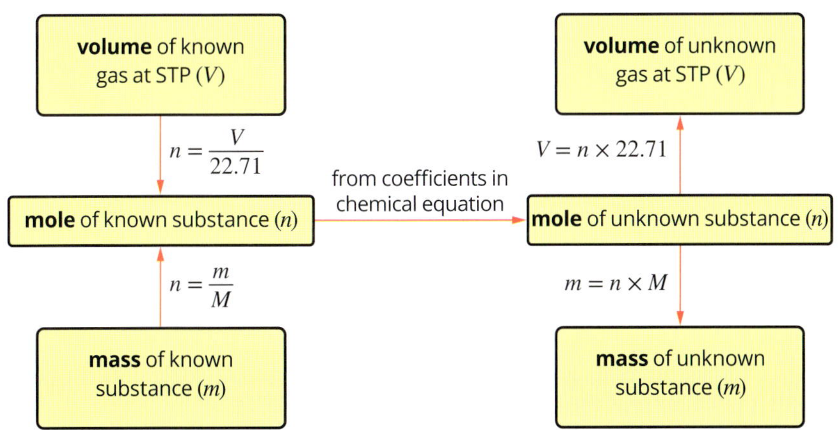

FIGURE 14.3.1 Most gas stoichiometric calculations follow the steps shown in the flow chart. Calculating the number of moles and using a mole ratio from a balanced chemical equation are always central to any stoichiometric calculation.

- To calculate the number of moles from the mass of a substance:

$$\text{moles}\,(n) = \frac{\text{mass in g}\,(m)}{\text{molar mass}\,(M)}$$

- To calculate the mass from the number of moles, rearrange this relationship as:

$$\text{mass in g}\,(m) = \text{moles}\,(n) \times \text{molar mass}\,(M)$$

When carrying out stoichiometric calculations, always keep the answer in your calculator and proceed to the next step. Only round to the correct number of significant figures when finished.

MASS–VOLUME STOICHIOMETRY

Some stoichiometric calculations require you to determine the volume of a gas that reacts with, or is produced from, a given mass of a substance. For these calculations, you need to determine the number of moles of the substance from its mass, and use the mole ratio from the balanced chemical equation.

The volume of the gas, in litres, is determined from its amount, in moles, at standard temperature of 0°C and 100 kPa (STP). The molar volume equation is used:

$$n = \frac{V}{V_m}$$

This formula can be rearranged to make volume the subject:

$$V = n \times V_m$$

Remember that, at STP, the accepted volume of 1 mole of any gas is 22.71 L.

These calculations are based on the mole ratios of reactants and products. For example, the reaction that occurs when calcium carbonate is heated can be represented by the equation:

$$CaCO_3(s) \rightarrow CaO(s) + CO_2(g)$$

Recall that the lack of a coefficient before a reactant or product in the equation indicates 1 mole. This balanced equation tells us that in any given reaction the amount, in mol, of calcium oxide produced is equal to the amount, in mol, of carbon dioxide produced. This is also equal to the amount, in mol, of calcium carbonate that decomposes.

Worked example 14.3.2

MASS–VOLUME STOICHIOMETRIC CALCULATIONS

A sample of calcium carbonate ($CaCO_3$) of mass 1.001 g, is heated until it has decomposed completely to form calcium oxide (CaO) and carbon dioxide (CO_2). Calculate the volume of carbon dioxide produced, measured at STP.

Thinking	Working
Write a balanced equation for the reaction.	$CaCO_3(s) \rightarrow CaO(s) + CO_2(g)$
Identify the known quantity and the unknown.	Known: mass of $CaCO_3$ = 1.001 g Unknown: What is the volume of CO_2 at STP?
Calculate the number of moles, *n*, of the known substance using: $n = \frac{m}{M}$	$n(CaCO_3) = \frac{1.001}{100.09}$ $= 0.01000$ mol
Find the mole ratio: $\frac{\text{coefficient of unknown}}{\text{coefficient of known}}$	$\frac{n(CO_2)}{n(CaCO_3)} = \frac{1}{1}$
Calculate the number of moles of the unknown substance using: *n*(unknown) = mole ratio × *n*(known)	$n(CO_2) = n(CaCO_3)$ $= 0.01000$ mol
Calculate the volume of the unknown substance using: $V = n \times 22.71$	$V(CO_2) = n(CO_2) \times V_m$ $= 0.01000 \text{ mol} \times 22.71 \text{ L}$ $= 0.2271$ L $= 227$ mL

Worked example: Try yourself 14.3.2

MASS–VOLUME STOICHIOMETRIC CALCULATIONS

A sample of calcium carbonate ($CaCO_3$) is heated until it has decomposed completely to form calcium oxide (CaO) and carbon dioxide (CO_2). A mass of 3.00 g of calcium oxide is produced. Calculate the volume of carbon dioxide produced, measured at STP. Give your answer to 3 significant figures.

GAS VOLUME-VOLUME CALCULATIONS

For chemical reactions where both the reactants and products are in the gaseous state, it is often convenient to measure volumes, rather than masses.

For example, the reaction between hydrogen and chlorine can be represented by the equation:

$$H_2(g) + Cl_2(g) \rightarrow 2HCl(g)$$

This equation tells us that when equal numbers of moles of hydrogen gas and chlorine gas react, twice as many moles of hydrogen chloride gas are produced.

Avogadro's law tells us that equal amounts, in moles, of all gases occupy equal volumes measured at the same temperature and pressure.

Therefore, the mole ratios in the balanced equation become volume ratios at the same temperature and pressure. In the above reaction, this means that when equal volumes of hydrogen gas and chlorine gas react, they will produce twice the volume of hydrogen chloride.

Worked example 14.3.3

GAS VOLUME–VOLUME CALCULATIONS

Methane gas (CH_4) is burned in a gas stove according to the following equation:

$$CH_4(g) + 2O_2(g) \rightarrow CO_2(g) + 2H2O(g)$$

If 50 mL of methane is burned, calculate the volume of O_2 gas required for complete combustion of the methane under constant temperature and pressure conditions.

Thinking	Working
Use the balanced equation to find the mole ratio of the two gases involved.	1 mol of CH_4 gas reacts with 2 mol of O_2 gas.
The temperature and pressure are constant, so volume ratios are the same as mole ratios.	1 volume of CH_4 reacts with 2 volumes of O_2 gas, so 50 mL of CH_4 reacts with 100 mL of O_2.

Worked example: Try yourself 14.3.3

GAS VOLUME–VOLUME CALCULATIONS

Methane gas (CH_4) is burned in a gas stove according to the following equation:

$$CH_4(g) + 2O_2(g) \rightarrow CO_2(g) + 2H_2O(g)$$

If 50 mL of methane is burned in air, calculate the volume of CO_2 gas produced under constant temperature and pressure conditions.

EXTENSION

Calculations involving excess reactants

Stoichiometry calculations become more complex if the reactants are not present in their stoichiometric ratio. In these cases, you must determine which reactant is completely consumed in the reaction, the limiting reactant, and which one is present in excess. The amount of limiting reactant determines how much product is formed.

Worked example 14.3.4 introduces a strategy that can be used to determine the limiting reactant in a reaction.

Worked example 14.3.4

EXCESS REACTANT CALCULATIONS

A gaseous mixture of 25.0 g of hydrogen gas and 100.0 g of oxygen gas are mixed and ignited. The water produced is collected and weighed. The equation for the reaction is:

$$2H_2(g) + O_2(g) \rightarrow 2H_2O(g)$$

a Which reactant is the limiting reactant? **b** What is the mass of water vapour formed?	
Thinking	**Working**
a Calculate the number of moles of each reactant using $n = \frac{m}{M}$ or $n = \frac{V}{V_m}$ as appropriate.	$n(H_2) = \frac{m}{M} = \frac{25.0}{2.0} = 12.5$ mol $n(O_2) = \frac{m}{M} = \frac{100.0}{32.0} = 3.13$ mol
Use the coefficients of the equation to find the limiting reactant.	The equation shows 2 mol of H_2 reacts with 1 mol of O_2. So to react all of the O_2 will require $2 \times n(O_2)$ of $H_2 = 2 \times 3.13 = 6.26$ mol As there is 12.5 mol of H_2, the H_2 is in excess. The O_2 is the limiting reactant (it will be completely consumed).
b Find the mole ratio using: $\frac{\text{coefficient of unknown}}{\text{coefficient of known}}$ The limiting reactant is the known substance.	$\frac{n(H_2O)}{n(O_2)} = \frac{2}{1}$
Calculate the number of moles of the unknown substance using: n(unknown) = mole ratio × n(known)	$n(H_2O) = 2 \times 3.13$ $= 6.26$ mol
Calculate the required quantity of the unknown using $m = n \times M$ or $V = n \times 22.71$ as appropriate.	$m(H_2O) = 6.26 \times 18.016$ $= 113$ g

Worked example: Try yourself 14.3.4

EXCESS REACTANT CALCULATIONS

Calculate the volume of carbon dioxide, in L, produced when 65.0 g of butane is burned completely in 200 L of oxygen. The gas volume is measured at STP. The equation for the reaction is:

$$2C_4H_{10}(g) + 13O_2(g) \rightarrow 8CO_2(g) + 10H_2O(l)$$

a Which reactant is the limiting reactant?

b What is the volume of carbon dioxide formed?

In these types of problems, remember to always use the number of moles of the limiting reactant to determine the amount of product that will be formed.

14.3 Review

SUMMARY

- A balanced equation shows the ratio of the amount, in moles, of reactants and products in the reaction.
- Stoichiometric calculations follow the general steps:
  1. Calculate the amount, in moles, of a known substance from the data given.
     Use $n = \frac{m}{M}$, or $n = \frac{V}{V_m}$
  2. Use the mole ratio from a balanced chemical equation to determine the amount, in moles, of the unknown substance.

$$\frac{n(\text{unknown chemical})}{n(\text{known chemical})} = \frac{\text{coefficient of unknown chemical}}{\text{coefficient of known chemical}}$$

  3. Find the desired quantity of the unknown substance from its amount, in moles, using $m = n \times M$ or $V = n \times 22.71$
- Stoichiometric calculations can be used to calculate the mass of substances involved in chemical reactions.
- The mole ratio in a balanced equation is also a volume ratio if all reactants and products are in the gaseous state and the temperature and pressure are kept constant.
- The volume of gases (at standard temperature and pressure) involved in a chemical reaction can also be calculated using the mole concept.

KEY QUESTIONS

1. Methanol burns in air according to the equation:
   $$2CH_3OH(g) + 3O_2(g) \rightarrow 2CO_2(g) + 4H_2O(g)$$
   Complete the following mathematical relationships.
   a $\frac{n(CH_3OH)}{n(O_2)} =$
   b $\frac{n(O_2)}{n(H_2O)} =$
   c $\frac{n(CH_3OH)}{n(CO_2)} =$
2. Create a flow chart for completing mass–mass stoichiometric calculations by placing these steps in the correct order.
   - Identify the known and unknown substances in the question.
   - Use mole ratios from the equation to calculate the amount of the unknown substance.
   - Calculate the mass of the unknown substance using $m = n \times M$.
   - Write a balanced equation for the reaction.
   - Calculate the amount, in mol, of the known substance using $n = \frac{m}{M}$.
3. Ammonia (NH_3) is produced by the reaction between hydrogen gas (H_2) and nitrogen gas (N_2) according to the reaction:
   $$N_2(g) + 3H_2(g) \rightarrow 2NH_3(g)$$
   Assuming complete reaction, what mass of ammonia would be produced when 12.0 L of hydrogen gas reacts with excess nitrogen gas at STP?
4. Glucose ($C_6H_{12}O_6$) ferments in an aqueous solution according to the equation:
   $$C_6H_{12}O_6(aq) \rightarrow 2C_2H_5OH(aq) + 2CO_2(g)$$
   Express your answers to the following questions to 3 significant figures.
   a If 3.604 g of glucose ferments, what is the amount in moles?
   b Determine the number of moles of CO_2 that are produced when this number of moles of glucose ferments.
   c Determine the volume of carbon dioxide (CO_2) that is produced, measured at STP.
5. Calculate the volume of oxygen gas (O_2), measured at STP, that is prepared when 7.35 g of potassium chlorate is decomposed by strong heat according to the equation:
   $$2KClO_3(s) \rightarrow 2KCl(s) + 3O_2(g)$$
   Express your answer to 3 significant figures.
6. Methane will burn in excess oxygen according to the equation:
   $$CH_4(g) + 2O_2(g) \rightarrow CO_2(g) + 2H_2O(g)$$
   This reaction produces 5.00 L of carbon dioxide (CO_2) at STP. Calculate the mass of water vapour produced. Express your answer to 3 significant figures.

Chapter review

KEY TERMS

Avogadro's law
bar
diffusion
elastic collision
ideal gas
kinetic energy
kinetic energy distribution diagram
kinetic molecular theory
Maxwell–Boltzmann distribution graph
molar volume
mole ratio
partial pressure
pascal
pressure
standard temperature and pressure (STP)
volume

Introducing gases

1 Which one of the following volumes is equal to 4.5 L?
- **A** 4.5×10^2 mL
- **B** 4.5×10^3 mL
- **C** 4.5×10^{-3} mL
- **D** 4.5×10 mL

2 Which one of the following best describes the effect of an increase in temperature on gas particles?
- **A** Both the average kinetic energy and the average speed of the particles increase.
- **B** Both the average kinetic energy and the average speed of the particles decrease.
- **C** The average kinetic energy of the particles increases and the average speed of the particles decreases.
- **D** The average kinetic energy of the particles decreases and the average speed of the particles increases.

3 Which one of the following would you expect to happen when a container of a smelly gas is opened in the corner of a small, closed room?
- **A** The smell of the gas will fill the room in a relatively short time.
- **B** The smell of the gas will be confined to the corner of the room were the container is found.
- **C** The smell of the gas will travel across the room and the corner where the container was opened no longer smells.
- **D** The smell of the gas will fill half of the room and will travel no further.

4 Select the correct answers from the pairs of words to complete the paragraph about gases.
The volume occupied by the atoms or molecules in a gas is much *smaller/larger* than the total volume occupied by the gas. The particles move in rapid, *straight-line/curved* paths and collide with each other and with the walls of the container. The forces between particles are extremely *weak/strong*. The collisions between particles are *elastic/rigid*. The average kinetic energy of the particles is *directly/inversely* proportional to the temperature of the gas, in units of K/°C.

5 Use the kinetic theory of gases to explain why:
- **a** the pressure of a gas increases if its volume is reduced at constant temperature
- **b** the pressure of a gas decreases if its temperature is lowered at a constant volume
- **c** in a mixture of gases, the total pressure is the sum of the partial pressure of each gas
- **d** the pressure of a gas, held at constant volume and temperature, will increase if more gas is added to the container.

6
- **a** If a container of gas is opened and some of the gas escapes, what happens to the pressure of the remaining gas in the container?
- **b** Use the kinetic molecular theory to explain what happens to the gas pressure in part **a**.

7 Which one of the following explains why an inflated balloon might pop when its temperature is increased?
- **A** The mass of gas inside the balloon increases.
- **B** The pressure of gas inside the balloon increases.
- **C** The number of molecules inside the balloon increases.
- **D** The pressure of gas outside the balloon decreases.

Molar volume of a gas

8 Determine the amount (in mol) of oxygen gas in a volume of 500 mL measured at STP. Assume the gas behaves like an ideal gas.

9 Determine the volume occupied by 0.1100 g of carbon dioxide gas (CO_2) at STP. Assume that carbon dioxide behaves like an ideal gas. Give your answer to 3 significant figures.

Calculations involving reactions with gases

10 Which one of the following is the mass of 50 L of oxygen gas (O_2) measured at STP?
- **A** 7.05 g
- **B** 2.20 g
- **C** 70.5 g
- **D** 32.0 g

11 Use the molar volume of a gas at STP to find the:

 a volume occupied by 8.0 g of oxygen

 b mass of nitrogen dioxide present in 10 L.

12 Carbon dioxide gas is a product of the complete combustion of fuels.

 a Calculate the mass of 1.00 mol of carbon dioxide.

 b What is the volume occupied by 1.00 mol of carbon dioxide at STP?

 c Given that density (d) is defined as $\frac{\text{mass}}{\text{volume}}$, calculate the density of carbon dioxide at STP in $g\,L^{-1}$.

13 A room has a volume of 220 m^3.

 a Calculate the amount, in moles, of air particles in the room at STP.

 b Assume that 20% of the molecules in the air are oxygen molecules and the remaining molecules are nitrogen. Calculate the mass of nitrogen in the room.

14 The ethanol produced by the fermentation of glucose is used as a biochemical fuel or biofuel. Fermentation of glucose produces ethanol and carbon dioxide according to the following equation:

$$C_6H_{12}O_6(aq) \xrightarrow{\text{yeast}} 2C_2H_5OH(aq) + 2CO_2(g)$$

Calculate the mass of ethanol produced by the fermentation of 80.0 g of glucose.

15 Large quantities of coal are burned in Australia to generate electricity, in the process generating significant amounts of the greenhouse gas carbon dioxide. The equation for this combustion reaction is:

$$C(s) + O_2(g) \rightarrow CO_2(g)$$

Determine the mass of carbon dioxide produced by the combustion of 1.0 tonne (10^6 g) of coal, assuming that the coal is pure carbon.

16 Propane (C_3H_8) burns in oxygen according to the equation:

$$C_3H_8(g) + 5O_2(g) \rightarrow 3CO_2(g) + 4H_2O(g)$$

Calculate the volume of:

 i oxygen at STP used

 ii carbon dioxide at STP produced

when the following masses of propane react completely with excess oxygen.

 a 22 g

 b 5.0 g

 c 0.145 g

Connecting the main ideas

17 Which of the following statements is not correct?

 A At the same temperature, a sample of CO_2 would exert a greater pressure than the same amount of helium molecules.

 B The measured air pressure is the sum of all the individual gas pressures in the sample of air.

 C The pressure of a fixed amount of gas is independent of the identity of the gas.

 D As the temperature of a sample of gas increases in a fixed volume container, the pressure of the sample will increase.

18 Suppose that 350 kg of iron(II) oxide (Fe_2O_3) reacts with excess carbon monoxide (CO) according to the equation:

$$Fe_2O_3(s) + 3CO(g) \rightarrow 2Fe(s) + 3CO_2(g)$$

 a Determine the number of moles of Fe_2O_3.

 b Determine the mole ratio of CO_2 to Fe_2O_3 and use this to calculate the number of moles of carbon dioxide produced.

 c Determine the volume of carbon dioxide (CO_2) measured at STP.

19 Consider two containers of equal size. One contains oxygen and the other carbon dioxide. Both containers are at 23°C and at a pressure of 1.0 atm. Answer each of the following questions about the two gases and give a reason for your answers.

 a Which of the two samples of gas contains more molecules?

 b Which of the two samples of gas contains the greater number of atoms?

 c Which of the two gases has the greater density?

20 Nitric acid (HNO_3) is decomposed by light according to the equation:

$$4HNO_3(aq) \rightarrow 4NO_2(g) + 2H_2O(l) + O_2(g)$$

 a From the equation, determine the molar ratio of NO_2 to HNO_3. If 0.500 mol of nitric acid is decomposed, how many moles of NO_2 gas are produced?

 b From the equation, determine the mole ratio of O_2 to HNO_3. If 0.500 mol of nitric acid is decomposed, how many moles of O_2 gas are produced?

 c Determine the total volume of gases produced at STP if 0.500 mol of NO_2 and 0.125 mol of O_2 are produced in the reaction.

21 Copper is extracted from chalcopyrite ($CuFeS_2$) according to the equation below.

$$2CuFeS_2(s) + 5O_2(g) + 2SiO_2(s) \rightarrow 2Cu(l) + 4SO_2(g) + 2FeSiO_3(l)$$

a If 350 L of oxygen gas measured at STP reacts, how many moles is this?

b Determine the number of moles of chalcopyrite, that would react with 350 L of O_2, measured at STP.

c Calculate the mass, in kilograms, of chalcopyrite that would react.

22 Hot-air balloons rise because hot air is less dense than cooler air.

a Explain why the density of a gas inside a hot-air balloon decreases as the temperature rises.

b A Chemistry student was explaining to a friend that as air is heated, the added heat causes the kinetic energy of all the air molecules to increase. Comment critically on this statement, identifying any inconsistencies and making appropriate corrections.

c Use the kinetic molecular theory of gases to explain why the pressure of a gas decreases if its temperature is lowered at a constant volume.

CHAPTER 15

Properties and uses of water

Think about the water you see and use every day. It falls from the sky as rain; it fills the rivers and oceans. You drink it, you cook with it, you bathe in it and you wash your clothes in it. Reactions in your body take place between chemicals that are dissolved in it. Without water, there is no life.

In this chapter, you will learn what makes water crucial for life. You will learn about the sources of water here on Earth and how water sampling and testing can be undertaken to check water quality.

You will learn about some of the unique properties of water and the way in which these properties support life on Earth. Lastly, you will also learn how each property can be explained by the structure and bonding present in water.

Science as a human endeavour

- The supply of potable drinking water is an extremely important issue for both Australia and countries in the Asian region. Water sourced from groundwater and seawater undergoes a number of purification and treatment processes (such as desalination, chlorination, fluoridation) before it is delivered into the supply system. Chemists regularly monitor drinking water quality to ensure that it meets the regulations for safe levels of solutes. Heavy metal contamination in groundwater is monitored to ensure that concentrations are at acceptable levels. Several methods can be used to reduce heavy metal contamination; the method used is influenced by economic and social factors.

Science understanding

- the shape and polarity of molecules can be used to explain and predict the nature and strength of intermolecular forces, including dispersion forces, dipole–dipole forces and hydrogen bonding
- the unique physical properties of water, including melting point, boiling point, density in solid and liquid phases, and surface tension, can be explained by its molecular shape and hydrogen bonding between molecules
- the solubility of substances in water, including ionic and polar and non-polar molecular substances, can be explained by the intermolecular forces, including ion–dipole interactions between species in the substances and water molecules, and is affected by changes in temperature

15.1 Essential water

Water is a special chemical. It moderates our weather, shapes our lands and is essential for the existence of life. Water is the most abundant liquid on Earth, covering more than 70% of our planet. The total amount of water is estimated to be more than 1.3 billion cubic kilometres.

This water on Earth constantly cycles, as shown in Figure 15.1.1. It is in continuous movement between land, ocean, rivers and creeks and the atmosphere. Warmth from the Sun causes the water in oceans, lakes and rivers to evaporate and become vapour in the air. As water vapour rises, the air temperature drops and the water vapour condenses to form droplets, which merge to form clouds. When the water droplets are large enough, water falls back to Earth as rain, ice or snow, depending on the air temperature.

Other processes also add water to the atmosphere. When fossil fuels undergo combustion, water is a product and is released into the air. Water also evaporates from the leaves, stems and flowers of plants.

FIGURE 15.1.1 The water cycle. The arrows show some of the ways water on Earth is redistributed.

Water is an exceptional solvent for a large variety of substances. However, the solvent properties of water do cause problems. Because water dissolves many chemicals readily, it can be easily contaminated. The quality of the water available for drinking, agriculture and other uses is vitally important to humans. In this section, you will learn about the main sources of water pollution and investigate the procedures used for taking water samples in order to test the levels of contaminants in water supplies and how to treat water to ensure it is safe for human consumption.

SOURCES OF WATER

Water is found on Earth in three states of matter (gas, liquid and solid), and it readily changes from one state into another. The **water cycle** shown in Figure 15.1.2 involves the continuous movement of water between the land, oceans, streams and atmosphere. Solar energy is the primary energy source for the cycle.

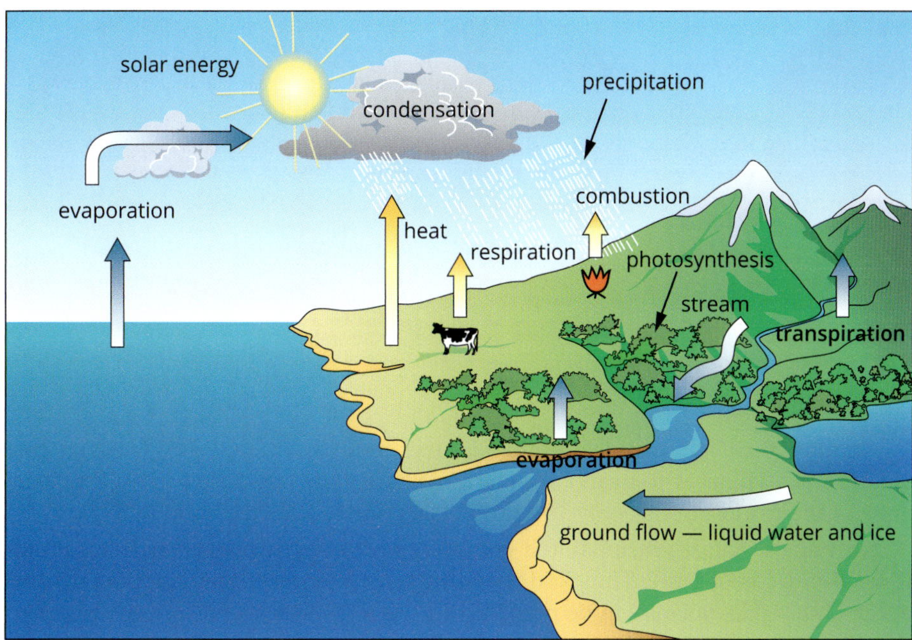

FIGURE 15.1.2 The water cycle illustrates how water is moved around the Earth through evaporation, condensation and precipitation.

The water cycle involves three main processes.

- Heat from the Sun causes water to evaporate from the oceans, lakes and streams.
- Water vapour in the air is transported around the globe until it condenses to form clouds.
- The water precipitates as rainwater, or occasionally as ice crystals in hail or snow, and falls to the ground.

Human activities such as the combustion of fossil fuels, which produce steam, also contribute to the water cycle.

Table 15.1.1 shows how water is distributed on Earth.

TABLE 15.1.1 The distribution of water on Earth

Location of water	State of matter	Volume (km^3)
oceans	liquid	1 300 000 000
ice caps and glaciers	solid	24 000 000
groundwater	liquid	23 000 000
ground ice and permafrost	solid	300 000
lakes	liquid	180 000
soil moisture	liquid	17 000
atmosphere as water vapour	gas	13 000
rivers	liquid	2100

The volumes in Table 15.1.1 are approximate but, from these figures, the total volume of water on Earth is more than 1.3 billion km^3.

Despite the large quantities of water, there is a limited supply of fresh water on our planet. Figure 15.1.3 shows that only 2.5% of the water on Earth is drinkable and most of this water is not accessible as it is locked in ice caps, glaciers or the soil.

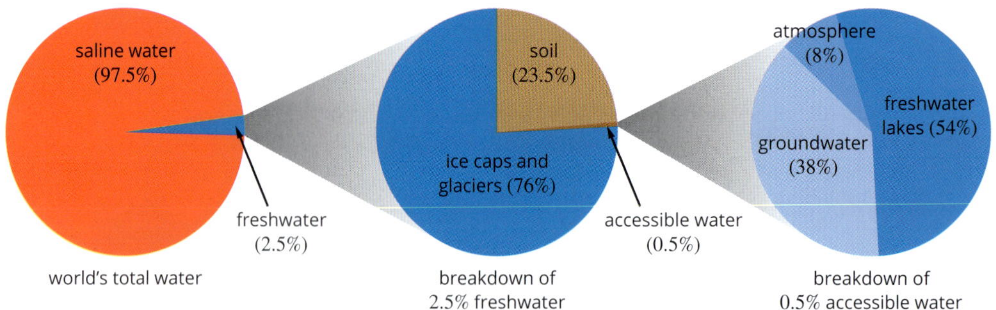

FIGURE 15.1.3 Distribution of water on Earth. Only 2.5% of the water on Earth is drinkable.

CHEMFILE

Great Artesian Basin

The **Great Artesian Basin** is located within Australia. It has the following key features.

- It is the largest **artesian basin** (underground water supply) in the world.
- It covers an area of more than 1 700 000 km^2. It underlies nearly a quarter of the Australian continent.
- In some places, the basin is up to 3000 m deep.
- The temperature of the water in the basin has been recorded to be anywhere from 30°C to 100°C.
- It provides a reliable source of **groundwater** for a very large part of inland Australia.

Traditionally, the water could be readily accessed as it flowed to the surface under natural pressure. However, in the last century, government bodies have set up initiatives to try to maintain stores of water within the basin.

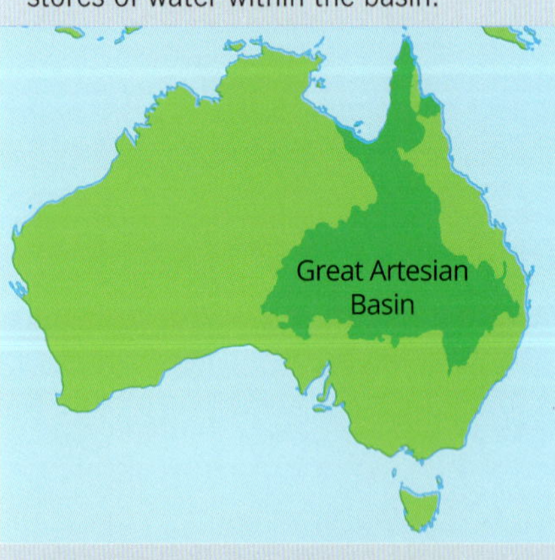

FIGURE 15.1.4 The Great Artesian Basin provides water for stock and the human population for a large area of inland Australia.

About 40% of the world's population lives in areas where water is scarce and that number will grow to nearly 50% by 2025. Australia has close to 5% of the land mass of the world but has only 1% of the water carried by the world's rivers.

Australia is the driest inhabited continent and has extremely variable rainfall. It has hundreds of large dams and thousands of farm dams that can hold a lot of water so that Australians can be prepared for periods of drought. Since Australia is also very flat and hot, most of the rain that falls evaporates before it can enter rivers and reservoirs.

 **Due to Australia's harsh climate and low rainfall, Australia has only 1% of the water carried by the world's rivers.**

POTABLE WATER

Obtaining clean drinking water, or **potable water**, has been a challenge for people living in communities for thousands of years.

Major sources of drinking water

Drinking water in Australia is obtained from a variety of sources. The main sources are:

- reservoirs filled by run-off from rivers and streams
- water obtained directly from rivers and lakes
- groundwater
- **recycled water**
- **desalinated seawater**.

In most of Australia's major cities, water comes from reservoirs built on rivers in protected areas. These reservoirs are able to provide the population with safe, clean water. Water from such sources needs only minor treatment before being released for consumption. Reservoirs and weirs like Mundaring Weir shown in Figure 15.1.5 are surrounded by protected land, classifying them as **protected catchments**. They are often located in national parks and forests with limited access to ensure the quality of the water is very high.

However, in some parts of inland Australia, and in some other countries in the Asia–Pacific region, water comes from sources other than protected catchments. In some cases, water is taken directly from rivers and lakes that may be subject to contamination from run-off from farms and urban areas. Drinking water may also be obtained from groundwater, which is often referred to as **bore water** in Australia. Such sources may contain levels of contamination that require a more complex purification process than for water from protected catchments.

FIGURE 15.1.5 The Mundaring Weir stores water from the Helena River, near Perth, Western Australia.

Table 15.1.2 shows the major sources of drinking water in some Australian towns and cities.

Potable water is suitable and safe for human consumption.

TABLE 15.1.2 Sources of drinking water in Australia

City or town	Main water source
Alice Springs, inland Northern Territory	groundwater
Adelaide, South Australia	Torrens and Murray rivers
Bourke, inland New South Wales	Darling River
Broome, coastal Western Australia	groundwater
Melbourne, Victoria	reservoirs
Mildura, inland Victoria	Murray River
Birdsville, inland Queensland	groundwater
Hobart, Tasmania	Derwent River
Perth, Western Australia	desalination water, groundwater and reservoirs

Sources of water in some other countries are shown in Table 15.1.3.

TABLE 15.1.3 Sources of drinking water in other regions of the world

Region	Main water source
Bali, Indonesia	reservoirs and groundwater
Dubai, United Arab Emirates	desalinated seawater (more than 98% sourced this way)
Hong Kong	Dong Jiang River (mainland China), reservoirs (some reclaimed from ocean inlets), and seawater (for flushing toilets)
Singapore	reservoirs with protected catchments, desalinated seawater, recycled water and water piped from Malaysia

WATER QUALITY

In 2012, Chinese newspapers reported that 54% of China's drinking water was too polluted to drink. Just two years later, the situation in the country had worsened as the percentage of polluted drinking water rose to 60%. About 60 000 people die each year in China from diseases caused by water pollution, such as cholera, dysentery and diarrhoea.

The quality of the drinking water around the world varies enormously. Most of the water in Australia is of high quality but there have been instances where this is not the case.

This section explores the **chemical contaminants** that can be found in drinking water and how samples are obtained to analyse water quality.

Chemical contaminants

Water is an excellent solvent. It can dissolve many different types of substances including minerals, nutrients, organic material and human-made wastes. This can cause problems for drinking water resources. As rainwater flows down a hillside or stream, or percolates through soil, it will dissolve many different materials on its journey. The dissolved materials may end up in rivers, reservoirs or groundwater, and therefore in Australia's water supplies.

As a result, water supplies may be subject to pollution caused by human activity. Although chemical contamination of water supplies is not common in Australia, it can happen (Figure 15.1.6). Contamination can occur because of:

- run-off from farms and cities
- run-off from industrial and mining wastes
- lead used in solder in copper water pipes.

FIGURE 15.1.6 Pollution can be caused by seepage of heavy metals and other contaminants into aquifers (water-bearing rock).

Chemical contaminants are elements or compounds that may be harmful if consumed. They can be naturally occurring or synthetic. The main types of chemical contaminants are:

- heavy metals
- pollutants from fertilisers
- organic pollutants.

Heavy metal contamination

Heavy metals, such as mercury and lead, can contaminate drinking water and have severe health effects. These effects include cancer, organ and nervous system damage, and, in the most extreme cases, death. Unfortunately, young people are even more susceptible to the effects of heavy metals, and permanent brain damage can be the consequence of exposure to heavy metals. Table 15.1.4 lists some of the common heavy metals and the effects they can have on human health.

TABLE 15.1.4 Heavy metal contaminants and the risks they pose to human health

Heavy metal	Source	Effects
copper	copper pipes	anaemia; liver and kidney damage; stomach and intestinal irritation
lead	lead pipes, batteries, leaded gasoline (petrol), paints, ammunition	restricts haemoglobin production; damage to kidneys, gastrointestinal tract, joints and reproductive system; can lower IQ levels in young children
cadmium	smelting, improper disposal of rechargeable batteries	kidney failure; liver disease; osteoporosis
nickel	power plants, waste incinerators, improper disposal of batteries	decreased body weight; damage to the heart and liver
zinc	mining, smelting, steel production	anaemia; damage to nervous system and pancreas
arsenic	mining, smelting, steel production	carcinogenic; stomach pain; numbness; blindness
mercury	improper disposal of batteries, various industrial processes	tremors; gingivitis; spontaneous abortion; damage to the brain and central nervous system

The quality of drinking water around Australia is governed by local water authorities and is generally high. However, some rural communities do not have access to the same quality of drinking water as city residents do. As a result, there are guidelines in place for the levels of heavy metals that are permissible in Australian drinking water, which are shown in Table 15.1.5.

TABLE 15.1.5 Australian Drinking Water Guidelines values for heavy metals

Heavy metal	Guideline value ($mg\,L^{-1}$)
arsenic	0.01
cadmium	0.002
copper	2
chromium	0.05 as chromium(VI)
lead	0.01
mercury	0.001

Example of contamination: mining

In 2013, it was discovered that one-third of Tasmania's town water supplies did not meet the national standards for heavy metals. The residents of five towns in the north-west of Tasmania were told not to drink the water due to contamination. In particular, the Tamar River (Figure 15.1.7) was polluted by effluent and the heavy metals zinc, cadmium and lead. The issue for most towns was unusually high concentrations of zinc, cadmium and lead resulting from nearby mining operations.

FIGURE 15.1.7 The Tamar River in Tasmania has been polluted by effluent, including the heavy metals zinc, cadmium and lead.

Example of contamination: water pipes

Lead-based solder was once used to join copper water pipes. As a result of corrosion, the lead can be released into water flowing through the pipes. In 1993 in Perth, Western Australia, a study showed that some households had higher than acceptable levels of heavy metals in their water. Table 15.1.6 shows the percentage of houses in Perth with unacceptable heavy metal levels.

To minimise this potential problem, Australian standards were introduced in 1998 to limit the amount of lead in solder to no more than 0.1% by weight.

TABLE 15.1.6 Heavy metal levels in Perth (1993)

Heavy metal	Households with unacceptable levels (%)
lead	5
copper	12
cadmium	2

One method of removing heavy metals from a contaminated water supply is by the use of precipitation reactions. This process involves the following steps.

1 The heavy metal cation that is to be removed is identified.
2 A soluble compound containing an anion that will form an insoluble compound with the cation of the heavy metal is selected.
3 A solution of the compound containing the anions is added to the water.
4 The anions combine with the heavy metal cations to form a precipitate.
5 The precipitate is filtered off, thereby removing the heavy metal from the water supply.

An example of this process is the addition of calcium hydroxide ($Ca(OH)_2$) to contaminated water to form a precipitate of a heavy metal hydroxide. Precipitation reactions like this will be looked at in more detail in Chapter 16.

MONITORING WATER QUALITY

To ensure high water quality and to test for contamination in water supplies, water samples are regularly collected for analysis. The results may be compared against water quality standards. This determines the water's suitability for its intended use and, if necessary, the treatment required to purify the water.

There are protocols for obtaining water samples to ensure that the samples are collected in a consistent manner and with the correct equipment. There are many different methods for obtaining water samples depending upon whether the samples are collected close to shore, from boats (Figure 15.1.8) or from bridges (Figure 15.1.9).

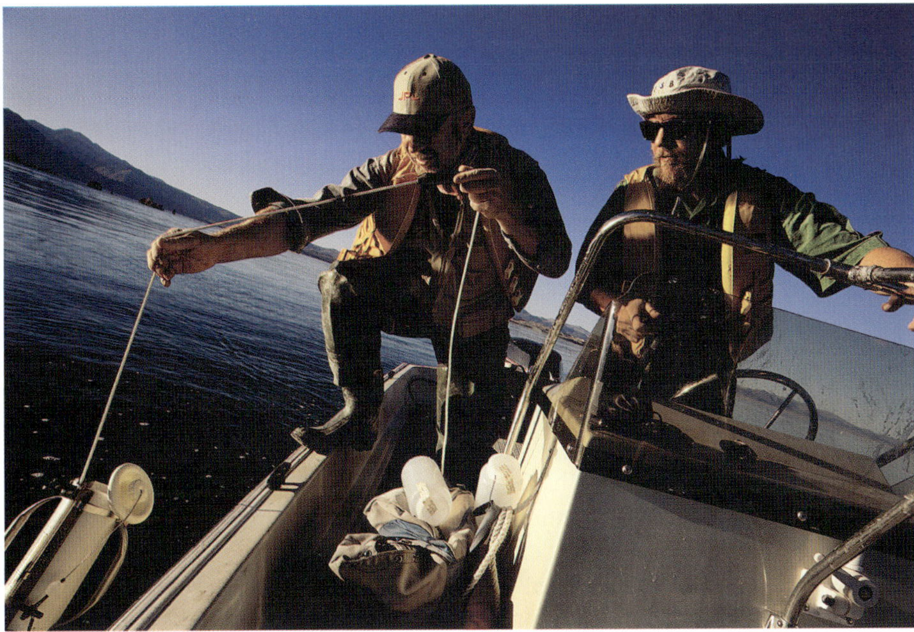

FIGURE 15.1.8 Sampling water from a boat

FIGURE 15.1.9 Sampling water from a bridge

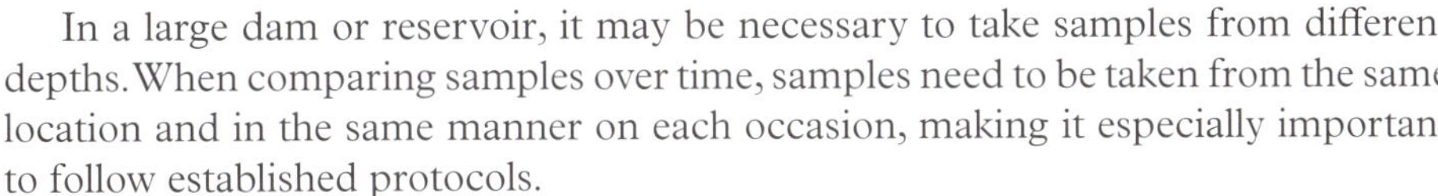

In a large dam or reservoir, it may be necessary to take samples from different depths. When comparing samples over time, samples need to be taken from the same location and in the same manner on each occasion, making it especially important to follow established protocols.

When collecting water samples for testing, it is important to consider the following questions.

- Which chemical is to be analysed and why is the testing to take place?
- What health risks are associated with the sampling?
- What equipment is required, and are **sterile** containers needed?
- What sample size is required for the selected test?
- What method is to be used to record the measurements?
- Where should the samples be taken and at what depths should the samples be taken?
- How will a representative sample be obtained?
- What are the labelling, storage and transport requirements for the sample?

Protocols for water sampling

The containers used for sampling should not react with the sample. To assure the quality of the sample, the following precautions are taken.

- The container can be rinsed with the sample before the final sample is taken.
- The container should be cleaned before taking another sample.
- The water used for cleaning the container can be tested to ensure there is no contamination between samples.

If the water is to be tested for bacteria and other microorganisms, it is important that the container is sterile.

Sampling methods and equipment

The method used to avoid contamination of a water sample depends on where the water is being sampled from, and the conditions under which the sample is taken. Some sampling methods are outlined in Table 15.1.7.

TABLE 15.1.7 Sampling methods for collecting water samples

Conditions	Sampling method
sample taken from a well-mixed body of water	Sampling near the surface is sufficient to obtain a representative sample. A sample should be taken about 10 cm below the surface and away from the water's edge.
sample taken from water contaminated by sediment (solid particles)	A sample should be drawn into the sample container by suction to avoid including the sediment.
sample taken from a river	A sample should be taken upstream from where the person taking the sample stands.

Sampling depths

To determine the depth at which a sample should be taken, the temperature is initially measured at every metre of depth.

- If the temperature is consistent, it can be assumed that the water is thoroughly mixed and a sample is taken halfway down.
- If there is a temperature variation, then a sample is taken in the middle of each temperature region.

When the water being sampled is more than 2 metres deep or samples are required from different depths, a Van Dorn sampler (Figure 15.1.10) can be used. This sampler can be lowered horizontally and pulled sideways before the sample is taken. Van Dorn samplers are made of a transparent polymer and are attached by a rope to the surface. After the sampler is raised to the surface, the water is drained and placed in a sterile sample bottle. The Van Dorn sampler is very useful when there are distinct layers in the water, such as in a deep lake or reservoir.

a

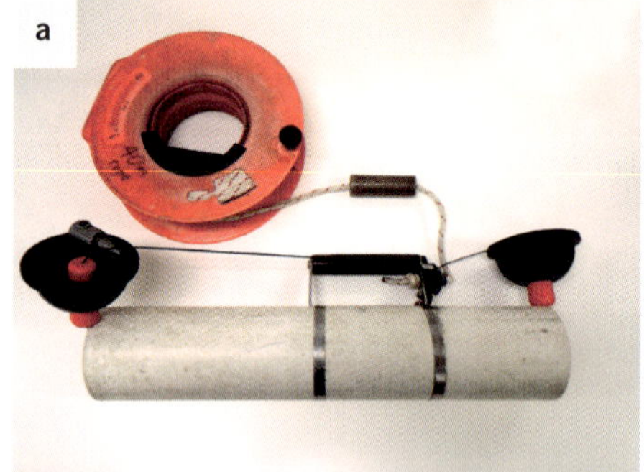

b

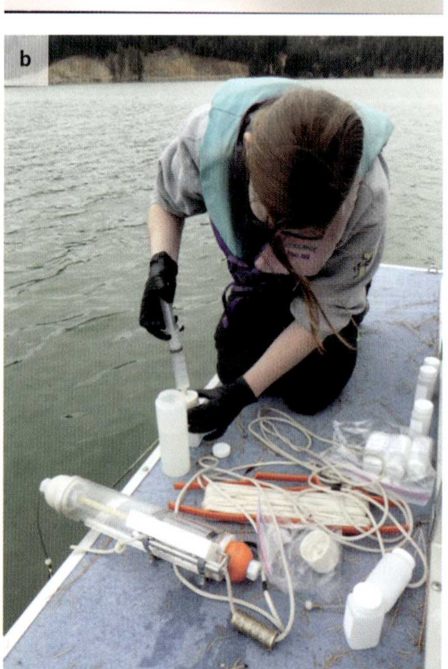

FIGURE 15.1.10 A Van Dorn sampler consists of a polymer tube with caps at each end.

Sampling locations

To develop a true picture of the water quality in a body of water, you need to take samples from many different locations.

For example, a water-quality monitoring program in the Tamar Estuary in Tasmania took samples from 20 sites divided into five zones, as shown in the map in Figure 15.1.11. The sites ranged from Launceston to where the lower estuary discharges into Bass Strait. The samples were taken monthly at specific depths and each set of 20 samples was collected on the same day.

Treatment of drinking water

In most Australian cities and countries in the Asian region, water is treated before being passed on to consumers through the water supply system (Figure 15.1.12). The amount of treatment required varies from city to city, but the usual purpose of the treatment is to remove suspended solids, bacteria, colour and odour from the water.

The steps involved in the purification of water for most cities are:

1 flocculation
2 settling of the floc
3 filtration
4 chlorination.

The water may also undergo fluoridation. Each of these processes is outlined in more detail.

Flocculation

Flocculation is the process by which small, suspended particles in the water are made to join together to form larger, heavier particles. The heavier particles then sink under their own weight and settle to the bottom of the water sample.

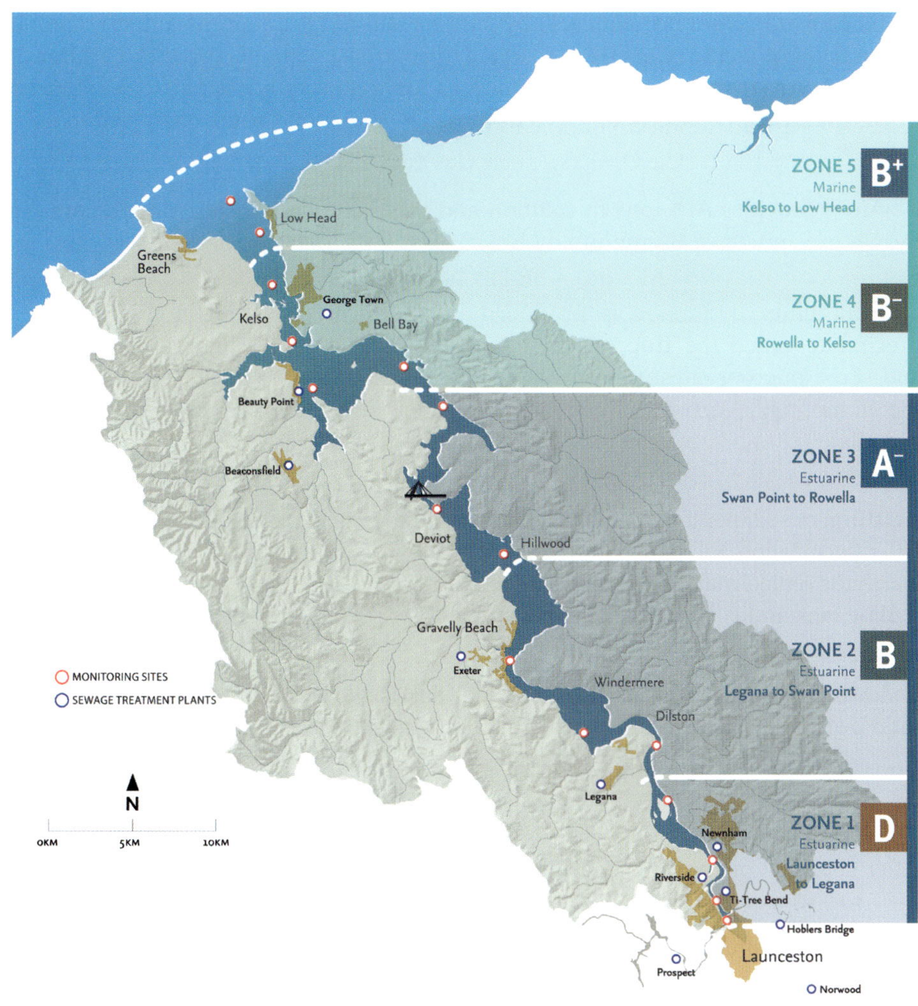

FIGURE 15.1.11 Sampling sites in the Tamar Estuary, Tasmania

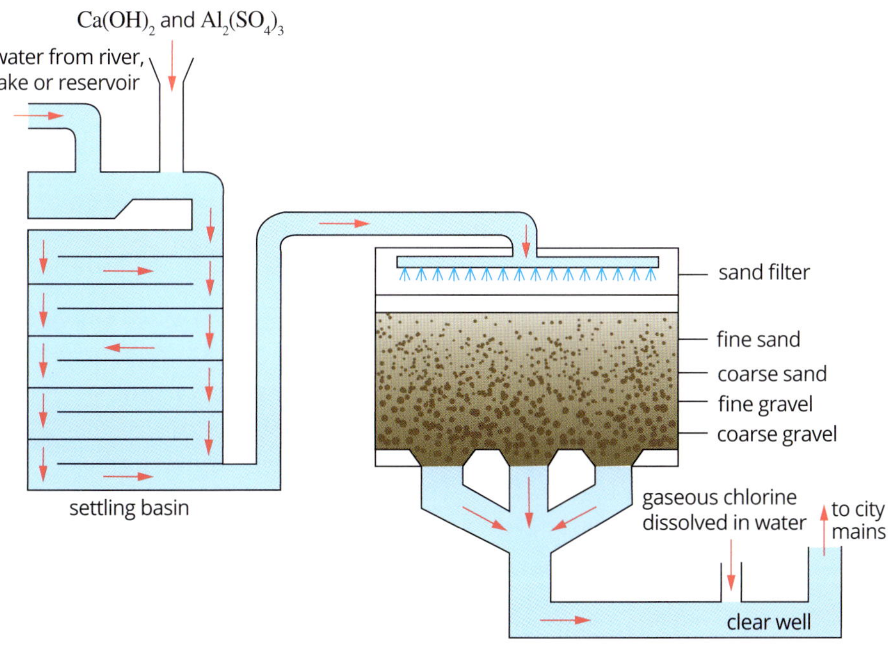

FIGURE 15.1.12 A typical system for treating drinking water

Flocculation can be achieved by adding alum (aluminium sulfate) and, if necessary, lime. Lime ($Ca(OH)_2$) is added to neutralise acids and provide a source of hydroxide ions in the water supply. Lime is an ionic compound that dissolves in water by **dissociating** (breaking up) into its ions. The equation for this dissociation is:

$$Ca(OH)_2(s) \xrightarrow{H_2O} Ca^{2+}(aq) + 2OH^-(aq)$$

Alum provides Al^{3+} ions in solution and these combine with hydroxide ions to form a precipitate of aluminium hydroxide:

$$Al^{3+}(aq) + 3OH^-(aq) \rightarrow Al(OH)_3(s)$$

Aluminium hydroxide is produced in the form of a gelatinous (jelly-like) precipitate called the **floc**, which traps other fine particles and removes colour and some microorganisms from the water. These other materials are adsorbed in the aluminium hydroxide precipitate. The finely divided precipitate particles then coagulate (clump together) to form heavier particles.

Settling, or sedimentation

Settling, or sedimentation, utilises gravity to remove suspended solids from the water. After flocculation, the water is left to stand to allow the floc to settle for a period, during which the settled materials form a sludge. The sludge accumulates at the bottom of the settling tank and is removed. The remaining water passes onto the filtering stage.

Filtration

Following settling, water from the settling tank is allowed to filter through a bed of sand over gravel. This removes any remaining suspended matter.

Chlorination

After filtering, the clear water is usually treated with gaseous chlorine to destroy any bacteria. The main purpose of **chlorination** is to remove biological contaminants. After chlorination, water is considered fit for human consumption.

Fluoridation

In all the capital cities and the majority of other cities in Australia, fluoride is added to drinking water in a process called **fluoridation** before it is released from storages. Fluoride in drinking water helps to reduce the incidence of tooth decay. It does this by interacting with tooth enamel ($Ca_{10}(PO_4)_6(OH)_2$). The fluoride ion replaces the hydroxyl ion in tooth enamel, forming fluorapatite ($Ca_{10}(PO_4)_6F_2$), which is stronger and more resistant to decay.

Fluoride is added to the water in the form of compounds including sodium hexafluorosilicate (Na_2SiF_6), fluorosilicic acid (H_2SiF_6) or sodium fluoride (NaF). Each of these compounds releases fluoride ions when it dissolves in water.

The water obtained from these procedures is suitable for drinking and is described as potable.

CHEMISTRY IN ACTION

Obtaining drinking water from seawater

Fresh water is a relatively scarce commodity on Earth, whereas seawater is plentiful. Seawater contains about 3.5% dissolved salts and so is not suitable for either drinking or agriculture. **Desalination** involves the removal of dissolved salts from seawater to obtain fresh water. Two main desalination processes are:

- distillation
- reverse osmosis.

Producing drinking water by distillation

Figure 15.1.13 illustrates simple laboratory distillation equipment that can be used to obtain fresh water from salt water.

- Salt water is placed in the round-bottomed flask on the left of the diagram, and then heated until it boils. Water vapour passes out from the flask and over into the condenser. Salt is left behind in the flask and so the salt and water vapour are separated.

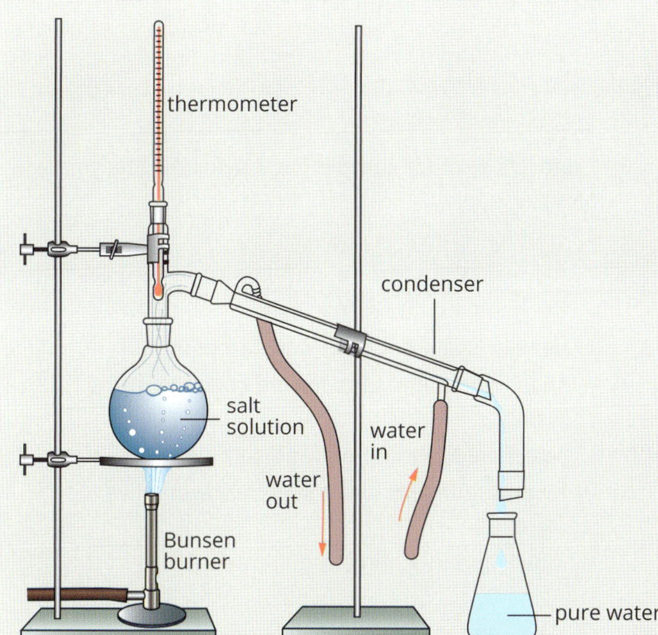

FIGURE 15.1.13 Distillation can be used to obtain fresh water from salt water.

- The condenser is a glass tube with a water-cooled 'jacket' surrounding it. As the water vapour passes down the condenser tube it is cooled and condenses back into water.
- The now pure water is collected in the conical flask at the end of the condenser.

Distillation is an expensive way to produce fresh water because of the cost of fuel to provide the heat required. In Australia, this is not a viable way to produce drinking water on a large scale.

The situation is different, however, in the United Arab Emirates, a country that has large supplies of natural gas. Dubai is a coastal city in the United Arab Emirates. The surrounding land is very dry and water is scarce. The city of Dubai has therefore turned to the desalination of seawater as a source of drinking water. In the Jebel Ali desalination plant near Dubai, shown in Figure 15.1.14, gas turbines generate electricity and the waste heat is used to produce fresh water from seawater in a distillation process. Over 98% of Dubai's water supply is provided by desalination.

Producing drinking water by reverse osmosis

Osmosis is the natural tendency of water to move from a region of low salt concentration to one of higher salt concentration. If salt water and fresh water are separated by a semipermeable membrane that allows water molecules, but not the dissolved ions, to pass through it, the water molecules will pass through the membrane from the fresh water to the salt water.

This natural tendency can be reversed if pressure is applied to the salt-water side of the semipermeable membrane. This process is called **reverse osmosis** and can be seen in Figure 15.1.15.

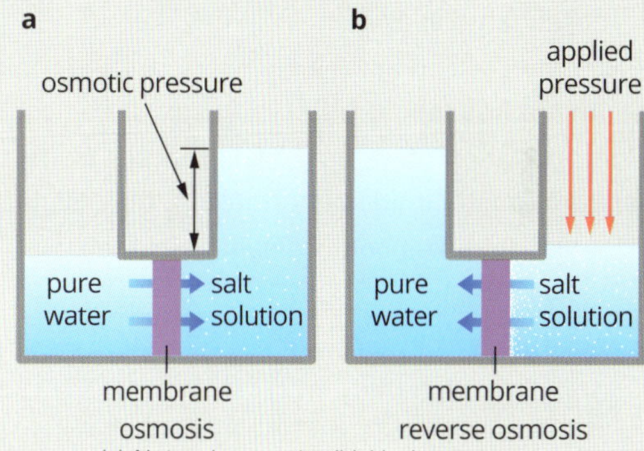

FIGURE 15.1.15 (a) Natural osmosis; (b) Under pressure, reverse osmosis can be used to obtain fresh water from seawater.

The difficulty and associated cost of reverse osmosis is in making suitable membranes that will not rupture under the high pressures used.

A number of reverse osmosis plants were built in Australia during the extended drought of the first decade of this century. They can be found in Western Australia, South Australia, Victoria, New South Wales and Queensland.

FIGURE 15.1.14 The Jebel Ali desalination plant has six gas turbines that generate power from burning natural gas.

FIGURE 15.1.16 Desalination plant in Kwinana, Western Australia

15.1 Review

SUMMARY

- Most of the water on Earth is in liquid form and contained in the oceans.
- Only 2.5% of all the water on Earth is drinkable, but just 0.5% of all the water on Earth is drinkable and accessible.
- Drinking water can be obtained from a variety of sources including:
  - rivers flowing through protected catchments
  - directly from rivers and lakes
  - groundwater
  - rainwater collected from a roof and stored in tanks
  - desalinated seawater.
- Reservoirs, which are fed by rivers, are the main source of household water in Australian cities.
- Contamination of water supplies comes from farms, industry, mines and through the soil.
- Types of chemical contaminants include:
  - heavy metals
  - pollutants from fertiliser
  - organic pollutants.
- Protocols exist for sampling water for the analysis of water quality.
- All protocols need to consider:
  - the reason for testing
  - the chemical to be analysed
  - health risks associated with sampling
  - specific requirements for the chemical being tested
  - selection of equipment and sample size
  - recording results and taking representative samples
  - labelling and storage requirements.
- Drinking water from reservoirs is usually purified using a sequence of steps including flocculation, settling, filtration and chlorination and fluoridation.
- Drinking water is often fluoridated to reduce dental decay.
- Elevated levels of heavy metals in drinking water can cause problems for human health.
- Heavy metals may be removed from waste water by chemical processes including precipitation.

KEY QUESTIONS

1. Which water source has the highest risk of contamination? Explain why.
2. Why is only 0.5% of water on Earth available for drinking?
3. Why does Australia store more water per person than any other country?
4. In a large body of water, the temperature of water should be measured at every metre of depth before taking samples. What is the reason for doing this?
5. Drinking water can be produced by distillation of salt water. Which of the following statements is correct? (There may be more than one.)
   A Salt water is placed in a flask and heated until it boils.
   B Water vapour passes out of the flask and over into a condenser.
   C The condenser is a tube with a water-cooled jacket surrounding it.
   D Pure water is collected into the flask that originally held the salt water.
6. Explain why fluoridation is an accepted additional treatment of drinking water in many Australian cities.

15.2 Properties of water

Water and life are so strongly linked that space scientists search the universe for water in their quest to discover possible life beyond our planet. It is the unique properties of water that allow it to support life. Some of the ways water supports life are summarised in Table 15.2.1.

TABLE 15.2.1 The properties of water allow it to sustain life.

Role water plays in sustaining life	Property of water that allows it to perform this role
provides an internal transport system in plants and animals	ability to dissolve many substances easily
provides a system to transport heat from cells to the body's surface of animals	high specific heat capacity
provides a cooling system in animals and plants by evaporation of water from skin and the surface of leaves	high latent heat of vaporisation
provides an environment for aquatic plants and animals	expands on freezing, so provides an insulating surface layer on ponds and rivers; ability to dissolve many substances easily
reactant for photosynthesis	ability to react with carbon dioxide to form glucose and oxygen

The unique properties of water can be explained by looking at the structure of a water molecule and the hydrogen bonding that occurs between the water molecules.

WATER AND HYDROGEN BONDING

Structure of water

Water has the chemical formula H_2O, which means that each water molecule contains two hydrogen atoms and one oxygen atom. Hydrogen and oxygen are both non-metals so they bond with each other by sharing electrons. Each hydrogen atom shares one electron with the oxygen atom and the oxygen atom shares one electron with each hydrogen atom. The sharing of electrons means these intramolecular bonds are covalent.

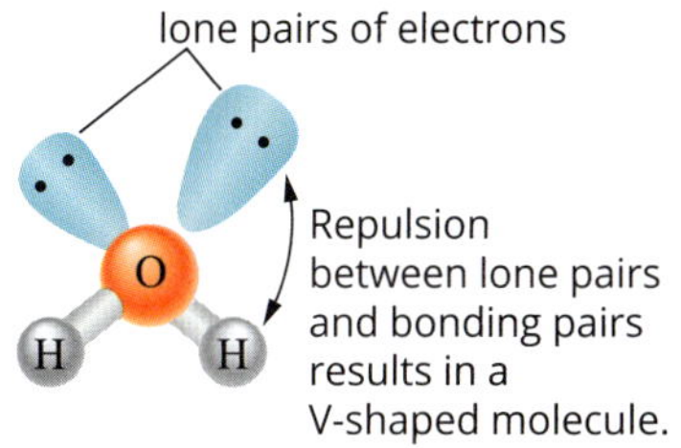

FIGURE 15.2.1 In this diagram of the structure of water, the two hydrogen atoms (white) form a V-shape with a central oxygen atom (red).

Each oxygen atom also has two lone pairs (non-bonding pairs) of electrons that contribute to the molecule's shape. A water molecule is V-shaped (or bent) as shown in Figure 15.2.1. The lone pairs of electrons on the oxygen atom have greater repulsion than the electrons forming the single bonds between oxygen and hydrogen. This pushes the hydrogen atoms closer together, giving the overall bent shape (the shape of molecules was explained in detail in Chapter 12).

The covalent bonds in a water molecule are polar. The oxygen atom has a higher electronegativity than the hydrogen atoms. This means that the electrons in the O–H bonds are more strongly attracted to the oxygen atom. Uneven sharing of the electrons gives the oxygen atom a partial negative (δ–) charge and each hydrogen atom carries a partial positive charge (δ+).

Since the bonding electrons are distributed unevenly across the water molecule, the molecule can be described as a dipole. Figure 15.2.2 shows how the individual dipoles of the O–H bonds in water add together, resulting in an overall molecular dipole.

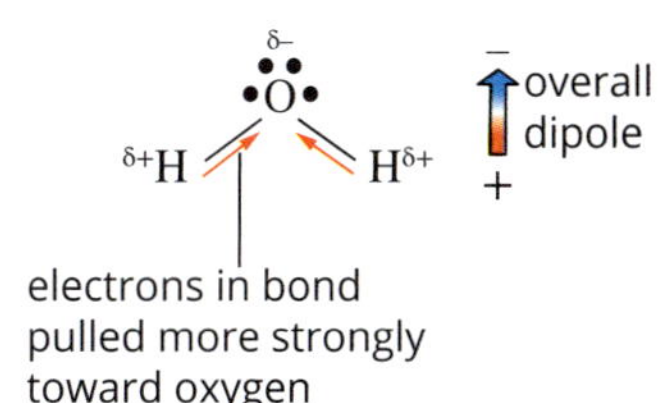

FIGURE 15.2.2 A water molecule has an overall dipole. The oxygen atom has a partial negative charge while the hydrogen atoms both have partial positive charges.

Polar bonds, polar molecules and dipoles were explained in detail in Chapter 12.

 Water molecules are polar and have a permanent overall dipole.

Hydrogen bonding

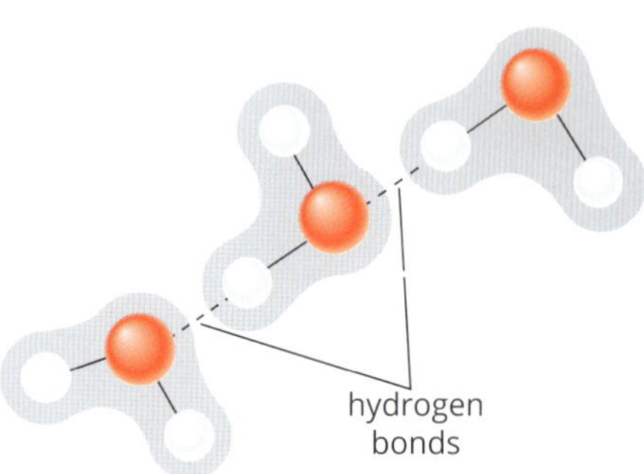

FIGURE 15.2.3 The hydrogen bonding between molecules of water

The main type of intermolecular force between molecules in water is a hydrogen bond. The hydrogen bonds are formed by an electrostatic attraction between the partial positive charge on a hydrogen atom on one water molecule and a lone pair of electrons on the oxygen atom of a neighbouring water molecule.

Figure 15.2.3 shows the hydrogen bonding between molecules of water. The two hydrogen atoms in each water molecule have a slight positive charge, while the central oxygen atom has a slight negative charge. The black dashed lines represent the hydrogen bonds that form between water molecules.

Because there is a large electronegativity difference between oxygen and hydrogen atoms, the partial charges on a water molecule are relatively large. The electrostatic attraction between these opposite partial charges makes the hydrogen bonds between water molecules quite strong.

As Figure 15.2.4 shows, each water molecule can form hydrogen bonds with up to four other water molecules.

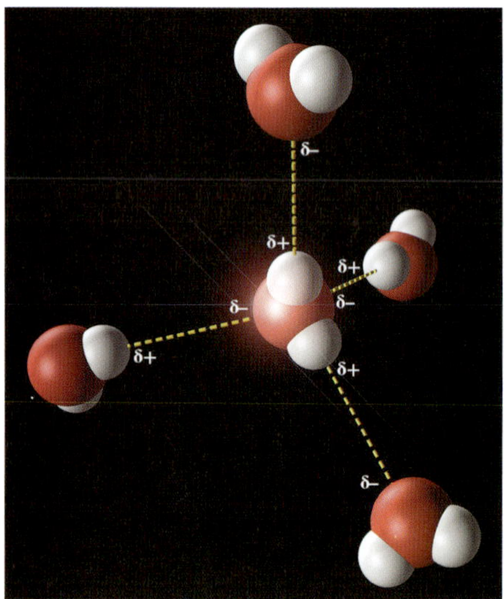

FIGURE 15.2.4 Each water molecule can form hydrogen bonds to four other water molecules.

WATER HAS RELATIVELY HIGH MELTING AND BOILING POINTS

Compared to other molecules of a similar size, water has high boiling and melting points. This can most easily be seen through the observed trends in these properties for the group 16 hydrides.

Trends in group 16 hydrides

The group 16 elements include oxygen (O), sulfur (S), selenium (Se), tellurium (Te) and polonium (Po). Each of these elements can bond with hydrogen to form compounds known as **hydrides**. Water is a group 16 hydride. Others are listed in Table 15.2.2.

TABLE 15.2.2 Names and formulae of the group 16 hydrides

Element	Name and formula of hydride	Formula
O	water	H_2O
S	hydrogen sulfide	H_2S
Se	hydrogen selenide	H_2Se
Te	hydrogen telluride	H_2Te
Po	hydrogen polonide	H_2Po

The group 16 hydrides are all molecular compounds, so their melting and boiling points reflect the size of the forces between their molecules. The higher the melting and boiling points, the stronger the intermolecular forces must be because more energy is required to overcome the forces and allow the molecules to move apart from each other.

The melting and boiling points of the group 16 hydrides are shown in Table 15.2.3 and Figure 15.2.5.

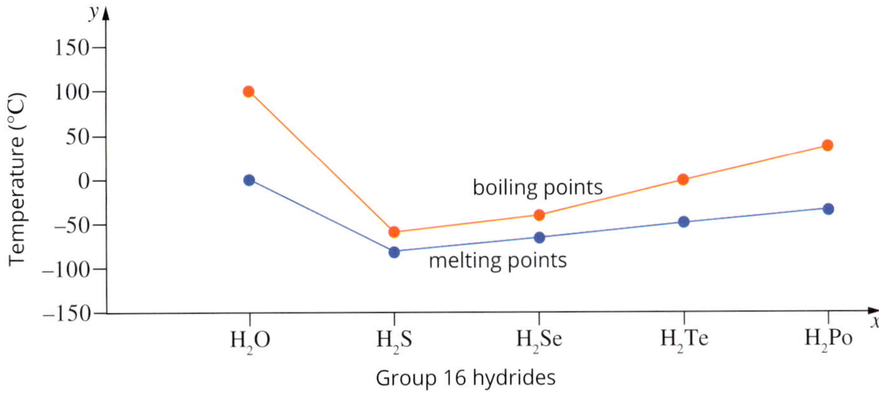

FIGURE 15.2.5 Graph showing the trend in melting and boiling points of water and other group 16 hydrides

You can see from Figure 15.2.5 that, apart from water, there are clear trends in the melting and boiling points of the group 16 hydrides. The melting and boiling points both increase going down the group. This increase indicates that the intermolecular forces are also getting stronger down the group. The intermolecular forces responsible for this trend are dispersion forces. Dispersion forces increase in strength with increasing mass. H_2Po has a larger mass than H_2S and H_2Se; hence it has stronger dispersion forces and higher melting and boiling points.

TABLE 15.2.3 Melting and boiling points of the group 16 hydrides

Hydride	Melting point (°C)	Boiling point (°C)
H_2O	0	100
H_2S	–82	–60.7
H_2Se	–66	–41.5
H_2Te	–49	–2.2
H_2Po	–35	36.1

Despite being the smallest hydride, water has significantly higher melting and boiling points than other group 16 hydrides.

The melting and boiling points of water are exceptional

Water has a melting point of 0°C and a boiling point of 100°C. Both of these values are significantly higher than those of other group 16 hydrides. The values are also significantly higher than those of other molecular substances of a similar size, as can be seen in Figure 15.2.6.

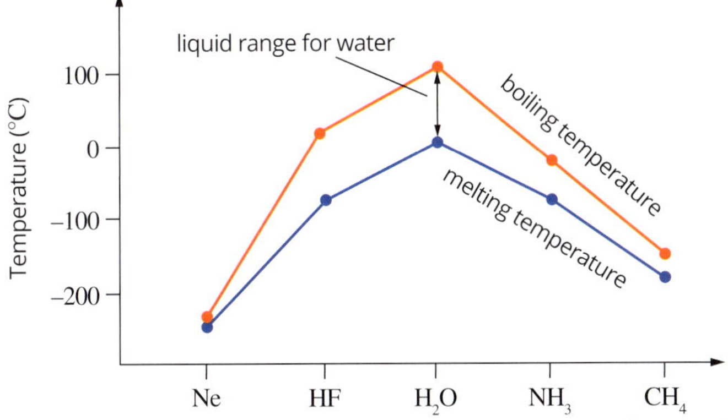

FIGURE 15.2.6 The melting and boiling points of water and other molecules of similar size

It is the relatively strong hydrogen bonds between its molecules that give water its exceptional properties.

Each water molecule has the potential to form four hydrogen bonds with surrounding water molecules. There are two partially charged hydrogen atoms and two lone pairs of electrons on the oxygen atom in each molecule so all the hydrogen atoms and lone pairs in the molecule can be involved in hydrogen bonding.

A significant amount of energy is needed to disrupt all of the hydrogen bonds between water molecules, resulting in the higher melting points and boiling points observed for water.

DENSITY IN THE LIQUID AND SOLID STATES

Water expands on freezing

When a sample of water freezes, the resulting ice crystal lattice has a greater volume than when the sample of water was a liquid. Water has the unusual property of having a lower density in the solid phase than in the liquid phase.

This property of water is important to life on Earth. When water at the surface of rivers or lakes freezes, the low density of the ice means it remains on the surface. This surface layer of ice insulates the water below from the cold air temperatures, reducing the possibility that the entire river or lake will freeze, and allowing aquatic life to survive.

As liquid water is cooled, the water molecules move more slowly. Upon approaching the freezing temperature of water, the molecules arrange in a way where each water molecule forms four hydrogen bonds to four neighbouring water molecules, as shown in Figure 15.2.7.

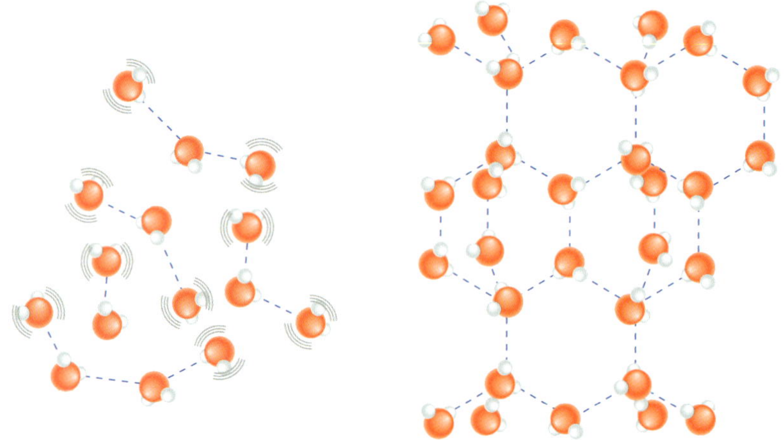

FIGURE 15.2.7 The arrangement of water molecules in liquid water (left) and ice (right)

This creates a very open arrangement of molecules, meaning that the water molecules in ice are more widely spaced than in liquid water. Therefore, ice is less dense than liquid water. When ice melts, the water molecules move more freely and move closer together.

High surface tension of water

The water molecules at the surface of a sample of water are not completely surrounded by other water molecules. As a result, they form hydrogen bonds with neighbouring molecules on the sides and below, but not above.

The sideways forces of attraction to neighbouring molecules are equal in all directions, so for molecules in the body of the liquid, there is no net force. However, the attraction of surface water molecules to the molecules below them has no opposite force, so the water molecules at the surface are pulled downwards.

In a beaker of water at room temperature, a water molecule will spend, on average, only 10^{-9} seconds or 1 nanosecond at the surface before being pulled back into the bulk of the liquid (Figure 15.2.8). The surface of the water is in a constant state of tension.

Water, with relatively strong hydrogen bonds between molecules, has a relatively high surface tension. The surface tension of a substance is a measure of the resistance of a liquid to increasing its surface area. The surface tension is why falling water tends to form droplets. It also explains why insects are able to walk across water.

In Figure 15.2.9, the surface of the water behaves like a stretched skin. The water molecules at the surface are so attracted to each other by strong hydrogen bonds that the paperclip floats, even though the density of the paperclip is higher than the density of water.

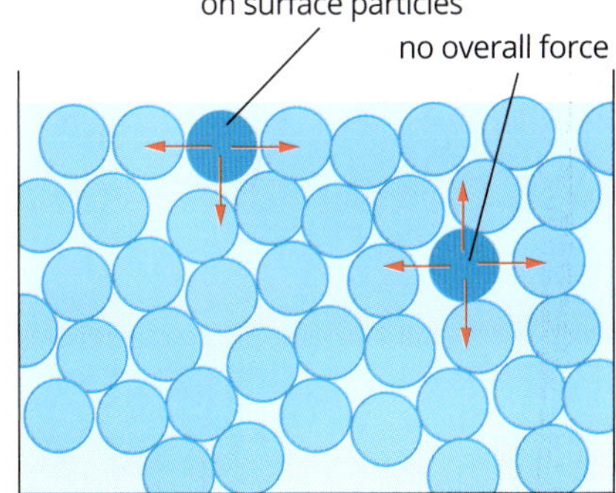

FIGURE 15.2.8 The forces of attraction on particles at the surface and in the body of a liquid. The particles in the body of the liquid experience no net force. The particles at the surface of the liquid have an overall downwards force of attraction.

FIGURE 15.2.9 The relatively high surface tension of water allows a paperclip to float.

HEAT CAPACITY

The **heat capacity** of a substance is a measure of the substance's capacity to absorb and store heat energy. It is a measure of how much energy the substance absorbs as its temperature increases. When the same quantity of heat energy is applied to two substances with different heat capacities, they will undergo different temperature changes.

This can be seen with water and ethanediol. Water has a higher specific heat capacity than ethanediol. When you heat equal amounts of the two liquids with the same amount of heat energy, the temperature of the two liquids will rise by different amounts. As shown in Figure 15.2.11, when 100 g of water and 100 g of ethanediol are heated with the same amount of energy, the beaker containing water will increase in temperature by a smaller amount than the ethanediol.

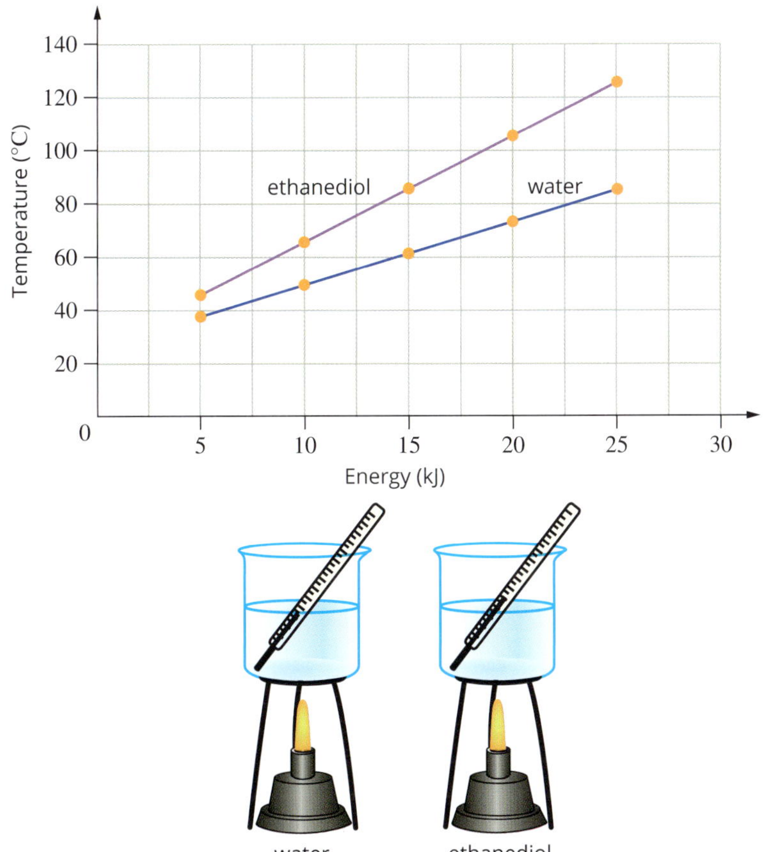

FIGURE 15.2.11 When 100 g of water and 100 g of ethanediol are heated with the same amount of energy, the temperature of the beaker containing water will increase less than the temperature of the ethanediol.

CHEMFILE

The deep blue sea

When light passes through water, the vibration of O–H bonds in water molecules, combined with the hydrogen bonds between molecules, causes a small amount of light to be absorbed in the red part of the visible colour spectrum. This leaves the remaining complementary colour, which is blue, for you to see. You cannot see this colour in small volumes of water. However, when light travels through several metres of water or ice, you can see the blue colour.

FIGURE 15.2.10 Large volumes of water or ice appear blue.

Specific heat capacity

The **specific heat capacity** of a substance measures the amount of energy (in joules) needed to increase the temperature of a certain amount (usually 1 gram) of that substance by 1°C.

Specific heat capacity is given the symbol *C* and is expressed in joules per grams per degrees Celsius, i.e. $J g^{-1} °C^{-1}$.

Specific heat capacity (*C*) is a measure of the energy required to raise 1 g of substance by 1°C. It is reported in units of $J g^{-1} °C^{-1}$.

The specific heat capacities of some common substances are listed in Table 15.2.4.

TABLE 15.2.4 Specific heat capacities of common substances

Substance	Specific heat capacity ($J g^{-1} °C^{-1}$)
water	4.18
ethanediol	2.42
ethanol	2.40
sand	0.48
copper	0.39
lead	0.16

The specific heat capacity of a substance is a reflection of the types of bonds holding the molecules, ions or atoms together in the substance. The relatively high specific heat capacity of water is related to the number and strength of the hydrogen bonds between water molecules.

The specific heat capacity of a substance reflects the type of bonding in that substance. For covalent molecules, this will depend on the strength of the intermolecular forces between molecules.

Specific heat capacity of water

Water has a specific heat capacity of $4.18 J g^{-1} °C^{-1}$. This means that 4.18 joules of heat energy are needed to increase the temperature of 1 gram of water by 1°C.

The specific heat capacity of water is relatively high. This is because of the presence of hydrogen bonds between water molecules. Hydrogen bonds are stronger than other intermolecular forces and they are able to absorb and store large amounts of heat energy before they break.

As water is a liquid, it is often convenient to measure a quantity of water as a volume, in mL or L, rather than as a mass measured in grams. Since 1 mL of water has a mass of 1 g, it is easy to convert a given volume of water to mass:

- 1 mL of water weighs 1 g.
- 100 mL of water weighs 100 g.
- 250 mL of water weighs 250 g.

The density of liquid water is $1 g mL^{-1}$, so volumes of water in millilitres (mL) are equal to the mass of water in grams (g).

EXTENSION

Calculations using specific heat capacity

The specific heat capacity of a substance can be used to calculate the heat energy in joules required to increase the temperature of a given mass of substance by a particular amount. Heat energy is given the symbol q.

A useful equation can be written:

heat energy = specific heat capacity × mass × temperature change

Using symbols, the equation can be written:

$$q = C \times m \times \Delta T$$

where q is the amount of heat energy (J), C is the specific heat capacity ($J\,g^{-1}\,°C^{-1}$), m is the mass (g) and ΔT is the temperature change (°C).

Worked example 15.2.1

CALCULATING THE AMOUNT OF ENERGY REQUIRED TO HEAT A SPECIFIED MASS OF A SUBSTANCE USING SPECIFIC HEAT CAPACITY

Calculate the heat energy, in kJ, needed to increase the temperature of 200 g of water by 15.0°C.	
Thinking	**Working**
Find the specific heat capacity (C) of the substance from Table 15.2.4.	The specific heat capacity of water is $4.18\,J\,g^{-1}\,°C^{-1}$.
To calculate the quantity of heat energy in joules, use the formula: $q = C \times m \times \Delta T$	$q = 4.18 \times 200 \times 15.0$ $= 1.25 \times 10^4\,J$
Express the quantity of energy in kJ. Remember that to convert from J to kJ, you multiply by 10^{-3}.	$q = 1.25 \times 10^4 \times 10^{-3}$ $= 12.5\,kJ$

Worked example: Try yourself 15.2.1

CALCULATING THE AMOUNT OF ENERGY REQUIRED TO HEAT A SPECIFIED MASS OF A SUBSTANCE USING SPECIFIC HEAT CAPACITY

Calculate the heat energy, in kJ, needed to increase the temperature of 375 g of water by 45.0°C.

CHEMFILE

Biological importance of water's high heat capacity

The high heat capacity of water means water will absorb a relatively large amount of energy before it heats up. This makes water resistant to sudden temperature changes, and plants and animals that live in water survive without experiencing wide temperature fluctuations.

Many living organisms are mainly composed of water. Water's high heat capacity allows organisms to maintain consistent internal body temperatures, regardless of the external environment.

FIGURE 15.2.12 Fish and coral in the sea can survive due to the relatively consistent temperature of the seawater. Water's high heat capacity makes it resistant to temperature fluctuations.

LATENT HEAT

When a solid is heated at constant pressure, the temperature of the solid increases until it reaches its melting point. The temperature then remains constant as the solid melts, even though further energy is being absorbed.

Similarly, when a liquid reaches its boiling point, its temperature remains constant until all of the liquid has evaporated. Think about a pot of boiling water. The temperature remains at 100°C until all of the water in the pot has evaporated.

Figure 15.2.13 shows how in both cases, at the melting point and boiling point of the substance, the temperature remains constant until the phase change is complete. Heat is still flowing into the substance, but the temperature does not change. Once a solid has completely melted, or a liquid has completely evaporated, if heating continues, then the temperature of the substance starts to increase again.

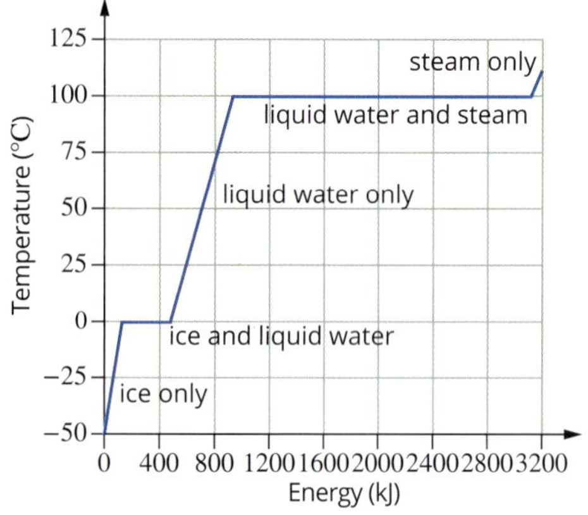

FIGURE 15.2.13 Latent heat is the energy absorbed by a substance in order to change state at its melting or boiling temperature. For a pure substance, such as water in this case, the horizontal regions of the heating curve show where latent heat is being absorbed and the temperature is not changing.

Even though the temperature does not change while the phase change is taking place, the substance is still absorbing energy. This energy is referred to as **latent heat** and is defined as the energy absorbed by a fixed amount of substance as it changes state from a solid to a liquid or a liquid to a gas at its melting point or boiling point respectively.

Latent heat is the energy required to change a fixed amount of substance, usually 1 mole, from either a solid to a liquid or a liquid to a gas. Over the period of time that latent heat is being absorbed, the temperature of a substance will not change.

Values of latent heat

Latent heat values are a measure of the quantity of heat energy required to melt or boil a given amount of a solid or liquid. Latent heat values have the unit kilojoules per mole, kJ mol^{-1}.

The **latent heat of fusion** of a substance is the heat needed to change 1 mole of the substance from a solid to a liquid at its melting point.

The **latent heat of vaporisation** of a substance is the heat needed to change 1 mole of the substance from a liquid to a gas at its boiling point.

Latent heat of water

The latent heat of fusion of water is 6.0 kJ mol^{-1}. This means that 6.0 kJ of energy is needed to change 1 mole of water from a solid to a liquid at 0°C. This energy is needed to disrupt the ice lattice by breaking some of the hydrogen bonds between water molecules.

The latent heat of vaporisation of water is 44.0 kJ mol^{-1}. This means that 44.0 kJ is needed to change the state of 1 mole of water from a liquid to a gas at 100°C. This relatively large quantity of energy is required to completely break the hydrogen bonds between the water molecules so they can separate and form a gas.

The latent heat values of water and some other substances are listed in Table 15.2.5. Much like specific heat capacity, latent heat depends on the strength of the intermolecular forces between molecules of the substance.

TABLE 15.2.5 Latent heat values for some common molecular substances

Substance	Latent heat of fusion (kJ mol^{-1})	Latent heat of vaporisation (kJ mol^{-1})
water	6.0	44.0
hydrogen	0.06	0.45
oxygen	0.22	3.4

As you can see, the latent heat values of water are relatively high. This is due to the strength and number of water's hydrogen bonds relative to its molecular size.

15.2 Review

SUMMARY

- A water molecule is V-shaped and contains polar covalent bonds.
- Water molecules are polar because an overall dipole exists as a result of the asymmetrical nature of the polar O–H bonds.
- As a result of the O–H bonds in water molecules, the forces that attract one water molecule to another are relatively strong hydrogen bonds.
- Each water molecule can form up to four hydrogen bonds with other water molecules.
- The strength of water's hydrogen bonds means relatively large amounts of energy are required to disrupt the bonds and separate the molecules from each other. This gives water relatively high melting and boiling points.
- Water has significantly higher melting and boiling points than the other group 16 hydrides.
- With the exception of water, the melting and boiling points of the group 16 hydrides increase going down the group. This is due to the increasing strength of dispersion forces.
- Ice is less dense than liquid water because of its unique geometric arrangement of water molecules as a result of hydrogen bonding.
- Heat capacity is a measure of a substance's capacity to absorb and store heat energy.
- The specific heat capacity of a substance measures the quantity of energy (in joules) needed to increase the temperature of a certain amount (usually 1 gram) of that substance by 1°C.
- The specific heat capacity of water is relatively high due to the ability of the hydrogen bonds between water molecules to absorb and store heat energy.
- The mass of a sample of liquid water in grams (g) is equal to its volume in millilitres (mL).
- Latent heat is the energy absorbed by a fixed amount of substance as it changes state from a solid to a liquid or a liquid to a gas.
- The latent heat of vaporisation of water is relatively high. This has significance for organisms and water supplies.

KEY QUESTIONS

1 **a** List the physical properties that make water unique.
b Explain the significance of polarity and hydrogen bonding in relation to these properties of water.

2 Explain why a water molecule can form up to four hydrogen bonds with other water molecules.

3 **a** Describe the forces that must be overcome in order for ice to melt.
b Sketch, or describe, a portion of a lattice of ice.
c Explain why ice is less dense than liquid water.

4 Which one of the following statements best describes the specific heat capacity of a substance?
A Specific heat capacity is a measure of the heat energy required to increase the temperature of a substance from its melting point to its boiling point.
B A substance with high specific heat capacity is less resistant to temperature change than a substance with low heat capacity.
C A substance with high specific heat capacity contains bonds that are unable to absorb a lot of energy.
D Specific heat capacity is a measure of the heat energy that can be absorbed by the bonds in a substance.

5 Select the overall force experienced by a water molecule at the surface of a glass of water.
A up
B down
C sideways
D no force at all

6 A heating curve for a substance is produced by plotting the temperature change against the energy input. When this is done from the melting point to the boiling point of a substance, two flat regions are observed. Explain what the flat regions of the graph represent.

7 What type of bonding is responsible for water's relatively high latent heat values?

15.3 Water as a solvent

FIGURE 15.3.1 When solid potassium permanganate is added to water, it dissolves. The particles disperse into the solution and move around freely.

Water is an excellent solvent. This is one of its most important properties. Almost all biological processes and many industrial processes occur in water. These systems are known as **aqueous** environments.

When substances are dissolved in water, the particles are free to move throughout the **solution**. When two aqueous reactants are combined in a reaction vessel, the dissolved reactant particles mix freely. This increases the chances of the reactants coming into contact. Because of the increased movement of the reactant particles, interactions between reactants are generally much more effective than if the same reactants were mixed as solids.

In Figure 15.3.1, you can see how the deep purple colour of potassium permanganate spreads through the water as the solid dissolves. Eventually the liquid will appear completely purple as the particles continue to mix and move.

This section looks at the importance of the solvent properties of water to living organisms and modern industry, as well as the process of dissolving. Sections 15.4 and 15.5 will explore in more detail the type of bonds that form when molecular and ionic substances dissolve in water.

FORCES BETWEEN SOLUTE AND SOLVENT PARTICLES

Aqueous solutions

An aqueous solution is formed when a solid, liquid or gas is dissolved in water. The substance being dissolved is called the **solute** and the liquid in which the substance is dissolved is the **solvent**.

The solute of an aqueous solution can change but the solvent will always be water. You can have an aqueous solution of salt (saline) or an aqueous solution of sugar. Table 15.3.1 lists some common aqueous solutions.

TABLE 15.3.1 Some everyday aqueous solutions

Solution	Solute/s	Solvent
saline solution (for use with contact lenses)	sodium chloride	water
soft drink	carbon dioxide, sugar, flavour, colour	water
coffee	coffee, sugar	water

All solutions have the following characteristics.

- The solute and the solvent cannot be distinguished from each other. This means the solution is **homogeneous**. (The opposite is a solution with visible distinguishable parts, otherwise known as a **heterogeneous** solution).
- The dissolved particles are too small to see.
- The amount of dissolved solute can vary from one solution to another.

EXTENSION

Suspensions and colloids

When a substance is mixed with water, it does not always dissolve to form a solution. It could also form a suspension or a colloid.

A suspension is a heterogeneous mixture (a mixture with visible distinguishable parts) that forms when a solute does not dissolve significantly in a solvent. Some particles will settle out over time and in many cases you can separate them from the solvent by using filter paper. Chalk in water and red blood cells in plasma are suspensions. Figures 15.3.2 and 15.3.3 show blood before and after it is separated into its component parts.

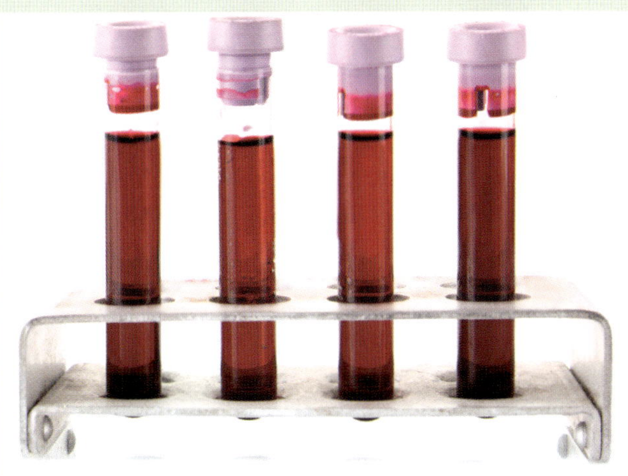

FIGURE 15.3.2 On first glance, blood looks like a homogeneous mixture.

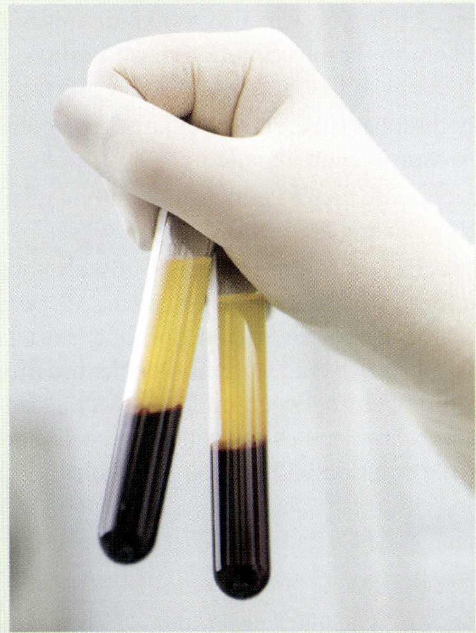

FIGURE 15.3.3 Centrifuging blood separates it into clear layers, demonstrating that blood is actually a heterogeneous suspension. Here you can see the plasma on top and the red blood cells on the bottom.

A colloid is a mixture of particles that consists of smaller clusters of ions or molecules. These are evenly dispersed throughout the solvent and do not settle on standing. For example, milk is a colloid of a fat and an aqueous solution (Figure 15.3.4). Milk solids and fat are finely dispersed throughout the more aqueous components of the milk. Other common colloids include mayonnaise, paint and ink.

FIGURE 15.3.4 Milk is a colloid. Small particles are dispersed throughout the liquid.

Process of dissolving

The process of a substance dissolving in another substance is called **dissolution**. When the two substances are liquids, you can say they are **miscible**.

During dissolution the following processes occur.

- The particles of the solute are separated from one another.
- The particles of the solvent are separated from one another.
- The solute and solvent particles are attracted to each other.

For a solution to form, the solute particles must interact with the solvent molecules. The solute particles are surrounded by solvent molecules and carried throughout the solution.

Forces involved in dissolving

For a substance to dissolve, there must be a change in the way particles in the solute and solvent interact. This means you need to look at the forces of attraction that occur between the particles.

It is useful to think of three different forces of attraction when considering if and how a substance will dissolve in water.

- Forces hold the particles of the substance (solute) together before it is added to the solvent.
- Forces hold the solvent molecules together. In water, these forces are hydrogen bonds.
- Forces can form between the solute particles and solvent molecules if the substance dissolves.

For a substance to dissolve, the attractive forces that form between the solute and solvent particles must be sufficient to overcome the forces between the particles in the solute and the forces between the solvent molecules.

For a substance to be soluble, the solute–solvent interactions must be stronger than the solute–solute and solvent–solvent interactions.

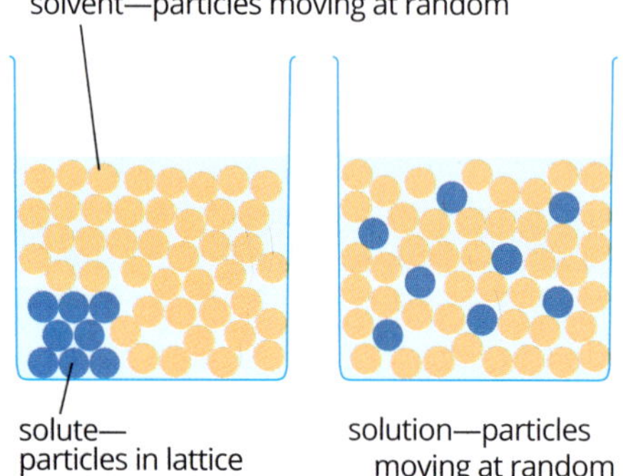

FIGURE 15.3.5 The rearrangement of particles when a solute dissolves in a solvent. The solute particles separate and become evenly distributed in the solvent.

Figure 15.3.5 shows that as a solute dissolves, the solute particles separate and become evenly distributed in the solvent. If the attraction between the solute and solvent particles is not strong enough, then the substance will not readily dissolve.

You will recall from Chapter 12 that you can use a 'like dissolves like' rule to predict whether a substance is likely to dissolve in another substance. The general statement that a solvent will only dissolve 'like' solutes tells you two things.

- **Polar solvents** will generally dissolve substances consisting of polar molecules or charged ions, but will not dissolve solutes made up of non-polar molecules.
- **Non-polar solvents** can dissolve substances consisting of non-polar molecules, but will not dissolve ones with polar molecules or ions.

Oil and other non-polar molecular substances do not dissolve well in water because the only intermolecular forces in these substances are dispersion forces, whereas the strong intermolecular forces of hydrogen bonding exist between the water molecules. These hydrogen bonds between water molecules are much stronger than the dispersion forces that could occur between molecules of oil and water. As a result, the attraction between the water molecules cannot be overcome and the water molecules do not separate to form a solution with the oil molecules.

Polar molecules have a permanent dipole and an uneven distribution of charge, and are asymmetrical.

Non-polar molecules have bonds with an equal distribution of valence (outer shell) electrons, therefore there is no charge on either end of the molecule and the molecule is symmetrical.

You can observe the 'like dissolves like' rule with the miscibility of different liquids. Figure 15.3.6 shows a solution of ethanol and water. The polar nature of the ethanol molecule means it readily dissolves in water, which is also polar. When mixed, a homogeneous solution is formed with no separation between solute and solvent.

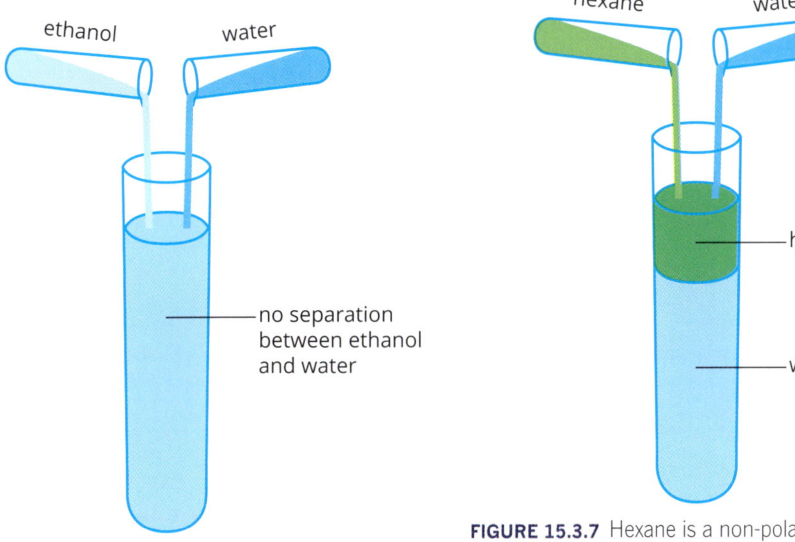

FIGURE 15.3.6 The polar compound ethanol is completely miscible in water. A homogeneous solution is formed on mixing.

FIGURE 15.3.7 Hexane is a non-polar molecular compound that is not miscible in water. Two layers are formed with the non-polar liquid sitting on top of the water layer.

In contrast, hexane is immiscible in water. Hexane is composed of non-polar molecules that will not interact with the polar water molecules. Figure 15.3.7 shows that when hexane and water are mixed, two layers form and the less dense hexane sits on top of the water.

Like hexane, olive oil is also composed of non-polar molecules. When olive oil and hexane are mixed with each other, the 'like dissolves like' rule applies and the two non-polar liquids readily mix with each other to form a homogeneous solution (Figure 15.3.8).

 **Use the 'like dissolves like' principle to predict if something will dissolve in a solvent.**

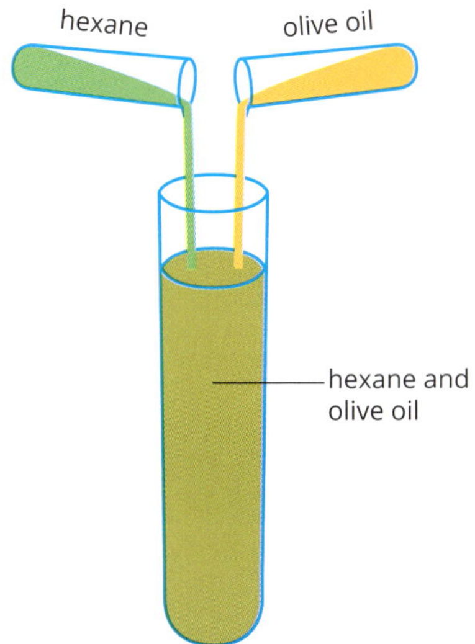

FIGURE 15.3.8 Hexane and olive oil are both non-polar molecular compounds and so are completely miscible.

CHEMFILE

Solubility of pharmaceutical drugs

Pharmaceutical drugs are usually carried through the blood to the place in a person's body where they need to act. The blood is mostly water so drugs need to dissolve in water to be transported. However, most drugs are organic molecules with low solubility in water. The force of attraction between water molecules and, often non-polar, drug molecules can be weak and so cannot overcome water–water molecular attractions and drug–drug molecular attractions.

To overcome this problem, a range of techniques are used to improve aqueous solubility of drugs, including physical and chemical modifications. Particle size reduction, a physical method of modification, involves producing the drugs in small particle size to increase their surface area, which leads to more attractions between the drug and water molecules. Although pharmaceutical drugs are often large organic molecules, they often contain groups of atoms within their structure that can be modified to become positively or negatively charged—that is, they have some ionic character. These charged parts of the organic molecule form ion–dipole attractions with water molecules and so improve the drug's solubility. When chemical modifications of a potential drug are made, there is testing to ensure it remains therapeutic and does not become toxic.

15.3 Review

SUMMARY

- A solution is a homogeneous mixture in which molecules or ions are evenly dispersed throughout a solvent.
- Solutions in which water is the solvent are called aqueous solutions.
- A solution forms when the bonds between the solute and solvent particles are sufficiently strong to compete with attractive forces between the solute particles and between the solvent particles.
- 'Like dissolves like' is a rule that predicts that polar solutes will dissolve in polar solvents and non-polar solutes will dissolve in non-polar solvents.

KEY QUESTIONS

1 Identify each of the following components of a glass of wine as solute, solvent or solution.
   a water
   b wine
   c ethanol
   d sugar

2 What is the one thing that all aqueous solutions have in common?

3 Classify the following substances as likely to be soluble or insoluble in water.
   a ammonia (NH_3)
   b oxygen gas (O_2)
   c hydrogen chloride (HCl)
   d methanol (CH_3OH)
   e methane (CH_4)
   f hydrogen fluoride (HF)
   g carbon dioxide (CO_2)

4 Sodium chloride is very soluble in water. Explain what can be concluded about the forces present between the solute and solvent particles.

5 Match each of the following terms with the correct definition: dissolution, solution, solvent, solute, soluble, insoluble.
   a capable of dissolving
   b incapable of dissolving
   c the minor component of a solution
   d the process by which a substance is dissolved in a solvent
   e a liquid mixture in which a solute is dispersed throughout a solvent
   f the major component of a solution

6 Which one of the following molecules is likely to dissolve best in hexane, a non-polar solvent: CH_4, NH_3, HF or H_2S? Explain your reasoning.

15.4 Water as a solvent of molecular substances

Most molecular substances are insoluble (or only very sparingly soluble) in water. However, some smaller molecules, such as ammonia (NH_3), hydrogen chloride (HCl) and sucrose ($C_{12}H_{22}O_{11}$), dissolve well in water. Solutions in which a molecular substance is at least one of the solutes include brick cleaner, various liquid fertilisers and drinks such as wine and cordial (Figure 15.4.1).

There are two main ways in which molecular compounds dissolve well in water. In this section, you will explore both of these ways in detail.

FIGURE 15.4.1 Lemon cordial is a solution that contains several dissolved molecular substances.

MOLECULAR COMPOUNDS THAT FORM HYDROGEN BONDS WITH WATER

One way a molecular compound might dissolve in water is if its molecules form hydrogen bonds with water molecules. An example of such a molecule is ethanol.

Ethanol (C_2H_5OH) is a liquid at room temperature. Its molecules contain the polar –OH group, with lone pairs of electrons on the oxygen atom. The presence of the hydrogen atoms bonded to the electronegative oxygen atom allow an ethanol molecule to form hydrogen bonds.

 Molecular compounds that can form hydrogen bonds are often soluble in water.

Figure 15.4.2 shows how hydrogen bonds form between molecules in pure ethanol.

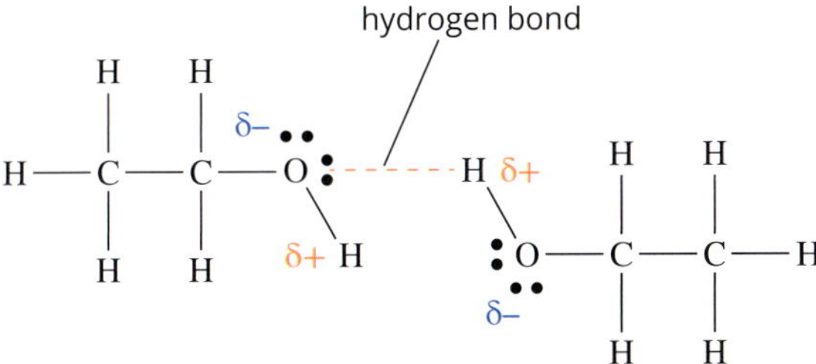

FIGURE 15.4.2 Hydrogen bonding in pure ethanol. The intermolecular hydrogen bond is formed between the lone pair of electrons on the oxygen atom of one ethanol molecule and the electron-deficient hydrogen atom of an adjacent ethanol molecule.

When ethanol is added to water, it dissolves. The two solutions are miscible. Figure 15.4.3 shows how hydrogen bonds form between the ethanol and surrounding water molecules.

Because the strength of the intermolecular forces between the solute and the solvent are similar, the two substances can readily interact with each other. Therefore, water and ethanol molecules mix freely with each other, held together in solution by hydrogen bonds.

In summary, when ethanol dissolves in water, hydrogen bonds between:

- water molecules break
- ethanol molecules break
- ethanol molecules and water molecules form.

An equation for the dissolution of ethanol can be written to represent this process:

$$C_2H_5OH(l) \xrightarrow{H_2O} C_2H_5OH(aq)$$

Note that the formula of water sits above the arrow. This is because there is no direct reaction between the water and the ethanol. The two substances simply mix together. No chemical change occurs; only the state symbol for ethanol is altered from (l) to (aq), indicating it is now dissolved in water.

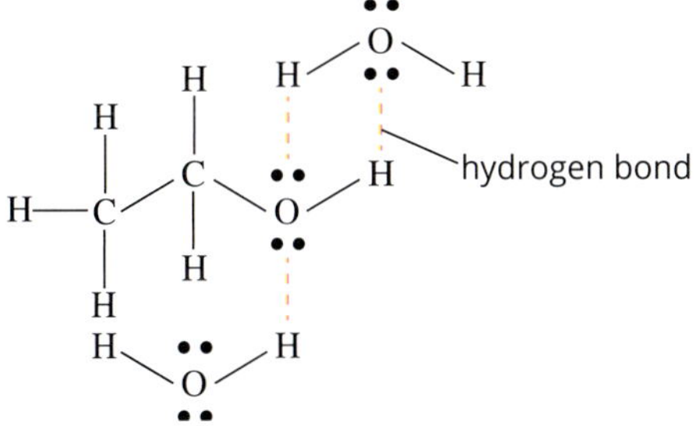

FIGURE 15.4.3 Hydrogen bonding between ethanol and water

When a polar molecular substance dissolves in water by forming hydrogen bonds, the equation for dissolution places water above the arrow.

Other polar compounds that form hydrogen bonds with water

Sugars, including glucose, fructose and sucrose, can all dissolve in water by forming hydrogen bonds. The structure of each of these sugars is shown in Figure 15.4.4. As with ethanol, their molecules also contain the polar –OH group.

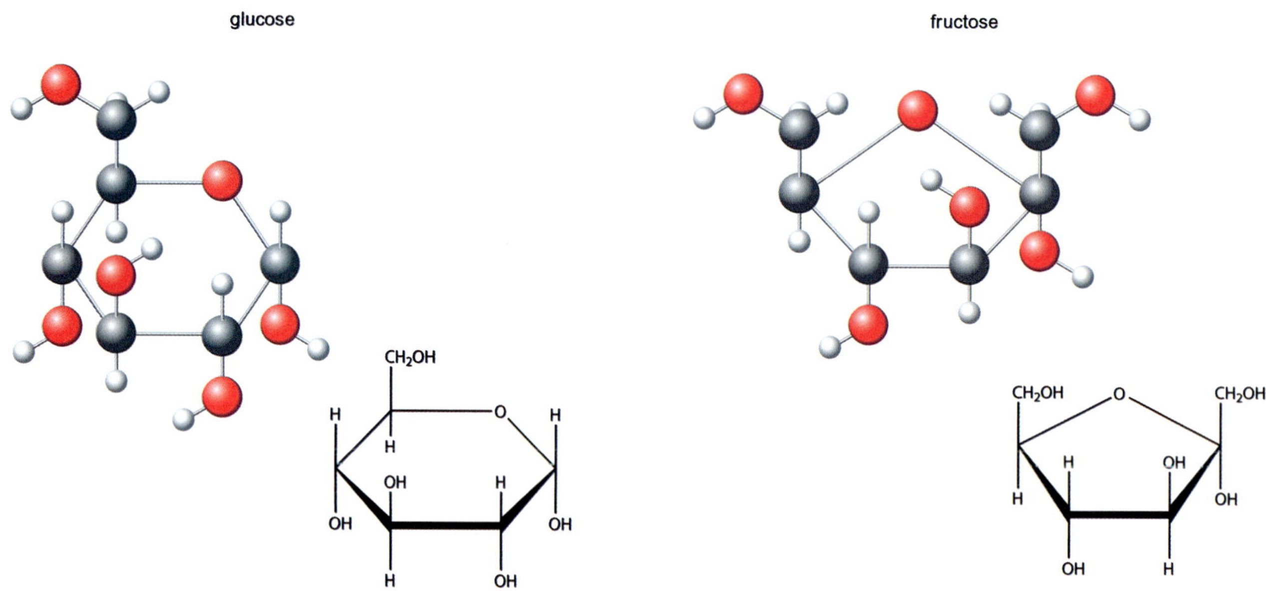

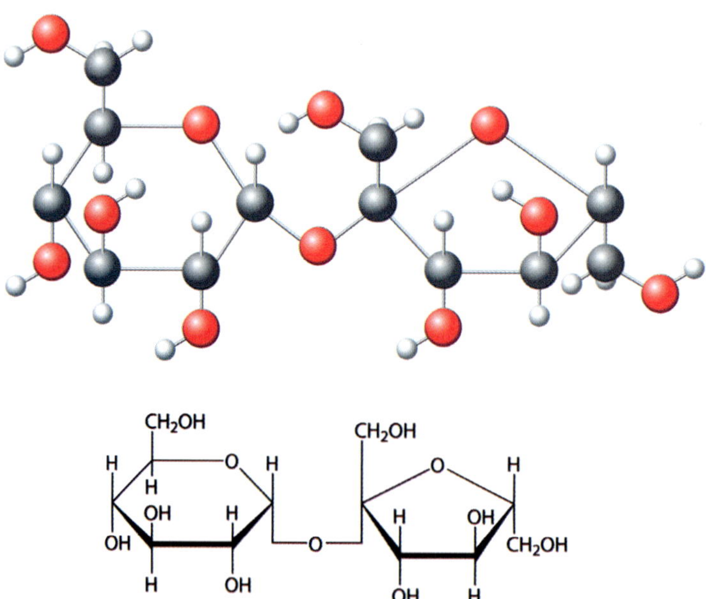

FIGURE 15.4.4 These are the molecular structures of common sugars: glucose (the sugar found in our blood), fructose (fruit sugar) and sucrose (table sugar—the sugar used to sweeten tea and coffee). Each of these sugars can dissolve in water because of the presence of polar –OH groups.

The more polar the molecules of a molecular compound are, the more likely the compound is to dissolve in water. Some molecules have a polar section and a non-polar section. In general, the larger the non-polar section of the molecule, the less soluble it is in water.

Non-polar molecular substances do not have charged ends so there is no significant attraction to water molecules. The only force of attraction that exists between non-polar substances and water are weak dispersion forces, which are not strong enough to overcome the relatively strong hydrogen bonding between water molecules.

Table 15.4.1 compares the structures and properties of vitamin C and vitamin A. Both molecules contain the polar –OH group, but only vitamin C is soluble in water. The higher proportion of polar –OH groups on the vitamin C molecule allows it to form sufficient hydrogen bonds with water to overcome the strong attraction between water molecules, therefore dissolving to form a solution.

TABLE 15.4.1 Comparison of the solubility and structures of vitamin C and vitamin A

	Vitamin C	Vitamin A
structure	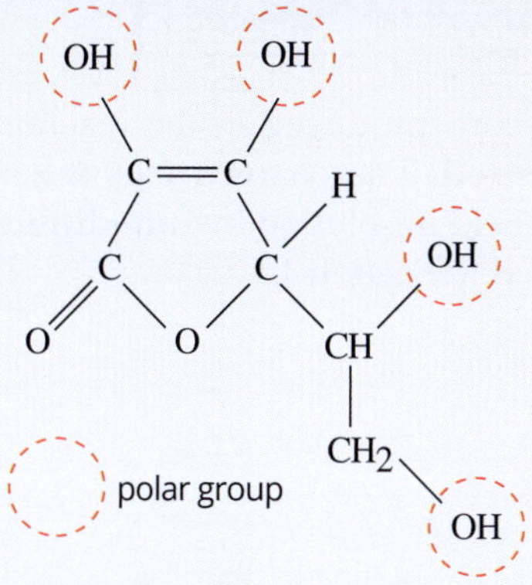 polar group Vitamin C: • contains four polar –OH groups • molecules are quite polar.	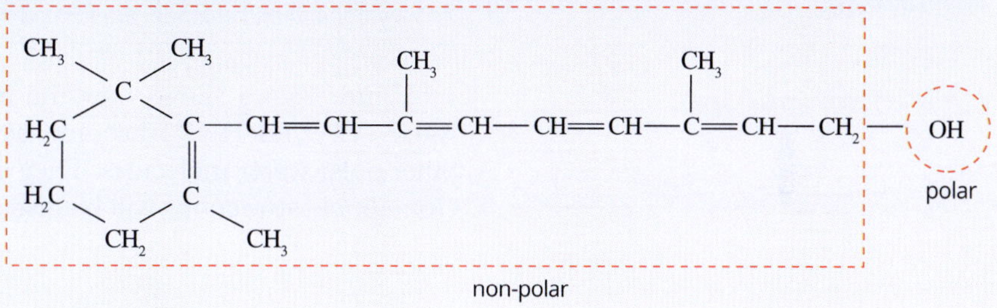 non-polar polar Vitamin A: • contains one polar –OH group • molecules are largely non-polar.
solubility	highly soluble in water, insoluble in fats	soluble in fats, insoluble in water
biological significance	The high solubility of vitamin C means it is excreted in urine so it must form a regular part of the diet.	Vitamin A is not excreted in urine but is stored in body fat. The body can tolerate a diet low in vitamin A for a limited period.

MOLECULAR COMPOUNDS THAT IONISE

Some compounds have molecules with one or more covalent bonds that are so polar they break when the compound is placed in water. Hydrogen chloride is such a compound. Hydrogen chloride (HCl) is a gas at room temperature. Chlorine is much more electronegative than hydrogen, so the H–Cl covalent bond is highly polar; the molecule forms a dipole.

When hydrogen chloride is added to water, the hydrogen atom in HCl forms such a strong attraction to the oxygen atom in a water molecule that the H–Cl covalent bond breaks. The two electrons that made up the H–Cl covalent bond remain with the more electronegative Cl atom and the newly formed hydrogen ion (H^+) joins the water molecule.

Figure 15.4.5 shows that when the hydrogen ion (H^+) bonds to the water molecule, it forms a new ion known as the **hydronium ion** (H_3O^+).

Since the Cl atom has gained an electron, it has a negative charge, forming a chloride ion (Cl^-).

Highly polar molecular compounds will form ions when dissolved in water.

The HCl molecule is split into smaller particles. The HCl is said to have become **ionised**; that is, it has undergone a process that has produced ions (Figure 15.4.5).

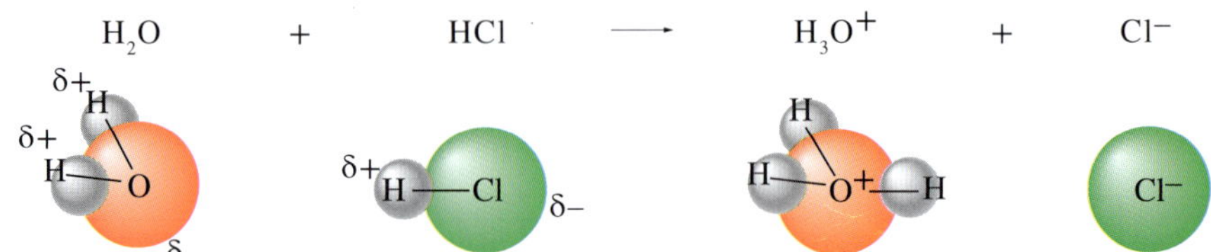

FIGURE 15.4.5 The dipole–dipole attraction between the molecules of water and hydrogen chloride leads to the breaking of the polar covalent bond between the hydrogen and chlorine atoms. New ions, hydronium and chloride, are formed in a process called ionisation.

Figure 15.4.6 shows how the two ions produced in the reaction of HCl with water (Cl^- and H_3O^+) become **hydrated**. The charged ions are surrounded by other polar water molecules. They are held in solution by **ion–dipole attractions**. (Ion–dipole attractions will be discussed in Section 15.5.)

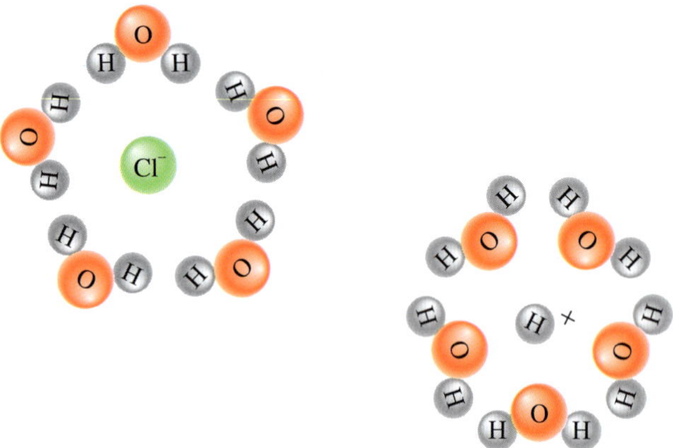

FIGURE 15.4.6 When HCl molecules ionise in water, the newly formed ions are surrounded by adjacent water molecules. The charges of the ions interact strongly with the polar water molecules. For simplicity, the hydronium ion (H_3O^+) is represented by the hydrogen ion (H^+).

In summary, this is what happens when hydrogen chloride dissolves in water.

- Polar covalent bonds within hydrogen chloride molecules break, producing hydrogen ions (H^+) and chloride ions (Cl^-).
- A covalent bond forms between each H^+ and H_2O molecule, forming H_3O^+ ions.
- Ion–dipole attractions form between the H_3O^+ and Cl^- ions and polar water molecules.

An equation can be written to represent this process:

$$HCl(g) + H_2O(l) \rightarrow H_3O^+(aq) + Cl^-(aq)$$

You should note two important points about this equation.

- Water is a reactant because there has been a rearrangement of atoms to form new substances.
- The aqueous state of the H_3O^+ and Cl^- ions tells you that they are hydrated in solution.

Other polar molecular compounds that ionise

Other compounds that dissolve in water by ionising include the common acids nitric acid (HNO_3), sulfuric acid (H_2SO_4) and ethanoic acid (CH_3COOH).

Ammonia (NH_3) is another polar molecule that can dissolve in water. Ammonia can dissolve by either forming hydrogen bonds with water or by reacting to form the ammonium ion (NH_4^+). The process by which ammonia forms the ammonium ion will be looked at in detail in Chapter 17.

CHEMFILE

Dissolved molecules in a fish tank

Many people keep fish at home in a small tank. Unfortunately, many pet fish do not live for quite as long as their owners hoped.

One of the biggest killers of aquarium fish is ammonia poisoning. Fish produce ammonia as a waste product. Ammonia is a polar molecule that readily dissolves in water, and is toxic to fish. Bacteria colonies living in a fish tank break down the ammonia produced by fish into other non-toxic products. In a healthy tank (Figure 15.4.7), a balance exists between ammonia production by fish and breakdown by bacteria.

FIGURE 15.4.7 Fish in this tank produce the molecular compound ammonia (NH_3) as a waste product. Bacteria living in the tank break down the toxic ammonia into harmless products.

15.4 Review

SUMMARY

- Water is a good solvent for some polar molecular compounds.
- Some polar molecular compounds dissolve by forming hydrogen bonds with water.
- Some covalent molecular compounds are so polar that they dissolve in water by ionising to form hydrated ions.
- The table below summarises the two ways that molecular compounds dissolve in water.

Type of solute	Example of solute	Bond broken in the solute	Bonds formed	Equation
polar covalent molecule that can hydrogen bond	ethanol, C_2H_5OH	hydrogen bonds broken between ethanol molecules	hydrogen bonds formed between ethanol and water molecules	$C_2H_5OH(l) \xrightarrow{H_2O(l)} C_2H_5OH(aq)$
polar molecules that ionise	hydrogen chloride, HCl	covalent bond broken between hydrogen and chlorine atoms in the HCl molecule	covalent bond formed between H^+ from HCl and oxygen atom in water molecule, forming H_3O^+ ions; H_3O^+ and Cl^- ions form ion–dipole bonds with water molecules	$HCl(g) + H_2O(l) \rightarrow H_3O^+(aq) + Cl^-(aq)$

KEY QUESTIONS

1 Which of the following substances are likely to dissolve in water by forming hydrogen bonds with water molecules.

A ammonium sulfate ($(NH_4)_2SO_4$)
B methanol (CH_3OH)
C diamond (C)
D hydrogen chloride (HCl)
E ethandiol ($HOCH_2CH_2OH$)
F sodium hydroxide (NaOH)
G hydrogen gas (H_2)

2 Methanol (CH_3OH) and glucose ($C_6H_{12}O_6$) form hydrogen bonds with water. They will dissolve in water without ionising. Write chemical equations to represent the dissolving process for each of these compounds.

3 With reference to the bonds that are broken and new bonds that are formed, describe how ethanol dissolves in water.

4 Hydrogen iodide (HI) will ionise when it dissolves in water. The ionisation reaction is similar to that of HCl. Write a chemical equation to represent the dissolving process for this compound.

5 Hydrogen chloride dissolves in water by ionisation. Using the terms provided, complete the following sentences to describe the process of how ionisation occurs. Some terms may be used more than once.

ion–dipole bonds	hydrogen bonds
covalent bonds	hydronium
ionises	H^+

__________ within hydrogen chloride molecules are broken.

__________ between water molecules are broken. The HCl _______ and produces Cl^- and H^+ ions. _________ form between ___ ions and water to produce _________ ions. ________ form between the Cl^- and H_3O^+ ions and polar water molecules.

15.5 Water as a solvent of ionic compounds

Sports drinks are advertised to athletes as a way to replace the electrolytes lost in sweat during exercise (Figure 15.5.1). The electrolytes are dissolved ionic substances such as sodium chloride and potassium phosphate.

Many ionic substances are soluble in water. However, not all ionic substances dissolve readily in water. In this section, you will look at how ionic substances dissolve and learn how to predict whether an ionic compound will be soluble in water.

FIGURE 15.5.1 Sports drinks are used by athletes to replace water and dissolved ionic solutes.

DISSOLUTION OF AN IONIC LATTICE IN WATER

Many ionic compounds dissolve readily in water. Sodium chloride is a typical ionic compound that exists as a solid at room temperature. In Figure 15.5.2, you can see the arrangement of sodium cations (Na^+) and chloride anions (Cl^-) in a three-dimensional ionic lattice. The ions are held together by strong electrostatic forces between the positive and negative charges of the ions.

When an ionic compound such as sodium chloride is added to water, the positive ends of the water molecules are attracted to the negatively charged chloride ions and the negative ends of the water molecules are attracted to the positively charged sodium ions (Figure 15.5.3). The attraction between an ion and a polar molecule such as water is described as an ion–dipole attraction.

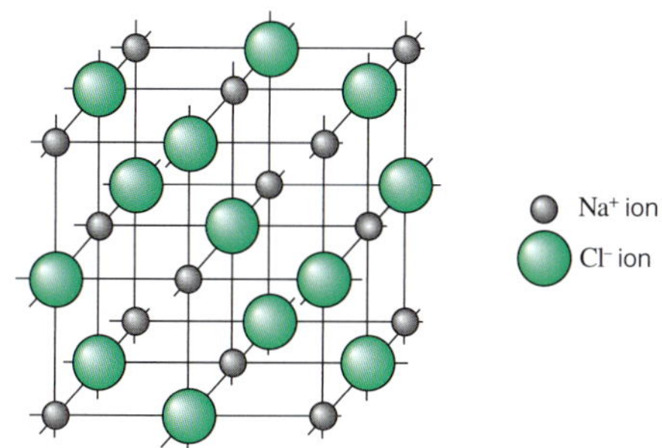

FIGURE 15.5.2 A representation of the crystal lattice of the ionic compound sodium chloride

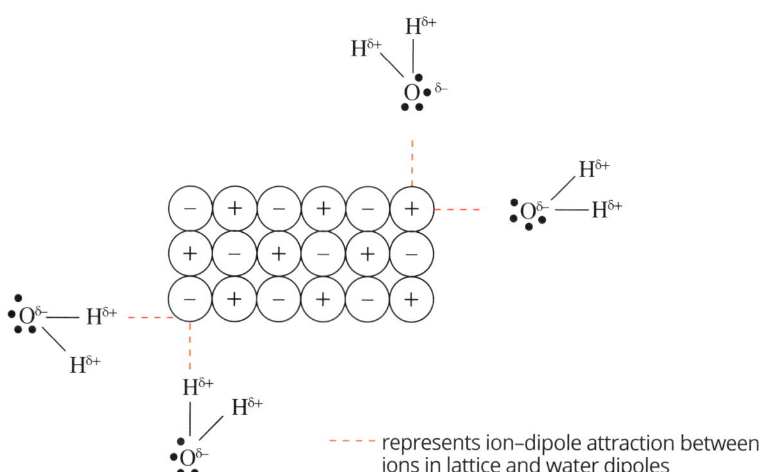

FIGURE 15.5.3 Electrostatic attraction occurs between the negative ions in a NaCl lattice and the hydrogen atoms in polar water molecules. Electrostatic attraction also occurs between the positive ions in the lattice and the oxygen atoms in water molecules.

Water molecules are in a continuous state of random motion. If the ion–dipole attractions between the ions and the water molecules are strong enough, the water molecules can pull the sodium and chloride ions on the outer part of the crystal out of the lattice and into the surrounding solution.

Sodium ions and chloride ions pulled out of the lattice become surrounded by water molecules. These ions are said to be hydrated. Water molecules are arranged around the ions as shown in Figure 15.5.4. Note the different arrangements of the water molecules around the positive and negative ions. The hydrogen atoms in the water molecule are more positive so they are orientated towards the negative chloride ion. The positive sodium ion is surrounded by the more negative oxygen atom of the water molecules.

The process of dissolving a solid ionic compound in water and separating positive and negative ions to form hydrated ions is called **dissociation**.

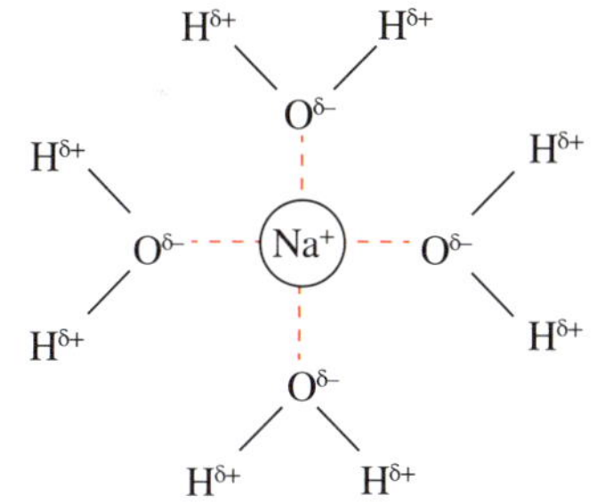

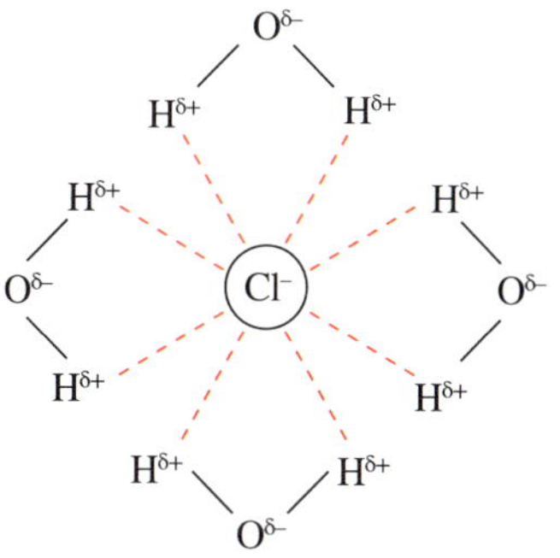

represents ion–dipole interaction

FIGURE 15.5.4 Ion–dipole attraction between the ions and adjacent water molecules form hydrated sodium and chloride ions.

Although the ionic bonds within the lattice are strong, the ions can be pulled away from the lattice by the interactions of many water molecules.

In summary, this is what happens when sodium chloride dissolves in water.

- Ionic bonds within the sodium chloride lattice are broken.
- Hydrogen bonds between water molecules are broken.
- Ion–dipole attractions form between ions and polar water molecules.

An equation can be written to represent the dissociation process:

$$NaCl(s) \xrightarrow{H_2O(l)} Na^+(aq) + Cl^-(aq)$$

Note that the formula of water sits above the arrow. This is because there is no direct reaction between the water and the sodium chloride. No chemical change occurs; only the state symbol for sodium chloride is altered from (s) to (aq), indicating it is now dissolved in water. (You may omit the H_2O from this equation if you wish.)

It is important to note that dissociation of ionic compounds is simply freeing ions from the lattice so that they can move throughout the solution. This is different from ionisation of molecular compounds where new ions are formed by the reaction of the molecule with water.

 Ionic substances dissolve by dissociation. Ion–dipole bonds are formed between the ions and water molecules.

INSOLUBLE IONIC COMPOUNDS

FIGURE 15.5.5 Limestone caves are formed by limestone (calcium carbonate) dissolving over very long periods of time. This photograph is of a limestone cave at Loch Ard Gorge, Victoria.

Not all ionic compounds are soluble in water. For example, limestone ($CaCO_3$) is almost completely insoluble in water. Limestone caves such as those seen in Figure 15.5.5 are formed over a long period of time as $CaCO_3$ is dissolved and redeposited. The ionic compound $Ca_3(PO_4)_2 \cdot Ca(OH)_2$, which gives strength to bones and teeth, is also (fortunately) insoluble in water.

Insoluble ionic compounds do not dissolve in water because the energy required to separate the ions from the lattice is greater than the energy released when the ions are hydrated.

Although substances are usually described as 'soluble' or 'insoluble', this is a generalisation. Substances that are described as 'insoluble' tend to dissolve very slightly. Those that are described as 'soluble' dissolve to varying extents.

 Substances are rarely ever completely 'soluble' or 'insoluble'. The solubility of substances can be considered as a scale.

The SNAPE rule

A handy way to remember many of the soluble salts is with the initials SNAPE. Salts that contain one or more of the following ions are soluble:

- **S**odium (Na^+)
- **N**itrate (NO_3^-)
- **A**mmonium (NH_4^+)
- **P**otassium (K^+)
- **E**thanoate (CH_3COO^-)

Solubility tables

A **solubility table** can be used to determine whether common ionic compounds are soluble in water. Anions and cations are listed with indications of their solubility in different compounds. Table 15.5.1 is a sample solubility table listing ionic compounds that are soluble in water. Note that for some ions there are exceptions. Insoluble compounds are given in Table 15.5.2.

TABLE 15.5.1 Relative solubilities of soluble ionic compounds

Soluble in water (>0.1 mol dissolves per L at 25°C)	Exceptions: insoluble (<0.1 mol dissolves per L at 25°C)	Exceptions: slightly soluble (0.01–0.1 mol dissolves per L at 25°C)
most chlorides (Cl^-), bromides (Br^-) and iodides (I^-)	AgCl, AgBr, AgI, PbI_2	$PbCl_2$, $PbBr_2$
all nitrates (NO_3^-)	no exceptions	no exceptions
all ammonium (NH_4^+) salts	no exceptions	no exceptions
all sodium (Na^+) and potassium (K^+) salts	no exceptions	no exceptions
all ethanoates (CH_3COO^-)	no exceptions	no exceptions
most sulfates (SO_4^{2-})	$SrSO_4$, $BaSO_4$, $PbSO_4$	$CaSO_4$, Ag_2SO_4

TABLE 15.5.2 Relative solubilities of insoluble ionic compounds

Insoluble in water	Exceptions: soluble	Exceptions: slightly soluble
most hydroxides (OH^-)	NaOH, KOH, $Ba(OH)_2$, NH_4OH*, AgOH†	$Ca(OH)_2$, $Sr(OH)_2$
most carbonates (CO_3^{2-})	Na_2CO_3, K_2CO_3, $(NH_4)_2CO_3$	no exceptions
most phosphates (PO_4^{3-})	Na_3PO_4, K_3PO_4, $(NH_4)_3PO_4$	no exceptions
most sulfides (S^{2-})	Na_2S, K_2S, $(NH_4)_2S$	no exceptions

*NH_4OH does not exist in significant amounts in an ammonia solution. Ammonium and hydroxide ions readily combine to form ammonia and water.

†AgOH readily decomposes to form a precipitate of silver oxide and water.

CHEMFILE

The Dead Sea

The Dead Sea, a salt lake in the Middle East, is one of the saltiest bodies of water on Earth. The concentration of dissolved ionic compounds, including salt, in the water makes the Dead Sea roughly 8.6 times saltier than the ocean. This prevents fish and aquatic plants from living in it. There is so much dissolved salt in the water that people can easily float in the Dead Sea due to natural buoyancy; the density of the salt water is greater than the density of the human body.

FIGURE 15.5.6 The Dead Sea contains so many dissolved ions it is easy to float.

Using a solubility table to predict solubility

The solubility tables provided as Tables 15.5.1 and 15.5.2 can be used to predict whether a particular ionic substance will be soluble, slightly soluble or insoluble in water.

Worked example 15.5.1 shows you a systematic way of using a solubility table to determine whether an ionic compound will be soluble or insoluble in water.

Worked example 15.5.1

DETERMINING IF IONIC COMPOUNDS ARE SOLUBLE OR INSOLUBLE IN WATER

Is barium sulfide (BaS) soluble or insoluble in water? You will need to refer to the solubility tables above to complete this question.

Thinking	Working
Identify the ions that are present in the ionic compound.	Barium (Ba^{2+}) and sulfide (S^{2-})
Check the solubility tables to see if compounds containing the cation are usually soluble or insoluble in water.	Barium ions do not appear in the table. This is no help, so look for the other ion.
Check the solubility tables to see if compounds containing the anion are usually soluble or insoluble in water.	Compounds containing sulfide ions are usually insoluble in water. So, barium sulfide will be insoluble in water.

Worked example: Try yourself 15.5.1

DETERMINING IF IONIC COMPOUNDS ARE SOLUBLE OR INSOLUBLE IN WATER

Is ammonium phosphate ($(NH_4)_3PO_4$) soluble or insoluble in water? You will need to refer to the solubility tables above to complete this question.

15.5 Review

SUMMARY

- Soluble ionic compounds dissociate in water to form hydrated ions.
- In a hydrated ion, water molecules are attracted to the central ion by ion–dipole attractions.
- The table below summarises the way soluble ionic compounds dissolve in water.
- A solubility table can be used to determine whether a particular ionic compound is likely to be soluble or insoluble in water.

Type of solute	Example	Bond broken in the solute	Bonds formed with water	Equation
soluble ionic compounds	sodium chloride (NaCl)	ionic bonds between Na^+ and Cl^- ions	ion–dipole attractions between dissociated ions and polar water molecules	$NaCl(s) \xrightarrow{H_2O(l)} Na^+(aq) + Cl^-(aq)$

KEY QUESTIONS

1 Sodium nitrate ($NaNO_3$) and calcium hydroxide ($Ca(OH)_2$) will both dissociate when they dissolve in water. Write chemical equations to represent the dissolving process for each of these compounds.

2 Redraw the representations of Na^+ and Cl^- ions below, then show the correct orientation of water molecules around hydrated sodium and chloride ions. Use a V-shape to represent the shape of water molecules.

3 Which of the following substances would you expect to be soluble in water? Refer to Table 15.5.1 and Table 15.5.2 on page 367 to complete this question.
- **A** sodium carbonate
- **B** lead(II) nitrate
- **C** magnesium carbonate
- **D** ammonium sulfate
- **E** iron(II) sulfate
- **F** magnesium phosphate
- **G** zinc carbonate
- **H** sodium sulfide
- **I** silver chloride
- **J** barium sulfate

4 Which of the following compounds would you expect to be insoluble in water? Refer to Table 15.5.1 and Table 15.5.2 on page 367 to complete this question.
- **A** silver carbonate
- **B** zinc nitrate
- **C** copper(II) carbonate
- **D** silver chloride
- **E** lead(II) bromide
- **F** magnesium hydroxide
- **G** barium nitrate
- **H** aluminium sulfide

5 Write the formulae for the ions produced when these compounds dissolve in water.
- **a** sodium carbonate
- **b** calcium nitrate
- **c** potassium bromide
- **d** iron(III) sulfate
- **e** copper(II) chloride

6 Suggest reasons for each of the following.
- **a** Concentrated deposits of nitrate compounds are only found in desert regions.
- **b** The sea is a rich source of sodium, chloride and sulfate ions.

15.6 Solubility

The solubility of a substance is a measure of how much of the substance will dissolve in a given amount of a solvent. For example, glucose is very soluble in water and is used by your body as a readily available energy source. Most rocks are made of minerals that are insoluble in water, but limestone (which contains calcium carbonate) is slightly soluble and so caves form in areas of limestone rock over long periods.

In this section, you will learn how to measure the solubility of some compounds in water and how to predict whether a compound will be soluble or not.

DEFINITION OF SOLUBILITY

In chemistry, the term **solubility** has a specific meaning. It refers to the maximum amount of a solute (substance that is dissolved) that can be dissolved in a given quantity of a solvent (the common quantity often used is 100g) at a certain temperature. Table 15.6.1 gives the solubility of some common substances in 100g of water at 18°C.

TABLE 15.6.1 Solubility of solutes at 18°C

Solute	Solubility (g per 100g of water)
table sugar (sucrose)	200
table salt (sodium chloride)	35
limestone (calcium carbonate)	0.0013

Different kinds of solutions

There are three terms that can be used to describe the different solutions that result from dissolving solutes in solvents.

- saturated
- unsaturated
- supersaturated.

A **saturated solution** is one in which no more solute can be dissolved at a particular temperature.

Saturated solutions are those in which no more solute can be dissolved in a solvent (usually water) at a particular temperature.

An **unsaturated solution** contains less solute than is needed to make the solution saturated. Unsaturated solutions can dissolve more solute.

A **supersaturated solution** is an unstable solution that contains more dissolved solute than a saturated solution. If this type of solution is disturbed, some of the solute will separate from the solvent as a solid.

Figure 15.6.1 shows a supersaturated solution of sodium ethanoate. Supersaturated sodium ethanoate is prepared by cooling a saturated solution very carefully so that solid crystals do not form. Adding a small **seed crystal** to the supersaturated solution causes the solute to **crystallise** (form solid crystals) so that a saturated solution remains.

FIGURE 15.6.1 Crystals of sodium ethanoate form after a seed crystal is added to a supersaturated solution of the compound.

SOLUBILITY CURVES

The solubility of many substances changes significantly as the temperature changes. For example, you can dissolve more chocolate powder in hot milk than in cold milk. For most solids, increasing the temperature increases the solubility in a liquid. This is because at higher temperatures, both the solute and solvent have more energy to overcome the forces of attraction that hold the particles together in the solid.

The relationship between solubility and temperature can be represented by a solubility curve, as shown in Figure 15.6.2.

Solubility curves show the solubility of a substance as a function of temperature. For the solubility curves featured in Figure 15.6.2, note the following.

- Each point on a curve represents the maximum amount of the solute that can be dissolved in 100 g of water at a particular temperature. Therefore, each point on the curve represents a saturated solution.
- Any point below a curve represents an unsaturated solution.
- Any point above a curve represents a supersaturated solution.
- For most solids, as temperature increases, the solubility increases.

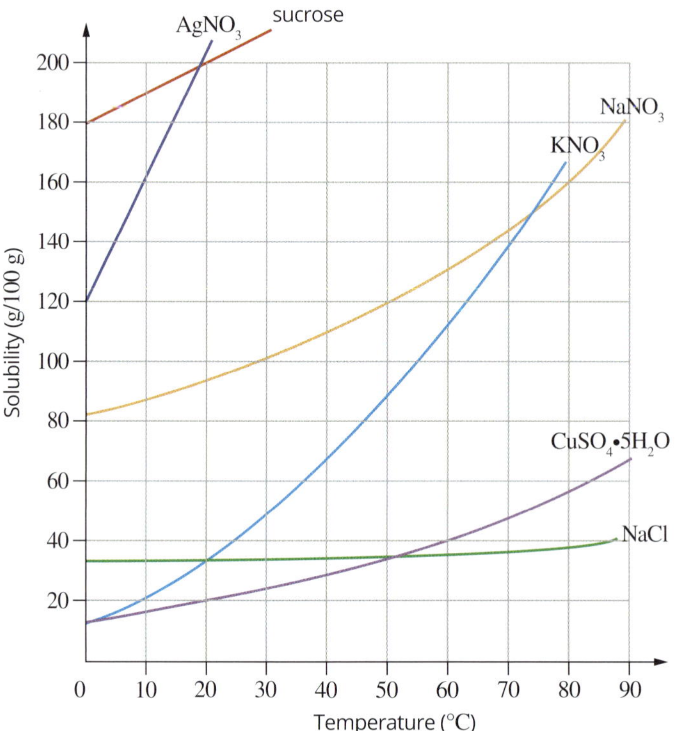

FIGURE 15.6.2 Solubility curves for some common chemicals

Worked example 15.6.1

SOLUBILITY CURVE CALCULATIONS

Use Figure 15.6.2 on page 370 to find how many grams of sodium nitrate ($NaNO_3$) will dissolve in 100 g of water at 50°C.	
Thinking	**Working**
Points on a solubility curve show the maximum amount of the solute that will dissolve in 100 g water at a given temperature. Draw a vertical line from the required temperature on the horizontal axis to intersect with the appropriate solubility curve.	Draw a vertical line from 50°C on the horizontal axis to intersect with the solubility curve for $NaNO_3$. 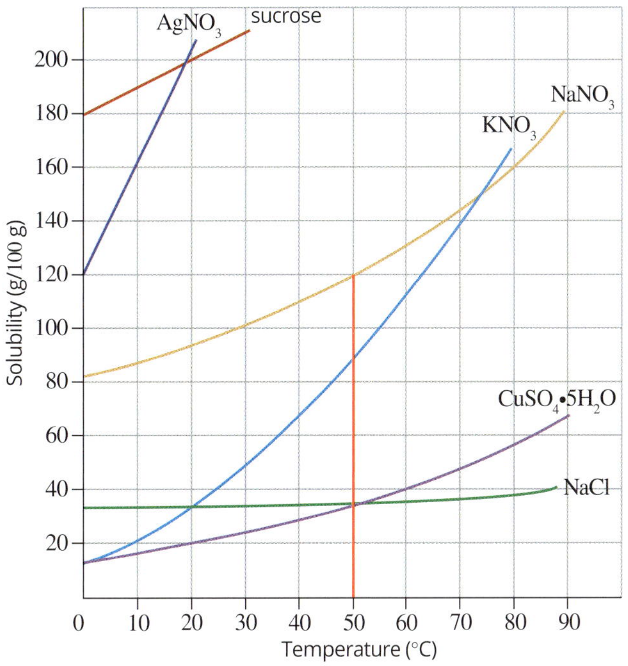
Draw a horizontal line from the intersection point of the vertical line drawn in the previous step. The point where this horizontal line intersects with the vertical axis will give the mass of dissolved substance.	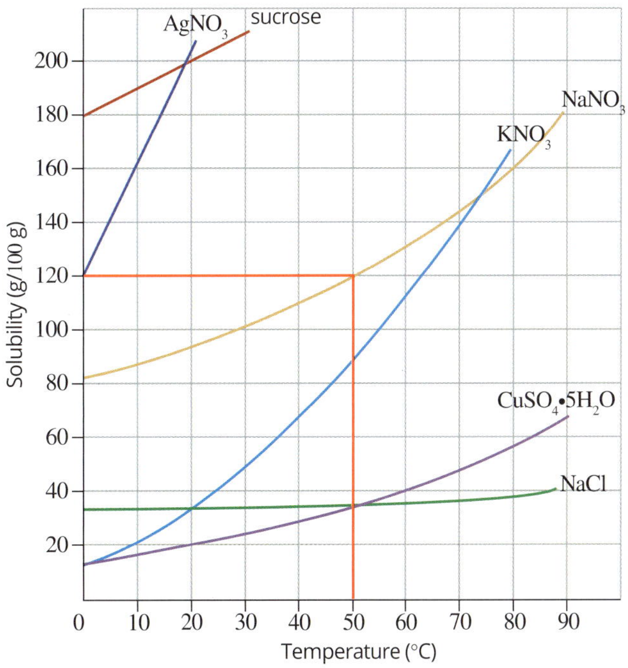
Read the solubility from the graph.	The horizontal line intersects with the vertical axis at 120 g. Therefore, the mass of $NaNO_3$ that will dissolve in 100 g of water at 50°C is 120 g.

Worked example: Try yourself 15.6.1

SOLUBILITY CURVE CALCULATIONS

Use the solubility curve below to find how many grams of potassium nitrate (KNO_3) will dissolve in 100 g of water at 70°C.
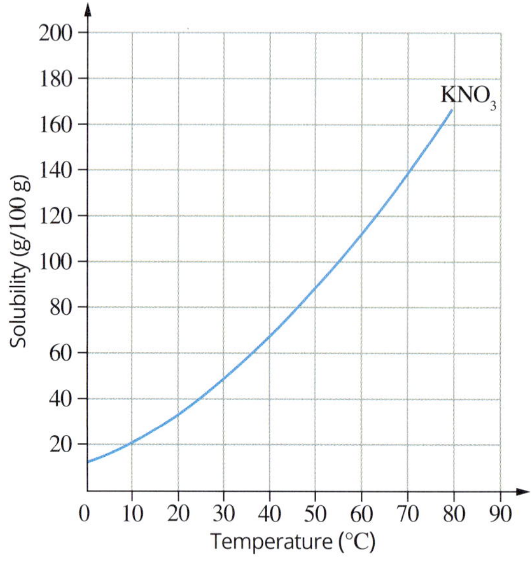

More calculations using solubility curves

Solubility curves allow you to directly calculate the mass of a compound that will dissolve in 100 g water at different temperatures. It is also possible to use the curves to calculate solubilities of compounds in quantities of water other than 100 g. Worked example 15.6.2 shows you how.

Worked example 15.6.2

SOLUBILITY CURVES

An 80 g sample of sodium nitrate ($NaNO_3$) is added to 200 mL of H_2O at 20°C. Use Figure 15.6.2 on page 370 to calculate how much more $NaNO_3$ must be added to make the solution saturated with $NaNO_3$ at 20°C.

Thinking	Working
Use the solubility curve to find the mass of solute in a saturated solution of 100 g of H_2O at the required temperature.	Draw a line from 20°C on the horizontal axis to the solubility curve for $NaNO_3$ and find the corresponding value on the vertical axis. The value is 92 g.

Use the amount of solute that will dissolve in 100 g of H_2O to find the mass of solute to make a saturated solution in the mass of H_2O for this question.	The density of water is $1.0\,g\,mL^{-1}$, so 200 mL of water will weigh 200 g. So twice the mass of solute can dissolve in 200 g of water as in 100 g. $m(NaNO_3) = 2 \times 92$ $= 184\,g$
To find out how much extra solute you need to add, find the difference between the mass of solute needed to make a saturated solution and how much has already been added.	80 g of $NaNO_3$ has already been added to 200 g H_2O. So the extra mass of $NaNO_3$ needed: $= 184 - 80$ $= 104\,g$

Worked example: Try yourself 15.6.2

SOLUBILITY CURVES

A 120 g sample of sodium nitrate ($NaNO_3$) is added to 300 mL of H_2O at 40°C. Use Figure 15.6.2 on page 370 to calculate how much more $NaNO_3$ must be added to make the solution saturated with $NaNO_3$ at 40°C.

Making your own solubility curve

Solubility curves can be developed relatively easily in a school laboratory. For example, a solubility curve for potassium nitrate in water can be derived in a group activity by carrying out the steps shown in Figure 15.6.3.

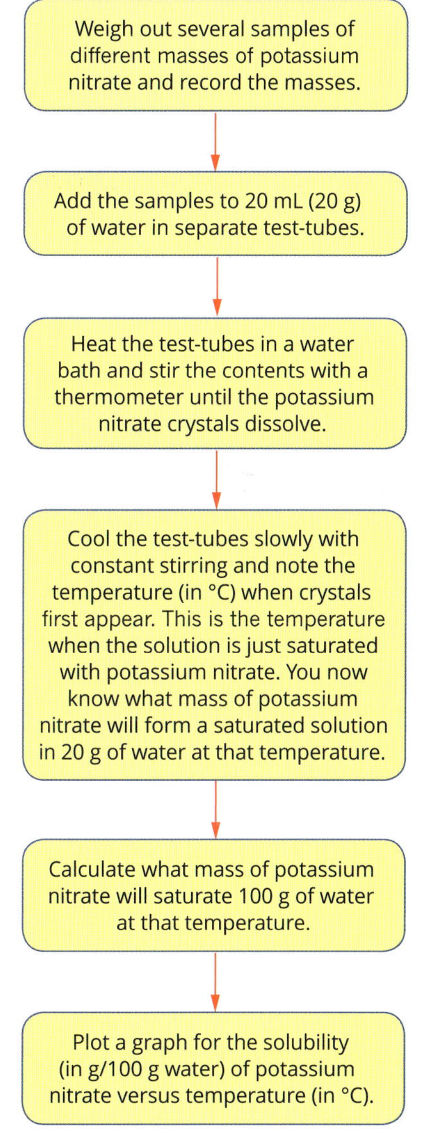

FIGURE 15.6.3 Steps for constructing a solubility curve for potassium nitrate in water

PROCESS OF CRYSTALLISATION

Methods of crystallisation

Crystallisation from solutions occurs when an unsaturated solution becomes saturated and crystals form. There are two main ways in which this can happen.

Cooling a solution may reduce the solubility of a dissolved solute to the point where not all of the substance present is soluble. An example of this occurs when copper(II) sulfate crystals form after a hot solution of the compound is cooled. At lower temperatures, less of the compound can dissolve in the water and a point will be reached when the solution becomes saturated. Further cooling will result in crystals being formed.

Alternatively, crystallisation can occur as a result of the evaporation of solvent from a solution. For example, salt forms as water evaporates from seawater. This process produced the crystals of salt seen in Figure 15.6.4.

FIGURE 15.6.4 White crystals of salt are left behind at the edge of a rock pool where seawater has evaporated.

CHEMFILE

Salt

In Australia, salt is made from seawater in salt pans (Figure 15.6.5). The concentration of sodium chloride in seawater is approximately 35 g L^{-1}. Seawater is released into shallow ponds and the water evaporates due to heat from the Sun. As the volume of water decreases, the concentration of salt increases. Eventually, the solution becomes saturated and the salt begins to crystallise out. Australia's hot climate makes it very suitable for producing salt by evaporation, and it is a major exporter of salt.

FIGURE 15.6.5 Salt is produced by the evaporation of water from seawater.

FIGURE 15.6.6 Large crystals of copper(II) sulfate formed by slowly cooling a saturated solution of the compound

Variables affecting crystal growth

Several factors affect the formation of crystals from solutions. Some of these factors can affect the size of crystals formed.

Rate of cooling of a solution or molten compound

The rate of temperature change during cooling can affect the size of the crystals that are formed.

- Slow cooling generally produces large crystals.
- Rapid cooling produces smaller crystals.

This is true whether crystals are formed from a solution or as a molten compound cools below its melting point.

You can investigate this phenomenon in a laboratory. For example, a saturated solution of copper(II) sulfate can be cooled at different rates. Some of the solution could be cooled quickly by placing it in a refrigerator. Another sample could be cooled slowly by allowing it to sit in an insulated dish. The slower cooling results in larger crystals, as shown in the Figure 15.6.6.

Igneous rocks illustrate the effects of cooling rates on crystal size. The molten magma from the Earth's interior that forms these rocks can cool down quickly on the Earth's surface during volcanic activity, or very slowly underground. Figure 15.6.7 shows a section of granite, a rock that forms when magma cools below the Earth's surface. As it cools below the surface, the rate of temperature change is slow so granite contains large crystals.

On the other hand, basalt is a rock that forms when magma flows out from a vent in the Earth and over surrounding land. This results in rapid cooling of the magma and formation of small crystals. The basalt plains around Melbourne and much of southwestern Victoria were formed in this way (Figure 15.6.8). The bluestone paving and kerb stones used for many streets and lanes in the city and older suburbs of Melbourne were cut from these basalt deposits.

FIGURE 15.6.7 Large crystals (or grains) in granite

FIGURE 15.6.8 The rock columns at the Organ Pipes National Park near Melbourne are basalt extrusions.

Rate of evaporation of solvent

If crystals are formed by evaporation, then the rate of evaporation of the solvent will affect the size of the crystals. Generally, a faster rate of evaporation results in smaller crystals. This can be investigated in the laboratory by placing solutions of a chemical in different situations where the rate of evaporation of the solvent varies, for example, in a cupboard and on a windowsill.

Nucleation

Crystallisation occurs more rapidly if a 'nucleus' is present in the solution on which the new crystals can form. A nucleus can be provided by adding a single seed crystal of the compound itself. You saw an example of this in Figure 15.6.1 on page 369. Even a small solid impurity in the solution, such as a speck of dust or a scratch on the container wall, can act as a nucleus.

Nature of the compound

Different compounds produce different-shaped crystals. Figure 15.6.9 shows two different-shaped crystals. Sodium chloride crystals are cubic in shape; the crystals of alum (potassium aluminium sulfate, $KAl(SO_4)_2$) are octahedral.

FIGURE 15.6.9 (a) Cubic crystals of table salt (sodium chloride) and (b) octahedral crystals of alum, $KAl(SO_4)_2$

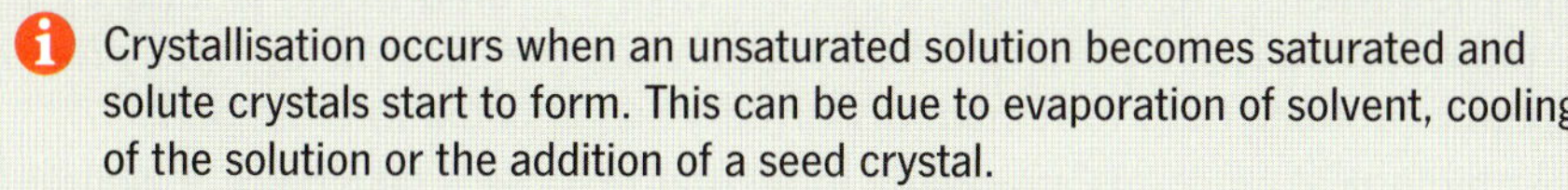

Crystallisation occurs when an unsaturated solution becomes saturated and solute crystals start to form. This can be due to evaporation of solvent, cooling of the solution or the addition of a seed crystal.

Calculations involving crystallisation

You can predict how much of a compound will crystallise out of a solution as temperature decreases by using a solubility curve. Worked example 15.6.3 shows how this can be done.

Worked example 15.6.3

CRYSTALLISATION

If 50 g of potassium nitrate (KNO_3) is dissolved in 100 g water at 40°C, what mass of KNO_3 crystals will form if the temperature is reduced to 20°C? You will need to refer to Figure 15.6.2 on page 370 to complete this question.

Thinking	Working
Identify the mass of solute dissolved in the original solution.	mass of KNO_3 in solution = 50 g
Find the maximum mass of solute that will remain dissolved in 100 g water at the final temperature.	The solubility curve shows that the maximum mass that will dissolve at 20°C is 32 g.
Calculate the mass of solute crystals that will form in the solution at the final temperature.	mass of crystals formed = original mass − remaining mass = 50 − 32 = 18 g

Worked example: Try yourself 15.6.3

CRYSTALLISATION

If 200 g of sucrose is dissolved in 100 g water at 20°C, what mass of sucrose crystals will form if the temperature is reduced to 10°C?

Uses of solubility curves in the workplace

Sometimes it is important to know how much of a solute will precipitate out of a solution as it cools to a particular temperature, or if all the solute will stay dissolved. Two examples, one from industry and one from biological research, illustrate how solubility curves can be used to help in situations like this.

Crystallising sugar from sugar cane syrup

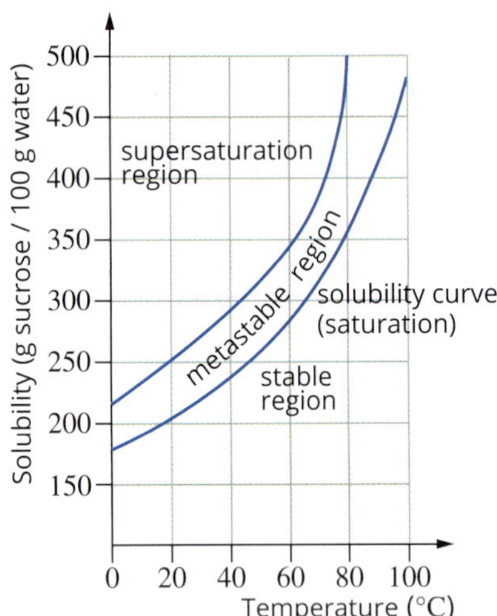

FIGURE 15.6.10 A solubility curve for sucrose used to predict the amount of sugar that will crystallise out under different conditions

In Australia, table sugar (sucrose) is prepared in large quantities from sugar cane. During processing, the cane is crushed to release juice. Impurities are removed and water is evaporated from the juice until the solution becomes saturated and the sugar begins to crystallise out. Chemists use solubility curves like the one in Figure 15.6.10 to predict the amount of sucrose that will crystallise at different temperatures and solubilities.

In this solubility curve, the area marked 'supersaturation region' indicates the conditions under which crystals will form spontaneously. In the 'metastable region', seed crystals must be added to induce crystallisation. In areas below the saturation curve, the solution is unsaturated and crystals will not form.

Research on therapeutic proteins

Therapeutic proteins are compounds that are manufactured in the laboratory for use as pharmaceuticals. Synthetic insulin is an example of a therapeutic protein. Therapeutic proteins need to remain dissolved when injected into the body.

Most biological reactions and many industrial processes take place in solution. Laboratory workers in medical, pharmaceutical and industrial fields need to know the temperature ranges at which compounds remain dissolved. If the temperature of the solutions they are using does not stay within a certain range, then reactants may crystallise out and interfere with the desired reactions.

Solubility curves allow researchers to make accurate predictions about the solubility at different temperatures of the compounds under investigation. They can then ensure that work will not be disrupted by unexpected and unwanted crystallisation.

SOLUBILITY OF LIQUIDS AND GASES IN WATER

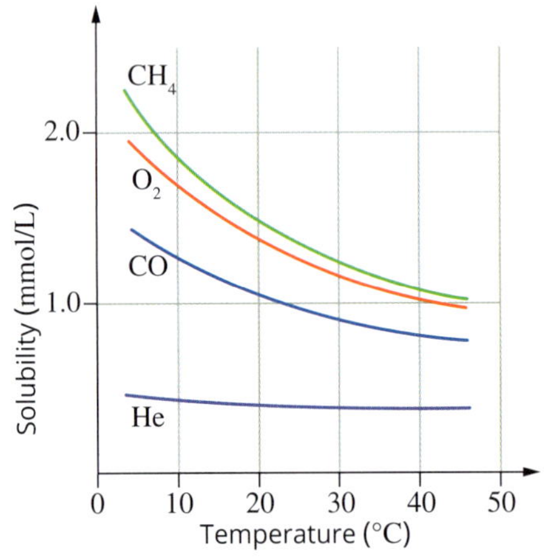

FIGURE 15.6.11 Solubility curves for some gases

It is generally true that as temperature increases, the solubility of solids in solution increases. However, the solubility of liquids in other liquids does not show clear trends in solubility as the temperature changes.

Gases generally become less soluble as the temperature increases. The graph in Figure 15.6.11 shows the solubility of different gases in water at varying temperatures.

Table 15.6.2 shows the solubility of some common gases in water at different temperatures. For each gas, it can be seen that as temperature increases, less gas is able to dissolve in water.

TABLE 15.6.2 Solubility of some gases in water at different temperatures

Gas	Solubility (g of gas per kg of water) at 0°C	Solubility (g of gas per kg of water) at 20°C	Solubility (g of gas per kg of water) at 60°C
oxygen	0.069	0.043	0.023
carbon dioxide	3.4	1.7	0.58
nitrogen	0.029	0.019	0.011
methane	0.040	0.023	0.011
ammonia	897	529	168

FIGURE 15.6.12 Bubbles of air appear as water is heated. This is because solubility of gases decreases at higher temperatures.

FIGURE 15.6.13 Soft drinks left standing in the sunshine go 'flat' quickly.

In Figure 15.6.12, you can see the bubbles that appear when water is first heated are not steam (water vapour) but air bubbles. Some of the air that was dissolved in the water comes out of solution as the temperature increases and the gas becomes less soluble.

You may have noticed that soft drinks will go 'flat' more quickly if they are left standing in the sunshine on a hot day than if they are in a cold refrigerator (Figure 15.6.13). The dissolved carbon dioxide in the drink comes out of the solution as the drink heats up.

The effect of temperature on the solubility of a gas can have environmental implications. If the temperature of the water in rivers, lakes and oceans increases even slightly, it will contain less dissolved oxygen. This can have serious consequences for oxygen-breathing aquatic organisms, such as the Australian Murray River cod.

FIGURE 15.6.14 Fish, such as this native Australian Murray River cod, are susceptible to lowered levels of oxygen in natural waters.

Therefore, hot water from power stations and other industries must be cooled before it can be discharged into waterways. Even a small increase in water temperature can cause the oxygen concentration in the water to drop below levels that are necessary for aquatic life to survive.

15.6 Review

SUMMARY

- Solubility is a measure of how much solute will dissolve in a given amount of solvent at a specified temperature.
- Solubility curves show how the solubility of a compound changes with temperature.
- Solubility curves can be used to calculate the amount of a substance that will dissolve in a given amount of solvent at a specified temperature.
- The solubility of most solids in water increases as the temperature increases.
- Crystallisation occurs when a solution is cooled to the point where it becomes saturated with solute and the solute begins to form solid crystals.
- As the temperature of a solution increases, the solubility of solids generally increases and the solubility of gases decreases. Liquids show no overall trend in solubility with temperature.
- Crystallisation can occur from a solution when:
  - the temperature of the solution decreases
  - some of the solvent evaporates.
- Factors that can affect crystallisation, including crystal size, are:
  - rate of cooling of solution
  - rate of evaporation of solvent
  - nucleation
  - nature of the compound.
- Solubility curves can be used to predict the amount of a compound that will precipitate from solution at a particular temperature.

KEY QUESTIONS

1 To form a saturated solution, 35 g of sodium chloride is dissolved into 100 g of water at 18°C. A number of different solutions were made by dissolving the following masses of sodium chloride in 50 g of water at 18°C. Which solution would be supersaturated?

A 9 g
B 3.5 g
C 20 g
D 17.5 g

2 Look at graph below and determine which coloured points correspond to saturated, unsaturated and supersaturated solutions of potassium nitrate.

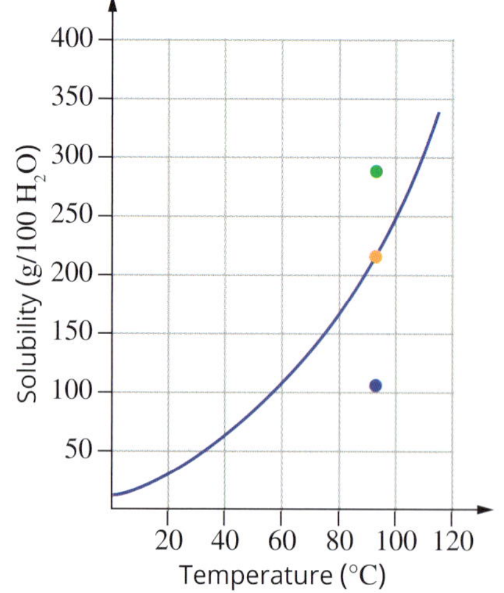

Refer to Figure 15.6.2 on page 370 to answer Questions **3**–**7**.

3 Find the mass of solute that could dissolve in the following situations.

a $CuSO_4{\cdot}5H_2O$ in 100 g of water at 60°C
b $AgNO_3$ in 100 g of water at 10°C
c KNO_3 in 50 g of water at 24°C

4 What mass of the following compounds will dissolve in 200 g of water at 30°C to form a saturated solution?

a $CuSO_4{\cdot}5H_2O$
b $NaNO_3$
c NaCl

5 A sample of 140 g of potassium nitrate (KNO_3) is dissolved in 100 g of water at 70°C. The temperature of the solution is then cooled to 55°C.
What mass of KNO_3 will crystallise out of solution?

A 33 g
B 40 g
C 100 g
D 140 g

6 Calculate the mass of crystals that will form when a solution containing 60 g of $CuSO_4{\cdot}5H_2O$ in 100 g of water at 90°C is cooled to 60°C.

7 A sample of 55 g of sodium nitrate is dissolved in 50 g of water at 80°C. What mass of sodium nitrate will remain dissolved if the solution is cooled to 50°C?

8 Explain why granite crystals are a different size from basalt crystals.

Chapter review

15

KEY TERMS

aqueous
artesian basin
bore water
chemical contaminant
chlorination
crystallisation
crystallise
desalinated seawater
desalination
dissociate
dissociation
dissolution
floc
flocculation
fluoridation
Great Artesian Basin
groundwater
heat capacity
heavy metal
heterogeneous
homogeneous
hydrated
hydride
hydronium ion
ion–dipole attraction
ionise
latent heat
latent heat of fusion
latent heat of vaporisation
miscible
non-polar solvent
osmosis
polar solvent
potable water
protected catchment
recycled water
reverse osmosis
saturated solution
seed crystal
settling
solubility
solubility curve
solubility table
solute
solution
solvent
specific heat capacity
sterile
supersaturated solution
unsaturated solution
water cycle

Essential water

1 Copy and complete the following diagram relating to drinking water and water quality.

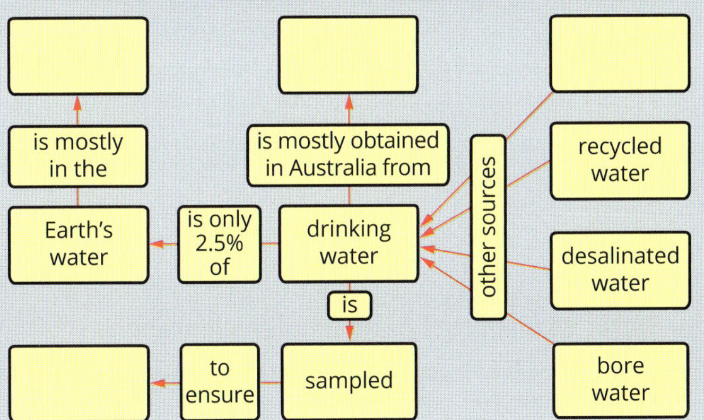

2 Where is most of the drinkable water found on Earth?
A ice
B rivers
C groundwater
D bore water

3 What is the drinking water that comes from groundwater called?

4 What is the source of drinking water for most Australians and why?

5 Arrange the following sources of water on Earth in order from highest volume to lowest volume: oceans, groundwater, rivers, soil moisture, ground ice and permafrost, lakes, ice caps and glaciers, atmosphere as water vapour.

6 Imagine that you are an environmental scientist and have been asked to develop a water-sampling program for a large reservoir. The quality of the water is to be tested over a month. What are some of the matters that should be addressed before starting your program?

7 Which one of the following factors does not need to be considered when sampling water for analysis?
A temperature
B depth
C distance from the shore
D humidity

8 A water sample was tested for heavy metals. The results are shown in the table below. Classify which of the concentrations are above and which are below the Australian Drinking Water Guidelines acceptable levels (Table 15.1.5 on page 337).

Heavy metal	Concentration ($mg\,L^{-1}$)
arsenic	0.009
cadmium	2.2×10^{-4}
copper	1.9
lead	0.1
mercury	0.9×10^{-2}

9 Give a possible source of error in each of the following steps in sampling water from a lake.
a Determine sampling location.
b Collect sample.
c Transfer sample to a container.
d Store sample.
e Transport sample.

Properties of water

10 Copy and complete the following sentences about water and its properties.

Water is a ____________ molecule. Within a single molecule, hydrogen and oxygen atoms are held together by strong ________ _________. Between different molecules, the most significant forces are __________ __________.

It is the relatively _______ strength of the intermolecular forces that gives water its unique properties of:

- relatively ________ boiling point, _______°C
- relatively _______ latent heat values, _____ kJ mol^{-1} and _____ kJ mol^{-1}
- relatively _____________ specific heat capacity, _______ J g^{-1} °C^{-1}.

11 The image below shows the arrangement of water molecules in ice. Identify the parts of the diagram labelled **a–d**.

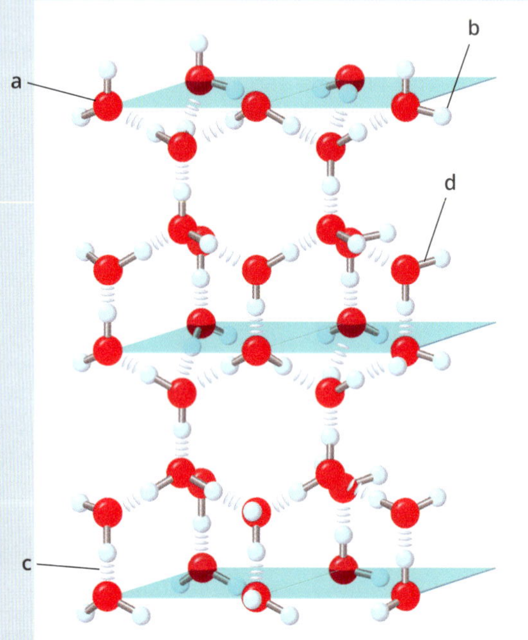

12 Water boils at 100°C. However, a much higher temperature (1000°C) is needed to decompose water molecules into hydrogen gas and oxygen gas.

a Using water as an example, explain the meaning of the terms 'intermolecular' and 'intramolecular' forces.

b Which of the two types of forces described in part **a** is stronger? Justify your answer by using the information at the beginning of the question.

13 Hydrogen sulfide and water are both group 16 hydrides. Explain why water exhibits a much higher boiling point than hydrogen sulfide.

14 In your own words, discuss the significance of polarity and hydrogen bonding in relation to the high boiling point of water.

15 An evaporative air conditioner is used to cool the air inside some buildings. The air conditioner lowers the air temperature by using heat from the air to evaporate water. Select the property that is most important when water is used in an evaporative air conditioner.

16 When heat energy is applied to a substance, the temperature increase of the substance depends on which of the following?

A the mass of the substance

B the types of bonds present in the substance

C the amount of heat energy transferred to the substance

D all of the above

Water as a solvent

17 Many of the solvent properties of water depend on its structure and bonding. List all the bonding present in a sample of water from weakest to strongest. Remember to include all intramolecular and intermolecular bonding.

18 Explain why water is such a good solvent for polar and ionic substances.

19 Nitrogen gas (N_2) and ethene (C_2H_6) are both insoluble in water, whereas ethanol (C_2H_5OH) is soluble. Refer to the bonding in each of the molecules to explain these differences.

20 Explain, with reference to its structure and bonding, why octane does not dissolve in water.

Water as a solvent of molecular substances

21 Classify the following substances according to the way they dissolve in water: $C_6H_{12}O_6$, HI, I_2, C_2H_4, C_3H_7OH, HNO_3, CH_4.

The structures of $C_6H_{12}O_6$ and C_3H_7OH are shown to assist you.

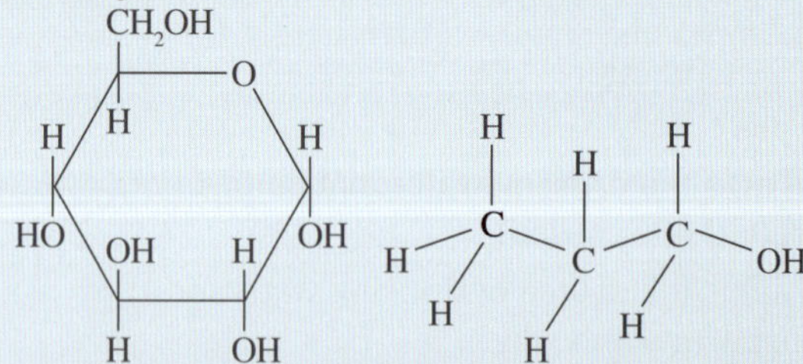

22 A student conducts an experiment to investigate the solubility of CH_4 and CH_3OH molecules in three different solvents, X, Y and Z. The student finds that the CH_3OH molecules dissolve well in solvent Y but do not dissolve in solvent X. The student also finds that CH_4 partially dissolves in solvent Z and completely dissolves in solvent X. List each of the solvents in order from most polar to least polar. Explain your reasoning.

23 DDT is a hazardous agricultural insecticide, which has been banned in many countries. It is only slightly soluble in water but is very soluble in fats and oils, so accumulates in the fat deposits of animals. What can you deduce about the polarity of the DDT molecule from its solubility characteristics?

Water as a solvent of ionic compounds

24 **a** What is the name given to the process that ionic solids undergo when dissolving in water?
b What ions will be produced when the following compounds are added to water?
i $Cu(NO_3)_2$
ii $ZnSO_4$
iii $(NH_4)_3PO_4$

25 What ions would be produced when the following compounds are added to water?
a potassium carbonate
b lead(II) nitrate
c sodium hydroxide
d sodium sulfate
e magnesium chloride
f zinc nitrate
g potassium sulfide
h iron(III) nitrate

26 Write equations to show the dissociation of the following compounds when they are added to water.
a magnesium sulfate
b sodium sulfide
c potassium hydroxide
d copper(II) ethanoate
e lithium sulfate

27 What particles are present in the solution that forms when HCl gas is bubbled through water?

28 Briefly explain why the arrangement of water molecules around dissolved magnesium ions is different from that around dissolved chloride ions.

29 Write the formulae of three carbonate compounds that are:
a soluble in water
b insoluble in water.

30 Write the formulae of three sulfate compounds that are:
a soluble in water
b insoluble in water.

31 Briefly describe what happens to the forces between solute and solvent substances when an ionic substance such as potassium bromide dissolves in water.

Solubility

32 What mass of the following compounds will dissolve in 50 g of water at 30°C to form a saturated solution?
a $NaNO_3$
b $CuSO_4{\cdot}5H_2O$
c KNO_3

33 What is the maximum mass of solute that will dissolve in the following situations?
a sucrose in 500 g of water at 20°C
b $CuSO_4{\cdot}5H_2O$ in 40 g of water at 60°C

34 If 160 g of KNO_3 were dissolved in 200 g water at 60°C, what mass of crystals would form if the temperature were decreased to 30°C?

35 What effect does increasing the temperature have on the concentration of gases available to aquatic animals and plants?

36 A gas forms a saturated solution when 0.50 mol dissolves in 500 mL water at 23°C. Use Figure 15.6.11 on page 376 to decide what the gas is most likely to be.

37 If 2.0 kg of water absorbed 1200 g of ammonia at 10°C, what mass of the gas would leave the solution if the temperature of the solution were increased to 20°C? (Refer to Table 15.6.2 on page 376.)

Connecting the main ideas

38 Match the type of compound with the way it is likely to behave in water.
i ionic compound
ii compound composed of polar molecules with –OH groups
iii compound composed of small polar molecules in which a hydrogen atom is covalently bonded to an atom of a group 17 element
iv non-polar molecular compound
v compound composed of covalent molecules with a large non-polar end and one –OH group.
a does not dissolve in water because a large proportion of the molecule is non-polar
b dissolves in water by ionising, then forming ion–dipole bonds with water
c does not dissolve in water
d dissolves in water by forming hydrogen bonds with water molecules
e dissolves in water by dissociating, then forming ion–dipole bonds with water

39 Give concise explanations for the following observations.
a Ammonia (NH_3) and methane (CH_4) are both covalent molecular substances. Ammonia is highly soluble in water, but methane is not.
b Glucose ($C_6H_{12}O_6$) and common salt (NaCl) are very different compounds. Glucose is a covalent molecular substance, whereas common salt is ionic, yet both of these substances are highly soluble in water.

CHAPTER

16 Aqueous solutions

Aqueous solutions are found all around you. The Earth's oceans, rivers and lakes are aqueous solutions. They contain dissolved minerals and gases. The plasma of human blood and the sap of plants are aqueous solutions carrying dissolved nutrients and wastes. Each body cell contains aqueous solutions. Even rain contains small quantities of dissolved gases and other materials.

In fields such as medicine, pharmaceutical manufacturing, and even food preparation, it is very important to know how much of a compound is present in a solution.

In this chapter, you will learn how to determine how much of a substance will dissolve in a particular solvent. You will also explore the nature of aqueous solutions and learn how to calculate the amount of a compound dissolved in a given solution.

Science understanding

- the presence of specific ions in solutions can be identified by observing the colour of the solution, flame tests and observing various chemical reactions, including precipitation and acid–base reactions
- solutions can be classified as saturated, unsaturated or supersaturated; the concentration of a solution is defined as the quantity of solute dissolved in a quantity of solution; this can be represented in a variety of ways, including by the number of moles of the solute per litre of solution ($mol\,L^{-1}$) and the mass of the solute per litre of solution ($g\,L^{-1}$) or parts per million (ppm)
- the mole concept can be used to calculate the mass of solute, and solution concentrations and volumes involved in a chemical reaction

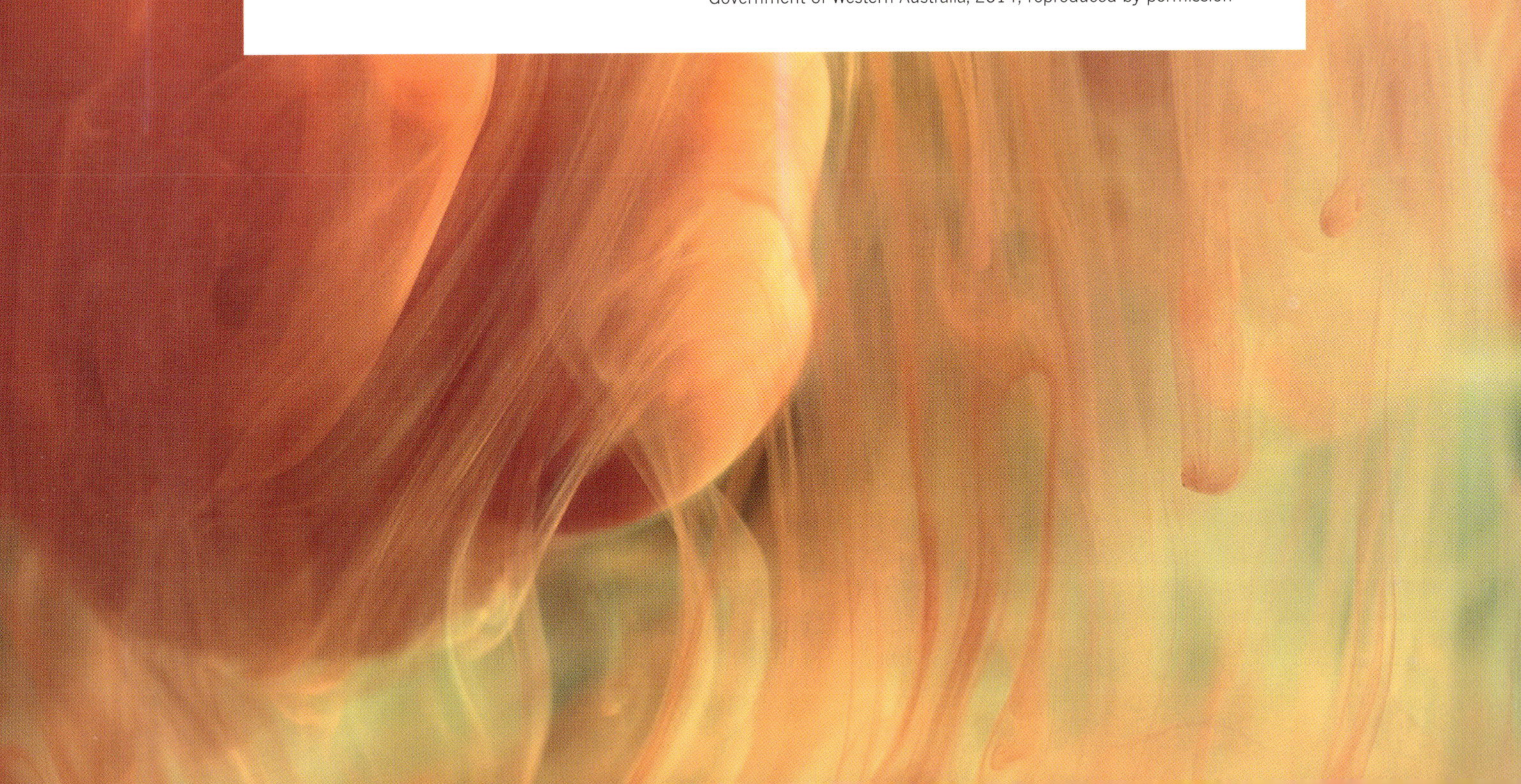

16.1 Precipitation reactions

A **precipitation reaction** occurs if ions in solution combine to form a new compound that is insoluble in water. The insoluble compound formed in such a reaction is called a **precipitate**.

Precipitation reactions occur naturally in undersea hydrothermal vents. The vents release superheated solutions containing sulfides, which then combine with metal ions to form precipitates of mineral sulfides, creating the chimney-like structures seen in Figure 16.1.1. The areas around these chimneys are biologically rich, often hosting complex communities fuelled by the chemicals dissolved in the vent fluids.

FIGURE 16.1.1 Undersea hydrothermal vents release superheated water containing sulfides, which form precipitates with metal ions.

FIGURE 16.1.2 Mixing aqueous solutions of sodium chloride and silver nitrate produces a solid, called a precipitate.

Precipitation reactions are used to remove minerals from drinking water, to remove heavy metals from waste water and in the purification plants of reservoirs.

In this section you will look at what takes place during precipitation reactions.

DEDUCING THE IDENTITY OF A PRECIPITATE

When a colourless solution of silver nitrate is mixed with a colourless solution of sodium chloride, a white solid is formed, as shown in Figure 16.1.2. The white solid is the precipitate, but what is its composition? What is the solid compound that forms?

To understand what happens in the reaction between silver nitrate and sodium chloride, you need to identify the ions present in the reactant solutions and how they interact with each other.

- In the silver nitrate solution, there are silver ions (Ag^+) and nitrate ions (NO_3^-).
- In the sodium chloride solution, there are sodium ions (Na^+) and chloride ions (Cl^-).
- When one solution is added to the other, the mixture formed will contain all of the ions.

In both of these solutions, all the ions are moving around independently. As the ions move in the solution, they will collide with one another. If cations (positive ions) and anions (negative ions) collide, they may join together to form a new, insoluble precipitate.

Two new combinations of cations and anions are possible:

- sodium and nitrate ions (to form sodium nitrate)
- silver and chloride ions (to form silver chloride).

Tables 15.5.1 and 15.5.2 on page 367 indicate that sodium nitrate is soluble in water, but that silver chloride is insoluble. Therefore, the precipitate must be silver chloride.

The process for the precipitation reaction between sodium chloride and silver nitrate is shown in Figure 16.1.3. When the hydrated Ag^+ and Cl^- ions come into contact, the attraction between the ions is greater than that of the ions for the water molecules. An ionic lattice of AgCl is formed.

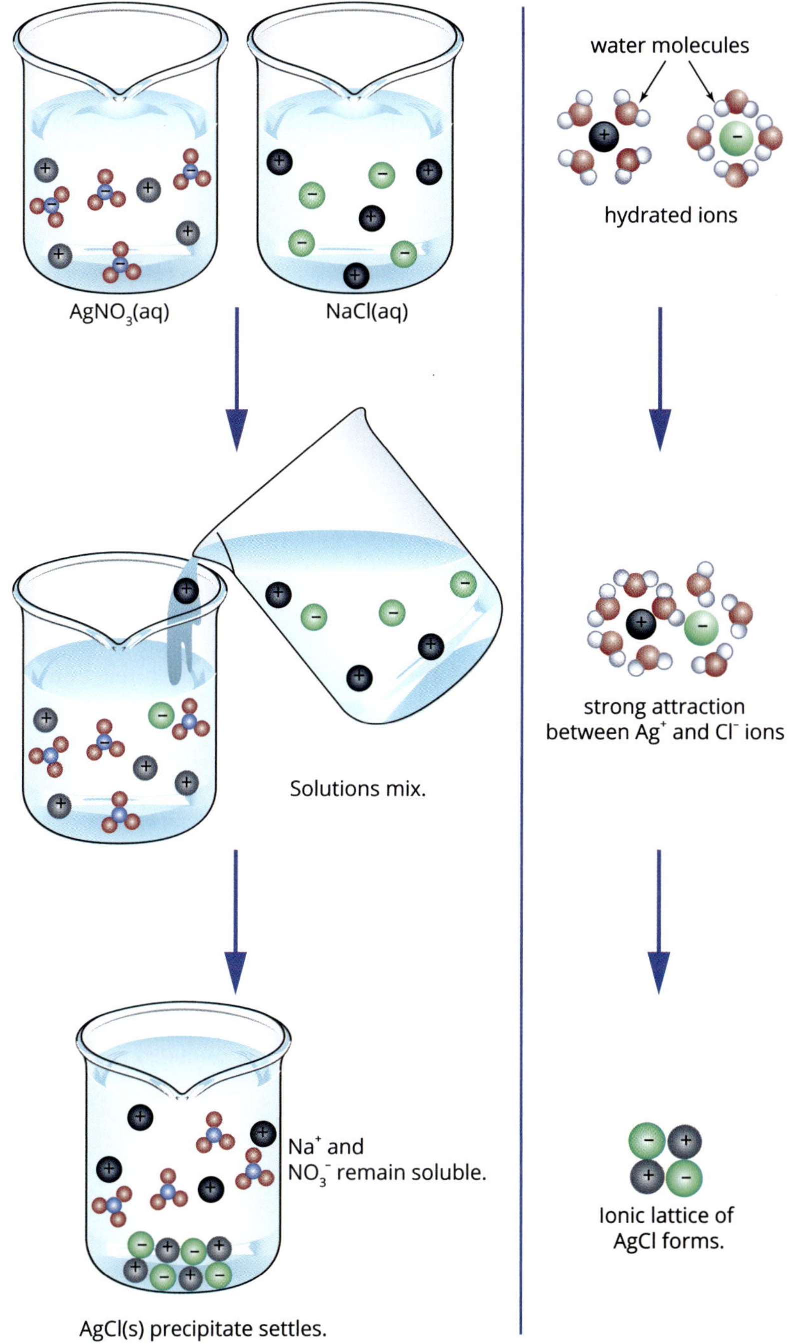

FIGURE 16.1.3 Pictorial representation of mixing aqueous solutions of sodium chloride and silver nitrate to produce a precipitate of silver chloride

There is a simple way that allows you to work out which compound will form the precipitate in a reaction between two ionic solutions, shown in column 2 of Figure 16.1.4.

1 Write down the formula for the cation of one of the compounds, followed by its anion. Repeat the process for the second compound. For example, Ag^+, NO_3^-, Na^+, Cl^-.
2 Then draw two lines. The first line joins the cation of the first solution to the anion of the second. The second line joins the anion of the first solution to the cation of the second.
3 Next, combine the ions to write the formulae for the two possible precipitates, if any.
4 Finally, use solubility tables to work out which of the two combinations of ions will result in an insoluble compound. This will be the precipitate. The other ions will remain in solution.

	Ions in solution	Possible precipitates	Actual precipitate, if any, from solubility tables
solution 1	Ag^+ and NO_3^-	AgCl or $NaNO_3$	AgCl
solution 2	Na^+ and Cl^-		

FIGURE 16.1.4 Process for identifying the precipitate, if any, when two aqueous ionic solutions are mixed

It is important for chemists to be able to predict whether a precipitate will form in a reaction and what this precipitate will be. Worked example 16.1.1 takes you through the process of predicting the products of a precipitation reaction.

Worked example 16.1.1

PREDICTING THE PRODUCTS OF A PRECIPITATION REACTION

What precipitate, if any, will be produced when solutions of potassium hydroxide and lead(II) nitrate are added together? You will need to refer to the solubility tables (Tables 15.5.1 and 15.5.2 on page 367 or in your data booklet) to complete this question.

Thinking	Working
Identify which ions are produced by each of the ionic compounds in the mixture.	(table below)

	Ions in solution	Possible precipitates	Actual precipitate, if any, from solubility tables
solution 1	K^+ and OH^-		
solution 2	Pb^{2+} and NO_3^-		

Thinking	Working
Identify which two new combinations of cations and anions are possible in the mixture of the solutions.	(table below)

	Ions in solution	Possible precipitates	Actual precipitate, if any, from solubility tables
solution 1	K^+ and OH^-	KNO_3 or $Pb(OH)_2$	
solution 2	Pb^{2+} and NO_3^-		

Use the solubility tables to check which, if any, of these combinations will produce an insoluble compound.	

	Ions in solution	Possible precipitates	Actual precipitate, if any, from solubility tables
solution 1	K^+ and OH^-	KNO_3 or $Pb(OH)_2$	$Pb(OH)_2$
solution 2	Pb^{2+} and NO_3^-		

Compounds containing potassium ions are usually soluble, so potassium nitrate will not form a precipitate.

Compounds containing hydroxide ions are usually insoluble, so lead(II) hydroxide will form a precipitate.

Worked example: Try yourself 16.1.1

PREDICTING THE PRODUCTS OF A PRECIPITATION REACTION

What precipitate, if any, will be produced when solutions of sodium sulfide (Na_2S) and copper(II) nitrate ($Cu(NO_3)_2$) are added together? You will need to refer to the solubility tables (Tables 15.5.1 and 15.5.2 on page 367 or in your data booklet) to complete this question.

CHEMFILE

Limescale accumulation

Have you ever wondered where that flaky white build-up on the element of your kettle comes from?

When you boil water in the kettle, ions present in the water can precipitate out, leaving a white coating called limescale on the element (Figure 16.1.5).

FIGURE 16.1.5 The accumulation of limescale on domestic kettles is the result of precipitation of calcium carbonate as water is repeatedly boiled.

The amount of build-up depends on the type of water treatment in your area. Areas that have hard water (i.e. high levels of dissolved ions) have a bigger problem with limescale.

Limescale mostly consists of calcium carbonate ($CaCO_3$) that precipitates out as a crystalline solid when the water is boiled. High levels of limescale can build up in pipes and eventually restrict or even block the flow of water (Figure 16.1.6).

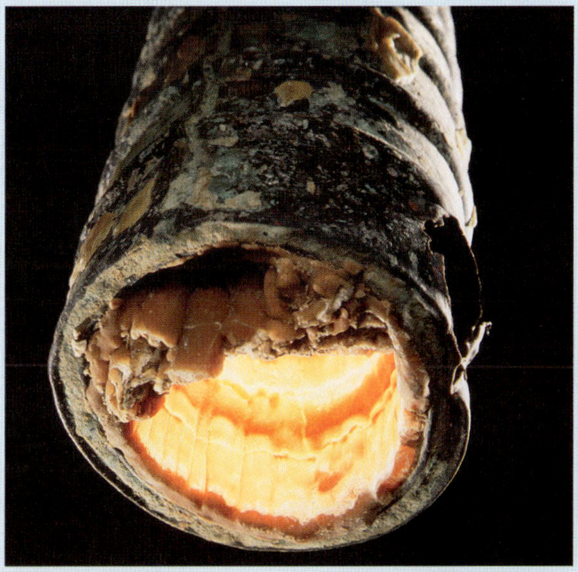

FIGURE 16.1.6 Accumulation of limescale in pipes is a result of precipitation of $CaCO_3$.

Limescale can be a big problem in the home. A coating as thin as 1.5 mm over a heating element can reduce its efficiency by as much as 12%.

Many modern houses use ion filters or water conditioners to remove dissolved ions from the water and reduce the accumulation of limescale.

WRITING IONIC EQUATIONS FOR PRECIPITATION REACTIONS

For any reaction, including precipitation reactions, the most accurate way to represent the reaction is to show only those species (ions or molecules) that participate in the reaction. For the reaction between silver nitrate and sodium chloride solutions, a **full equation** with the complete formulae of all the substances shown in the reaction can be written. The state symbols for each of the species in the chemical reaction must be shown:

$$AgNO_3(aq) + NaCl(aq) \rightarrow AgCl(s) + NaNO_3(aq)$$

However, the essential feature of the reaction between silver nitrate and sodium chloride is the combination of silver ions and chloride ions to form a precipitate. This reaction is more accurately summarised in an **ionic equation**:

$$Ag^+(aq) + Cl^-(aq) \rightarrow AgCl(s)$$

Note that spectator ions are not included in an ionic equation. Only the species that change are included. Following the process outlined in Figure 16.1.4, writing the ionic equation requires writing on the reactant side the formulae of the ions joined by the line between the ions in the precipitate and the formula for the precipitate on the product side, and then balancing the equation if needed.

Another way to write an ionic equation for a precipitation reaction is shown in the following example of the reactions between solutions of aluminium nitrate and sodium sulfide. The full equation for the precipitation reaction is:

$$2Al(NO_3)_3(aq) + 3Na_2S(aq) \rightarrow Al_2S_3(s) + 6NaNO_3(aq)$$

To write an ionic equation, use the following steps.

1 Write down the formula of the precipitate on the right-hand side of the page. Include a symbol of state. Place an arrow to the left of it.

$$\rightarrow Al_2S_3(s)$$

2 To the left of this formula, add the formulae of the ions that form the precipitate, in the ratio shown by the formula of the precipitate. Include state symbols.

$$2Al^{3+}(aq) + 3S^{2-}(aq) \rightarrow Al_2S_3(s)$$

3 Check that the equation is balanced.

Worked example 16.1.2

WRITING EQUATIONS FOR PRECIPITATION REACTIONS

Write a balanced ionic equation for the reaction between iron(III) nitrate and sodium sulfide, in which the precipitate is iron(III) sulfide. Identify the spectator ions in this reaction.	
Thinking	**Working**
Identify the ions present in each solution.	Iron(III) nitrate solution has $Fe^{3+}(aq)$ and $NO_3^-(aq)$ ions. Sodium sulfide solution has $Na^+(aq)$ and $S^{2-}(aq)$ ions.
Write down the formula of the precipitate on the right-hand side. Include the state symbol (s). Place an arrow to the left of it.	$\rightarrow Fe_2S_3(s)$
To the left of this formula, add the formulae of the ions that form the precipitate, in the ratio shown by the formula of the precipitate. Include state symbols. Check the equation is balanced.	$2Fe^{3+}(aq) + 3S^{2-}(aq) \rightarrow Fe_2S_3(s)$
Write the formulae of the ions that do not form a precipitate in the reaction. These are the spectator ions.	$Na^+(aq)$ and $NO_3^-(aq)$ are spectator ions.

Worked example: Try yourself 16.1.2

WRITING EQUATIONS FOR PRECIPITATION REACTIONS

Write a balanced ionic equation for the reaction between copper(II) sulfate and sodium hydroxide, in which the precipitate is copper(II) hydroxide. Identify the spectator ions in this reaction.

Worked example 16.1.3

IDENTIFYING PRECIPITATES AND WRITING EQUATIONS FOR PRECIPITATION REACTIONS

Identify the precipitate, if any, when solutions of copper(II) chloride and sodium hydroxide are mixed. Write the balanced ionic equation for the formation of any precipitate. Identify the spectator ions in any reaction.

Thinking	Working
Identify the ions present in each solution.	Copper(II) chloride solution has Cu^{2+}(aq) and Cl^-(aq) ions. Sodium hydroxide solution has Na^+(aq) and OH^-(aq) ions.
Identify the possible products.	copper(II) hydroxide and sodium chloride
Use the solubility rules to determine if one of the products is insoluble.	Yes: copper(II) hydroxide
Write down the formula of the precipitate on the right-hand side. Include the state symbol (s). Place an arrow to the left of it.	$\rightarrow Cu(OH)_2(s)$
To the left of this formula, add the formulae of the ions that form the precipitate, in the ratio shown by the formula of the precipitate. Include state symbols. Check the equation is balanced.	$Cu^{2+}(aq) + 2OH^-(aq) \rightarrow Cu(OH)_2(s)$
Write the formulae of the ions that do not form a precipitate in the reaction. These are the spectator ions.	Na^+(aq) and Cl^-(aq) are spectator ions.

Worked example: Try yourself 16.1.3

IDENTIFYING PRECIPITATES AND WRITING EQUATIONS FOR PRECIPITATION REACTIONS

Identify the precipitate, if any, when solutions of potassium carbonate and copper(II) sulfate are mixed. Write the balanced ionic equation for the formation of any precipitate. Identify the spectator ions in any reaction.

CHEMISTRY IN ACTION

The chemistry of colour

If you have ever walked through an art supplies store, you will have noticed displays selling paints with names that sound as though they belong in a chemistry laboratory. Names such as lead yellow, titanium white and cobalt blue are just a few of the colours still in use today (Figure 16.1.7).

The names are not just for show. The colours that they represent have their basis in chemistry. All paints consist of a pigment (the colour) and a binder that holds the pigments in a suspension. Historically, paints were purchased as powdered pigments like those in Figure 16.1.8.

Artists would mix these pigments with their own binders, such as linseed oil, to produce the required paints. Ancient civilisations such as the Egyptians would grind minerals like lapis lazuli and ochre from the earth to make pigments. Early alchemists were able to manufacture synthetic pigments using precipitation reactions, collecting the coloured precipitates through filtration.

Prussian blue was a very popular colour through the 19th century and is made from a precipitate of iron(III) hexacyanoferrate(II). It has been used extensively in ceramics and painting, including the famous Japanese block print from the 1830s *The Great Wave off Kanagawa* (Figure 16.1.9).

FIGURE 16.1.7 Paint colours are often named after the pigments that were once used to make them. Many of these pigments were the product of precipitation reactions.

FIGURE 16.1.8 An array of different coloured pigments. Historically, pigments for paints were collected from minerals and other sources. However, in the 19th century a better knowledge of chemistry was used to manufacture pigments more reliably and in a larger variety of colours.

FIGURE 16.1.9 *The Great Wave off Kanagawa* (c. 1830–1833). This woodblock print is by the Japanese artist Hokusai (1760–1849). The deep blue tones were painted with a paint known as Prussian blue made from iron(III) hexacyanoferrate(II).

Pigments based on ionic precipitates provided a much wider array of colours to artists. This paved the way for the more flamboyant use of colour seen in the work of the French impressionists. Cadmium sulfide was used to make red pigments and cadmium red is still used widely today. Viridian, a deep green pigment, is manufactured from chromium(III) oxide dihydrate.

So the next time you visit an art gallery, take a moment to think about the complex chemistry that went into bringing those colours to life.

16.1 Review

SUMMARY

- A precipitation reaction occurs when two solutions of compounds are mixed and a solid product is formed. The solid product is called a precipitate.
- Solubility tables can be used to predict which compound, if any, will precipitate in a reaction.
- Ions that are not directly involved in the formation of the precipitate are called spectator ions.
- Full and ionic equations can be written for precipitation reactions.
- Precipitation reactions are most accurately represented by ionic equations.
- Ionic equations do not include spectator ions.

KEY QUESTIONS

1 Use the solubility tables to identify the precipitate formed, if any, when the following solutions are mixed.
  a silver nitrate and potassium carbonate
  b potassium hydroxide and lead(II) nitrate
  c magnesium chloride and sodium sulfide
  d sodium chloride and iron(II) sulfate

2 **a** Name the precipitate formed when aqueous solutions of the following compounds are mixed.
    i K_2S and $MgCl_2$
    ii $CuCl_2$ and $AgNO_3$
    iii KOH and $AlCl_3$
    iv $MgSO_4$ and NaOH
  b Write a balanced ionic equation for each reaction.

3 Write an ionic equation for any precipitate formed when the following aqueous soutions are mixed.
  a $AgNO_3(aq) + NaI(aq) \rightarrow$
  b $CuSO_4(aq) + NaCl(aq) \rightarrow$
  c $(NH_4)_2SO_4(aq) + BaCl_2(aq) \rightarrow$
  d $K_2SO_4(aq) + Pb(NO_3)_2(aq) \rightarrow$
  e $CaCl_2(aq) + Na_3PO_4(aq) \rightarrow$
  f $NaOH(aq) + Pb(NO_3)_2(aq) \rightarrow$

4 Where a precipitate formed in Question **3**, identify the spectator ions.

16.2 Concentration of solutions

The **concentration** of a solution describes the relative amount of solute and solvent present in the solution. A solution in which the ratio of solute to solvent is high is said to be concentrated. Cordial that has not had any water added to it is an example of a **concentrated solution**. A solution in which the ratio of solute to solvent is low is said to be a **dilute solution**. A quarter of a teaspoon of sugar dissolved in a litre of water will produce a dilute solution that tastes slightly sweet.

'Concentrated' and 'dilute' are general terms. However, sometimes you need to know the actual concentration of a solution—the exact ratio of solute to solvent. The use of exact solution concentrations is important in the prescription of medicines, chemical manufacturing and chemical analysis.

UNITS OF CONCENTRATION

Chemists use different measures of concentration depending on the situation. The most common measures describe the amount of solute in a given amount of solution.

Some examples of units of concentration are:

- mass of solute per litre of solution (grams per litre, $g\,L^{-1}$)
- moles of solute per litre of solution (moles per litre, $mol\,L^{-1}$). (You will look more closely at this unit of concentration in Section 16.3.)
- parts per million (ppm)
- percentage by mass.

Units of concentration measured in this way have two parts.

- The first part gives information about how much solute there is.
- The second part gives information about how much solution there is.

For example, if a solution contains sodium chloride (NaCl) with a concentration of $17\,g\,L^{-1}$, then (first part) 17 g of NaCl is dissolved in every (second part) 1 L of the solution.

You will now look at how to perform calculations of concentration using different units.

Concentration in grams per litre ($g\,L^{-1}$)

The concentration of a solution in grams per litre ($g\,L^{-1}$) indicates the mass, in grams, of solute dissolved in 1 litre of the solution. For example, if the concentration of sodium chloride in seawater is $20\,g\,L^{-1}$, this means that in 1 L of seawater there is 20 g of sodium chloride. A formula used to calculate the concentration in $g\,L^{-1}$ is:

$$\text{concentration } (g\,L^{-1}) = \frac{\text{mass of solute (in g)}}{\text{volume of solution (in L)}}$$

Worked example 16.2.1

CALCULATING CONCENTRATION IN $g\,L^{-1}$ (GRAMS PER LITRE)

Calculate the concentration, in $g\,L^{-1}$, of a solution containing 8.00 g of sodium chloride in 500 mL of solution.	
Thinking	**Working**
Change the volume of solution so it is expressed in litres.	$500\,mL = \frac{500}{1000} = 0.500\,L$
Calculate the concentration in $g\,L^{-1}$.	$c = \frac{\text{mass of solute (in g)}}{\text{volume of solution (in L)}} = \frac{8.00}{0.500} = 16.0\,g\,L^{-1}$

Worked example: Try yourself 16.2.1

CALCULATING CONCENTRATION IN $g\,L^{-1}$ (GRAMS PER LITRE)

Calculate the concentration, in $g\,L^{-1}$, of a solution containing 5.00 g of glucose in 250 mL of solution.

Concentration in parts per million (ppm)

When very small quantities of solute are dissolved to form a solution, the concentration is often measured in **parts per million (ppm)**. For example, the concentration of mercury in fish that is safe for consumption is usually expressed in parts per million. The maximum concentration allowed for sale in Australia is 1 ppm (Figure 16.2.1).

FIGURE 16.2.1 Fish sold in Australia must have no more than 1 ppm of mercury (Australia New Zealand Food Standards Code).

In simple terms, the concentration in parts per million can be thought of as the mass in grams of solute dissolved in 1 000 000 g of solution. This can also be expressed as the mass in milligrams of solute dissolved in 1 kg of solution, because there are 1 million milligrams in a kilogram.

For example, a solution of sodium chloride that has a concentration of 154 ppm contains 154 mg of sodium chloride dissolved in 1 kg of solution. A formula used to calculate the concentration of a solution in ppm is:

$$\text{concentration (ppm)} = \frac{\text{mass of solute (in mg)}}{\text{mass of solution (in kg)}}$$

Worked example 16.2.2

CALCULATING CONCENTRATION IN PARTS PER MILLION (PPM)

A saturated solution of calcium carbonate was found to contain 0.0198 g of calcium carbonate dissolved in 2000 g of solution. Calculate the concentration, in ppm, of calcium carbonate in the saturated solution.

Thinking	Working
Calculate the mass of solute in mg. Remember: mass (in mg) = mass (in g) × 1000	mass of solute (calcium carbonate) in mg = 0.0198 × 1000 = 19.8 mg
Calculate the mass of solution in kg. Remember: $\text{mass (in kg)} = \frac{\text{mass (in g)}}{1000}$	$\text{mass of solution in kg} = \frac{2000}{1000}$ = 2.000 kg
Calculate the concentration of the solution in $mg\,kg^{-1}$. This is the same as concentration in ppm.	concentration of calcium carbonate in ppm $= \frac{\text{mass of solute (in mg)}}{\text{mass of solution (in kg)}}$ $= \frac{19.8}{2.000}$ $= 9.90\,mg\,kg^{-1}$ = 9.90 ppm

Worked example: Try yourself 16.2.2

CALCULATING CONCENTRATION IN PARTS PER MILLION (PPM)

A sample of tap water was found to contain 0.0537 g of NaCl per 250.0 g of solution. Calculate the concentration of NaCl in parts per million (ppm).

EXTENSION

Other units of concentration

You might have noticed symbols such as w/w, v/v or w/v on the labels of some foods, drinks and pharmaceuticals. These symbols represent other concentration units based on masses and volumes of solutes and solutions. These are useful in practical situations because people are familiar with these quantities (Figure 16.2.2).

FIGURE 16.2.2 Consumer products from hardware stores and supermarkets show a wide range of concentration units on their labels.

Percentage by mass (w/w)

Percentage by mass describes the mass of solute, measured in grams, present in 100 g of the solution.

Normal saline solution for washing contact lenses has a concentration of 0.9%(w/w). This can also be written as 0.9%(m/m). The abbreviation w/w indicates that the percentage is based on the weights or, more correctly, masses of both solute and solution. A concentration of 0.9% indicates that there is 0.9 g of sodium chloride dissolved in 100 g of solution.

Percentage by volume (v/v)

The abbreviation v/v indicates that the percentage is based on volumes of both solute and solution. The same units must be used to record both volumes. Percentage by volume is a more convenient unit to use than w/w when the solute is a liquid.

Just like percentage by mass, percentage by volume is frequently expressed as volume per 100 mL of solution. For example, the wine label in Figure 16.2.3 shows 13.5% alc./vol. This means the wine contains 13.5% alcohol (ethanol) by volume (13.5%(v/v)). There will be 13.5 mL of alcohol in 100 mL of the wine.

For example, a 200 mL glass of champagne contains 28 mL of alcohol. The concentration as %(v/v) of alcohol in this solution can be calculated as follows:

$$\text{concentration (\% v/v)} = \frac{\text{volume of solute (in mL)}}{\text{volume of solution (in mL)}} \times 100\%$$

$$= \frac{28}{200} \times 100\%$$

$$= 14\%(\text{v/v})$$

FIGURE 16.2.3 The alcohol content of this wine is 13.5% v/v.

Percentage mass/volume (w/v)

Percentage mass/volume describes the mass of solute, measured in grams, present in 100 mL of the solution.

For example, if a solution of plant food contains a particular potassium compound at a concentration of 3%(w/v), this indicates that there is 3 g of potassium in 100 mL of solution.

CHEMFILE

Salinity in Western Australia

The south-west region of Western Australia, like other parts of Australia, has increasing salt levels in both surface and groundwater. Salt has accumulated in south-west soils over hundreds of thousands of years due to wind and rain carrying it from the ocean. Prior to European settlement, the salt was mostly undissolved in the soil but with clearing of trees, which used much of the rainfall, more rain now enters the groundwater causing a rise in the water table, which in turn dissolves salt in the soil bringing it closer to the surface. As groundwater rises, some enters the surface water bringing its salt with it. For example, in the 1940s the Warren River had about 300 mg L^{-1} of salt but now the level is around 800–900 mg L^{-1}, and the Blackwood River has gone from about 500 mg L^{-1} to 2000 mg L^{-1} over the past 50 years.

Water is classified as fresh if the salt concentration is less than 500 mg L^{-1}, marginal if it is 500–1000 mg L^{-1} and saline above 2000 mg L^{-1}. Marginal water can be used for most irrigation purposes but the ecosystem begins to suffer, and saline water may be suitable for some livestock but it is no longer suitable for irrigation or cropping.

CHEMFILE

Saline drip

A 'saline drip' is sometimes used during medical procedures to replenish body fluids in patients. The solution used contains sodium chloride, commonly with a concentration of 0.9%(m/v). That means that 100 mL of the solution contains 0.9 g of sodium chloride.

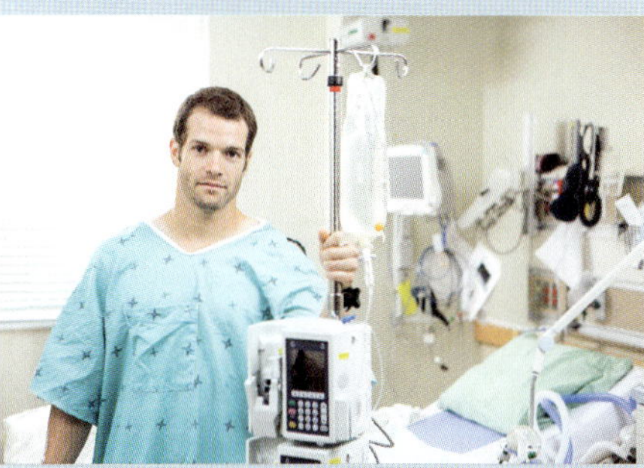

FIGURE 16.2.4 A saline drip is a solution of sodium chloride.

16.2 Review

SUMMARY

- The concentration of a solution is defined as the quantity of solute dissolved in a quantity of solution.
- Concentration units have two parts. The first part provides information about the quantity of solute. The second part provides information about the quantity of solution.
- The concentrations of solutions can be expressed in different units including g L^{-1}, mol L^{-1} and ppm.

KEY QUESTIONS

1 What is the concentration, in g L^{-1}, of a 60 mL solution that contains 5.0 g of sugar?
 A 300
 B 0.83
 C 12
 D 83

2 Calculate the concentration of the following solutions in ppm.
 a 25 mg of $CaCl_2$ dissolved in 5.0 kg of solution
 b 1.25 g of lead nitrate dissolved in 2000 g of solution
 c 4.0×10^{-3} g of $MgSO_4$ dissolved in 150 g of solution

3 The nutrition panel on a flavoured milk drink states that it contains 35.0 g of sugar and 7.5 g of fat in every 250 mL serving. What are the concentrations of sugar and fat expressed as %(w/v)?

4 Fish sold in Australia can have a maximum concentration of mercury of 1 ppm. A fish catch is sampled and a fish of mass 1.26 kg is found to contain 1.092 mg of Hg. Can the fish catch be sold in Australia?

16.3 Molar concentration

In the previous section you saw that units of concentration such as $g\,L^{-1}$, %(w/w) and %(v/v) are commonly used on the labels of consumer products found in supermarkets, pharmacies and hardware stores. Another unit of concentration that is commonly used by chemists is $mol\,L^{-1}$.

In this section, you will learn how to carry out calculations using this unit of concentration.

CONCENTRATION IN MOLES PER LITRE

Concentrations expressed in moles per litre ($mol\,L^{-1}$) allow chemists to compare the amount, in moles, of atoms, molecules or ions present in a given volume of solution.

For example, a 0.1 $mol\,L^{-1}$ solution of hydrochloric acid would contain 0.1 moles of HCl in 1 L of the solution. (A bottle of HCl of this concentration is shown in Figure 16.3.1.)

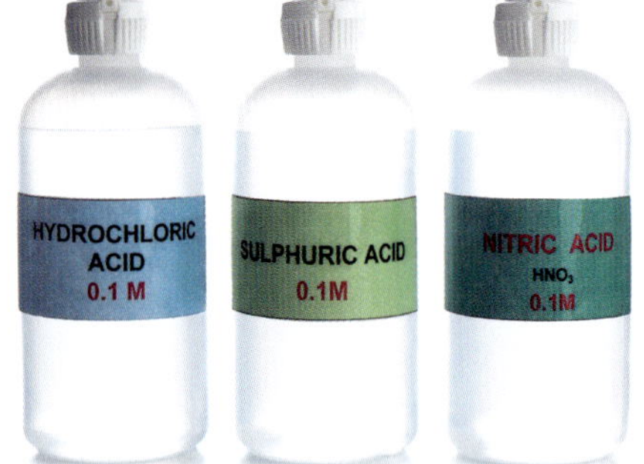

FIGURE 16.3.1 These containers contain solutions with concentrations shown in units of $mol\,L^{-1}$. The $mol\,L^{-1}$ unit is often written as M.

The concentration, in $mol\,L^{-1}$, of a solution can be calculated as follows:

$$\text{concentration (mol L}^{-1}) = \frac{\text{amount of solute (in mol)}}{\text{volume of solution (in L)}}$$

$$c = \frac{n}{V}$$

where c is the concentration ($mol\,L^{-1}$), n is the amount (mol) and V is the volume (L).

The concentration of a solution in moles per litre is often referred to as the **molarity**, or molar concentration of the solution. Molarity has the unit $mol\,L^{-1}$, which is sometimes represented as M.

A solution containing 1 mole of solute dissolved in 1 litre of solution can therefore be described in several different ways. You can say that the solution:

- has a concentration of 1 mole per litre
- has a concentration of 1 $mol\,L^{-1}$
- is 1 molar
- is 1 M
- has a molarity of 1 M.

> The symbol *M* was used earlier to represent molar mass. Take care not to confuse the two different uses, one is a quantity symbol, the other a unit symbol.

> Molarity is calculated using the formula $c = \frac{n}{V}$, where c is the concentration in $mol\,L^{-1}$, n is the amount in moles and V is the volume in litres.

Worked example 16.3.1

CALCULATING MOLAR CONCENTRATIONS (MOLARITY)

Calculate the molar concentration of a solution that contains 0.105 mol of potassium nitrate dissolved in 200 mL of solution.	
Thinking	**Working**
Convert the given volume to litres.	$V(KNO_3) = \frac{200}{1000}$ $= 0.200$ L
Calculate the molar concentration using the formula: $c = \frac{n}{V}$	$c(KNO_3) = \frac{n}{V}$ $= \frac{0.105}{0.200}$ $= 0.525\ mol\,L^{-1}$

Worked example: Try yourself 16.3.1

CALCULATING MOLAR CONCENTRATIONS (MOLARITY)

Calculate the molar concentration of a solution that contains 0.24 mol of glucose dissolved in 500 mL of solution.

Calculating molarity given the mass of solute

Sometimes you will know the mass of a solute and volume of solution and want to calculate the molarity. Two main calculations are involved.

1 Calculate the number of moles of solute from its mass.
2 Calculate the concentration using the number of moles and the volume (in litres).

Worked example 16.3.2

CALCULATING MOLARITY GIVEN THE MASS OF SOLUTE

Calculate the concentration, in mol L^{-1}, of a solution that contains 16.8 mg of silver nitrate ($AgNO_3$) dissolved in 150 mL of solution.	
Thinking	**Working**
Convert the volume to litres.	$V(AgNO_3) = \frac{150}{1000}$ $= 0.150$ L
Convert the mass to grams.	$m(AgNO_3) = \frac{16.8}{1000}$ $= 0.0168$ g
Calculate the molar mass of the solute. To do this, add up the atomic masses of all the atoms in the compound.	$M(AgNO_3) = 107.9 + 14.0 + (3 \times 16.0)$ $= 169.9$ g mol^{-1}
Calculate the number of mol of solute using the formula: $n = \frac{m}{M}$	$n(AgNO_3) = \frac{m}{M}$ $= \frac{0.0168}{169.9}$ $= 9.89 \times 10^{-5}$ mol
Calculate the molar concentration using the formula: $c = \frac{n}{V}$	$c(AgNO_3) = \frac{n}{V}$ $= \frac{9.89 \times 10^{-5}}{0.150}$ $= 6.59 \times 10^{-4}$ mol L^{-1}

Worked example: Try yourself 16.3.2

CALCULATING MOLARITY GIVEN THE MASS OF SOLUTE

Calculate the concentration, in mol L^{-1}, of a solution that contains 4000 mg of ethanoic (acetic) acid (CH_3COOH) dissolved in 100 mL of solution.

The formula used to calculate the molarity of a solution is $c = \frac{n}{V}$. If the formula is rearranged, it can be used to calculate the number of moles of solute in a solution of given concentration and volume:

$$n = c \times V$$

where n is the amount (mol), c is the concentration (mol L^{-1}) and V is the volume (L).

CHEMFILE

Molarity—the chemist's unit of concentration

If you ever look at the labels on household and commercial products you are very unlikely to see the concentrations of the active ingredients given in units of molarity (mol L^{-1}). The units typically used will include g/L (grams per litre), mg/g (milligrams per gram), %(w/v) (weight/volume per cent) or %(w/w) (weight/weight per cent). Chemists however most commonly use molarity—the number of moles in one litre of solution.

When working with solutions, chemists are often dealing with the reaction of a substance in the solution with another substance (which may also be in solution). Substances react at the molecular or particle level so chemists are more likely to be interested in the number of particles present so that the quantity of the substances in the reaction can be easily related to each other. It is simplest to relate the quantity of substances in a chemical reaction through their number of moles, and molarity allows this more readily for solutions than do other units of concentration.

Worked example 16.3.3

CALCULATING THE NUMBER OF MOLES OF SOLUTE IN A SOLUTION

Calculate the amount, in moles, of ammonia (NH_3) in 25.0 mL of a 0.3277 mol L^{-1} ammonia solution.	
Thinking	**Working**
Convert the given volume to litres.	$V(NH_3) = \frac{25.0}{1000}$ $= 0.0250$ L
Calculate the amount of compound, in moles, using the formula: $n = c \times V$	$n(NH_3) = c \times V$ $= 0.3277 \times 0.0250$ $= 8.19 \times 10^{-3}$ mol

Worked example: Try yourself 16.3.3

CALCULATING THE NUMBER OF MOLES OF SOLUTE IN A SOLUTION

Calculate the amount, in moles, of potassium permanganate ($KMnO_4$) in 100 mL of a 0.0250 mol L^{-1} solution of the compound.

16.3 Review

SUMMARY

- Molarity is defined as the number of moles of solute per litre of solution.
- Molarity is calculated using the formula $c = \frac{n}{V}$, where c is the concentration in mol L^{-1}, n is the amount in moles and V is the volume in litres.
- The formula $n = c \times V$ can be used to calculate the number of moles of solute in a given volume of solution.

KEY QUESTIONS

1 Hydrogen peroxide solutions for hair bleaching are sold as 120 mL solutions containing 5.00 g of H_2O_2 dissolved in water. What is the molar concentration of hydrogen peroxide in this bleaching product?
  - **A** 0.81 %(w/v)
  - **B** 1.23 mol L^{-1}
  - **C** 41.7 g L^{-1}
  - **D** 1.23×10^{-3} mol L^{-1}

2 Calculate the molar concentration of the following solutions.
  - **a** a 25.0 mL solution that contains 2.0×10^{-3} mol of NaCl
  - **b** a 4.1 L solution that contains 1.23 mol of CH_3COOH
  - **c** a 9.3×10^3 L solution that contains 1.8×10^3 mol of KCl

3 Calculate the molar concentration of the following solutions.
  - **a** an 8.0 L solution that contains 2.0 mol of NaCl
  - **b** a 500 mL solution that contains 0.25 mol of $MgCl_2$
  - **c** a 200 mL solution that contains 0.0876 mol of sucrose

4 Calculate the molar concentration of a 250 mL solution that contains 5.09 g of $AgNO_3$.

5 Calculate the molar concentration of a 1.55 L solution that contains 1.223 g of $CaCl_2$.

6 Calculate the amount, in mol, of solute in each of the following solutions.
  - **a** 0.10 L of 0.22 mol L^{-1} KOH solution
  - **b** 10 mL of 0.64 mol L^{-1} NaI solution
  - **c** 15.6 mL of 0.0150 mol L^{-1} $CuSO_4$ solution
  - **d** 1.5×10^{-1} mL of 5.2 mol L^{-1} HCl solution

16.4 Dilution

Many commercially available domestic and industrial products come in the form of concentrated solutions. Examples are pesticides (Figure 16.4.1), fertilisers, detergents, fruit juices, acids and other chemicals. A major reason for using concentrates is to save on transportation costs. Diluted solutions contain a lot of water and that extra mass has to be transported, which increases costs. It is also more convenient to buy concentrated products and dilute with water at home or in the workplace.

Everyday examples of dilution are:

- adding water to cordial
- a laboratory technician making a 1 mol L^{-1} solution of hydrochloric acid from a bottle of concentrated hydrochloric acid
- a home gardener diluting fertiliser concentrate to spray on the lawn
- a farmer diluting weedkiller concentrate to spray on a wheat crop
- an assistant in a commercial kitchen diluting concentrated detergent solution before using it to wash dishes.

In this section, you will learn how to calculate the concentrations of solutions before and after they have been diluted. As well, you will learn how to calculate the volume of water that needs to be added to a concentrated solution to give a diluted solution of a required concentration, and the volume of concentrated solution needed to produce a diluted solution of known volume and concentration.

FIGURE 16.4.1 Pesticide solutions used in aerial spraying are prepared by diluting a concentrated solution of the pesticide compound.

CALCULATING CONCENTRATIONS WHEN SOLUTIONS ARE DILUTED

The process of adding more solvent to a solution is known as **dilution**. When a solution is diluted, its concentration is decreased.

For example, if 50 mL of water is added to 50 mL of 0.10 mol L^{-1} sugar solution, the amount of sugar remains unchanged but the volume of the solution in which it is dissolved doubles. As Figure 16.4.2 shows, this means the sugar molecules are spread further apart during the dilution process, and so the concentration of the sugar solution is decreased (it will become 0.050 mol L^{-1} in this instance).

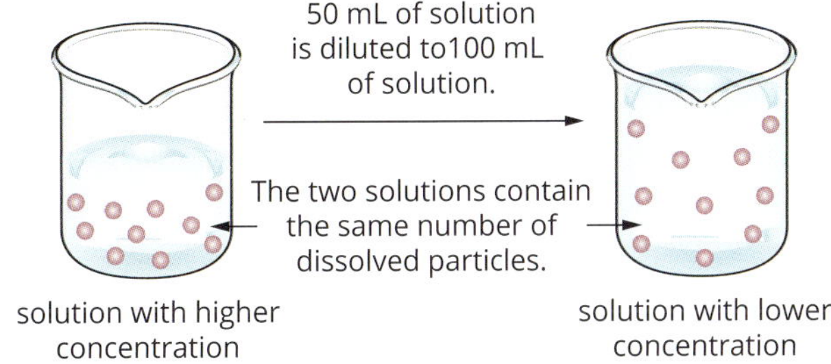

FIGURE 16.4.2 Dilution does not change the number of solute particles but the concentration of the solute decreases.

It is important to recognise that diluting a solution (by adding more solvent) does not change the amount of solute present.

Suppose you had V_1 litres of a solution and the concentration was c_1 mol L^{-1}. The amount of solute, in mole, is given by:

$$n_1 = c_1 \times V_1$$

Suppose water was added to make a new volume, V_2, and change the concentration to c_2. The amount of solute, n_2, in this diluted solution is given by:

$$n_2 = c_2 \times V_2$$

Since the number of moles of solute has not changed, $n_1 = n_2$; therefore:

$$c_1V_1 = c_2V_2$$

Numerical problems involving dilution can be solved using the formula: $c_1V_1 = c_2V_2$

This formula is useful when solving problems involving diluted solutions. (Note that this formula can also be used with different concentration and volume units, as long as they are the same on both sides.)

Worked example 16.4.1

QUESTIONS INVOLVING DILUTION

Calculate the concentration of the solution formed when 10.0 mL of water is added to 5.00 mL of 1.2 mol L^{-1} HCl.

Thinking	Working
Write down the value of c_1 and V_1. Note: c_1 and V_1 refer to the original solution, before water is added.	$c_1 = 1.2$ mol L^{-1} $V_1 = 5.00$ mL
Write down the value of V_2. Note: V_2 is the total volume of the original solution plus the added water.	$V_2 = 10.0 + 5.00$ $= 15.0$ mL
Transpose the equation $c_1V_1 = c_2V_2$ to allow calculation of the concentration, c_2, of the new solution.	$c_1V_1 = c_2V_2$ $c_2 = \frac{c_1V_1}{V_2}$
Calculate the concentration of the diluted solution.	$c_2 = \frac{1.2 \times 5.00}{15.0}$ $= 0.40$ mol L^{-1}

Worked example: Try yourself 16.4.1

QUESTIONS INVOLVING DILUTION

Calculate the concentration of the solution formed when 95.0 mL of water is added to 5.00 mL of 0.500 mol L^{-1} HCl.

CALCULATING VOLUMES WHEN SOLUTIONS ARE DILUTED

When a laboratory technician is preparing a diluted solution from a concentrated solution to use in a laboratory procedure, it is usual that they know the concentration and volume of the diluted solution they want to prepare. As well, they will know the concentration of the concentrated solution that is being used to make the dilute solution. Thus to make their diluted solution they need to calculate how much water to add to the concentrated solution.

For example, sulfuric acid is typically provided to laboratories as a 98% solution which has a molar concentration of 18.4 mol L^{-1}. The concentrations at which the acid is more typically used in laboratory work are around the 1 mol L^{-1} range.

Worked example 16.4.2

QUESTIONS INVOLVING DILUTION

Calculate the volume of 18.4 mol L^{-1} sulfuric acid solution required to prepare 5.00 L of a 1.50 mol L^{-1} solution. What volume of water must be added to the 18.4 mol L^{-1} sulfuric acid solution to give the required volume of the diluted solution?

Thinking	Working
Write down the value of c_1 and V_1. Note: c_1 and V_1 refer to the diluted solution, after water is added.	$c_1 = 1.50$ mol L^{-1} $V_1 = 5.00$ L

Write down the value of c_2. Note: c_2 is the concentration of the concentrated solution.	$c_2 = 18.4\,\text{mol L}^{-1}$
Transpose the equation $c_1V_1 = c_2V_2$ to allow calculation of the volume, V_2, of the concentrated solution.	$c_1V_1 = c_2V_2$ $V_2 = \frac{c_1V_1}{c_2}$
Calculate the volume of the concentrated solution.	$V_2 = \frac{1.50 \times 5.00}{18.4}$ $= 0.408\,\text{L}$
Calculate the volume of water required by finding the difference between the final volume of the diluted solution and the volume of concentrated solution.	$V(\text{water}) = 5.00 - 0.408$ $= 4.59\,\text{L}$

Worked example: Try yourself 16.4.2

QUESTIONS INVOLVING DILUTION

Calculate the volume of $11.5\,\text{mol L}^{-1}$ hydrochloric acid solution required to prepare 2.5 L of a $0.50\,\text{mol L}^{-1}$ solution. What volume of water must be added to the $11.5\,\text{mol L}^{-1}$ hydrochloric acid solution to give the required volume of the diluted solution?

CHANGING THE UNITS OF CONCENTRATION

At times it is useful to change, or 'convert', concentration values from one unit to another unit. One way to do this is to assume that you have 1 litre of the solution. Worked example 16.4.3 shows how this can be done.

Worked example 16.4.3

CONCENTRATION UNIT CONVERSIONS

Calculate the concentration, in ppm, of a $0.00200\,\text{mol L}^{-1}$ solution of NaCl. Note that concentration in ppm is the same as mg L^{-1}, if the density of the solution is $1\,\text{kg L}^{-1}$, which can be assumed for solutions of low concentrations.

Thinking	Working
Calculate the number of moles of solute in 1.00 L of the solution.	$n(\text{NaCl}) = c \times V$ $= 0.00200 \times 1.00$ $= 0.00200\,\text{mol}$
Calculate the mass, in grams, of solute in 1.00 L of the solution.	$M(\text{NaCl}) = 35.5 + 23.0$ $= 58.5\,\text{g mol}^{-1}$ $m(\text{NaCl}) = n \times M$ $= 0.00200 \times 58.5$ $= 0.117\,\text{g}$
Calculate the mass, in mg, of solute in 1.00 L of the solution.	$m(\text{NaCl}) = 0.117 \times 1000$ $= 117\,\text{mg}$
Express the concentration of the solute in ppm.	$c(\text{NaCl}) = 117\,\text{ppm}$

Worked example: Try yourself 16.4.3

CONCENTRATION UNIT CONVERSIONS

Calculate the concentration, in ppm, of a $0.0100\,\text{mol L}^{-1}$ solution of NaOH. Note that concentration in ppm is the same as mg L^{-1}, if the density of the solution is $1\,\text{kg L}^{-1}$, which can be assumed for solutions of low concentrations.

16.4 Review

SUMMARY

- Numerical problems involving dilution can be solved using the formula:

$$c_1V_1 = c_2V_2$$

where c is concentration and V is volume.

- When using this formula, c_1 and c_2 must be in the same units of concentration and V_1 and V_2 must be in the same units of volume.
- One type of concentration unit can be converted to another unit in simple steps. For example, units of %(w/w) can be converted to units of mol L^{-1}.

KEY QUESTIONS

1. Calculate the concentration of each of the following diluted solutions.
   - **a** 10.0 mL of water added to 5.0 mL of 1.2 mol L^{-1} HCl
   - **b** 1.0 L of water added to 3.0 L of 0.10 mol L^{-1} HCl
   - **c** 5.0 mL of 0.50 mol L^{-1} HCl added to 95.0 mL of water
2. What volume of 10 mol L^{-1} hydrochloric acid would be required to prepare 250 mL of a 0.30 mol L^{-1} HCl solution?
   - **A** 7.5 L
   - **B** 133 L
   - **C** 0.133 L
   - **D** 0.0075 L
3. The concentration of a solution of ammonia (NH_3) is 1.5%(w/v). What is the molar concentration of a solution produced by diluting 25.0 mL of this solution with 250 mL of water?
4. The concentration of a solution of NaOH is 17.0%(w/v). What is the concentration of this solution in mol L^{-1}?

16.5 Calculations involving reactions in solutions

As you learnt earlier in this chapter, water is an exceptional solvent for a large variety of substances. In this section, you will learn to calculate amounts of the reactants or products involved in chemical reactions that occur in solution. These reactions can involve precipitation reactions as well as other types of reactions. Using your knowledge of the mole concept, you will be able to calculate the mass or concentration of products formed or the amounts of reactants consumed.

FIGURE 16.5.1 The reaction of a solution of mercury(II) ethanoate with a solution of sodium iodide. When colourless aqueous solutions of each reactant are mixed, they produce a red precipitate, mercury(II) iodide.

CALCULATING THE MASS OF A PRECIPITATE IN A REACTION

Several steps are involved in calculating the mass of a precipitate produced in a precipitation reaction:

1 Write a balanced equation for the reaction.

2 Calculate the number of moles of any reactant that has reacted completely, using the information supplied in the question, and the formula $n = c \times V$ or $n = \frac{m}{M}$, depending on the information.

3 Using the mole ratios in the equation, calculate the number of moles of the precipitate.

4 Calculate the mass of the precipitate using $m = n \times M$.

Figure 16.5.2 summarises this process. The worked example below will help you to understand these steps.

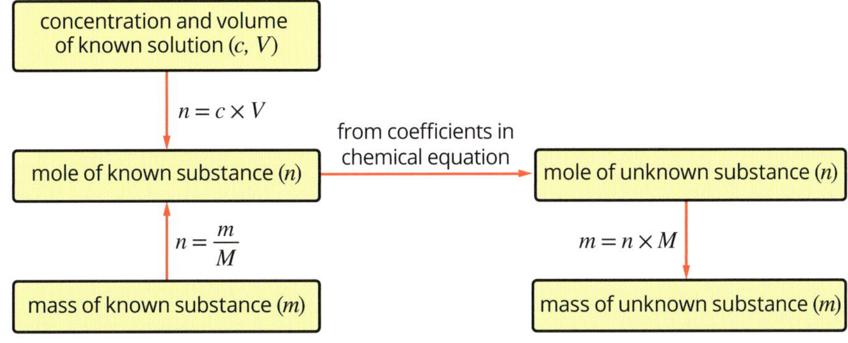

FIGURE 16.5.2 A flow chart for mass–mass or solution volume–mass stoichiometric calculations is helpful when trying to solve these problems.

Worked example 16.5.1

STOICHIOMETRY PROBLEMS INVOLVING SOLUTIONS

Calculate the mass of silver chloride precipitated when 25.0 mL of a 0.450 mol L^{-1} solution of calcium chloride reacts completely with a solution of silver nitrate.

Thinking	Working
Write a balanced equation for the reaction.	$CaCl_2(aq) + 2AgNO_3(aq) \rightarrow 2AgCl(s) + Ca(NO_3)_2(aq)$
Calculate the number of moles of the known substance using $n = c \times V$. The 'known substance' is the one you are provided information about in the question.	$V(CaCl_2) = 0.0250\,L$ $n(CaCl_2) = c \times V$ $= 0.450 \times 0.0250$ $= 0.0113\,mol$
Write the mole ratio for the known and unknown substances.	mole ratio $= \frac{n(AgCl)}{n(CaCl_2)} = \frac{2}{1}$
Calculate the number of moles of the unknown substance using: n(unknown) = n(known) × mole ratio The 'unknown substance' is the one whose mass you are required to calculate.	$n(AgCl) = 0.0113 \times \frac{2}{1}$ $= 0.0225$ mol
Calculate the mass of the unknown substance using: m(unknown) = n(unknown) × molar mass	$M(AgCl) = 143.4\,g\,mol^{-1}$ $m(AgCl) = 0.0225 \times 143.4$ $= 3.23\,g$

Worked example: Try yourself 16.5.1

STOICHIOMETRY PROBLEMS INVOLVING SOLUTIONS

Calculate the mass of aluminium hydroxide precipitated when 45.0 mL of a 0.200 mol L^{-1} solution of sodium hydroxide reacts completely with a solution of aluminium chloride.

CALCULATING OTHER QUANTITIES IN A REACTION

Not only can the mass of the precipitate formed from mixing two aqueous solutions be calculated, but the quantities of other products formed or of reactants consumed can also be found.

1 Calculate the number of moles of the known substance, either a reactant that has reacted completely, or the product, using the formula $n = c \times V$ or $n = \frac{m}{M}$, depending on the information supplied in the question.
2 Using the mole ratios in the equation, calculate the number of moles of the unknown.
3 Calculate the quantity of the unknown by rearranging $n = c \times V$ or $n = \frac{m}{M}$ as needed.

Figure 16.5.3, an extension of Figure 16.5.2, summarises this process. Worked example 16.5.2 will help you to understand these steps.

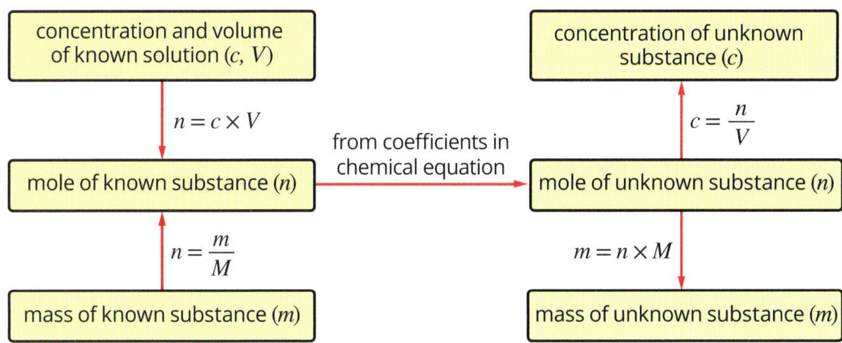

FIGURE 16.5.3 A flow chart for determining the quantity of a reactant or product by stoichiometric calculations is helpful when trying to solve these problems.

Worked example 16.5.2

STOICHIOMETRY PROBLEMS INVOLVING SOLUTIONS

Calculate the concentration of a sodium carbonate solution if 2.36 g of copper(II) carbonate is precipitated when 25.0 mL of the sodium carbonate solution is reacted with excess copper(II) sulfate solution.

Thinking	Working
Write a balanced equation for the reaction.	$Na_2CO_3(aq) + CuSO_4(aq) \rightarrow CuCO_3(s) + Na_2SO_4(aq)$
Calculate the number of moles of the known substance using $n = \frac{m}{M}$. The 'known substance' is the one you are provided information about in the question.	$M(CuCO_3) = 123.5\ g\ mol^{-1}$ $n(CuCO_3) = \frac{m(CuCO_3)}{M(CuCO_3)}$ $= \frac{2.36}{123.5}$ $= 0.0191\ mol$
Write the mole ratio for the known and unknown substances.	mole ratio $= \frac{n(Na_2CO_3)}{n(CuCO_3)} = \frac{1}{1}$
Calculate the number of moles of the unknown substance using: $n(\text{unknown}) = n(\text{known}) \times \text{mole ratio}$ The 'unknown substance' is the one whose mass you are required to calculate.	$n(Na_2CO_3) = 0.0191 \times \frac{1}{1}$ $= 0.0191\ mol$
Calculate the quantity (volume in this example) of the unknown substance by rearranging: $n = c \times V$	$c(Na_2CO_3) = \frac{n}{V}$ $= \frac{0.0191}{0.025}$ $= 0.764\ mol\ L^{-1}$

Worked example: Try yourself 16.5.2

STOICHIOMETRY PROBLEMS INVOLVING SOLUTIONS

Calculate the concentration of a potassium sulfate solution if 7.88 g of barium sulfate is precipitated when 55.0 mL of the potassium sulfate solution is reacted with excess barium nitrate solution.

16.5 Review

SUMMARY

- A balanced equation shows the ratio of the amount (in moles) of reactants and products formed in the reaction.
- Given the quantity of one of the reactants or products of a chemical reaction, such as in a precipitation reaction, the quantity of all other reactants and products can be predicted by working through the following steps:
  1. Write a balanced equation for the reaction.
  2. Calculate the amount (in moles) of the given substance.
  3. Use the mole ratios of reactants and products in the balanced equation to calculate the amount (in moles) of the required substance.
  4. Use the appropriate formula to determine the required answer. Formulae are:

     $n = \frac{m}{M}$

     $n = c \times V$

KEY QUESTIONS

1. Calculate the mass of precipitate produced when 120.0 mL of 0.075 mol L^{-1} calcium chloride solution reacts completely with a sodium sulfide solution.
2. Calculate the volume of 0.050 mol L^{-1} lead(II) nitrate solution that reacts with sodium iodide solution to produce 0.984 g of lead(II) iodide precipitate.
3. Silver nitrate solution will react with sodium chloride solution to give a precipitate of silver chloride according to the equation:

   $AgNO_3(aq) + NaCl(aq) \rightarrow AgCl(s) + NaNO_3(aq)$

   Calculate the mass of silver chloride produced when 2.40 g of silver nitrate in aqueous solution is added to excess sodium chloride.
4. What volume of 0.125 mol L^{-1} calcium nitrate solution will be needed to precipitate all the phosphate from 30.0 mL of a 0.155 mol L^{-1} sodium phosphate solution? The balanced equation for the reaction is:

   $3Ca(NO_3)_2(aq) + 2Na_3PO_4(aq) \rightarrow Ca_3(PO_4)_2(s) + 6NaNO_3(aq)$

Chapter review

16

KEY TERMS

concentrated solution
concentration
dilute solution
dilution
full equation
ionic equation
molarity
parts per million (ppm)
precipitate
precipitation reaction

Precipitation reactions

1 Consider the reaction represented by the following equation:

$$NaCl(aq) + AgNO_3(aq) \rightarrow AgCl(s) + NaNO_3(aq)$$

Indicate whether the following statements about the reaction and equation are true or false.

- **a** The equation represents a precipitation reaction.
- **b** Ionic compounds that contain sodium or nitrate ions are always soluble.
- **c** The precipitate in this reaction is $AgNO_3$.
- **d** The equation is an example of an ionic equation.
- **e** Silver chloride is insoluble in water.
- **f** The sodium and chloride ions can be described as spectator ions.

2 Copy and complete the following table. Identify which reaction mixtures will produce precipitates and write their formulae.

	NaOH	KBr	NaI	$MgSO_4$	$BaCl_2$
$Pb(NO_3)_2$					
KI					
$CaCl_2$					
Na_2CO_3					
Na_2S					

3 What precipitate will be formed (if any) when the following solutions are mixed?

- **a** barium nitrate and sodium sulfate
- **b** sodium chloride and copper(II) sulfate
- **c** magnesium sulfate and lead(II) nitrate
- **d** potassium chloride and barium nitrate

4 Write ionic equations for each of the following precipitation reactions.

- **a** $NH_4Cl(aq) + AgNO_3(aq) \rightarrow$
- **b** $FeCl_2(aq) + Na_2S(aq) \rightarrow$
- **c** $Fe(NO_3)_3(aq) + KOH(aq) \rightarrow$
- **d** $CuSO_4(aq) + NaOH(aq) \rightarrow$
- **e** $Ba(NO_3)_2(aq) + Na_2SO_4(aq) \rightarrow$

5 Write an ionic equation for the reaction that takes place when solutions of the following compounds are mixed. In each case, also give the spectator ions. Refer to the solubility tables on page 367 to help identify the precipitate for each reaction.

- **a** copper(II) sulfate and sodium carbonate
- **b** silver nitrate and potassium chloride
- **c** sodium sulfide and lead(II) nitrate
- **d** iron(III) chloride and sodium phosphate
- **e** iron(III) sulfate and potassium hydroxide

Concentration of solutions

6 **a** What is the concentration, in ppm, of lead in a 6.0 kg solution that contains 12 mg of lead?

- **b** What is the concentration as percentage by mass?

7 The label on a laundry bleach lists the concentration of the active ingredient, sodium hypochlorite, as 42 g/L (42 g L^{-1}) What mass of sodium hypochlorite will be in a 750 mL bottle of the bleach?

Molar concentration

8 What is the molar concentration of a 2.0 L solution containing 30 g of NaOH?

9 How many moles of solute are present in the following solutions?

- **a** 12 mL of 0.22 mol L^{-1} NaI
- **b** 150 mL of 0.0250 mol L^{-1} $KMnO_4$
- **c** 7.2 L of 3.15×10^{-3} mol L^{-1} KBr

10 What is the mass, in grams, of solute present in the following solutions?

- **a** 100 mL of 1.20 mol L^{-1} NH_3
- **b** 20 mL of 0.50 mol L^{-1} $AgNO_3$

Dilution

11 How much water must be added to 25 mL of a 4.0 mol L^{-1} solution of potassium carbonate solution to dilute it to a concentration of 1.6 mol L^{-1}?

12 A sample of 1.50 mL of a 0.0500 mol L^{-1} solution of calcium chloride ($CaCl_2$) was diluted with water to a volume of 10.0 L. What was the concentration of Cl^- ions in the diluted solution in ppm? Assume 1.0 L of the diluted solution has a mass of 1.0 kg.

Calculations involving reactions in solutions

13 What mass of precipitate is produced when 25.6 mL of 0.550 mol L^{-1} copper(II) chloride solution reacts with excess potassium carbonate solution?

14 What volume of 0.752 mol L^{-1} sodium iodide solution is needed to precipitate all the lead(II) ions from 150.0 mL of 0.085 mol L^{-1} lead(II) nitrate solution?

Connecting the main ideas

15 Copy and complete the diagram below by inserting into the white boxes the processes required to convert between the quantities in the yellow boxes. The answer to the top left box is given to you.

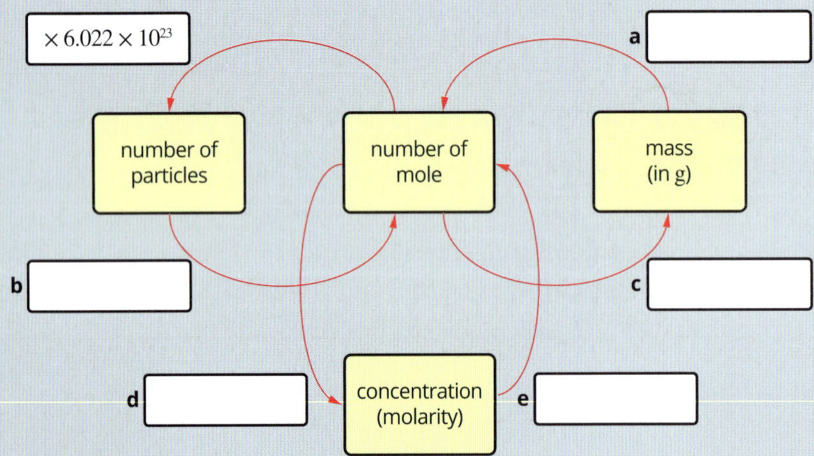

16 A sample of 100.0 mL of 0.150 mol L^{-1} sodium sulfate solution reacts completely with 25.0 mL of barium nitrate solution.

a What mass of precipitate is produced from the reaction?

b What is the molar concentration of the sodium ions after the reaction?

CHAPTER 17

Acids and bases

Acids and bases have an important and diverse role. They are found in homes and are used extensively in industry and agriculture. Acids and bases are also the reactants and products of many chemical reactions that take place in environmental and biological systems.

In this chapter, you will study a theory that explains the characteristic reactions of acids and bases. You will learn to represent the common reactions of acids and bases using ionic equations and investigate the distinction between strong and weak acids, and concentrated and dilute acids. Finally, you will learn how to calculate the pH of an acid and determine the concentration of unknown acids and bases using stoichiometry.

Science understanding

- indicator colour and the pH scale are used to classify aqueous solutions as acidic, basic or neutral
- pH is used as a measure of the acidity of solutions and is dependent on the concentration of hydrogen ions in the solution
- the Arrhenius model can be used to explain the behaviour of strong and weak acids and bases in aqueous solutions
- patterns of the reactions of acids and bases, including reactions of acids with bases, metals and carbonates and the reactions of bases with acids and ammonium salts, allow products and observations to be predicted from reactants; ionic equations represent the reacting species and products in these reactions
- the presence of specific ions in solutions can be identified by observing the colour of the solution, flame tests and observing various chemical reactions, including precipitation and acid-base reactions
- the mole concept can be used to calculate the mass of solute, and solution concentrations and volumes involved in a chemical reaction

17.1 Properties of acids and bases

Acids and **bases** make up some of the household products in your kitchen and laundry (Figure 17.1.1). In this section, you will be introduced to a theory that explains the chemical properties of acids and bases, helping you to explain their usefulness within the home and industry. You will also look at how the **acidity** of a solution can be measured, therefore defining a solution as a strong or weak acid.

FIGURE 17.1.1 Common household products that contain acids, bases or salts

ACIDS AND BASES

Table 17.1.1 gives the names, chemical formulae and uses of some common acids.

TABLE 17.1.1 Common acids and their everyday uses

Name	Formula	Uses
hydrochloric acid	HCl	present in stomach acid to help break down proteins
sulfuric acid	H_2SO_4	used in car batteries and in the manufacture of fertilisers and detergents
nitric acid	HNO_3	used in the manufacture of fertilisers, dyes and explosives
ethanoic acid (acetic acid)	CH_3COOH	found in vinegar; used as a preservative
carbonic acid	H_2CO_3	found in carbonated soft drinks
citric acid	$H_8C_6O_7$	found in the juice of citrus fruits, such as lemons

Many cleaning agents used in the home, such as washing powders and oven cleaners, contain bases. Solutions of ammonia are used as floor cleaners, and sodium hydroxide is the major active ingredient in oven cleaner. Bases are effective cleaners because they react with fats or oils to produce water-soluble soaps. A soluble base is referred to as an **alkali**.

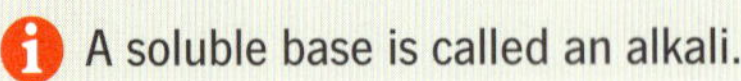

A soluble base is called an alkali.

Table 17.1.2 gives the names, chemical formulae and uses of some common bases.

TABLE 17.1.2 Common bases and their uses

Name	Formula	Uses
sodium hydroxide	$NaOH$	used in drain and oven cleaners, and soap making
ammonia	NH_3	used in household cleaners, fertilisers and explosives
calcium hydroxide	$Ca(OH)_2$	found in cement and mortar; used in garden lime to adjust soil pH

PROPERTIES OF ACIDS AND BASES

All acids have some properties in common. Bases also have common properties. The properties of acids and bases are summarised in Table 17.1.3.

TABLE 17.1.3 Properties of acids and bases

Properties of acids	Properties of bases
turn litmus indicator red	turn litmus indicator blue
tend to be corrosive	are caustic and feel slippery
taste sour	taste bitter
react with bases	react with acids
solutions have a pH of less than 7	solutions have a pH of greater than 7
solutions conduct an electric current	solutions conduct an electric current

INDICATORS

One of the characteristic properties of acids and bases is their ability to change the colour of certain plant extracts. Litmus is a purple dye obtained from lichen. In the presence of acids, litmus turns red. The colouring of rose petals, blackberries and red cabbage is also altered by acids and bases. Such plant extracts are called **indicators**.

Universal indicator and the pH scale

Universal indicator (Figure 17.1.2) is widely used to estimate the **pH** of a solution. It is a mixture of several indicators and changes through a range of colours, from red through yellow, green and blue, to violet, depending on the pH of the solution. If a more accurate measurement of pH is needed, a pH meter can be used instead of universal indicator.

FIGURE 17.1.2 Universal indicator pH scale. When universal indicator is added to a solution, it changes colour depending on the solution's pH. The tubes contain solutions of pH 0 to 14 from left to right. The green tube (centre) is neutral, pH 7.

17.1 Review

SUMMARY

- Acids have a range of uses and are often found in food and drink products.
- Bases are often used in cleaning products.
- A soluble base is called an alkali.
- Indicators are chemicals that change colour, dependent on the acidity of a solution.
- The level of acidity or alkalinity of a solution can be measured using the pH scale.
- Colours shown by universal indicator can be used to determine the approximate pH of solutions.
- Acidic solutions have a pH of less than 7 and alkaline solutions have a pH of greater than 7.

KEY QUESTIONS

1 Which one of the following acids is not usually found in food or drinks?

A hydrochloric acid
B ethanoic acid
C carbonic acid
D citric acid

2 Which one of the following is not a property of an alkaline solution?

A pH of greater than 7
B tastes sour
C conducts an electric current
D turns litmus blue

3 Explain why universal indicator tells us more about an acidic solution than litmus indicator.

4 Copper(II) oxide (CuO) is a base. A few grams of copper(II) oxide were added to a beaker of distilled water and the mixture stirred. When universal indicator solution was added to the water in the beaker, the indicator was green, showing that the water was still neutral with a pH of 7. Explain this observation.

17.2 The Arrhenius model of acids and bases

There have been many attempts to define acids and bases. At first, acids and bases were just defined in terms of their observed properties such as taste, effect on an indicator and reactions with other substances. In 1884, the Swedish scientist Svante Arrhenius passed an electric current through various solutions, including acids and bases. As a result of his observations, he developed a theory that explained some of the common properties of acids and bases.

THE ARRHENIUS MODEL OF ACIDS

In the **Arrhenius model**, an acid is defined as a substance that is ionised in water to produce hydrogen ions. For example, when hydrogen chloride gas is bubbled through water, the HCl molecules break apart (a process called dissociation) and form ions (a process called ionisation) to form hydrochloric acid, which consists of hydrogen ions and chloride ions in the aqueous solution:

$$HCl(g) \rightarrow H^+(aq) + Cl^-(aq)$$

In water, the H^+ ion exists as a hydronium ion (H_3O^+), so the equation above may also be written as:

$$HCl(g) + H_2O(l) \rightarrow H_3O^+(aq) + Cl^-(aq)$$

In this chapter, the hydrogen ion in water will be represented as $H^+(aq)$.

The Arrhenius model is useful because it explains the similarity in the reactions of different acids. The hydrogen ions that are present in acid solutions account for the common properties of acids. For example, as you will see later in this chapter, the hydrogen ions present in all acidic solutions form hydrogen gas when the acid reacts with metals such as magnesium:

$$2H^+(aq) + Mg(s) \rightarrow H_2(g) + Mg^{2+}(aq)$$

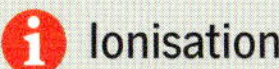

Ionisation is:

- the removal of one or more electrons from an atom or ion
- the reaction of a molecular substance with a solvent to form ions in solution.

Polyprotic acids

Some acids can react with water to form more than one hydrogen ion per molecule. The molecules of these acids have more than one proton that can be ionised and are said to be **polyprotic acids**.

The number of hydrogen ions an acid can donate depends on the structure of the acid molecule. It is difficult to judge how many hydrogen ions can be produced by an acid by looking at its formula alone.

Ethanoic (acetic) acid, CH_3COOH, contains four hydrogen atoms, yet each molecule can only form one hydrogen ion and it is therefore a **monoprotic acid**. Only the hydrogen that is part of the highly polar O–H bond is ionised in water. In general, hydrogen atoms bonded to a highly electronegative atom, thus forming polar bonds, are ionised in solution.

Diprotic acids

Diprotic acids, such as sulfuric acid (H_2SO_4) and carbonic acid (H_2CO_3), can donate two protons.

Overall, according to the Arrhenius model, the ionisation can be represented as shown here:

$$H_2SO_4(l) \rightarrow SO_4^{2-}(aq) + 2H^+(aq)$$

In reality, a diprotic acid such as sulfuric acid ionises in two stages. You will learn more about this process in Year 12.

Triprotic acids

Triprotic acids can donate three protons. These include phosphoric acid (H_3PO_4) and boric acid (H_3BO_3). Each molecule can provide three hydrogen ions when dissolved in water as shown here:

$$H_3PO_4(aq) \rightleftharpoons PO_4^{3-}(aq) + 3H^+(aq)$$

Note that in this equation, a reversible arrow ($\rightleftharpoons$) is used. This indicates that phosphoric acid is a weak acid. The differences between weak and strong acids are discussed in the next section.

The structure and bonding exhibited by ethanoic acid, sulfuric acid and phosphoric are shown in Figure 17.2.1.

FIGURE 17.2.1 The bonds between a hydrogen atom attached to an electronegative oxygen atom are broken when these molecules ionise in water. (a) ethanoic acid, (b) sulfuric acid, (c) phosphoric acid. The hydrogen atoms that become ions are circled.

Ionic substances dissolve by dissociation. Ion–dipole bonds are formed between the ions and water molecules.

THE ARRHENIUS MODEL OF BASES

Arrhenius defined a base as a substance that dissociates in water to form **hydroxide ions**. Sodium hydroxide is an ionic compound, containing sodium ions and hydroxide ions. When it is added to water it dissociates:

$$NaOH(s) \rightarrow Na^{+}(aq) + OH^{-}(aq)$$

Some bases dissociate to form more than one hydroxide ion:

$$Ca(OH)_2(s) \rightarrow Ca^{2+}(aq) + 2OH^{-}(aq)$$

The presence of the hydroxide ions in solutions of bases accounts for the common properties of bases. For example, the hydroxide ion reacts with the litmus indicator to form a blue-coloured compound.

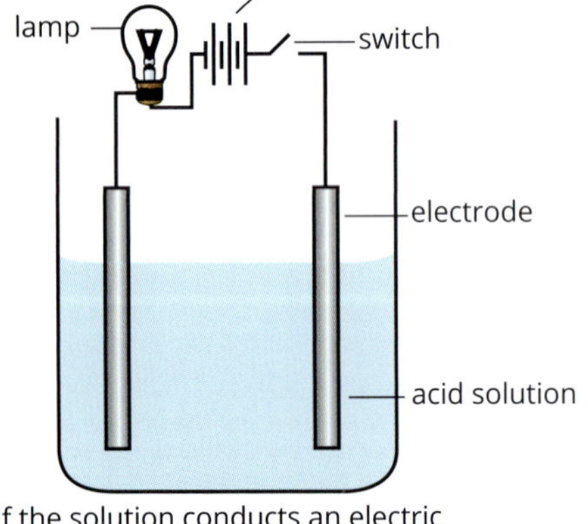

FIGURE 17.2.2 Apparatus used to test the conductivity of acid solutions

ACIDS AND BASES AS ELECTROLYTES

Hydrochloric acid is formed when hydrogen chloride ionises in water to form a solution containing hydrogen ions and chloride ions. When a solution of hydrochloric acid is tested with the conductivity apparatus shown in Figure 17.2.2, the light globe will glow. The chloride ions are attracted to the positive electrode of the apparatus and the hydrogen ions to the negative electrode. The flow of current in the solution is a result of the movement of these ions.

Because sodium hydroxide dissociates into sodium and hydroxide ions in aqueous solution, a sodium hydroxide solution will also conduct an electric current.

According to the Arrhenius model, acids ionise in water to produce hydrogen ions (H^{+}), whereas bases ionise in water to produce hydroxide ions (OH^{-}). The production of these charged particles in solution allows acids and bases to conduct electricity.

Neutralisation reactions

According to the Arrhenius model, when acids and bases react together, the hydrogen and hydroxide ions combine to form water according to the equation:

$$H^{+}(aq) + OH^{-}(aq) \rightarrow H_2O(l)$$

Arrhenius called this a **neutralisation reaction**. You will look at neutralisation reactions in more detail in the Section 17.3.

Limitations of the Arrhenius model of acids and bases

The Arrhenius model explains the properties of acids in terms of the formation of hydrogen ions by ionisation of the acid molecule. However, the model does have some limitations.

The model does not explain why some substances that do not contain hydrogen form acidic solutions when mixed with water. For example, an acidic solution is formed when carbon dioxide (CO_2) or sulfur dioxide (SO_2) is dissolved in water. Also, the model does not explain why some substances such as ammonia (NH_3) or sodium hydrogen carbonate ($NaHCO_3$) form basic solutions when mixed with water, even though they do not contain hydroxide ions.

The Arrhenius model is also restricted to acids and bases that dissolve in water. The model does not explain acid–base behaviour in non-aqueous solutions.

An improved model for acid–base behaviour is the Brønsted–Lowry theory, which describes an acid as a substance that can donate a hydrogen ion (proton) to a base, which accepts the proton. You will learn more about the Brønsted–Lowry theory in Year 12.

ACID AND BASE STRENGTH

The acidic solution in the left-hand beaker shown in Figure 17.2.3 is more acidic, as can be seen from the more vigorous reaction with the zinc metal. If the solutions in the two beakers are of the same concentration, both contain zinc and the temperature of the acidic solutions are the same, the acid in the left-hand beaker must be a stronger acid than the acid in the beaker on the right.

FIGURE 17.2.3 Zinc reacts more vigorously with a strong acid (left) than with a weak acid (right). The acid solutions are of equal concentration and volume.

Experiments show that different acid solutions of the same concentration do not have the same concentrations of hydrogen ions.

In the Arrhenius model, an acid is defined as a substance that is ionised in water to produce hydrogen ions. Some acids can be ionised more readily than others, which makes them stronger acids. Likewise, some bases can be ionised more readily than others.

Table 17.2.1 gives the names and chemical formulae of some strong and weak acids and bases.

TABLE 17.2.1 Examples of common strong and weak acids and bases

Strong acids	Weak acids	Strong bases	Weak bases
hydrochloric acid (HCl)	ethanoic acid (CH_3COOH)	sodium hydroxide (NaOH)	ammonia (NH_3)
sulfuric acid (H_2SO_4)	carbonic acid (H_2CO_3)	potassium hydroxide (KOH)	
nitric acid (HNO_3)	phosphoric acid (H_3PO_4)	calcium hydroxide ($Ca(OH)_2$)	

Strong acids

As you saw previously, when hydrogen chloride gas (HCl) is bubbled through water, it ionises completely—virtually no HCl molecules remain in the solution (Figure 17.2.4a). Similarly, pure HNO_3 which is a covalent molecular compound, also ionises completely in water:

$$HCl(g) \rightarrow H^+(aq) + Cl^-(aq)$$
$$HNO_3(l) \rightarrow H^+(aq) + NO_3^-(aq)$$

The single reaction arrow → in each equation above indicates that the ionisation reaction is complete.

Acids that are completely ionised are called **strong acids**. Therefore, solutions of strong acids contain ions, with virtually no unreacted acid molecules remaining. Hydrochloric acid, sulfuric acid and nitric acid are the most common strong acids.

Weak acids

Vinegar is a solution of ethanoic acid. Pure ethanoic acid is a polar covalent molecular compound that ionises in water to produce hydrogen ions and ethanoate (acetate) ions. In a 1.0 mol L^{-1} solution of ethanoic acid (CH_3COOH), only a small proportion of ethanoic acid molecules are ionised at any one time (Figure 17.2.4b). A 1.0 mol L^{-1} solution of ethanoic acid contains a high proportion of CH_3COOH molecules and only some hydronium ions and ethanoate ions. At 25°C, in a 1.0 mol L^{-1} solution of ethanoic acid, the concentrations of $CH_3COO^-(aq)$ and H_3O^+ are only about 0.004 mol L^{-1}.

The partial ionisation of a **weak acid** is shown in an equation using reversible (double) arrows:

$$CH_3COOH(aq) + H_2O(l) \rightleftharpoons CH_3COO^-(aq) + H_3O^+(aq)$$

Therefore, ethanoic acid is described as a weak acid in water.

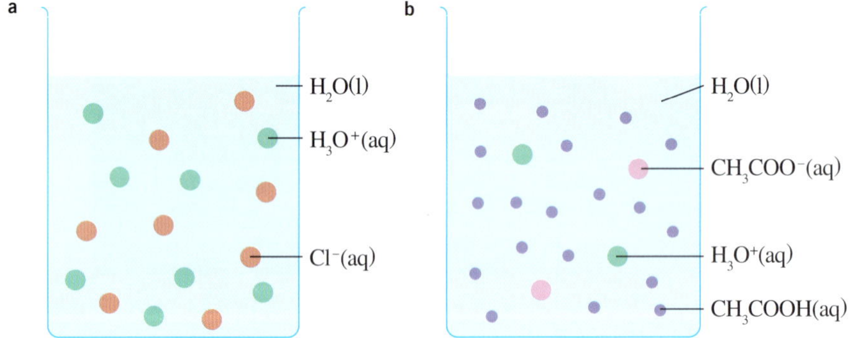

FIGURE 17.2.4 (a) In a 1.0 mol L^{-1} solution of a hydrochloric acid the acid molecules are completely ionised in water. (b) However, in a 1.0 mol L^{-1} solution of ethanoic acid only a small proportion of the ethanoic acid molecules are ionised.

Strong bases

Sodium hydroxide is a **strong base** as it is completely dissociated into Na^+ and OH^- ions in water. This can be shown by the equation:

$$NaOH(s) \rightarrow Na^+(aq) + OH^-(aq)$$

Note that because sodium hydroxide is an ionic substance, this process is just described as dissociation. It is not ionisation because the original solid is already comprised of ions, which separate from each other when dissolved in water.

Weak bases

Ammonia is a covalent molecular compound that ionises in water to produce hydroxide ions. This ionisation can be represented by the equation:

$$NH_3(aq) + H_2O(l) \rightleftharpoons NH_4^+(aq) + OH^-(aq)$$

Only a small proportion of ammonia molecules are ionised at any instant, so a 1.0 mol L^{-1} solution of ammonia contains mostly ammonia molecules together with a smaller number of ammonium ions and hydroxide ions. This is shown by the double arrow in the equation. Ammonia is a **weak base** in water.

CHEMFILE

Superacids

Fluorosulfuric acid (HSO_3F) is one of the strongest acids known. It has a similar geometry to the sulfuric acid molecule (Figure 17.2.5). The highly electronegative fluorine atom causes the oxygen–hydrogen bond in fluorosulfuric acid to be more polarised than the oxygen–hydrogen bond in sulfuric acid. The acidic proton is easily transferred to a base.

FIGURE 17.2.5 Structure of sulfuric acid (left) and fluorosulfuric acid (right) molecules

Fluorosulfuric acid is classified as a superacid. Superacids are defined as acids that have acidity greater than the acidity of pure sulfuric acid.

Superacids such as fluorosulfuric acid and triflic acid (CF_3SO_3H) are about 1000 times stronger than sulfuric acid. Carborane acid ($H(CHB_{11}Cl_{11})$) is one million times stronger than sulfuric acid. The strongest known superacid is fluoroantimonic acid (H_2FSbF_6), which is 10^{16} times stronger than 100% sulfuric acid.

Superacids are used in the production of plastics and high-octane petrol, in coal gasification and in research.

Strength versus concentration

When referring to solutions of acids and bases, it is important not to confuse the terms 'strong' and 'weak' with 'concentrated' and 'dilute'. Concentrated and dilute describe the amount of acid or base dissolved in a given volume of solution. Hydrochloric acid is a strong acid because the HCl molecules totally ionise in solution. A concentrated solution of hydrochloric acid can be prepared by bubbling a large amount of hydrogen chloride into a given volume of water. By using only a small amount of hydrogen chloride, a dilute solution of hydrochloric acid would be produced.

However, in both cases, the hydrogen chloride is completely ionised—it is a strong acid. Similarly, a solution of ethanoic acid may be concentrated or dilute. However, as it is partially ionised, it is a weak acid (Figure 17.2.6).

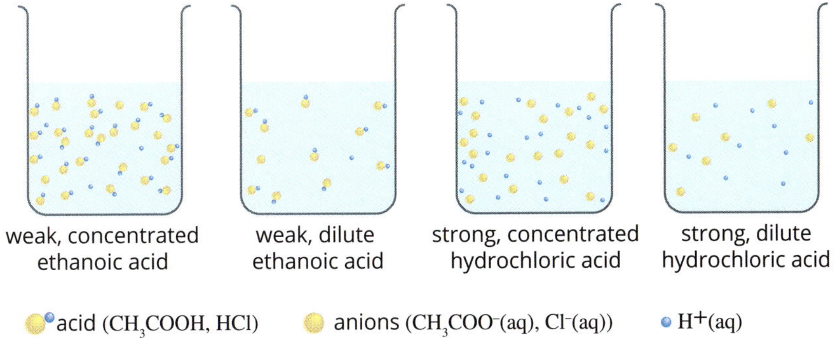

FIGURE 17.2.6 The concentration of ions in an acid solution depends on both the concentration and strength of the acid.

Terms such as 'weak' or 'strong' acids, or solutions classified as 'dilute' or 'concentrated' are qualitative (or descriptive) terms. Solutions can be given accurate, quantitative descriptions by stating concentrations in mol L^{-1} or g L^{-1}.

pH AND HYDROGEN ION CONCENTRATION

Definition of pH

The range of H^+ concentrations in solutions is so great that a convenient scale, called the **pH scale**, has been developed to measure acidity. The pH scale was first proposed by the Danish scientist Sören Sörenson in 1909 as a way of expressing levels of acidity. The pH of a solution is defined as:

$$pH = -\log_{10}[H^+]$$

where $[H^+]$ represents the concentration of $H^+(aq)$ ions in the solution, measured in $mol\,L^{-1}$.

Alternatively, this expression can be rearranged to give:

$$[H^+] = 10^{-pH}$$

The pH scale eliminates the need to write cumbersome powers of 10 to describe hydrogen ion concentration. The use of pH greatly simplifies the measurement and calculation of acidity. Since the scale is based upon the negative logarithm of the hydrogen ion concentration, the pH of a solution *decreases* as the concentration of hydrogen ions *increases*.

pH of acidic and basic solutions

Figure 17.2.7 shows a pH meter, which is used to accurately measure the pH of a solution.

Acidic, basic and neutral solutions can be defined in terms of their pH at 25°C.

- **Neutral solutions** have a pH equal to 7.
- **Acidic solutions** have a pH less than 7.
- **Basic solutions** have a pH greater than 7.

On the pH scale, the most acidic solutions have pH values slightly less than 0 and the most basic solutions have values of about 14. The pH values of some common substances are provided in Table 17.2.2.

FIGURE 17.2.7 A pH meter is used to measure the acidity of a solution.

TABLE 17.2.2 pH values of some common substances at 25°C

Solution	pH	$[H^+]$ ($mol\,L^{-1}$)
1.0 $mol\,L^{-1}$ HCl	0.0	1
lemon juice	3.0	10^{-3}
vinegar	4.0	10^{-4}
tomatoes	5.0	10^{-5}
rain water	6.0	10^{-6}
pure water	7.0	10^{-7}
seawater	8.0	10^{-8}
soap	9.0	10^{-9}
oven cleaner	13.0	10^{-13}
1.0 $mol\,L^{-1}$ NaOH	14.0	10^{-14}

A solution with pH 2 has 10 times the concentration of hydrogen ions as one of pH 3.

Measurement and understanding of pH are important in a large variety of everyday applications. For example, there is a complex system of pH control in your blood because even small deviations from the normal pH range of 7.35–7.45 for any length of time can lead to serious illness and death.

Calculations involving pH

pH can be calculated using the formula $pH = -\log_{10}[H^+]$. Your scientific calculator has a logarithm function that will simplify pH calculations.

Worked example 17.2.1

CALCULATING THE pH OF AN AQUEOUS SOLUTION FROM $[H^+]$

Calculate the pH of a solution in which the concentration of H^+ is $0.14\,mol\,L^{-1}$. Express your answer to 2 decimal places.	
Thinking	**Working**
Write down the concentration of H^+ ions in the solution.	$[H^+] = 0.14\,mol\,L^{-1}$
Substitute the value of $[H^+]$ into: $pH = -\log_{10}[H^+]$ Use the logarithm function on your calculator to determine the answer.	$pH = -\log_{10}[H^+]$ $= -\log_{10}(0.14)$ (use your calculator) $= 0.85$

Worked example: Try yourself 17.2.1

CALCULATING THE pH OF AN AQUEOUS SOLUTION FROM $[H^+]$

Calculate the pH of a solution in which the concentration of H^+ is $6.0 \times 10^{-9}\,mol\,L^{-1}$. Express your answer to 2 significant figures.

You will learn more about pH, including how to calculate the pH of basic solutions, in the Year 12 course.

17.2 Review

SUMMARY

- An Arrhenius acid ionises in water to produce hydrogen ions.
- In an ionisation reaction, a molecular substance reacts with a solvent such as water to form ions in solution.
- A molecule of a polyprotic acid can lose more than one hydrogen ion.
- The common properties of acids are due to their ability to form the hydrogen ion, H^+, in solution.
- An Arrhenius base dissociates in water, producing hydroxide ions, OH^-.
- In a dissociation process, a molecule or ionic compound separates or splits into smaller particles.
- The common properties of bases are due to the presence of the hydroxide ion, OH^-, in solution.
- Strong acids are completely ionised; weak acids are partially ionised.
- Strong bases are ionic compounds that are completely dissociated. Weak bases are usually molecular compounds that are partially ionised.
- A concentrated acid or base contains more moles of solute per litre than a dilute acid or base.
- The concentration of H^+ is measured using the pH scale:
  $pH = -\log_{10}[H^+]$
- At 25°C the pH of a neutral solution is 7. The pH of an acidic solution is less than 7 and the pH of a basic solution is greater than 7.

KEY QUESTIONS

1 An acid solution is formed when hydrogen bromide gas is bubbled into water. Write a balanced ionic equation for this reaction.

2 Write a balanced ionic equation for the reaction that occurs when H_2SO_4 is completely ionised in water.

3 Which one of the following statements correctly describes an Arrhenius base?

- **A** a molecular substance that ionises in water, forming H^+ ions
- **B** a molecular substance that ionises in water, forming OH^- ions
- **C** an ionic compound that dissociates in water, forming OH^- ions
- **D** an ionic compound that ionises in water, forming H^+ ions

4 Perchloric acid is a stronger acid than ethanoic acid. If you had $1.0\ mol\ L^{-1}$ solutions of each acid, which acid would you expect to be a better conductor of electricity? Explain why.

5 Calculate the pH of a solution in which $[H^+] = 0.01\ mol\ L^{-1}$.

6 Calculate the pH of a $0.001\ mol\ L^{-1}$ solution of nitric acid (HNO_3).

17.3 Reactions of acids and bases

Acids were originally grouped together because of their similar chemical behaviour. Chemists use indicators, such as litmus, to identify acidic solutions. Acids and bases react readily with many other chemicals, and some of the early definitions of acids and bases were derived from these reactions.

In this section, you will learn to use the patterns in the reactions of acids and bases to predict the products that are formed.

GENERAL REACTION TYPES INVOLVING ACIDS AND BASES

Acids and bases react in many ways. However, it is possible to group some reactions together according to the similarity of the reactants involved and products formed. While the identification of products should be based on experimental data, these groups, or reaction types, can be useful. The reaction types you will be studying are the reaction of acids with:

- metal hydroxides
- metal carbonates and hydrogen carbonates
- reactive metals.

You will also study the reaction of bases with ammonium salts.

ACIDS AND METAL HYDROXIDES

Soluble metal hydroxides, such as NaOH, dissociate in water to form metal cations and hydroxide ions, $OH^-(aq)$. The products of a reaction of an acid with a metal hydroxide are an ionic compound, called a **salt**, and water.

The general rule for the reaction between acids and metal hydroxides can be expressed as:

$$\text{acid} + \text{metal hydroxide} \rightarrow \text{salt} + \text{water}$$

For example, solutions of sulfuric acid and sodium hydroxide react to form sodium sulfate and water. This can be represented by the full (or overall) equation:

$$H_2SO_4(aq) + 2NaOH(aq) \rightarrow Na_2SO_4(aq) + 2H_2O(l)$$

The salt formed in the reaction between sulfuric acid and sodium hydroxide is sodium sulfate. If water were evaporated from the reaction mixture, solid sodium sulfate would be left behind.

Salts

Salts consist of the positive ion or cation from the base and the negative ion or anion from the acid. In Chapter 4, you saw that the names of anions that contain oxygen often end in '-ate'; for example, sulfate, phosphate. The names of anions that do not contain oxygen end in '-ide'; for example, chloride, bromide.

Table 17.3.1 lists the names of salts formed from some neutralisation reactions of acids with metal hydroxides. Note that the name of the positive ion is listed first and the name of the negative ion from the acid is listed second.

TABLE 17.3.1 Salts formed from some common neutralisation reactions

Reactants (acid + metal hydroxide)	Name of salt formed	Formulae of ions present in the salt solution
hydrochloric acid + potassium hydroxide	potassium chloride	$K^+(aq) + Cl^-(aq)$
hydrochloric acid + magnesium hydroxide	magnesium chloride	$Mg^{2+}(aq) + Cl^-(aq)$
nitric acid + sodium hydroxide	sodium nitrate	$Na^+(aq) + NO_3^-(aq)$
sulfuric acid + zinc hydroxide	zinc sulfate	$Zn^{2+}(aq) + SO_4^{2-}(aq)$
phosphoric acid + potassium hydroxide	potassium phosphate	$K^+(aq) + PO_4^{3-}(aq)$
ethanoic acid + calcium hydroxide	calcium ethanoate	$Ca^{2+}(aq) + CH_3COO^-(aq)$

Salts is the general name given to an ionic compound. They are therefore formed from a positive metal cation and a negative non-metal anion. Salts are often formed from reactions involving acids and bases.

IONIC EQUATIONS

The hydroxide ions from metal hydroxides, such as sodium hydroxide (NaOH), calcium hydroxide ($Ca(OH)_2$) and magnesium hydroxide ($Mg(OH)_2$), react readily with the hydrogen ion ($H^+(aq)$) from acids.

The reaction between an acid and a metal hydroxide can be represented by an ionic equation as well as by an overall equation.

When writing ionic equations, there are a few points to remember.

- Ionic equations are balanced with respect to both the number of atoms of each element and charge.
- **Spectator ions** are ions that are dissolved in the solution and are present as ions before and after the reaction, but are not involved in the reaction. These are omitted from the ionic equation.
- If a reactant or product is a solid, liquid or a gas, it cannot be written as ions and it must be present in the ionic equation.

You also need to remember when writing ionic equations for neutralisation reactions.

- Strong acids ionise in solution and are written as ions; for example, HCl in solution is written as $H^+(aq) + Cl^-(aq)$.
- Metal hydroxides and salts are ionic and, if soluble, dissociate in solution and are written as ions; for example, KOH dissolving in water is written as:

$$KOH(s) \rightarrow K^+(aq) + OH^-(aq)$$

- Water is a covalent molecular substance that does not ionise to any significant extent. It is written as $H_2O(l)$.

Figure 17.3.1 is a diagrammatic representation of the ions in solutions of HCl and NaOH when mixed in a neutralisation reaction.

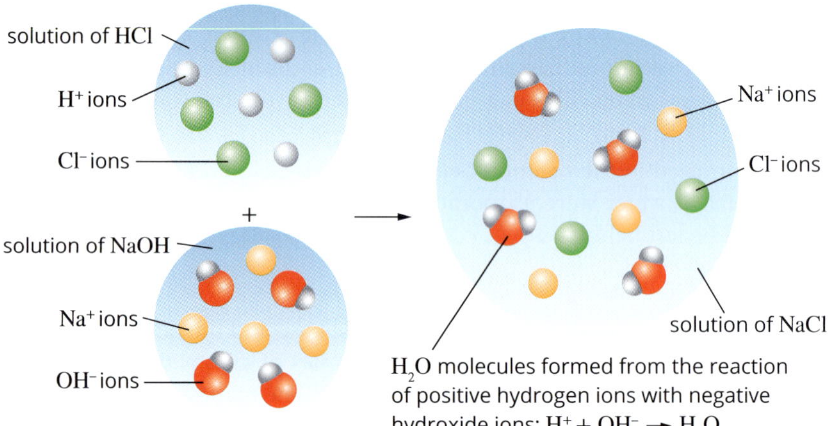

FIGURE 17.3.1 A representation of the reaction between $H^+(aq)$ and $OH^-(aq)$ ions that occurs when solutions of HCl and NaOH are mixed

Worked example 17.3.1 indicates the steps to follow when writing ionic equations for the reactions of acids with bases: **acid–base reactions**.

Worked example 17.3.1

WRITING AN IONIC EQUATION FOR AN ACID–BASE REACTION

Write an ionic equation for the reaction that occurs when hydrochloric acid is added to sodium hydroxide solution. A representation of this reaction is shown in Figure 17.3.1.

Thinking	Working
What is the general reaction? Identify the products formed.	acid + metal hydroxide → salt + water A solution of sodium chloride and water is formed.
Identify the reactants and products. Indicate the state of each, i.e. (aq), (s), (l) or (g).	Reactants: HCl(aq) is ionised in solution, forming $H^+(aq)$ and $Cl^-(aq)$. NaOH(aq) is dissociated in solution, forming $Na^+(aq)$ and $OH^-(aq)$. Products: Sodium chloride is dissociated and exists as $Na^+(aq)$ and $Cl^-(aq)$. Water is a molecular compound and its formula is $H_2O(l)$.
Write the equation showing all reactants and products, in ionised form where possible. (There is no need to balance the equation yet.)	$H^+(aq) + Cl^-(aq) + Na^+(aq) + OH^-(aq) \rightarrow Na^+(aq) + Cl^-(aq) + H_2O(l)$
Identify the spectator ions: the ions that have an (aq) state both as a reactant and as a product.	$Na^+(aq)$ and $Cl^-(aq)$
Rewrite the equation without the spectator ions. Balance the equation with respect to number of atoms of each element and charge.	$H^+(aq) + OH^-(aq) \rightarrow H_2O(l)$

Worked example: Try yourself 17.3.1

WRITING AN IONIC EQUATION FOR AN ACID–BASE REACTION

Write an ionic equation for the reaction that occurs when sulfuric acid is added to potassium hydroxide solution.

When writing ionic equations, if a reactant or product is a solid, liquid or a gas, it cannot be written as ions and it must be present in the ionic equation.

NEUTRALISATION REACTIONS

If a solution of a metal hydroxide is added to a solution of an acid, the hydroxide ions will react with the hydronium ions. The acid and base are said to have been **neutralised** at the point when all the hydroxide ions have reacted with all the hydrogen ions, forming water (H_2O).

CHEMISTRY IN ACTION

Benefits of neutralisation

A neutralisation reaction enables the adjustment of the acidity of a solution. If excess acid is harmful, it can be neutralised by adding a base. Conversely, an environment that is too alkaline can be neutralised by adding an acid. The following are some common examples.

- Methanoic acid (otherwise known as formic acid) is released from the sting of ants, bees and nettles. If affected skin is rinsed with limewater or a dilute solution of ammonia, the acid becomes neutralised and is no longer painful to the person affected.
- The venom from wasps is alkaline. A common treatment for wasp bites is to rinse the site of the bite with vinegar (ethanoic acid, CH_3COOH) as this neutralises the base within the venom.
- Excess acid in the stomach is the cause of indigestion. This condition is treated with substances that neutralise acids. Antacid tablets are bases that neutralise the excess acid in the stomach.
- The bacteria occurring on tooth enamel feed on the sugars present in food. The products of their metabolism are acids that attack the enamel, which leads to tooth decay. This is why toothpastes are weak bases.

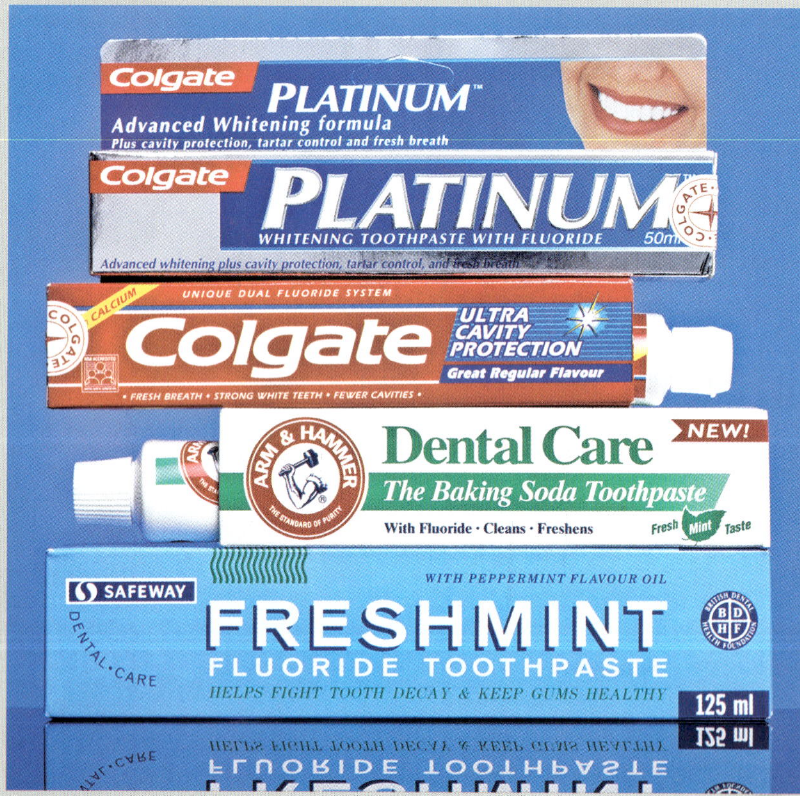

FIGURE 17.3.3 A selection of toothpastes. These products are advertised as providing protection against tooth cavities, gum disease and tooth decay.

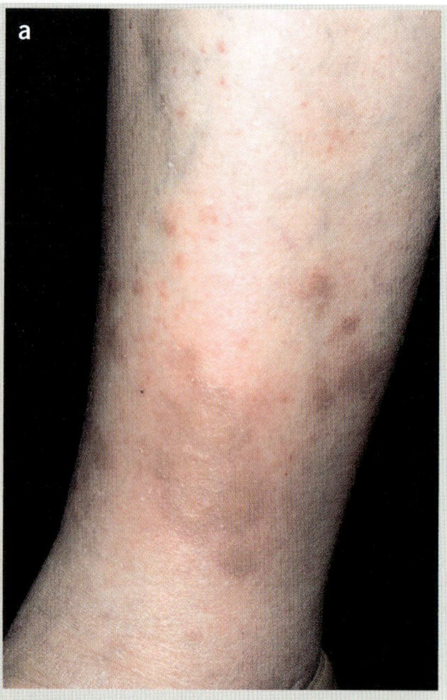

FIGURE 17.3.2 (a) Irritation on the leg of a person bitten by ants. (b) Scanning electron microscope image of a nettle (*Urtica* sp.) surface showing stinging, hair-like structures (colourised)

ACIDS AND METAL CARBONATES

The weathering of buildings and statues (Figure 17.3.4) is due in part to the reaction between acid rain and the carbonate minerals in the stone.

FIGURE 17.3.4 This statue has been heavily eroded by acid rain, which reacts with carbonate salts in limestone. Acid rain is formed when gases, such as sulfur dioxide and nitrogen oxides, dissolve in water to form acidic solutions.

Acids reacting with metal carbonates and metal hydrogencarbonates produce carbon dioxide gas, together with a salt and water. Metal carbonates include sodium carbonate (Na_2CO_3), magnesium carbonate ($MgCO_3$) and calcium carbonate ($CaCO_3$).

The general reaction for metal carbonates with acids can be summarised as:

$$\text{acid} + \text{metal carbonate} \rightarrow \text{salt} + \text{water} + \text{carbon dioxide}$$

For example, a solution of hydrochloric acid reacting with sodium carbonate solution produces a solution of sodium chloride, water and carbon dioxide gas. The reaction is represented by the equation:

$$2HCl(aq) + Na_2CO_3(aq) \rightarrow 2NaCl(aq) + CO_2(g) + H_2O(l)$$

The reaction between hydrochloric acid and sodium carbonate is represented in Figure 17.3.5.

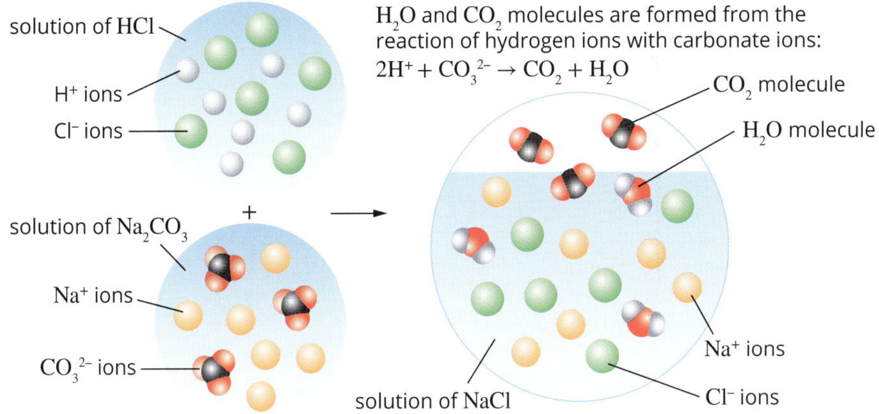

FIGURE 17.3.5 The reaction between solutions of sodium carbonate and hydrochloric acid

Metal hydrogencarbonates (also known as bicarbonates) include sodium hydrogencarbonate ($NaHCO_3$), potassium hydrogencarbonate ($KHCO_3$) and calcium hydrogencarbonate ($Ca(HCO_3)_2$). Acids added to metal hydrogencarbonates also produce carbon dioxide together with a salt and water. The general reaction is:

$$\text{acid} + \text{metal hydrogencarbonate} \rightarrow \text{salt} + \text{water} + \text{carbon dioxide}$$

For example, when a solution of hydrochloric acid and solid sodium hydrogencarbonate are mixed, the following reaction occurs:

$$HCl(aq) + NaHCO_3(s) \rightarrow NaCl(aq) + H_2O(l) + CO_2(g)$$

The reactions between acids and metal carbonates, and reactions between acids and metal hydrogencarbonates, can also be represented as ionic equations by following steps similar to the steps for writing reactions between acids and bases.

CHEMFILE

Bicarbonate of soda

Self-raising flour contains tartaric acid and some sodium hydrogencarbonate (bicarbonate of soda). It is used in baking cakes because on heating in the oven, the acid and hydrogencarbonate react. Carbon dioxide is released, which causes the cake mixture to rise.

FIGURE 17.3.6 Bicarbonate of soda leads to the production of carbon dioxide, causing these muffins to rise.

Worked example 17.3.2

WRITING IONIC EQUATIONS FOR REACTIONS BETWEEN ACIDS AND METAL CARBONATES

What products are formed when a dilute solution of nitric acid is added to solid magnesium carbonate? Write an ionic equation for this reaction.	
Thinking	**Working**
What is the general reaction? Identify the products.	acid + metal carbonate → salt + water + carbon dioxide Products of this reaction are magnesium nitrate in solution, water and carbon dioxide gas.
Identify the reactants and products. Indicate the state of each, i.e. (aq), (s), (l) or (g).	Reactants: Nitric acid is ionised in solution, forming $H^+(aq)$ and $NO_3^-(aq)$ ions. Magnesium carbonate is an ionic solid, $MgCO_3(s)$. Products: Magnesium nitrate is dissociated into $Mg^{2+}(aq)$ and $NO_3^-(aq)$ ions. Water has the formula $H_2O(l)$. Carbon dioxide has the formula $CO_2(g)$.
Write the equation showing all reactants and products. (There is no need to balance the equation yet.)	$H^+(aq) + NO_3^-(aq) + MgCO_3(s) \rightarrow Mg^{2+}(aq) + NO_3^-(aq) + H_2O(l) + CO_2(g)$
Identify the spectator ions.	$NO_3^-(aq)$
Rewrite the equation without the spectator ions. Balance the equation with respect to number of atoms of each element and charge.	$2H^+(aq) + MgCO_3(s) \rightarrow Mg^{2+}(aq) + H_2O^+(l) + CO_2(g)$

Worked example: Try yourself 17.3.2

WRITING IONIC EQUATIONS FOR REACTIONS BETWEEN ACIDS AND METAL CARBONATES

What products are formed when a solution of hydrochloric acid is added to a solution of sodium hydrogencarbonate? Write an ionic equation for this reaction.

FIGURE 17.3.7 In the limewater test, carbon dioxide is bubbled through limewater, turning the limewater cloudy.

Testing for carbonate salts

Acids can be used to detect the presence of carbonate salts. Carbon dioxide is produced when an acid is added to a carbonate.

The **limewater test** is a simple laboratory test used to confirm the presence of carbon dioxide gas. Limewater is a saturated solution of calcium hydroxide ($Ca(OH)_2$). When carbon dioxide is bubbled through this solution, it will turn 'milky' or 'cloudy' in appearance (Figure 17.3.7) due to the precipitation of calcium carbonate:

$$Ca(OH)_2(aq) + CO_2(g) \rightarrow CaCO_3(s) + H_2O(l)$$

ACIDS AND REACTIVE METALS

When dilute acids are added to main group metals and some transition metals, bubbles of hydrogen gas are released and a salt is formed. In general, the equation for the reaction is:

$$\text{acid} + \text{reactive metal} \rightarrow \text{salt} + \text{hydrogen}$$

Reactive metals include calcium, magnesium, iron and zinc, but *not* copper, silver or gold. For example, the reaction between dilute hydrochloric acid and zinc metal can be seen in Figure 17.3.8 and represented by the equation:

$$2HCl(aq) + Zn(s) \rightarrow ZnCl_2(aq) + H_2(g)$$

This reaction can also be represented by an ionic equation. In an aqueous solution, the hydrochloric acid is ionised and the salt, zinc chloride (a soluble ionic compound), is dissociated. Worked example 17.3.3 shows how the ionic equation can be determined.

FIGURE 17.3.8 Hydrogen gas is produced from the reaction of zinc with dilute hydrochloric acid.

Worked example 17.3.3

WRITING IONIC EQUATIONS FOR REACTIONS BETWEEN ACIDS AND REACTIVE METALS

Write an ionic equation for the reaction that occurs when dilute hydrochloric acid is added to a sample of zinc metal.	
Thinking	**Working**
What is the general reaction? Identify the products formed.	acid + reactive metal → salt + hydrogen Hydrogen gas and zinc chloride solution are produced.
Identify the reactants and products. Indicate the state of each, i.e. (aq), (s), (l) or (g).	Reactants: Zinc is a solid, $Zn(s)$. Hydrochloric acid is ionised, forming $H^+(aq)$ and $Cl^-(aq)$ ions. Products: Hydrogen gas, $H_2(g)$, forms. Zinc chloride is dissociated into $Zn^{2+}(aq)$ and $Cl^-(aq)$ ions.
Write the equation showing all reactants and products. (There is no need to balance the equation yet.)	$H^+(aq) + Cl^-(aq) + Zn(s) \rightarrow Zn^{2+}(aq) + Cl^-(aq) + H_2(g)$
Identify the spectator ions.	$Cl^-(aq)$
Rewrite the equation without the spectator ions. Balance the equation with respect to number of atoms of each element and charge.	$2H^+(aq) + Zn(s) \rightarrow Zn^{2+}(aq) + H_2(g)$

Worked example: Try yourself 17.3.3

WRITING IONIC EQUATIONS FOR REACTIONS BETWEEN ACIDS AND REACTIVE METALS

Write an ionic equation for the reaction that occurs when aluminium is added to a dilute solution of hydrochloric acid.

BASES WITH AMMONIUM SALTS

Ammonium salts are similar to metal salts except that the metal cation is replaced by the ammonium ion (NH_4^+). Some examples of ammonium salts are ammonium chloride (NH_4Cl) ammonium nitrate (NH_4NO_3) and ammonium sulfate ($(NH_4)_2SO_4$). Ammonium salts are highly soluble in water and dissociate into their constituent ions.

Ammonia gas (NH_3) and water are formed when bases react with ammonium salts:

$$\text{ammonium salt} + \text{base} \rightarrow \text{salt} + \text{ammonia} + \text{water}$$

For example, when a solution of sodium hydroxide is added to a solution of ammonium sulfate, sodium sulfate, ammonia gas and water are produced. The ammonia can be detected by its characteristic odour or by reaction with moist red litmus paper, which turns blue in the presence of ammonia.

The equation for this reaction is:

$$2NaOH(aq) + (NH_4)_2SO_4(aq) \rightarrow Na_2SO_4(aq) + 2NH_3(g) + 2H_2O(l)$$

The ionic equation can be written using the steps outlined earlier in this chapter. Remember that NaOH, $(NH_4)_2SO_4$ and Na_2SO_4 are ionic solids that dissociate in water and that NH_3 and H_2O are molecular compounds. Na^+ and SO_4^{2-} are spectator ions and take no part in this reaction. The ionic equation simplifies to:

$$NH_4^+(aq) + OH^-(aq) \rightarrow NH_3(g) + H_2O(l)$$

17.3 Review

SUMMARY

- Generalisations can be made about the likely products of reactions involving acids and bases:
  - acid + metal hydroxide → salt + water
  - acid + metal carbonate → salt + water + carbon dioxide
  - acid + metal hydrogencarbonate → salt + water + carbon dioxide
  - acid + reactive metal → salt + hydrogen
  - base + ammonium salt → salt + ammonia gas + water
- Each of these reactions can be represented by full and ionic equations.
- An ionic equation only shows those ions, atoms or molecules that take part in the reaction.
- Spectator ions (ions that do not take part in the reaction) are not included in ionic equations.
- Ionic equations are balanced with respect to both the number of atoms of each element and charge.

KEY QUESTIONS

1 Write full and ionic chemical equations for the following reactions.
   a magnesium and sulfuric acid
   b calcium and hydrochloric acid
   c zinc and ethanoic acid
   d aluminium and hydrochloric acid

2 Name the salt produced in each of the reactions in Question **1**.

3 For each of the following reactions, write:
   i a full chemical equation to represent the reaction (remember to include states)
   ii an ionic equation.
   a solid zinc oxide and sulfuric acid
   b solid calcium and nitric acid
   c solid copper(II) hydroxide and nitric acid
   d solid magnesium hydrogencarbonate and hydrochloric acid
   e solid tin(II) carbonate and sulfuric acid

4 Predict the products of the following reactions and write full and ionic chemical equations for each.
   a A solution of sulfuric acid is added to a solution of potassium hydroxide.
   b Nitric acid solution is mixed with sodium hydroxide solution.
   c Hydrochloric acid solution is poured onto some solid magnesium hydroxide.
   d Green copper(II) carbonate powder is added to dilute sulfuric acid.
   e Dilute hydrochloric acid is mixed with a solution of potassium hydrogencarbonate.
   f Dilute nitric acid is added to a spoon coated with solid zinc.
   g Hydrochloric acid solution is added to some marble chips (calcium carbonate).
   h Solid bicarbonate of soda (sodium hydrogencarbonate) is mixed with vinegar (a dilute solution of ethanoic acid).
   i A solution of ammonium carbonate is added to a solution of calcium hydroxide. The mixture is then warmed.

17.4 Calculations involving acids and bases

Although acids are frequently purchased as concentrated solutions, you will often need to use them in a more diluted form. For example, a bricklayer uses a 10% solution of hydrochloric acid to remove mortar splashes from bricks used to build a house. The brick-cleaning solution is prepared by diluting concentrated hydrochloric acid by a factor of 10.

In this section, you will learn how to calculate the concentration and pH of acids and bases once they have been diluted.

CONCENTRATION OF ACIDS AND BASES

The concentration of acids and bases is usually expressed in units of $mol\,L^{-1}$ (or M). This is also referred to as molar concentration or molarity.

You will recall from Chapter 16 that the molar concentration of a solution, in $mol\,L^{-1}$, is given by the expression:

$$c = \frac{n}{V}$$

where c is the molar concentration ($mol\,L^{-1}$), n is the amount of solute (mol) and V is the volume of the solution (L).

The most convenient way of preparing a solution of a dilute acid is by mixing concentrated acid with water, as shown in Figure 17.4.1. This is known as a dilution.

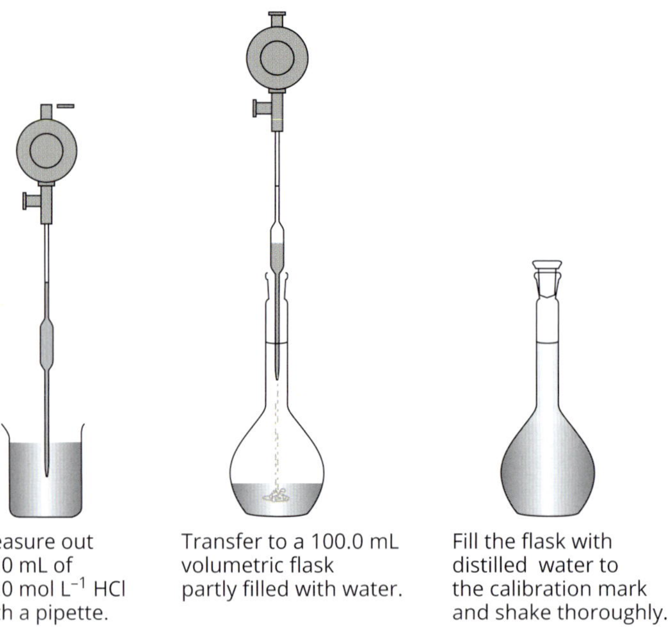

FIGURE 17.4.1 Preparing a 1.00 $mol\,L^{-1}$ HCl solution by diluting a 10.0 $mol\,L^{-1}$ solution. Heat is released when a concentrated acid is added to water, so the volumetric flask is partly filled with water before the acid is added. (Extra safety precautions would be required for diluting concentrated sulfuric acid.)

The amount of solute (in moles) in a solution does not change when a solution is diluted; the volume of the solution increases and the concentration decreases. The change in concentration or volume can be calculated using the formula:

$$c_1 V_1 = c_2 V_2$$

where c_1 and V_1 are the initial concentration and volume, and c_2 and V_2 are the concentration and volume after dilution.

The concentration of a dilute acid can be calculated if the following are known:

- volume of the concentrated solution
- concentration (molarity) of the concentrated solution
- total volume of water added.

The molar concentration of some concentrated acids are shown in Table 17.4.1.

TABLE 17.4.1 Molar concentrations of some concentrated acids

Concentrated acid (% by mass)	Formula	Molar concentrations (mol L^{-1})
ethanoic acid (99.5%)	CH_3COOH	17
hydrochloric acid (36%)	HCl	12
nitric acid (70%)	HNO_3	16
phosphoric acid (85%)	H_3PO_4	15
sulfuric acid (98%)	H_2SO_4	18

In the laboratory, you can prepare solutions of a base of a required concentration by diluting a more concentrated solution or dissolving a weighed amount of the base in a measured volume of water, as shown in Figure 17.4.2.

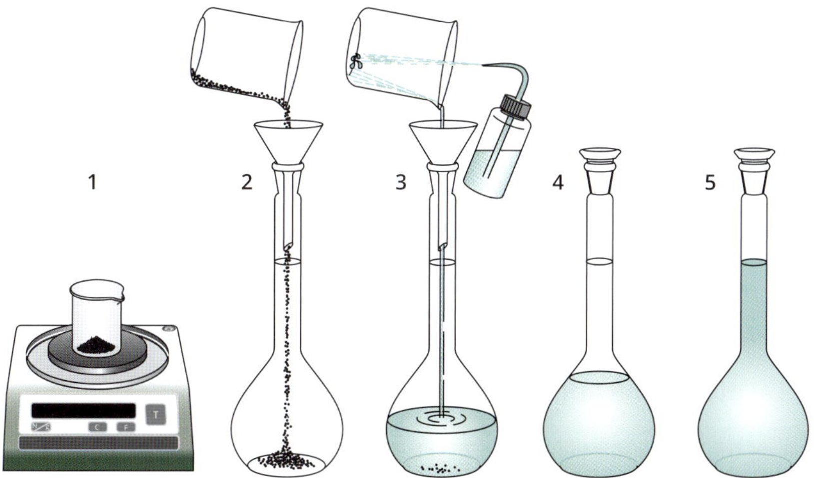

1 Accurately weigh out a mass of the base.
2 Transfer the base to a volumetric flask.
3 Ensure complete transfer of the base by washing with water.
4 Dissolve the base in water.
5 Add water to make the solution up to the calibration mark and shake thoroughly.

FIGURE 17.4.2 Preparing a solution by dissolving a weighed amount of base in a measured volume of water

Worked example 17.4.1

CALCULATING MOLAR CONCENTRATION AFTER DILUTION

Calculate the molar concentration of hydrochloric acid when 10.0 mL of water is added to 5.0 mL of 1.2 mol L^{-1} HCl.

Thinking	Working
The number of moles of solute does not change during a dilution. So $c_1V_1 = c_2V_2$, where *c* is the concentration in mol L^{-1} and *V* is the volume of the solution. (Each of the volume units must be the same, although not necessarily litres.)	$c_1V_1 = c_2V_2$
Identify given values for concentrations and volumes before and after dilution. Identify the unknown.	10.0 mL was added to 5.0 mL, the final volume is 15.0 mL. (In practice, small volume changes can occur when solutions are mixed; assume no volume changes occur.) $c_1 = 1.2$ mol L^{-1} $V_1 = 5.0$ mL $V_2 = 15.0$ mL You are required to calculate, c_2, the concentration after dilution.
Transpose the equation and substitute the known values into the equation to find the required value.	$c_2 = \frac{c_1 \times V_1}{V_2}$ $= 1.2 \times \frac{5.0}{15.0}$ $= 0.40$ mol L^{-1}

Worked example: Try yourself 17.4.1

CALCULATING MOLAR CONCENTRATION AFTER DILUTION

Calculate the molar concentration of nitric acid when 80.0 mL of water is added to 20.0 mL of 5.00 mol L^{-1} HNO_3.

Worked example 17.4.2

CALCULATING THE VOLUME OF WATER TO BE ADDED IN A DILUTION

How much water must be added to 30.0 mL of 2.50 mol L^{-1} HCl to dilute the solution to 1.00 mol L^{-1}?

Thinking	Working
The number of moles of solute does not change during a dilution. So $c_1V_1 = c_2V_2$, where *c* is the concentration in mol L^{-1} and *V* is the volume of the solution. (Each of the volume units must be the same, although not necessarily litres.)	$c_1V_1 = c_2V_2$
Identify given values for concentrations and volumes before and after dilution. Identify the unknown.	$c_1 = 2.50$ mol L^{-1} $V_1 = 30.0$ mL $c_2 = 1.00$ mol L^{-1} You are required to calculate V_2, the volume of the diluted solution.

Transpose the equation and substitute the known values into the equation to find the required value.	$V_2 = \frac{c_1 \times V_1}{c_2}$ $= \frac{2.50 \times 30.0}{1.00}$ $= 75.0\,mL$
Calculate the volume of water to be added.	volume of dilute solution = 75.0 mL initial volume of acid = 30.0 mL so 75.0 – 30.0 = 45.0 mL of water must be added.

Worked example: Try yourself 17.4.2

CALCULATING THE VOLUME OF WATER TO BE ADDED IN A DILUTION

How much water must be added to 15.0 mL of 10.0 mol L^{-1} NaOH to dilute the solution to 2.00 mol L^{-1}?

EFFECT OF DILUTION ON pH OF STRONG ACIDS AND BASES

Consider a 0.1 mol L^{-1} solution of hydrochloric acid. HCl is a strong acid, so the concentration of H^+ ions in this solution is 0.1 mol L^{-1}. Since $pH = -\log_{10}[H^+]$, the pH of this solution is 1.0 (see Section 17.2 if you need to review this formula).

If a 1.0 mL solution is diluted by a factor of 10 to 10.0 mL by the addition of 9.0 mL of water, the concentration of H^+ ions decreases to 0.01 mol L^{-1} and the pH increases to 2.0. A further dilution by a factor of 10 to 100 mL will increase pH to 3.0. However, note that when acids are repeatedly diluted, the pH cannot increase above 7.

We will now look at how to calculate the pH of solutions of strong acids after dilution. (A discussion of the effect of dilution on the pH of solutions of weak acids and bases is complex and beyond the scope of this course.)

Worked example 17.4.3

CALCULATING THE pH OF A DILUTED ACID

If 5.0 mL of 0.010 mol L^{-1} HNO_3 is diluted to 100.0 mL, what is the pH of the diluted solution?

Thinking	Working
Identify given values for concentrations and volumes before and after dilution.	$c_1 = 0.010\,mol\,L^{-1}$ $V_1 = 5.0\,mL$ $V_2 = 100.0\,mL$ $c_2 = ?$
Calculate c_2, which is the concentration of H^+ after dilution, by transposing the formula: $c_1V_1 = c_2V_2$	$c_2 = \frac{c_1 \times V_1}{V_2}$ $= \frac{0.010 \times 5.0}{100.0}$ $= 0.00050\,mol\,L^{-1}$
Calculate pH using: $pH = -\log_{10}[H^+]$ Use the logarithm function on your calculator to determine pH.	$pH = -\log_{10}[H^+]$ $= -\log_{10}(0.00050)$ $= 3.30$

Worked example: Try yourself 17.4.3

CALCULATING THE pH OF A DILUTED ACID

If 10.0 mL of 0.1 mol L^{-1} HCl is diluted to 30.0 mL, what is the pH of the diluted solution?

REACTING QUANTITIES OF ACIDS AND BASES

Calculations based on the reactions of acids usually involve determining the number of mole of a substance.

You will recall the following from earlier chapters.

- The coefficients in a balanced chemical equation gives the ratio in which substances react.
- The amount of solid, liquid or gas (in mole) can be calculated from the expression $n = \frac{m}{M}$, where m is the mass in gram and M is the molar mass in g mol^{-1}.
- The amount of solute in a solution is given by $n = cV$, where c is the concentration in mol L^{-1} and V is the volume in litres.
- The amount, in mol, of a gas at STP is given by $n = \frac{V}{22.71}$, where V is the volume of the gas in litres.

The major steps in solving calculation problems involving acids are bases are as follows.

1 Write a balanced equation for the reaction.

2 Calculate the amount, in mol, of the substance with the known volume and concentration.

3 Use the mole ratio from the equation to calculate the amount, in mol, of the required substance.

The steps in a stoichiometric calculation are summarised in Figure 17.4.3.

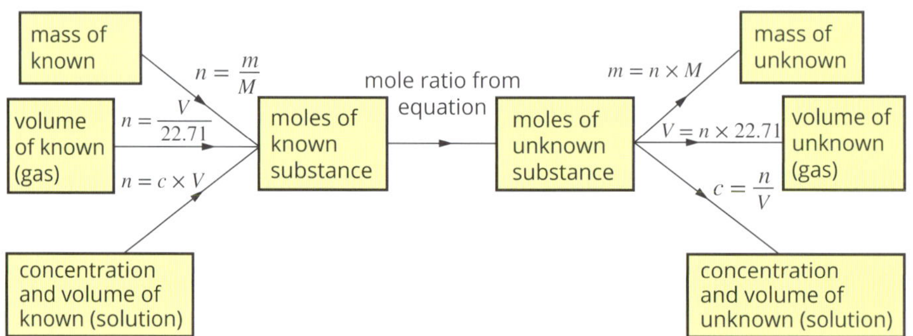

FIGURE 17.4.3 Flow chart for mass, solution and gas calculations involving reactions

Worked example 17.4.4

DETERMINING THE VOLUME OF AN UNKNOWN SOLUTION FROM A NEUTRALISATION REACTION

What volume of 0.100 mol L^{-1} sulfuric acid reacts completely with 17.8 mL of 0.150 mol L^{-1} potassium hydroxide?	
Thinking	**Working**
Write a balanced chemical equation.	$2KOH(aq) + H_2SO_4(aq) \rightarrow K_2SO_4(aq) + 2H_2O(l)$
Calculate the amount, in mol, of the solution of known concentration and volume.	$n(KOH) = cV = 0.150 \times 0.0178 = 0.00267$ mol
Determine the mole ratio as given by the balanced chemical equation.	mole ratio $= \frac{n(\text{unknown})}{n(\text{known})} = \frac{1}{2}$
Determine the amount, in mol, of the unknown solution.	$n(H_2SO_4) = 0.00267 \times \frac{1}{2} = 0.00134$ mol
Determine the volume of the unknown solution needed to react.	$V(H_2SO_4) = \frac{n}{c} = \frac{0.00134}{0.100} = 0.0134$ L $= 13.4$ mL Therefore, 13.4 mL of H_2SO_4 is required to react with 17.8 mL of 0.150 mol L^{-1} KOH.

Worked example: Try yourself 17.4.4

DETERMINING THE VOLUME OF AN UNKNOWN SOLUTION FROM A NEUTRALISATION REACTION

Calculate the volume of 0.500 mol L^{-1} hydrochloric acid (HCl) that reacts completely with 25.0 mL of 0.100 mol L^{-1} calcium hydroxide ($Ca(OH)_2$) solution.

STOICHIOMETRY PROBLEMS INVOLVING EXCESS REACTANTS

Reactants are not always mixed in stoichiometric amounts in acid–base reactions. There are occasions when one of the reactants, an acid or a base, is in excess. Follow these steps to determine the **limiting reagent** and the amount of the unknown substance.

- Calculate the number of moles of each reactant.
- Determine which reactant is in excess and which is the limiting reactant.
- Use the amount of limiting reactant to work out the amount of product formed or the amount of reactant in excess.

Worked example 17.4.5

A LIMITING REAGENT ACID–BASE PROBLEM

A sample of 20.0 mL of a 1.00 mol L^{-1} LiOH solution is added to 30.0 mL of a 0.500 mol L^{-1} HNO_3 solution.

a Which reactant is the limiting reagent?

b What mass of $LiNO_3$ will the reaction mixture contain when the reaction is complete?

Thinking	Working
Write a balanced chemical equation.	$LiOH(aq) + HNO_3(aq) \rightarrow LiNO_3(aq) + H_2O(l)$
Calculate the amount, in mol, of the first reactant.	$n(LiOH) = cV = 1.00 \times 0.0200 = 0.0200$ mol
Calculate the amount, in mol, of the second reactant.	$n(HNO_3) = cV = 0.500 \times 0.0300 = 0.0150$ mol
Use the coefficients of the reaction to determine the limiting reagent.	The equation shows that 1 mole of HNO_3 reacts with 1 mole of LiOH. So the HNO_3 is the limiting reagent.
Find the mole ratio of the unknown substance to the limiting reagent from the equation coefficients.	mole ratio $= \frac{n(\text{unknown})}{n(\text{known})} = \frac{1}{1}$
Determine the amount, in mol, of the unknown substance.	$n(LiNO_3) = 0.00150$ mol
Determine the mass, in g, of the unknown substance.	$(LiNO_3) = nM = 0.00150 \times 68.98 = 1.035$ g $= 1.04$ g

Worked example: Try yourself 17.4.5

A LIMITING REAGENT ACID–BASE PROBLEM

A sample of 30.0 mL of a 0.100 mol L^{-1} H_2SO_4 solution is mixed with 40.0 mL of a 0.200 mol L^{-1} KOH solution.

a Which reactant is the limiting reagent?

b What will be the mass of K_2SO_4 produced by this reaction?

17.4 Review

SUMMARY

- Amounts of acid or base in solution do not change during a dilution. The volume of the solution increases and its concentration decreases.
- Solutions of acids and bases of a required concentration can be prepared by diluting more concentrated solutions using the formula:

$$c_1V_1 = c_2V_2$$

where c_1 and V_1 are the initial concentration and volume, and c_2 and V_2 are the final concentration and volume after dilution.
- The pH increases when a solution of an acid is diluted.
- The pH of a diluted acid can be determined by calculating the concentration of hydrogen ions in the diluted solution.
- The pH decreases when a solution of a base is diluted.
- Given the quantity of one of the reactants or products in a chemical reaction, the quantity of all other reactants and products can be predicted by working through the following steps.
  1. Write a balanced equation for the reaction.
  2. Calculate the amount (in mol) of the given substance.
  3. Use the mole ratios of reactants and products in the balanced chemical equation to calculate the amount (in mol) of the required substance.
  4. Convert the amount (in mol) of the required substance to the units required in the question.

KEY QUESTIONS

1. If 1.0 L of water is added to 3.0 L of 0.10 mol L^{-1} HCl, what is the concentration of the diluted acid?
2. How much water must be added to 10 mL of a 2.0 mol L^{-1} sulfuric acid solution to dilute it to 0.50 mol L^{-1}?
3. What volume of water must be added to dilute a 20.0 mL volume of 0.600 mol L^{-1} HCl to 0.100 mol L^{-1}?
4. Describe the effect on the pH of a monoprotic acid solution of pH 1.0 when it is diluted by a factor of 10.
5. A sample of 18.26 mL of dilute nitric acid reacts completely with 20.00 mL of 0.09927 mol L^{-1} potassium hydroxide solution.
   a. Write a balanced chemical equation for the reaction between nitric acid and potassium hydroxide.
   b. Calculate the amount, in mol, of potassium hydroxide consumed in this reaction.
   c. What amount, in mol, of nitric acid reacted with the potassium hydroxide in this reaction?
   d. Calculate the concentration of the nitric acid.
6. A sample of 10.0 mL of a 0.200 mol L^{-1} sulfuric acid solution is added to 16.0 mL of a 0.100 mol L^{-1} sodium carbonate solution. The equation for the reaction is:

   $Na_2CO_3(aq) + H_2SO_4(aq) \rightarrow Na_2SO_4(aq) + CO_2(g) + H_2O(l)$

   a. Calculate the number of moles of H_2SO_4 in the sulfuric solution.
   b. Calculate the number of moles of Na_2CO_3 in the sodium carbonate solution.
   c. Identify the limiting reagent.

Chapter review

17

KEY TERMS

acid
acid–base reaction
acidic solution
acidity
alkali
Arrhenius model
base
basic solution
diprotic acid
dissociation
hydroxide ion
indicator
ionisation
limewater test
limiting reagent
monoprotic acid
neutral solution
neutralisation reaction
neutralise
pH
pH scale
polyprotic acid
salt
spectator ions
strong acid
strong base
triprotic acid
weak acid
weak base

Properties of acids and bases

1 Describe how you could use red litmus paper to distinguish between an acidic and an alkaline solution.

2 Using suitable examples, distinguish between:

a a diprotic and a monoprotic substance

b a strong acid and a concentrated acid.

3 Draw a structural formula of the monoprotic ethanoic acid molecule. Identify which proton is donated in an acid–base reaction.

The Arrhenius model of acids and bases

4 Describe what is meant by a 'scientific model'.

5 Using the Arrhenius model of acids and bases, explain the difference between strong and weak acids.

6 Write an equation to show perchloric acid ($HClO_4$) acting as a strong acid.

7 Write an equation to show hypochlorous acid ($HClO_3$) acting as a weak acid.

8 Write an equation to show ammonia (NH_3) acting as a weak base in water.

9 Human blood has a pH of 7.4. Is blood acidic, basic or neutral?

10 The pH of a cola drink is 3 and of black coffee is 5. How many more times acidic is the cola than black coffee?

11 Calculate the pH of each of the following solutions at 25°C.

a $0.01\ mol\ L^{-1}$ solution of nitric acid (HNO_3)

b $0.20\ mol\ L^{-1}$ solution of hydrochloric acid (HCl)

c $2.0\ mol\ L^{-1}$ solution of sulfuric acid (H_2SO_4)

Reactions of acids and bases

12 Complete, and balance, the following chemical equations.

a $HNO_3(aq) + KOH(aq) \rightarrow$

b $H_2SO_4(aq) + K_2CO_3(aq) \rightarrow$

c $H_3PO_4(aq) + Ca(HCO_3)_2(s) \rightarrow$

d $HF(aq) + Zn(OH)_2(s) \rightarrow$

13 Which one of the following correctly identifies all the products formed when magnesium hydroxide reacts with hydrochloric acid?

A water

B chloride ions

C magnesium ions

D magnesium chloride precipitate

E water, magnesium ions and chloride ions

14 Acids can be represented by a general formula HA. Write a balanced ionic equation for the reaction between an acid HA and a soluble metal carbonate MCO_3.

15 Dilute hydrochloric acid is added to a white solid in a test-tube. A colourless gas is produced. The gas turns limewater cloudy. What is a possible identity of the white solid?

16 Hydrogen is produced when dilute sulfuric acid reacts with aluminium metal.

a Write a balanced full equation for this reaction.

b Write a balanced ionic equation for this reaction.

Calculations involving acids and bases

17 A solution of hydrochloric acid has a pH of 2.

a What is the molar concentration of hydrogen ions in the solution?

b What amount of hydrogen ion, in mol, would be present in 500 mL of this solution?

18 Calculate the pH of each of the following mixtures at 25°C.

a 10 mL of $0.25\ mol\ L^{-1}$ HCl diluted to 50 mL of solution

b 10 mL of $0.15\ mol\ L^{-1}$ HCl diluted to 1.5 L of solution

19 The molarity of concentrated sulfuric acid is $18.0\ mol\ L^{-1}$. What volume of concentrated sulfuric acid is required to prepare a 1.00 L solution of $2.00\ mol\ L^{-1}$ H_2SO_4?

20 A sample of 40.0 mL of $0.10\ mol\ L^{-1}$ HNO_3 is diluted to 500.0 mL. Calculate whether the pH will increase or decrease.

Connecting the main ideas

21 A laboratory assistant forgot to label 0.1 mol L^{-1} solutions of sodium hydroxide (NaOH), hydrochloric acid (HCl), glucose ($C_6H_{12}O_6$), ammonia (NH_3) and ethanoic acid (CH_3COOH). In order to identify them, temporary labels A–E were placed on the bottles and the electrical conductivity and pH of each solution were measured. The results are shown in the table below. Identify each solution and briefly explain your reasoning.

Solution	Electrical conductivity	pH
A	poor	11
B	zero	7
C	good	13
D	good	1
E	poor	3

22 Write concise definitions for:

a an Arrhenius base
b strong acid
c molarity.

CHAPTER

18 Rates of chemical reactions

Chemical reactions occur at many different rates. The explosion of gunpowder and the combustion of petrol in a car's engine occur very quickly. On the other hand, the chemical weathering of limestone buildings and statues, the ripening of fruit and the rusting of iron all occur quite slowly.

During chemical reactions, particles such as atoms, molecules and ions collide with each other and undergo a rearrangement to produce new substances. It is important to appreciate that collisions between reactant particles do not always result in a chemical reaction. For example, while a car's fuel tank is being filled with petrol, the hydrocarbon molecules in the petrol are colliding with oxygen molecules in the air without a reaction occurring.

By the end of this chapter, you will understand how rates of chemical reactions can be measured. You will also be able to describe how varying the conditions of chemical reactions can affect the rate of a reaction.

Using collision theory, you will learn to predict the effects of concentration of solutions, temperature, surface area, gas pressure and the presence of a catalyst on the rate of chemical reactions, and explain these effects.

Science understanding

- the rate of chemical reactions can be quantified by measuring the rate of formation of products or the depletion of reactants
- varying the conditions under which chemical reactions occur can affect the rate of the reaction
- collision theory can be used to explain and predict the effects of concentration, temperature, pressure, the presence of catalysts and surface area on the rate of chemical reactions
- the activation energy is the minimum energy required for a chemical reaction to occur and is related to the strength and number of the existing chemical bonds; the magnitude of the activation energy influences the rate of a chemical reaction
- energy profile diagrams, which can include the transition state and catalysed and uncatalysed pathways, can be used to represent the enthalpy changes and activation energy associated with a chemical reaction

18.1 Investigating the rate of chemical reactions

FIGURE 18.1.1 How quickly do the chemical reactions involved in baking occur?

The time it takes for a batch of Anzac biscuits to cook in the oven (Figure 18.1.1) and the time taken for a fibreglass patch on a surfboard to set are processes related to the rate of chemical reactions.

In this section, you will learn how changes to reaction conditions affect reaction rates. You will also learn how chemists measure the rate at which a chemical reaction occurs.

FAST AND SLOW REACTIONS

Chemical reactions are taking place all around us:

- in the soil and rocks beneath our feet
- in the air around and above us
- inside every plant and animal
- in our homes, schools and workplaces.

Some of these reactions are over in a flash. In a car accident when a car's airbag needs to be inflated, the chemical reactions producing the gas that expands the airbag need to happen extremely quickly. On the other hand, if the car's painted surface is scratched to expose the metal beneath, the rusting reactions take place at a very slow rate.

FACTORS THAT AFFECT REACTION RATES

Experimental investigations have shown that five main factors can change the rate of a chemical reaction:

- surface area of solid reactants
- concentration of reactants in a solution
- gas pressure
- temperature
- the presence of catalysts.

You can probably think of some examples of situations where one or more of these conditions is changed and a reaction becomes noticeably faster or slower.

FIGURE 18.1.2 Particle size can be used to control the rate of reaction and create different effects during fireworks displays.

Surface area

The surface area of solid reactants can have a significant effect on reaction rate. Smaller particles have a much larger surface area than the same mass of large particles. As a result, the smaller particles react much faster.

Manufacturers of fireworks modify the surface area of solid reactants to control the rate at which fuels in the fireworks burn and create different effects (Figure 18.1.2). For example, very small pieces of aluminium confined in a shell explode violently. If larger pieces of aluminium are used, the reaction is slower and sparks are seen as pieces of burning metal being ejected.

FIGURE 18.1.3 This limestone statue has become pitted in recent years. Limestone (calcium carbonate) reacts more rapidly with the increased concentration of hydrogen ions in rainwater.

Concentration

The concentration of solutes dissolved in a solution can influence the rate of their reactions: higher concentrations usually lead to increased reaction rates.

Pollutants, such as sulfur dioxide and nitrogen dioxide, are released by cars and many industrial processes. When these compounds react with rainwater, acids such as sulfurous acid and nitric acid are formed. This causes the rainwater's hydrogen ion concentration to increase significantly and become acid rain. The increasing acidity of rain over the past 200 years has caused many famous marble buildings and statues to deteriorate much more rapidly due to the reaction between marble and the acids (Figure 18.1.3).

CHEMISTRY IN ACTION

Saved by a very fast chemical reaction

Imagine the scene. An 18-year-old borrows his parents' car to take his girlfriend for a drive to celebrate gaining his driver's licence. Roof down, enjoying the beautiful afternoon and the countryside, the driver rounds a corner to find the road wet. The car begins to slide on the wet surface. In his inexperience, the driver brakes; the car starts to spin. Suddenly, the car is leaving the road and heading straight for a large tree. Then, bang!

Later, the car was estimated to have been travelling at 60 km/h when it hit the tree. The collision was a 'head-on', with the front and passenger side taking most of the impact. Yet the girl in the passenger seat suffered just a chipped tooth, and her boyfriend sustained only minor bruising.

This is the true story of a lucky escape, thanks to a very rapid chemical reaction. As the collision took place, airbags were inflating and then deflating as the travellers were slammed forward towards the windscreen. The driver described it as being 'all over in a flash' and had no clear recollection of the airbags going off.

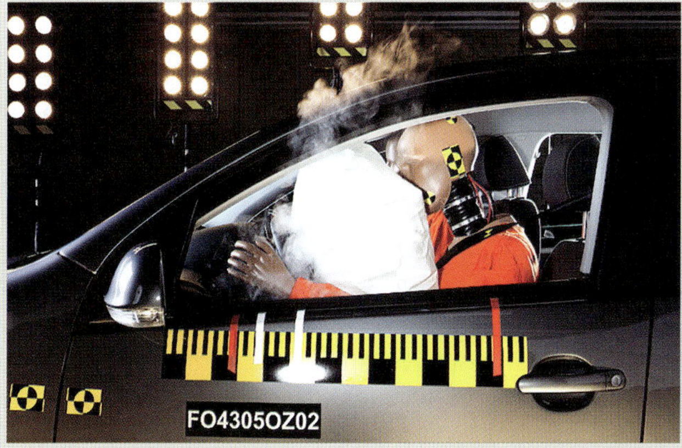

FIGURE 18.1.4 Airbags are deployed within 30 milliseconds of impact.

Hidden in the car's steering wheel, dashboard and windscreen pillars, special nylon bags fill with gas within 30 milliseconds of impact (Figure 18.1.4). As a consequence, the car's occupants are prevented from smashing their heads against the steering wheel, dashboard, windscreen or front pillars, all within the blink of an eye. As the head and body strike the airbags, the cushion of gas is forced out of the bag through tiny vents, and within 100 milliseconds the bag has completely deflated.

Air bags contain a mixture of crystalline solids—sodium azide (NaN_3), potassium nitrate (KNO_3) and silica (SiO_2)—stored in a canister. Sensors in the front of the car detect the difference between a bump and life-threatening impact. When a response is required, an electronic impulse initiates a series of three separate reactions. The electronic impulse 'ignites' the sodium azide. Sodium metal and hot nitrogen gas are the products of this energy-releasing, exothermic reaction:

$$2NaN_3(s) \rightarrow 2Na(l) + 3N_2(g)$$

The pulse of hot nitrogen gas released from this reaction starts to inflate the nylon bag. The molten sodium metal immediately reacts with the potassium nitrate, generating more nitrogen gas, as well as sodium oxide and potassium oxide, which are white powdery solids.

The equation for this second reaction is:

$$10Na(l) + 2KNO_3(s) \rightarrow K_2O(s) + 5Na_2O(s) + N_2(g)$$

A filtration system prevents any of the oxides from entering the nylon bag, while a third reaction 'captures' them to produce a harmless glassy solid.

In this third reaction the metal oxides combine with silica:

$$K_2O(s) + Na_2O(s) + SiO_2(s) \rightarrow \text{alkaline silicate ('glass')}$$

Chemical reactions do save lives!

Pressure

In reactions involving gases, increasing the pressure on a reaction increases the rate at which the reaction takes place. Increasing the pressure at constant temperature will result (on average) in the reactant particles becoming closer together.

This will increase the chance of the gas molecules colliding, and therefore increase the rate of reaction. Increasing the pressure of a reacting gas is the same as increasing the concentration because you have the same mass in a smaller volume.

For this reason, engineers often employ high gas pressures in their design of chemical processes that use gas-phase reactions. An example is the production of ammonia gas by reacting hydrogen gas and nitrogen gas. Increasing the pressure ensures a faster rate of reaction.

Temperature

As every cook knows, the temperature of an oven affects the rate of the chemical reactions during baking. The higher the temperature, the more rapidly the reactions occur.

FIGURE 18.1.5 Food is stored at low temperatures in a refrigerator to slow down the rate of reactions that cause food spoilage.

On the other hand, in hot weather it is wise to store fruit and vegetables in the refrigerator so that the chemical reactions that cause them to over-ripen and then spoil will be slowed down at lower temperatures (Figure 18.1.5).

Catalysts

Some chemical reactions occur much more rapidly if another substance is added to the reaction mixture. Such substances are called **catalysts**. Catalysts allow the reaction to follow a more energetically favourable pathway.

For example, if you chew a piece of dry biscuit or bread for several minutes, you may notice it tasting much sweeter. This happens because there is a catalyst present in your saliva that speeds up the breakdown of starch into sweet-tasting sugars.

In the following sections, you will learn how each of these factors can cause these changes in reaction rate.

MEASURING RATES OF REACTION EXPERIMENTALLY

The rate of reaction is defined as the change in concentration of a reactant or product per unit time.

The **rate of reaction** is defined as the change in concentration of a reactant or product per unit time.

To experimentally determine the rate of reaction, either directly or indirectly, you need to measure how much of a reactant is being used up or how much of a product is being formed in a given time period.

When a reaction involves gaseous products, this might involve measuring changes in mass or gas volume with time.

The graphs shown in Figure 18.1.7 were obtained by measuring the volume of carbon dioxide produced in the reaction between marble chips and hydrochloric acid. The experiment was performed twice, first with large marble chips, then with small marble chips. The rate of the reactions is shown by the gradient of the graphs. The graphs demonstrate two phenomena.

Firstly, the reaction rate of both reactions slows as the reactants are used up. The concentration of the acid will always be highest at the start of the reaction so it makes sense that the initial rate will be the fastest. In both of these examples, the graph levels out, indicating that the reaction that is producing the gas has finished, and one of the reactants has been totally consumed.

Secondly, the steeper initial gradient of the graph with small marble chips indicates that the rate of production of carbon dioxide gas is faster with the marble chips that have a higher surface area. Note that in comparing the two situations, in this case the size of the marble chips, a valid comparison can only be made of the initial rate. This is because as soon as the reaction starts, other variables such as concentration of reactants, and possibly temperature will be changing.

Colour changes and pH changes can also be used to follow the rate of some reactions, depending on the colour and acidity of reactants and products.

CHEMFILE

Fireworks and nanoparticles

Studies have shown that the fireworks (Figure 18.1.6) associated with celebrations such as New Year's Eve and the Lantern Festival in China can significantly increase air pollution levels. Sulfur dioxide and particles of metals are released when fireworks explode. These can cause breathing problems and lung disease.

Recent research aimed at lowering the environmental impact of fireworks has focused on reducing the particle size of chemicals used as fuels in the fireworks. By using smaller nanoparticles, reaction rates are increased and smaller amounts of chemicals are needed for the same performance. Thus, fewer pollutants are released into the atmosphere.

However, this new approach carries increased risks because fireworks made of such small particles could be even more explosive.

FIGURE 18.1.6 Fireworks on New Year's Eve, Sydney

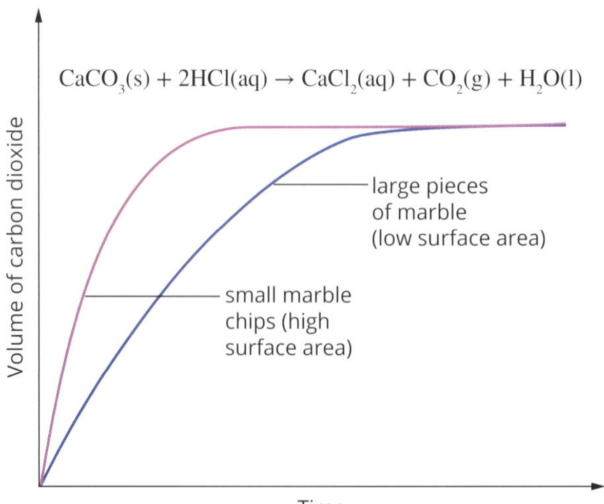

FIGURE 18.1.7 As carbon dioxide is produced from the reaction between marble chips and hydrochloric acid, it is collected in a gas syringe and its volume is recorded.

18.1 Review

SUMMARY

- The rate of a reaction is the change in concentration of reactants or products over time.
- The rate of a reaction may be increased by:
  - increasing surface area
  - increasing the concentration of a reactant
  - increasing the pressure of a gaseous reactant
  - increasing temperature
  - adding a catalyst.
- A range of experimental methods can be used to measure the rate of a reaction, including measuring the following during specific time intervals:
  - mass loss
  - volume of gas produced
  - colour change
  - concentration changes
  - pH change.

KEY QUESTIONS

1 Which one of the following changes would *decrease* the rate of the reaction between zinc metal and dilute hydrochloric acid?

A increasing the temperature of the hydrochloric acid
B decreasing the size of the pieces of zinc
C decreasing the concentration of the hydrochloric acid
D decreasing the volume of hydrochloric acid used

2 Select the correct response in the statements about the five main ways in which reaction rates can be increased.

a *increasing/decreasing* surface area of solid reactants
b *increasing/decreasing* the temperature of a reaction mixture
c *increasing/decreasing* the concentration of a reactant in solution
d *increasing/decreasing* the pressure of gaseous reactants
e *adding/removing* a catalyst

3 The following graph shows the mass of carbon dioxide gas produced during a four minute period of a reaction between marble chips (calcium carbonate) and 1.0 mol L^{-1} nitric acid.

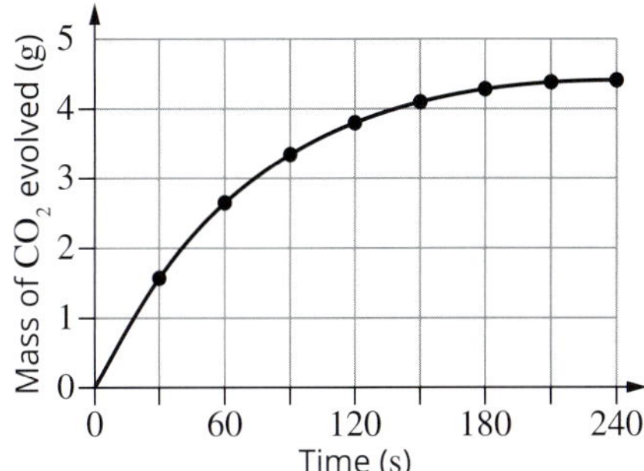

a Write a full chemical equation for this reaction.
b Explain whether the rate of this chemical reaction is increasing or decreasing over time.

4 Describe one way of increasing the rate of each of the following reactions.

a wood burning on a camp fire
b removing excess mortar from between bricks using brick cleaner
c baking a cake in the oven

18.2 Collision theory

The chemical equation for a reaction shows the nature of the reactants and products. It provides no information about the way in which the reaction proceeds.

Look at the equation for the reaction of zinc metal with hydrochloric acid:

$$2HCl(aq) + Zn(s) \rightarrow ZnCl_2(aq) + H_2(g)$$

This equation gives no indication about whether the reaction proceeds quickly or slowly; nor can you tell how the products have been formed.

In fact, the rate of this reaction can vary dramatically by changing the size of the zinc pieces or the concentration of acid used. If concentrated acid is used, for example, you see vigorous bubbling as hydrogen is rapidly produced. When you reduce the concentration of the acid, the bubbling slows as the reaction occurs more slowly (Figure 18.2.1).

Chemical reactions occur as a result of collisions between the reacting particles. This idea is part of the **collision theory** of reaction rates, which will be discussed in this section.

FIGURE 18.2.1 Metals reacting in concentrated (left) and then more dilute (centre and right) solutions of hydrochloric acid.

COLLISION THEORY AND ACTIVATION ENERGY

During chemical reactions, particles (atoms, molecules or ions) collide and are rearranged to produce new particles. Consider the decomposition reaction of hydrogen peroxide:

$$2H_2O_2(l) \rightarrow 2H_2O(l) + O_2(g)$$

The collision that forms the first step of the reaction occurs between the two hydrogen peroxide molecules. If this collision is to result in the formation of molecules of water and oxygen, the collision must occur in such a way that the covalent bonds in the hydrogen peroxide break. To break bonds, energy is required.

The collision theory of reactions explains why some collisions result in reactions and others do not. According to collision theory, for a reaction to occur, the reactant particles must:

- collide with each other
- collide with sufficient energy to break the bonds within the reactants
- collide with the correct orientation to break the bonds within the reactants and so allow the formation of new products.

If a collision does not meet all of these requirements, then no reaction occurs. In fact, most collisions do not result in a chemical reaction. Collision theory helps us to explain why this is the case.

Activation energy

For a reaction to occur between reactant molecules, the molecules must collide with a certain minimum amount of energy. Unless this minimum amount of energy is met or exceeded, the colliding molecules will rebound and simply move away from each other without reacting.

The minimum energy that a collision must possess for a reaction to occur is called the activation energy, E_a. When the energy of a collision is equal to or greater than the activation energy, a reaction can occur.

Activation energy can be represented on an energy profile diagram. An energy profile diagram represents the enthalpy (chemical potential energy) of the reactants and the products over the course of the reaction.

Energy profile diagrams for both exothermic (Figure 18.2.2) and endothermic (Figure 18.2.3) reactions have a peak that represents the activation energy. This is sometimes referred to as the activation energy barrier, and represents the minimum energy that must be absorbed in order to break the bonds of reactants so that a chemical reaction can progress. The activation energy is measured from the enthalpy of the reactants to the top of the peak.

You will recall from Chapter 10 that an exothermic reaction releases more heat energy during the reaction than it absorbs. An endothermic reaction absorbs more heat energy during the reaction than it releases. This is represented on the diagram as ΔH and is the difference in enthalpy between the reactants and the products (Figures 18.2.2 and 18.2.3).

> Reactant particles must have energy equal to or greater than the activation energy before a reaction can occur.

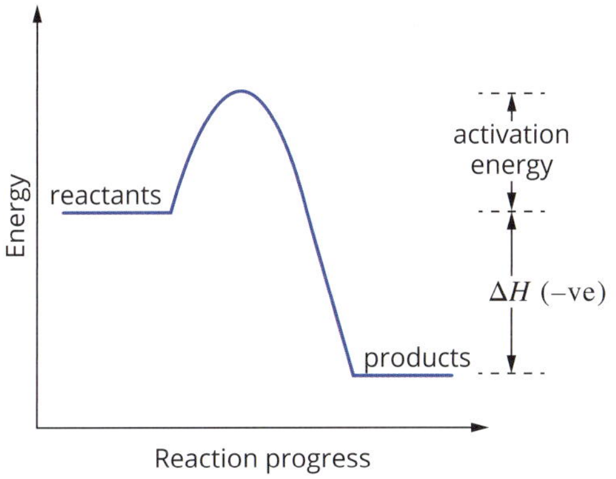

FIGURE 18.2.2 Energy profile diagram for an exothermic reaction showing E_a and ΔH. The ΔH value is negative as overall energy is released during the reaction.

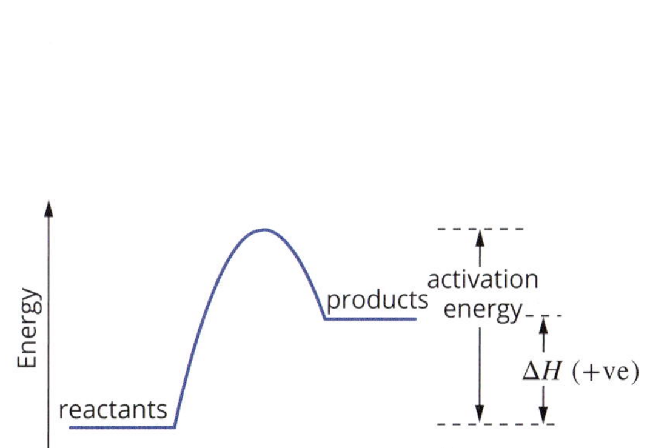

FIGURE 18.2.3 Energy profile diagram for an endothermic reaction showing E_a and ΔH. The ΔH value is positive as overall energy is absorbed during the reaction.

Transition state

When the activation energy is absorbed, a new arrangement of the atoms known as the **transition state** occurs. The transition state occurs at the stage of maximum potential energy in the reaction: the activation energy (Figure 18.2.4). Bond breaking and bond forming are both occurring at this stage, and the arrangement of atoms is unstable. The atoms in the transition state rearrange into the products as the reaction progresses.

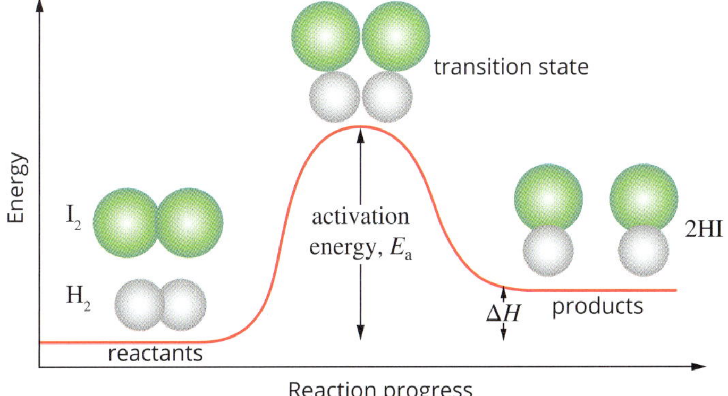

FIGURE 18.2.4 Energy profile diagram for the endothermic reaction $H_2(g) + I_2(g) \rightarrow 2HI(g)$

Activation energy and reaction rate

The magnitude of the activation energy determines how easy it is for a reaction to occur and therefore what proportion of collisions results in a successful reaction. For this reason, the reaction rate is dependent upon the activation energy.

The existence of an activation energy for a reaction means that collisions between reactants do not always result in a chemical change. For example, nitrogen (N_2) and oxygen (O_2) molecules collide frequently in the air around us at room temperature. However, it is only when energy is provided by a spark, such as in car engines or during a lightning strike, that the energy of the collisions is increased enough to overcome the activation energy barrier. This allows nitrogen monoxide to be produced. The nitrogen monoxide formed in this process can then react to form brown nitrogen dioxide (NO_2), a poisonous gas that is a major contributor to the formation of the **photochemical smog** seen in Figure 18.2.5.

FIGURE 18.2.5 Photochemical smog, such as seen here over Barcelona, Spain, in December 2013, is caused by NO_2, a poisonous gas.

Orientation of colliding molecules

For a reaction to occur, reactants need to collide with enough energy to provide the activation energy. Reacting molecules must also collide with each other in the correct orientation in such a way that particular bonds in the reactants are broken and new bonds are formed in the products.

Figure 18.2.6 shows the importance of collision orientation. In the decomposition of hydrogen iodide gas into hydrogen gas and iodine gas, two hydrogen iodide molecules must collide with hydrogen and iodine atoms orientated towards each other for a reaction to possibly occur. If the collision orientation is incorrect, the particles simply bounce off each other, and no reaction occurs.

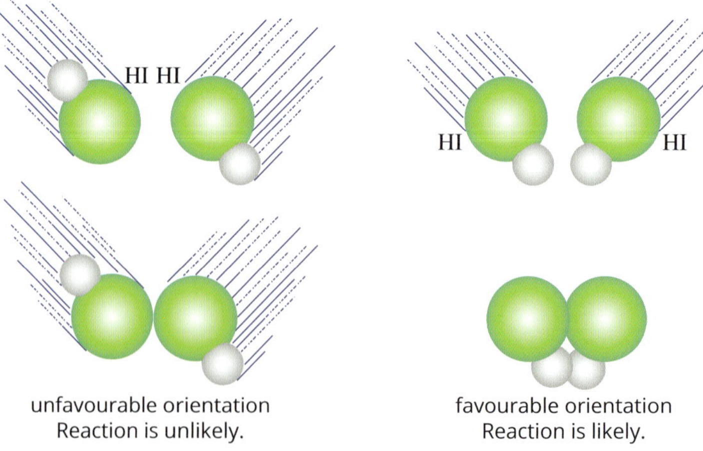

$$2HI(g) \rightarrow H_2(g) + I_2(g)$$

FIGURE 18.2.6 A reaction between colliding molecules is more likely to occur if the orientation of the collision is favourable.

18.2 Review

SUMMARY

- The activation energy of a reaction is the minimum amount of energy required to break reactant bonds to allow a reaction to proceed.
- The transition state is an unstable state in which bonds in the reactants are being broken and bonds in the products are starting to form.
- An energy profile diagram shows the activation energy as the difference between the enthalpy of the reactants and the maximum energy reached.
- Collision theory is a theoretical model that explains rates of chemical reactions in terms of collisions between particles during a chemical reaction.
- According to collision theory, for a reaction to occur, the reactant particles must:
  - collide with each other
  - collide with sufficient energy
  - collide with the correct orientation.

KEY QUESTIONS

1 According to the collision theory, which one of the following is *not* essential for a reaction to occur?

A Molecules must collide to react.

B The reactant particles should collide with the correct orientation.

C The reactant particles should collide with enough energy to overcome the activation energy barrier.

D The reactant particles should collide with double the energy of the activation energy.

2 Which one of the following is the energy required to produce the transition state in a reaction?

A activation energy

B difference in energy between the products and reactants

C difference in energy between the products and the activation energy

D transition state energy

3 When 1 mol of methane gas burns completely in oxygen, the process of bond breaking uses 3380 kJ of energy and 4270 kJ of energy is released as new bonds form.

a Write a balanced chemical equation for the reaction.

b Calculate the value of the heat of reaction, ΔH, for the reaction.

c Draw and label a diagram to show the changes in energy during the course of the reaction.

4 The formation of hydrogen iodide from its elements is represented by the equation:

$$H_2(g) + I_2(g) \rightarrow 2HI(g)$$

This reaction has an activation energy of 167 kJ mol^{-1} and the heat of reaction, ΔH, is +28 kJ mol^{-1}.

a Draw and label an energy profile diagram for the reaction.

b Is the reaction as described, the forward reaction, endothermic or exothermic? Explain.

Consider the reverse reaction.

c What is the ΔH for the reverse reaction?

d Calculate the activation energy for the reverse reaction, the decomposition of 2 mol of hydrogen iodide.

18.3 Applying collision theory

Earlier in this chapter you saw that the rate of a reaction can be changed by the:

- surface area of a solid reactant
- concentration of reactants in a solution
- pressure of any gaseous reactants
- temperature of the reaction
- presence of a catalyst.

In any given reaction mixture, only a certain proportion of the collisions are successful. To increase a reaction rate, you can increase the proportion of successful collisions by:

- increasing the number of collisions that can occur in a given time
- increasing the proportion of collisions with an energy equal to or greater than the activation energy.

In this section, you will consider how various changes to conditions can affect the proportion of successful collisions that occur between reactant particles and hence change the rate of reaction.

INCREASING THE FREQUENCY OF COLLISIONS

The rate of a reaction increases as the frequency of collisions increases. The frequency of collisions between reactants can be increased by:

- increasing the concentration of the reactants, as collisions occur more frequently when particles are closer together
- increasing the surface area of a solid reactant.

Changing concentration or pressure

In Section 18.1, you learnt that the rate of a reaction can be increased by increasing the concentration of a reacting solution or the pressure of a reacting gas. This can be explained by collision theory.

The rate of a reaction increases when the frequency of collisions between reactants increases. When the concentration of a solution increases, there are more reactant particles moving randomly in a given volume of solution (Figure 18.3.1). More particles increases the frequency of collisions, which increases the number of successful collisions in a given time.

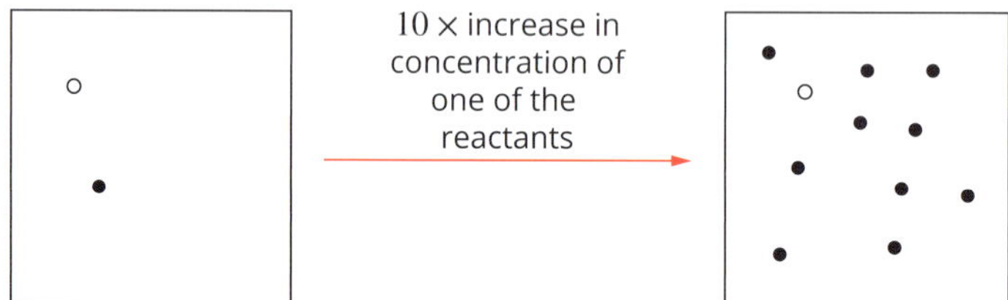

FIGURE 18.3.1 On the left, the concentration of both reactants is low. On the right, the concentration of one of the reactants has been increased ten-fold, resulting in an increase in collision frequency.

For a reaction in the gas phase, the pressure of the gases can be increased either by adding more gas to a fixed volume container or by decreasing the volume of a container with a variable volume, such as a gas syringe. Increasing the pressure increases the concentration of gas molecules, causing more frequent collisions and increasing the number of successful collisions in a given time.

Changing surface area

When a solid is involved in a reaction, only the particles at the surface of the solid are involved in the reaction. The number of particles at the surface depends on the **surface area** of the substance. As you can see in Figure 18.3.2, breaking a solid into smaller parts means that more particles are present at the surface and available to react. The surface area has increased.

As a consequence of the greater number of exposed particles, the frequency of collisions between these particles and the other reactant particles increases, and so the reaction occurs more rapidly.

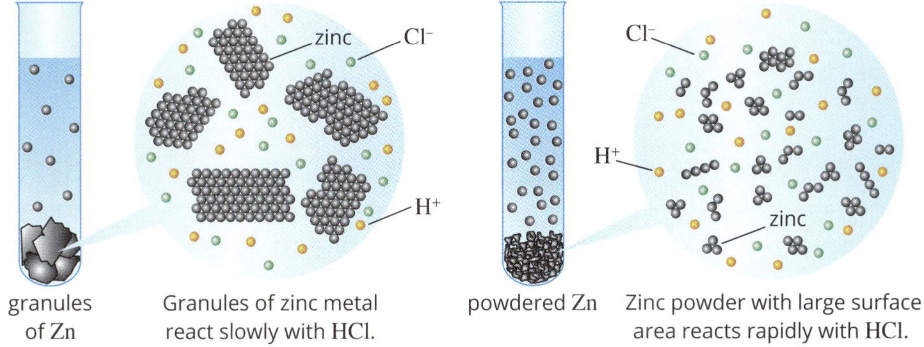

FIGURE 18.3.2 The reaction of hydrochloric acid and zinc. As the surface area of zinc increases, the rate of reaction with hydrochloric acid increases. (Hydrochloric acid exists as Cl^- ions and H^+ ions in solution.)

The effect of increasing surface area can be seen when setting up an open fire at home or on a camp. It is best to first light a pile of kindling rather than trying to directly light large logs. The kindling has a larger surface area than the logs, so it catches fire more easily and will then burn rapidly, providing enough sustained heat energy to make the logs catch fire as well.

When using collision theory to explain the effect of concentration, pressure and surface area on the rate of a chemical reaction, you must discuss the effect on either collision frequency or the number of successful collisions per unit time.

Worked example 18.3.1

USING COLLISION THEORY TO EXPLAIN CHANGES IN RATES OF REACTIONS

The most common fuel sources for explosions in underground mines are flammable gases and explosive dust. Explain, in terms of collision theory, why the presence of coal dust can cause explosions in underground coal mines.	
Thinking	**Working**
Note that this answer uses an explanation structure called the Premise, Reasoning, Outcome (PRO) approach. The premise is the knowledge, theory or the law that is relevant to the situation. The reasoning is where you apply this knowledge or understanding to the situation, and the outcome is the observed result. You can try to use this approach in the Try yourself example.	
Premise: What is the key knowledge, law or theory that you are using?	More successful collisions mean a greater rate of reaction.
Reasoning: Consider the state of the reactants. Relate the state of the reactant to the factor that affects the reaction rate and explain in terms of collision theory.	Coal is a solid. In the mine, there would be lumps of coal and also powdered coal. The surface area of powdered coal is greater than that of solid coal. When the surface area increases, the frequency of collisions increases and so the rate of reaction increases.
Outcome: Return to the question to complete your answer.	The very large surface area of the coal dust allows for an increase in the frequency of collisions with reacting particles, which increases the reaction rate so much that explosions occur.

Worked example: Try yourself 18.3.1

USING COLLISION THEORY TO EXPLAIN CHANGES IN RATES OF REACTIONS

Iron anchors recovered from shipwrecks at considerable depths can show little corrosion after years in the sea, whereas anchors recovered from shallow water are badly corroded. Explain this observation in terms of collision theory.

INCREASING THE ENERGY OF COLLISIONS

As you have seen, a reaction can be made to occur more rapidly by increasing the concentration of the reactants or, for solids, increasing the surface area. A change in temperature can also have a major effect on the rate of a reaction. An increase in temperature not only increases the frequency of collisions, it also increases the kinetic energy of the particles and hence the energy of their collisions.

Maxwell–Boltzmann distribution

At any particular temperature, the particles in a substance have a range of kinetic energies. Although most of the particles have similar kinetic energies, there are always some particles with a high energy or a low energy. This range of energies is shown on a graph called a **Maxwell–Boltzmann distribution graph,** also known as a kinetic energy distribution diagram. Figure 18.3.3 shows how the distribution of energies is represented in a Maxwell–Boltzmann distribution graph.

The vertical axis gives the number of particles with a particular energy, while the horizontal axis gives the kinetic energies of the particles. The area under the graph represents the sum of all the reactant particles. So a **kinetic energy distribution diagram** considers all reacting particles and represents their probable energy distribution, while an energy profile diagram represents the energy of individual particles during the course of a reaction.

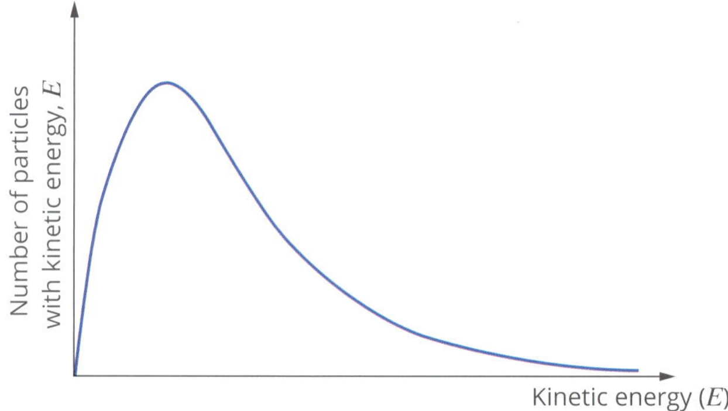

FIGURE 18.3.3 This Maxwell–Boltzmann graph shows the distribution of energies of particles in a sample at a particular temperature. The peak of the graph corresponds to the energy of the greatest number of particles.

During a reaction at a given temperature, only a small proportion of the reactant particles have kinetic energy that is equal to or greater than the activation energy and so are able to react. You can see this in Figure 18.3.4 as a small shaded area to the right of the activation energy, E_a.

Effect of temperature and rate of reaction

The relationship between kinetic energy and velocity is given by the formula $KE = \frac{1}{2}mv^2$. As the temperature of a reaction system increases, the average kinetic energy of the particles increases and the average speed of the particles in the system increases as well.

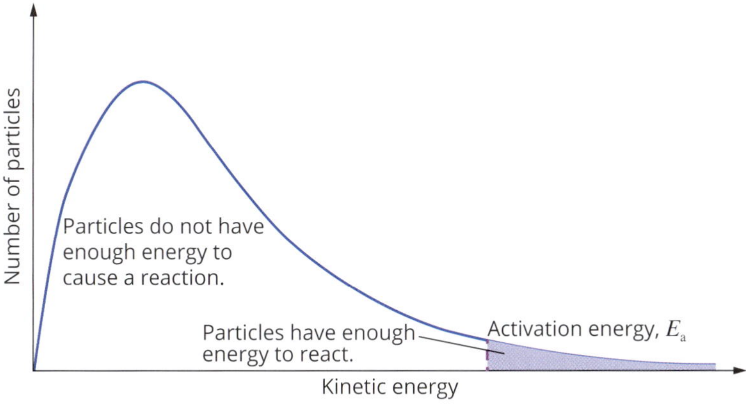

FIGURE 18.3.4 Only a small number of higher energy particles (represented by the shaded area) have sufficient energy to overcome the activation energy barrier.

This is illustrated in Figure 18.3.5 in which the range of kinetic energies for a gas at three different temperatures is shown. Note that the area under the graph, which is equal to the total number of particles in the sample, stays constant when the temperature is changed. As the temperature increases, the increasing average kinetic energy of the particles can be seen by the movement to the right of the peak in the Maxwell–Boltzmann graph.

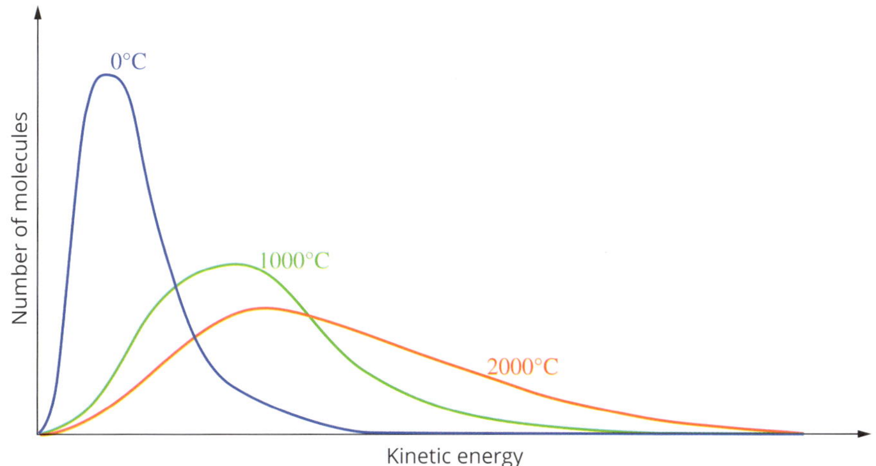

FIGURE 18.3.5 Kinetic energy distribution for a gas at a range of temperatures

As the temperature of a reaction increases, the increased speed of particles causes more collisions, increasing the frequency of successful collisions and the rate of reaction.

The increased kinetic energy of the particles also means that collisions occurring at higher temperatures have greater energy than those at lower temperatures. More particles will have energies that are greater than or equal to the activation energy (Figure 18.3.6) and so the proportion of successful collisions also increases.

When the temperature increases, the increase in reaction rate due to the increased energy of the collisions significantly outweighs the increase in reaction rate due to the increased frequency of collisions.

A temperature increase of just 10°C causes the rate of many reactions to double, but it can be shown that this is not just due to the increased frequency of collisions. The frequency of collisions only increases by about 3% when the temperature increases by 10°C. The main reason why the reaction rate increases is that more particles have sufficient energy to overcome the activation energy barrier of the reaction.

The effect of increasing temperature on the rate of reaction is mainly through increasing the proportion of reacting particles that have energies equal to or greater than the activation energy for the reaction. This increases the proportion of successful collisions.

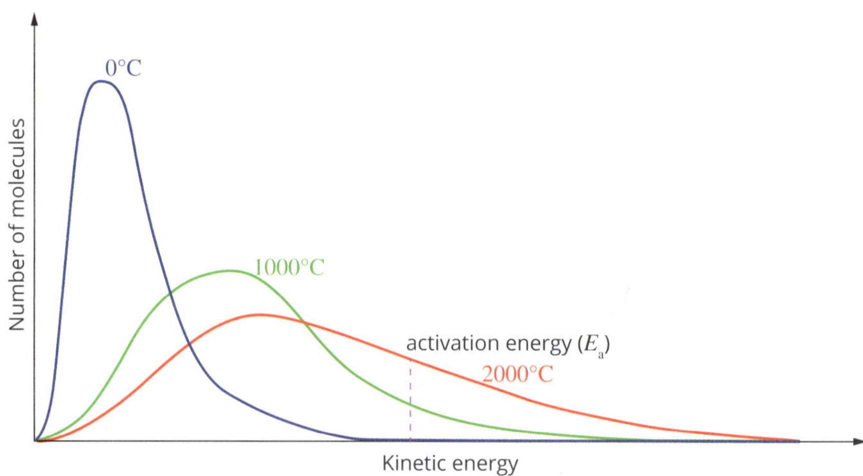

FIGURE 18.3.6 Kinetic energy distribution for a gas at a range of temperatures showing the activation energy for the combustion of this gas. The higher energy particles (those to the right of the E_a) have enough energy to react. Note that there are more particles at the higher temperatures with enough energy to react. This is shown by an increase in the area under the section of the graph to the right of the activation energy as the temperature increases.

FIGURE 18.3.7 The decomposition of hydrogen peroxide is usually very slow, but the addition of crystals of potassium permanganate results in the rapid evolution of oxygen gas and water vapour.

Effect of a catalyst on rate of reaction

Look at the equation for the decomposition of hydrogen peroxide:

$$2H_2O_2(l) \rightarrow 2H_2O(l) + O_2(g)$$

This reaction normally occurs very slowly. When a catalyst, such as crystals of potassium permanganate, are added to the hydrogen peroxide, the reaction occurs rapidly, producing so much oxygen gas and heat that the reaction mixture foams and some of the liquid water vaporises (Figure 18.3.7).

The change in reaction rate when a catalyst is present is often very substantial. The action of a catalyst can be understood using collision theory and the changes in energy that occur during a chemical reaction. Catalysts speed up a reaction by reducing the activation energy barrier for the reaction. You will study the effects of catalysts in more detail in Chapter 19.

 **Catalysts lower the activation energy for a reaction.**

CHEMFILE

Ötzi the Iceman

In September 1991, Erika and Helmut Simon were walking in the Ötztal Alps near the border between Austria and Italy when they discovered the body of what they thought was a dead mountaineer. It was known that, in this region, bodies decompose very slowly because they are enclosed in ice. Closer examination of the body, and the Bronze Age items with it, eventually established that he had died approximately 5300 years ago.

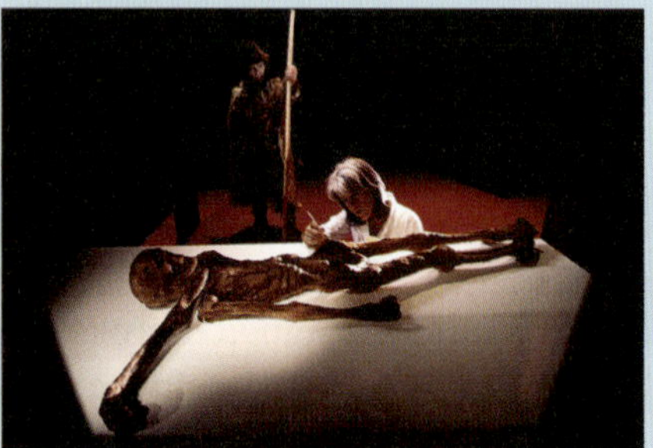

FIGURE 18.3.8 Ötzi the Iceman was so well preserved for 5300 years in the ice that his stomach contents could be identified and pollen of a spring plant was found on his clothes.

18.3 Review

SUMMARY

- Collision theory can be used to explain the increase in rate of reaction by:
  - an increase in concentration of a reactant solution
  - an increase in pressure of a gaseous reactant
  - an increase in surface area of a solid reactant
  - an increase in temperature
  - use of a catalyst.
- Increase in concentration, pressure or surface area results in an increase in:
  - the frequency of collisions between reactants
  - the number of successful collisions in a given time.
- Increase in temperature results in an increase in:
  - the frequency of collisions between reactants
  - the number of successful collisions in a given time
  - the energy of collisions, so an increased proportion of collisions has an energy larger than the activation energy for the reaction.
- A catalyst speeds up a reaction by reducing the activation energy barrier for the reaction.

KEY QUESTIONS

1 Which one of the following alternatives correctly explains why a sample of magnesium reacts more rapidly with 1 mol L^{-1} HCl than with 0.1 mol L^{-1} HCl?

A The energy of collisions between reactant particles is greater for the reaction containing 1 mol L^{-1} HCl.

B The rate of collisions between reactant particles is greater for the reaction containing 0.1 mol L^{-1} HCl.

C There are more collisions between the magnesium and 1 mol L^{-1} hydrochloric acid.

D The frequency of collisions between reactant particles is greater for the reaction containing 1 mol L^{-1} HCl.

2 Which one or more of the following may be true if a reaction is observed to proceed very slowly?

A The activation energy may be very large.

B The temperature may be low.

C Few collisions may be occurring with the correct orientation.

3 A number of experiments involving the reaction between 100 mL of hydrochloric acid and 5 g of calcium carbonate were carried out. Rearrange experiments A–F to place them in increasing order of rate of reaction (slowest to fastest).

A Powdered $CaCO_3$ and 1 mol L^{-1} HCl are mixed at 40°C.

B Small pieces of $CaCO_3$ and 1 mol L^{-1} HCl are mixed at 15°C.

C Powdered $CaCO_3$ and 1 mol L^{-1} HCl are mixed at 25°C.

D Large pieces of $CaCO_3$ and 0.5 mol L^{-1} HCl are mixed at 15°C.

E Powdered $CaCO_3$ and 1 mol L^{-1} HCl are mixed at 15°C.

F Small pieces of $CaCO_3$ and 0.5 mol L^{-1} HCl are mixed at 15°C.

4 Account for the following observations with reference to the collision model of particle behaviour.

a Surfboard manufacturers find that fibreglass plastics set within hours in summer but may remain tacky for days in winter.

b A bottle of fine aluminium powder has a caution sticker warning ‘Highly flammable, dust explosion possible’.

c A potato cooks much more slowly in a pot of boiling water on a trekking holiday in Nepal than a potato boiled in a similar way in the Australian bush.

5 Consider the reaction between solutions V and W that produces X and Z according to the equation:

$$V(aq) + W(aq) \rightarrow X(aq) + Z(aq)$$

The energy profile diagram for this process is shown below.

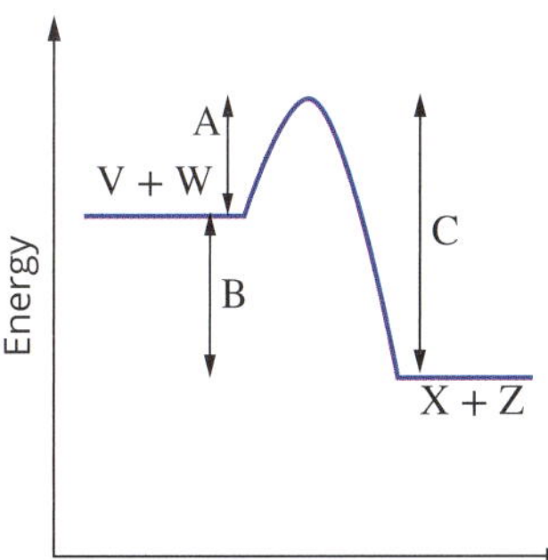

a What do A, B and C represent?

b What effect would the use of a catalyst have on the values of A, B and C?

Chapter review

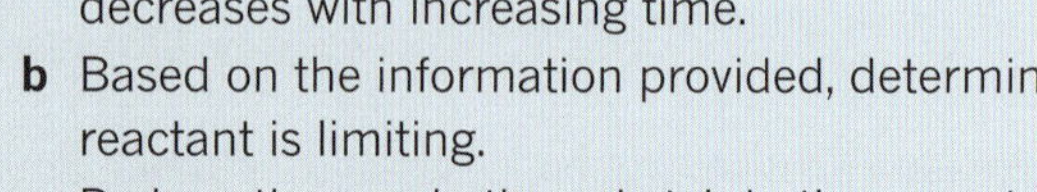

KEY TERMS

catalyst
collision theory
photochemical smog
rate of reaction
surface area
transition state

Investigating the rate of chemical reactions

1 Which one of the following is the correct definition of rate of reaction?
- **A** the time it takes for all of a reactant to be used up
- **B** how fast a reaction is going at the end of one minute
- **C** how much a reaction is bubbling
- **D** the change in concentration of reactants or products over time

2 Which of the following combinations of reactants will produce the greatest initial reaction rate?
$2HCl(aq) + CaCO_3(s) \rightarrow CaCl_2(aq) + H_2O(l) + CO_2(g)$
- **A** $CaCO_3$ chips and 1 mol L^{-1} HCl
- **B** $CaCO_3$ chips and 2 mol L^{-1} HCl
- **C** $CaCO_3$ powder and 2 mol L^{-1} HCl
- **D** $CaCO_3$ powder and 1 mol L^{-1} HCl

3 The following changes are made to a reaction mixture. Which change will lead to a decrease in reaction rate?
- **A** Smaller solid particles are used.
- **B** The temperature is decreased.
- **C** A catalyst is added.
- **D** The concentration of an aqueous reactant is increased.

4 Which of the following is the correct unit for measuring the rate of a reaction?
- **A** $mol^{-1} L s^{-1}$
- **B** $mol L^{-1} s^{-1}$
- **C** $mol^{-1} L^{-1} s$
- **D** $mol L s^{-1}$

5 A 5.00 g piece of copper was dissolved in a beaker containing 500 mL of 2.00 mol L^{-1} nitric acid. The equation for the reaction that occurred is:
$3Cu(s) + 8HNO_3(aq) \rightarrow 3Cu(NO_3)_2(aq) + 2NO(g) + 4H_2O(l)$
The changing mass of the mixture was observed for a period of time, and the following graph was obtained.

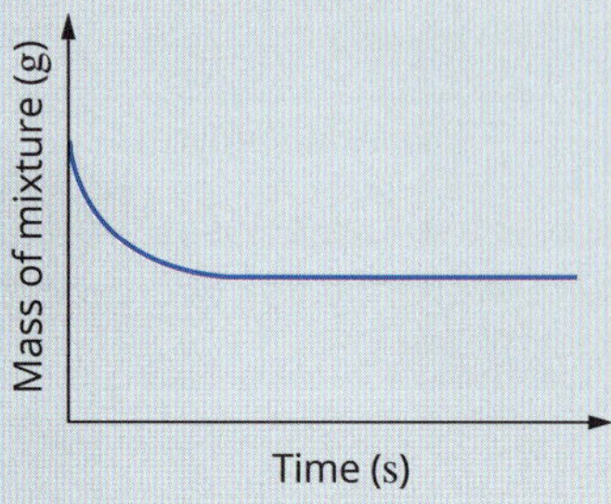

- **a** Explain why the mass of the mixture initially decreases with increasing time.
- **b** Based on the information provided, determine which reactant is limiting.
- **c** Redraw the graph, then sketch in the expected curve if 500 mL of 1.00 mol L^{-1} nitric acid had been used instead. Label your new graph line. Explain the difference in shape.
- **d** Redraw the graph, then sketch in the expected curve if 5.00 g of powdered copper was used instead. Label this new graph line. Explain the difference in shape.

Collision theory

6 According to collision theory, what must happen for a reaction to occur?

7 The following figure shows the reaction between hydrogen gas and fluorine gas to make hydrogen fluoride. The equation for the reaction is:

$$H_2(g) + F_2(g) \rightarrow 2HF(g)$$

Using collision theory, explain why collision 1 might be successful while collisions 2–4 will not be successful.

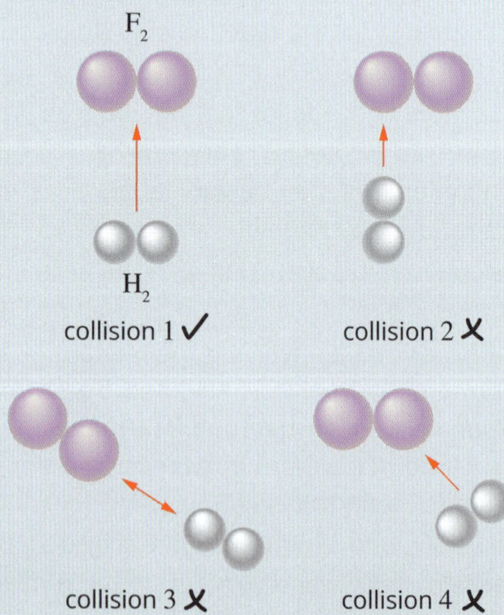

8 The figure below is an energy profile diagram for the substitution reaction between fluorine gas and hydrogen gas.

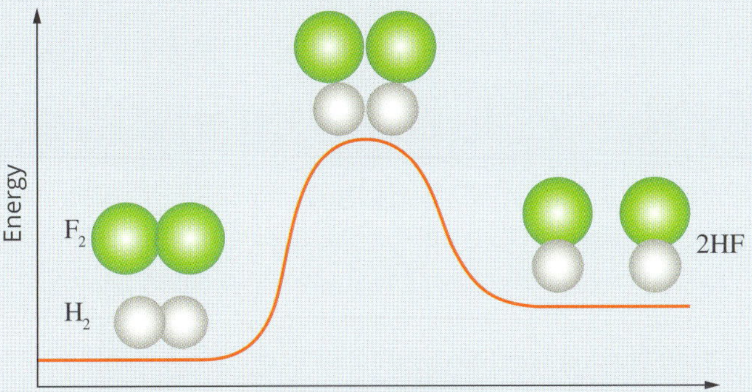

a Copy this diagram and label ΔH and activation energy.

b Explain what is meant by the term 'activation energy'.

c Is the reaction endothermic or exothermic?

d Label the transition state in this reaction on your diagram.

e What bonds are beginning to be broken and formed to produce the transition state?

9 Hydrogen reacts explosively with oxygen to form water.

a What chemical bonds are broken in the reaction?

b What chemical bonds are formed?

c Explain how the energy changes during bond-breaking and bond-forming affect the overall energy change for the reaction.

d Why is there no reaction until the reaction mixture is ignited?

Applying collision theory

10 Which one of the following alternatives correctly explains why the rate of reaction between 1 mol L^{-1} $CuSO_4$ and powdered zinc is greater than with an equal amount of large zinc pieces.

A The energy of collisions of the Cu^{2+} ions with powdered zinc is greater than with the large zinc pieces.

B The frequency of collisions of the Cu^{2+} ions with powdered zinc is greater than with the large zinc pieces.

C The energy of collisions of the Cu^{2+} ions with large zinc pieces is greater than with the powdered zinc.

D The frequency of collisions of the Cu^{2+} ions with large zinc pieces is greater than with the powdered zinc.

11 Which one of the following statements correctly describes what must occur when reactant particles collide and react?

A Colliding particles must have an equal amount of kinetic energy.

B Colliding particles must have different amounts of kinetic energy.

C Colliding particles must have kinetic energy equal to or greater than the average kinetic energy.

D Colliding particles must have kinetic energy equal to or greater than the activation energy of the reaction.

12 Account for the following observations with reference to the collision model of particle behaviour.

a Refrigeration slows down the browning of sliced apples.

b Hydrogen gas burns in air to produce water vapour. Using pure oxygen gas instead of air increases the rate of this reaction.

13 Which one of the following factors would *not* increase the rate of decomposition of hydrogen peroxide?

$$2H_2O_2(aq) \rightarrow 2H_2O(l) + O_2(g)$$

A increasing the pressure of oxygen gas

B increasing the concentration of hydrogen peroxide

C increasing the temperature of hydrogen peroxide

D adding a potassium iodide catalyst

14 Which pair of statements is correct for the effects of adding a catalyst and increasing the concentration on the rate of reaction?

	Adding a catalyst	Increasing the concentration
A	Collision frequency increases.	Collision frequency increases.
B	Activation energy decreases.	Activation energy decreases.
C	Activation energy decreases.	Collision frequency increases.
D	Collision frequency increases.	Activation energy decreases.

15 a What are the five factors that influence the rate of a reaction?

b Classify the five factors from part **a** according to whether they increase the proportion of successful collisions by:

i increasing collision frequency

ii increasing the proportion of collisions that have energy equal to or greater than the activation energy.

Connecting the main ideas

16 Lumps of limestone, calcium carbonate, react readily with dilute hydrochloric acid. Four large lumps of limestone, mass 10.0 g, were reacted with 100 mL 0.100 mol L^{-1} acid.

- **a** Write a balanced equation to describe the reaction.
- **b** Which reactant is in excess? Use a calculation to support your answer.
- **c** Describe a technique that you could use in a school laboratory to measure the rate of the reaction.
- **d** Small lumps of limestone will react at a different rate from four large lumps. Will the rate of reaction with the smaller lumps be faster or slower? Explain your answer in terms of collision theory.
- **e** List two other ways in which the rate of this reaction can be altered. Explain your answer in terms of collision theory.

17 Chemical reactions in the body normally take place at 37°C. Explain how the rate of chemical reactions in the body can account for the following facts.

- **a** The body often responds to illness by an increase in temperature, accompanied by a higher pulse rate and faster breathing.
- **b** People rescued from drowning after 20–30 minutes in freezing water can sometimes survive and recover with no brain damage.

18 The first step in most toffee recipes is to dissolve about 3 cups of sugar in 1 cup of water. Although sugar is quite soluble in water, this step could be time-consuming. Use your knowledge of reaction rates to suggest at least three things you could do to increase the rate of dissolution without ruining the toffee.

CHAPTER

19 Catalysts

As you have seen in Chapter 18, chemical reactions occur at many different rates. The explosion of gunpowder and combustion of petrol in a car's engine occur very quickly. On the other hand, weathering of buildings, ripening of fruit and rusting of iron all occur quite slowly.

In chemical industry, the rate of a chemical reaction must be high enough to make products at an economical rate. Catalysts increase the rate of a chemical reaction that might otherwise be uneconomical. As well, nearly all reactions in living organisms depend on biological catalysts called enzymes.

By the end of this chapter, you will understand what a catalyst is and its role in changing the reaction rate in relation to reaction pathways and the energy changes occurring during reactions.

Using collision theory, you will learn the effect of a catalyst on the rate of chemical reactions, and explain these effects.

Science as a human endeavour

- Catalysts are used in many industrial processes in order to increase the rates of reactions that would otherwise be uneconomically slow. Catalysts are also used to reduce the emission of pollutants produced by car engines. Motor vehicles have catalytic converters which are used to catalyse reactions that reduce the amount of carbon monoxide, unburnt petrol and nitrogen oxides that are emitted.

Science understanding

- catalysts, including enzymes and metal nanoparticles, affect the rate of certain reactions by providing an alternative reaction pathway with a reduced activation energy, hence increasing the proportion of collisions that lead to a chemical change

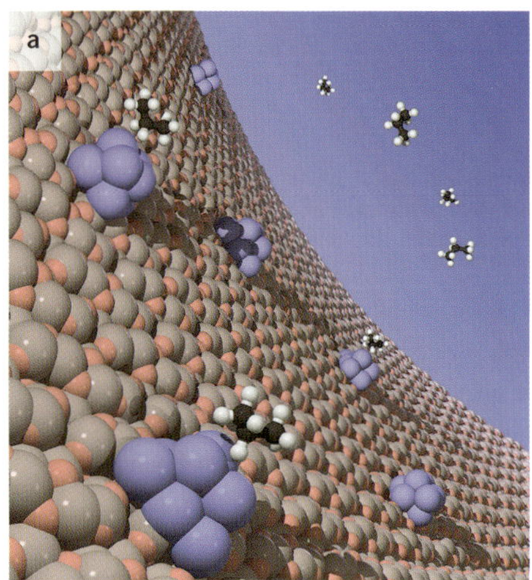

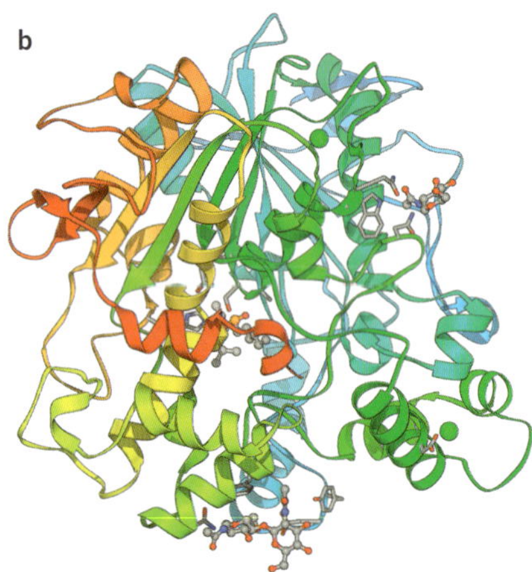

FIGURE 19.1.1 (a) A representation of the structure of a synthetic zeolite catalyst that is widely used in petroleum refineries to break down large hydrocarbon molecules into smaller, more useful molecules. Zeolite is a silica–alumina mineral. (b) A model of a lipase enzyme. Lipase is a catalyst that breaks down fats in the digestive system of the human body.

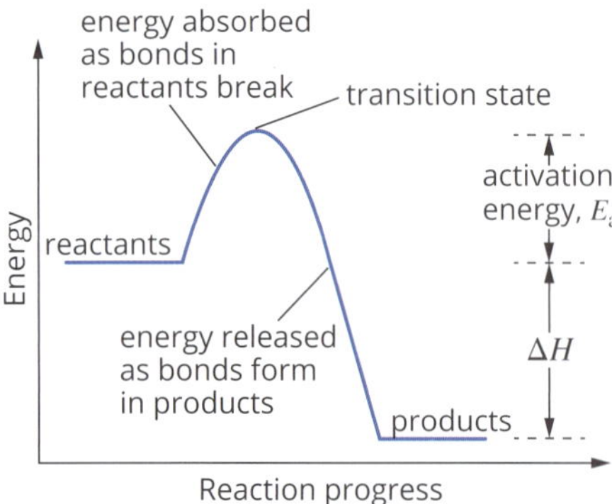

FIGURE 19.1.2 Energy profile diagram of an exothermic reaction such as burning natural gas

19.1 Catalysts

In this section, you will learn how catalysts increase the rate of reaction.

The change in reaction rate when a catalyst is present is often very substantial. The action of a catalyst can be understood using collision theory and the changes in energy that occur during a chemical reaction. Catalysts play an important role in industrial chemistry, in limiting air pollution and in controlling biochemical processes, as shown in the examples in Figure 19.1.1.

CATALYSTS AND ACTIVATION ENERGY

You have already learnt that the potential energy changes associated with a reaction can be represented as an energy profile diagram of the reaction. An energy profile diagram for an exothermic reaction is shown in Figure 19.1.2.

The activation energy is the minimum amount of energy required for a reaction to take place. On the energy profile diagram, the activation energy is measured from the energy of the reactants to the peak of the energy profile diagram.

The enthalpy change, ΔH, can also be represented on an energy profile diagram and is equal to the difference in energy between the products and the reactants.

Some reactions occur very readily because they have very small activation energies, E_a. These reactions need only a small amount of energy to be absorbed for bonds in the reactants to be broken.

Many reactions occur much more rapidly in the presence of a particular element or compound. These substances are known as catalysts. Catalysts are not consumed during the reactions they speed up, and so do not appear as either reactants or products in reaction equations.

Catalysts are able to cause a reaction to occur more quickly because they provide an alternative **reaction pathway**, which causes the activation energy barrier of the overall reaction to be dramatically reduced, as seen in the energy profile diagram in Figure 19.1.3. It is important to note that the original higher energy reaction pathway is still available but with the catalyst present a second, lower energy pathway is also available to the reactants.

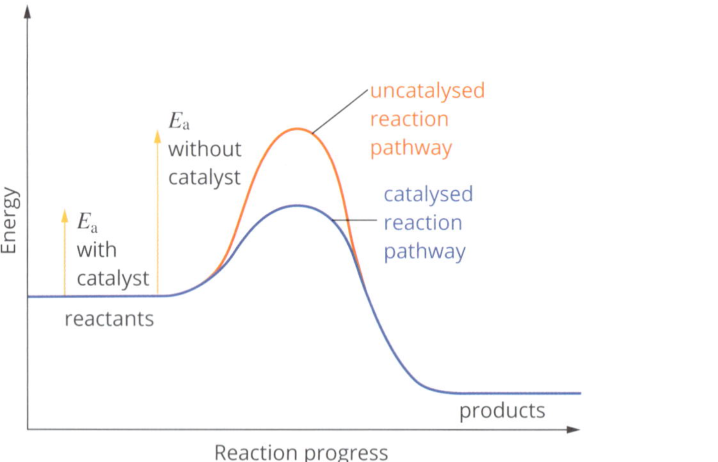

FIGURE 19.1.3 Energy profile diagrams of a catalysed and an uncatalysed reaction

With the catalyst present and a lower activation energy, the colliding particles are more likely to have energies that exceed this lower barrier, causing the bonds in the reactants to be broken more frequently. As a result, a greater proportion of collisions are 'successful'; that is, they lead to the formation of products. Thus, the reaction rate is increased.

The Maxwell–Boltzmann graph in Figure 19.1.4 shows the smaller number of particles with energies that exceed the activation energy in an uncatalysed reaction (shaded in red) compared with the number of particles with energies that exceed the activation energy in a catalysed reaction (regions shaded in red and blue).

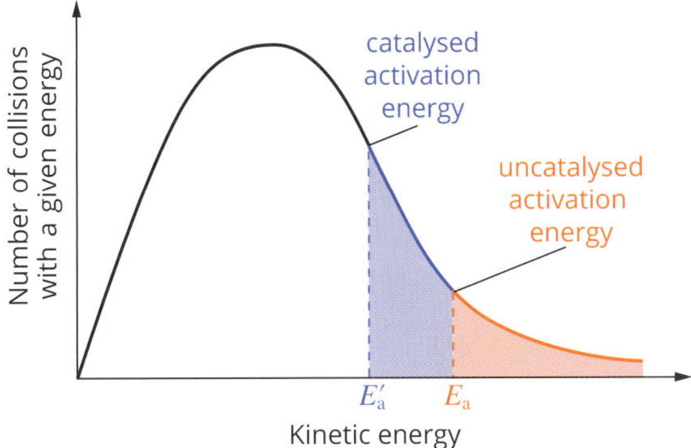

FIGURE 19.1.4 A catalyst provides a reaction pathway with a lower activation energy, increasing the proportion of collisions that exceed the activation energy and lead to a reaction.

> Catalysts lower the activation energy by providing an alternative reaction pathway for the reaction. This increases the proportion of reactant particles with energies greater than the lower activation energy. This increases the proportion of successful collisions.

CHEMFILE

Routes from Melbourne to Canberra

Figure 19.1.5 shows two groups of tourists travelling from Melbourne to Canberra by two different routes. The slower route can be likened to the progress of an uncatalysed reaction; the faster route can be likened to the progress of a catalysed reaction.

FIGURE 19.1.5 An analogy for the role of a catalyst in a chemical reaction

TYPES OF CATALYSTS

Depending on the physical state of the chemicals involved in the reaction and the catalyst, catalysts can be divided into two groups.

- **Homogeneous catalysts** are in the same physical state as the reactants and products of the reaction.
- **Heterogeneous catalysts** are in a different physical state from the reactants and products of the reaction.

An example of homogeneous **catalysis** occurs in the upper atmosphere and has contributed to the depletion of the ozone layer. Chlorine atoms in the gaseous state act as catalysts in the decomposition of ozone gas into oxygen gas. The chlorine atoms may have come from chlorofluorocarbons (CFCs) released into the atmosphere from refrigerators or air conditioners.

You may be familiar with the catalysed decomposition of a hydrogen peroxide solution using the black powder manganese(IV) oxide (MnO_2). This is an example of the use of a heterogeneous catalyst.

Catalysis is the increase in the rate of a chemical reaction due to the presence of a catalyst.

CATALYSTS IN INDUSTRY

The chemical industry uses catalysts extensively. Chemists prefer to use heterogeneous catalysts for industrial processes because they are:

- more easily separated from the products of a reaction
- much easier to reuse
- able to be used at high temperatures.

Particles at the surface of some solids of high surface area tend to adsorb (form a bond with) gas molecules that strike the surface. Adsorption distorts bonds in the gas molecules or may even break them completely. This allows a reaction to proceed more easily than it would if the solid were absent.

These solid surfaces provide a new way for the reaction to occur (a new reaction pathway) that has a significantly lower activation energy.

A powdered or sponge-like form of a solid catalyst is often used to provide the greatest possible surface area. With a larger surface area, more reactant molecules can be adsorbed and the reaction is even faster.

Catalysts enable products to be made very rapidly at significantly lower temperatures than would otherwise be needed. Operating at lower temperatures reduces the energy requirements for the process. Every year, millions of tons of chemicals are produced using processes that include the most cost-effective catalysts available. Some examples of these products are shown in Table 19.1.1.

Adsorption is the attraction and binding of molecules or particles of one substance to the surface of another.

TABLE 19.1.1 Some common products and the catalysts used in their manufacture

Catalyst	Reactants	Product
titanium/aluminium compound	ethene	poly(ethene)
iron	nitrogen, hydrogen	ammonia
nickel	edible vegetable oils, hydrogen	margarine
vanadium(v) oxide	sulfur dioxide, oxygen	sulfuric acid
platinum/rhodium	ammonia, oxygen	nitric acid

ENZYMES

Thousands of chemical reactions are involved in sustaining living things. In the same way that most industrial processes involve catalysts, so do almost all chemical reactions happening in our bodies. These biological catalysts are called enzymes.

Enzymes are highly efficient catalysts that can increase reaction rates by as much as a factor of 10^{10}. This is like taking one second to complete a task that normally takes 300 years.

Digestive enzymes catalyse the breakdown of large food molecules into smaller molecules. Two enzymes involved in the digestion of a potato chip are shown in Table 19.1.2. You will learn more about the specific uses of enzymes in Year 12.

TABLE 19.1.2 Some digestive enzymes and their roles

Enzyme	Role	Type of bond broken (shown circled)
amylase	breaks the ether link of starch molecules (in the mouth)	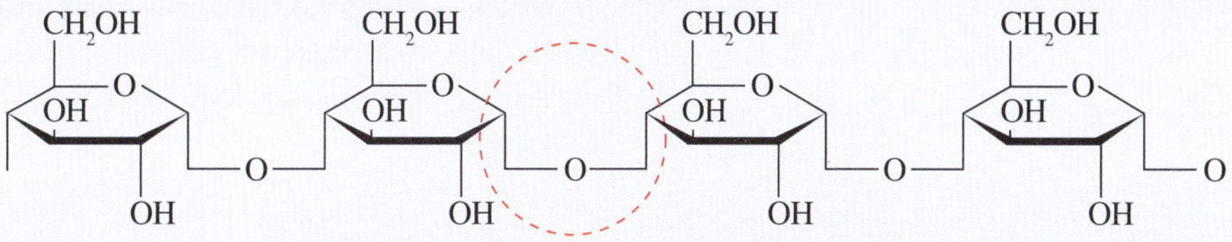
lipase	breaks the ester link of molecules of fats and oils (in the small intestine)	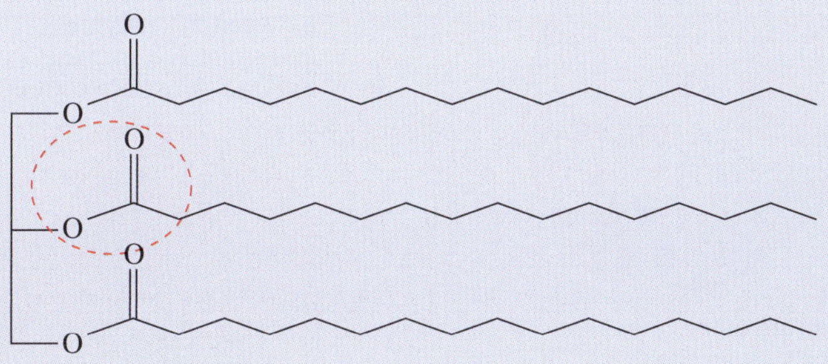

EXTENSION

Heterogeneous catalysis and the Haber process

The Haber process is a very important commercial reaction that produces ammonia gas, which is used to make fertilisers, nylon, explosives and some pharmaceuticals. In the Haber process, hydrogen and nitrogen gases are converted to ammonia (NH_3), using a catalyst of powdered iron.

The reaction is represented by the equation:

$$N_2(g) + 3H_2(g) \rightarrow 2NH_3(g) \quad \Delta H = -92\,kJ\,mol^{-1}$$

Hydrogen and nitrogen molecules both adsorb onto the iron surface (Figure 19.1.6a). As they attach themselves to the surface, the covalent bonds within their molecules break (Figure 19.1.6b).

The hydrogen and nitrogen atoms now readily combine to form ammonia molecules and move away from the iron surface (Figure 19.1.6c). The catalyst remains unaltered by the reaction.

Without a catalyst, temperatures over 3000°C are needed for a significant reaction to occur. The catalyst allows the manufacture of ammonia to proceed at an economical rate using a temperature of about 500°C.

The iron catalyst provides an alternative reaction pathway that dramatically reduces the activation energy 'barrier'—the energy needed to break the covalent bonds in the nitrogen and hydrogen molecules. Even though collisions are less frequent at 500°C than at 3000°C, a greater proportion of colliding particles have sufficient energy to successfully react.

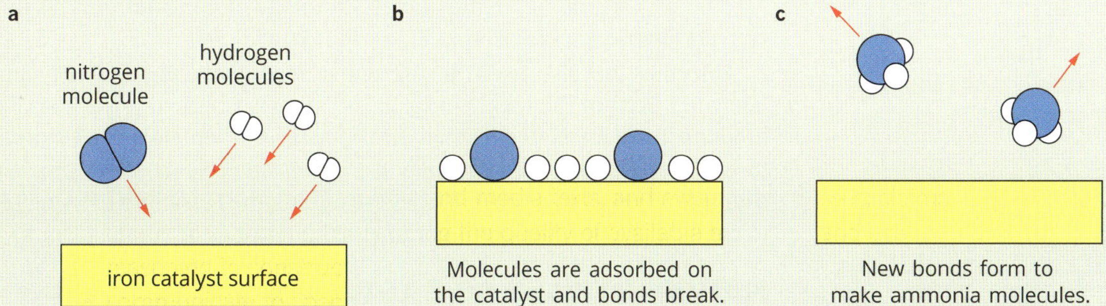

FIGURE 19.1.6 Behaviour of a catalyst in ammonia production. (a) Nitrogen and hydrogen molecules approach the iron catalyst surface. (b) The nitrogen and hydrogen molecules adsorb on the surface of the catalyst and their covalent bonds are broken. (c) The hydrogen and nitrogen atoms readily combine to form ammonia molecules. The molecules then leave the catalyst surface, and the catalyst remains unaltered by the reaction.

Enzymes are very large organic molecules, called proteins, and have specifically shaped sections that are described as active sites. These active sites can interact with particular reactants.

Figure 19.1.7 depicts the reaction pathway for an enzyme catalysing the breakdown of a reactant (the substrate) into two smaller molecules. The specific part of the enzyme molecule with which the reactant can interact is known as the **active site**.

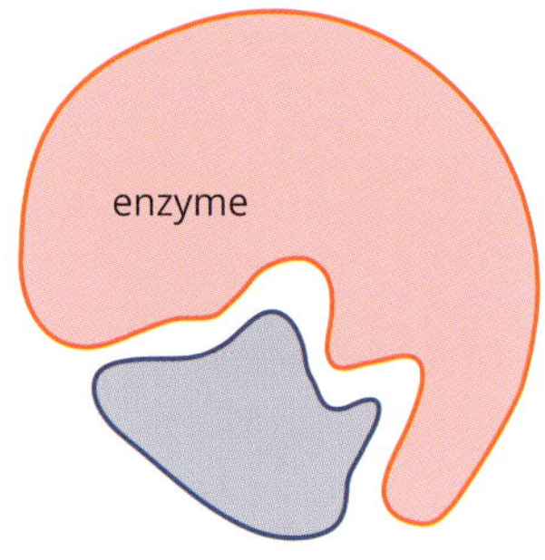

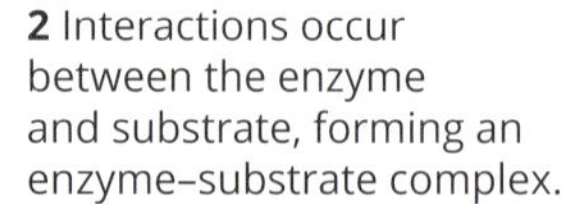

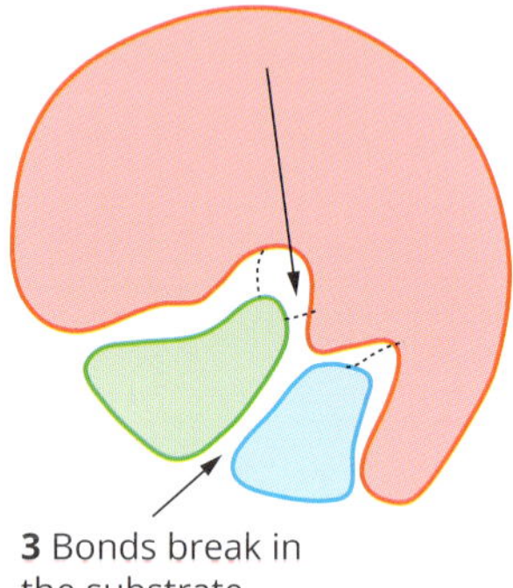

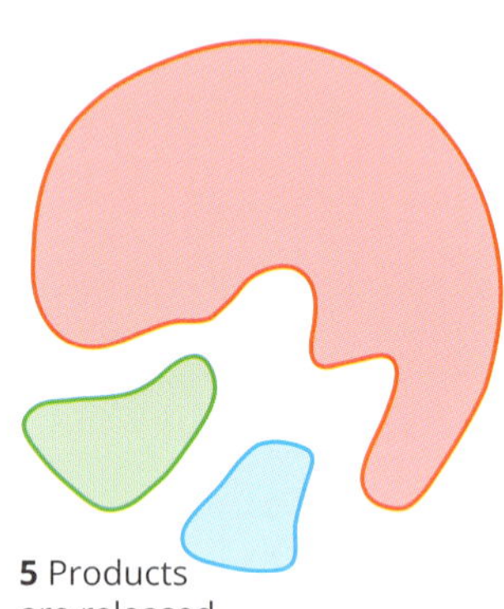

FIGURE 19.1.7 Steps in the action of an enzyme

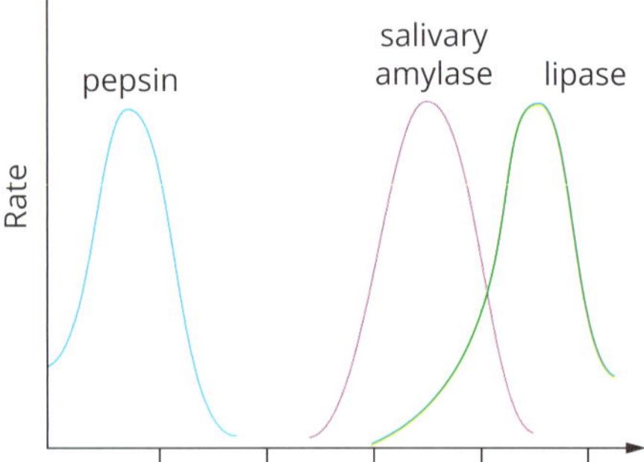

FIGURE 19.1.8 Enzymes can only operate effectively within a narrow pH range. Pepsin is a protein-digesting enzyme secreted into the stomach. Salivary amylase is the enzyme in human saliva. Lipase is the enzyme that breaks down fats. These enzymes are most effective at different pH values.

Compared with inorganic catalysts, enzymes:

- can produce much faster reaction rates
- are considerably more selective, often catalysing only one specific biochemical reaction. An inorganic catalyst, such as platinum, can be used to catalyse many different reactions, often using a variety of reactants
- are much more sensitive to pH changes and temperature changes because the intricate three-dimensional structure of their active site is destroyed at high temperatures and altered as pH changes.

Figure 19.1.8 shows the narrow pH ranges in which some common enzymes work and Figure 19.1.9 shows the effect of temperature on enzyme action.

Enzymes have numerous domestic and industrial applications. Some detergents contain enzymes that enable protein stains or greasy fat residues to be removed effectively. Fabric softeners may contain enzymes to reduce pilling of cotton fabrics.

The manufacture of vinegar, wine, beer and bread all take advantage of enzymes present in yeast to catalyse reactions such as the fermentation of sugar to ethanol:

$$C_6H_{12}O_6(aq) \rightarrow 2C_2H_5OH(aq) + 2CO_2(g)$$

The production of yoghurt takes advantage of enzymes present in bacteria that convert lactose, the sugar in milk, to lactic acid.

Enzymes are also used during paper manufacturing and recycling processes. They are used in the textile industry to 'stonewash' denim jeans and in the leather industry to make softer leather. In the future, when fossil fuel supplies are likely to be limited, enzymes may play an important role in the biofuel industry. For example, one group of chemists is researching the feasibility of using plant enzymes to catalyse the production of commercial quantities of ethene from biomass.

Rate

10° 20° 30° 40° 50° 60°

Temperature

FIGURE 19.1.9 The effect of temperature on the action of an enzyme. Enzymes are only effective in a narrow range of temperatures.

NANOPARTICLES AS CATALYSTS

Since catalysis often involves a reaction of gases passing across a metal or metal oxide surface, it is not surprising that chemists try to optimise the total surface area of a solid catalyst. Metal nanoparticles, which have a very large surface area, are therefore an exciting prospect for catalyst development.

With their much greater surface area, contact between reactant particles and catalyst is increased, which means more catalytic reactions can occur in any given time. This makes the reaction occur even faster.

Considerable research effort around the world is currently devoted to identifying the best size for nanocatalysts, the most suitable design and support materials for them, and the variety of reactions they can be applied to.

One car manufacturer recently announced a new pollution-control catalyst that features nano-sized metal particles embedded in fixed positions on a ceramic base, as seen in Figure 19.1.10. The innovation is claimed to use 70% less metal and to overcome the problem of small metal particles sticking together when heated (agglomeration), which lowers the effectiveness of the catalyst.

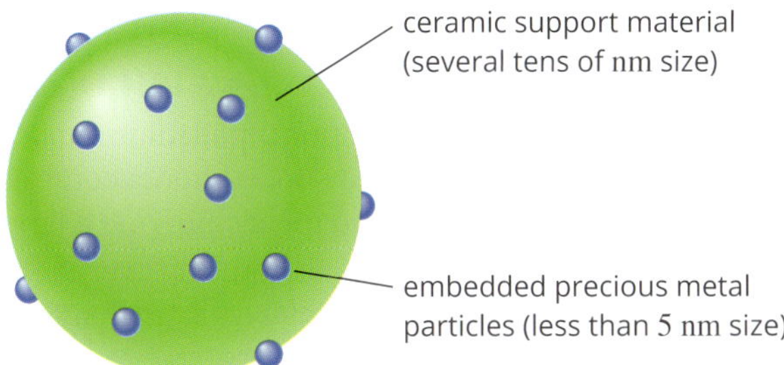

FIGURE 19.1.10 Catalyst nanoparticles on the surface of a ceramic support improve the efficiency of pollution-control systems in cars.

Another research group, also focused on pollution prevention, has designed a 'nanocage' for a catalyst to remove unwanted sulfur from syngas fuel—a fuel that is produced in the petrochemical industry. This nanocage uses a cage-like three-dimensional structure made of silica to protect nickel and copper nanocatalysts from heat damage.

CHEMISTRY IN ACTION

Small size, big potential

Fuel cells, like batteries, convert chemical energy into electrical energy but unlike batteries, where all the reactants are eventually reacted to make the electrical energy, fuel cells have a continuous supply of reactants to provide the electrical energy. Hydrogen fuel cells use hydrogen gas or methanol to provide hydrogen ions, which react with oxygen to produce water. Catalysts speed the production of hydrogen ions, and the breaking of the bonds in oxygen molecules, so the reaction occurs at rates that make the hydrogen fuel cells practical. Platinum metal, which is very expensive, is the catalyst most often used for these reactions. Platinum catalysts are now being used in the form of nanoparticles, which means less metal is required as well as giving high catalytic activity. This is contributing to lower costs for fuel cells. Other nano-based catalysts are incorporating less expensive metals to further reduce the need for expensive platinum.

Researchers have made nanowires with a nickel oxide core surrounded by a platinum shell. The nickel was leached out leaving a rough surface with increased surface area. The catalytic activity shown is one of the highest reported.

The reaction involving oxygen is very slow due to the strong double bond in an oxygen molecule. A nanoplate catalyst with a platinum/lead core surrounded by a platinum shell has shown high catalytic activity for this oxygen reaction. As well as improved catalytic activity, these nanoplate catalysts have better durability. In the acid environment of the fuel cell, in other shell-structure catalysts, acid leaches through the thin outer shell to react with the metal in the core, reducing catalyst stability and its lifetime. This new nanoplate catalyst has shown improved durability.

One of the requirements for the operation of a fuel cell is a semipermeable membrane that allows only hydrogen ions to pass through but no other atoms such as oxygen. Although not giving any catalytic effect, nanomaterials are also being used to improve the semipermeable membrane, allowing for lighter weight and more efficient fuel cells.

CATALYTIC CONVERTERS IN CARS

All new cars sold in Australia have a catalyst fitted between the engine and the exhaust pipe. The purpose of the catalyst is to clean the exhaust gases from the engine and reduce the air pollution that could be caused if these gases entered the atmosphere.

Catalysts in cars convert carbon monoxide and nitrogen oxide, formed in the engine, to the non-toxic gases carbon dioxide and nitrogen. Several reactions are involved in this process, including:

$$2NO(g) \rightarrow N_2(g) + O_2(g)$$

$$2CO(g) + O_2(g) \rightarrow 2CO_2(g)$$

Unburnt hydrocarbons are also converted by the catalyst to carbon dioxide and water:

$$2C_8H_{18}(g) + 25O_2(g) \rightarrow 16CO_2(g) + 18H_2O(g)$$

The catalyst is usually a mixture of platinum, palladium and rhodium metals and aluminium oxide, and it is mounted on a honeycomb-shaped support (Figure 19.1.11). Millions of tiny pores in the metals provide a large surface area.

Exhaust gases enter the catalyst chamber, pass quickly over the metals, and leave the exhaust pipe in a purified condition. The catalyst is unchanged by the reaction and can be used without replacement for many years.

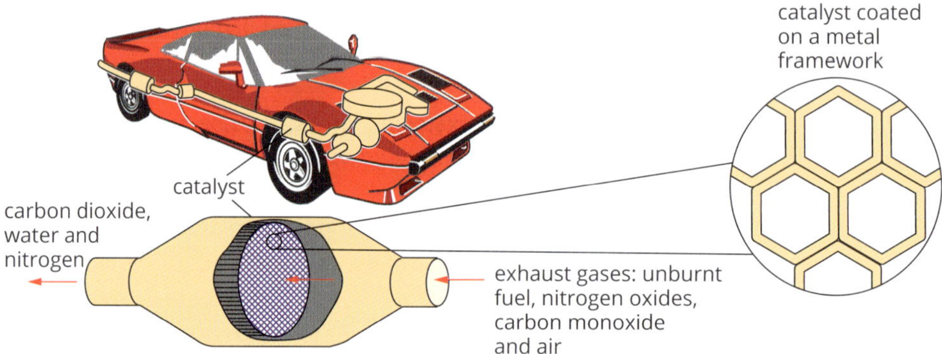

FIGURE 19.1.11 A catalyst fitted in the exhaust system of a car reduces air pollution.

Catalytic converters could not be used until unleaded petrol was available in Australia. Unleaded petrol was introduced in Australia in 1985, with the introduction of catalytic converters following in 1986 (since 2002 all petrol in Australia is unleaded). Lead, in the form of tetraethyl lead, was previously added to petrol to make engines run more smoothly and improve fuel efficiency. Catalytic converters cannot be used with leaded petrol because lead in exhaust gases would coat the surface of the metal catalysts and this would prevent the other components in exhaust gases coming into contact with the catalyst, thus making the catalyst ineffective. The phasing out of leaded petrol had a two-fold advantage in terms of air quality. Lead, like many heavy metals, is a neurotoxin (harmful to the nervous system), particularly in children, so its removal from car exhaust gases has reduced health problems. As well, its removal from petrol allowed the use of catalytic converters, which reduce the concentration of the toxic gases carbon monoxide and nitrogen oxides coming from car exhausts.

19.1 Review

SUMMARY

- A catalyst provides an alternative reaction pathway that has a lower activation energy.
- Energy profile diagrams (Figure 19.1.12), which can include catalysed and uncatalysed pathways, may be used to represent the enthalpy changes and activation energy associated with a chemical reaction.

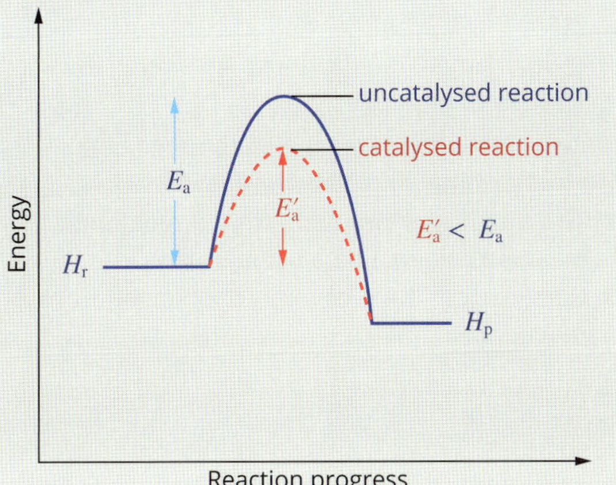

FIGURE 19.1.12 Energy profile diagram of a reaction showing the effect of a catalyst

- A catalyst provides a new reaction pathway and it is not used up in the reaction.
- When a catalyst is present, a greater proportion of the collisions between particles exceed the activation energy barrier of the reaction and therefore lead to a chemical change.
- Homogeneous catalysts are in the same state as the reactants and products of the reaction, whereas heterogeneous catalysts are in a different physical state from the reactants and products of the reaction.
- Catalysts are important in industry to enable faster production of chemicals, and to lower production costs.
- Enzymes are proteins that act as catalysts in biological systems.
- Metal nanoparticles have very large surface area to mass ratios and can be cost-efficient catalysts for commercial applications.
- Catalytic converters are used to remove pollutant gases from automobile exhaust.

KEY QUESTIONS

1 Consider the reaction between solutions V and W that produces X and Z according to the equation:

$$V(aq) + W(aq) \rightarrow X(aq) + Z(aq)$$

The energy profile diagram for this process is shown in in the figure below.

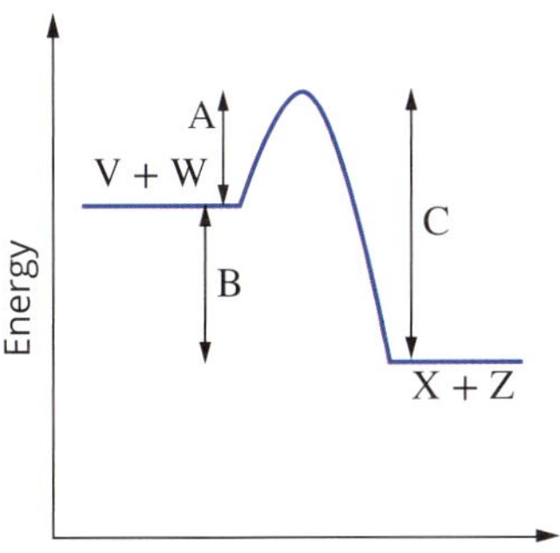

Which one of the following alternatives describes the change that a catalyst produces to increase the reaction rate?

A B only is decreased.
B A only is decreased.
C A, B and C are decreased.
D A and C only are decreased.

2 Explain the meaning of:
a catalyst
b activation energy.

3 If a sugar cube is held in the flame of a candle, the sugar melts and browns but does not burn. However, the cube burns if salt is first rubbed into it, even though the salt does not react. Explain the effect of the salt on the activation energy of this combustion reaction.

4 **a** Explain why surface properties are important to the operation of catalysts.
b Many industrial catalysts are made into porous pellets. What is the reason for this?

Chapter review

19

KEY TERMS

active site
catalysis
heterogeneous catalyst
homogeneous catalyst
reaction pathway

Catalysts

1 The Haber process involves the reaction of nitrogen gas and hydrogen gas to make ammonia gas.

a Describe two ways the rate of this reaction could be increased, at constant temperature.

b Using collision theory, explain why rate is increased.

2 Many major car makers have plans for hydrogen-powered cars. In the fuel cells of these cars, hydrogen reacts with oxygen from the air to produce water.

$$2H_2(g) + O_2(g) \rightarrow 2H_2O(g)$$

Energy changes for the reaction are shown in the graph below.

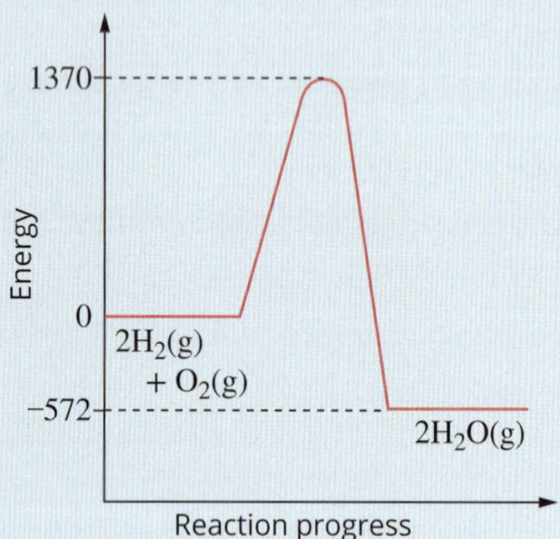

a What is the magnitude of the activation energy of this reaction?

b What is ΔH for this reaction?

Several groups of scientists have claimed to have split water into hydrogen and oxygen using a molybdenum catalyst:

$$2H_2O(g) \xrightarrow{Mo} 2H_2(g) + O_2(g)$$

c Sketch energy profile diagrams for this reaction with and without the presence of a catalyst.

d What is the value of ΔH for this water-splitting equation?

3 For the following reactions, state whether the catalyst is homogeneous or heterogeneous. Give a reason for your answer.

a Hydrogen peroxide solution is decomposed to oxygen gas and water in the presence of aqueous dichromate ions.

b Sulfur dioxide gas is reacted with oxygen gas to produce sulfur trioxide gas in the presence of solid vanadium pentoxide (V_2O_5).

c Methane (CH_4) and steam react to produce hydrogen gas and carbon monoxide gas in the presence of nickel metal.

d Conversion of aqueous ethanol to acetic acid in the presence of an enzyme.

4 Why do chemists prefer to use heterogeneous catalysts for industrial processes?

5 State whether the following statements about catalysts are true or false. Explain your answer.

a The quantity of products from a reaction is increased when a catalyst is used.

b Catalysts can be recovered unchanged from a reaction.

c Catalysts increase the rate of a reaction by increasing the energy with which reactant particles collide so that a higher proportion of collisions exceed the activation energy.

d The enthalpy of a reaction is decreased in the presence of a catalyst.

6 A finely divided nickel catalyst is selected to catalyse the hydrogenation of vegetable oils to produce margarine. Which one of the following would *not* be a reason for using this catalyst?

A Nickel is less toxic than other metals that are more effective catalysts.

B Finely divided nickel will be a more effective catalyst than nickel plate.

C Nickel lowers the energy change of the reaction involved more than other metals do.

D The finely divided nickel is readily removed from the reaction mixture.

7 The reaction of tartrate ion ($C_4H_4O_6^{2-}$) and hydrogen peroxide (H_2O_2) is shown below.

$$C_4H_4O_6^{2-}(aq) + 5H_2O_2(aq) \rightarrow 4CO_2(g) + 6H_2O(l) + 2OH^-(aq)$$

At room temperature, little reaction occurs but the addition of pink cobalt(II) ions (Co^{2+}) to the reaction mixture causes rapid effervescence and a colour change from pink to green and back to pink again. What evidence is there that the cobalt(II) ions are acting as a catalyst? Account for the colour change in your answer.

8 In the production of lactose-free milk, the enzyme lactase is used to catalyse the conversion of lactose into galactose and glucose. Which of the following statements about this catalysed process are correct? (More than one answer is possible.)

A Lactase is a protein.

B Lactose is described as the substrate.

C Lactose has an 'active site' where bond breaking occurs.

D Glucose is described as the substrate.

9 The graph below shows the range of activity for the enzymes pepsin, salivary amylase and lipase. Why are these enzymes active only over a narrow pH range? In your answer refer to the ability of the substrate to interact with the enzymes.

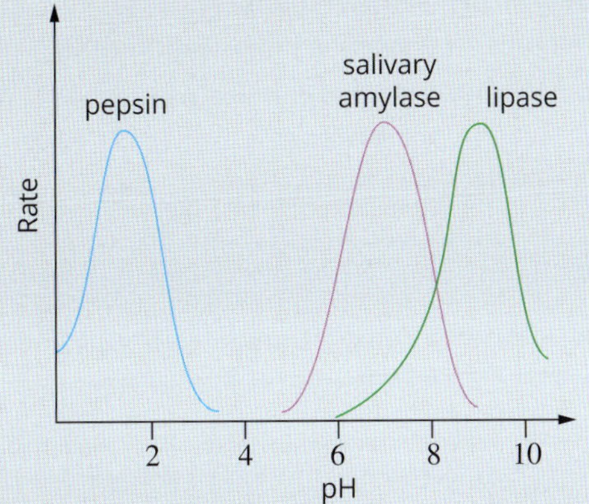

10 Describe how catalytic converters in motor vehicles are able to limit the potential for atmospheric pollution.

11 In the catalytic converter of a car, carbon monoxide gas is removed from exhaust gases by passing the gas mixture over a platinum, Pt, catalyst. Sort the sequence of steps W–Z below to describe the pathway, in order, for this catalysed reaction.

W Adsorbed O atoms combine with adsorbed CO molecules.

X O_2 and CO molecules adsorb on to the Pt surface.

Y Adsorbed CO_2 molecules leave the Pt surface.

Z Covalent bonds within O_2 molecules break.

12 When a mixture of hydrogen and oxygen gases is passed over a platinum gauze, the mixture spontaneously ignites. State the role of the platinum gauze and explain how it causes the hydrogen and oxygen gases to ignite.

13 Give two reasons that nanomaterials are seen as having great potential as catalysts.

14 The figure below shows the interactions at the active site of an enzyme, called ACE, responsible for catalysing a biochemical reaction in which a small section is chopped off a short polypeptide molecule. The product of this reaction is important in controlling blood pressure.

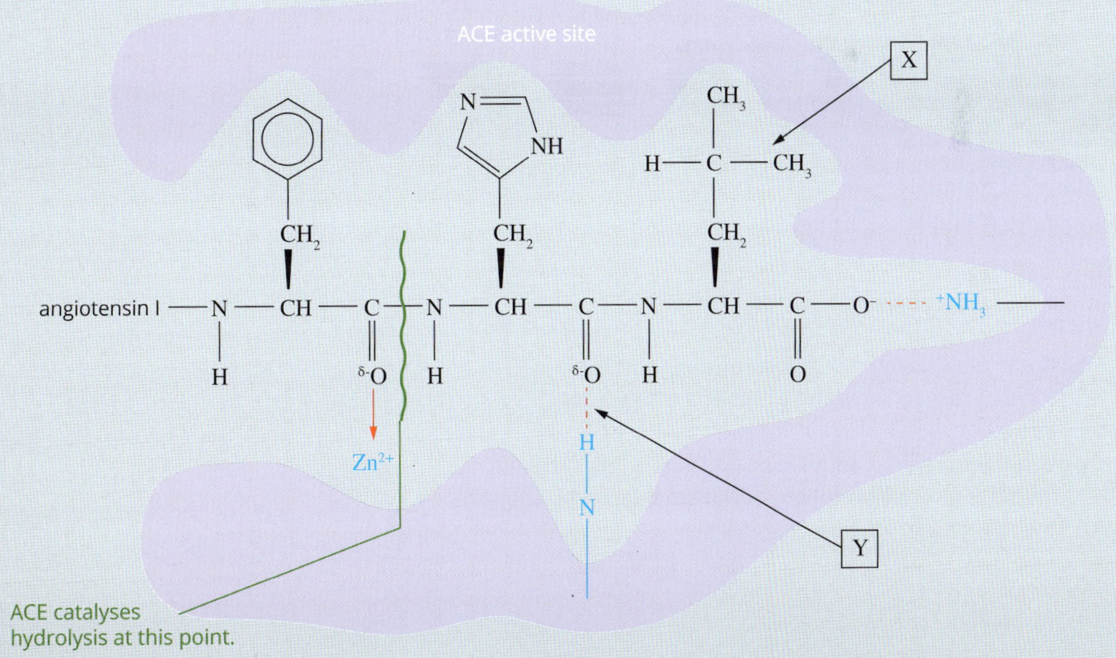

ACE active site

X

Y

angiotensin I

Zn^{2+}

ACE catalyses hydrolysis at this point.

Which one of the following statements about this enzyme-catalysed reaction is correct?

A ACE would be described as the substrate in this process.

B The part of the molecule labelled X interacts with the active site because it is polar.

C The interaction labelled Y (the dashed line) represents a hydrogen bond.

D This biochemical reaction would have an even faster rate if the temperature were lowered from 37°C to 20°C.

15 Use the word or phrases from the following list to describe what happens to nitrogen and hydrogen molecules when their reaction to produce ammonia is catalysed at an iron surface.

absorb	dispersion	molecules
adsorb	form covalent bonds	react with iron
atoms		strengthened
broken	hydrogen	strong
covalent	increasing	weak
decreasing	ions	

A balanced equation for the reaction is:

$$N_2(g) + 3H_2(g) \rightarrow 2NH_3(g)$$

Nitrogen molecules, N_2, consist of two nitrogen ___ held together by a triple covalent bond. This is a particularly ___ covalent bond and so the reaction has a large activation energy in the absence of a catalyst. Iron is a suitable catalyst for this reaction, capable of providing a new reaction pathway and so ___ the activation energy.

Nitrogen and hydrogen molecules ___ onto the iron surface and form bonds with the iron surface. The ___ bonds in the N_2 and H_2 molecules are ___ and individual neighbouring nitrogen and hydrogen atoms ___ and become ammonia molecules. The ammonia molecules then leave the iron surface.

16 **a** The figure below shows the distribution of energies of particles in a substance at two different temperatures, 40°C and 60°C. Indicate which temperatures are represented by graphs A and B.

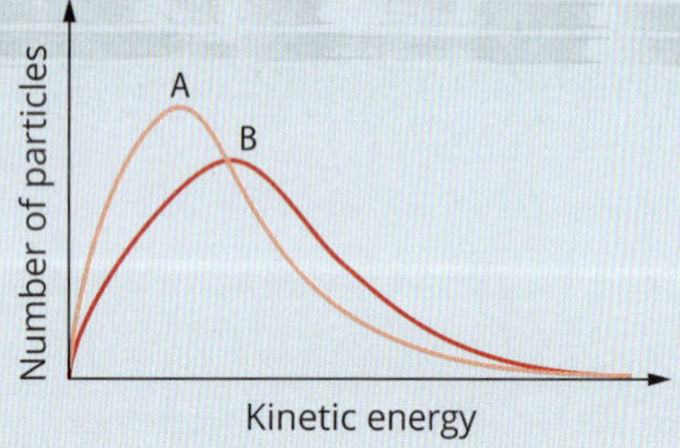

b Copy this diagram for temperature B and use the diagram to show the effect of a catalyst on a reaction.

c Use the diagram you have drawn in part **b** to explain in terms of collision theory how a catalyst increases the rate of a reaction.

17 The graph below shows the energy profile diagram for the reaction of hydrogen and iodine to form hydrogen iodide:

$$H_2(g) + I_2(s) \rightarrow 2HI(g)$$

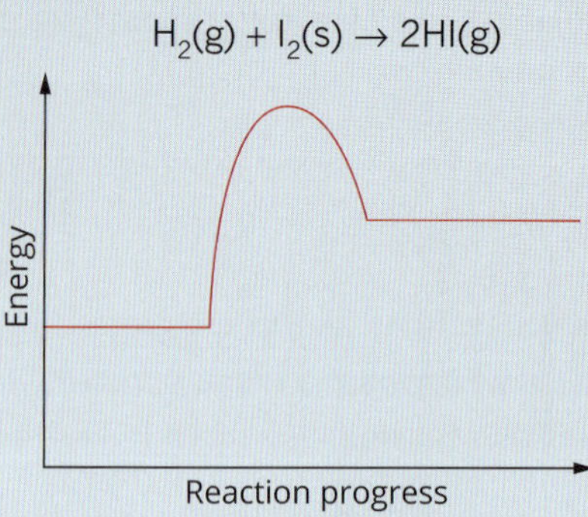

a Copy the diagram and label the following: $H_2(g)$ and $I_2(s)$, HI(g), ΔH, activation energy.

b Is the reaction endothermic or exothermic?

c On the diagram, draw the energy profile that would result if a catalyst were used in the reaction.

CHAPTER

20 Science inquiry skills in chemistry

This chapter covers the skills needed to successfully plan, conduct and evaluate results from experiments and investigations in chemistry.

For investigations, you will learn how to develop inquiry questions and hypotheses. There are many aspects to designing an investigation and you will consider risk assessments and ethical concerns, and the identification, measurement and control of variables which allow for reliable collection of data. Uncertainty and error are common to all experimental work, and you will learn to identify causes of random and systematic error, and importantly, how appropriate materials, technology and procedures can be used to accurately collect and record data to reduce uncertainty in experimental data.

The chapter describes how best to represent your data, how to identify trends and patterns and explains how to analyse your results, including using mathematical models. Many investigations in chemistry involve solving problems, such as the identification of an unknown or a determination of a quantity. Strategies for utilising critical thinking that will assist in problem-solving tasks are discussed.

Finally, the sharing of findings is vital, and in this chapter you will learn how to effectively communicate conclusions using appropriate scientific language, nomenclature and scientific notation.

Science inquiry skills

- identify, research and refine questions for investigation; propose hypotheses; and predict possible outcomes
- design investigations, including the procedure(s) to be followed, the materials required, and the type and amount of primary and/or secondary data to be collected; conduct risk assessments; and consider research ethics
- conduct investigations safely, competently and methodically for the collection of valid and reliable data, including: the use of devices to accurately measure temperature change and mass, flame tests, separation techniques and heat of reaction
- conduct investigations safely, competently and methodically for the collection of valid and reliable data, including: chromatography, measuring pH, rate of reaction, identification of the products of reactions, and determination of solubilities of ionic compounds to recognise patterns in solubility
- represent data in meaningful and useful ways, including using appropriate graphic representations and correct units and symbols; organise and process data to identify trends, patterns and relationships; identify sources of random and systematic error and estimate their effect on measurement results; and select, synthesise and use evidence to make and justify conclusions
- interpret a range of scientific and media texts, and evaluate processes, claims and conclusions by considering the quality of available evidence; and use reasoning to construct scientific arguments
- communicate to specific audiences and for specific purposes using appropriate language, nomenclature and formats, including scientific reports

20.1 Questioning

TYPES OF INVESTIGATIONS

There are many different types of investigation that can be conducted in chemistry. You are probably most familiar with practical investigations or experiments. Other types of investigations involve researching and evaluating primary and/or secondary sources of information to answer an inquiry question. Examples of the types of different investigations are listed in Table 20.1.1 and Table 20.1.2.

TABLE 20.1.1 Examples of primary research

Primary research	Example tasks
conducting experiments in a laboratory	• planning a valid experiment • conducting a risk assessment • working safely • recording observations and results • analysing and evaluating data and information
conducting field work	• conducting a risk assessment • working safely • recording observations and results • analysing and evaluating data and information
conducting surveys	• writing questions • collecting data • analysing data and information
designing a model	• identifying a problem to be modelled • summarise key findings • identify advantages and limitations of the model

TABLE 20.1.2 Examples of secondary research

Secondary research	Example tasks
researching published data from primary and secondary sources	• finding published information in scientific magazines and journals, books, databases, media texts, labels of commercially available products • analysing and evaluating data and information

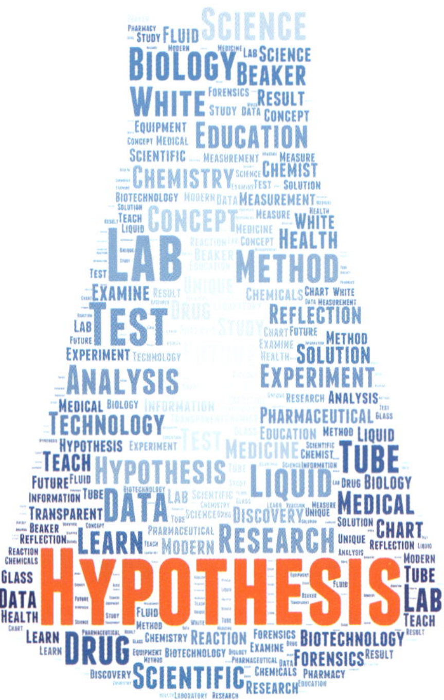

FIGURE 20.1.1 There may be many elements to an investigation. Taking a step-by-step approach will help the process and assist in completing a solid and worthwhile investigation.

Before you are able to start an investigation, you first need to understand the theory behind it. This section is a guide to some of the key steps that should be taken when first developing inquiry questions and hypotheses.

DEVELOPING AIMS AND QUESTIONS, FORMULATING HYPOTHESES AND MAKING PREDICTIONS

The inquiry question, aim and hypothesis are interlinked (Figure 20.1.1). It is important to note that each of these can be refined as the planning of the investigation continues.

Inquiry questions and hypotheses

An inquiry question defines what is being investigated, for example: 'What is the relationship between concentration and rate of reaction?'

A **hypothesis** is a prediction based on previous knowledge and evidence or observations that attempts to answer the inquiry question. For example: 'Increasing the concentration of the acid will increase the rate of the reaction of hydrochloric acid solution with marble chips.'

It is important that you can interpret what an inquiry question is asking you to do. These steps will help you to do this.

- Identify a 'guiding' word, such as *who*, *what*, *where*, *why*.
- Link the guiding word to command verbs that are often used in senior high school, such as *identify*, *describe*, *compare*, *contrast*, *distinguish*, *analyse*, *evaluate* and *create*.

Table 20.1.3 gives examples of inquiry questions that could be investigated.

TABLE 20.1.3 How to interpret an inquiry question

Guiding word	Example inquiry questions	What are you being asked to do? What are the command verbs?
what	**What** difference can nanomaterials make to society and the environment? **What** are electrons, protons and neutrons made of?	Identify and describe specific examples, evidence, reasons and analogies from a variety of possibilities. *Identify* and *describe*.
where	**Where** would an element with an atomic number of 130 be placed in the modern periodic table, what properties would it have and how likely is it to be discovered?	Identify and describe giving reasons for a place or location. *Identify* and *describe*.
how	**How** are atoms 'seen'? **How** can metal ores be transformed into metals? **How** do different crude oil extraction methods compare in terms of their ease of extraction and environmental impact?	Identify and describe in detail a process or mechanism. Give examples using evidence and reasons. *Identify* and *describe*.
why	**Why** are the 10 most abundant elements in the universe not the same as the 10 most abundant elements on Earth? **Why** does the composition of crude oil vary between different oil wells?	Explain in detail the causes, reasons, mechanisms and evidence for and against. *Identify* and *describe*.
would	**Would** there be life if elements did not form compounds?	Evaluate. Justify, giving reasons for and against (using evidence, analogies, comparisons). *Evaluate*, *assess*, *justify*.
is/are	**Are** there more elements to be discovered? **Is** it an advantage or a disadvantage to speed up chemical reactions used in cooking?	Evaluate. Justify, giving reasons and evidence. *Evaluate*, *assess*, *justify*.
on what basis	**On what basis** are alternative forms of the periodic table constructed?	Evaluate. Justify using reasons and evidence. *Evaluate*, *assess*, *justify*.
can	**Can** we live without the lanthanide and actinides elements?	Evaluate and assess. Is it possible? Give reasons, suggest possible alternatives. *Evaluate*, *assess*, *justify*, *create*.
do/does	**Do** the lanthanide and actinides rust or corrode? **Do** we need crude oil?	Evaluate. Justify using reasons and evidence for and against. *Evaluate*, *assess*, *justify*.
should	**Should** cars be made from shape memory metals?	Evaluate pros and cons, implications and limitations. Make a judgement. *Evaluate*, *assess*, *justify*, *create*.
might	What **might** we do if crude oil supplies become too expensive?	Evaluate. Justify, giving reasons for and against (using evidence, analogies, comparisons). *Evaluate*, *assess*, *justify*, *compare*, *contrast*, *create*.

Formulating a question

If you are required to formulate a question, it is good practice to research the topic to be investigated. You should become familiar with the relevant scientific concepts and key terms.

During this research, write down questions or correlations as they arise.

Compile a list of possible ideas. Do not reject ideas that initially might seem improbable. Use these ideas to generate questions that are answerable.

Before constructing a hypothesis, formulate a question that needs an answer. This question will lead to a hypothesis when:

- the question is confined to measurable variables
- a prediction is made based on knowledge and experience.

Evaluating your question

Once a question has been chosen, stop to evaluate the question before progressing. The question may need further refinement or even further investigation before it is suitable as a basis for investigation. A major planning point is to not attempt something that is not possible to complete in the time available or with the resources on hand. For example, it might be difficult to create a specific piece of specialist equipment in the school laboratory of if experiments take too long it may be impossible to conduct sufficient trials in the time allocated.

To evaluate the question, consider the following.

- Relevance: Is the question related to the appropriate course content?
- Clarity and measurability: Can the question be framed as a clear hypothesis?

If the question cannot be stated as a specific hypothesis, then it is going to be very difficult to complete the investigation.

- Time frame: Can the question be answered within a reasonable period of time? Is the question too broad?
- Knowledge and skills: Do you have a level of knowledge and the laboratory skills that will allow the question to be explored?
- Practicality: Are resources, such as laboratory equipment and materials, likely to be readily available?
- Safety and ethics: Consider the safety and ethical issues associated with the question you will be investigating. If there are issues, can they be addressed?

Sourcing information

Once you have selected a topic, the next step is to source reliable information. Some of the steps involved in sourcing information are:

- identifying key terms
- evaluating the credibility of sources
- evaluating experimental data/evidence.

Sources can be:

- **primary sources**—original sources of data and evidence; for example, articles containing research findings that have been published in peer-reviewed scientific journals or research presented at a scientific conference
- **secondary sources**—analyses and interpretations of primary sources; for example, textbooks, magazine articles and newspaper articles.

Some of the sources that may contain useful information are listed in Table 20.1.4.

TABLE 20.1.4 Useful sources of primary and secondary information

Primary sources of information	Secondary sources of information
journal articles from peer-reviewed journals, such as *Nature*	newspaper articles and opinion pieces
global databases, statistics and surveys	magazine articles; examples of reputable science magazines: • *Cosmos* • *Double Helix* (CSIRO) • *New Scientist* • *Popular Science* • *Scientific American*
laboratory work	government reports
computer simulations and modelling	

Hypothesis

A hypothesis is a prediction that is based on evidence and prior knowledge. A hypothesis often takes the form of a proposed relationship between two or more variables in a cause and effect relationship; or in other words, 'If X is done, then Y will occur.'

Here are some examples of hypotheses.

- If the number of carbon atoms per molecule of a hydrocarbon fuel increases, the energy released per kilogram of fuel during combustion will increase.
- If the concentration of sulfuric acid in a reaction with solid copper(II) oxide is increased, the rate of reaction will increase proportionally to the increase in concentration.
- If the structure of hydrocarbons with the same molecular formula varies (isomers), the boiling point of the isomers will also vary.

Variables

An investigation can test a good scientific hypothesis and the hypothesis will be supported or refuted. To be a testable hypothesis, it should be possible to measure both what is changed or carried out and what will happen. The factors that are monitored during an experiment or investigation are called the **variables**. An experiment or investigation determines the relationship between variables by measuring the results.

There are three categories of variables.

- The **independent variable** is the variable that is determined by the researcher (the one that is selected and changed).
- The **dependent variable** is the variable that may change in response to a change in the independent variable. This is the variable that will be measured or observed.
- **Controlled variables** are all the variables that must be kept constant during the investigation.

Note that you should only test one variable at a time, otherwise it cannot be stated that the changes in the dependent variable are the result of changes in the independent variable.

Qualitative and quantitative variables

Variables are either qualitative or quantitative, with further subsets in each category.

- **Qualitative** variables can be observed but not measured. They can only be sorted into groups or categories such as brightness, type of material, colour of flame.
- **Quantitative** variables can be measured. Mass, volume, temperature, pH and time are all examples of quantitative data.
  - Discrete variables consist of only integer numerical values, not fractions; for example, the number of protons in an atom, the number of atoms of each element in a compound and the number of isotopes of a particular element.
  - Continuous variables allow for any numerical value within a given range; for example, the measurement of temperature, volume, mass, pH and conductivity.

Formulating a hypothesis

Once the inquiry question is confirmed, formulating a hypothesis comes next. A hypothesis requires a proposed relationship between two variables. It should predict that a relationship exists or does not exist.

Identify the two variables in your question. State the independent and dependent variables.

For example: If I do/change this (independent variable), then this (dependent variable) will happen.

A good hypothesis should:

- be a statement
- be based on information contained in the **research question** (purpose)
- be worded so that it can be tested in the experiment
- include an independent and a dependent variable
- include variables that are measurable.

The hypothesis should also be falsifiable. This means that a negative outcome would disprove it. For example, the hypothesis that all apples are round cannot be proved beyond doubt, but it can be disproved—in other words, it is falsifiable. In fact, only one square apple is needed to disprove this hypothesis. Unfalsifiable hypotheses cannot be proved by science. These include hypotheses on ethical, moral and other subjective judgements.

Worked example 20.1.1

DEVELOPING A HYPOTHESIS

Develop a hypothesis and identify variables for investigating the rate of the reaction between marble chips and hydrochloric acid.	
Thinking	**Working**
Write an inquiry question.	What is the relationship between the concentration of the acid and reaction rate?
Identify the independent variable.	concentration of hydrochloric acid solution
Identify the dependent variable.	the mass lost during the first 30 seconds of the reaction
Identify the controlled variables.	starting temperature of hydrochloric acid volume of hydrochloric acid mass of the marble chips surface area of the marble chips
Write a potential hypothesis.	For a constant particle size of marble chips and volume of hydrochloric acid, if the concentration of the hydrochloric acid solution is halved, the rate of reaction will halve.

Worked example: Try yourself 20.1.1

DEVELOPING A HYPOTHESIS

- In this example, the heat given off in the combustion reactions of a range of hydrocarbons is measured by burning a certain amount of hydrocarbon and measuring the heat produced by measuring the temperature change of water. A possible set-up for the experiment is shown in Figure 20.1.2.

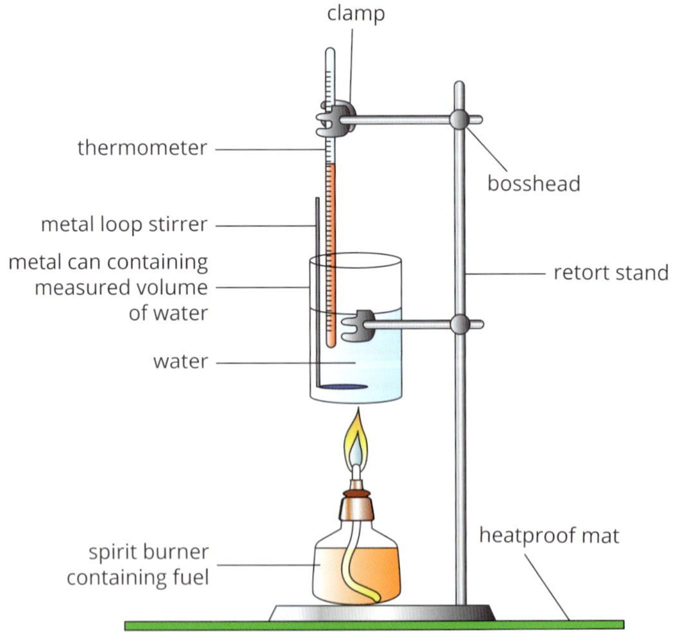

FIGURE 20.1.2 Apparatus for measuring heat energy released when a fuel burns

Develop a hypothesis and identify variables for investigating how the molecular mass of a hydrocarbon affects the energy released per gram of the hydrocarbon.

20.1 Review

SUMMARY

- By utilising data from primary and/or secondary sources, you will better understand the context of your investigation to create an informed inquiry question.
- A hypothesis is a prediction based on previous knowledge and evidence or observations that attempts to answer the research question. For example: 'If the temperature of the acid is increased, the rate of reaction of calcium carbonate and the hydrochloric acid will increase.'
- Once a question has been chosen, evaluate the question before progressing. The question may need further refinement or even further investigation before it is suitable as a basis for an achievable and worthwhile investigation.
- There are three categories of variables.
  - The independent variable is the variable that is controlled by the researcher.
  - The dependent variable is the variable that may change in response to a change in the independent variable. This is the variable that will be measured or observed.
  - Controlled variables are all the variables that must be kept constant during the investigation.

KEY QUESTIONS

1 Scientists make observations from which a hypothesis is stated and this is then experimentally tested.

a Define 'hypothesis'.

b How are theories and principles different from a hypothesis?

2 Which of the following is an example of an inquiry question?

A How are chemicals in solutions measured?

B A compound consists of two or more elements.

C Decreasing the volume of a container of gas will increase the pressure.

D The mass of the reactants equalled the mass of the products.

3 For each of the following hypotheses, select the dependent variable.

a If water is filtered through a domestic water purifier, then the electrical conductivity of the water will decrease.

b The concentration of lead in water will be higher in storm water close to an industrial site than in drinking water.

c The electrical conductivity of water from Fremantle Harbour will be greatest where ocean water can mix with fresh water.

d The pH of commercially available sparkling mineral water will be lower than commercially available non-sparkling mineral water.

4 In describing an experiment, a student uses the following range of values to record flame tests of ionic compounds: yellow, lilac, red and green. What type of measurement is the variable 'colour'?

5 Which of the following methods is likely to be the most accurate quantitative method for measuring the pH of water?

A using pH paper (e.g. litmus paper)

B using universal indicator and a colour chart

C using a calibrated pH meter at a particular temperature

D using a conductivity meter

6 Select the best hypothesis from the three options below. Give two reasons for your choice.

A Hypothesis 1: If volume and temperature of a gas are changed, then this will affect the pressure of the gas.

B Hypothesis 2: Concentration of solutions can be expressed using different units.

C Hypothesis 3: If water is filtered through a domestic water purifier, then the electrical conductivity of the water will decrease.

20.2 Planning investigations

Once you have formulated your hypothesis, you will need to plan and design your investigation. Taking the time to carefully plan and design a practical investigation before beginning will help you to maintain a clear and concise focus throughout. Preparation is essential. This section is a guide to some of the key steps that should be taken when planning and designing an investigation.

DEVELOPING THE METHOD

The method of your investigation is a step-by-step procedure. When detailing the method, ensure it complies as a valid, reliable and accurate investigation.

Validity

Validity refers to whether an experiment or investigation is in fact testing the set hypothesis and aims. Is the investigation obtaining data that is relevant to the question?

To ensure an investigation is valid, it should be designed so that only one variable is being changed at a time. The remaining variables must remain constant so that meaningful conclusions can be drawn about the effect of each variable in turn.

To ensure validity, carefully determine:

- the independent variable: the variable that will be changed, and how it will change
- the dependent variable: the variable that will be measured
- the controlled variables: the variables that must remain constant, and how they will be maintained.

Reliability

Reliability refers to the notion that if the experiment is repeated many times, the results obtained will be consistent.

It is important to determine how many times the experiment needs to be replicated. Many scientific investigations lack sufficient repetition to ensure that the results can be considered reliable and repeatable.

- Repeat readings: repeat each reading (at least three times if possible), record each measurement and then average the measurements. This reduces systematic errors and allows random errors to be identified. If a reading differs too much from the rest (known as an **outlier**), discard it before averaging (Figure 20.2.1).
- Sample size: where there might be differences in construction or manufacture of a sample, there should be various samples with the same conditions in the same experiment. The greater the sample size, the more reliable the data.
- Repeats: if possible repeat the experiment on a different day. Don't change anything. If the results are not the same, think about what could have happened. For example, was the equipment faulty, or were all the controlled variables correctly identified? Repeat the experiment a third time to confirm which run was correct. Repeat an experiment as many times as possible; three is a good number but, if time and resources allow, aim for five.

FIGURE 20.2.1 Replication increases the reliability of your investigation. It ensures that if anyone repeats the investigation they will obtain similar data. Outliers should be discarded before averaging measurements.

In some situations, it may be appropriate to conduct a **control experiment**. The control is an identical experiment carried out at the same time, except that in the control experiment the independent variable is not changed.

When the controls do not behave as expected, the data obtained from an experiment or observation is not reliable.

Accuracy and precision

In science and statistics, the terms 'accuracy' and 'precision' have very specific and different meanings:

- **Accuracy** is the ability to obtain the correct measurement. To obtain accurate results, you must minimise systematic errors.

- **Precision** is the ability to consistently obtain the same measurement. To obtain precise results, you must minimise random errors. You will learn more about types of errors in Section 20.3.

To understand more clearly the difference between accuracy and precision, think about firing arrows at an archery target (Figure 20.2.2). Accuracy is being able to hit the bullseye, whereas precision is being able to hit the same spot every time you shoot. If you hit the bullseye every time you shoot, you are both accurate and precise (Figure 20.2.2a). If you hit the same area of the target every time but not the bullseye, you are precise but not accurate (Figure 20.2.2b). If you hit the area around the bullseye each time but don't always hit the bullseye, you are accurate but not precise (Figure 20.2.2c). If you hit a different part of the target every time you shoot, you are neither accurate nor precise (Figure 20.2.2d).

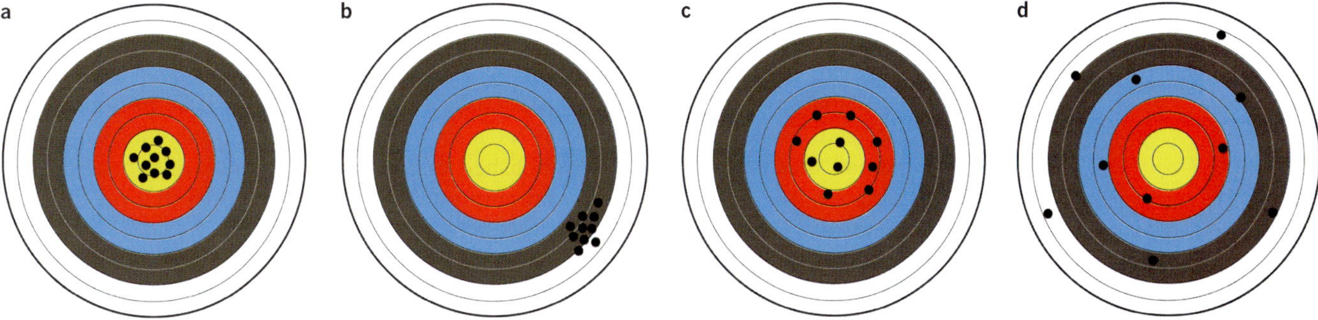

FIGURE 20.2.2 Examples of accuracy and precision: (a) both accurate and precise, (b) precise but not accurate, (c) accurate but not precise, and (d) neither accurate nor precise

Applying this idea to a chemistry situation, if you are timing the rate of reaction by stopping a stopwatch when 100 mL of gas has been released, and the you get consistent results but you are always stopping the stopwatch too early, your results are precise but not accurate.

Uncertainty

When scientists take a measurement, the measurement is always subject to variations in accuracy and precision. The goal when taking measurements is to get as close as possible to the 'true' or 'correct value. The **uncertainty** is how far an experimental quantity might be from the true value. It can be calculated, or estimated and is often represented as a percentage. For example, if the uncertainty was described as ±1%, it means that measurement is likely to be no more than 1% above or below the 'true' value of the quantity that is being measured.

In this course you are not required to calculate uncertainties, but it is a good idea to consider uncertainty when planning experiments, and later when you come to evaluate your results.

Recording numerical data

Reasonable steps to ensure the accuracy of the investigation include considering:

- the unit in which the independent and dependent variables will be measured
- the instrument that will be used to measure the independent and dependent variables.

Select and use appropriate equipment, materials and methods. For example, select equipment that measures to greater precision to reduce uncertainty and repeat the measurements to confirm them.

When using measuring instruments, the number of significant figures and decimal places you use is determined by how precise your measurements are.

This depends on the scale, accuracy and precision of the instrument and technique you are using (Figure 20.2.3). For example, a beaker is only used to measure volumes approximately and has a relatively high uncertainty, for example ±5%. A graduated pipette is more accurate, with uncertainties of ±0.1% or ±0.2%. Your pipette may be accurate but if your technique using the pipette is variable, the overall accuracy and precision will be limited.

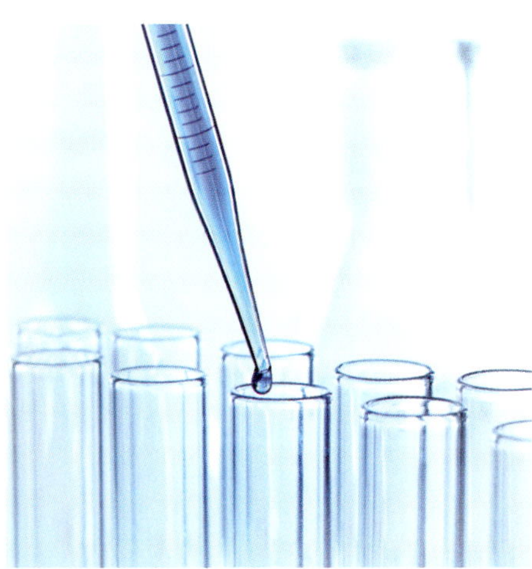

FIGURE 20.2.3 A 5 mL graduated pipette can measure volumes to an accuracy of one hundredth of a millilitre, or 5 mL ± 0.01 mL. The pipette has major divisions of 1 mL and minor divisions of 0.1 mL. You can estimate to 0.01 mL and record volumes to 2 decimal places, for example 3.80 mL or 4.52 mL.

When you record **raw data** and report processed data, use the number of significant figures available from your equipment or observation. Using either a greater or smaller number of significant figures can be misleading. For example, Table 20.2.1 shows measurements of a chemical taken using an electronic balance accurate to 2 decimal places. The data was entered into a spreadsheet to calculate the mean, which was displayed with 4 decimal places. You would record the mean as 20.83 g, not 20.8260 g, because 2 decimal places is the precision limit of the instrument. Recording 20.8260 g would be an example of false precision.

TABLE 20.2.1 An example of false precision in a data calculation

Sample	1	2	3	4	5	Mean
mass (g)	20.13	20.62	21.22	20.99	21.17	20.8260

Data analysis

Data analysis is part of the method. Consider how the data will be presented and analysed. A wide range of analysis tools are available. For example, tables organise data so that patterns can be established and graphs can show relationships and comparisons. In fact, preparing an empty table showing the data that needs to be obtained will help in the planning of the investigation.

Sourcing appropriate equipment and materials

When designing your investigation, you will need to decide on the materials, technology and instrumentation that will be used to carry out your investigation. It is important to find the right balance between items that are easily accessible and those that will give you accurate results. When conducting your investigation, it will be important to take note of the precision of your chosen instrumentation and how this affects the accuracy and validity of your results.

Modifying the method

The method may need modifying as the investigation is carried out. The following actions will help to determine any issues in the method and how to modify them.

- Record everything.
- Be prepared to make changes to the approach.
- Note any difficulties encountered and the ways they were overcome. What were the failures and successes? Every test carried out can contribute to the understanding of the investigation as a whole, no matter how much of a disaster it may first appear.

If the expected data is not obtained, don't worry. As long as it can be critically and objectively evaluated, the limitations of the investigation are identified and further investigations proposed, the work is worthwhile.

COMPLYING WITH ETHICAL AND SAFETY GUIDELINES

Ethical considerations

When planning an investigation, researchers should always identify all possible ethical considerations and evaluate their necessity or ways that can reduce or mitigate them.

Ethical issues might include the following.

- How can this affect wider society?
- Does one party benefit over another; for example, one individual, a group of individuals or a community?
- Is there a risk of harm (physical or mental) to people involved in the research?
- Does it prevent anyone from gaining their basic needs?
- How can this impact on future ethical decisions or issues?

- Does the research cause damage to the environment?
- Does the research cause harm to other living things?

In reality, school chemistry investigations generally will have minor ethical issues, if any, but these should be considered in your planning.

Safety in practical experiments and investigations

While planning for an investigation, it is important for your safety and the safety of others that the potential risks are considered (Figure 20.2.4).

Everything we do has some risk involved. **Risk assessments** are performed to identify, assess and control hazards. Always identify the risks and control them to keep everyone safe.

To identify risks think about:

- the activity that will be carried out
- the equipment or chemicals that will be used.

The following hierarchy of risk controls is organised from the most effective to least effective:

1. *Elimination:* Eliminate dangerous equipment, procedures or substances.
2. *Substitution:* Find different equipment, procedures or substances to use that will achieve the same result, but have less risk associated.
3. *Isolation:* Ensure there is a barrier between the person and the hazard. Examples include physical barriers such as guards in machines, or fume hoods to work with volatile substances.
4. *Administrative controls:* Provide guidelines, special procedures, warning signs and safe behaviours for any participants.
5. *Personal protective equipment (PPE):* Wear safety glasses, lab coats, gloves and respirators etc. where appropriate, and provide these to other participants.

Figure 20.2.5 shows a flow chart of how to consider and assess the risks involved in a research investigation.

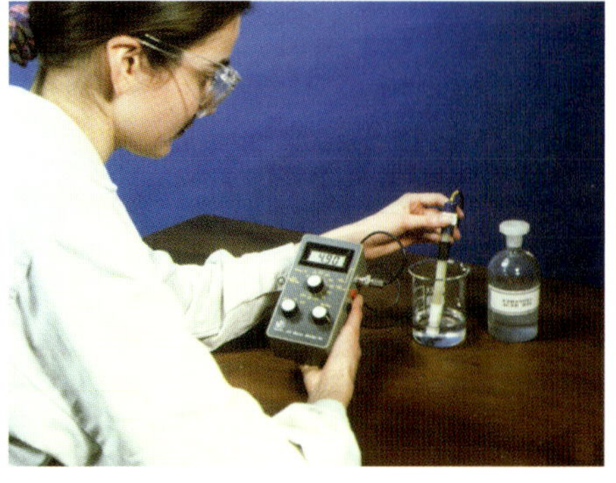

FIGURE 20.2.4 When planning an investigation, you need to identify, assess and control hazards.

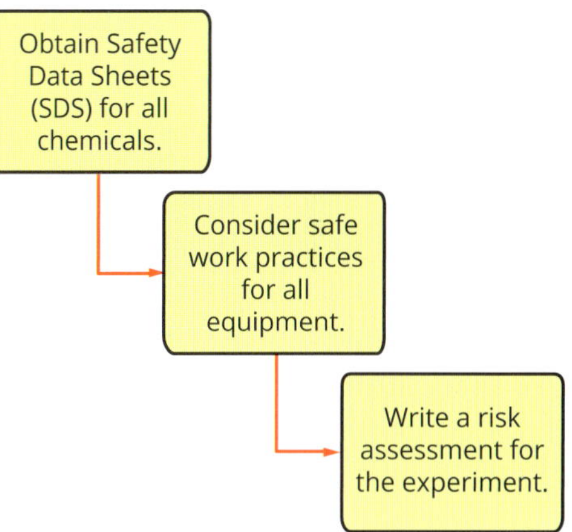

FIGURE 20.2.5 Steps involved in identifying risks

Chemical codes

The chemicals at school or at the hardware shop have a warning symbol on the label. These are **chemical (HAZCHEM) codes**. Some common codes and their meanings are shown in Figure 20.2.6. Trucks that carry chemicals display hazard symbols, as shown in Figure 20.2.7.

HAZCHEM INTERPRETATION

NUMBER			
1		Water Jets	
2		Water Fog	
3		Foam	
4		Dry Agent	
FIRST LETTER			
P	V	Full Protective Clothing*	DILUTE
R		Full Protective Clothing*	DILUTE
S	V	Breathing Apparatus	DILUTE
S	V	Breathing Apparatus for Fire Only	DILUTE
T		Breathing Apparatus	DILUTE
T		Breathing Apparatus for Fire Only	DILUTE
W	V	Full Protective Clothing*	CONTAIN
X		Full Protective Clothing*	CONTAIN
Y	V	Breathing Apparatus	CONTAIN
Y	V	Breathing Apparatus for Fire Only	CONTAIN
Z		Breathing Apparatus	CONTAIN
Z		Breathing Apparatus for Fire Only	CONTAIN
SECOND LETTER			
E		Consider Evacuation	

Note V: Danger of violent reaction or explosion
* Full Protective Clothing includes Breathing Apparatus

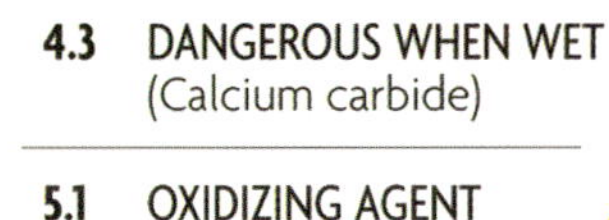

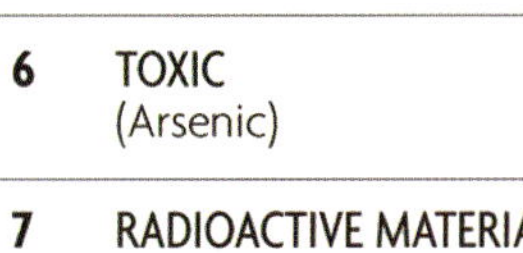

1 EXPLOSIVE (Gunpowder, flares)

2.1 FLAMMABLE GAS (LP gas, acetylene)

2.2 NON-FLAMMABLE NON-TOXIC GAS (Carbon dioxide)

2.3 TOXIC GAS (Chlorine gas)

3 FLAMMABLE LIQUID (Petrol, kerosene)

4.1 FLAMMABLE SOLID (Firelighters, matches)

4.2 SPONTANEOUSLY COMBUSTIBLE (Carbon, white phosphorus)

4.3 DANGEROUS WHEN WET (Calcium carbide)

5.1 OXIDIZING AGENT (Calcium hypochlorite)

5.2 ORGANIC PEROXIDE

6 TOXIC (Arsenic)

7 RADIOACTIVE MATERIAL (Uranium)

8 CORROSIVE (Hydrochloric acid)

9 MISCELLANEOUS DANGEROUS GOODS (Dry ice, asbestos)

MIXED CLASS LABEL (For road transport)

FIGURE 20.2.6 HAZCHEM signs

FIGURE 20.2.7 Trucks transporting hazardous substances, such as flammable liquids, have hazard symbols attached.

CHEMFILE

A new system of chemical classification

In all states (apart from Western Australia and Victoria) chemicals in laboratories must now be labelled according to the Globally Harmonised System of Classification and Labelling of Chemicals (GHS). This ruling forms part of the labelling provisions of the *Work Health and Safety Act*, which became fully effective on 1 January 2017. The GHS replaces national systems, such as Hazchem symbols, in order to create a global system for hazard identification. A selection of GHS labels is shown in Figure 19.2.7.

Schools that store, handle, generate, use or dispose of chemicals classified under the new GHS system will be required to provide staff with new training in relation to chemical and hazard identification and risk minimisation. Western Australia has not yet adopted the use of the GHS, but is likely to do so in the future.

FIGURE 20.2.8 New labels used in the 'Globally harmonised system of classification and labelling of chemicals' (GHS)

Safety Data Sheets

Each chemical substance has an accompanying document called a **Safety Data Sheet (SDS)** (Figure 20.2.9), which was previously called a Material Safety Data Sheet (MSDS). An SDS contains important safety and first aid information about each chemical you commonly use in the laboratory. If the products of a reaction are toxic to the environment, you must pour your waste into a special container (not down the sink).

1. IDENTIFICATION OF THE MATERIAL AND SUPPLIER

Product Name: **HYDROCHLORIC ACID - 20% OR GREATER**

Recommended use of the chemical and restrictions on use: Precursor for generation of chlorine dioxide gas used in water treatment.

Supplier: Ixom Operations Pty Ltd
ABN: 51 600 546 512
Street Address: Level 8, 1 Nicholson Street
Melbourne 3000
Australia

Telephone Number: +61 3 9665 7111
Facsimile: +61 3 9665 7937
Emergency Telephone: **1 800 033 111 (ALL HOURS)**

Please ensure you refer to the limitations of this Safety Data Sheet as set out in the "Other Information" section at the end of this Data Sheet.

2. HAZARDS IDENTIFICATION

Classified as Dangerous Goods by the criteria of the Australian Dangerous Goods Code (ADG Code) for Transport by Road and Rail; DANGEROUS GOODS.

This material is hazardous according to Safe Work Australia; HAZARDOUS SUBSTANCE.

Classification of the substance or mixture:
Corrosive to Metals - Category 1
Skin Corrosion - Sub-category 1B
Eye Damage - Category 1
Specific target organ toxicity (single exposure) - Category 3

SIGNAL WORD: DANGER

Hazard Statement(s):
H290 May be corrosive to metals.
H314 Causes severe skin burns and eye damage.
H335 May cause respiratory irritation.

Precautionary Statement(s):

Prevention:
P234 Keep only in original container.
P260 Do not breathe mist / vapours / spray.
P264 Wash hands thoroughly after handling.
P271 Use only outdoors or in a well-ventilated area.
P280 Wear protective gloves / protective clothing / eye protection / face protection.

FIGURE 20.2.9 Extracts of a Safety Data Sheet (SDS) for concentrated hydrochloric acid. The SDS alerts the reader to any potential hazards when using a substance, including appropriate measures to reduce risk of harm.

The SDS provides employers, workers and health and safety representatives with the necessary information to safely manage the risk of hazardous substance exposure.

Personal protective equipment

Everyone who works in a laboratory wears items that help keep them safe. This is called **personal protective equipment (PPE)** and includes:

- safety glasses
- shoes with covered tops
- disposable gloves when handling certain chemicals
- a disposable apron or a lab coat if there is risk of damage to clothing.

20.2 Review

SUMMARY

- The method of your investigation is a step-by-step procedure. When detailing the method, ensure it complies as a valid, reliable and accurate investigation.
- It is also important to determine how many times the experiment needs to be replicated. Many scientific investigations lack sufficient repetition to ensure that the results can be considered reliable and repeatable.
- It is important to choose appropriate equipment for your experiment. Select instrumentation that will give you accurate results.
- Risk assessments must be carried out before conducting an investigation to make sure that when you carry out your investigation you and others are kept safe. If you have elements of your investigation that are high risk, you will need to revise your design.
- Appropriate personal protective equipment (PPE) that will help keep you safe should be used according to the risk assessment.

KEY QUESTIONS

1 A journal article reported the materials and method used to conduct an experiment. The experiment was repeated three times, and all values were reported in the results section of the article.
What does repeating an experiment and reporting results support?
   A validity
   B reliability
   C credibility
   D systematic errors

2 **a** Explain what is meant by the term 'control experiment'.
   b Using an example, distinguish between independent and dependent variables.

3 Give the correct term that describes an experiment with each of the following conditions.
   a The experiment addresses the hypothesis and aims.
   b The experiment is repeated and consistent results are obtained.
   c Appropriate equipment is chosen for the desired measurements.

4 You are conducting an experiment to determine the pH of various soft drinks. Identify:
   a the independent variable
   b the dependent variable
   c at least one controlled variable.

5 You are conducting an experiment to determine the pH of a solution. Discuss the accuracy of your results if you are:
   a using litmus paper or universal indicator
   b recording the pH using a calibrated pH meter.

20.3 Uncertainty and error in data

During the planning and conducting of an experiment or investigation, minimising potential error is an important consideration.

IDENTIFYING ERRORS

Most practical experiments have errors associated with them. Errors can occur for a variety of reasons. Being aware of potential errors helps you to avoid or minimise them. For an investigation to be accurate, it is important to identify and record any errors.

There are two types of errors:

- systematic errors
- random errors.

When considering whether an error is random or systematic, you have to consider the measurement that is made compared to the true value of the value being measured.

Systematic errors

A **systematic error** will cause the measured error to be always above, or always below, the true value. It is an error that is consistent and will occur again if the investigation is repeated in the same way.

An example of a systematic error is if you were measuring the mass of a substance on a piece of filter paper and you didn't take into account the mass of the filter paper. The mass recorded will always be higher than the true mass of just the substance being weighed. Another example would be if you were measuring the volume of a solution, and you transferred the solution from a beaker to a measuring cylinder to measure the volume and didn't ensure that all the solution was transferred. In this case the volume measured would always be less than the true volume of the solution in the beaker.

Systematic errors are usually a result of instruments that are not **calibrated** correctly or methods that are flawed.

Random errors

Random errors occur in an unpredictable manner and are generally small. They could give a value higher or lower than the true value. Whenever a measuring device, whether it is a measuring cylinder or an electric balance is used, random error will occur.

Mistakes

Mistakes are not classified as errors. Mistakes could include misreading the numbers on a scale or spilling a portion of a sample. A measurement that involves a mistake must be rejected and not included in any calculations or averaged with other measurements of the same quantity.

TECHNIQUES FOR REDUCING ERROR

Designing the method carefully, including selection and use of equipment, will help reduce errors.

Appropriate equipment

Use the equipment that is best suited to the data to be collected to validate the hypothesis. Determining the units of the data being collected and at what scale will help to select the correct equipment. Using the right unit and scale will ensure that measurements are more accurate and precise (with smaller systematic errors).

Significant figures

Significant figures are the numbers that convey meaning and precision. The number of significant figures used depends on the scale of the instrument. It is important to record data to the number of significant figures available from the equipment or observation. Using either a greater or smaller number of significant figures can be misleading.

Review the following examples to learn more about significant figures.

- 15 has two significant figures.
- 3.5 has two significant figures.
- 3.50 has three significant figures.
- 0.037 has two significant figures.
- 1401 has four significant figures.

When adding or subtracting, the sum or difference of a set of numbers should have as many numbers after the decimal as the least accurate in the set.

Adding: $5.0 + 3.21 + 1.2345 = 9.445 = 9.4$ (because 5.0 has only one figure after the decimal point)

Subtracting: $19.8765 - 1.23 = 18.6465 = 18.64$ (because 1.23 has two figures after the decimal point)

When multiplying or dividing, the answer has no more significant figures than the number with the fewest significant figures.

Multiplying: $1.234 \times 2.34 = 2.88756 = 2.89$ (because 2.34 has only 3 significant figures)

Dividing: $34.56 \div 6.7 = 5.158208 = 5.2$ (because 6.7 has only 2 significant figures).

Calibrated equipment

Some equipment, such as pH meters, need to be calibrated before use. Before carrying out the investigation, make sure the instruments or measuring devices are properly calibrated and are, in general, functioning correctly. Record the precision of glassware that you intend to use. If you are preparing a solution of known concentration, you might have access to a pipette, which has much less uncertainty associated with measurements compared to using a beaker.

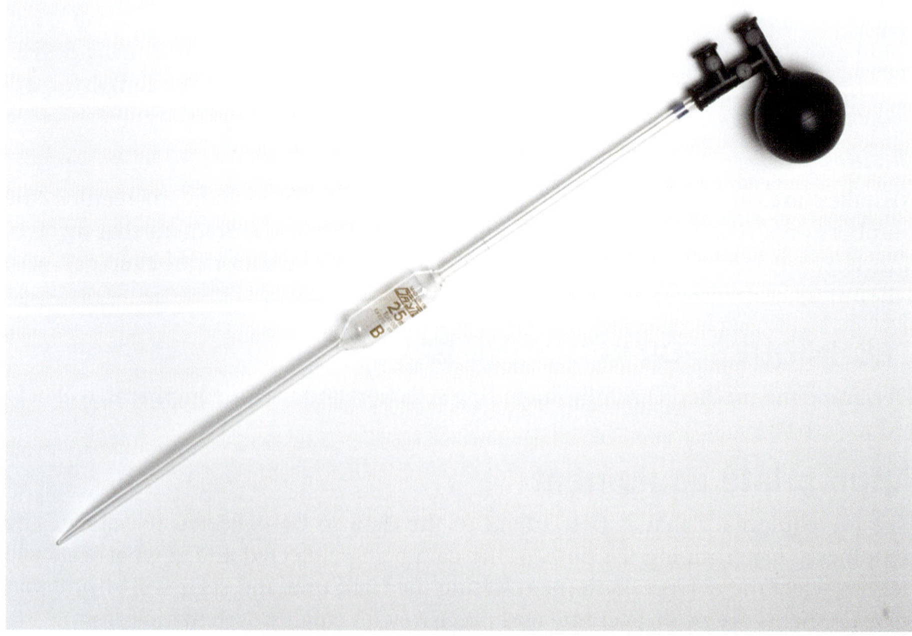

FIGURE 20.3.1 Record the uncertainty for glassware and instruments. This pipette can dispense a volume (aliquot) of 25.00 ±0.03 mL. When used correctly, the volume dispensed will be between 24.97 mL and 25.03 mL.

Correct use of equipment

Use the equipment properly. Ensure any necessary training has been done to use the equipment and that you have had an opportunity to practise using the equipment before beginning the investigation. Improper use of equipment can result in inaccurate, imprecise data with large errors, and the validity of the data can be compromised.

Incorrect reading of measurements is a common misuse of equipment. Make sure all of the equipment required in the investigation can be used correctly and record the instructions in detail so they can be referred back to if the data doesn't appear correct.

Repeat trials

As discussed in Section 20.2, reliability is improved by repeating your experiment. Note that repeating trials will only reduce the effect of random errors. Repeating a flawed method will not reduce the effect of the systematic error within that method. Therefore modifications to your procedure may need to be looked at before repeating the investigation to ensure all variables are being tested under the same conditions.

20.3 Review

SUMMARY

- It is essential that during the investigation, the following are recorded in the logbook:
  - all quantitative and qualitative data collected
  - the methods used to collect the data
  - any incident, feature or unexpected event that may have affected the quality or validity of the data.
- A systematic error will cause the measured error to always be above or always below the true value.
- Random errors could give a value higher or lower than the true value.
- The number of significant figures used depends on the scale of the instrument used. It is important to record data to the number of significant figures available from the equipment or observation.

KEY QUESTIONS

1 What type of error is associated with:
   a inaccurate measurements?
   b imprecise measurements?

2 Identify whether each of the situations below is a mistake, a systematic error or a random error.
   a A student measured the mass of samples of octane in beakers and left the beakers uncovered in the lab before using them in an experiment.
   b A student was measuring the mass of approximately 5 g samples of solid sodium chloride using a balance that was accurate to the nearest 1 g.
   c A student sometimes forgot to re-zero the balance between weighing samples of solid sodium carbonate.

3 A scientist carries out a set of experiments, analyses the results and publishes them in a scientific journal. Other scientists in different laboratories repeat the experiment, but do not get the same results as the original scientist. Suggest three reasons that could explain this.

20.4 Processing data and information

Once you have conducted your investigation and collected data, you will need to find the best way of collating it. This section is a guide to the different forms of representation that will help you to better understand your data.

Raw data is unlikely to be used directly to validate the hypothesis. However, raw data is essential to the investigation and plans for collecting the raw data should be made carefully. Consider the formulae or graphs that will be used to analyse the data at the end of the investigation. This will help to determine the type of raw data that needs to be collected in order to validate the hypothesis.

Once you have determined the data that needs to be collected, prepare a table in which to record the data.

The raw data that has been obtained should be presented in a way that is clear, concise and accurate. There are a number of ways of presenting data, including tables, graphs, flow charts and diagrams.

PRESENTING RAW AND PROCESSED DATA IN TABLES

Tables organise data into rows and columns and can vary in complexity according to the nature of the data. Tables can be used to organise raw data and processed data or to summarise results.

The simplest form of a table is a two-column format. In a two-column table, the first column should contain the independent variable (the one being changed) and the second column should contain the dependent variable (the one that may change in response to a change in the independent variable).

Tables should have the following features:

- a descriptive title
- column headings (including the unit)
- the independent variable placed in the left column
- the dependent variable placed in the right column.

Look at Table 20.4.1, which has been used to organise raw and processed data about the water near the desalination plant in Cockburn Sound.

TABLE 20.4.1 Analysis of seawater from Cockburn Sound and how the data could be recorded

Location	Temperature (°C)	pH	Conductivity (mS m^{-1})	Turbidity (NTU)	Dissolved oxygen (mg L^{-1})	Biological oxygen demand (mg L^{-1})
A	10.4	6.5	72	5	11.8	0.9
B	9.2	6.6	73	10	11.4	0.7
C	9.5	6.4	77	10	10.9	0.9
D	9.9	6.5	75	10	11.3	1.0

A table of processed data usually presents the average values of replicates: the **mean**.

GRAPHS

In general, tables provide more detailed data than graphs, but it is easier to observe **trends** and patterns in data in graphical form than in tabular form.

Graphs are used when two variables are being considered and one variable is dependent on the other. The graph shows the relationship between the variables.

There are several types of graphs that can be used, including line graphs, bar graphs and pie charts. The best one to use will depend on the nature of the data.

General rules to follow when making a graph (Figure 20.4.1) include the following.

- Keep the graph simple and uncluttered.
- Use a descriptive title.
- Represent the independent variable on the x-axis and the dependent variable on the y-axis.
- Make axes proportionate to the data.
- Clearly label axes with both the variable and the unit in which it is measured.

Rate of reaction of magnesium ribbon with different concentrations of hydrochloric acid

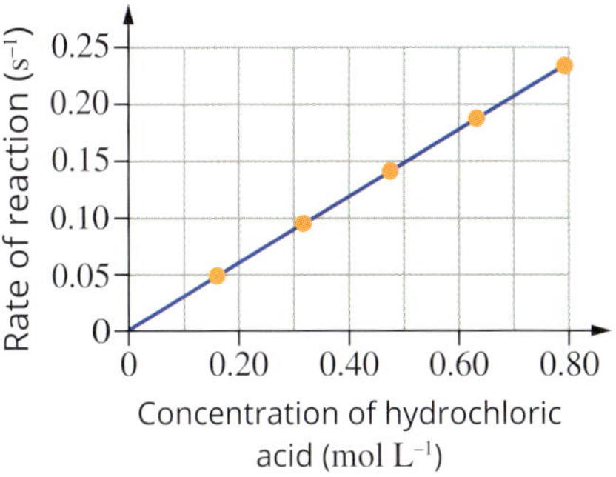

FIGURE 20.4.1 A graph is a better way to observe trends and patterns in the data.

Line graphs

Line graphs are a good way of representing continuous quantitative data. In a line graph, the values are plotted as a series of points on the graph. There are two ways of joining these points.

- A line can be ruled from each point to the next. It shows the overall trend; it is not meant to predict the value of the points between the plotted data.
- The points can be joined with a single smooth straight or curved line (Figure 20.4.2). This creates a **trend line**, also known as a **line of best fit**. The line of best fit does not have to pass through every point but should go close to as many points as possible. It is used when there is an obvious trend between the variables.

Sometimes when the data is collected, there may be one point that does not fit with the trend and is clearly an error. This is called an outlier. An outlier is often caused by a mistake made in measuring or recording data. If there is an outlier, include it on the graph, but ignore it when adding a line of best fit. As you can see in Figure 20.4.2, the point (0.48, 0.08) is an outlier because it does not fit the trend of the other data points, therefore it was not used in drawing the line of best fit.

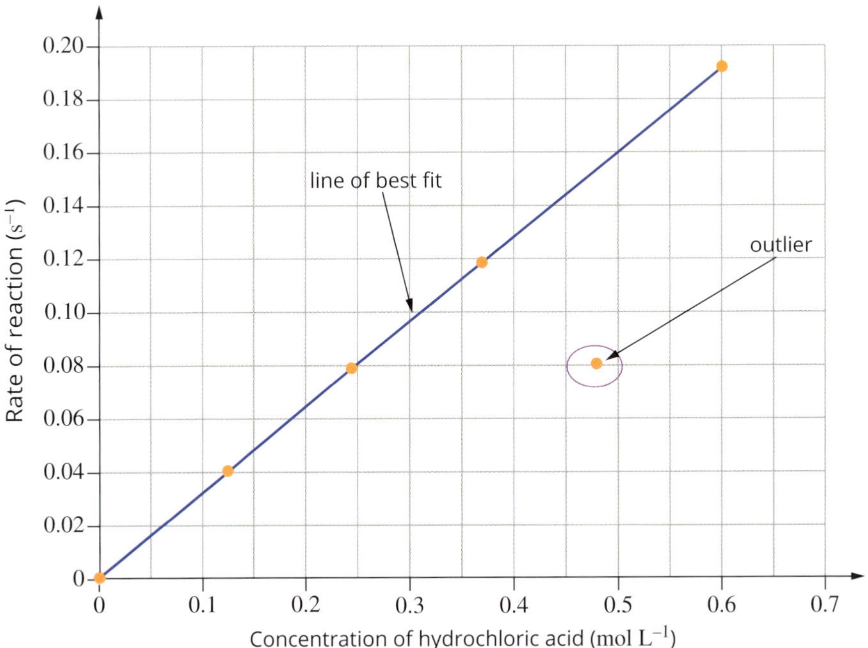

FIGURE 20.4.2 The graph shows an example of an outlier ignored in forming a line of best fit.

20.4 Review

SUMMARY

- Consider how the data will be presented and analysed. A wide range of analysis tools could be used. For example, tables organise data so that patterns can be established and graphs can show relationships and comparisons.
- The simplest form of a table is a two-column format in which the first column contains the independent variable and the second column contains the dependent variable
- General rules to follow when making a graph include the following.
  - Keep the graph simple and uncluttered.
  - Use a descriptive title.
  - Represent the independent variable on the *x*-axis (horizontal) and the dependent variable on the *y*-axis (vertical).
  - Make axes proportionate to the data.
  - Clearly label axes with both the variable and the unit in which it is measured.

KEY QUESTIONS

1 Determine the mean for the data set: {21, 28, 19, 19, 25, 24}.

2 How can the general pattern (trend) of a graph be represented once the points are plotted?

3 **a** Use the values in the table to plot a graph showing the rate of reaction between sodium hydrogencarbonate and solutions of citric acid with a range of concentrations. The rate of reaction was measured by recording the volume of carbon dioxide produced in the first three seconds of the experiment.

Concentration of citric acid ($mol\ L^{-1}$)	Volume of CO_2 produced (mL)
0.10	5
0.20	10
0.30	14.5
0.40	20.5
0.50	21.5
0.60	30

b From the graph you have drawn, select the data point that is an outlier.

c Define the term 'outlier'.

20.5 Analysing data and information

Now that the investigation has been conducted and data collected, it is time to draw it all together. You will now analyse your results to better understand the chemical processes behind them.

FIGURE 20.5.1 To discuss and conclude your investigation, utilise the raw and processed data.

ANALYSING AND EVALUATING DATA

In the discussion, the findings of the investigation need to be analysed and interpreted.

- State whether a pattern, trend or relationship was observed between the independent and dependent variables. Describe what kind of pattern it was and specify under what conditions it was observed.
- Were there discrepancies, deviations or anomalies in the data? If so, these should be acknowledged and explained.
- Identify any limitations in the data you have collected. Perhaps a larger sample or further variations in the independent variable would lead to a stronger conclusion.

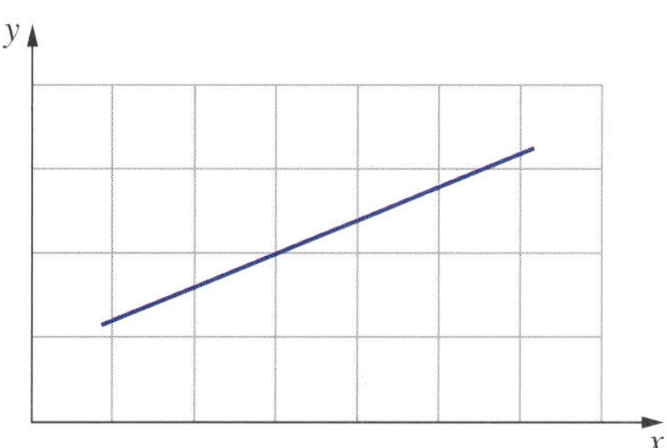

FIGURE 20.5.2 Trend line for variables that change in direct response to each other

Trends in line graphs

Graphs are drawn to show the relationship, or trend, between two variables.

- Variables that change in linear or direct proportion to each other produce a straight, sloping trend line (Figure 20.5.2).
- Variables that change exponentially in proportion to each other produce a curved trend line (Figure 20.5.3).
- When there is an inverse relationship, one variable increases as the other variable decreases (Figure 20.5.4).
- When there is no relationship between two variables, one variable will not change even if the other changes (Figure 20.5.5).

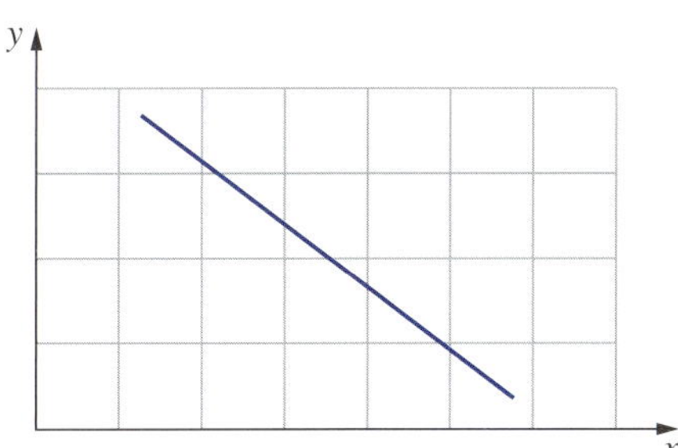

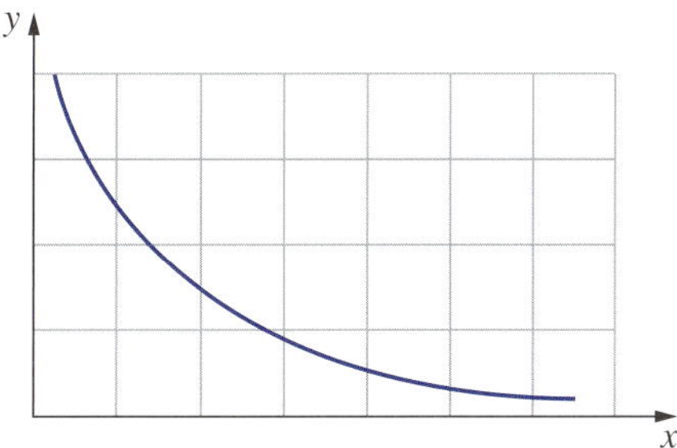

FIGURE 20.5.4 An inverse relationship in which one variable decreases in response to the other variable increasing. It may be (a) direct or (b) non-linear.

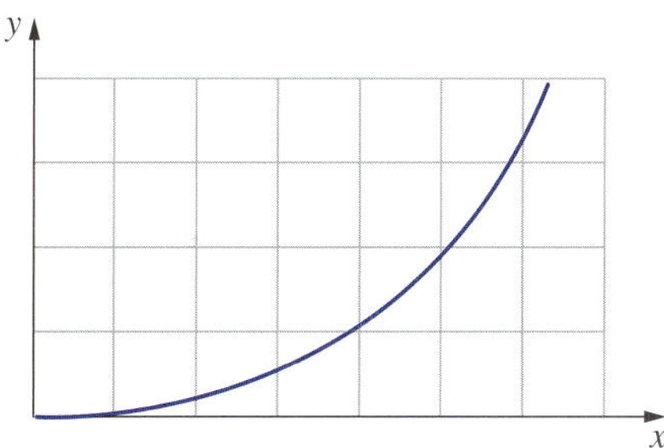

FIGURE 20.5.3 Variables that change in response to each other in a non-linear way

Remember that the results might be unexpected. This does not make the investigation a failure. However, the findings must be related to the hypothesis, aims and method.

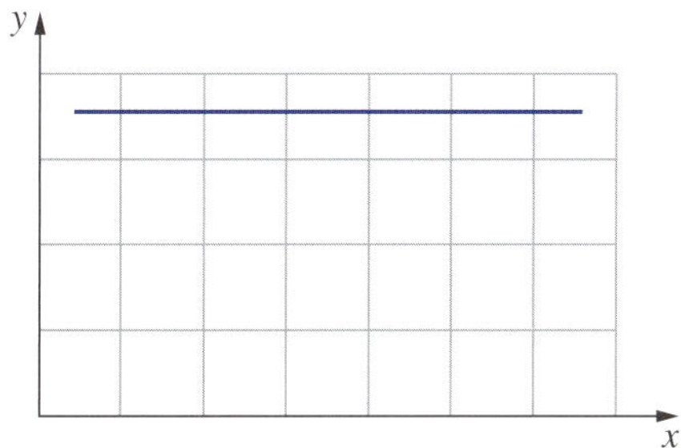

FIGURE 20.5.5 When two variables show no relationship, there is no trend in the data.

Table 20.5.1 lists the types of graphs used for various examples.

TABLE 20.5.1 Examples of the types of graphs that could be used in your report

Type of graph	When to use	Example
scatter graph	when showing quantitative data where one variable is dependent on another variable; draw a line of best fit to show the relationship between the two variables	showing the relationship between the rate of a reaction and the concentration of acid
line graph	with continuous quantitative data	concentration of dissolved oxygen at a particular location of a creek over a period of time
bar graphs	when comparing data in an investigation with a qualitative independent variable	turbidity of water at various locations, as shown in Figure 20.5.6
pie diagrams	when summarising qualitative data; to display proportions	relative proportion of different pesticides in a sample of water

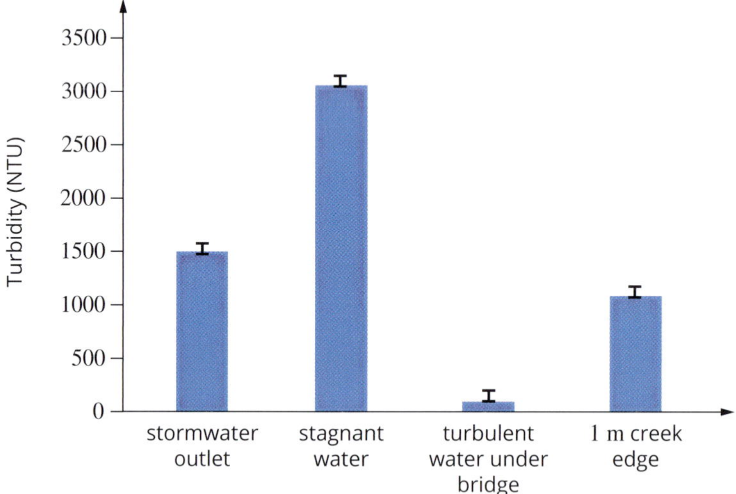

FIGURE 20.5.6 A bar graph can be used to compare data. This graph shows a measure of water quality, called turbidity, of water samples taken from various locations along Sunnyside Creek.

EVALUATING THE METHOD

It is important to discuss the limitations of the investigation method. Evaluate the method and identify any issues that could have affected the validity, accuracy, precision or reliability of the data. Sources of errors and uncertainty must also be stated in the discussion.

Once any limitations or problems in the methodology have been identified, recommend improvements on how the investigation could be conducted if repeated; for example, suggest how a journal article reported the materials could be minimised or eliminated.

Bias

Bias may occur in any part of the investigation method, including sampling and measurements.

Bias is a form of systematic error resulting from the researcher's personal preferences or motivations. There are many types of bias, including:

- poor definitions of both concepts and variables (for example, not defining pH)

- incorrect assumptions (for example, measuring pH of a solution without also measuring the temperature of the solution)
- errors in the investigation design and methodology (for example, taking water samples from more than one site).

Some biases cannot be eliminated, but should be addressed in the discussion.

Accuracy and precision

In the discussion, evaluate the degree of accuracy and precision of the measurements for each variable of the hypothesis. Comment on the uncertainties obtained.

When relevant, compare the chosen method with any other methods that might have been selected, evaluating the advantages and disadvantages of the selected method and the effect on the results.

Reliability

When discussing the results, indicate the **range** of the data obtained from replicates. Explain how the sample size was selected. Larger samples are usually more reliable, but short time frames and scarce resources might have prevented multiple sets of data from being collected. Discuss whether the results of the investigation have been limited by the sample size.

The control group is important to the reliability of the investigation. A control group helps determine if a variable that should have been controlled has been overlooked and might explain any unexpected results.

Error

Discuss any source of systematic or random error and suggest ways of improving the investigation.

CRITICALLY EVALUATING SOURCES OF INFORMATION

Not all sources are **credible**. When you have used primary or secondary data from information sources it is essential to critically evaluate the content and its origin. Questions you should always ask when evaluating a source include the following.

- Who created this source? What are the qualifications, **expertise**, **reputation** and affiliation of the authors?
- Why was it written?
- Where was the information published?
- When was the information published?
- Are conclusions supported by data or evidence?
- What is implied?
- What is omitted?
- Are any opinions or values being presented in the piece?
- Is the writing objectively and accurately describing a scientific concept or **phenomenon**?
- How might other people understand or interpret this message differently from me?
- When evaluating the validity or bias of a website, consider its domain extension:
  - .gov stands for government
  - .edu stands for education
  - .org stands for non-profit organisations
  - .com stands for commercial/business.

20.5 Review

SUMMARY

- After completing your investigation you will need to analyse and interpret your data. A discussion of your results is required where the findings of the investigation need to be analysed and interpreted.
- State whether a pattern, trend or relationship was observed between the independent and dependent variables. Describe what kind of pattern it was and specify under what conditions it was observed.
- Were there discrepancies, deviations or anomalies in the data? If so, these should be acknowledged and explained.
- Identify any limitations in the data collected. Perhaps a larger sample or further variations in the independent variable would lead to a stronger conclusion.
- It is important to discuss the limitations of the investigation method. Evaluate the method and identify any issues that could have affected the validity, accuracy, precision or reliability of the data. Sources of errors and uncertainty must also be stated in the discussion.
- When discussing the results, indicate the range of the data obtained from replicates. Explain how the sample size was selected. Larger samples are usually more reliable, but time and resources might have been scarce. Discuss whether the results of the investigation have been limited by the sample size.

KEY QUESTIONS

1. What relationship between the variables is indicated by a sloping linear graph?
2. What relationship exists if one variable decreases as the other increases?
3. What relationship exists if both variables increase or both decrease at the same rate?
4. What might cause a sample size to be limited in an investigation?

20.6 Conclusions

Having analysed your results you can then apply them to chemical concepts in order to evaluate your conclusions. In this section you will learn how analysing your investigation leads to a better understanding of the underlying scientific principles of your research.

DISCUSSING RELEVANT CHEMICAL CONCEPTS

To make the investigation more meaningful, it should be explained within the right context, meaning the related chemical ideas, concepts, theories and models. Within this context, explain the basis for the hypothesis. For example, if you were studying the impact of dissolved carbon dioxide on the pH of sparkling mineral water, you could include the information in Table 20.6.1 in your discussion.

TABLE 20.6.1 Examples of how to include chemical concepts in your discussion

Key ideas	Example
definitions of key terms	'pH', 'dissolved carbon dioxide' and 'sparkling' mineral water
the function of added carbon dioxide	in order to create 'sparkling' water
relationship between variables	dissolved carbon dioxide and pH of water, temperature was controlled in the experiment
chemical principles	dissolved carbon dioxide and formation of carbonic acid (H_2CO_3), including relevant equations
sources of error	reducing random error, by repeating measurements and calculating average

Relating your findings to a chemical concept

In Section 20.5 you learnt that during the analysing stage of your investigation, you should look for trends, patterns and mathematical models in your results. This is the framework needed to discuss whether the data supported or refuted the hypothesis. You should ask questions such as the following.

- Was the hypothesis supported?
- Has the hypothesis been fully answered? (If not, give an explanation of why this is so and suggest what could be done to either improve or complement the investigation.)
- Do the results contradict the hypothesis? If so, why? (The explanation must be plausible and must be based on the results and previous evidence.)

Be sure to discuss the broader implications of the findings. Implications are the bigger picture. Outlining them for the audience is an important part of the investigation. Ask questions like these.

- Do the findings lead to further questions?
- Can the findings be extended to another situation?
- Are there any practical applications for the findings?

DRAWING EVIDENCE-BASED CONCLUSIONS

A conclusion is usually written as a paragraph that links the collected evidence to the hypothesis and provides a justified response to the research question.

Indicate whether the hypothesis was supported or refuted and the evidence on which this decision is based (that is, the results). Do not provide irrelevant information. Only refer to the specifics of the hypothesis and the research question and do not make generalisations.

Examples of poor and better conclusions for the hypothesis and research question given are listed in Table 20.6.2 and Table 20.6.3.

TABLE 20.6.2 Examples of strong and weak conclusions to a hypothesis

Hypothesis: A decrease in the concentration of solution of hydrochloric acid will result in an increase in the measured pH of the solution.

Strong conclusion	Weak conclusion
A decrease in the concentration of the acid by a factor of 10 resulted in a decrease in the pH of the acid solution from 1.5 to 2.5.	The pH of the solution increased as the hydrochloric acid concentration decreased.

TABLE 20.6.3 Examples of strong and weak conclusions in response to a research question

Research question: How is the rate of the reaction between magnesium ribbon and an acid related to the temperature of the acid?

Strong conclusion	Weak conclusion
Analysis of the results on the effect of increasing the temperature of the acid suggests that each increase of 10°C results in an approximate doubling of the rate of the reaction. This supports the collision theory, which states that higher temperature increases both the number of collisions and the proportion of successful collisions.	The results show that the reaction is faster at higher temperatures.

20.6 Review

SUMMARY

- To make the investigation more meaningful, it should be explained within the right context, meaning the related chemical ideas, concepts, theories and models. Within this context, explain the basis for the hypothesis.
- Indicate whether the hypothesis was supported or refuted and on what evidence this decision is based (that is, the results). Do not provide irrelevant information. Only refer to the specifics of the hypothesis and the research question and do not make generalisations.

KEY QUESTIONS

1 Which of the following would **not** support a strong conclusion to a report?

A The concluding paragraphs are relevant and provide evidence.

B The concluding paragraphs are written in emotive language.

C The concluding paragraphs include reference to limitations of the research.

D The concluding paragraphs include suggestions for further avenues of research.

2 A procedure was repeated 30 times. How should the following statement be rewritten?

Many repeats of the procedure were conducted.

20.7 Communicating

How you approach communicating your results will depend on the audience you want to reach. If you are communicating with a general audience you may want to discuss your investigation in the style of a news article or blog post. These types of communication use simple language and analogies, and don't use too much scientific language as you need to assume that your audience does not have a science background.

Throughout this course you will need to present your research using appropriate scientific language and notation. There are many different presentation formats that you can use, such as posters, oral presentations and reports. This section covers the main characteristics of effective science communication and report writing, including objectivity, clarity, conciseness and coherence.

STRUCTURING A REPORT

Your report should have a clear, logical structure.

Introduction:

- The first paragraph should introduce your topic and define key terms.

Body paragraphs:

- Each subsequent paragraph should cover one main idea.
- Use evidence to support statements.
- Avoid very long or very short paragraphs.

Conclusion:

- The final section should summarise the main findings.
- It should relate to the aim of the investigation.
- The conclusion should include limitations and should discuss implications and applications of the research and potential future research.

Analysing information relevant to your research investigation

Scientific research should always be objective and neutral. Any premise presented must be supported with facts and evidence to allow the audience to make its own decision. Identify the evidence supporting or contradicting each point you want to make. It is also important to explain connections between ideas, concepts, theories and models. Figure 20.7.1 lists the questions you need to consider when writing your investigation report.

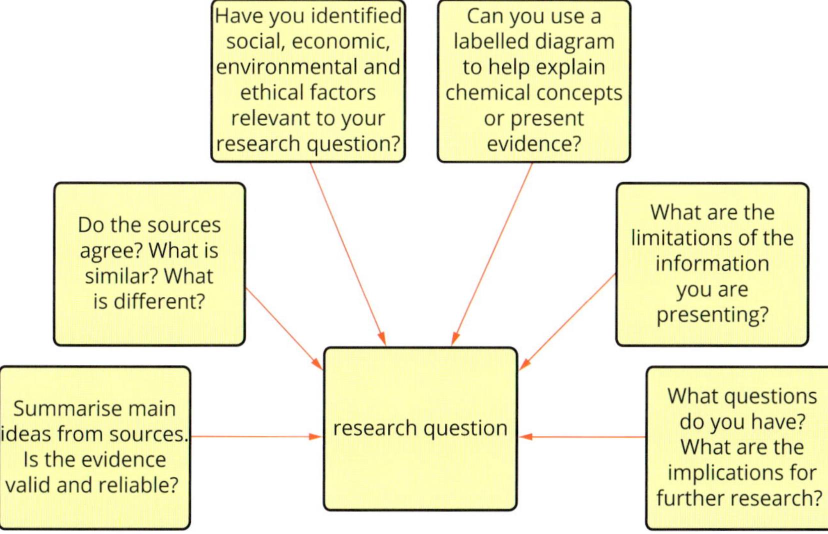

FIGURE 20.7.1 Discuss relevant information, ideas, concepts and implications, and make sure your discussion is relevant to the question under investigation.

WRITING FOR SCIENCE

Scientific reports are usually written in an objective or unbiased style. This is in contrast with English writing that most often uses the subjective techniques of rhetoric or persuasion. Table 20.7.1 contrasts persuasive and scientific writing styles.

TABLE 20.7.1 Persuasive writing versus scientific writing styles

Persuasive writing examples	Scientific writing equivalent examples
Use of biased and subjective language The results were extremely bad, atrocious, wonderful etc. This is terrible because ... This produced a disgusting odour. Health crisis	**Use of unbiased and objective language** The results showed ... The implications of these results suggest ... The results imply ... This produced a foul odour. Health issue
Use of exaggeration The object weighed a colossal amount, like an elephant.	**Use of non-emotive language** The object weighed 256 kg.
Use of everyday or colloquial language The subject passed away. The results don't ... The researchers had a sneaking suspicion ...	**Use of formal language** The subject died. The results do not ... The researchers predicted/hypothesised/theorised ...

Consistent reporting narrative

Scientific writing can be written either in first-person or in third-person narrative. Your teacher may advise you on which to select. In either case, ensure that you keep the narrative point of view consistent. Table 20.7.2 gives examples of first-person and third-person narrative.

TABLE 20.7.2 Examples of first-person and third-person narrative

First person	Third person
I put 50.0 g marble chips in a conical flask and then added 10.0 mL of 2 mol L^{-1} hydrochloric acid.	First, 50.0 g of marble chips was weighed into the conical flask and then 10.0 mL of 2.0 mol L^{-1} hydrochloric acid was added.
After I observed the reaction, I found that ...	After the reaction was completed, the results showed ...
My colleagues and I found ...	Researchers found ...

Qualified writing

Be careful of words that are absolute, such as *always*, *never*, *shall*, *will* and *proven*. Sometimes it may be more accurate and appropriate to use qualifying words, such as *may*, *might*, *possible*, *probably*, *likely*, *suggests*, *indicates*, *appears*, *tends*, *can* and *could*.

Concise writing

It is important to write concisely, particularly if you want to engage and maintain the interest of your audience. Use shorter sentences. Table 20.7.3 shows some examples of concise wording.

TABLE 20.7.3 Examples of verbose and concise language

Verbose example	Concise example
Due to the fact that ...	Because ...
Smith undertook an investigation into ...	Smith investigated
It is possible that the cause could be ...	The cause might be ...
a total of five experiments	five experiments
end result	result
in the event that	if
shorter in length	shorter

Visual support

Identify concepts that can be explained using visual models and information that can be presented in graphs or diagrams. This will not only reduce the word count of your work but will also make it more accessible for your audience.

For example, flow charts convey the steps in a process or method. The flow chart in Figure 20.7.2 shows how polymers used in the production of a consumer item can be decomposed. A limitation of flow charts is that the details of the process are omitted. Of course, simplification is often a benefit of visual models.

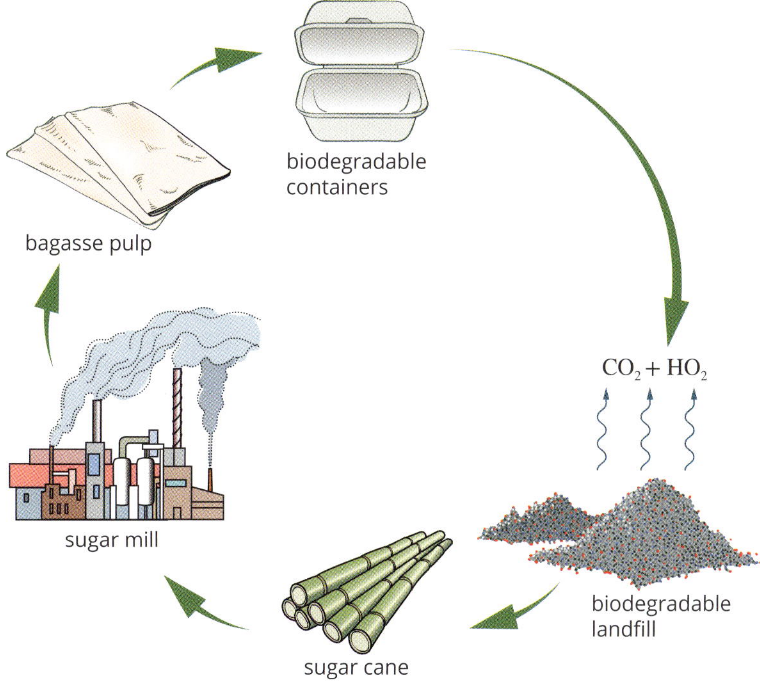

FIGURE 20.7.2 This flow chart shows how polymers used in the production of a consumer item can be decomposed. A limitation of this diagram is that it does not indicate the timeline or details involved in this process.

EDITING YOUR REPORT

Editing your report is an important part of the process. After editing your report, save new drafts with a different file name and always back up your files in another location.

Pretend you are reading your report for the first time when editing. Once you have completed a draft, it is a good idea to put it aside and return to it with 'fresh eyes' a day later. This will help you find areas that need further work and give you the opportunity to improve them. Look for content that is:

- ambiguous or unclear
- repetitive

- awkwardly phrased
- too lengthy
- not relevant to your research question
- poorly structured
- lacking evidence
- lacking a reference (if it is another researcher's work).

Use a spellchecker tool to help you identify typographical errors, but first, check that your spellchecker is set to Australian English.

REFERENCES AND ACKNOWLEDGEMENTS

All the quotations, documents, publications and ideas used in the investigation need to be acknowledged in the 'references and acknowledgments' section in order to avoid plagiarism and to ensure authors are credited for their work. References and acknowledgements also give credibility to the study and allow the audience to locate information sources should they wish to study it further. The standard referencing style used is the American Psychological Association (APA) academic referencing style.

When referencing a book, include in this order:

- author's surname and initials
- date of publication
- title
- publisher's name
- place of publication.

For example: Rickard, Greg, et al. (2016), *Pearson Science SB9* (2nd ed.), Pearson Education, Melbourne, Australia.

When referencing a website, include in this order:

- author's surname and initials, or name of organisation, or title
- year website was written or last revised
- title of webpage
- date website was accessed
- website address.

For example: National Geographic (2015), 'Killer fungus that's devastating bats may have met its match', accessed 29 May 2015, from http://news.nationalgeographic.com/2015/05/150527-bats-white-nose-syndrome-treatment-conservation-animals-science.

In-text citations

Each time you write about the findings of other people or organisations, you need to provide an in-text citation and provide full details of the source in a reference list. In the APA style, in-text citations include the first author's last name and date in brackets (author, date). List the full details in your list of references.

The following examples show the use of in-text citation.

> It was reported that in testing of five pro-oxidant additives added to commonly manufactured polymers, none resulted in significant biodegradation after three years (Selke et al., 2015).

Or

> Selke et al. (2015) reported that in testing of five pro-oxidant additives added to commonly manufactured polymers, none resulted in significant biodegradation after three years.

The bibliographic details of the example above would be:

Selke, S., Auras, R., Nguyen, T.A., Aguirre, E.C., Cheruvathur, R., & Liu, Y. (2015). Evaluation of biodegradation—promoting additives for plastics. *Environmental Science & Technology,* 49(6), 3769–3777.

CHEMICAL NOMENCLATURE

Using appropriate chemical nomenclature, scientific notations and units is also important. Table 20.7.4 has examples of the most common terms used in chemistry.

TABLE 20.7.4 The most common terms used in chemistry

Chemical term	Definition	Examples
element	substance whose atoms have the same atomic number; atoms of different elements have different atomic numbers	sodium, chlorine, tin
compound	substance consisting of two or more elements combined in fixed proportions; a chemical formula can be written for a pure compound	H_2O, NaCl
substance	element, compound	H_2, diamond, H_2O
particle	an atom, ion, molecule, proton, neutron, electron	Ne, Na^+, H_2O, H^+, ^{12}C, e^-
atom	building block of matter; the smallest unit of an element	Na, He, C, Sn
molecule	two or more atoms covalently bonded together	H_2O, water, C_4H_{10}, butane, $C_6H_{12}O_6$ glucose
ion	a positively charged or negatively charged atom or group of atoms resulting from the loss or gain of one or more electrons.	Na^+, Cl^-, NO_3^-
cation	positively charged ion	Na^+, Mg^{2+}, Al^{3+}, NH_4^+
anion	negatively charged ion	Cl^-, O^{2-}, PO_4^{3-}
network layer lattice	covalently bonded layer lattice	graphite
network lattice	covalently bonded lattice in which the bonds extend in three dimensions	diamond, silicon dioxide

Table 20.7.5 describes some chemical nomenclature (naming conventions) commonly used in chemistry.

TABLE 20.7.5 Chemical nomenclature

Application	Convention	Examples
naming elements	the first letter is not capitalised unless at the start of a sentence	carbon, hydrogen
element symbols	first letter is capitalised, subsequent letter (if present) is lower case	N, Na, Ne, Ni
naming metals that can form ions of different charges (oxidation states)	write the name of the metal, followed by a Roman numeral in brackets	Cu^+, name is copper(I) ion Cu^{2+}, name is copper(II) ion
naming ionic compounds	name the cation before the anion	sodium chloride tin(IV) chloride
writing ionic formulae	write the cation before the anion	NaCl $SnCl_4$
use of brackets in chemical formulae	to indicate atoms that need to be considered together	$Al_2(CO_3)_3$
use of brackets in condensed structural (semistructural) formulae	can indicate which groups of atoms: • are attached to a central atom (such as a carbon atom); • is repeated, for example repeating CH_2 groups	 $CH_3CH(CH_3)CH_3$ $CH_3(CH_2)_5CH_3$

Examples of uses and limitations of some commonly used chemical conventions and representations are presented in Table 20.7.6.

TABLE 20.7.6 Examples of uses and limitations of some commonly used chemical conventions and representations

Representation	Use	Limitations	Example
molecular formula	gives the actual number of atoms of each element in a molecule of a compound	does not show the arrangement of atoms in a molecule	C_4H_{10}, H_2O
structural formula	shows the relative location of atoms within a molecule in two dimensions	does not show the arrangement of atoms in three dimensions	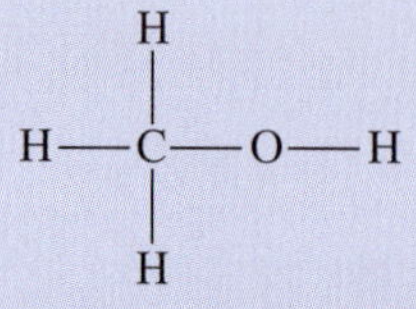 *methanol* 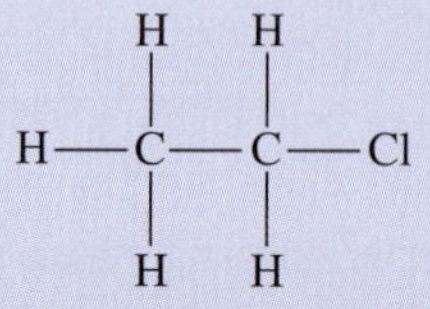 *chloroethane* 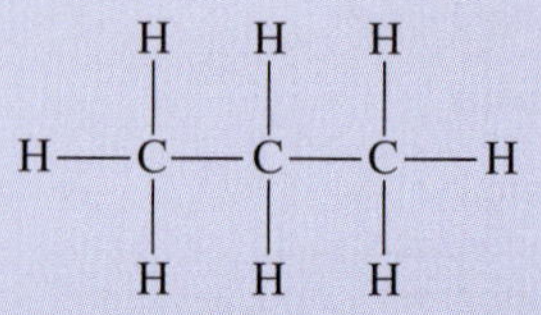 *propane* 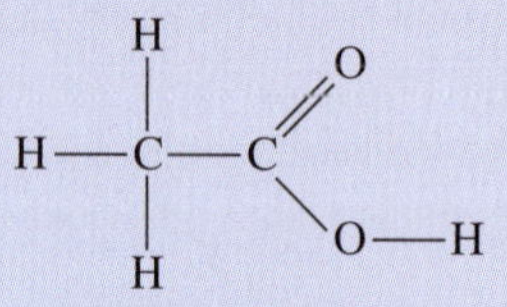 *ethanoic acid* 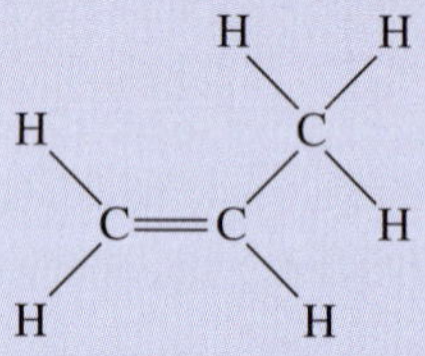 *propene*
condensed structural (semi-structural) formula	enables a structural formula to be drawn	does not show the arrangement of atoms in three dimensions	$CH_3CH_2CH_2CH_3$ or $CH_3(CH_2)_2CH_3$

MEASUREMENT AND UNITS

In every area of chemistry, we have attempted to quantify the phenomena we study. In investigations we generally make measurements and process those measurements in order to come to some conclusions. Scientists have a number of conventional ways of interpreting and analysing data from their investigations. There are also conventional ways of writing numerical measurements and their units.

Table 20.7.7 presents the main quantities and units used in chemistry.

TABLE 20.7.7 Quantities and units used in chemistry

Quantity	Symbol for quantity	Unit
mass	m	gram (g)
volume	V	litre (L)
amount of substance	n	mole (mol)
molar mass	M	grams per mole (g mol^{-1})
relative molecular or formula mass	M_r	no units, relative to one atom of ^{12}C exactly
relative atomic mass	A_r	no units, relative to one atom of ^{12}C exactly
density	d	grams or kilograms per litre (g mL^{-1}, kg L^{-1})
molarity	c	moles per litre (mol L^{-1})

Correct use of unit symbols

The correct use of unit symbols removes ambiguity, as symbols are recognised internationally. The symbols for units are not abbreviations and should not be followed by a full stop unless they are at the end of a sentence.

Upper-case letters are only used for the *symbols* of the units that are named after people. For example, the unit of energy is joule and the symbol is J. The joule was named after James Joule who was famous for studies into energy conversions. The exception to this rule is 'L' for litre. We do this because a lower-case 'l' looks like the numeral '1'.

The product of a number of units is shown by separating the symbol for each unit with a space. The division or ratio of two or more units can be shown in fraction form, using a slash, or using negative indices. Prefixes should not be separated by a space.

Table 20.7.8 provides some examples of correct and incorrect use of symbols for derived units.

TABLE 20.7.8 Some examples of the use of symbols for derived units

Preferred	Incorrect
$g\ mol^{-1}$	$gmol^{-1}$
kPa	K Pa

Units named after people can take the plural form by adding an 's' when used with numbers greater than one. *Never* do this with the unit symbols. It is acceptable to say 'two kilojoules' but it is wrong to write 2 kJs.

Numbers and symbols should not be mixed with words for units and numbers. For example, thirty grams and 30 g are correct while 30 grams and thirty g are incorrect.

Scientific notation

To overcome confusion or ambiguity, measurements are often written in scientific notation. Quantities are written as a number between one and 10 and then multiplied by an appropriate power of 10. Note that 'scientific notation', 'standard notation' and 'standard form' all have the same meaning.

Examples of some values written in scientific notation:

6.022×10^{23} particles

$25.25\ mL = 2.525 \times 10^{-2}\ L$

$0.00302\ mol = 3.02 \times 10^{-3}\ mol$

You should be routinely using scientific notation to express numbers. This also involves learning to use your calculator intelligently. Scientific and graphics calculators (Figure 20.7.3) can be put into a mode whereby all numbers are displayed in scientific notation. It is useful when doing calculations to use this mode rather than frequently attempting to convert to scientific notation by counting digits on the calculator display.

FIGURE 20.7.3 A scientific calculator

An important reason for using scientific notation is that it removes ambiguity about the precision of some measurements. For example, a measurement recorded as 240 g could be a measurement to the nearest gram; that is, somewhere between 239.5 g and 240.5 g, or it could have been rounded up to the nearest 10 g. By writing this quantity as 2.40×10^2 g it makes clear the accuracy of the value (in this case, 3 significant figures).

Prefixes and conversion factors

Conversion factors should be used carefully. You should be familiar with the prefixes and conversion factors in Table 20.7.9. Note that the table gives all conversions as a multiplying factor.

TABLE 20.7.9 Prefixes and conversion factors

Multiplying factor		Prefix	Symbol
1 000 000 000 000	10^{12}	tera	T
1 000 000 000	10^{9}	giga	G
1 000 000	10^{6}	mega	M
1 000	10^{3}	kilo	k
0.01	10^{-2}	centi	c
0.001	10^{-3}	milli	m
0.000 001	10^{-6}	micro	μ
0.000 000 001	10^{-9}	nano	n
0.000 000 000 001	10^{-12}	pico	p

Do not put spaces between prefixes and unit symbols.

It is important to give the symbol the correct case (upper or lower case). There is a big difference between 1 mm and 1 Mm.

20.7 Review

SUMMARY

- A scientific report must include an introduction, body paragraphs and conclusion.
- The conclusion should include a summary of the main findings, a conclusion related to the issue being investigated, limitations of the research, implications and applications of the research, and potential future research.
- Scientific writing uses unbiased, objective, accurate, formal language. Scientific writing should also be concise and qualified.
- Visual support can assist in conveying scientific concepts and processes efficiently.
- Ensure you edit your final report.
- Scientific notation needs to be used when communicating your results.

KEY QUESTIONS

1 Which of the following statements is written in scientific style?
  - **A** The results were fantastic ...
  - **B** The data in Table 2 indicates ...
  - **C** The researchers felt ...
  - **D** The smell was awful ...

2 Which of the following statements is written in first-person narrative?
  - **A** The researchers reported ...
  - **B** Samples were analysed using ...
  - **C** The experiment was repeated three times ...
  - **D** I reported ...

3 The variables molar mass and molarity each have different units. Write the units for each of the following in correct notation.
  - **a** molar mass; grams per mole
  - **b** molarity; moles per litre

4 Show how to convert grams into kilograms.

5 Discuss why you might need to convert between different multiplying factors, for example, litres to millilitres.

Chapter review

20

KEY TERMS

accuracy
bias
calibrate
chemical (HAZCHEM) code
control experiment
controlled variable
credible
dependent variable
expertise
hypothesis
independent variable
line of best fit
mean
outlier
personal protective equipment (PPE)
phenomenon
precision
primary source
qualitative
quantitative
random error
range
raw data
reliability
reputation
research question
risk assessment
Safety Data Sheet (SDS)
secondary source
significant figures
systematic error
trend
trend line
uncertainty
validity
variable

1 Consider the following research question:
'Is the concentration of lead in water sampled from Esperance Harbour within acceptable limits?'
For each of the following, decide whether it is the independent, dependent and controlled variable.
- **a** concentration of lead
- **b** analytical technique, temperature of water sample, type of sampling container
- **c** source and location of water

2 Match the following command verbs with their definitions: describe, analyse, apply, create, identify, reflect, investigate.
- **a** think deeply about
- **b** produce or make new
- **c** identify connections and relationships, interpret to reach a conclusion
- **d** observe, study, examine, inquire systematically in order to establish facts or derive conclusions
- **e** use knowledge and understanding in a new situation
- **f** recognise or indicate what or who
- **g** give a detailed account.

3 Consider the following hypothesis:
'The concentration of heavy metals in industrial waste water will be greater than that of drinking water.'
Name the independent, dependent and controlled variables for an experiment with this hypothesis.

4 Which graph from the following list would be best to use with each set of data listed here?
Graph types: pie diagram, scatter graph (with line of best fit), bar graph, line graph
- **a** the levels of a pesticide detected in drinking water at various locations
- **b** the temperature of water sampled at the same time of day over a period of a month
- **c** a calibration curve showing conductivity of a solution compared to the concentration of dissolved ions
- **d** the proportion of specific contaminants detected in water

5 What is the meaning of each of the following chemical codes?

6 Explain the terms 'accuracy' and 'validity'.

7 Identify whether the following are mistakes, systematic errors or random errors.
- **a** A student spills some solution during an experiment.
- **b** The reported measurements are above and below the true value.
- **c** A weighing balance has not been calibrated.

8 What relationship between variables could be indicated by a curved trend line?

9 A scientist designed and completed an experiment to test the following hypothesis: 'Increasing the temperature of water would result in an increase in the measured electrical conductivity of water.'
The discussion section of the scientist's report included comments on the reliability, validity, accuracy and precision of the investigation.
Determine if the following sentences comment on reliability, validity, accuracy or precision.
- **a** Three water samples from the same source were examined at each temperature. Each water sample was analysed and the measurements were recorded.
- **b** The temperature and the electrical conductivity of the water samples were recorded using data-logging equipment. The temperatures of some of the water samples were measured using a glass thermometer.
- **c** The data-logging equipment was calibrated for electrical conductivity against a known standard. The equipment was calibrated before measurements were taken.

- **d** The temperature probe (data logger) measured temperature to the nearest 0.1°C. The glass thermometer measured temperature to the nearest 1°C.

10 What factors should you consider to ensure you discuss the limitations of your method?

11 Explain the meaning of the term 'trend' in a scientific investigation and describe the types of trends that might exist.

12 What is meant by the 'limitations' of the investigation method?

13 What is 'bias' in an investigation?

14 Which of the following consists only of secondary sources of information?

- **A** an interactive periodic table, an article published in a science magazine, a science documentary, a practical report written by a Year 11 student
- **B** an article published in a peer-reviewed science journal, an article published in a science magazine, a science documentary
- **C** an interactive periodic table, an article published in a science magazine, a science documentary, this Year 11 textbook
- **D** a science article published in a newspaper, an article published in a science magazine, a science documentary, a practical report written by a Year 11

15 What is the purpose of referencing and acknowledging documents, ideas and quotations in your investigation?

16 For the following source, what is the correct way to make an in-text citation in APA style?

Selke, S., Auras, R., Nguyen, T.A., Aguirre, E.C., Cheruvathur, R., & Liu, Y. (2015). Evaluation of biodegradation—promoting additives for plastics. *Environmental Science & Technology,* 49(6), 3796–3777.

- **A** However, Selke et al. (2015) did not find any significant difference in biodegradability.
- **B** However, Selke et al. did not find any significant difference in biodegradability[1].
- **C** However, Selke et al. did not find any significant difference in biodegradability (Selke, S., Auras, R., Nguyen, T.A., Aguirre, E.C., Cheruvathur, R., & Liu, Y. (2015). Evaluation of biodegradation—promoting additives for plastics. *Environmental Science & Technology, 49*(6), 3769–3777).
- **D** However, Selke et al. (2015) did not find any significant difference in biodegradability (Evaluation of biodegradation-promoting additives for plastics. *Environmental Science & Technology*).

17
- **a** Convert 30.0 mL into L.
- **b** Convert 34.5 kg into g.

18 A scientist designed and completed an experiment to test the following hypothesis:

'Increasing the temperature of water would result in an increase in the measured electrical conductivity of water.'

- **a** Write a possible aim for this scientist's experiment.
- **b** What would be the independent, dependent and controlled variables in this investigation?
- **c** What kind of data would be collected? Would it be qualitative or quantitative?
- **d** List the equipment that could be used and the precision expected for each item.
- **e** Explain the difference between raw data and processed data, using this hypothesis as an example. What would you expect the graph of the results to look like if the scientist's hypothesis is correct?

UNIT 2 • MOLECULAR INTERACTIONS AND REACTIONS

REVIEW QUESTIONS

Section 1: Multiple choice

1 Which of the following statements about water is correct?

A Water molecules are non-polar and so water is a good solvent.

B Water has a high latent heat value and so it is a good coolant.

C Water contracts on freezing and so lakes are covered with a layer of ice in very cold climates.

D Water has a low specific heat capacity and so large bodies of water moderate temperatures on Earth.

2 Which one of the equations below best represents table sugar ($C_{12}H_{22}O_{11}$) dissolving in water?

A $C_{12}H_{22}O_{11}(s) + H_2O(l) \rightarrow C_{12}H_{23}O_{11}^{+}(aq) + OH^-$

B $C_{12}H_{22}O_{11}(s) + H_2O(l) \rightarrow C_{12}H_{21}O_{11}^{-}(aq) + H_3O^+(aq)$

C $C_{12}H_{22}O_{11}(s) \xrightarrow{H_2O(l)} C_{12}H_{22}O_{11}(aq)$

D $C_{12}H_{22}O_{11}(s) \xrightarrow{H_2O(l)} C_{12}H_{22}O_{11}(l)$

3 In which one of the following reactions is the substance in bold acting as an acid?

A $\mathbf{C_6H_5COO^-(aq)} + H_2O(aq) \rightarrow OH^-(aq) + CH_3COOH(aq)$

B $\mathbf{H_2(g)} + C_2H_4(g) \rightarrow C_2H_6(g)$

C $\mathbf{HSO_4^-(aq)} + PH_3(g) \rightarrow SO_4^{2-}(aq) + PH_4^+(aq)$

D $\mathbf{NH_3(aq)} + H_2O(l) \rightarrow NH_4^+(aq) + OH^-(aq)$

4 Which one of the following aqueous solutions would have the lowest pH?

A 0.1 mol L^{-1} H_2SO_4

B 0.1 mol L^{-1} HNO_3

C 0.1 mol L^{-1} H_3PO_4

D 0.1 mol L^{-1} $Ca(OH)_2$

5 If the acidity of a solution were increased, how would the hydrogen ion concentration, $[H^+]$, and the pH be affected?

A $[H^+]$ and the pH would increase.

B $[H^+]$ and the pH would decrease.

C $[H^+]$ would increase and the pH would decrease.

D $[H^+]$ would decrease and the pH would increase.

6 In which chemical reaction will an increase in pressure increase the rate of reaction?

A $CH_4(g) + H_2O(g) \rightarrow CO(g) + 3H_2(g)$

B $H(g) + Br_2(g) \rightarrow 2HB(g)$

C $N_2(g) + 3F_2(g) \rightarrow 2NF_6(g)$

D all of the above

7 A saturated solution of calcium hydroxide was prepared in a flask using an excess of solid calcium hydroxide so that some remained undissolved.

$$Ca(OH)_2(s) \rightleftharpoons Ca^{2+}(aq) + 2OH^-(aq)$$

A small amount of 10 mol L^{-1} hydrochloric acid was added to the flask. What would you most likely observe?

A No change is observed.

B $CaCl_2$ is precipitated.

C More solid $Ca(OH)_2$ is precipitated.

D Some solid $Ca(OH)_2$ dissolves.

8 When a solution of KBr is placed in a beaker connected to a power supply, a current flows through the circuit and a reaction occurs at the electrodes. The set-up is shown below.

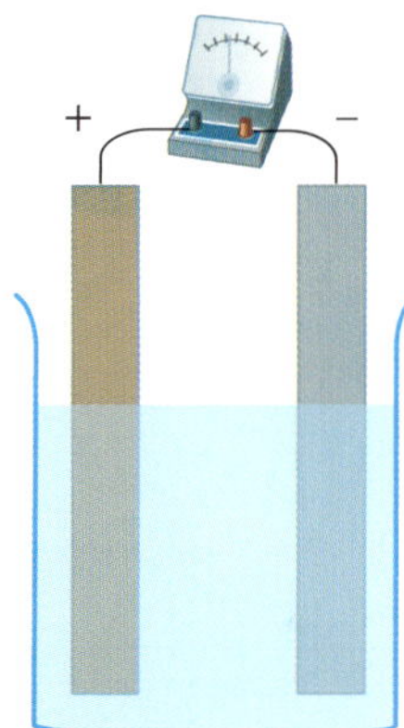

Which one of the following statements best describes how the current is moving in the KBr solution?

A The water molecules in the solution carry the current.

B KBr molecules move through the solution to each electrode.

C There are freely moving ions in the solution.

D Electrons flow through the solution to the electrodes.

9 Which of the following compounds would most readily dissolve in water?

A NH_4I

B $AlPO_4$

C $BaCO_3$

D Ag_2SO_4

10 Which of the following molecules are polar?

i CH_4 ii CH_3Cl iii CH_2Cl_2 iv $CHCl_3$ v CCl_4

A all of the above

B none of the above

C ii and iv

D ii, iii and iv

UNIT 2 • REVIEW

11 Which compound is most soluble in water?

A $Al(OH)_3$
B CH_3OH
C CHF_3
D $CF_2 = CF_2$

12 If a bottle of concentrated sulfuric acid were spilt, the most appropriate action would be to:

A dilute the acid with water.
B add sodium hydroxide to neutralise the acid.
C cover it with powered sodium carbonate.
D soak it up with paper towels.

Section 2: Short answer

1 **a** Copy the grid below and draw the energy profile diagram for the following data for the reaction between gaseous H_2 and gaseous I_2.

$H_2(g) + I_2(g) \rightarrow 2HI(g)$

$H_r = 250\,kJ \quad H_p = 450\,kJ \quad E_a = 300\,kJ$

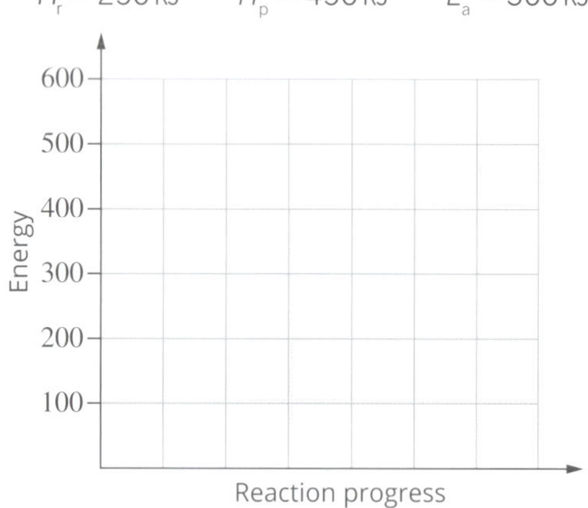

b On your graph, show the $\Delta H_{reaction}$.

c To catalyse the above reaction, some sulfuric acid is added. The E_a of the forward reaction is lowered by 75 kJ. Draw the resulting reaction path on your graph.

2 Hydrogen peroxide, H_2O_2, *exothermically* decomposes into water and oxygen.

a Write an equation for this reaction, including the enthalpy as a reactant or product in the equation.

b At room temperature, the reaction is very slow in an open beaker.

Sketch the enthalpy diagram, showing the activation energy, E_a, and the change in the enthalpy of reaction, ΔH.

c When a catalyst is added a vigorous reaction occurs.

Sketch the enthalpy diagram, labelling E_a and ΔH. Use the same scale (size) as you did for the graph in part **b**.

3 Write the balanced equation for each of the following reactions.

a the molecular equation for the reaction of a sulfuric acid solution with a potassium hydroxide solution

b the ionic equation for the reaction of a sulfuric acid solution with a potassium hydroxide solution

c the molecular equation for the reaction of a hydrochloric acid solution with aluminium metal

d the ionic equation for the reaction of a hydrochloric acid solution with aluminum metal

4 **a** A 20.0 g sample of calcium bromide is dissolved in 400.0 mL of water.

Calculate the concentration of the solution in mol L^{-1}.

b A 1000.0 L sample of seawater is evaporated. If the concentration of sodium chloride in seawater is 0.600 mol L^{-1}, what mass of NaCl is obtained?

5 The boiling points (BP) and the molar masses (*M*) for the hydrides (hydrogen compounds) of the halogens are given in the following table.

Molecule	BP (°C)	*M* (g)
HF	19	20
HCl	–85	36.5
HBr	–67	81
HI	–35	128

The boiling points increase as the molar mass increases; however, HF does not fit this trend in terms of molecular bonding. Explain. You can draw a graph to help explain your answer.

6 **a** In 1660, Robert Boyle observed 'that for a fixed mass of gas the volume of the gas decreases as the pressure increases provided the temperature remains constant'. (This is known as Boyle's law.)

Explain Boyle's law in terms of the kinetic theory of gases.

b In 1780, Jacques Charles observed 'that for a fixed mass of gas the volume of the gas increases as the temperature increases provided the pressure remains constant'. (Now known as Charles' law)

Explain Charles' law In terms of the kinetic theory of gases.

Section 3: Extended answer

1 The production of the fertiliser and mining explosive, ammonium nitrate, is produced via a series of chemical reactions.

In the first stage, ammonia and oxygen from the air are heated together to produce nitrogen monoxide and water.

a For every 1.00 tonne (1.00×10^6 g) of ammonia calculate the:

i mass of oxygen required

ii volume of nitrogen monoxide (NO) produced at STP.

The reaction is extremely slow and uneconomical at room temperature and pressure, therefore conditions of high concentrations of reactants at temperatures of 800°C and 10 times atmospheric pressure in the presence of a finely divided platinum catalyst are required to increase the rate of reaction.

b Discuss, using collision theory, how each of the following conditions increases the rate of this reaction.

i high concentrations of the reactants ($NH_3(g)$ and $O_2(g)$)

ii high temperature (800°C)

iii high pressure (10 times atmospheric pressure)

iv the presence of a catalyst (finely divided platinum)

2 The solubility of compounds in water is determined experimentally and can be measured in grams of solute per 100 g of water.

The solubility of salt (sodium chloride, NaCl) in water is 36 g at 25°C.

a Based on the data above, what other way could the solubility can be expressed?

b Calculate the solubility of NaCl expressed as moles per 100 g of solution.

c Calculate is the solubility of NaCl in moles per litre ($mol\ L^{-1}$). (Assume the mass of 1.0 L is 1000.0 g.)

A sample of 15.0 g of NaCl was added to 25 g of water in a beaker at 25°C, stirred vigorously for a few minutes then left to stand.

d Would the solution be unsaturated or saturated?

e Why is the temperature given 'at 25°C' when reporting the solubility?

f Describe the appearance of the contents of the beaker. You may include a diagram in your answer.

g Based on your understanding of the properties of ionic solids, explain why a solid crystal of NaCl is a non-conductor of electricity, however, when dissolved in water, the aqueous solution is a conductor of electricity. Include an equation in your answer.

h The solubility of sugar (sucrose, $C_{12}H_{22}O_{11}$) is 68.0 g per 100.0 g of water at 25°C.

Although sugar has a higher solubility in water than sodium chloride, the solution has almost zero electrical conductivity. Based on your understanding of the properties of materials, explain the difference between the conductivity of a salt solution and the conductivity of a sugar solution.

APPENDIX A Symbols, units and fundamental constants

TABLE 1 Units and symbols based on the SI system. Units listed in red are the arbitrarily defined fundamental units of the SI system.

Quantity	Symbol for physical quantity	Corresponding SI unit	Symbol for SI unit	Definition of SI unit
Mechanics				
volume	V	cubic metre	m^3	—
mass	m	kilogram	kg	fundamental unit
density	d	—	$kg\ m^{-3}$	—
time	t	second	s	fundamental unit
force	F	newton	N	$kg\ m\ s^{-2}$
pressure	P	pascal	Pa	$N\ m^{-2}$
energy	E	joule	J	N m
Electricity				
electric potential difference	V	volt	V	$J\ A^{-1}\ s^{-1}$
Nuclear and chemical quantities				
atomic number	Z	—	—	—
neutron number	N	—	—	—
mass number	A	—	—	$Z + N$
amount of substance	n	mole	mol	fundamental unit
relative atomic mass	A_r	—	—	—
relative molecular mass	M_r	—	—	—
molar mass	M	—	—	$kg\ mol^{-1}$
molar volume	V_m	—	—	$m^3\ mol^{-1}$
concentration	c	—	—	$mol\ m^{-3}$
Thermal quantities				
temperature	T	kelvin	K	fundamental unit
specific heat capacity	C	—	—	$J\ K^{-1}\ kg^{-1}$

TABLE 2 SI prefixes, their symbols and values

SI prefix	Symbol	Value
pico	p	10^{-12}
nano	n	10^{-9}
micro	μ	10^{-6}
milli	m	10^{-3}
centi	c	10^{-2}
deci	d	10^{-1}
kilo	k	10^{3}
mega	M	10^{6}
giga	G	10^{9}
tera	T	10^{12}

TABLE 3 Some physical constants

Description	Symbol	Value
Avogadro's constant	N_A	$6.022 \times 10^{23}\ mol^{-1}$
charge of an electron	e	-1.60×10^{-19} C
mass of electron	m_e	9.109×10^{-31} kg
mass of proton	m_p	1.673×10^{-27} kg
mass of neutron	m_n	1.675×10^{-27} kg
gas constant	R	$8.31\ J\ K^{-1}\ mol^{-1}$
ionic product for water	K_w	1.00×10^{-14} at 298 K
molar volume of an ideal gas at 273 K, 100 kPa	V_m	$22.71\ L\ mol^{-1}$
specific heat capacity of water	c	$4.18\ J^{-1}\ g^{-1}\ K^{-1}$
density of water	d	$1.00\ g\ mL^{-1}$ at 298 K

Significant figures and standard form

SIGNIFICANT FIGURES

The number of significant figures a piece of data has indicates the precision of a measurement. For example, compare the following data:

- A jogger takes 20 minutes to cover 4 kilometres.
- A sprinter takes 10.21 seconds to cover 100.0 metres.

The sprinter's data has been measured more precisely than that of the jogger. This is indicated by the greater number of significant figures in the second set of data.

Which figures are significant?

A significant figure is an integer or a zero that follows an integer.

In the data above:

- the distance '4 kilometres' has one significant figure
- the time '20 minutes' has two significant figures (the zero follows the integer 2)
- the 10.21 seconds and 100.0 metres each have four significant figures.

A zero that comes before any integers, however, is not significant. For example, the value 0.0004 has only one significant figure, whereas the value 0.0400 has three significant figures. The zeros that come before the integer 4 are not significant, whereas those that follow the integer are significant.

Using significant figures

In chemistry, you will often need to calculate a value from a set of data. It is important to remember that the final value you calculate is only as precise as your *least precise piece of data*.

Addition and subtraction

When adding or subtracting values, the answer should have no more digits to the right of the decimal place than the value with the least number of digits to the right of the decimal place.

Example

A 12.78 mL sample of water was added to 10.0 mL of water. What is the total volume of water?

12.78 mL + 10.0 mL = 22.78 = 22.8 mL

Because one of the values (10.0 mL) has only one digit to the right of the decimal place, the answer will need to be adjusted so that it too has only one digit to the right of the decimal place.

Multiplication and division

When multiplying and dividing values, the answer should have no more significant figures than the value with the least number of significant figures.

Example

An athlete takes 3.5 minutes to complete four laps of an oval. What is the average time taken for one lap?

$$\textit{Average time} = \frac{3.5 \text{ minutes}}{4} = 0.875 = 0.88 \text{ minutes}$$

Because the data (3.5 minutes) has only two significant figures, the answer will need to be adjusted to two significant figures so that it has the same degree of precision as the data. (Note: The 'four' is taken to indicate a precise number of laps and so is considered to have as many significant figures as the calculation requires. This applies to values that describe *quantities* rather than *measurements*.)

Rounding off

When adjusting the number of significant figures, if the integer after the last significant figure is equal to or greater than '5', then the last significant integer is rounded up. Otherwise, it is rounded down.

STANDARD FORM

A value written in standard form is expressed as a number equal to or greater than 1 and less than 10 multiplied by 10^x, where x is an integer. For example, when written in standard form:

- 360 becomes 3.6×10^2
- 0.360 becomes 3.60×10^{-1}
- 0.000456 becomes 4.56×10^{-5}.

Sometimes you will need to use standard form to indicate the precision of a value.

APPENDIX C Table of relative atomic masses*

TABLE 1 Table of relative atomic masses*

Element name	Symbol	Atomic number	Relative atomic mass	Element name	Symbol	Atomic number	Relative atomic mass	Element name	Symbol	Atomic number	Relative atomic mass
actinium	Ac	89	—	hafnium	Hf	72	178.49	praseodymium	Pr	59	140.9077
aluminium	Al	13	26.9815	hassium	Hs	108	—	promethium	Pm	61	—
americium	Am	95	—	helium	He	2	4.00260	protactinium	Pa	91	231.0359
antimony	Sb	51	121.76	holmium	Ho	67	164.9303	radium	Ra	88	—
argon	Ar	18	39.948	hydrogen	H	1	1.0080	radon	Rn	86	—
arsenic	As	33	74.9216	indium	In	49	114.82	rhenium	Re	75	186.21
astatine	At	85	—	iodine	I	53	126.9045	rhodium	Rh	45	102.9055
barium	Ba	56	137.33	iridium	Ir	77	192.22	roentgenium	Rg	111	—
berkelium	Bk	97	—	iron	Fe	26	55.845	rubidium	Rb	37	85.468
beryllium	Be	4	9.01218	krypton	Kr	36	83.80	ruthenium	Ru	44	101.07
bismuth	Bi	83	208.9804	lanthanum	La	57	138.9055	rutherfordium	Rf	104	—
bohrium	Bh	107	—	lawrencium	Lr	103	—	samarium	Sm	62	150.4
boron	B	5	10.81	lead	Pb	82	207.2	scandium	Sc	21	44.9559
bromine	Br	35	79.904	lithium	Li	3	6.94	seaborgium	Sg	106	—
cadmium	Cd	48	112.41	livermorium	Lv	116	—	selenium	Se	34	78.97
caesium	Cs	55	132.9055	lutetium	Lu	71	174.967	silicon	Si	14	28.086
calcium	Ca	20	40.08	magnesium	Mg	12	24.305	silver	Ag	47	107.868
californium	Cf	98	—	manganese	Mn	25	54.9380	sodium	Na	11	22.9898
carbon	C	6	12.011	meitnerium	Mt	109	—	strontium	Sr	38	87.62
cerium	Ce	58	140.12	mendelevium	Md	101	—	sulfur	S	16	32.06
chlorine	Cl	17	35.453	mercury	Hg	80	200.59	tantalum	Ta	73	180.9479
chromium	Cr	24	51.996	molybdenum	Mo	42	95.95	technetium	Tc	43	—
cobalt	Co	27	58.9332	moscovium	Mc	115	—	tellurium	Te	52	127.60
copernicium	Cn	112	—	neodymium	Nd	60	144.24	tennessine	Ts	117	—
copper	Cu	29	63.55	neon	Ne	10	20.180	terbium	Tb	65	158.9254
curium	Cm	96	—	neptunium	Np	93	—	thallium	Tl	81	204.384
darmstadtium	Ds	110	—	nickel	Ni	28	58.693	thorium	Th	90	232.038
dubnium	Db	105	—	nihonium	Nh	113	—	thulium	Tm	69	168.9342
dysprosium	Dy	66	162.50	niobium	Nb	41	92.9064	tin	Sn	50	118.71
einsteinium	Es	99	—	nitrogen	N	7	14.0067	titanium	Ti	22	47.87
erbium	Er	68	167.26	nobelium	No	102	—	tungsten	W	74	183.84
europium	Eu	63	151.96	oganesson	Og	118	—	uranium	U	92	238.0289
fermium	Fm	100	—	osmium	Os	76	190.2	vanadium	V	23	50.942
flerovium	Fl	114	—	oxygen	O	8	15.9994	xenon	Xe	54	131.29
fluorine	F	9	18.9984	palladium	Pd	46	106.4	ytterbium	Yb	70	173.05
francium	Fr	87	—	phosphorus	P	15	30.9738	yttrium	Y	39	88.9058
gadolinium	Gd	64	157.25	platinum	Pt	78	195.08	zinc	Zn	30	65.38
gallium	Ga	31	69.72	plutonium	Pu	94	—	zirconium	Zr	40	91.22
germanium	Ge	32	72.63	polonium	Po	84	—				
gold	Au	79	196.9666	potassium	K	19	39.098				

* Based on the atomic mass of $^{12}C = 12$.
The values for relative atomic masses given in the table apply to elements as they exist in nature, without artificial alteration of their isotopic composition, and, further, to natural mixtures that do not include isotopes of radiogenic origin.

APPENDIX D Common positive and negative ions and solubility of common ionic compounds in water

TABLE 1 Names and formulae of some common positive and negative ions

Positive ions (cations)						Negative ions (anions)			
+1		**+2**		**+3**		**−1**		**−2**	
caesium	Cs^+	barium	Ba^{2+}	aluminium	Al^{3+}	bromide	Br^-	carbonate	CO_3^{2-}
copper(I)	Cu^+	cadmium(II)	Cd^{2+}	chromium(III)	Cr^{3+}	chloride	Cl^-	chromate	CrO_4^{2-}
gold(I)	Au^+	calcium	Ca^{2+}	gold(III)	Au^{3+}	cyanide	CN^-	dichromate	$Cr_2O_7^{2-}$
lithium	Li^+	cobalt(II)	Co^{2+}	iron(III)	Fe^{3+}	dihydrogen phosphate	$H_2PO_4^-$	hydrogen phosphate	HPO_4^{2-}
potassium	K^+	copper(II)	Cu^{2+}			ethanoate	CH_3COO^-	oxalate	$C_2O_4^{2-}$
rubidium	Rb^+	iron(II)	Fe^{2+}	**+4**		fluoride	F^-	oxide	O^{2-}
silver	Ag^+	lead(II)	Pb^{2+}	lead(IV)	Pb^{4+}	hydrogen carbonate	HCO_3^-	sulfide	S^{2-}
sodium	Na^+	magnesium	Mg^{2+}	tin(IV)	Sn^{4+}	hydrogen sulfide	HS^-	sulfite	SO_3^{2-}
		manganese(II)	Mn^{2+}			hydrogen sulfite	HSO_3^-	sulfate	SO_4^{2-}
		mercury(II)	Hg^{2+}			hydrogen sulfate	HSO_4^-		
		nickel	Ni^{2+}			hydroxide	OH^-	**−3**	
		strontium	Sr^{2+}			iodide	I^-	nitride	N^{3-}
		tin(II)	Sn^{2+}			nitrite	NO_2^-	phosphate	PO_4^{3-}
		zinc	Zn^{2+}			nitrate	NO_3^-		
						permanganate	MnO_4^-		

TABLE 2 Solubility of common ionic compounds in water

Soluble ionic compounds		
Soluble in water (>0.1 mol dissolves per L at 25°C)	**Exceptions: insoluble (<0.01 mol dissolves per L at 25°C)**	**Exceptions: slightly soluble (0.01–0.1 mol dissolves per L at 25°C)**
most chlorides (Cl^-), bromides (Br^-) and iodides (I^-)	AgCl, AgBr, AgI, PbI_2	$PbCl_2$
all nitrates (NO_3^-)	no exceptions	no exceptions
all ammonium (NH_4^+) salts	no exceptions	no exceptions
all sodium (Na^+) and potassium (K^+) salts	no exceptions	no exceptions
all ethanoates (CH_3COO^-)	no exceptions	no exceptions
most sulfates (SO_4^{2-})	$SrSO_4$, $BaSO_4$, $PbSO_4$	$CaSO_4$, Ag_2SO_4
Insoluble ionic compounds		
Insoluble in water	**Exceptions: soluble**	**Exceptions: slightly soluble**
most hydroxides (OH^-)	NaOH, KOH, $Ba(OH)_2$, NH_4OH*, AgOH**	$Ca(OH)_2$, $Sr(OH)_2$
most carbonates (CO_3^{2-})	Na_2CO_3, K_2CO_3, $(NH_4)_2CO_3$	no exceptions
most phosphates (PO_4^{3-})	Na_3PO_4, K_3PO_4, $(NH_4)_3PO_4$	no exceptions
most sulfides (S^{2-})	Na_2S, K_2S, $(NH_4)_2S$	no exceptions

*NH_4OH does not exist in significant amounts in an ammonia solution. Ammonium and hydroxide ions readily combine to form ammonia and water.
**AgOH readily decomposes to form a precipitate of silver oxide and water.

APPENDIX E Mathematical skills for Chemistry ATAR

1 Transforming decimal notation to scientific notation

Scientists use scientific notation to handle very large and very small numbers.

For example, instead of writing 0.000 000 035, scientists would write 3.5×10^{-8}.

A number in *scientific notation* (also called standard form or power of ten notation) is written as:

$$a \times 10^n$$

where a is a number equal to or greater than 1 and less than 10, that is $1 \leq a < 10$

n is an integer (a positive or negative whole number).

n is the power that 10 is raised to and is called the index value.

Use the following steps to transform a very large or very small number into scientific notation.

1 Write the original number as a decimal number greater than or equal to 1 but less than 10.

2 Multiply the decimal number by the appropriate power of 10.

The index value is determined by counting the number of places the decimal point needs to be moved to form the original number again.

If the decimal point is moved n places to the right, n will be a positive number. For example:

$$51 = 5.1 \times 10^1$$

If the decimal point is moved n places to the left, n will be a negative number. For example:

$$0.51 = 5.1 \times 10^{-1}$$

You will notice from these examples that when large numbers are written in scientific notation, the 10 has a positive index value. When very small numbers are written in scientific notation, the 10 has a negative index value.

Practice questions

1 Match each number with its correct scientific notation.

Number	Scientific notation
0.002	2×10^3
2000	1.234×10^{-1}
0.1234	2×10^{-3}
12.34	1.234×10^1
123.4	1.234×10^2

2 Write 7.009×10^{-4} using decimal notation.

2 Identifying significant figures

When giving an answer to a calculation it is important to take note of the number of significant figures that you use.

You should give an answer that is as accurate as possible. However, an answer can't be more accurate than the data or the measuring device used to calculate it. For example, if a set of scales that measures to the nearest gram shows that an object has a mass of 56 g, then the mass should be recorded as 56 g, not 56.0 g. This is because you do not know whether it is 56.0 g, 56.1 g, 56.2 g or 55.8 g.

The number 56 has 2 significant figures. Recording to 3 significant figures (e.g. 56.0 g or 55.8 g) would not be scientifically 'honest'.

If this mass of 56 g is used to calculate another value it would also not be 'honest' to give an answer that has more than 2 significant figures.

Determining to what number of significant figures to give an answer to depends on what kind of calculation you are doing.

If you are multiplying or dividing, use the smallest number of significant figures provided in the initial values.

If you are adding or subtracting, use the smallest number of decimal places provided in the initial values.

Working out the number of significant figures

The following rules should be followed to avoid confusion in determining how many significant figures are in a number.

1. All non-zero digits are always significant. For example, 21.7 has 3 significant figures.
2. All zeroes between two non-zero digits are significant. For example, 3015 has 4 significant figures.
3. A zero to the right of a decimal point and following a non-zero digit is significant. For example, 0.5700 has 4 significant figures.
4. Any other zero is not significant, as it will be used only for locating decimal places. For example, 0.005 has just 1 significant figure.

Practice questions

1. Which of the following is written to 2 significant figures?
   A 30.1
   B 0.000 40
   C 0.5
   D 5.12

2. George multiplied 1.22 by 1.364. Which of the options below shows the result of this multiplication with the correct number of significant figures?
   A 1.66
   B 1.664
   C 1.65
   D 1.7

3. How can 41 be written to 4 significant figures?
   A 00.41
   B 4100
   C 41.00
   D 4.100

4. Alex is getting ready to go for a bike ride. Alex's mass is 65.3 kg. The bicycle has a mass of 12.92 kg.
   a Calculate the combined mass of Alex and the bicycle. Give your answer to the correct number of significant figures.
      A 78
      B 78.2
      C 78.22
      D 78.3
   b Using the combined mass calculated in part **a** above, and the formula $F_{net} = ma$, calculate the force Alex needs to apply to achieve an acceleration of 1.250 m s^{-2}. Give your answer to the correct number of significant figures.

3 Calculating percentages

Scientists use percentages to express a ratio or fraction of a quantity.

To express one quantity as a percentage of another, use the second quantity to represent 100%.

For example, expressing 6 as a percentage of 24 is like saying '6 is to 24 as x is to 100':

$$\frac{6}{24} = \frac{x}{100}$$
$$x = \frac{6}{24} \times 100$$
$$= 25\%$$

To calculate a percentage of a quantity, the percentage is expressed as a decimal then multiplied by the quantity.

For example, to calculate 40% of 20:

$$x = \frac{40}{100} \times 20$$
$$= 0.4 \times 20$$
$$= 8$$

Practice questions

1 What is 9 as a percentage of 12?
A 25%
B 50%
C 75%
D 30%

2 What is 25% of 24?
A 6.6
B 6
C 5
D 0.5

3 Which of the following values expresses 15 as a percentage of 120?
A 8%
B 5%
C 12.5%
D 0.125%

4 Converting between percentages and fractions

To write a percentage as a fraction, divide the percentage by 100.

For example:

$$25\% = \frac{25}{100}$$

But $\frac{25}{100}$ is not the simplest form of this fraction. If you divide both the numerator and the denominator by 25 (their highest common factor) then the fraction simplifies to $\frac{1}{4}$.

Whenever you give a fraction as an answer, always try to simplify it by dividing the numerator and denominator by the highest common factor.

To write a fraction as a percentage, multiply the fraction by 100%. In many cases, it is easier to convert the fraction to a decimal number first.

For example:

$$\frac{1}{4} = 0.25 \times 100 = 25\%$$

The value of the fraction or percentage has not changed. It is just being represented in a different way.

Practice questions

1 Choose the option that expresses $\frac{1}{5}$ as a percentage.
A 25%
B 20%
C 30%
D 50%

2 Match each percentage with its corresponding fraction.

Percentage	Fraction
0.2%	$\frac{7}{20}$
2.5%	$\frac{7}{40}$
17.5%	$\frac{1}{40}$
35%	$\frac{111}{250}$
44.4%	$\frac{1}{500}$

5 Changing the subject of an equation

Scientists use equations to represent relationships between variables. In an equation like $A = \pi r^2$, A is called the subject of the equation.

Sometimes the subject of the equation has to be changed in order to express the relationship in a more useful way. For example, if you need to find the radius of a circle, you would want r to be the subject of the equation above.

To change the subject of a simple equation, transpose the equation to leave the new subject on its own. In the example above, the equation needs to read

$r = \ldots$

Keep the equation balanced by performing the same operation to both sides of the equation to cancel operations being performed on the desired subject. Inverse operations (the opposite operation; for example, dividing is the inverse of multiplying) will allow cancelling.

For example, make r the subject of the equation $A = \pi r^2$.

1 Divide both sides of the equation by π.

$$A = \pi r^2$$

$$\frac{A}{\pi} = \frac{\pi r^2}{\pi}$$

The π in the numerator and denominator on the right-hand side of the equation cancel out, giving

$$\frac{A}{\pi} = r^2$$

2 To cancel the squaring operation of r, take the square root of both sides of the equation.

$$\frac{A}{\pi} = r^2$$

$$\sqrt{\frac{A}{\pi}} = \sqrt{r^2}$$

The square and square root on the right side of the equation cancel out, giving

$$\sqrt{\frac{A}{\pi}} = r$$

3 r is now the subject of the equation.

$$r = \sqrt{\frac{A}{\pi}}$$

Practice questions

1 Rearrange the formula $A = \frac{2}{3}R$ to make R the subject.

A $R = \frac{2A}{3}$

B $R = \frac{A}{3}$

C $R = \frac{3A}{2}$

D $R = 6A$

2 Rearrange the formula $y = 3\sqrt{\frac{p}{q}}$ to make p the subject.

6 Interpreting the slope of a linear graph

Scientists often represent a relationship between two variables as a graph. For directly proportional relationships, the variables are connected by a straight line, where the slope (or gradient) of the line represents the constant of proportionality between the two variables.

The slope or gradient of the line is defined as the ratio of change between two points in the vertical axis (ΔY), divided by the change between two points in the horizontal axis (ΔX). In other words, it measures the rate at which one variable (the dependent variable) changes with respect to the other (the independent variable).

The graph below has two straight lines with different slopes. The steeper slope (blue line) indicates that the rate of change is higher. This means the change is happening more quickly. On the other hand, the flatter slope (red line) indicates that the rate of change is lower. This means the change is happening more slowly.

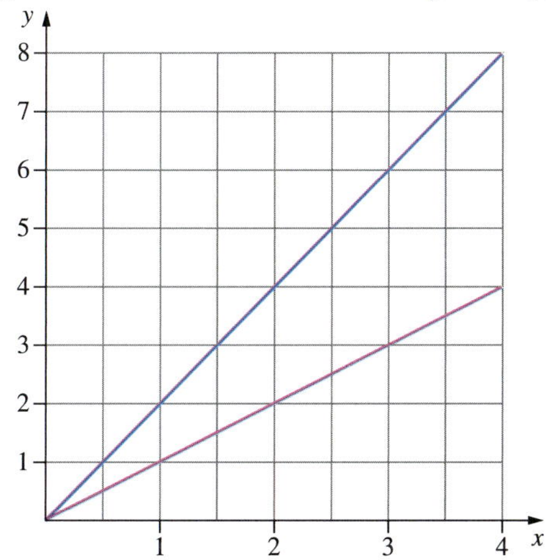

Practice questions

1 On a graph with two sloped lines, what does the steeper sloped line indicate?
A a faster rate of change
B a slower rate of change
C the same rate of change
D a much slower rate of change

2 What gives rate of change of a straight line on a graph?
A y-intercept
B x-intercept
C gradient
D area under the graph

7 Understanding mathematical symbols

Part of the language of science is using symbols to represent quantities or to give meanings. For example, the four symbols $<$, $>$, $\leq$ and $\geq$ are known as 'inequalities'.

The following mathematical symbols are commonly used in science.

Symbol	Meaning	Example	Explanation
$<$	less than	$2 < 3$	2 is less than 3
$>$	greater than	$6 > 1$	6 is greater than 1
$\leq$	less than or equal to	$2x \leq 10$	$2x$ is less than or equal to 10
$\geq$	greater than or equal to	$3y \geq 12$	$3y$ is greater than or equal to 12
$\sqrt{\ }$	square root	$\sqrt{4} = 2$	The square root of 4 is 2
Δ	change in (difference between)	Δt	change in t (time)
$\approx$	approximately equal to	$\pi \approx 3.14$	π is approximately equal to 3.14
Σ	summation	$\sum_{i=1}^{4} i$	The sum of consecutive integers from 1 to 4, i.e. $1 + 2 + 3 + 4 = 10$

Practice questions

1 The symbol that means 'less than' is:
A $<$
B $>$
C $\leq$
D $\geq$

2 Which of these symbols is an inequality?
A $\approx$
B Δ
C $\sqrt{\ }$
D $\leq$

8 Understanding the difference between discrete and continuous data

Quantitative data forms the backbone of science. Scientists are constantly working with data—measuring, recording, analysing and interpreting it.

Quantitative data consists of numerical values that can either be discrete or continuous.

Discrete data is data that has a set of clearly defined values. For example, the number of students in a class would have a discrete set of possible data.

Continuous data is usually data that is measured in some way and can have an infinite number of values. For example, your height or weight would have a continuous set of possible data.

The easiest way to distinguish between the two types of quantitative data is to ask, 'Is the data measured or counted?' If it is counted, the data set is discrete. If it is measured, the data set is continuous.

Practice questions

1 Which one of these data sets is continuous?
  A the number of cars parked in a street
  B the temperature of the air over a 24-hour period
  C the number of students at a school
  D the number of nails used to build a fence

2 Which one of these data sets is discrete?
  A the number of cars parked in a street
  B the temperature of the air over 24 hours
  C the heights of a team of footballers
  D the mass of a team of netballers

9 Calculating the mean, median and range of a data set

When handling data, scientists often look for ways to describe patterns in the data. Common terms used when analysing a set of data include the mean, median and range.

Mean: the *average* value in the data set. To calculate the mean, sum all the values in the data set and then divide this by the number of data values.

Median: the *middle* value in an ordered data set. To calculate the median, arrange the data set in ascending order and then count the number of data values. If the number of values is odd, the median is the middle value. If the number of values is even, calculate the median by adding the two middle values and dividing by 2, i.e. by calculating the average of the two middle numbers.

Range: the *spread* of values in the data set. To calculate the range, take the largest data value and then subtract the smallest data value.

Practice questions

1 The following set of data is recorded:
  44, 17, 21, 26, 42, 18
  Find the:
  a mean
  b median
  c range.

2 The mass in kilograms of each student in a class of 25 students is recorded below. The combined mass of all the students is 1340 kg.
  Find the:
  a mean
  b median
  c range.
  Students' weights: 67, 60, 41, 52, 39, 60, 42, 55, 55, 50, 46, 62, 48, 48, 56, 64, 55, 56, 59, 61, 41, 63, 53, 62, 45

10 Solving simple algebraic equations

To solve an equation means to find the values that make the equation true. Scientists manipulate equations and substitute in known variables in order to solve for the variable required.

For example, you can solve

$$a = \frac{F_{net}}{m}$$

where F_{net} is the net force on the car, which is 2400 N
m is the mass of the car, which is 1200 kg
a is the acceleration of the car in m s^{-2}, which is unknown.

$$a = \frac{2400}{1200}$$
$$= 2\,\text{m s}^{-2}$$

Practice questions

1 Solve the equation $V = IR$ if $I = 3$ and $R = 9$.
A $V = 3$
B $V = 6$
C $V = 12$
D $V = 27$

2 Solve the equation and find the value of Q if $Q = mc\Delta T$, $m = 1.2$, $c = 4200$ and $\Delta T = 30$.
A $Q = 840$
B $Q = 25\,200$
C $Q = 126\,000$
D $Q = 6$

11 Completing calculations with more than one operation

Scientists often deal with complex calculations that can involve numerous operations within the one calculation. The order in which these operations are performed can affect the result of the calculation.

For example, the calculation $2 + 3 \times 4$ will give an incorrect answer of 20 if you calculate the $2 + 3$ part first, but it will give the correct answer of 14 if you calculate the 3×4 part first.

Scientists and mathematicians have agreed on a set order in which operations are carried out so that calculations are consistent. You can remember this order using the acronym 'BIDMAS':

- brackets
- indices (powers, square roots, etc.)
- division and multiplication
- addition and subtraction.

The operations present in a calculation are performed in the order shown in the list. If there are multiple instances of division and multiplication, or addition and subtraction, work from left to right.

For example, the 3×4 part in the original example would always be performed first, since multiplication is higher in the list than addition.

When dealing with scientific notation, it is important to keep each individual number complete. Use brackets to help do this, especially when dividing.

For example, when dividing 3.01×10^{21} by 6.02×10^{23}, brackets are used to keep the second number together.

If you used a calculator and entered $\frac{3.01 \times 10^{21}}{6.02 \times 10^{23}}$ (i.e. $3.01 \times 10^{21} \div 6.02 \times 10^{23}$), the answer would come out as 5.00×10^{43}. This is not the correct answer.

The correct answer is only obtained by entering $\frac{3.01 \times 10^{21}}{(6.02 \times 10^{23})}$. This time, the answer comes out correctly as 5.00×10^{-3}.

Using the calculator's *EXP* button or the $x10^x$ button keeps the number and power of 10 together as one number and avoids the problems of using the number multiplied by 10^x. If your calculator does not have an *EXP* or $x10^x$ button, check the user manual. It may be that it is just labelled differently on your calculator. Alternatively, remember to always use brackets to keep the terms in the denominator together.

Practice questions

1 What is 3.4×10^{-4} divided by 1.7×10^{-3}?
A 2×10^{-7}
B 2
C 20
D 0.2

2 Substitute $m = 1.4$, $d = 3.9$ and $c = 2.7$ into $W = 6m - 4(d + c)$ and solve for W.
A −4.5
B −18
C 34.8
D 26.7

12 Understanding the relationship between data, graphs and algebraic rules

Scientists use graphs to analyse the data they collect from experiments. All graphs tell a story. The shape of the graph shows the relationship between the variables, and this relationship can be written algebraically and numerically. The horizontal axis is known as the *x*-axis and the vertical axis is known as the *y*-axis.

Once the algebraic rule is known, the values for one variable can be substituted and the values for the other variable can be calculated. These values can also be determined by reading them from the graph.

For example, when investigating how current and voltage vary across a light bulb, the following data was collected:

Current, I (A)	Voltage, V (V)
0.14	0.6
0.22	1.0
0.33	1.4
0.47	2.0
0.66	2.8
0.80	3.6
1.20	5.0

Graphing this data produced:

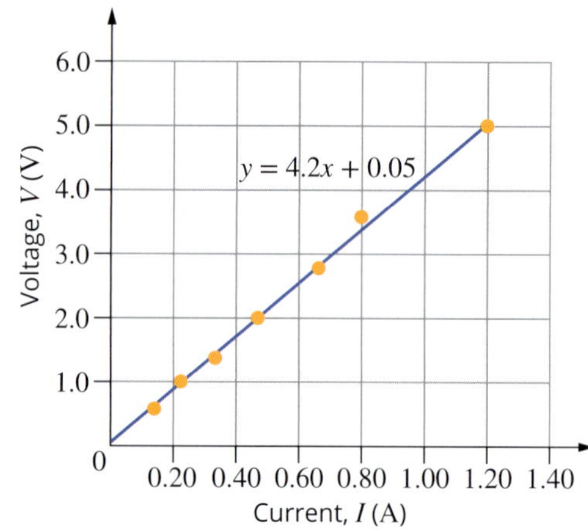

The numerical values from the experiment are listed in the table and plotted on the graph. The algebraic relationship between the variables is given by the equation of the line:

$$y = 4.2x + 0.05$$

The value of the y-intercept is approximately zero, so assuming that the y-intercept is zero, and labelling the x-axis as current and the y-axis as voltage, the relationship can be written as:

$$y = 4.2 \times \text{current}$$

Using the appropriate symbols, this can also be written as: $V = 4.2I$

Practice questions

1 If a graph had L on the y-axis and B on the x-axis and the equation of the straight line was $y = 3.7x$, what is the algebraic form of the graph?
A $L = 3.7x$
B $y = 3.7x$
C $y = 3.7B$
D $L = 3.7B$

2 If a relationship were written as $m = 5.9L$, what shape would the graph be and which variable would be plotted on which axis?
A The graph would be non-linear, with m on the y-axis and L on the x-axis.
B The graph would be non-linear, with m on the x-axis and L on the y-axis.
C The graph would be linear, with m on the y-axis and L on the x-axis.
D The graph would be linear, with m on the x-axis and L on the y-axis.

13 Recognising and using ratios

A ratio is the relationship between two numbers of the same kind. It could be the quantities in a recipe, the division of profits from a sale, or the number of different types of the same thing.

Scientists use ratios to compare quantities. This might be the numbers of atoms of different elements in a compound, or the number of primary and secondary windings of a transformer.

You can also use the principle of ratios to solve problems. For example, if one reaction produces 2.5 MeV of energy, then how much energy does 34 reactions produce?

The reaction-to-energy ratio of 1:2.5 should remain constant as the number of reactions increases. So you need to find the factor that 1 needs to be multiplied by to give 34:

$$1 \times 34 = 34$$

You then multiply the energy amount by the same factor:

$$2.5 \times 34 = 85\,\text{MeV}$$

You may also see ratios expressed as fractions.

Practice questions

1 If the primary coil of a transformer has 2400 windings and the secondary coil has 600 windings, what is the simplest ratio of primary to secondary windings?

A 2400:600
B 24:6
C 12:3
D 4:1

14 Understanding pie charts, frequency graphs, and histograms

It is essential in science to collect data and arrange it in an orderly way. Tables are often used to organise data, which can then be displayed in a graph.

Pie charts

A pie chart is a circle that is divided into sectors. Each sector represents one item in the data set and is shown as a percentage or fraction of the total data set.

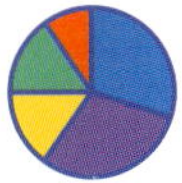

Frequency graphs and histograms

Frequency graphs and histograms are another way of representing data visually.

If data is discrete (i.e. can be counted), each column in a column graph will represent one category, e.g. apples or strawberries. Often these columns have a gap between them.

If the data is continuous (i.e. can be measured), such as the heights of the students in that class, each column will represent a range of possible heights, e.g. 140 cm to 160 cm, and there will be no gaps between the columns. These are called histograms.

Practice questions

1 Choose the pie chart that correctly shows the data from Emmanuel's poll of soccer fans.

Number of matches watched	10	12	15	18	21
Relative frequency (%)	25	35	15	15	10

Matches watched:

- (blue) = 10
- (green) = 18
- (purple) = 12
- (red) = 21
- (yellow) = 15

A 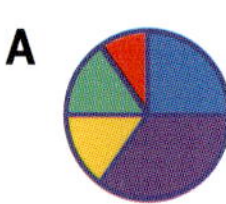

B 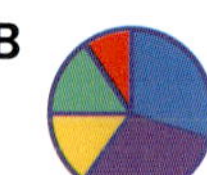

C 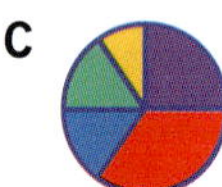

D 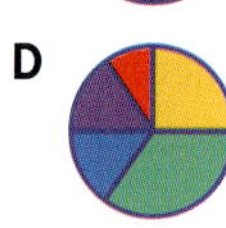

2 A class of Year 11 Chemistry students had a test. A histogram of their test scores is shown below.

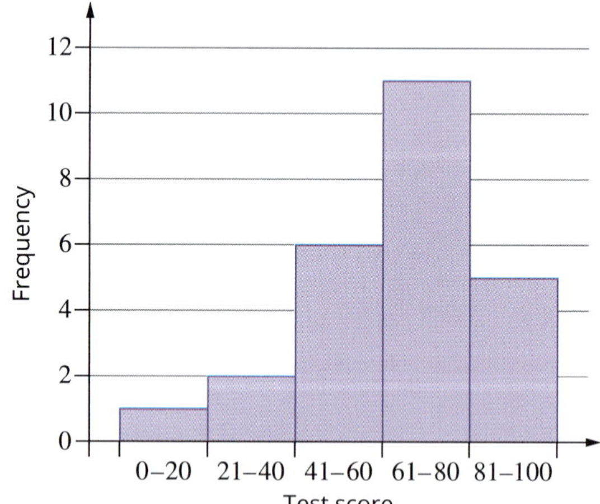

a How many students scored over 80?
b How many students scored below 40?
c How many student are in this class?

15 Using units in an equation to check for dimensional consistency

Scientists know that each term in an equation represents a quantity. The units used to measure that quantity are not used in the calculations. Units are only indicated on the final line of the solved equation.

For example, this is the equation for the area (A) of a rectangle of length (L) and width (W):

$$A = L \times W$$

If L has a value of 7 m and W has a value of 4 m, it is written:

$$\begin{aligned} A &= L \times W \\ &= 7 \times 4 \\ &= 28\,\text{m}^2 \end{aligned}$$

Note that the units are left out of the actual calculation on the second line and only included at the end, after the numerical answer.

You can use units to check the dimensional consistency of the answer. In the example above, the two quantities of L (length) and W (width) both have to be expressed in consistent units, in this case metres (m), to give an answer that is expressed in square metres ($\text{m} \times \text{m} = \text{m}^2$).

If you had made a mistake, and used the formula $A = L + W$ instead, the answer would be expressed in metres only. This is not the correct unit to express area, so you would know that was wrong.

Practice questions

1 Which formula has the correct dimensions for calculating volume in m^3?

A $\text{m} \times \text{m}$
B $\text{m} \times \text{s}$
C $\text{m} \times \text{m} \times \text{s}$
D $\text{m} \times \text{m} \times \text{m}$

2 Which of these shows the correct substitution into $P = 2L + 2W$ using consistent units for $L = 3.5\,\text{m}$ and $W = 240\,\text{cm}$?

A $P = 2 \times 3.5 + 2 \times 240$
B $P = 2 \times 3.5 + 2 \times 24$
C $P = 2 \times 35 + 2 \times 240$
D $P = 2 \times 3.5 + 2 \times 2.4$

3 By using the equation $p = mv$ as a guide, select the correct units in which to measure momentum, p.

A m s
B m s^{-1}
C kg m s^{-1}
D kg m s^{-2}

16 Understanding lines of best fit

When scientists observe that the points on a graph seem to form a straight line, then a line of best fit (or trendline) can be drawn through them. Computer programs such as Excel can fit a trendline to a data set; lines of best fit can also be drawn by hand onto a printed or drawn graph.

The line of best fit should pass as close as possible to as many of the points as possible (i.e. it should 'fit' the data closely). It may not pass exactly through any of the points, but once the line of best fit is drawn, the points should be spaced equally on each side, above and below the line. There should be no points very far away from the line, unless they are considered to be unreliable. Unreliable points are called 'outliers' and can be disregarded for the purposes of creating a line of best fit.

The gradient and y-intercept of the line (not the points) can be determined to find the relationship between the variables using the general equation for a straight line: $y = mx + c$.

For example:

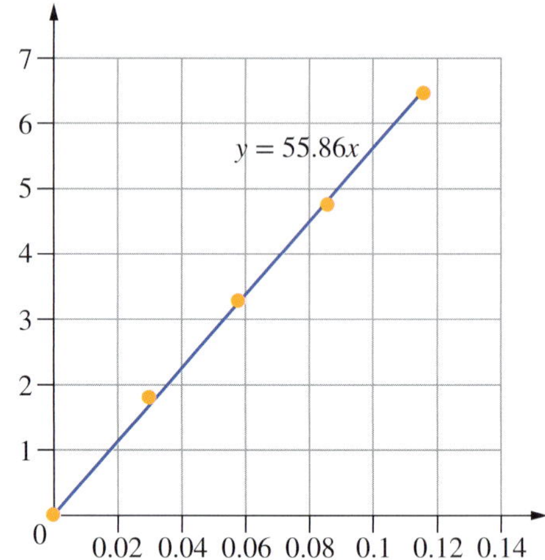

Practice questions

1 Are the following statements true or false?

- **a** All the points on a scatter plot must lie on the line of best fit.
- **b** The line of best fit may pass through none of the points.
- **c** The points of the scatter plot should lie close to the line of best fit.

2 Are the following statements true or false?

- **a** A trendline is the same as a line of best fit.
- **b** A line of best fit can only be drawn using a computer program such as Excel.
- **c** Outliers should be included as normal points when considering where to draw a line of best fit.

Answers to practice questions

1 Transforming decimal notation to scientific notation

1

0.002	2×10^{-3}
2000	2×10^{3}
0.1234	1.234×10^{-1}
12.34	1.234×10^{1}
123.4	1.234×10^{2}

0.002: Move the decimal point 3 places to the right, which gives an index of –3, so 0.002 is written as 2×10^{-3}.
2000: Move the decimal point 3 places to the left, which gives an index of +3, so 2000 is written as 2×10^{3}.
0.1234: Move the decimal point 1 place to the right, which gives an index of –1, so 0.1234 is written as 1.234×10^{-1}.
12.34: Move the decimal point 1 place to the left, which gives an index of +1, so 12.34 is written as 1.234×10^{1}.
123.4: Move the decimal point 2 places to the left, which gives an index of +2, so 123.4 is written as 1.234×10^{2}.

2 The –4 index shows the decimal point has been moved 4 places to the right. Move it back 4 places to the left to give 0.000 7009.

2 Identifying significant figures

1 B

2 A
When two numbers are multiplied, use the smallest number of significant figures in the initial values to give your answer. In George's multiplication, the answer is 1.66408, but as 1.22 has 3 significant figures, the correct answer is 1.66.

3 C
The zeros to the right of the decimal point are significant. When writing an answer to a correct number of significant figures, you may need to use rounding if the initial answer has more figures (digits) than you need. These extra figures are called 'non-significant figures.' If the first non-significant figure is ≥5, you round up. If the first non-significant figure is <5, then do not round up. For example, if the initial answer for a calculation was 2.1259 but you only needed an answer to 3 significant figures, you would round it to 2.13. If the initial answer was 2.1241, then 3 significant figures becomes 2.12.

4 **a** B
When 65.3 is added to 12.92, the answer is 78.22, but only if it is assumed that 65.3 is actually 65.30. When adding or subtracting with significant figures, use the smallest number of significant figures provided in the initial values. As there is no way of knowing the accuracy of beyond 65.3, the answer should only be given to 3 significant figures, which is 1 decimal place, 78.2.

b When multiplying or dividing, use the smallest number of significant figures in the initial values to give your answer.

$$\begin{aligned} F_{net} &= ma \\ &= 78.2 \times 1.250 \\ &= 97.8\,\text{kg m s}^{-2} \end{aligned}$$

Since the mass value only has 3 significant figures, the answer should also only have 3 significant figures. Therefore, even though the calculated answer is 97.75, the answer to 3 significant figures is $97.8\,\text{kg m s}^{-2}$.

3 Calculating percentages

1 C

$$\frac{9}{12} = \frac{x}{100}$$
$$x = \frac{9}{12} \times 100 = 75\%$$

2 B

$$x = \frac{25}{100} \times 24 = 0.25 \times 24 = 6$$

3 C

$$\frac{15}{120} = \frac{x}{100}$$
$$x = \frac{15}{120} \times 100 = 12.5\%$$

4 Converting between percentages and fractions

1 B
To write a fraction as a percentage, multiply the fraction by 100.

$$\frac{1}{5} = 0.2 \times 100 = 20\%$$

2

0.2%	$\frac{1}{500}$
2.5%	$\frac{1}{40}$
17.5%	$\frac{7}{40}$
35%	$\frac{7}{20}$
44.4%	$\frac{111}{250}$

To change a percentage to a fraction, divide by 100.

$$0.2\% = \frac{0.2}{100} = \frac{2}{1000} = \frac{1}{500}$$
$$2.5\% = \frac{2.5}{100} = \frac{25}{1000} = \frac{1}{40}$$
$$17.5\% = \frac{17.5}{100} = \frac{175}{1000} = \frac{35}{200} = \frac{7}{40}$$
$$35\% = \frac{35}{100} = \frac{7}{20}$$
$$44.4\% = \frac{44.4}{100} = \frac{444}{1000} = \frac{222}{500} = \frac{111}{250}$$

5 Changing the subject of an equation

1 C

Multiply both sides of the equation by 3.

$$3A = 3 \times \frac{2}{3}R$$
$$= 2R$$

Divide both sides of the equation by 2.

$$\frac{3A}{2} = R$$

Rewrite the equation.

$$R = \frac{3A}{2}$$

2 $p = \frac{y^2 q}{9}$

Divide both sides of the equation by 3.

$$\frac{y}{3} = \sqrt{\frac{p}{q}}$$

Square both sides.

$$\frac{y}{3}^{2} = \frac{p}{q}$$

Expand the brackets on the left.

$$\frac{y^2}{9} = \frac{p}{q}$$

Multiply both sides by q.

$$\frac{y^2 q}{9} = p$$

Rewrite the equation.

$$p = \frac{y^2 q}{9}$$

6 Interpreting the slope of a linear graph

1 A

The steepness of the slope indicates the rate of change. A line with a steeper slope indicates a faster rate of change.

2 C

The gradient or slope of a linear graph indicates the rate of change.

7 Understanding mathematical symbols

1 A

One way is to remember that the smaller end of the shape points toward the smaller number. For example, $3 < 6$ means 3 is less than 6.

2 D

The symbols $<$, $>$, $\le$ and $\ge$ are all inequalities.

8 Understanding the difference between discrete and continuous data

1 B

If the data can be counted, the data set is discrete. If the data can be measured, the data set is continuous. The temperature of the air can be measured with a thermometer, so the data set is continuous.

2 A

If the data can be counted, the data set is discrete. If the data can be measured, the data set is continuous. The number of cars parked in a street can be counted, so the data set is discrete.

9 Calculating the mean, median and range of a data set

1 **a** $44 + 17 + 21 + 26 + 42 + 18 = 168$

$$\frac{168}{6} = 28$$

b Place the numbers in ascending order: 17, 18, 21, 26, 42, 44
As there is an even number of values, add the two middle values and divide by 2

$$\frac{21 + 26}{2} = 23.5$$

c $44 - 17 = 27$

2 **a** $\frac{1340}{25} = 53.6$ kg

b 55 kg

c $67 - 39 = 28$ kg

10 Solving simple algebraic equations

1 D

Substitute the values and solve the equation.

$$V = 3 \times 9$$
$$= 27$$

2 B

Substitute the values and solve the equation.

$$Q = 0.2 \times 4200 \times 30$$
$$= 25\,200$$

11 Completing calculations with more than one operation

1 D

The correct calculation, using brackets, is:

$$\frac{3.4 \times 10^{-4}}{(1.7 \times 10^{-3})} = 0.2$$

2 B

$$W = 6 \times 1.4 - 4 \times (3.9 + 2.7)$$
$$= 6 \times 1.4 - 4 \times 6.6$$
$$= 8.4 - 26.4$$
$$= -18$$

12 Understanding the relationship between data, graphs and algebraic rules

1 D

Substituting L for y and B for x gives $L = 3.7B$.

2 C

The graph would be linear as the equation is written in the form $y = mx + 0$ with m on the y-axis and L on the x-axis.

13 Recognising and using ratios

1 D

The primary to secondary ratio is 2400:600. Dividing both sides of the ratio by 600 simplifies it to 4:1.

14 Understanding pie charts, frequency graphs, and histograms

1 A

The largest sector of the pie chart is purple, representing the percentage of people who watched 12 matches (35%). The next in size is blue (10 matches, 25%), then green (18 matches, 15%) and yellow (15 matches, 15%). The smallest sector is red (21 matches, 10%).

2 **a** 5

b 3

c 25

The height of the 81–100 column is 5. This shows five students scored over 80.

The heights of the 0–20 and 21–40 columns are 1 and 2. This shows that three students scored less than 40.

If you add up the heights of all the columns ($1 + 2 + 6 + 11 + 5 = 25$), it tells you that there are 25 students in the class.

15 Using units in an equation to check for dimensional consistency

1 D

2 D

Both units must be consistent.

Since 100 cm = 1 m the consistent units must be either

$L = 3.5$ m and $W = 2.4$ m

or

$L = 350$ cm and $W = 240$ cm

The only correct combination is $P = 2 \times 3.5 + 2 \times 2.4$.

3 C

The unit for momentum is taken from the unit for mass (kg) multiplied by the unit for velocity ($m\,s^{-1}$). Therefore, momentum is measured in $kg\,m\,s^{-1}$.

16 Understanding lines of best fit

1 **a** false

b true

c true

2 **a** true

b false

c false

Answers

Comprehensive answers and fully worked solutions for all section review questions, try yourself worked examples, chapter review questions and unit review questions are available via Pearson Chemistry 11 Western Australia Reader+.

Chapter 1 Materials in our world

1.1 Materials science

1 A material is a substance used to construct an object.
2 Metals generally have high tensile strength, are ductile, are malleable, have a shiny lustre, have high melting points and are good thermal and electrical conductors.
3 An alloy is a mixture of a metal with other metals or small amounts of non-metals.
4 Polymers generally are less dense than metals, are corrosion resistant, possess excellent electrical resistance and some are compatible with human tissue.
5 Ceramics generally are hard, demonstrate high compressive strength and are able to withstand high temperatures.

1.2 Nanotechnology

1 **a** 8.35×10^7 nm
b 1.35×10^7 nm
c 4.2×10^6 nm
2 Due to the size of nanoparticles, they can travel freely throughout the body, even into cells themselves.
3 At the nanoscale, the ratio of surface area to volume of the particles becomes such that surface area effects, or quantum effects, begin to significantly influence the properties of the material.

1.3 Purifying materials

1 The particle size of salt and pepper is too similar and will likely be either passed or blocked by a sieve.
2 No, you cannot. The solute in a solution will not settle out of the solution no matter how long you wait.
3 No. Water is not a mixture of hydrogen and oxygen, but a pure compound composed of molecules of hydrogen chemically bonded to oxygen.
4 Refer to Figure 1.3.10.
5 **a** Use a separating funnel or distillation.
b Use evaporation or distillation.
c Use sieving.
d Add water then filter or decant.
e Add water then decant (the wood chips float).
6 pentane, benzene, heptane, octane

Chapter 1 review

1 a combination of two or more distinct materials that creates a substance that has a unique combination of properties of its components
2 paper, glue
3 Metals = aluminium and stainless steel. Ceramics = porcelain. Polymers = cotton and nylon.
4 There are two possible reasons. Pure gold is very soft, so alloys of gold are used in jewellery as they are more durable than pure gold. Alloys of gold are also cheaper, using less gold in total but still maintaining the attractiveness of pure gold.
5 The properties demonstrated by elements and compounds depend on the arrangement of the atoms and molecules in the substance. Diamond, graphite and carbon nanotubes must have different arrangements of particles/atoms.
6 Reinforced concrete gains the benefit of steel reinforcing bars with high tensile strength, while maintaining the high compressive strength of concrete.
7 Iron corrodes quite readily and would rust in a relatively short period of time. Stainless steel is much more corrosion-resistant and durable.
8 **a** copper and tin
b tin and copper, antimony, bismuth or lead
c gold and nickel, manganese, platinum or palladium
9 **a** 5.0×10^7 nm
b 1.2×10^7 nm
c 2.0×10^4 nm
10 Zinc oxide nanoparticles are colourless.
11 **a** Top-down fabrication involves the removal of material from particles until the desired material is produced, whereas bottom-up fabrication is the building up of the material, one atom or molecule at a time.
b Advantages of top-down: large quantities can be produced relatively cheaply, with good uniformity and with existing technology. Disadvantages: limited to relatively simple structure and by the scale of the tools available.
Advantages of bottom-up: far more complicated materials can be manufactured. Disadvantages: only small amounts of material can be produced economically.
12 A colloid is a mixture of an insoluble substance dispersed throughout another substance. Unlike suspensions, the two components in a colloid will not separate (based on density) over a period of time as a suspension will.
13 In distillation, all components of the mixture are retained, whereas in evaporation the solvent is lost to the atmosphere as vapour.
14 The heat from the Sun causes water to evaporate, forming a vapour. This water vapour then condenses on the plastic and drips into the container located at the centre of the hole. Any salt or contaminants present in the original water are left behind.
15 Evaporation of the solution will only remove the solvent from the solution, leaving the multiple different salts still mixed but as a solid instead of a solution.
16 **a** Add water, decant off saw dust then filter out sand as the residue, then evaporate off water to leave sugar.
b Use a separating funnel to remove oil (less dense and floats) then use fractional distillation to separate ethanol and water.
c Dissolve the instant coffee in water, filter to collect tea leaves as a residue, evaporate off water to obtain instant coffee.
17 Carbon fibres are usually used to reinforce polymer or plastic and produce a material that is very strong but still light.
18 The ease with which nanoparticles enter the body and the lack of evidence for their long-term safety are causes for concern.
19 Grey goo is an apocalyptic scenario where self-assembling/self-replicating nanobots consume all the available resources on Earth by uncontrolled replication of themselves.

Chapter 2 Atoms: structure and mass

2.1 Atomic theory

1 proton, neutron and electron
2 protons and neutrons; found in the nucleus
3 The electrostatic attraction of the protons; the negative electrons are attracted to the positive protons and pulled towards them.

2.2 Describing atoms

WE 2.2.1 92
1 **a** Na **b** Cu **c** Ne **d** K **e** Be **f** Ca
2 atomic number
3 mass number
4 protons and electrons
5 $Z = 15$: protons = 15, electrons = 15, neutrons = 16
6 nitrogen

2.3 Isotopes

WE 2.3.1 107.9 **WE 2.3.2** 29.50%

1 an atom of an element that has a different number of neutrons and therefore a different mass number
2 Isotopes of the same element have the same atomic number and therefore the same number of protons and electrons. The number of neutrons is different between isotopes of the same element, therefore they have different mass numbers.
3 two
4 **a** 15.999 **b** 1.008
5 8%

2.4 Mass spectrometry

WE 2.4.1 % abundance ^{206}Pb = 28%; % abundance ^{207}Pb = 22%; % abundance ^{208}Pb = 50%

1 1 The sample is vaporised and then ionised using high-energy electrons.
  2 The ions are separated and accelerated according to their mass-to-charge ($\frac{m}{z}$) ratios in an electric/magnetic field.
  3 The ions that have a particular mass-to-charge ratio are detected by a device that counts the number of ions that strike it.
2 the number of different isotopes in the sample
3 the mass-to-charge ratio, which equates to the relative isotopic mass
4 the relative abundance of the isotopes
5 **a** % abundance ^{90}Zr = 51%
  % abundance ^{91}Zr = 11%
  % abundance ^{92}Zr = 17%
  % abundance ^{94}Zr = 17%
  % abundance ^{96}Zr = 4%
  b A_r(Zr) = 91

Chapter 2 review

1 Atoms are hard, indivisible structures.
2 the nucleus
3 The protons and neutrons form the nucleus. The electrons are grouped in shells and occupy the space around the nucleus.
4 The mass of a proton is approximately equal to the mass of a neutron and is about 1840 times the mass of an electron. The proton and electron have equal but opposite charges and the neutron has no charge.
5 Thomson discovered the electron, Rutherford discovered the proton and Chadwick discovered the neutron.
6 The periodic table is organised by atomic number. Although it appears to also be organised by mass number, it is not consistent throughout the whole table.
7 **a** atomic number is 24; mass number is 52
  b 24 electrons, 24 protons, 28 neutrons
8 **a** atomic number is 25; mass number is 55
  b 25 electrons, 25 protons, 30 neutrons
9 Atoms are electrically neutral. The positive charge on one proton balances the negative charge on one electron. Therefore, for electrical neutrality, there must be an equal number of protons and electrons.
10 Most elements have more than one isotope, so they will have more than one mass number. All bromine atoms have 35 protons in their nuclei. No other type of atom has 35 protons in its nucleus (i.e. no other atom has an atomic number of 35). Isotopes of bromine, however, differ in their mass numbers, so mass number is not fixed for an element (except for those elements such as sodium, which have only one naturally occurring isotope). In addition, an isotope of one element may have the same mass number as an isotope of another element.
11 No. Isotopes have the same number of protons (atomic number) but different numbers of neutrons (and therefore different mass numbers). These atoms have different atomic numbers and different mass numbers.
12 The relative atomic mass of carbon is the weighted average of the isotopic masses of all carbon isotopes (i.e. ^{12}C, ^{13}C and ^{14}C). Small amounts of ^{13}C and ^{14}C make this average slightly greater than 12, the relative isotopic mass of the ^{12}C isotope.
13 **a** proportion of the lighter isotope = 61.5%
  b 20.2%
14 The isotopes of an element all have the same chemical properties, they only differ by the physical properties.
15 Nothing. Argon has 18 protons, 18 electrons and 22 neutrons whereas calcium has 20 protons, 20 electrons and 20 neutrons.
16 106.4
17 **a** A_r(Ar) = 39.96; A_r(K) = 39.11
  b Ar: close to 40. K: close to 39
18 **a** peak heights ^{50}Cr = 0.3 units, ^{52}Cr = 12 units, ^{53}Cr = 1 unit, ^{54}Cr = 0.2 units, Total height = 13.5 units
  percentages ^{50}Cr = 2.2%, ^{52}Cr = 88.9%, ^{53}Cr = 7.4%, ^{54}Cr = 1.5%
  b A_r(Cr) = 52
19 48.0% and 52.0%
20 close to 20
21 **a** The mass number is the sum of the protons and neutrons present in the nucleus of an atom.
  b The relative isotopic mass of an isotope is the mass of an atom of that isotope relative to the mass of an atom of ^{12}C, taken as 12 units exactly.
  c The relative atomic mass of an element is the weighted average of the relative masses of the isotopes of the element, based on a natural sample, on the ^{12}C scale.
22 **a** 0.011 70% **b** 39.09

Chapter 3 Electrons and the periodic table

3.1 Electronic structure of atoms

1 Each line in an emission spectrum corresponds to a specific amount of energy. This energy is emitted when electrons from higher energy electron shells transition to a lower energy shell. Different lines indicate that there are differences in energy between shells. This is evidence that electrons are found in shells with discrete energy levels.
2 Electrons revolve around the nucleus in fixed, circular orbits.
  Electron orbits correspond to specific energy levels in the atom.
  Electrons can only occupy fixed energy levels and cannot exist between two energy levels.
  Orbits of larger radii correspond to higher energy levels.
3 Energy is emitted as light or electromagnetic radiation.

3.2 Electron arrangement in the periodic table

WE 3.2.1 2,8,8

1 **a** 2,3 **b** 2,8,2 **c** 2,8,8,2
2 **a** 2,2 **b** 2,8,6 **c** 2,8,8 **d** 2,8,2 **e** 2,8
3 **a** helium, He **b** fluorine, F **c** aluminium, Al
  d nitrogen, N **e** chlorine, Cl

3.3 Trends in the periodic table

1 +4
2 The atomic radii can be considered the distance from the centre of the nucleus to the valence shell. As the number of protons increases, the attraction on the valence electrons increases, thus pulling the electrons closer to the nucleus.
3 **a** Ionisation energy is the least amount of energy needed to remove an electron from an atom or ion in the gas phase.
  b The factors that affect ionisation energy across a period are the size of the atom (i.e. the distance of the outermost (highest energy) electron from the nucleus), and the charge on the nucleus.

4 Across a period, the number of occupied shells in the atoms remains constant but core charge increases. The valence electrons become more strongly attracted to the nucleus, so more energy is required to remove an electron from an atom. Therefore, the first ionisation energy increases across a period.

5 a i fluorine, F ii francium, Fr
b i group 17 ii group 1
c Elements in group 18, the noble gases, have a very stable electron configuration and so are unreactive.

6 As core charge increases, electronegativity increases.

3.4 Quantisation of energy

1 Each line in an emission spectrum corresponds to a specific amount of energy. This energy is emitted when electrons from higher-energy electron shells transition to a lower-energy shell. Different lines indicate that there are differences in energy between shells. This is evidence that electrons are found in shells with discrete energy levels.

2 Different wavelengths and frequencies correspond to different colours of visible light.

3 A 4 C

5 No. Although they share the same noble gas electron configuration, they are different in terms of the arrangement of electrical charges both in the electron shells and the nucleus, therefore they will demonstrate different emission spectra.

6 No. Although they share the same electron configuration, differences in the mass of the nucleus (due to different number of neutrons) affects the system of electrical charges in the atom and therefore will alter their emission spectra slightly.

3.5 Spectroscopy

1 Spectroscopy is the study of the interaction between matter and electromagnetic radiation.

2 Advantage: it is a quick and easy test for metal atoms.
Disadvantages: it only provides qualitative data; only a small range of metals are detectable; metals in low concentrations may be difficult to observe; mixtures of metals will produce confusing results.

3 The light source that produces the light characteristic of the metal being analysed is the source of energy for the excitation of the atoms. The flame merely vaporises the sample to produce free gaseous atoms in their ground state.

4 Atomic absorption spectroscopy (AAS) provides qualitative data whereas the flame test does not. A larger range of elements can be analysed by AAS whereas those suitable to be analysed by the flame test are fewer. AAS is very sensitive and can detect low concentrations whereas the flame test is not as sensitive. AAS is highly selective and can analyse mixtures well whereas the flame test is only applicable to pure samples.

5 Maritime flares traditionally use strontium or calcium compounds to produce their characteristic red colour.

Chapter 3 review

1 Bohr's idea was that the energy possessed by electrons is quantised, that is, can only exist as certain discrete values.

2 Nothing is between those electrons and the nucleus.

3 a i group 13 ii group 17 iii group 1
iv group 18 v group 14 vi group 14
b i 4 ii 2 iii 1
iv 1 v 7 vi 3
c i silicon, Si, 2,8,4 ii beryllium, Be, 2,2
iii argon, Ar, 2,8,8

4 a beryllium b fluorine
c silicon d calcium

5 a silicon or any other group 14 element b calcium or any other group 2 element
c bromine or any other group 17 element d nitrogen or any other group 15 element

6 groups 1, 2 and 13–18

7 a 1 b $15 - 10 = 5$
c $17 - 10 = 7$ d 2

8 Moving from left to right across groups 1, 2 and 13–17, the charge on the nucleus increases. Each time the atomic number increases by one, the electrons are attracted to an increasingly more positive nucleus. Within a period, the outer electrons are in the same shell—that is, they have the same number of inner-shell electrons shielding them from the nucleus. Therefore, the additional nuclear charge attracts the electrons more strongly, drawing them closer to the nucleus and so decreasing the size of the atom.

9 a Magnesium and phosphorus, with outer electrons in the third shell, are in the same period. Magnesium has a nuclear charge of +12 but, with two completed inner shells, the outer electrons experience the attraction of a core charge of +2. The outer-shell electrons of phosphorus, which has a nuclear charge of +15 and the same number of inner shells as magnesium, are attracted by a core charge of +5. The stronger attraction of the phosphorus electrons to the core means that more energy is required to remove an electron from a phosphorus atom than from a magnesium atom.
b Both fluorine and iodine are in group 17, so the outer electrons of each atom experience the attraction of the same core charge. Because the outer-shell electrons of a fluorine atom are closer to the nucleus than those of an iodine atom, they are attracted more strongly and so more energy is needed to remove one.

10 a The radius of the atoms decreases as the core charge increases.
b There is a trend from metals (lithium, beryllium) to non-metals (boron, carbon, nitrogen, oxygen and fluorine).
c Electronegativity increases as the core charge increases and size of the atoms decreases.

11 a nitrogen b chlorine c chlorine

12 Fluorine. Both oxygen and fluorine will have higher first ionisation energy than chlorine due the reduced distance between the valence shell and the nucleus, and the additional distance between the nucleus and chlorine's valence (third) shell. Out of oxygen and fluorine, fluorine will be highest due to the increased core charge.

13 so that the energy value is a true indication of only the energy required to ionise the atoms (remove electrons) rather than the energy required to separate atoms during the melting and/or vaporisation of the element.

14 Smaller. Sodium, with an electron configuration of 2,8,1 has a single valence electron in the third shell that is lost when it becomes a cation. The resultant sodium ion, Na^+, has an electron configuration of 2,8. With only two shells now, the sodium ion is significantly smaller.

15 Although the electron that fluorine gains to become an ion enters the same valence shell as fluorine's existing seven valence electrons, the increased repulsion within the (now crowded) valence shell causes the shell to enlarge.

16 $n = 1$

17 The ionisation of an atom requires more energy as this involves the removal of an electron from the atom, completely overcoming the electrostatic attraction to the nucleus, whereas the excitation of an atom does not completely remove an electron, but only raises it temporarily to a higher energy level.

18 The blue colour is a result of the emission spectrum of various chemicals generated during the combustion of the gas that have been excited due to the temperature of the flame.

19 The absorption lines characteristic of oxygen that Fraunhofer measured were the result of absorption taking place within Earth's atmosphere.

20 When atoms absorb energy, it is often possible for electrons to be promoted to various higher energy levels. Electrons can return to the ground state from these excited states by undergoing a number of transitions of different energy. Each transition results in a line of specific energy in the emission spectrum.

21 Using a luminous flame will obscure any colour produced by the metal atoms with the orange colour of the flame itself. Also, an orange flame may not be hot enough to achieve excitation of the metal atoms.

22 Excite a pure sample of the element itself.

23 The excitation and emission of light via the flame test will result in a mixture of different colours that will be hard to identify individually. In atomic absorption spectroscopy, the process relies on using the specific characteristic radiation of the element being tested for, and as no other elements will absorb the same frequencies of light, they should not interfere with the determination.

24 a Until Rutherford's work, the plum pudding model of the atom was widely accepted. However, his discovery that a beam of alpha particles directed at thin gold foil caused a few particles to be deflected through high angles led to the development of a new atomic model.
b Although Rutherford's atomic model accounted for a number of atomic properties, it was not able to account for the characteristic emission spectrum of each element.

25 Possible answer: Periodic table **a** shows the atoms arranged in increasing atomic number (starting from the centre of the 'spiral') but also aligned with other atoms with the same number of valence electrons moving out from this centre. The transition metals and the lanthanide and actinide series are shown as an expansion of the periodic table, corresponding to an expansion of the electron shells within the atoms themselves.
Periodic table **b** shows the expanding shells of the atoms as more electrons are added to atoms of increasing atomic number, but keeping the 'main group' elements in the eight columns to the right of the table.

Chapter 4 Metals

4.1 Properties of metals

WE 4.1.1 +3

1 a Li +1 b Mg +2 c Al +3

2 a Both potassium and gold have good thermal and electrical conductivity. However, gold has a higher density, and higher melting and boiling temperatures than potassium.
b sodium
c silver
d Sodium and potassium are in group 1. Gold and silver are transition metals.

3 a silver, copper, gold, aluminium
b Availability and cost need to be considered; also properties such as malleability and ductility.

4 Sodium belongs to the alkali metals, which have relatively low melting and boiling points, relatively low density and are relatively soft. Iron is a transition metal, which have relatively high melting and boiling points, relatively high density and are relatively hard.

5 tensile strength, cost, availability

4.2 Metallic bonding

WE 4.2.1 If the Mg is part of an electric circuit, the delocalised electrons are able to move through the lattice towards a positively charged electrode.

1 a

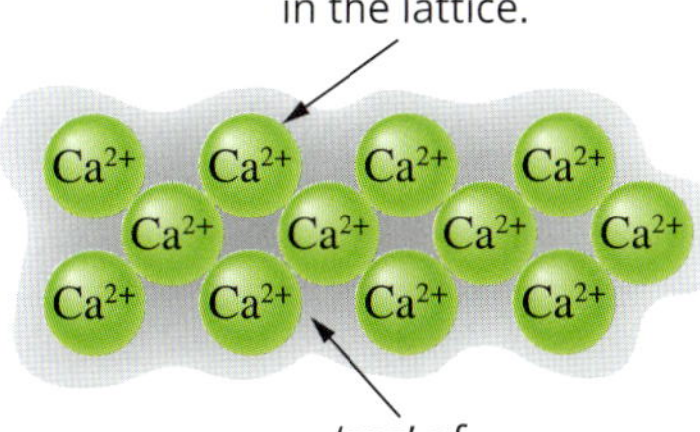

b strong electrostatic forces of attraction between Ca^{2+} ions and the delocalised valence electrons

2 Barium has a high melting temperature because there are strong attractive forces between the positive ions and the delocalised electrons. Barium conducts electricity because the delocalised electrons from the outer shell are free to move.

3 a Graphite and other metals are lustrous and conduct heat and electricity.
b These properties are explained by the presence of free-moving electrons. Graphite and other metals must contain delocalised electrons.

4.3 Reactivity of metals

1 $2K(s) + 2H_2O(l) \rightarrow 2KOH(aq) + H_2(g)$
potassium + water → potassium hydroxide + hydrogen

2 a Metals react more energetically further down the group.
b Metal atoms further down the group have more electron shells, which provide more shielding of the valence electrons from the core charge, reducing the electronegativity and allowing more spontaneous reaction.

3 zinc > iron > gold 4 calcium

4.4 Modifying metals

1 a A 20-cent coin contains copper and nickel. High-carbon steel contains iron and carbon.
b The 20-cent coin is a substitutional alloy similar to Figure 4.4.3. High-carbon steel is an interstitial alloy, similar to Figure 4.4.2.

2 The metal in the hooks becomes work hardened and brittle.

3 a Aluminium contains small areas of regular metallic lattice called crystals. When the aluminium is annealed, the crystal structure is changed to contain more large crystals. Larger crystals are more flexible and easier to shape than smaller crystals and the metal is less likely to break along crystal boundaries during shaping.
b Quenching the aluminium by heating to the critical temperature and then rapidly cooling causes the growth of small crystals. These crystals make the metal stiffer and do not allow the metal to deform as easily.

4 An alloy with properties appropriate to the task is selected. The metal is annealed to allow shaping of the chisel. The chisel is tempered to make the shaft flexible and strong. The tip of the chisel could then be hardened either by work hardening or local quenching.

5 A: nanoparticle. B: nanowire. C: nanorod. D: not a nanoparticle.

6 iron + oxygen → iron oxide

Chapter 4 review

1 Mg, Ca, Sr

2 a silver
b It is too expensive and tarnishes readily.
c aluminium, copper (combined with stainless steel)

3 electrical conductivity

4 a low density
b high electrical conductivity
c high tensile strength

5 20 in Ca, 18 in Ca^{2+}

6 Al: 2,8,3; Al^{3+}: 2,8

7 It is able to be drawn into a wire.

8 a When a current is applied to the copper wire, the free-moving, delocalised electrons move from one end to the other and so the copper wire conducts electricity.
b The delocalised electrons in the metal spoon obtain energy from the boiling mixture and move more quickly. These electrons move freely throughout the spoon, colliding with other electrons and metal ions, transferring energy so that the spoon becomes warmer and, eventually, too hot to hold.
c A lot of energy is required to overcome the strong forces of attraction between the iron ions and the delocalised electrons in the metal lattice, so that the iron changes from a solid to a liquid.
d Because of the strong forces of attraction between them, the lead ions and the delocalised electrons form a closely packed three-dimensional structure. Also, the lead atom itself has a higher mass-to-volume ratio than the sulfur atom. This means that the density—the mass per volume—is high.
e As the copper is drawn out, the copper ions are forced apart and the delocalised electrons rearrange themselves around these ions and re-establish strong forces of attraction.

9 a i valence electrons not restricted to a region between two atoms
ii a regular three-dimensional arrangement of a very large number of positive ions or cations
iii the electrostatic attraction between a lattice of cations and delocalised electrons
b valence electrons

10 A metal wire contains an extended lattice of metal cations surrounded by a 'sea' of delocalised electrons. The electrons are charged and free to move and so can conduct electricity.

11 In a metal lattice, metal cations are in a regular three-dimensional arrangement and have a positive charge. The positive cations are surrounded by a mobile sea of delocalised electrons.

12 a any of the group 2 metals, e.g. magnesium
b Magnetic metals are found in the transition metals.

13 aluminium

14 The bubbles are hydrogen gas, which is produced when a reactive metal reacts with water.

15 The reaction on the left is more vigorous and the metal must be more reactive. Iron is a more reactive metal than silver and so iron must be on the left. Silver is less reactive than iron and so silver must be on the right.

16 Metal A is copper. Metal B is sodium. Metal C is aluminium.

17 a false b true c false d false e true

18 magnesium + oxygen → magnesium oxide

19 a copper and nickel; harder, more corrosion-resistant and a silver colour
b tin and lead; lower melting temperature
c gold, silver and copper; harder
d iron, nickel and chromium; resists rusting, stronger
e mercury and zinc (sometimes a little silver is added); harder, non-toxic

20 needle 2 < needle 3 < needle 1
Needle 2 has been quenched, producing a hard but brittle metal, so it is the least malleable.
Needle 3 has been tempered, producing a hard but more malleable metal.
Needle 1 has been annealed, producing a soft, malleable metal.

21 Steel is used instead of iron as this alloy is stronger, more flexible and resistant to corrosion. Heating the horseshoe during the process allows the worker to change the crystal structure through heat treatment. The final shaping and hammering is an example of work hardening, which aligns the crystals and increases strength.

22 nanowire

23 8.34×10^{-7} m is 834 nm. This particle would be too big to be classified as a nanoparticle.

24 The width of a nanowire is in the range of standard nanomaterials. This changes the properties of the metal atoms as they are not exposed to delocalised electrons in the same way as in bulk metals. The long length of a nanowire is not enough to give it electrons that behave in the same way as electrons in bulk metals.

25 a i iron (steel) or aluminium
ii Iron and steel are strong. Aluminium has a low density (light) and can be easily coloured.
iii Iron rusts easily. Aluminium is soft and lacks strength.
b i copper
ii It is a good conductor of electricity and is ductile.
c i gold, silver and platinum
ii They are non-reactive, malleable, ductile and lustrous.

26 A variety of answers is possible.

27 a aluminium Al; copper Cu; gold Au; iron Fe; silver Ag
b aluminium: group 13, period 3
copper: 1st transition series, period 4
gold: 3rd transition series, period 6
iron: 1st transition series, period 4
silver: 2nd transition series, period 5
c gold and silver
d copper, gold, iron and silver
e gold

28 a The positive ions are arranged in a regular, three-dimensional lattice.
b Stress corrosion cracking can occur between the crystal grains.

29 a Na: group 1, period 3
K: group 1, period 4
Ca: group 2, period 4
b Na: 2,8,1
K: 2,8,8,1
Ca: 2,8,8,2
c i The atoms of Na are smaller than those of K, so the delocalised valence electrons of Na are closer to the positive nuclear charge than those of K. The electrostatic forces of attraction between delocalised electrons and cations are stronger in Na, so Na requires more energy to overcome the metallic bonding to boil the metal.
ii Valence electrons are in the fourth shell in the atoms of both Ca and K. However, there are twice as many valence electrons in the atoms of Ca. Also, the charge on a calcium cation is +2 as opposed to +1 on the potassium cation. So the electrostatic forces of attraction between delocalised electrons and cations are stronger in Ca and so it requires more energy to overcome the metallic bonding to boil the metal.

30 Aluminium is extracted from its ore by electrolysis. There was no process available until 1886.

Chapter 5 Ionic bonding

5.1 Properties and structures of ionic compounds

1 B

2 a The diagram shows that in solid sodium chloride, the sodium and chloride ions are held in fixed positions in the crystal lattice and are not free to move and conduct electricity.
b Molten sodium chloride contains sodium and chloride ions that are free to move and, therefore, it can conduct electricity.

3 Aluminium is a metal. It has an electronic configuration of 2,8,3. It would lose three electrons to have a valence shell with eight electrons and therefore become a cation (positive ion).

4 The negative ions are slightly further away from each other than they are from the positive ions in the lattice and the attractive force of the oppositely charged ions outweighs the repulsive force of two positively charged or two negatively charged ions near each other.

5.2 Using the ionic bonding model to explain properties

1 The electrostatic forces of attraction between the positive and negative ions holding the lattice together are very strong and a lot of energy is required to break them apart.

2 a When hit with a hammer or hard object, the ions move within the lattice so that like-charged ions line up adjacent to each other.
b When like-charged particles are near each other, they repel due to electrostatic repulsion and this causes the ionic compound to shatter.

3 In solid form, the ionic compound forms a crystal lattice. This is a very strong structure as the strong electrostatic forces of attraction between the positively charged cations and negatively charged anions means that the ions are not free to move. For a substance to be able to conduct electricity, the particles not only need to be charged but also free to move. In solid form, these particles cannot move, but when heated so the ionic compound is now molten, they are able to conduct electricity.

4 salad; salami; saline; expressions such as 'salt of the Earth', 'take with a pinch of salt', 'worth one's salt'; and superstitions, such as throwing salt over one's left shoulder to keep away evil spirits

5.3 Formation of ionic compounds

WE 5.3.1 3Ca (2,8,8,2) + 2P (2,8,5) → $3Ca^{2+}$ (2,8,8) + $2P^{3-}$ (2,8,8)

1 Cations: calcium, aluminium. Anions: nitrogen, fluorine and phosphorus. Metals form cations and non-metals form anions. Metals have lower ionisation energies than non-metals so they tend to lose valence electrons more readily. Non-metals have higher electronegativity than metals, so they tend to gain electrons more readily.

2 a K (2,8,8,1) and F (2,7) → K^+ (2,8,8) and F^- (2,8)

b Mg (2,8,2) and S (2,8,6) → Mg^{2+} (2,8) and S^{2-} (2,8,8)

c Al (2,8,3) and F (2,7), F (2,7), F (2,7) → Al^{3+} (2,8) and F^- (2,8), F^- (2,8), F^- (2,8)

d Na (2,8,1), Na (2,8,1) and O (2,6) → Na^+ (2,8), Na^+ (2,8) and O^{2-} (2,8)

e Al (2,8,3), Al (2,8,3) and O (2,6), O (2,6), O (2,6) → Al^{3+} (2,8), Al^{3+} (2,8) and O^{2-} (2,8), O^{2-} (2,8), O^{2-} (2,8)

3 Group 2 metals have two electrons in their valence shell. They lose these two electrons and therefore become positively charged, as they still have the original number of protons but have lost two electrons. Cations are ions with a positive charge.

4 KCl; $CaCl_2$

5 a Na (2,8,1) + Cl (2,8,7) → Na^+ (2,8) + Cl^- (2,8,8)
b Mg (2,8,2) + O (2,6) → Mg^{2+} (2,8) + O^{2-} (2,8)
c 2Al (2,8,3) + 3S (2,8,6) → $2Al^{3+}$ (2,8) + $3S^{2-}$ (2,8,8)

5.4 Chemical formulae of simple ionic compounds

WE 5.4.1 BaF_2

1 B

2 a 2:1 b 1:3 c 3:2 d 1:1 e 1:2

3 a NaCl b KBr c $ZnCl_2$ d K_2O e $BaBr_2$
f AlI_3 g AgBr h ZnO i BaO j Al_2S_3

4 a potassium chloride b calcium oxide
c magnesium sulfide d potassium oxide
e sodium fluoride

5.5 Writing formulae of more complex ionic compounds

1 a Na_2CO_3 b $Ba(NO_3)_2$ c $Al(NO_3)_3$ d $Ca(OH)_2$ e $Zn(SO_4)_2$
f KOH g KNO_3 h $ZnCO_3$ i K_2SO_4 j $Ba(OH)_2$

2 a CsCl b Fe_2O_3 c CuO d $Cr_2(SO_4)_3$ e FeO
f Al_2O_3 g $Ca(NO_3)_2$

3 a magnesium hydroxide b sodium carbonate
c iron(II) sulfate d copper(II) sulfate
e barium nitrate f copper(I) sulfate
g iron(III) sulfate h ammonium nitrate
i sodium hydrogenphosphate

Chapter 5 review

1 a Assemble equipment to test conductivity. Add a globe to the circuit. When the electrodes are touching the solid magnesium chloride, the globe will not light up.
b Using the same equipment with molten sodium chloride, the globe will glow.
c If a crystal of sodium chloride was hit firmly with a hammer, it would shatter.

2 a both metallic and ionic lattices
b both metallic and ionic lattices
c both metallic and ionic lattices
d ionic lattices only
e both metallic and ionic lattices

3 top left: non-metals; gain
top right: metals: lose
bottom left: an ionic lattice
bottom right: electrostatic forces of attraction

4 a The electrostatic forces of attraction between the positive and negative ions are strong and will be overcome only at high temperatures.
b The strong electrostatic forces of attraction between the ions mean that a strong force is needed to break up the lattice, giving the ionic crystals the property of hardness. However, the crystal lattice will shatter when a strong force is applied, suddenly causing ions of like charge to become adjacent to each other and be repelled.
c In the solid state, the ions are not free to move. However, when the solid melts or dissolves in water, the ions are free to move and conduct electricity.

5 a Na^+, Cl^-; Mg^{2+}, O^{2-}
b MgO. More energy is required to overcome the stronger forces. The higher melting temperature therefore reflects the solid with stronger forces between particles.
c The strength of electrostatic attraction between ions will depend on the size of the ions and on their charge. The Mg^{2+} ion is slightly smaller than the Na^+ ion, and the O^{2-} ion is much smaller than the Cl^- ion. More importantly, the Mg^{2+} ion and the O^{2-} ion each have twice the charge of the Na^+ ion and the Cl^- ion. The attraction between the ions in MgO is therefore much stronger than in NaCl. Magnesium oxide therefore has a much higher melting temperature.

6 The strength of the forces remains unchanged, but the kinetic energy of the ions increases until the forces can no longer hold the ions in the solid lattice, and the lattice breaks up as the solid melts.

7 a Solid ionic compounds do not conduct electricity.
b Ionic compounds are hard.
c In solution, ionic compounds conduct electricity.

8 a

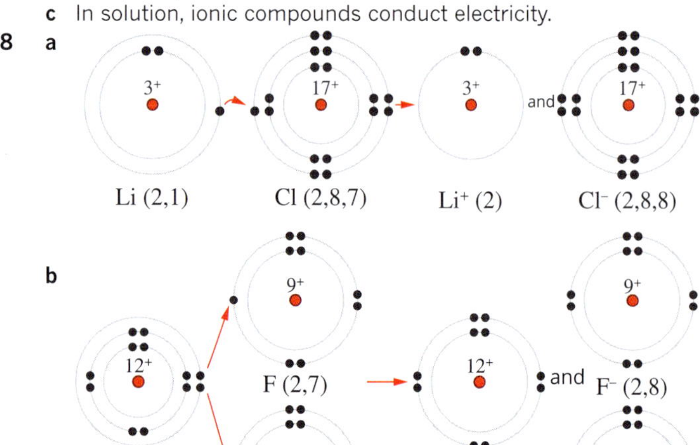

b

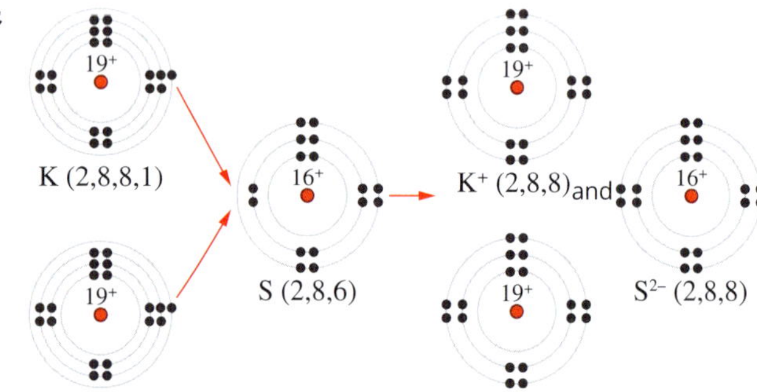

c

K (2,8,8,1)
K (2,8,8,1)

d

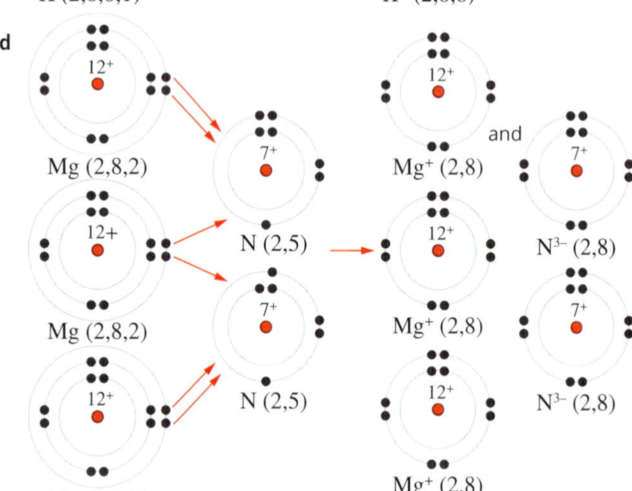

9 a 2,8 b 2,8 c 2,8 d 2,8

10 b CD_3 c EF d G_3H e KL

11 a $MgCl_2$ or MgF_2 b NaCl or CaS c Na_2O or K_2S
d Na_3N or Li_3N e $AlCl_3$ or AlF_3 f Mg_3N_2

12 They have seven electrons in their outer shell so only need to gain one electron to satisfy the octet rule. This means they become negative by gaining one electron.

13 a Mg (2,8,2) + 2Cl (2,8,7) → Mg^{2+} (2,8) + $2Cl^-$ (2,8,8)
b 2Al (2,8,3) + 3O (2,6) → $2Al^{3+}$ (2,8) + $3O^{2-}$ (2,8)

14 a KBr. Potassium ion has a charge of +1, bromide ion has −1.
b MgI_2. Magnesium ion has a charge of +2, iodide ion has −1.
c CaO. Calcium ion has a charge of +2, oxide ion has −2.
d AlF_3. Aluminium ion has a charge of +3, fluoride ion has −1.
e Ca_3N_2. Calcium ion has a charge of +2, nitride ion has −3.

15 a CuCl b Ag_2O c Li_3N d KI

16 The subscripts represent the ratio of metal to non-metal ions in the ionic compound.

17 a i −3 ii +1 iii −2
b i Y_2SO_4 ii K_2Z iii Y_3X iv Y_2Z

18 a $CuNO_3$ b CrF_2 c K_2CO_3
d $Mg(HCO_3)_2$ e $Ni_3(PO_4)_2$

19 a ammonium carbonate
b copper(II) nitrate
c chromium(III) bromide

20 a agree b agree c agree
d agree e disagree

21 Possible answer:

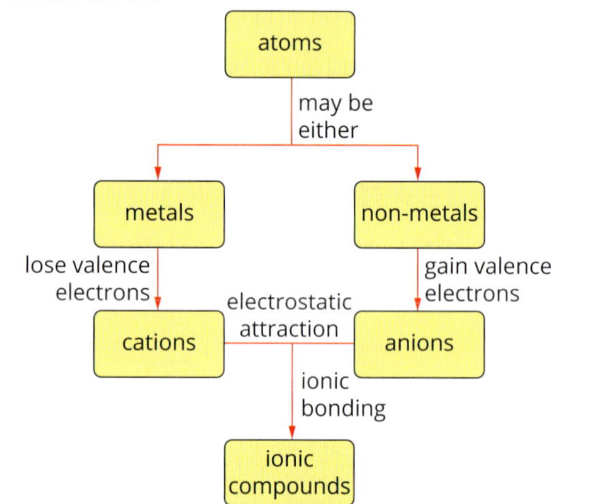

Chapter 6 Materials made of molecules

6.1 Properties of non-metallic substances

1 Non-metallic elements: H_2 and Br_2 as they consist of only one type of atom. Non-metallic compounds: NO_2 as it consists of more than one type of atom (nitrogen and oxygen).

2 A molecule is a discrete group of atoms of known formula, bonded together.

3 a Non-metals do not conduct electricity because they do not contain free-moving charged particles (neither delocalised electrons nor ions).
b Non-metals have low melting and boiling temperatures because they have weak intermolecular forces between molecules.

4 When sugar turns to a liquid, it is melting; the intermolecular forces between sugar molecules are broken. When the liquid turns black and a gas is produced, the bonds between atoms in the sugar molecules are broken, allowing new substances to be produced.

6.2 Covalent bonding

WE 6.2.1 H : N : H (H above and below)

1 a one b three c two d one

2 a fluorine (F_2)

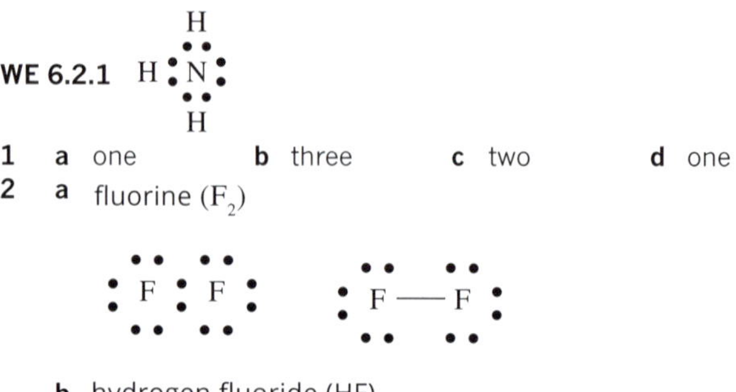

b hydrogen fluoride (HF)

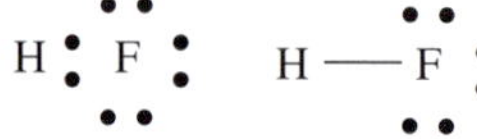

c water (H_2O)

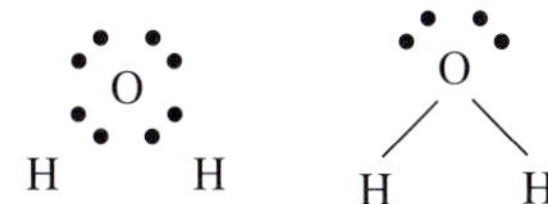

d tetrachloromethane (CCl_4)

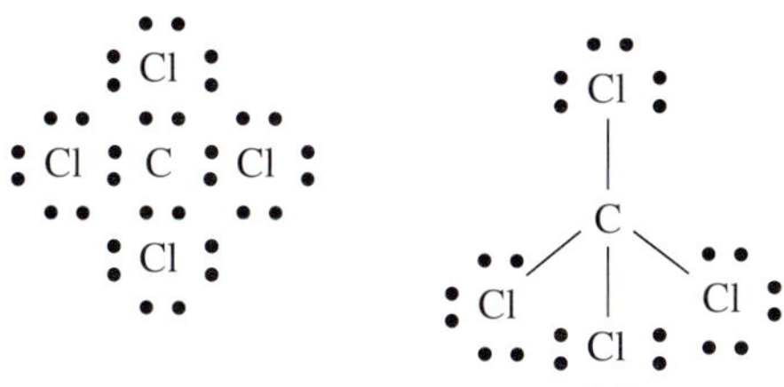

e phosphine (PH_3)

H : P : H
H

P
H H H

f butane (C_4H_{10})

H H H H
H : C : C : C : C : H
H H H H

H H H H H C C C C H H H H H

g carbon dioxide (CO_2)

O :: C :: O   O = C = O

3 a one b two c three d four
e one f zero

4 To complete its valence shell, the oxygen atom uses two of its valence-shell electrons to form two single bonds or a double bond with suitable non-metal atoms. The remaining four electrons in the valence shell are not required for bonding, as the valence shell is now complete, and they arrange themselves as two lone pairs around the oxygen atom.

5 a CCl_4 b NBr_3 c SiO_2 d HF e PF_3

6 a true b false c true d false

7 a ball-and-stick model
b Lewis structure (or electron dot diagram)
c space-filling model

Chapter 6 review

1 The strength of the intermolecular forces in pure hydrogen chloride must be relatively weak. Since the pure hydrogen chloride exists as a gas at room temperature, it must have a low boiling temperature, which indicates that not much energy is required to break the intermolecular forces between molecules.

2 Non-metallic substances do not conduct electricity.

3 Bonds are the forces that hold the atoms within a molecule together. In carbon dioxide molecules they are the bonds between the carbon atoms and the oxygen atoms. Intermolecular forces are the forces between one molecule and its neighbouring molecules. It is the intermolecular forces that are broken when carbon dioxide sublimes.

4 For an odour to be detected, gas molecules need to enter our nose. Because covalent molecular substances have weak forces of attraction between their molecules some molecules at the surface of a solid or liquid will have enough energy to leave the surface and become gaseous.

5 A is naphthalene. It is a soft, covalent molecular compound.
B is magnesium chloride. As it is a conductor in aqueous solution, brittle and has a high melting point it is an ionic compound.
C is copper. As it is not soluble in water, has a high melting point and is hard it is a metal.
D is phosphorus trihydride. As it is a non-conductor in aqueous solution, has a low melting point and is soft as a solid, it is a covalent molecular compound.

6 D

7 a six b three
c four when both atoms are counted

8 B 9 XY_2

10 X will have seven valence electrons. If X has seven valence electrons, each X atom can share one electron with 1 of the four Si valence electrons so giving Si its octet. As well, each X atom will have an octet.

11 Neon will not form bonds to other atoms as it has a stable valence shell.

12 a zero b one c three d four

13 a Two bonding electrons, six non-bonding electrons H : Br :

b Four bonding electrons, four non-bonding electrons H : O : H

c Eight bonding electrons, 24 non-bonding electrons

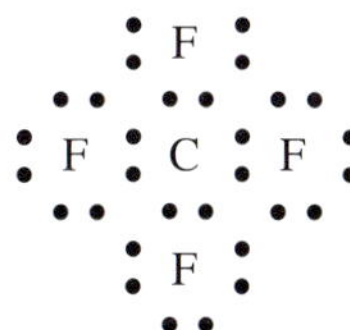

d 14 bonding electrons, zero non-bonding electrons

H H
H : C : C : H
H H

e Six bonding electrons, 20 non-bonding electrons

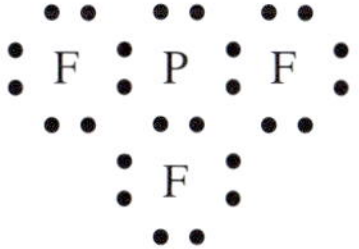

f Four bonding electrons, 16 non-bonding electrons

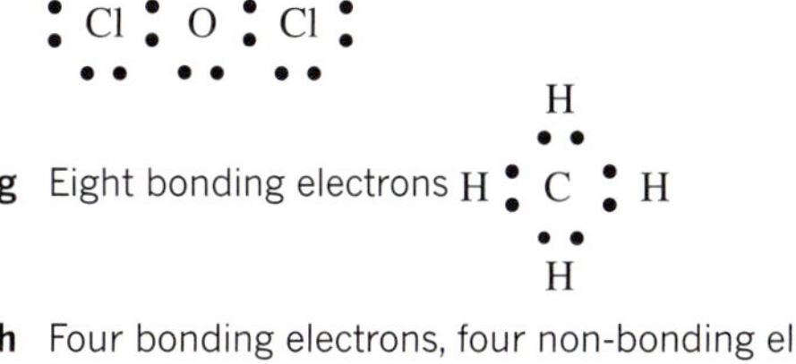

g Eight bonding electrons

h Four bonding electrons, four non-bonding electrons : S : H
H

14 a The bonds are similar in that they all involve the sharing of electron pairs between two atoms; that is, they are covalent bonds.

b They differ in the number of electron pairs shared: one pair (fluorine), two pairs (oxygen) and three pairs (nitrogen).

15 a

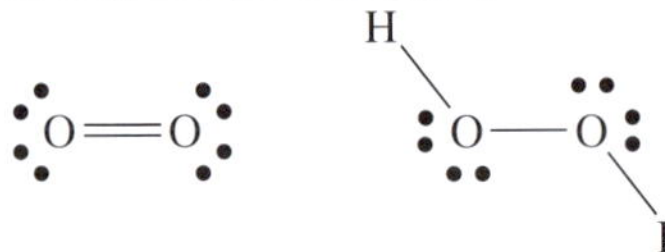

b In a single bond there are two electrons that are shared between the atoms, whereas in a double bond there are four electrons. The net attraction is stronger and more energy is needed to break the double bond than the single bond.

c Oxygen has six valence-shell electrons and completes its valence shell by forming two single covalent bonds or a double bond with another non-metal. There is no need for the oxygen to form a triple bond or three single covalent bonds as this exceeds its requirement of two electrons to complete its valence shell.

16 Metallic: Ag, Cu. Ionic: $CuCl_2$, CaS. Molecular: NH_3, HCl, H_2O. Ag and Cu are both metals. $CuCl_2$ and CaS are both formed from a combination of a metal and non-metal(s). NH_3, HCl, H_2O are compounds made from non-metals.

17 Dissolve a small quantity of both substances in water and test for electrical conductivity. The solution that conducts electricity will be calcium chloride and the non-conductor will be glucose.

18 The nuclei of the two atoms involved in the bond each have an (electrostatic) attraction for the shared electrons due to their opposite charges. This holds the atoms together in the molecule.

19 a true b true c false d true
e false f true

Chapter 7 Carbon

Section 7.1 Carbon lattices

1 It can form four covalent bonds. Carbon atoms can form single, double and triple covalent bonds. Carbon atoms can bond to each other.

2 a to turn from a solid directly into a gas

b Diamond and graphite contain extended networks of strong covalent bonds, which must be overcome to allow the material to sublime.

3 a Diamond is hard because it has strong covalent bonds throughout the lattice, with all atoms being held in fixed positions. Graphite is soft because there are weak dispersion forces between the layers in graphite, so layers can be made to slide over each other easily.

b Diamond is a non-conductor of electricity because all of its electrons are localised in covalent bonds and are not free to move.
Graphite is able to conduct electricity because it has delocalised electrons between its layers of carbon atoms.

4 a Graphite is used as a dry lubricant because the dispersion forces between the layers in graphite enable the layers to slide over each other easily and to reduce the friction between moving parts.

b The strong covalent bonding throughout the lattice means that the carbon atoms are fixed in place. This makes diamond very hard and suitable as a material for cutting softer materials.

7.2 Carbon nanomaterials

1 Fullerenes are similar to graphite in that they are allotropes of carbon in which each carbon atom has bonds to 3 other carbon atoms. Fullerenes conduct electricity and heat, similar to graphite. Fullerenes differ from graphite in that they are nanomaterials, which give them different physical properties.

2 Each carbon atom in a buckyball is covalently bonded to three other carbon atoms. Each carbon atom has one free electron, which is shared between the other carbon atoms in the sphere. The structure consists of hexagonal and pentagonal rings of atoms.

3 Each carbon atom in a graphene sheet is covalently bonded to 3 other carbon atoms. Each carbon atom has one free electron that is shared with the other carbon atoms in the sheet. The structure consists of hexagonal rings of atoms. The sheet is one atom thick but may be of any length and width.

4 Carbon nanotubes are immensely strong for their mass. They can theoretically be produced with any length, allowing for very long cables to be produced. A space elevator would require an extremely strong, long and uniform cable to be produced.

Chapter 7 review

1 Carbon exists in different forms with different arrangements of atoms.

2 Refer to Figure 7.1.10. Carbon atoms bond covalently to three other carbon atoms in graphite to form layers. The strong covalent bonds between the atoms in each layer explain graphite's resistance to heat and, hence, why graphite sublimes at high temperature. These layers consist of hexagonal rings connected to each other. The fourth electron in each carbon atom is delocalised, which explains its electrical conductivity. There are weak dispersion forces between layers of graphite, allowing the layers to slide over each other and enabling it to act as a lubricant.

3 Charcoal has many pores and pockets giving it a very large surface area. The surface of charcoal may form dispersion forces with the coloured contaminants. The coloured contaminants may also be trapped in the internal pores of the charcoal.

4 Graphite can be used under very low temperature conditions because it does not freeze (unlike the potential for oil-based lubricants). It can be used under very high temperature conditions because it has a very high sublimation temperature.

5 Diamond consists of an extended three-dimensional network of strong covalent bonds, which must be overcome for sublimation to occur.

6 Silicon dioxide forms a covalent network lattice with each silicon atom bonded tetrahedrally to four oxygen atoms and each oxygen atom bonded to two silicon atoms. The high melting point and hardness are due to these strong covalent bonds. Its non-conductivity is due to the absence of any mobile charged particles in the structure. All valence electrons are involved in forming covalent bonds and so are not free to move through the structure.

7 a Methane is an example of a molecular substance. It has strong, covalent intramolecular bonds and very weak intermolecular bonds. Diamond is an example of a covalent network lattice. It has strong covalent bonds throughout its structure.

b Due to the weak dispersion forces between molecules, methane will have extremely low boiling and melting points. If it were a solid it would be crystalline, brittle and soft. Due to its extended covalent lattice, diamond is extremely hard, does not exist as a liquid and has a very high sublimation point.

8 Carbon atoms in diamond have a tetrahedral bond geometry; carbon atoms in graphene are arranged with one atom at the centre and three atoms at the corners of an equilateral triangle.

9 It is composed of single layers and the covalent bonding in each layer is very strong.

10 Fullerenes are an allotrope of carbon and carbon atoms have four valence electrons. In fullerenes, each carbon atom forms single covalent bonds with three neighbouring carbon atoms. This leaves one delocalised valence electron in each carbon atom that can move throughout the fullerene, allowing it to conduct electricity.

11 A carbon nanotube is like a cylinder of graphene with half a buckyball on each end.

12 Carbon nanotubes are strong due to their chain of unbroken covalent carbon–carbon bonds. As well, each nanotube is a single large molecule which means it does not have any weak points such as boundaries between crystalline grains.

13 Carbon nanotubes have a very high surface area. Thus coating a nanotube with a catalyst would give a very high surface area of catalyst for reactants to come into contact with.

14 O_2 and O_3; S_6 and S_8

15

allotrope name	diamond	graphite	carbon nanotube	graphene
bonding feature	Each carbon is bonded to four others in a network lattice structure.	Each carbon is bonded to three others in a layered structure.	Each carbon is bonded to three others in a cylindrical shape.	Each carbon is bonded to three others in a layer one atom thick.
properties	hard, brittle	conducts electricity, soft	very strong, conducts electricity	very strong, conducts electricity

16 Common structural feature: layered structure. Both contain layers of carbon atoms with each carbon bonded to three neighbouring atoms, with three covalent bonds. Each carbon atom still has another valence electron that is delocalised in the layer, explaining the electrical conductivity of both allotropes. The bonding between the atoms in the layer is strong. Graphene is just a single layer but graphite contains layer upon layer.
Properties: Graphene is very strong. There are dispersion forces between layers in graphite, allowing the layers to slide past each other. Graphite can therefore act as a lubricant. Both allotropes have a high sublimation point.

17 **a** An electric circuit is used that contains a light globe and a power supply. When the power supply is switched on, the light globe will glow if the circuit is complete. The diamond must conduct electricity for the circuit to be complete and the light globe to glow. For this to occur, the diamond must contain charged particles that are free to move. Since the light globe does not glow, the diamond is not an electrical conductor and therefore does not contain free-moving charged particles.
b Graphite, graphene and nanotubes all conduct electricity. Diamond does not.

18 Germanium, like carbon, is a group 14 element so has four valence electrons. This enables germanium atoms to bond to other germanium atoms to form covalent network bonds in the same way as carbon.

19 **a** Silicon and carbon are both group 14 elements so have four valence electrons. This enables them to form four covalent bonds to other carbon and/or silicon atoms to form a covalent network lattice.
b Silicon carbide would be expected to have a high melting point, to be hard, insoluble in water and most solvents, and a non-conductor of electricity.

20 All models have mobile charged particles enabling them to conduct an electrical current. In an aqueous solution of sodium chloride, it is sodium ions and chloride ions that carry the charge, while in graphite and solid copper it is delocalised valence electrons. In solid copper, the valence electrons are not fixed to one atom but can move freely within the structure of the copper. The copper atoms are described as existing in a 'sea' of delocalised valence electrons. In graphite, each carbon atom has three of its four valence electrons covalently bonded to three other carbon atoms with the fourth valence electron delocalised, enabling an electric current to be carried by these delocalised electrons.

Chapter 8 Organic compounds

8.1 Alkanes

WE 8.1.1 3-methylpentane

1 A, D

2 **a** CH_4
b It is a compound of carbon and hydrogen.
c Carbon has the electron configuration of 2,4. Each carbon atom needs four electrons to complete its outer shell. Hydrogen has an electron configuration of 1. Each hydrogen atom needs one electron to complete its outer shell. Because both atoms need electrons, they will share electrons, that is, form covalent bonds. It will take four hydrogen atoms to provide the four electrons required by each carbon atom.
d

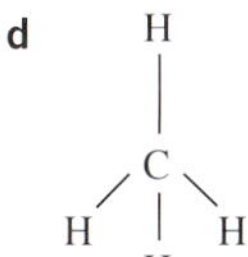

The hydrogen atoms are arranged around the central carbon atom in a tetrahedral configuration. This tetrahedral arrangement gives minimum electrostatic repulsion between the four pairs of bonding electrons.

3 **a** propane
b

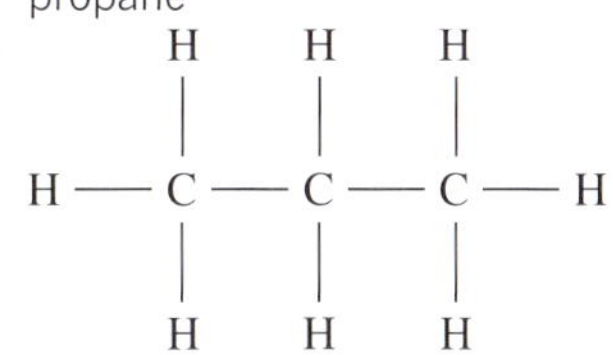

c $CH_3CH_2CH_3$

4 **a** butane **b** heptane
c 2-methylpentane **d** 2-methylbutane
e 2,4-dimethylpentane

5 **a** butane **b** methylpropane
c 2,4-dimethylpentane **d** 3-methylhexane

6 **a** hexane

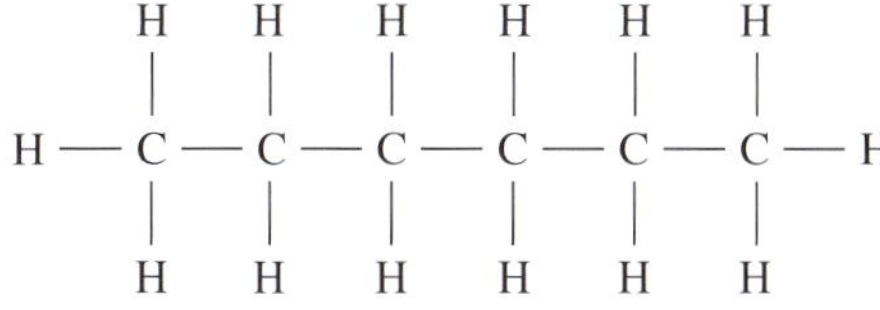

b 3-methylhexane

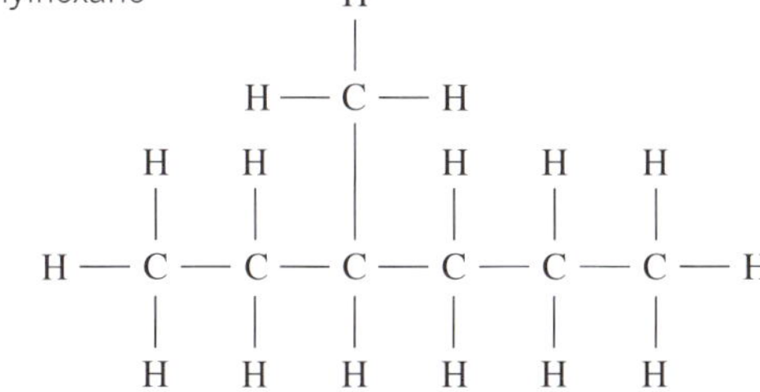

c 3,3-dimethylpentane

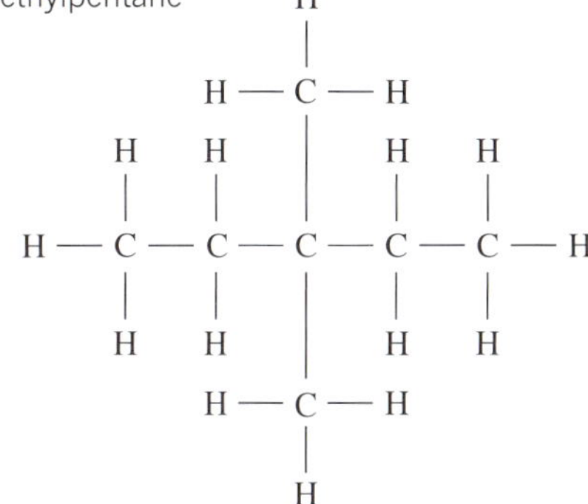

d 3-ethyl-2-methylpentane

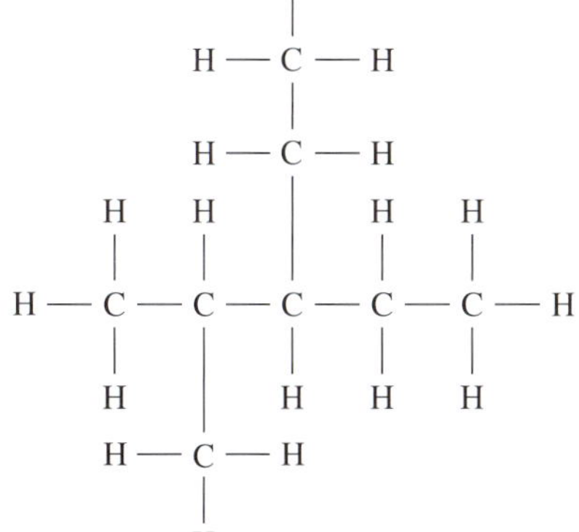

8.2 Alkenes

WE 8.2.1 2,3-dimethylbut-1-ene

1 ethene (C_2H_4), propene (C_3H_6), methylpropene (C_4H_8), pentene (C_5H_{10}), octene (C_8H_{16})

2 **a** four **b** eight
c Methylpropene is an isomer of butene (C_4H_8).

3 B

4 **a** 4-methylpent-1-ene **b** pent-2-ene
c 2-methylbut-1-ene

5 **a**

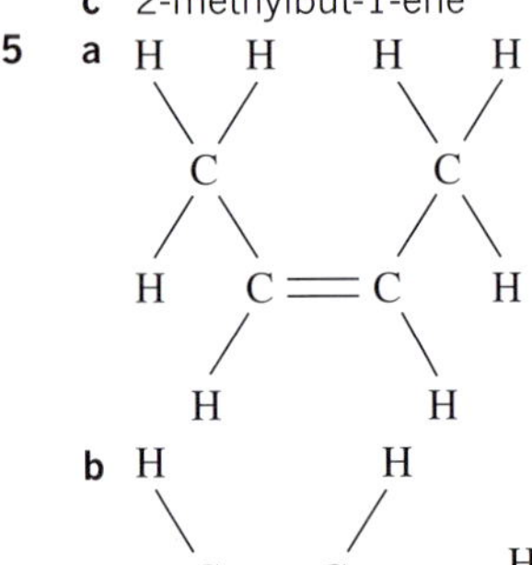

b

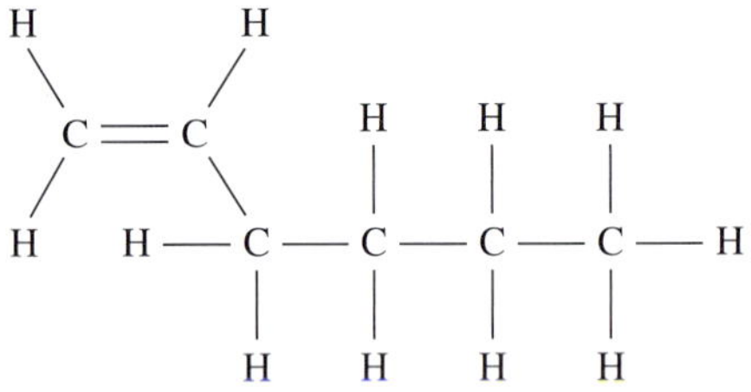

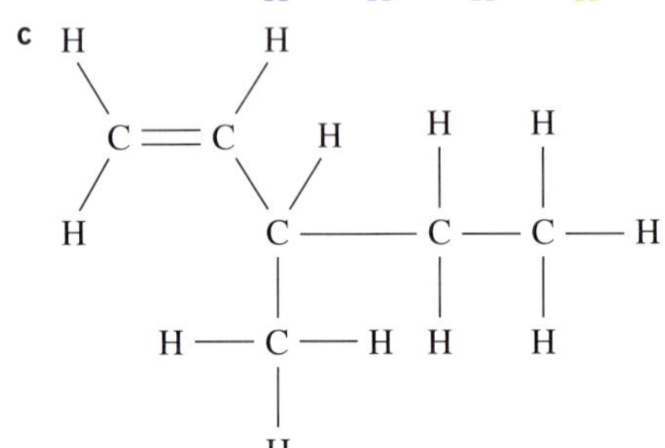

c

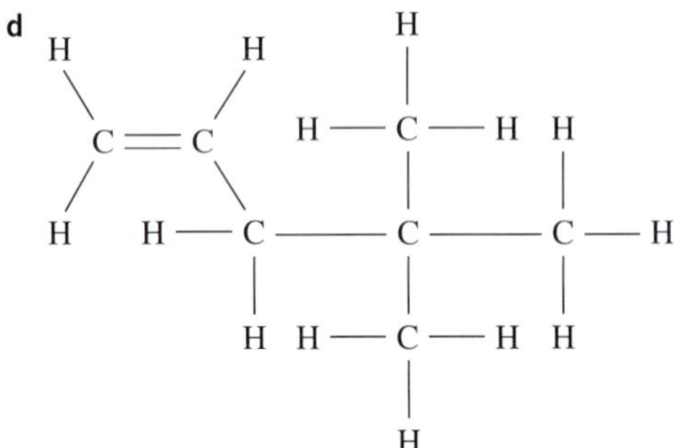

d

8.3 Benzene

1 D 2 C

8.4 Reactions of hydrocarbons

1 B

2 $CH_4 + Br_2 \rightarrow CH_3Br + HBr$

3 $CH_2CHCH_2CH_3 + Cl_2 \rightarrow CH_2ClCHClCH_2CH_3$

4 $C_6H_6 + Br_2 \xrightarrow{AlBr_3} C_6H_5Br + HBr$

5 B

Chapter 8 review

1 Carbon atoms can use two electrons each to bond to form very long chains. This leaves the other two valence electrons able to bond other non-metal atoms onto the chain. In addition, there can be double and triple bonds between carbon atoms as well as ring structures.

2 A

3 **a** alkanes **b** $C_{17}H_{36}$ **c** $C_{15}H_{32}$

4 **a** ethane

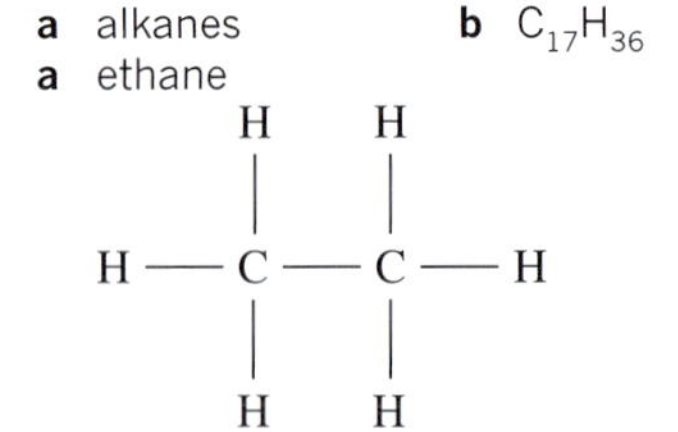

b methylpropane

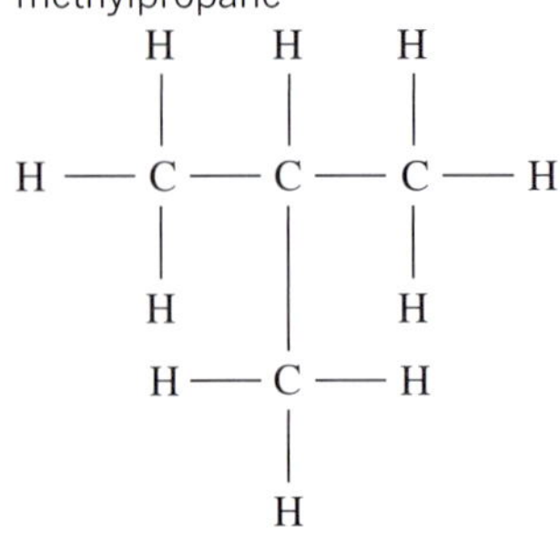

c 3-methylpentane

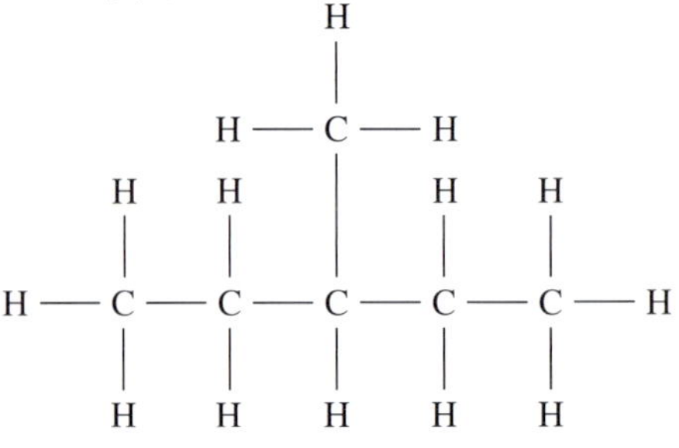

d pentane

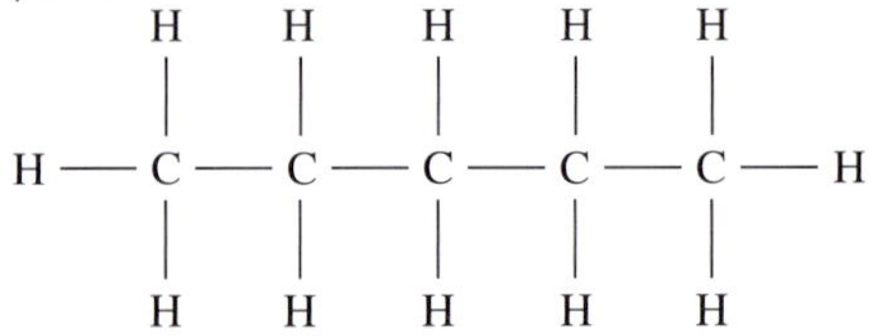

5 **a**

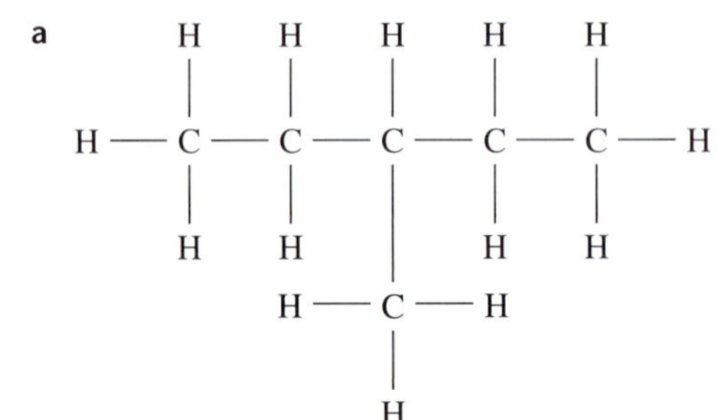

b

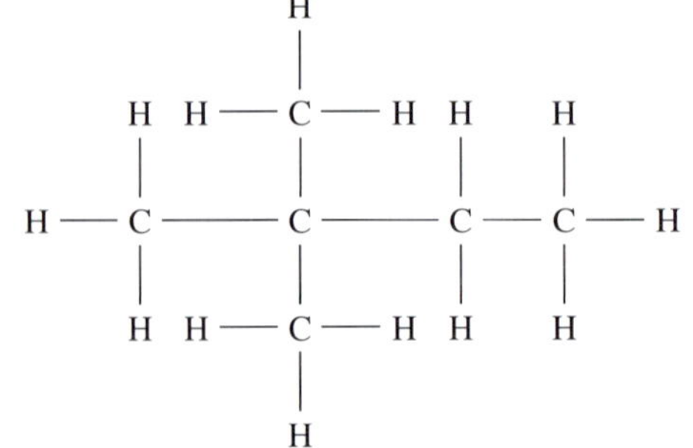

c

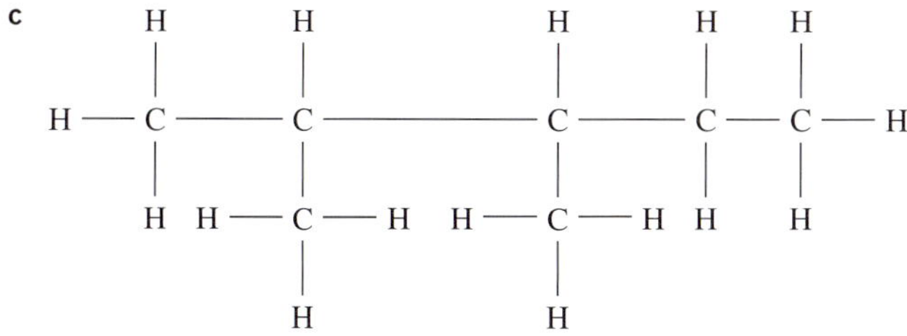

6 Hexane molecules are larger than propane molecules, so hexane has a higher boiling point than propane.

7 a alkane b alkene c alkane d alkene e alkane

8

but-1-ene

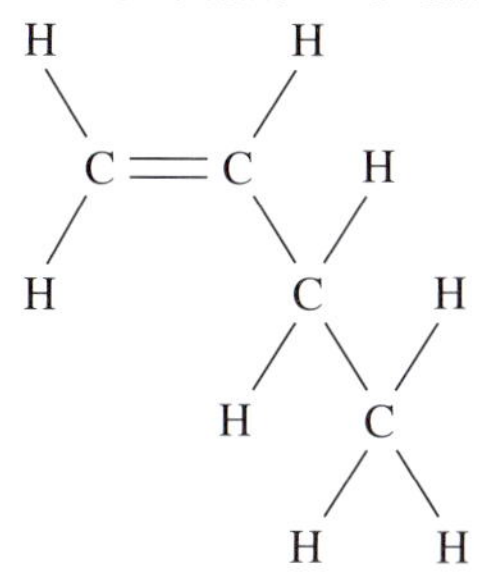

cis-but-2-ene

trans-but-2-ene

methylpropene

9 a *cis*-hex-2-ene b 4-methylpent-2-ene c propene

10

The Kekulé structure of benzene depicts the six-carbon ring as containing alternating double and single bonds. This is incorrect as experimental results show that each carbon–carbon bond is in fact of equal length. Furthermore, benzene does not readily undergo addition reactions, as it would if it did contain three double bonds. Instead, it has a similar reactivity to saturated compounds. The benzene ring in fact contains six single bonds and six delocalised electrons (one per atom).

11 B

12 a $CH_4(g) + 2O_2(g) \rightarrow CO_2(g) + 2H_2O(g)$
b $2C_6H_{14}(l) + 19O_2(g) \rightarrow 12CO_2(g) + 14H_2O(g)$
c $2C_4H_{10}(g) + 13O_2(g) \rightarrow 8CO_2(g) + 10H_2O(g)$
d $C_{31}H_{64}(s) + 47O_2(g) \rightarrow 31CO_2(g) + 32H_2O(g)$

13 a $C_2H_4(g) + 3O_2(g) \rightarrow 2CO_2(g) + 2H_2O(g)$
b $C_3H_6(g) + Br_2(aq) \rightarrow C_3H_6Br_2(aq)$
c $C_4H_8(g) + H_2O(l) \rightarrow C_4H_9OH(l)$
d $C_2H_4(g) + HCl(g) \rightarrow C_2H_5Cl(l)$

14 You could distinguish between them by adding a few drops of orange bromine water to each. The hex-2-ene will rapidly decolourise the bromine water, whereas the hexane will not readily react.

15 Substitution: $C_2H_6(g) + Cl_2(g) \rightarrow C_2H_5Cl(g) + HCl(g)$
Addition: $C_2H_4(g) + HCl(aq) \rightarrow C_2H_5Cl(g)$

16 10 Br_2 molecules

17 $C_2H_4(g) + H_2O(g) \rightarrow CH_3CH_2OH(g)$

18 a An alkene contains one double carbon–carbon bond, which requires two carbon atoms. The first alkene is therefore ethene.
b The carbon atom has four electrons in the outer shell, which are available for sharing with other atoms to produce four covalent bonds.

19 Polyunsaturated: contains many double bonds; monounsaturated: contains one double bond; saturated: contains only single bonds between carbon atoms.

20 The compound would be an alkene, as it has the general formula C_nH_{2n}. Alkenes are unsaturated and react readily with bromine water via an addition reaction. The orange bromine water would rapidly decolourise.

Chapter 9 The mole

9.1 Masses of particles

WE 9.1.1 63.02 **WE 9.1.2** 187.6

1 a 98.09 b 17.03 c 30.07

2 a 74.55 b 106.0 c 342.2

9.2 Introducing the mole

WE 9.2.1 1.5 mol **WE 9.2.2** 9.6×10^{23} molecules

WE 9.2.3 8.4×10^{23} atoms **WE 9.2.4** 0.0013 mol or 1.3×10^{-3} mol

1 a 1.2×10^{24} atoms b 6.0×10^{22} molecules
c 1.20×10^{25} atoms d 2.5×10^{24} molecules
e 6.02×10^{21} atoms f 2.78×10^{19} molecules

2 a 0.50 mol b 0.25 mol c 70 mol d 70 mol

3 a 1.7×10^{-4} mol b 1.7×10^{-4} mol
c 1.7×10^{-4} mol

4 a 0.8 mol b 4.8 mol c 0.72 mol d 6.0 mol

9.3 Molar mass

WE 9.3.1 496 g **WE 9.3.2** 7.4×10^{21} molecules

1 a 28.0 b 17.0 c 98.1 d 241.9
e 60.1 f 32.1 g 176.1 h 249.7

2 a 23.0 g b 64.0 g c 1.60 g d 25.5 g

3 a 4.96 mol b 2.48 mol c 0.10 mol d 0.025 mol
e 0.0031 mol f 0.0063 mol g 9.7×10^{-6} mol
h 3.87×10^{-5} mol

4 a 6.02×10^{23} atoms b 6.03×10^{22} atoms
c 6.02×10^{21} atoms d 3.05×10^{22} atoms

5 a i 3.01×10^{23} molecules ii 6.02×10^{22} molecules
b 6.02×10^{22} atoms. c 4.07×10^{25} atoms

9.4 Percentage composition

WE 9.4.1 35.0%

1 a 69.9% b 84.8% c 26.2% d 51.2%

2 a urea: percentage nitrogen = 46.6%
b ammonium nitrate: percentage nitrogen = 35.0%
c ammonium sulfate: percentage nitrogen = 21.2%
Therefore urea has the highest nitrogen content.

3 a silicon carbide (SiC): Si = 70.0%, C = 30.0%
b water (H_2O): H = 11.2%, O = 88.8%
c potassium phosphate (K_3PO_4): K = 55.3%, P = 14.6%, O = 30.1%
d sucrose ($C_{12}H_{22}O_{11}$): C = 42.1%, H = 6.50%, O = 51.4%
e gold sulfate ($Au_2(SO_4)_3$): Au = 57.8%, S = 14.1%, O = 28.1%

Chapter 9 review

1 a 18.02 b 123.9 c 28.01

2 a 225.2 b 171.3 c 291.7

3 a 0.747 mol b 14.9 mol c 3.8×10^4 mol
d 1.7×10^{-24} mol

4 a i 8.73×10^{23} molecules ii 3.49×10^{24} atoms
b i 3.47×10^{23} molecules ii 1.04×10^{24} atoms
c i 9.21×10^{21} molecules ii 4.61×10^{22} atoms
d i 1.5×10^{24} molecules ii 6.8×10^{25} atoms

5 The molar mass, M, has the same numerical value as the relative molecular mass, M_r, which is the sum of the relative atomic masses, A_r, of the elements in the compound. The molar mass, M, is the actual mass of 1 mole and so has the unit g mol^{-1}.

6 a Fe : 55.85 g mol^{-1} b H_2SO_4 : 98.09 g mol^{-1}
c Na_2O : 61.98 g mol^{-1} d $Zn(NO_3)_2$:189.4 g mol^{-1}
e H_2NCH_2COOH : 75.07 g mol^{-1}
f $Al_2(SO_4)_3$: 342.2 g mol^{-1}
g $FeCl_3{\cdot}6H_2O$: 270.3 g mol^{-1}

7 a 1.8 g b 58 g c 0.41 g d 389 g

8 a 0.10 mol b 0.0390 mol c 1.25 mol
d 0.00167 mol e 3.4×10^4 mol

9 a 6.66×10^{-23} g b 2.99×10^{-23} g c 7.31×10^{-23} g

10 a i 0.034 mol ii 2.05×10^{22} molecules
iii 8.2×10^{22} atoms
b i 0.292 mol ii 1.76×10^{23} molecules
iii 1.41×10^{24} atoms
c i 0.0088 mol ii 5.3×10^{21} molecules
iii 1.1×10^{22} atoms
d i 1.22×10^{-4} mol ii 7.3×10^{19} molecules
iii 1.8×10^{21} atoms

11 62.0 g

12 a i $n(NaCl) = 0.100$ mol
ii $n(Na^+) = 0.100$ mol; $n(Cl^-) = 0.100$ mol
b i $n(CaCl_2) = 0.405$ mol
ii $n(Ca^{2+}) = 0.405$ mol; $n(Cl^-) = 0.810$ mol
c i $n(Fe_2(SO_4)_3) = 0.00420$ mol
ii $n(Fe^{3+}) = 0.00840$ mol; $n(SO_4^{2-}) = 0.0126$ mol

13 a 144 g mol^{-1} b 100 g mol^{-1}

14 a 40 g mol^{-1} b 98 g mol^{-1}
c 44 g mol^{-1} d 106 g mol^{-1}

15 B

16 a 1.25×10^4 g mol^{-1} b 1.6×10^{-7} mol
c 9.6×10^{16} molecules

17 a %(Al) = 52.9%
%(O) = 47.1%
b Cu 65.1%; O 32.8%; H 2.1%
c %(Mg) = 12.0%
%(Cl) = 34.9%
%(H) = 5.9%
%(O) = 47.2%
d Fe 27.9%; S 24.1%; O 48.0%
e H 1.0%; Cl 35.3%; O 63.7%

18 a %C = 93.7% b %C = 40.0%
c %C = 19.9% d %C = 60.0%

19 a 3.0×10^{24} water molecules b 4.0 mol
c 4.04×10^{-23} g d 2.99×10^{-23} g

20 a i 2 mol Au ii 3 mol S iii 12 mol O
b 682.2 g c 0.403 mol
d 450 g e 57.8%

21 a 295.7 g b 0.101 mol c 36.5%

22 Kr

23 a 35.2 g
b Some gases, such as neon, are monatomic (composed of single atoms), and some gases, such as oxygen, are diatomic (formed from two atoms). This will cause results to be incorrect if the scientists did not know whether the gases being tested were monatomic or diatomic.

Chapter 10 Energy changes in chemical reactions

10.1 Exothermic and endothermic reactions

1 D

2 a 180 kJ
b 1.5×10^3 kJ
c 0.0100 kJ or 1.00×10^{-2} kJ
d 2.0×10^{-6} kJ

3 In chemistry, the system is usually the chemical reaction, whereas the surroundings refer to everything else; for example, the beaker or test-tube in which the reaction takes place.

4 In any reaction, the total amount of chemical energy of the reactants is made up of the bonds between atoms within the reactants. If the total amount of chemical energy within the reactants is less than the total amount of chemical energy within the products, energy must be supplied to the system, and the reaction is said to be endothermic.

10.2 Thermochemical equations, energy profile diagrams and enthalpy

WE 10.2.1 $6CO_2(g) + 6H_2O(g) + 2803 \text{ kJ mol}^{-1} \rightarrow C_6H_{12}O_6(aq) + 6O_2(g)$

WE 10.2.1 $4C_3H_5N_3O_9(l) \rightarrow 12CO_2(g) + 10H_2O(g) + 6N_2(g) + O_2(g)$; $\Delta H = -1456 \text{ kJ mol}^{-1}$

1 A negative ΔH value indicates a reaction is *exothermic*. This is because the enthalpy of the reactants is *greater than* the enthalpy of the products. Energy is being *released to* the surroundings.

2 $CH_4(g) + 2O_2(g) \rightarrow CO_2(g) + 2H_2O(l)$ $\Delta H = -890 \text{ kJ mol}^{-1}$

3 It would be lower because the change of state of the H_2O from liquid to gas will require energy to be absorbed.

4 a endothermic
b The total enthalpy of the product (HI) is greater than that of the reactants (hydrogen gas and iodine gas).
c The activation energy is greater than the ΔH value.

5 $6CO_2(g) + 6H_2O(l) \rightarrow C_6H_{12}O_6(aq) + 6O_2(g)$ $\Delta H = +2803 \text{ kJ mol}^{-1}$

Chapter 10 review

1 a 2.21 kJ b 152 J c 1.89 MJ d 12.5 kJ

2 a Exothermic, because heat and light energy are released to the surrounding environment by the combustion of wood.
b Endothermic, because thermal energy is absorbed from the surrounding environment to melt the ice.
c Endothermic, because electrical energy is consumed from a power supply as the battery is recharged.
d Exothermic, because heat energy is released to the surrounding environment as organisms in the compost heap decompose the plant material. The temperature of the heap rises as a consequence.

3 During chemiluminescence, energy is given off mainly in the form of light. Although the temperature of the surroundings may not increase, energy is still being lost to the surroundings, and the enthalpy of the chemicals decreases. Therefore it is an exothermic process, with a negative enthalpy change.

4 C

5 **a** true **b** false **c** false **d** true

6 If a chemical equation is written for an endothermic reaction, ΔH is positive, telling you that energy is absorbed as the reaction proceeds. The enthalpy of the products must be higher than the enthalpy of the reactants.
If this reaction is reversed, the enthalpy of the reactants is now higher than the enthalpy of the products. The reaction releases energy as the reaction proceeds, so is exothermic. ΔH becomes negative.

7 **a** The energy in the bonds of the reactants is higher than the energy in the bonds of the products.

b

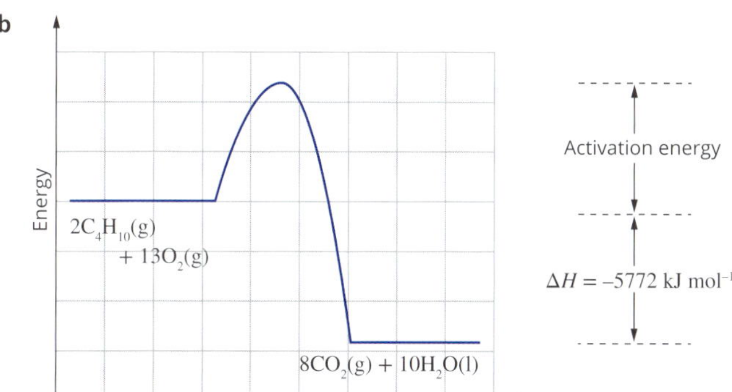

8 $H_2(g) + \frac{1}{2}O_2(g) \rightarrow H_2O(l)$ $\Delta H = -286$ kJ mol^{-1} or
$2H_2(g) + O_2(g) \rightarrow 2H_2O(l)$ $\Delta H = -572$ kJ mol^{-1}

9 **a** 1 mol of CO(g) and 0.5 mol of $O_2(g)$
b i $\Delta H = -566$ kJ mol^{-1} **ii** $\Delta H = +566$ kJ mol^{-1}.

Chapter 11 Fuels and introduction to stoichiometry

11.1 Types of fuels

1 A non-renewable fuel cannot be replenished at the rate at which it is consumed. Renewable fuels are those that can be replenished at a rate similar to that at which they are consumed.

2 Renewable: bioethanol, biogas, biodiesel. Non-renewable: coal, oil, LPG, natural gas, coal seam gas.

3 Coal has the longest expected lifespan based on reserves and current rates of use. Carbon dioxide emission restrictions may reduce long-term dependence on coal.

4 **a** The rate of global energy use is more than can be supplied by wood. Wood has a relatively low energy density and is unsuitable for many portable/transport applications.
b Using a non-renewable energy source cannot be sustained indefinitely, but moderate and careful use now can increase the likelihood that it will meet the needs of future generations.

5 Crude oil consists of a range of hydrocarbons with different boiling points. The use to which a fraction is put is influenced by its boiling point and so fractional distillation is needed to separate them.

6 Some CO_2 is consumed in the production of the plant materials that biodiesel is made from.

11.2 Combustion reactions

WE 11.2.1 $2C_6H_{14}(l) + 19O_2(g) \rightarrow 12CO_2(g) + 14H_2O(g)$
WE 11.2.2 $2CH_3OH(l) + 3O_2(g) \rightarrow 2CO_2(g) + 4H_2O(g)$
WE 11.2.3 $CH_3OH(l) + O_2(g) \rightarrow CO(g) + 2H_2O(g)$

1 $2C_6H_6(l) + 15O_2(g) \rightarrow 12CO_2(g) + 6H_2O(g)$

2 $C_2H_5OH(l) + 2O_2(g) \rightarrow 2CO(g) + 3H_2O(g)$

3 $2H_2S(g) + 3O_2(g) \rightarrow 2SO_2(g) + 2H_2O(g)$

4 11

5 **a** Burning biofuels: produces carbon dioxide, however the net effect is less than the burning of fossil fuels.
b Burning petrol: produces carbon dioxide.
c Carbon dioxide will increase over time, as less photosynthesis can take place, which will reduce the amount of CO_2 being absorbed from the atmosphere.
d Carbon dioxide deceases as the process of growth requires photosynthesis, which uses atmospheric carbon dioxide.
e Using windpower rather than fossil fuels reduces the amount of carbon dioxide produced.
f Increased industrialization increases carbon dioxide as industrial processes generally create carbon dioxide.

6 **a** Carbon dioxide is a greenhouse gas that makes rainwater naturally acidic.
b Carbon monoxide is a toxic gas.
c Methane is a greenhouse gas.
d Water vapour is a greenhouse gas that does not affect water acidity.
e Sulfur dioxide is responsible for high levels of acidity in rainwater.

11.3 Calculations involving fuels

WE 11.3.1 1170 tonnes **WE 11.3.2** 16.2 kg
WE 11.3.3 2.72×10^5 kJ **WE 11.3.4** 2.0 mol
WE 11.3.5 10.9 kg

1 250 mol

2 Propane: heat of combustion = -50.5 kJ g^{-1}
Octane: heat of combustion = -47.8 kJ g^{-1}
Propane produces more energy per kilogram.

3 826 g

4 **a** 1.39×10^4 kJ **b** 4.86×10^5 kJ
c 1.20×10^7 kJ **d** 1.7×10^{10} kJ

5 **a** 702 g **b** 618 g

6 **a** 3.59 kg **b** 3.03 kg

7 **a** advantage **b** disadvantage **c** advantage
d advantage **e** disadvantage

8 **a** Energy will be absorbed by the evaporation of the water. The energy content per gram will be less than that for coal with a lower water content.
b removal of some water by drying and/or crushing
c Use of off-peak (e.g. night-time) electricity to dry the coal would reduce the cost.
d Potential atmospheric pollution from oxides of sulfur require either pre-treatment and/or post-treatment to remove or reduce sulfur from the brown coal, and post-treatment to remove or reduce it from exhaust gases.

Chapter 11 review

1 Apart from the difficulties in daily travel, lack of crude oil and natural gas will make transport of manufactured goods difficult and costly. It will also stop the production of all the products that are derived from crude oil—plastics, synthetic fibres, dyes, paints, solvents, detergents and pharmaceuticals.

2 **a** oil
b coal, oil, natural gas, biofuels, nuclear energy, new technologies, hydroelectricity

3 The formation of fossil fuels is a process that occurs over millions of years. The organic matter produced by plants and animals undergoes complex changes as it is subjected to heat and pressure under tonnes of mud and sand. Once the current reserves of fossil fuels have been used, they will not be replaced in the foreseeable future.

4 **a** black coal **b** peat **c** black coal

5 Natural gas. Australia has a large supply of natural gas, but globally other fuel resources will last longer based on current patterns of usage.

6 Photosynthesis uses six molecules of carbon dioxide to produce one molecule of glucose:
$6CO_2(g) + 6H_2O(l) \rightarrow C_6H_{12}O_6(aq) + 6O_2(g)$
Fermentation of glucose produces ethanol and two molecules of CO_2:
$C_6H_{12}O_6(aq) \rightarrow 2CO_2(g) + 2CH_3CH_2OH(aq)$
Combustion of ethanol releases heat and four molecules of CO_2:
$2CH_3CH_2OH(l) + 3O_2(g) \rightarrow 3H_2O(g) + 2CO_2(g)$
Therefore, all of the carbon dioxide released into the atmosphere by the fermentation and combustion of ethanol has been taken in by plants in the process of photosynthesis.

7 Biogas is formed through the action of anaerobic bacteria on organic matter. Anaerobic bacteria operate in the absence of oxygen. Methane is the most abundant component of biogas, followed by carbon dioxide and then a mix of other gases in low percentages.

8 $C_4H_9OH(g) + 6O_2(g) \rightarrow 4CO_2(g) + 5H_2O(l)$

9 $2C_4H_{10}(g) + 9O_2(g) \rightarrow 8CO(g) + 10H_2O(l)$

10 All of the points are interrelated. Many other valid responses are possible.
 a Polar ice caps are shrinking, causing sea levels to rise worldwide. This has local effects on the wildlife adapted to polar conditions. Melting ice adds cold, fresh water to the ocean with the possibility of altering the major ocean currents, resulting in climate changes, e.g. to Western Europe.
 b Changed weather patterns causing droughts, floods, hurricanes and resulting in changed growth patterns in plants that might lead to both plant and animal extinctions.
 c The changed weather patterns can result in crops failing, affecting the economy and driving some people to starvation.
 d Some plants and animals that are dependent on particular weather conditions for propagation or for key parts of their life cycle may become extinct.

11 278 MJ

12 a 20.1 g b 24.4 g c 11.0 g

13 a $CH_3CH_2OH(g) + 3O_2(g) \rightarrow 2CO_2(g) + 3H_2O(g)$
 b 148 kJ c 9.55 g

14 methane: 4.95 kg; propane: 5.95 kg. A greater mass of propane is required to produce the 100 000 kJ of energy and a higher mass of carbon dioxide is released in the process.

15 a a mixture of 10% ethanol and 90% conventional petrol
 b extends the availability of petrol as a fuel; will allow more of the larger fractions from crude oil to be used as a feedstock for other uses rather than being burnt as a fuel

16 heat losses to the surroundings, friction, release of light and sound

17 a 1 mol of CO(g) and 0.5 mol of $O_2(g)$
 b i $\Delta H = -566\ kJ\ mol^{-1}$ ii $\Delta H = +566\ kJ\ mol^{-1}$

18 42.0 mg

19 a 46.1 g
 b Incomplete combustion of the propane may occur. This will likely result in the formation of carbon monoxide instead of carbon dioxide. Note that carbon monoxide is a highly toxic gas that can cause death if inhaled in large enough quantities.

20 a Biochemical fuels are derived from renewable resources such as plants. Ethene, the feedstock used for the industrial production of ethanol, is derived from the distillation and cracking of crude oil, a non-renewable resource.
 b Ethanol can be produced from the fermentation of sugar by yeasts. Sugar may be derived from sugar cane or by the hydrolysis of starch from grains such as maize.

21 a methane, octane, biodiesel, ethanol
 b B

22 Individual student response required.

23 a Burning coal to generate electricity produces greenhouse gas emissions, which would not help Australia meet its Paris Agreement obligations. However, the cost of electricity could rise if other energy sources were required to replace coal, which is relatively abundant and cheap.
 b Biofuels could potentially be carbon neutral—the carbon dioxide produced from the combustion of the fuels is offset by the carbon dioxide absorbed to grow the crops that are used to make the fuels. This would help Australia meet its Paris Agreement obligations.

Unit 1 review: Chemical fundamentals: Structure, properties and reactions

Section 1: Multiple choice

1 B 2 C 3 D 4 D
5 C 6 A 7 C 8 A
9 D 10 C 11 B 12 B

Section 2: Short answer

1 a i Au_2S_3 ii $Sr(OH)_2$ iii $(NH_4)_3PO_4$
 b i iron(III) sulfate ii diphosphorus pentoxide

2 a 180.2 g b 3.0×10^{-22} g
 c hydrogen bonds and dispersion forces
 d hydrogen bonds and dispersion forces

3 a i water (H_2O)

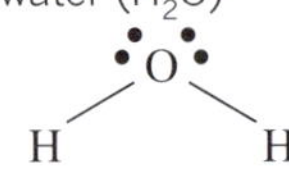

 ii dichlorodifluoromethane (CCl_2F_2)

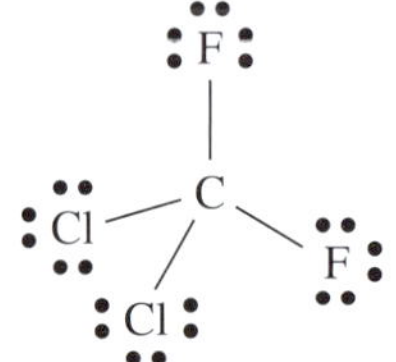

 b i *cis*–dibromoethene ($C_2H_2Br_2$)
Br Br
C=C
H H
 ii carbon disulfide (CS_2)
S=C=S

4 a i methylpropene ii *trans*-difluoroethene
 iii 3-ethyl-5,5-trimethylheptane
 b i and ii only as they are unsaturated
 c $C_{12}H_{26}$

5 a 33.7% b 3.11 mol

6

Reaction	$\Delta H_{reaction}$	Exothermic or endothermic
$2H_2(g) + O_2(g) \rightarrow 2H_2O + 287\ kJ\ mol\ L^{-1}$	$-287\ kJ\ mol\ L^{-1}$	exothermic
$H_2S + 90\ kJ\ mol\ L^{-1} \rightarrow H_2 + S(s)$	$+90\ kJ\ mol\ L^{-1}$	endothermic

Section 3: Extended answer

1 a metallic bond; a lattice of positive ions surrounded by a 'sea' of mobile delocalised electrons; conductor
 b ionic bonds between positive sodium ions and negative chloride ions; no free ions or electrons; non-conductor
 c covalent bonds between the hydrogen atoms and the oxygen atoms within the H_2O molecules; weak intermolecular bonds (hydrogen bonds) between the molecules; no free charged particles; non-conductor
 d two-dimensional covalent network with free mobile electrons; conductor
 e three-dimensional covalent network in which all electrons are covalently bonded; no free electrons; non-conductor

2 a

b

methylpropane

c $2C_4H_{10}(g) + 13O_2(g) \rightarrow 8CO_2(g) + 10H_2O(g)$

d exothermic

e $2C_4H_{10}(g) + 13O_2(g) \rightarrow 8CO_2(g) + 10H_2O(g) + 5750\,kJ$

f 3.01×10^3 g (3 sig. figs)

g $CH_3–CH_2–CH_2–CH_3 + F–F \rightarrow CH_3–CH_2–CH_2–CH_2F + H–F$

h fluorobutane

i

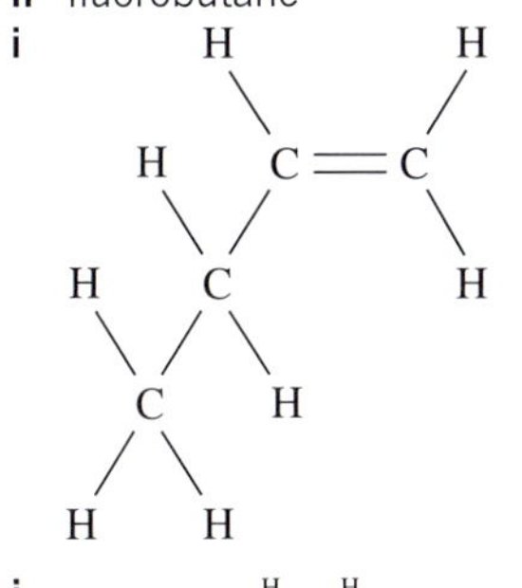

j

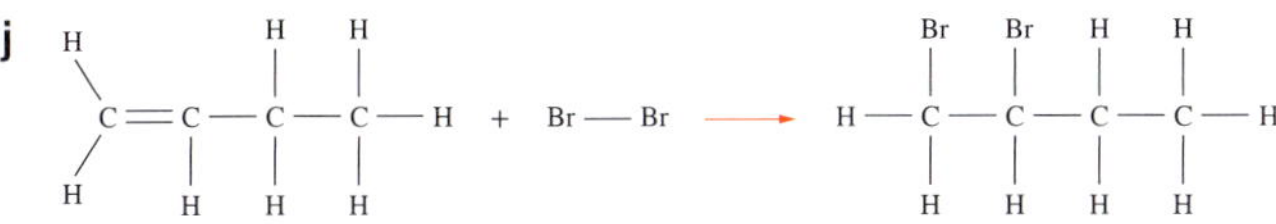

k 1,2-dibromobutane

Chapter 12 Intermolecular forces

12.1 Shapes of molecules

WE 12.1.1 bent or V-shaped

1 The VSEPR theory is based on the principle that negatively charged electron pairs in the outer shell of an atom repel each other. As a consequence, these electron pairs are arranged as far away from each other as possible.

2 Four pairs (one bonding pair and three lone pairs)

3 a

b

c

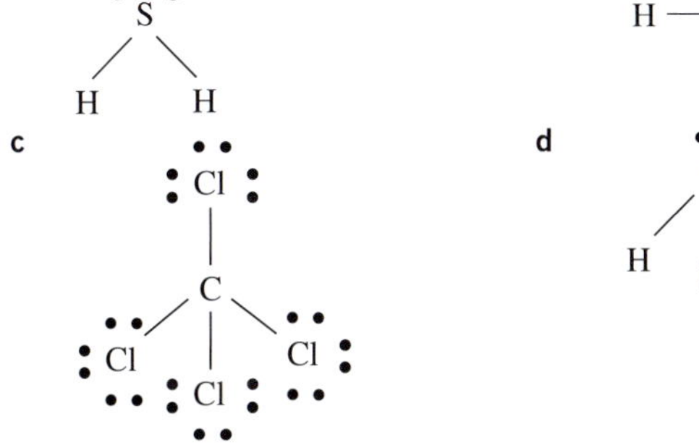

d

e

4 a V-shaped b linear c tetrahedral
d pyramidal e linear

5 a tetrahedral b pyramidal c V-shaped

12.2 Properties of covalent molecular substances

WE 12.2.1 HCl is more polar than NO.

1 a O b C c N d N e F f F

2 a P–F b C–H

3 hydrogen–nitrogen bond, carbon–nitrogen bond, sulfur–nitrogen bond, oxygen–nitrogen bond, nitrogen–nitrogen bond

4

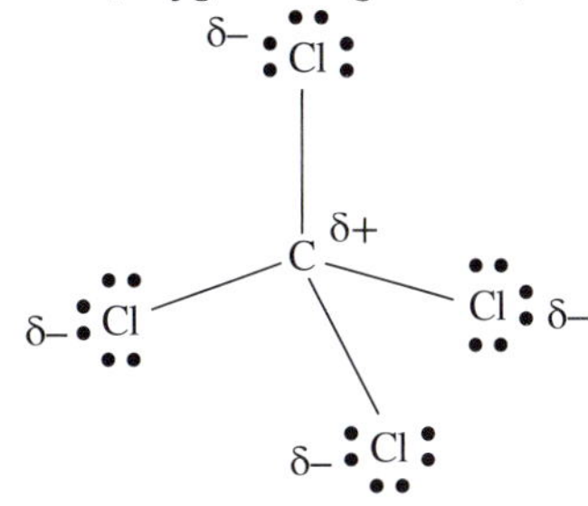

5 a polar b polar c polar d polar e non–polar

6 C

12.3 Types of intermolecular forces

1 Hydrogen chloride and chloromethane would form dipole–dipole forces.

2 C

3 dipole–dipole forces: a to h; hydrogen bonds: a, g, h

4

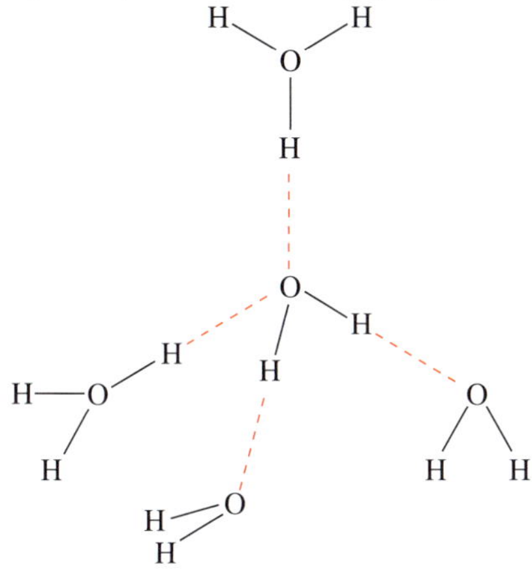

5

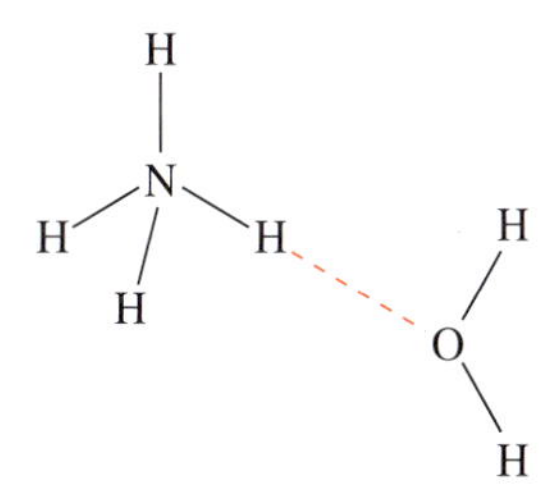

6 a dispersion forces only
b dispersion forces and hydrogen bonding
c dispersion and dipole–dipole forces
d dispersion forces and hydrogen bonding
e dispersion forces only

Chapter 12 review

1 PCl_3 – pyramidal, HOCl – V-shaped, $CHCl_3$ – tetrahedral, HF – linear

2 V-shaped

3 Beryllium does not obey the octet rule in this molecule and forms only two single bonds. Each bond has two electrons, so there are four bonding electrons involved in bonding.

4 a tetrahedral b pyramidal c tetrahedral
d V-shaped e pyramidal

5 a non-polar b polar c non-polar
d polar e non-polar f non-polar

6 B

7 a non-polar b polar c non-polar
d polar e polar

8 Si–O, H–Br, N–O, O–Cl, F–F

9 a SO_3 i non-polar ii dispersion forces
b $SiCl_4$ i non-polar ii dispersion forces
c CF_4 i non-polar ii dispersion forces
d NF_3 i polar ii dipole–dipole attraction
e CH_3NH_2 i polar ii hydrogen bonding

10 A and C

11 Melting temperatures increase down the table because the molecules increase in mass and size and there are more electrons in the molecules; therefore, the strength of the dispersion forces increases.

12 CF_4 has a slightly higher boiling temperature (–128°C) than OF_2 (–145°C), indicating that the forces between molecules in CF_4 are stronger. OF_2 is slightly polar; CF_4 is non-polar. OF_2 molecules are held together by forces of dipole–dipole attraction and dispersion forces. Although CF_4 molecules are attracted by dispersion forces only, the much larger size of CF_4 molecules makes the dispersion forces stronger than the sum of the dipole–dipole forces and the dispersion forces between OF_2 molecules.

13 Neon exists as single atoms, with the only forces of attraction being dispersion forces. As Ne atoms have very few electrons, the dispersion forces are extremely weak. Neon therefore has a very low boiling temperature. Hydrogen fluoride molecules, however, are very polar and so are held together by electrostatic attraction between permanent dipoles. Because hydrogen is bonded to the very electronegative fluorine, the forces between molecules are hydrogen bonds. These are relatively strong intermolecular bonds and HF, therefore, has a much higher boiling temperature than Ne. (The dispersion forces operating between HF molecules are extremely weak.)

14 a CCl_4
b CH_4 and CCl_4 are both non-polar and so are held together in a lattice only by dispersion forces. CCl_4 is the larger of these two molecules and has more electrons, so the dispersion forces between CCl_4 molecules will be greater than those between CH_4 molecules. As there are stronger dispersion forces between molecules of CCl_4 than for CH_4, it takes more energy to vaporise CCl_4.

15 Iodine is a much larger molecule, so it has more electrons and therefore the dispersion forces are much stronger in iodine than in fluorine.

16 A permanent dipole is formed if there is a difference in electronegativity between the two atoms that form a bond. The more electronegative atom has a partial negative charge and the less electronegative atom has a partial positive charge. In larger molecules, individual bonds may be polar but if the molecule is symmetrical the molecule will not have a permanent dipole overall and is non-polar. Asymmetry in the molecule causes an asymmetry in the electron distribution around the molecule, causing one end of the molecule to develop a partial negative charge while the other end develops a partial positive charge. The positive and negative ends of neighbouring molecules attract each other, forming dipole–dipole bonds.
A temporary dipole is caused by random fluctuations in the electron distributions around the molecule. The electrons are constantly moving and can occasionally concentrate at one end of the molecule, causing that end to develop a temporary negative charge while the other end develops a temporary positive charge. This temporary dipole can then induce dipoles in the neighbouring molecules. The induced dipoles attract each other. Such attractions are known as dispersion forces and are present between all molecules.

17 a i N_2 :N≡N:
ii O_2 O=O (with lone pairs)
iii CO_2 O=C=O (with lone pairs)
b i NH_3, HCl, H_2O, $CHCl_3$
ii N_2, Cl_2, O_2, CH_4, CO_2, CCl_4
iii NH_3, H_2O

18 If water were a linear molecule, the polarity of the two O–H bonds would cancel each other out and make the molecule non-polar. As water *is* polar, it cannot be a linear molecule.

19 The intermolecular bonds that hold molecules together in covalent molecular substances are much weaker (100 times) than the chemical bonds holding the atoms together in ionic, metallic and covalent network substances. As a result, it takes much less heat energy to break the intermolecular bonds holding covalent molecular solids and liquids together and these substances have relatively low melting and boiling points.

20 A 21 C 22 D

Chapter 13 Chromatography

13.1 Principles of chromatography

WE 13.1.1 $R_f = 0.3$

1 B

2

Term	Description
adsorption	the attraction of one substance to the surface of another
desorption	the breaking of the attraction between a substance and the surface to which the substance is adsorbed
components	the different compounds in the mixture, which can be separated by chromatography
polar molecule	a molecule that acts as a dipole; it has one or more polar covalent bonds, with the charge being distributed asymmetrically
mobile phase	the solvent that moves over the stationary phase in chromatography
stationary phase	the components of a mixture undergo adsorption to this phase

3

Band	a Distance from origin (mm)	b R_f	c Compound
light green	20	0.33	chlorophyll b
dark green	27	0.45	chlorophyll a
orange	40	0.67	xanthophyll
yellow	50	0.83	β-carotene
solvent front	60	–	–

d The chromatogram would be likely to be different because separation of components depends on their solubility in the mobile phase (as well as strength of adsorption to the stationary phase). The polarity of the solvent used in TLC and paper chromatography will affect the R_f of the sample components. A polar solvent will dissolve polar samples readily; a non-polar solvent will dissolve nonpolar samples readily.

4
1 Dissolve a sample of pure phenacetin in a volume of chloroform. This is the standard solution.
2 Dissolve a tablet of the analgesic in chloroform. This is the sample solution.
3 Place a small spot of the sample solution near the bottom of a thin-layer plate.
Place a spot of the standard solution next to it, at the same distance from the bottom of the plate.
4 When the spots are dry, place the plate in a container with a small volume of solvent, such as chloroform. The lower edge of the plate, but not the spots, should be immersed.
5 Allow the solvent to rise until it almost reaches the top of the plate and then remove the plate from the container.
6 Let the plate dry and examine it under ultraviolet light. If a spot from the sample appears at the same distance from the origin as the spot from the standard solution, the tablet is likely to contain phenacetin.

5 During column chromatography, the components of the sample *adsorb* onto the stationary phase and *desorb* into the liquid mobile phase. A component that adsorbs most strongly to the *stationary* phase and is least soluble in the *mobile* phase would be expected to take the *longest* time to pass through the column.

13.2 Advanced applications of chromatography

WE 13.2.1 0.75 mg kg^{-1}

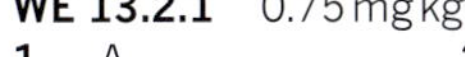
1 A 2 A 3 B, A, C

4 a

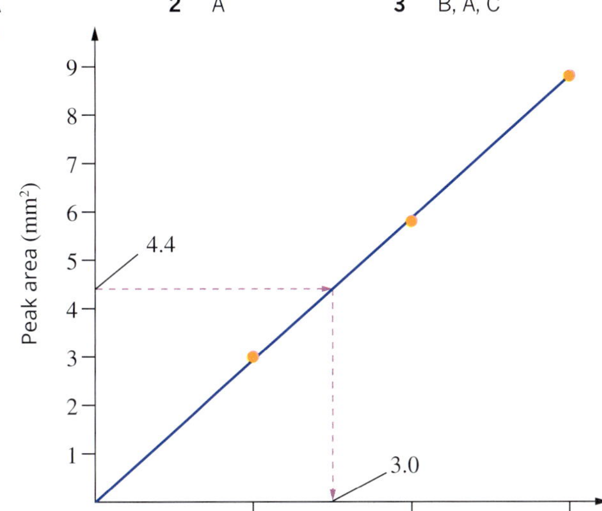

b 3.0 μg mL^{-1}

5 a

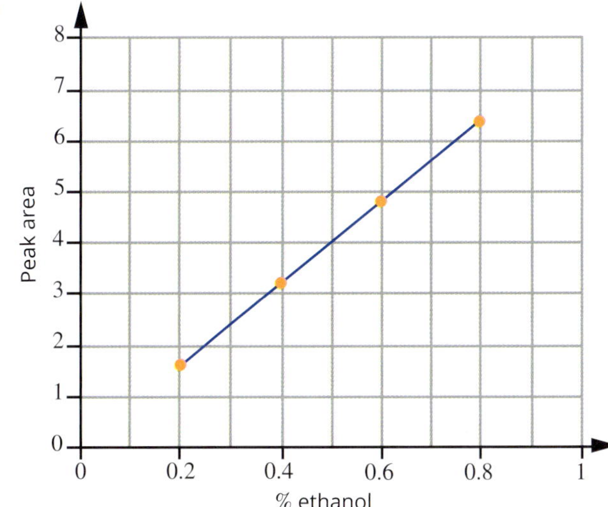

b 0.7% (using the calibration graph)

Chapter 13 review

1 In paper chromatography, the paper acts as the *stationary* phase. A small spot of a solution is placed at one end of the paper, called the *origin*. A small spot of a solution is placed at one end of the paper, called the *sample*. The sample solution contains a number of different coloured compounds, the *components*.
The paper is suspended so that the end with the spot is *above* the surface of the solvent. The solvent or *mobile* phase moves up the *stationary* phase. Different coloured spots are observed at various places on the paper, due to the separation of different *components*.

2 a Water was *absorbed* by the towel as the wet swimmer dried himself.
A thin layer of grease *adsorbed* onto the cup when it was washed in the dirty water.
b Absorb: Atoms or molecules are taken *into* the material.
Adsorb: Atoms or molecules accumulate and bond weakly to the *surface* of a solid or liquid.

3 The component at the top of the chromatogram has a greater rate of adsorption and desorption compared to a component at the bottom of the chromatogram.

4 D

5 R_f(blue) = 0.8
R_f(purple) = 0.6
R_f(yellow) = 0.2

6 R_f(blue) = 0.83
R_f(red) = 0.58

7 a If the solvent were above the level of the origin, the compounds under test would dissolve and disperse throughout the solvent.
b Components in a mixture undergoing chromatography cannot move faster than the solvent that is carrying them over the stationary phase. R_f values must therefore be less than one.
c three
d B: blue; C: green. They can be identified on the basis of their colour and R_f values.
e blue
f 0.63; 0.13

8 a 3.2 cm b 15 cm
c

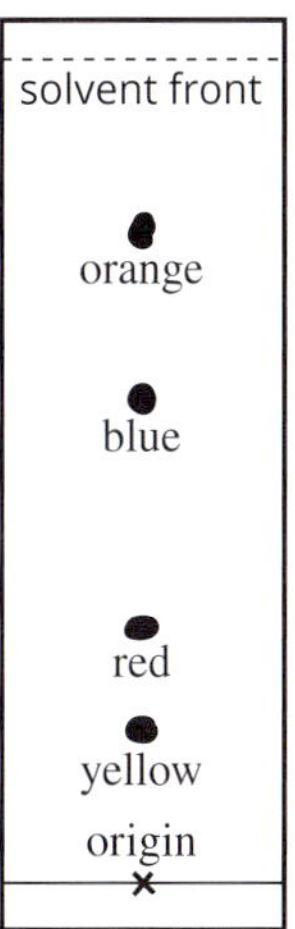

9 a taurine, glycine and an unknown
b View the chromatogram under UV light. Spray the finished chromatogram with a compound that causes the amino acids to fluoresce. Spray the finished chromatogram with sulfuric acid that causes a brown spot for organic compounds.
c R_f = 0.12
d leucine

10 The component most strongly adsorbed to the stationary phase is band A.
As band B begins to emerge from the column, it appears to separate into two bands. These two components would be more readily separated if you *increased the column length*.
HPLC differs from column chromatography because the particles in the stationary phase are *smaller* and *high* pressure is applied to the mobile phase.

11 A 12 A 13 A 14 B, D, A, C, E

15 a Chemists can determine the number of components and the absorption of components, from which concentration can be determined.
b four
c Solid samples are dissolved in a suitable solvent. The liquid sample is injected into the top of an HPLC column. The stationary and mobile phases are chosen to achieve a good separation of the components in the sample. The sample components alternately adsorb onto the stationary phase and then desorb into the solvent as they are swept forward. The time taken to exit the column increases if the component strongly absorbs onto the stationary phase and has a low solubility in the mobile phase.

16 a

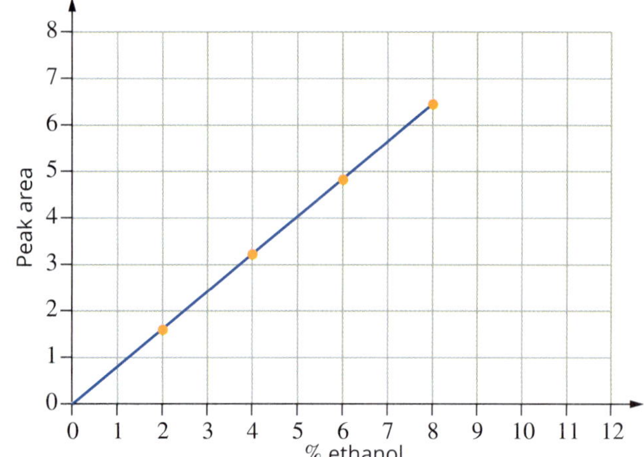

b 7.0%

17 0.6% **18** B

19

Technique	Problem
paper or thin-layer chromatography	B It is difficult or impossible to obtain quantitative data. D Samples must be able to be dissolved in solvent.
high performance chromatography (HPLC)	A It requires large amounts of solvents to operate. C Expensive equipment is needed. D Samples must be able to be dissolved in solvent.
gas chromatography (GC)	C Expensive equipment is needed.

20 a

Standard	Peak height (cm)
Standard 0 ppm parathion	0.2
Standard 10 ppm parathion	1.5
Standard 20 ppm parathion	3.1
Standard 30 ppm parathion	4.4
Reservoir water	2.0

b

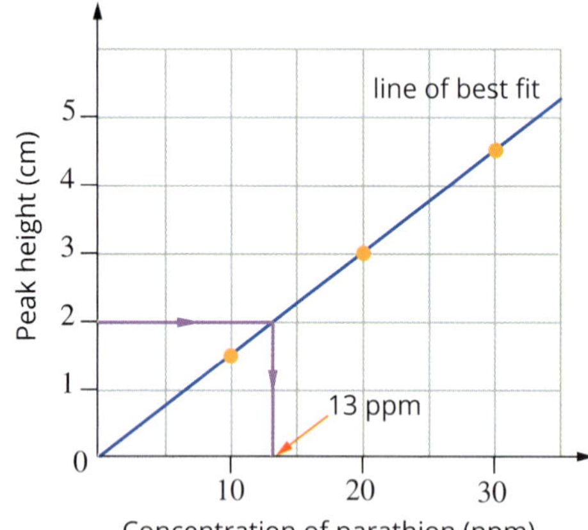

c concentration of parathion in water sample = 13 ppm

d 13 ppm ≡ 13 μg/mL ≡ 13 mg/L

No, this is more than 1000 times the allowable concentration.

e LD_{50} of 8 mg/kg = 1.2 mg. A mouse of 150 g has a 50% chance of being killed by a 1.2 mg dose of parathion.

volume of water that contains 1.2 mg parathion = $\frac{1.2}{13}$ L = 0.092 L = 92 mL

Chapter 14 Gases

14.1 Introducing gases

WE 14.1.1 **a** 700 cm^3 **b** 0.700 L **c** 7.00×10^{-4} m^3

WE 14.1.2 **a** 90.2 kPa **b** 0.891 atm **c** 90.2 kPa **d** 0.902 bar

1 **a** Molecules of gases are in constant, rapid, random motion and the forces between molecules are negligible. They continue to move outwards until stopped by the walls of the container, filling all the space available.

b Most of the volume occupied by a gas is space, so compression can be achieved by reducing the space between the particles.

c The molecules in a gas are spread much further apart than those of a liquid. A given mass of gas would occupy a much greater volume than the same mass of the liquid phase. Therefore, the density of the gas is less.

d Gases mix easily together because of the large amount of space between the molecules.

e The pressure exerted by a gas depends on the number of collisions of gas particles and the wall of the container. The pressure is independent of the type of gas involved. The total pressure exerted by a mixture of gases will depend on the total number of collisions each gas has with the container.

2 **a** Tyres have a recommended maximum pressure to give a comfortable ride as well as good traction on the road. If the pressure in a tyre is too high, the gas inside cannot be compressed as easily and passengers will be more aware of bumps on the road.

b During a long journey on a hot day, the air in a tyre warms up. This means the air molecules have increased kinetic energy, and collisions with the walls of the tyres will increase in frequency and exert more force, and so the pressure will increase.

c Particles from the cooking food escape the pot and move randomly through the house. If the food has an odour, and if there are enough particles in the air, you will detect the odour as you enter the house.

d As air is pumped into a balloon, air molecules collide with the rubber of the balloon, forcing it to expand. If too much air is pumped in, the balloon reaches a stage where it cannot stretch any further. If the number of collisions by molecules per given surface area is increased still further, the rubber will break.

3 The order should be from gases with the smallest molecular mass to largest molecular mass. He < Ne < Ar < Kr

This is because, at the same temperature, gases have the same kinetic energies. As the mass of the gas increases, its velocity decreases. Therefore, it will diffuse more slowly.

4 **a** As temperature increases, the average kinetic energy of gas molecules in the can will increase. This will lead to an increase in the frequency and force of collisions of gas molecules with the inside walls of the aerosol cans. This will cause an increase in pressure.

b As the syringe is compressed, the inside surface area of the syringe will decrease. The number of collisions of molecules per unit area per second with the inside walls of the syringe will increase. This will cause a pressure increase.

5 **a** 1.40×10^5 Pa **b** 92 kPa

c 3.22×10^3 mmHg, 4.30×10^5 Pa

d 900 mmHg, 1.18 atm, 1.20 bar

6 **a** 2×10^3 mL **b** 4.5×10^{-3} m^3

c 2.250 L **d** 0.120 L

14.2 Molar volume of a gas

WE 14.2.1 $V_{(STP)}$ = 79.5 L **WE 14.2.2** 0.15 mol

1 **a** 32 L **b** 23 mL **c** 1.1 L

2 **a** 0.123 mol **b** 2.20 mol **c** 6.16×10^{-3} mol

3 D

14.3 Calculations involving reactions with gases

WE 14.3.1 1.02×10^2 mol **WE 14.3.2** 1.21 L

WE 14.3.3 1 volume of CH_4 produces 1 volume of CO_2 gas, so 50 mL of CH_4 reacts with 50 mL of CO_2.

WE 14.3.4 **a** As there is 1.12 mol of C_4H_{10}, the O_2 is in excess. The C_4H_{10} is the limiting reactant (it will be completely consumed).
b 101.7 L

1 **a** $\frac{2}{3}$ **b** $\frac{3}{4}$ **c** 1

2
- Write a balanced equation for the reaction.
- Identify the known and unknown substances in the question.
- Calculate the amount, in mol, of the known substance using $n = \frac{m}{M}$.
- Use mole ratios from the equation to calculate the amount of the unknown.
- Calculate the mass of the unknown substance using $m = n \times M$.

3 6.00 g of NH_3

4 **a** 0.0200 mol **b** 0.0400 mol **c** 0.909 L

5 2.04 L of O_2 **6** 7.93 g of H_2O

Chapter 14 review

1 B **2** A **3** A

4 smaller, straight-line, weak, elastic, directly

5 **a** As volume is reduced, there is an increase in the frequency of molecular collisions per unit wall area. This is measured as an increase in pressure.
b When the temperature of a gas is lowered, the average kinetic energy of the particles decreases. The rate of collisions between particles and the walls of the container decreases and particles collide with less force. As pressure is a measure of the force of molecular collisions per unit wall area of the container, pressure is found to decrease.
c In a mixture of gases, the particles of each gas are moving and colliding with the walls of the container, independently of each other. Each gas therefore exerts a pressure. As the gases behave independently of each other, total pressure is simply the sum of the individual gas (or partial) pressures.
d When more gas is added to a container, the total number of particles in the container increases. Provided that the volume of the container and the temperature have not changed, the collisions of these additional particles means that the total pressure in the container has increased.

6 **a** The pressure inside the container is reduced when some of the gas escapes.
b There are fewer gas molecules to collide with each other and the walls of the container. Pressure is the force exerted by the molecules over a defined area, so this will decrease.

7 B **8** 0.022 mol **9** 0.0568 L **10** C

11 **a** 5.7 L **b** 20 g

12 **a** 44.0 g **b** 22.7 L **c** 1.94 g L^{-1}

13 **a** 9.69×10^3 mol **b** 108 kg

14 40.9 g **15** 3.7 tonne

16 **a** **i** 57 L **ii** 34 L
b **i** 13 L **ii** 7.7 L
c **i** 0.374 L **ii** 0.225 L

17 A

18 **a** 2.19×10^3 mol
b $\frac{n(CO_2)}{n(Fe_2O_3)} = \frac{3}{1}$, 6.58×10^3 mol
c 1.49×10^5 L

19 **a** Equal. With pressure, volume and temperature the same, n will be the same.
b Carbon dioxide. Each CO_2 molecule contains three atoms and each O_2 molecule contains two atoms. As there is an equal number of molecules of each gas, there are more atoms in the CO_2 sample.
c Carbon dioxide. Density = mass ÷ volume. The volume is the same for each gas, but the mass of CO_2 is greater, so it has the greater density.

20 **a** $\frac{n(NO_2)}{n(HNO_3)} = \frac{4}{4}$, 0.500 mol
b $\frac{n(NO_2)}{n(HNO_3)} = \frac{1}{4}$, 0.125
c 14.2 L

21 **a** 15.4 mol **b** 6.16 mol **c** 1.13 kg

22 **a** As the air is heated, the molecules move faster. Because a hot-air balloon has an opening at the bottom, some of the gas molecules escape. The balloon filled with warmer air has fewer molecules in the same volume as a balloon filled with cooler air. Fewer molecules have less mass. Therefore, the warmer balloon has less mass per unit volume, or density, than the cooler balloon.
b The molecules in a sample of gas at any temperature have a range of kinetic energies. Some have low energy and some have high energy. As the temperature of a gas increases, the average kinetic energy of molecules increases. However, as the temperature increases, there will still be some molecules with low kinetic energy. So the student is mistaken, as not all of the kinetic energies will have increased.
c When the temperature of a gas is lowered, the average kinetic energy of the particles decreases. The frequency of collisions between particles and the walls of the container decreases and particles collide with less force. As the pressure is a measure of the force of molecular collisions per unit surface area of the container, the pressure of the gas will decrease.

Chapter 15 Properties and uses of water

15.1 Essential water

1 Surface water has the highest risk of contamination since it can be easily polluted. It is not a protected water source like mains water where the water is tested and treated if needed.

2 Most of the freshwater on Earth is locked in ice caps, glaciers or the soil.

3 Australia is the driest inhabited continent. Rainfall in Australia is extremely variable. Most of the rain that falls evaporates before it can enter rivers and reservoirs.

4 To determine if the water is mixed and therefore what depth to take the samples. If the temperature is consistent, it can be assumed that the water is thoroughly mixed and samples can be taken halfway down. If there is a temperature variation, then accurate information can only be obtained if a sample is taken in the middle of each temperature region.

5 A, B and C

6 Fluoride in drinking water helps to reduce the incidence of tooth decay. It does this by interacting with tooth enamel. The fluoride ion replaces the hydroxyl ion in tooth enamel, forming fluorapatite, which is stronger and more resistant to decay.

15.2 Properties of water

WE 15.2.1 70.5 kJ

1 **a**
- relatively high melting and boiling temperatures for its molecular size
- decrease in density on freezing
- high heat capacity
- high latent heat of fusion and evaporation for a substance of its molecular size.

b The bond between H and O atoms in water is highly polar. As a result, hydrogen bonds exist between water molecules. Hydrogen bonds are stronger than other intermolecular bonds (although still weaker than the covalent intramolecular bonds) and so require more energy to break. Thus, water has relatively high melting and boiling temperatures. Hydrogen bonding between water molecules in ice results in a very open arrangement of molecules so ice is less dense than liquid water.

2 Each water molecule has two hydrogen atoms and one oxygen atom. The oxygen atom has two pairs of non-bonding electrons, each of which can form one hydrogen bond.
So, the maximum number of water molecules with which one water molecule can form hydrogen bonds is four: up to two hydrogen bonds involving the two hydrogen atoms and up to two hydrogen bonds involving the two pairs of non-bonding electrons on the oxygen atom.

3 a In ice, the distribution of hydrogen bonds in water is almost tetrahedral in shape. In order for these hydrogen bonds to form as the water solidifies, the water molecules must move into these positions, so occupying more volume than in liquid water.

b

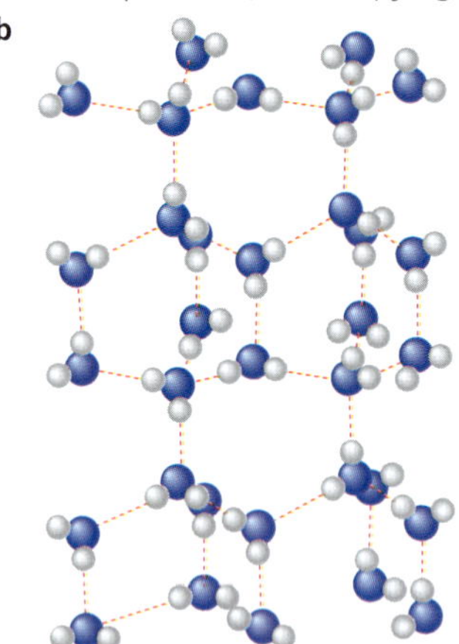

c As it freezes, water expands, unlike most liquids. This is because of hydrogen bonding. Each water molecule is surrounded by four others in what is almost a crystal-type situation. Therefore, ice is less dense than liquid water, and it floats on liquid water. (For most liquids, the solid is denser than the liquid.)

4 D 5 B

6 At the melting point the flat region of the graph represents the substance *changing from a solid to a liquid.* The energy change is equal to *the latent heat of fusion.* At the boiling point the flat region of the graph represents the substance *changing from a liquid to a gas.* The energy change is equal to *the latent heat of vaporisation.*

7 hydrogen bonding

15.3 Water as a solvent

1 a solvent b solution c solute d solute

2 They all have water as the solvent.

3 a soluble b likely to be insoluble
c soluble d soluble
e likely to be insoluble f soluble
g likely to be insoluble

4 The solute–solvent forces are stronger than the solute–solute and solvent–solvent forces.

5 a soluble b insoluble c solute
d dissolution e solution f solvent

6 CH_4 (methane). Methane is a non-polar molecule. According to the principle of 'like dissolves like' a non-polar substance will dissolve in a non-polar solvent.

15.4 Water as a solvent of molecular substances

1 B and E

2 $CH_3OH(l) \xrightarrow{H_2O(l)} CH_3OH(aq)$
$C_6H_{12}O_6(s) \xrightarrow{H_2O(l)} C_6H_{12}O_6(aq)$

3
- Hydrogen bonds between water molecules are broken.
- Hydrogen bonds between ethanol molecules are broken.
- New hydrogen bonds form between ethanol molecules and water molecules.

4 $HI(s) + H_2O(l) \rightarrow H_3O^+(aq) + I^-(aq)$

5 *Covalent bonds* within hydrogen chloride molecules are broken. *Hydrogen bonds* between water molecules are broken. The HCl *ionises* and produces Cl^- and H^+ ions. *Covalent bonds* form between H^+ ions and water to produce *hydronium* ions. *Ion–dipole bonds* form between the Cl^- and H_3O^+ ions and polar water molecules.

15.5 Water as a solvent of ionic compounds

WE 15.5.1 soluble

1 $NaNO_3(s) \xrightarrow{H_2O(l)} Na^+(aq) + NO_3^-(aq)$
$Ca(OH)_2(s) \xrightarrow{H_2O(l)} Ca^{2+}(aq) + 2OH^-(aq)$

2

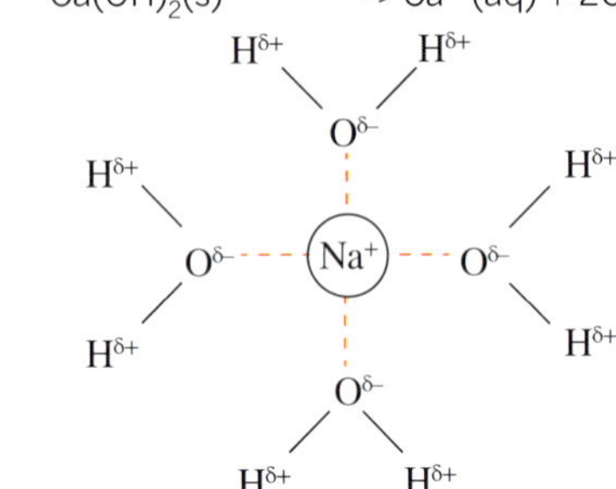

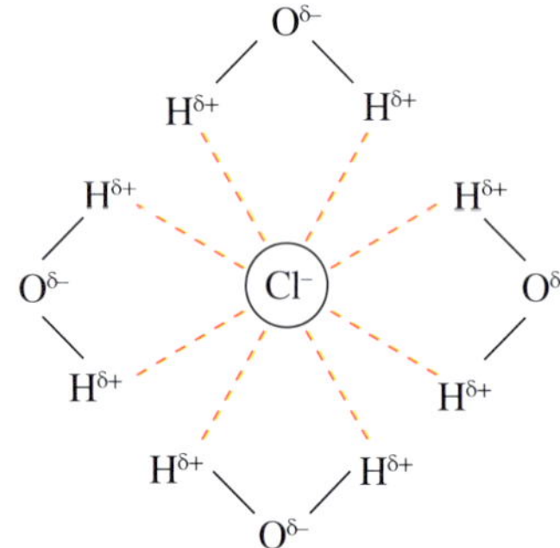

---- represents ion–dipole interaction

The positive sodium ion attracts the partial negative charges on the oxygen atoms in the water molecule. The negative chloride ion attracts the partial positive charges on the hydrogen atoms in the water molecule.

3 A, B, D, E, H

4 A, C, D, E, F, H

5 a Na^+/CO_3^{2-} b Ca^{2+}/NO_3^- c K^+/Br^-
d Fe^{3+}/SO_4^{2-} e Cu^{2+}/Cl^-

6 a Nitrates are highly soluble in water. If found on Earth, they would dissolve in rainwater and wash into the oceans. Therefore, they are found only in areas of low rainfall.
b The high solubility of sodium, chloride and sulfate ions results in them dissolving and flowing into the world's oceans.

15.6 Solubility

WE 15.6.1 140 g (from the graph) **WE 15.6.2** 210 g
WE 15.6.3 10 g

1 C

2 The solubility curve represents the maximum mass of solute that can be dissolved at a set temperature. The green dot represents a supersaturated solution as it is above the solubility curve.
The orange dot represents a saturated solution as it lies on the solubility curve. The blue dot represents an unsaturated solution as it lies below the solubility curve.

3 a 40 g b 160 g c 20 g

4 a 48 g b 200 g c 70 g

5 B 6 20 g 7 55 g

8 Granite is a rock that is formed from magma cooling beneath the Earth's surface. This means the crystals are formed slowly. Crystals formed from slow cooling are larger than those formed from cooling quickly. Magma that cools above the surface forms basalt, the cooling process is faster, and therefore the crystals in basalt are smaller than those in granite.

Chapter 15 review

1 a

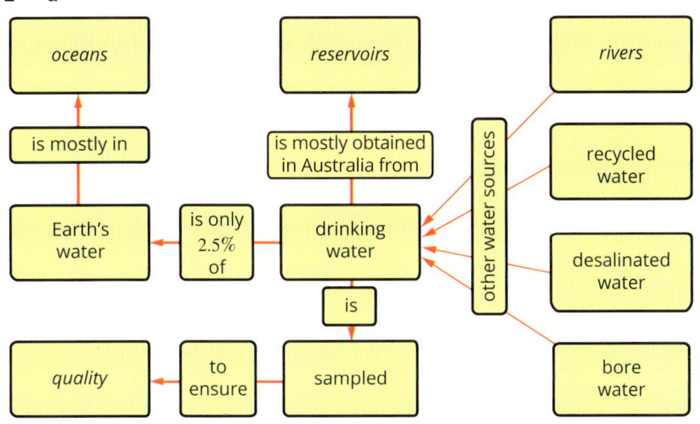

2 A

3 Bore water is water collected in aquifers (underground water-bearing rock) below Earth's surface.

4 Most Australians live in the capital cities and their drinking water is supplied from reservoirs, as cities in Australia obtain their water from protected sources (although Perth sources a large amount of its water from desalination).

5 oceans, ice caps and glaciers, groundwater, ground ice and permafrost, lakes, soil moisture, atmosphere as water vapour, rivers

6 Considerations include:
- which chemical is to be analysed
- why the testing is to take place
- what are the health risks associated with the sampling
- which equipment is required
- what is the sample size required for the selected test
- what method is to be used to record the measurements
- where and at what depths should the samples should be taken
- how to obtain a representative sample
- what are the labelling, storage and transport requirements for the sample

7 D

8 Levels of arsenic, cadmium and copper are below the Australian Drinking Water Guideline levels and therefore acceptable. Lead and mercury are above the Guideline levels and therefore unacceptable.

9 **a** unsafe position
b sample taken from surface of groundwater
c container not sterile
d container not protected from light
e water sample not secured safely

10 Water is a *polar* molecule. Within a single molecule, hydrogen and oxygen atoms are held together by strong *covalent bonds*. Between different molecules, the most significant forces are *hydrogen bonds*. It is the relatively *high* strength of the intermolecular forces that give water its unique properties of:
- relatively *high* boiling point, *100*°C
- relatively *high* latent heat values *6.0* kJ mol^{-1} and *44.0* kJ mol^{-1}
- relatively *high* specific heat capacity *4.18* J g^{-1} °C^{-1}.

11 **a** oxygen atom
b hydrogen atom
c hydrogen bond (and dispersion forces)
d covalent bond

12 **a** Intermolecular forces are those between one molecule and other molecules. For water, these are hydrogen bonds. Intramolecular forces are those holding the atoms together within a molecule. For water, these are covalent bonds.
b Covalent bonds are stronger. Evidence for this is the high temperatures required to break the bonds between the oxygen and hydrogen atoms inside the water molecule and so decompose it into its constituent gases. Changing liquid water into gaseous water involves breaking hydrogen bonds to separate one molecule from another. The lower temperatures needed to do so indicate that hydrogen bonds are weaker.

13 Water has a significantly higher melting point than hydrogen sulfide due to the hydrogen bonds between water molecules. Hydrogen sulfide cannot form hydrogen bonds.

14 It is the high polarity of the water molecule that allows relatively strong hydrogen bonding to occur between molecules. As a consequence, a relatively large quantity of energy is required to break the hydrogen bonds between water molecules when water changes from a liquid to a gas. This gives water a high boiling point.

15 Water has a high latent heat of vaporisation. Water is effective as a coolant because it absorbs a relatively large amount of energy when it evaporates, giving it a high latent heat of vaporisation.

16 D

17 dispersion forces, hydrogen bonds, covalent bonds

18 A solution is most likely to form when the polarity of bonding of the solute is similar to that of the solvent. The bonds formed between solute and solvent are then similar to those that existed between solute particles and between solvent particles. Water, being polar, is therefore a good solvent for ionic and polar substances.

19 Using the 'like dissolves like' rule, only polar substances will dissolve in water. Nitrogen is a non-polar molecule so will not dissolve well. Ethene is a non-polar hydrocarbon that will not be expected to dissolve in water. Ethanol, however, isa small polar molecule that can be expected to dissolve in water.

20 Carbon is slightly more electronegative than hydrogen, so each C–H bond is slightly polar. However, the resulting partial charges are distributed symmetrically across the octane molecule, making the molecule non-polar overall.
The energy released in the formation of solute–solvent bonds is not enough to overcome the intermolecular bonds between solute molecules and the intermolecular bonds between solvent molecules, so octane will not dissolve in water.

21 $C_6H_{12}O_6$ and C_3H_7OH are polar molecules. They contain the polar –OH group and so are able to form hydrogen bonds with water.
HI and HNO_3 contain polar molecules but they are unable to form hydrogen bonds with water. They dissolve by ionising.
I_2, CH_4 and C_2H_4 are non-polar covalent molecules. They do not dissolve well in polar water.

22 Y, Z, X. As the CH_3OH molecule is polar, it will dissolve readily in polar solvents. Because CH_3OH dissolves in Y, it can be ascertained that Y is a polar solvent. The non-polar nature of a CH_4 molecule means CH_4 will dissolve readily in non-polar solvents. It can be concluded that X is a non-polar solvent because CH_4 completely dissolves in it. CH_3OH molecules and CH_4 molecules partially dissolve in Z, which indicates it is more polar than X. No information is given about methanol dissolving in Z, although it can be implied.

23 DDT is most likely non-polar because it is soluble in fats, which are non-polar, and insoluble in polar water.

24 **a** dissociation
b **i** $Cu^{2+}(aq)$, $NO_3^-(aq)$
ii $Zn^{2+}(aq)$, $SO_4^{2-}(aq)$
iii $NH_4^+(aq)$, $PO_4^{3-}(aq)$

25 **a** K^+/CO_3^{2-} **b** Pb^{2+}/NO_3^- **c** Na^+/OH^- **d** Na^+/SO_4^{2-}
e Mg^{2+}/Cl^- **f** Zn^{2+}/NO_3^- **g** K^+/S^{2-} **h** Fe^{3+}/NO_3^-

26 **a** $MgSO_4(s) \xrightarrow{H_2O(l)} Mg^{2+}(aq) + SO_4^{2-}(aq)$
b $Na_2S(s) \xrightarrow{H_2O(l)} 2Na^+(aq) + S^{2-}(aq)$
c $KOH(s) \xrightarrow{H_2O(l)} K^+(aq) + OH^-(aq)$
d $(CH_3COO)_2Cu(s) \xrightarrow{H_2O(l)} 2CH_3COO^-(aq) + Cu^{2+}(aq)$
e $Li_2SO_4(s) \xrightarrow{H_2O(l)} 2Li^+(aq) + SO_4^{2-}(aq)$

27 Hydrated hydronium ions and chloride ions. Hydrogen chloride is a polar molecule that cannot form hydrogen bonds. When added to water, the molecule ionises to form a hydronium ion, H_3O+, and a chloride ion, Cl^-. These ions become hydrated by ion-dipole attractions. There will also be water molecules and a very, very few HCl molecules.

28 Magnesium ions are cations; they have a positive charge. This means the negatively charged non-bonding electron pairs on the oxygen atoms in the water molecules are attracted to them. The water molecules arrange around the magnesium ion with their oxygen atoms, rather than the hydrogen atoms, closest to the chloride ion.

Chloride ions are anions; they have a negative charge. This means the hydrogen atoms in the water molecules, which have a partial positive charge, are attracted to them. The water molecules arrange around the chloride ion with their hydrogen atoms, rather than the oxygen atoms, closest to the chloride ion.

29 Possible answers: Na_2CO_3, Li_2CO_3 and K_2CO_3 are soluble, whereas $CaCO_3$, $MgCO_3$ and Ag_2CO_3 are insoluble.

30 Possible answers: Na_2SO_4, K_2SO_4 and $(NH_4)_2SO_4$ are soluble, whereas $CaSO_4$, $BaSO_4$ and $PbSO_4$ are insoluble.

31 Potassium ions and bromide ions are held in the ionic lattice by ionic bonds that are based on electrostatic attraction. These ionic bonds in the solute break when it dissolves in water.

Water, the solvent, has hydrogen bonds between water molecules.

When potassium bromide is added to water, the hydrogen atoms of the water molecules are attracted to the negative bromide ions, and the oxygen atoms of the water molecules are attracted to the positive potassium ions.

Ion–dipole bonds form between the ions and water molecules and the surface ions are pulled into solution. Gradually the ionic lattice dissociates and a solution is formed.

32 **a** 50g **b** 10.5g **c** 25g

33 **a** 1000g **b** 16g

34 60g

35 As the temperature increases, the solubility of most gases in water decreases. This means that less oxygen and other gases will be available to aquatic life.

36 carbon monoxide (CO) **37** 142g

38 **a** v **b** i **c** iv **d** ii **e** iii

39 **a** Ammonia is a highly polar molecule and forms hydrogen bonds with water. It is therefore very soluble in water. Methane, however, is non-polar. Weak (dispersion) forces would occur between methane and water, and these are unable to disrupt the stronger hydrogen bonds between water molecules. Therefore, methane does not dissolve in water.

b Glucose dissolves in water because it has very polar –OH groups that can form hydrogen bonds with water molecules. Sodium chloride is ionic; hence, there are ion–dipole attraction between the ions and water. This is strong enough to overcome the attraction between the sodium ions and chloride ions in the solid NaCl lattice.

Chapter 16 Aqueous solutions

16.1 Precipitation reactions

WE 16.1.1 Compounds containing sodium ions are usually soluble, so sodium nitrate will not form a precipitate. Compounds containing sulfide ions are usually insoluble, so copper(II) sulfide will form a precipitate.

WE 16.1.2 $Na^+(aq)$ and $SO_4^{2-}(aq)$

WE 16.1.3 $K^+(aq)$ and $SO_4^{2-}(aq)$

1 **a** a precipitate of silver carbonate

b a precipitate of lead(II) hydroxide

c a precipitate of magnesium sulfide

d no precipitate

Silver, lead and magnesium ions are not found in the solubility table. However, the anions that each of these are combined with in these questions all form compounds that are usually insoluble. It is also worth noting that compounds containing sodium and nitrate ions are usually soluble.

The iron(II) ion is also not found in the solubility table. However, iron(II) does not appear in the 'exceptions' column of the table for either nitrates or sulfates, both of which generally form soluble compounds.

2 Chemicals containing the ions Na^+, K^+, NH_4^+ and NO_3^- almost never form a precipitate.

a **i** magnesium sulfide **ii** silver chloride **iii** aluminium hydroxide **iv** magnesium hydroxide

b **i** $Mg^{2+}(aq) + S^{2-}(aq) \rightarrow MgS(s)$

ii $Ag^+(aq) + Cl^-(aq) \rightarrow AgCl(s)$

iii $Al^{3+}(aq) + 3OH^-(aq) \rightarrow Al(OH)_3(s)$

iv $Mg^{2+}(aq) + 2OH^-(aq) \rightarrow Mg(OH)_2(s)$

3 **a** $Ag^+(aq) + I^-(aq) \rightarrow AgI(s)$

b No precipitate is formed.

c $Ba^{2+}(aq) + SO_4^{2-}(aq) \rightarrow BaSO_4(s)$

d $Pb^{2+}(aq) + SO_4^{2-}(aq) \rightarrow PbSO_4(s)$

e $3Ca^{2+}(aq) + 2PO_4^{3-}(aq) \rightarrow Ca_3(PO_4)_2(s)$

f $Pb^{2+}(aq) + 2OH^-(aq) \rightarrow Pb(OH)_2(s)$

4 **a** $Na^+(aq)$ and $NO_3^-(aq)$ **b** no spectator ions

c $NH_4^+(aq)$ and $Cl^-(aq)$ **d** $K^+(aq)$ and $NO_3^-(aq)$

e $Na^+(aq)$ and $Cl^-(aq)$ **f** $Na^+(aq)$ and $NO_3^-(aq)$

16.2 Concentration of solutions

WE 16.2.1 20.0g L^{-1} **WE 16.2.2** 215 ppm

1 D

2 **a** 5.0 ppm **b** 625 ppm **c** 27 ppm

3 sugar 14.0% (m/v); fat 3.0% (m/v)

4 yes, 0.870 ppm

16.3 Molar concentration

WE 16.3.1 0.48 mol L^{-1} **WE 16.3.2** 6.6×10^{-1} mol L^{-1}

WE 16.3.3 2.50×10^{-3} mol

1 B

2 **a** 8.0×10^{-2} mol L^{-1} **b** 0.30 mol L^{-1} **c** 0.19 mol L^{-1}

3 **a** 0.25 mol L^{-1} **b** 0.50 mol L^{-1} **c** 0.438 mol L^{-1}

4 0.120 mol L^{-1} **5** 7.10×10^{-3} mol L^{-1}

6 **a** 2.2×10^{-2} mol **b** 6.4×10^{-3} mol

c 2.34×10^{-4} mol **d** 7.8×10^{-4} mol

16.4 Dilution

WE 16.4.1 0.0250 mol L^{-1} **WE 16.4.2** 2.4 L

WE 16.4.3 400 ppm

1 **a** 0.4 mol L^{-1} **b** 0.075 mol L^{-1} **c** 0.025 mol L^{-1}

2 D **3** 8.0×10^{-2} mol L^{-1} **4** 4.25 mol L^{-1}

16.5 Calculations involving reactions in solutions

WE 16.5.1 0.234g **WE 16.5.2** 0.615 mol L^{-1}

1 0.65g **2** 4.27×10^{-2} L

3 2.03g **4** 5.58×10^{-2} L

Chapter 16 review

1 **a** true **b** true **c** false **d** false **e** true **f** false

2

	NaOH	KBr	NaI	$MgSO_4$	$BaCl_2$
$Pb(NO_3)_2$	$Pb(OH)_2$	$PbBr_2$	PbI_2	$PbSO_4$	$PbCl_2$
KI					
$CaCl_2$	$Ca(OH)_2$			$CaSO_4$	
Na_2CO_3				$MgCO_3$	$BaCO_3$
Na_2S				MgS	BaS

3 **a** barium sulfate **b** none

c lead(II) sulfate **d** none

4 **a** $Ag^+(aq) + Cl^-(aq) \rightarrow AgCl(s)$

b $Fe^{2+}(aq) + S^{2-}(aq) \rightarrow FeS(s)$

c $Fe^{3+}(aq) + 3OH^-(aq) \rightarrow Fe(OH)_3(s)$

d $Cu^{2+}(aq) + 2OH^-(aq) \rightarrow Cu(OH)_2(s)$

e $Ba^{2+}(aq) + SO_4^{2-}(aq) \rightarrow BaSO_4(s)$

5 **a** $Cu^{2+}(aq) + CO_3^{2-}(aq) \rightarrow CuCO_3(s)$

spectator ions: Na^+, SO_4^{2-}

b $Ag^+(aq) + Cl^-(aq) \rightarrow AgCl(s)$

spectator ions: K^+, NO_3^-

c $Pb^{2+}(aq) + S^{2-}(aq) \rightarrow PbS(s)$
spectator ions: Na^+, NO_3^-
d $Fe^{3+}(aq) + PO_4^{3-}(aq) \rightarrow FePO_4(s)$
spectator ions: Na^+, Cl^-
e $Fe^{3+}(aq) + 3OH^-(aq) \rightarrow Fe(OH)_3(s)$
spectator ions: K^+, SO_4^{2-}

6 a 2 ppm b 2.0×10^{-4} %
7 31.5 g
8 0.38 mol L^{-1}
9 a 2.6×10^{-3} mol b 3.75×10^{-3} mol c 2.3×10^{-2} mol
10 a 2.04 g b 1.7 g
11 37.5 mL
12 0.533 ppm
13 1.74 g
14 34 mL
15

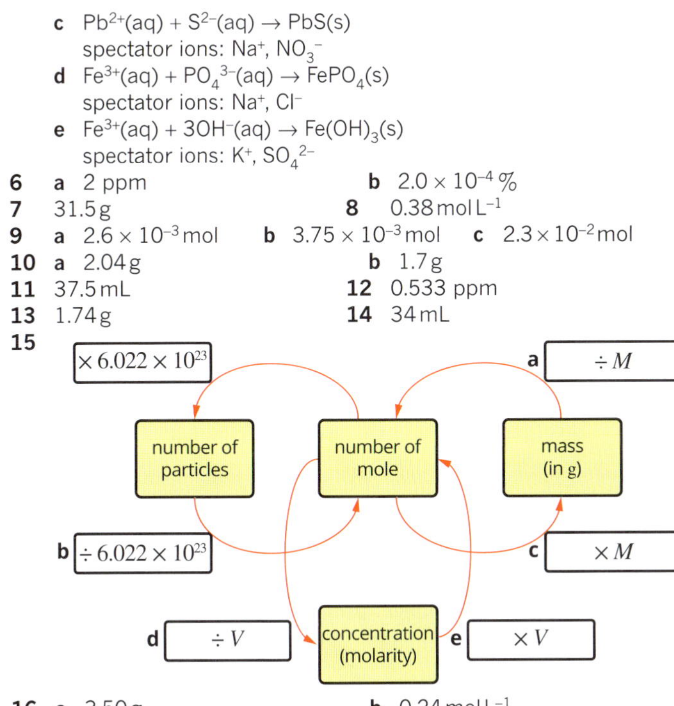

16 a 3.50 g b 0.24 mol L^{-1}

Chapter 17 Acids and bases

17.1 Properties of acids and bases

1 A
2 B
3 Litmus just tells you whether the solution is acidic or basic. Universal indicator will give an indication of the level of acidity.
4 Although CuO is a base, it is insoluble, so will not affect the pH of the water, i.e. copper oxide is a base, but not an alkali.

17.2 The Arrhenius theory of acids and bases

WE 17.2.1 pH = 8.2

1 $HBr(g) \rightarrow H^+(aq) + Br^-(aq)$
2 $H_2SO_4(aq) \rightarrow 2H^+(aq) + SO_4^{2-}(aq)$
3 C
4 Stronger acids more readily ionise, forming ions in solution. As perchloric acid is stronger, more hydronium ions would be present in solution than in a solution of ethanoic acid, making it a better conductor of electricity.
5 pH = 2
6 pH = 3

17.3 Reactions of acids and bases

WE 17.3.1 spectator ions $K^+(aq)$ and $SO_4^{2-}(aq)$;
$H^+(aq) + OH^-(aq) \rightarrow H_2O(l)$

WE 17.3.2 spectator ions $Na^+(aq) + Cl^-(aq)$;
$H^+(aq) + HCO_3^-(aq) \rightarrow H_2O(l) + CO_2(g)$

Note that if hydronium ions are represented as $H_3O^+(aq)$, rather than as $H^+(aq)$, this reaction would be written as: $H_3O^+(aq) + HCO_3^-(aq) \rightarrow 2H_2O(l) + CO_2(g)$

WE 17.3.3 spectator ion $Cl^-(aq)$;
$6H^+(aq) + 2Al(s) \rightarrow 2Al^{3+}(aq) + 3H_2(g)$

1 a $Mg(s) + H_2SO_4(aq) \rightarrow MgSO_4(aq) + H_2(g)$
$Mg(s) + 2H^+(aq) \rightarrow Mg^{2+}(aq) + H_2(g)$
b $Ca(s) + 2HCl(aq) \rightarrow CaCl_2(aq) + H_2(g)$
$Ca(s) + 2H^+(aq) \rightarrow Ca^{2+}(aq) + H_2(g)$
c $Zn(s) + 2CH_3COOH(aq) \rightarrow Zn(CH_3COO)_2 + H_2(g)$
$Zn(s) + 2H^+(aq) \rightarrow Zn^{2+}(aq) + H_2(g)$
d $2Al(s) + 6HCl(aq) \rightarrow 2AlCl_3(aq) + 3H_2(g)$
$2Al(s) + 6H^+(aq) \rightarrow 2Al^{3+}(aq) + 3H_2(g)$

2 a magnesium sulfate b calcium chloride
c zinc acetate or ethanoate d aluminium chloride
3 a i $ZnO(s) + H_2SO_4(aq) \rightarrow ZnSO_4(aq) + H_2O(l)$
ii $ZnO(s) + 2H^+(aq) \rightarrow Zn^{2+}(aq) + H_2O(l)$
b i $Ca(s) + 2HNO_3(aq) \rightarrow Ca(NO_3)_2(aq) + H_2(g)$
ii $Ca(s) + 2H^+(aq) \rightarrow Ca^{2+}(aq) + H_2(g)$
c i $Cu(OH)_2(s) + 2HNO_3(aq) \rightarrow Cu(NO_3)_2(aq) + 2H_2O(l)$
ii $Cu(OH)_2(s) + 2H^+(aq) \rightarrow Cu^{2+}(aq) + 2H_2O(l)$
d i $Mg(HCO_3)_2(s) + 2HCl(aq) \rightarrow MgCl_2(s) + 2H_2O(l) + 2CO_2(g)$
ii $Mg(HCO_3)_2(s) + 2H^+(aq) \rightarrow Mg^{2+}(aq) + H_2O(l)$
e i $SnCO_3(s) + H_2SO_4(aq) \rightarrow SnSO_4(s) + H_2O(l) + CO_2(g)$
ii $SnCO_3(s) + 2H^+(aq) \rightarrow Sn^{2+}(s) + H_2O(l) + CO_2(g)$
4 a $2KOH(aq) + H_2SO_4(aq) \rightarrow K_2SO_4(aq) + 2H_2O(l)$
$OH^-(aq) + H^+(aq) \rightarrow H_2O(l)$
b $NaOH(aq) + HNO_3(aq) \rightarrow NaNO_3(aq) + H_2O(l)$
$OH^-(aq) + H^+(aq) \rightarrow H_2O(l)$
c $Mg(OH)_2(s) + 2HCl(aq) \rightarrow MgCl_2(aq) + 2H_2O(l)$
$Mg(OH)_2(s) + 2H^+(aq) \rightarrow Mg^{2+}(aq) + 2H_2O(l)$
d $CuCO_3(s) + H_2SO_4(aq) \rightarrow CuSO_4(aq) + H_2O(l) + CO_2(g)$
$CuCO_3(s) + 2H^+(aq) \rightarrow Cu^{2+}(aq) + H_2O(l) + CO_2(g)$
e $KHCO_3(aq) + HCl(aq) \rightarrow KCl(aq) + H_2O(l) + CO_2(g)$
$HCO_3^-(aq) + H^+(aq) \rightarrow H_2O(l) + CO_2(g)$
f $Zn(s) + 2HNO_3(aq) \rightarrow Zn(NO_3)_2(aq) + H_2(g)$
$Zn(s) + 2H^+(aq) \rightarrow Zn^{2+}(aq) + H_2(g)$
g $CaCO_3(s) + 2HCl(aq) \rightarrow CaCl_2(aq) + H_2O(l) + CO_2(g)$
$CaCO_3(s) + 2H^+(aq) \rightarrow Ca^{2+}(aq) + H_2O(l) + CO_2(g)$
h $NaHCO_3(s) + CH_3COOH(aq) \rightarrow CH_3COONa(aq) + H_2O(l) + CO_2(g)$
$NaHCO_3(s) + H^+(aq) \rightarrow Na^+(aq) + H_2O(l) + CO_2(g)$
i $(NH_4)_2CO_3(aq) + Ca(OH)_2(aq) \rightarrow 2NH_3(g) + CaCO_3(s) + 2H_2O(l)$
$2NH_4^+(aq) + Ca^{2+}(aq) + CO_3^{2-} + 2OH^-(aq) \rightarrow CaCO_3(s) + 2NH_3(g) + 2H_2O(l)$

17.4 Calculations involving acids and bases

WE 17.4.1 1.00 mol L^{-1} **WE 17.4.2** 60.0 mL
WE 17.4.3 pH = 1.5 **WE 17.4.4** 10.0 mL
WE 17.4.5 0.523 g

1 0.075 mol L^{-1} 2 30 mL 3 100 mL
4 This increases the pH by one unit.
5 a $HNO_3(aq) + KOH(aq) \rightarrow KNO_3(aq) + H_2O(l)$
b 0.001985 mol c 0.001985 mol d 0.1087 mol L^{-1}
6 a 0.00200 mol b 0.00160 mol
c Na_2CO_3

Chapter 17 review

1 Dip pieces of the red litmus paper into the two solutions. The paper will remain red in the acidic solution and turn blue in the alkaline solution.
2 a Sulfuric acid (H_2SO_4) is a diprotic acid because each molecule can produce two hydrogen ions when it ionises:
$H_2SO_4(aq) \rightarrow 2H^+(aq) + SO_4^{2-}(aq)$
Hydrochloric acid (HCl) is a monoprotic acid as each molecule can only produce one hydrogen ion when the molecule ionises:
$HCl(g) \rightarrow H^+(aq) + Cl^-(aq)$
b A strong acid is one that ionises completely in solution (e.g. HCl). A concentrated acid is one in which there is a large amount of acid dissolved in a given volume of solution; for example, 5 mol L^{-1} HCl and 5 mol L^{-1} CH_3COOH are concentrated acids.
3

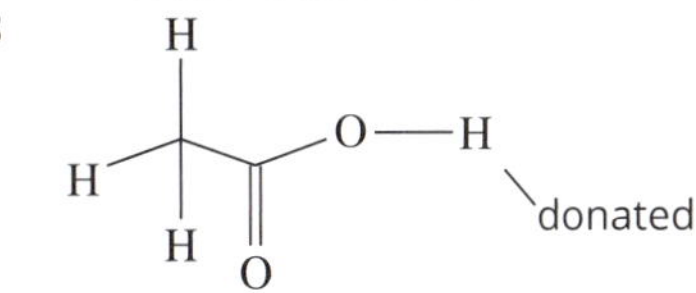

4 A scientific model is a representation that can be used to describe and explain phenomena or observations.
5 Strong acids are completely ionised; weak acids are partially ionised. Lower levels of ionisation mean there are fewer charged hydrogen ions in solution.

6 $HClO_4(aq) \rightarrow ClO_4^-(aq) + H^+(aq)$
7 $HClO_3(aq) \rightleftharpoons ClO_3^-(aq) + H^+(aq)$
8 $NH_3(aq) + H_2O(l) \rightleftharpoons NH_4^+(aq) + OH^-(aq)$
9 basic
10 The difference is a factor of 100.
11 **a** 2 **b** 0.7 **c** −0.6
12 **a** $HNO_3(aq) + KOH(aq) \rightarrow KNO_3(aq) + H_2O(l)$
b $H_2SO_4(aq) + K_2CO_3(aq) \rightarrow K_2SO_4(aq) + H_2O(l) + CO_2(g)$
c $2H_3PO_4(aq) + 3Ca(HCO_3)_2(s) \rightarrow Ca_3(PO_4)_2(s) + 6H_2O(l) + 6CO_2(g)$
d $2HF(aq) + Zn(OH)_2(s) \rightarrow ZnF_2(aq) + 2H_2O(l)$
13 E
14 $2H^+(aq) + CO_3^{2-}(aq) \rightarrow H_2O(l) + CO_2(g)$
15 a carbonate or hydrogencarbonate
16 **a** $2Al(s) + 3H_2SO_4(aq) \rightarrow Al_2(SO_4)_3(aq) + 3H_2(g)$
b $2Al(s) + 6H^+(aq) \rightarrow 2Al^{3+}(aq) + 3H_2(g)$
17 **a** $[H^+] = 10^{-2}$ or 0.01 mol L^{-1} **b** 0.005 mol
18 **a** pH = 1.3 **b** pH = 3
19 111 mL
20 The pH will increase.
21 solution A: weaker base, few freely moving charged particles—ammonia
solution B: neutral, no freely moving charged particles—glucose
solution C: strong base, many freely moving charged particles—sodium hydroxide
solution D: strong acid, many freely moving charged particles—hydrochloric acid
solution E: weaker acid, few freely moving charged particles—ethanoic acid
22 **a** a substance that dissociates in water to form hydroxide (OH^-) ions
b a substance that completely ionises (in water) to form hydrogen ions (H^+) ions
c a measure of concentration expressed in moles per litre of solution (mol L^{-1})

Chapter 18 Rates of chemical reactions

18.1 Investigating the rate of chemical reactions

1 C
2 **a** increasing **b** increasing **c** increasing
d increasing **e** adding
3 **a** $CaCO_3(s) + 2HNO_3(aq) \rightarrow Ca(NO_3)_2(aq) + CO_2(g) + H_2O(l)$
b Because the gradient of the graph is decreasing, the reaction rate is decreasing over time.
4 **a** using smaller pieces of wood with a larger surface area
b using a brick cleaner with a higher concentration
c increasing the temperature of the oven

18.2 Collision theory

1 D **2** A
3 **a** $CH_4(g) + 2O_2(g) \rightarrow CO_2(g) + 2H_2O(g)$
b −890 kJ mol^{-1}
c

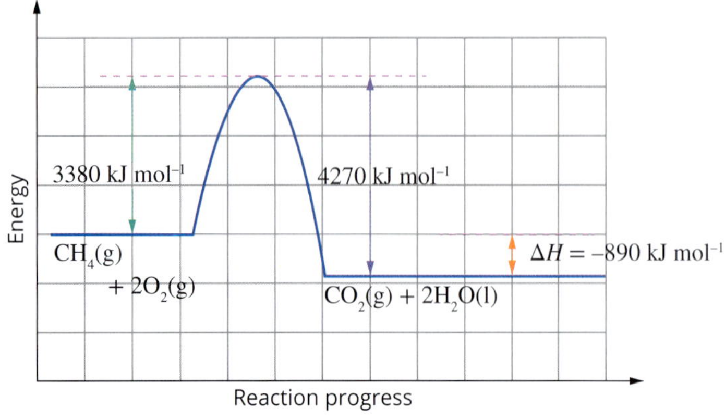

4 **a**

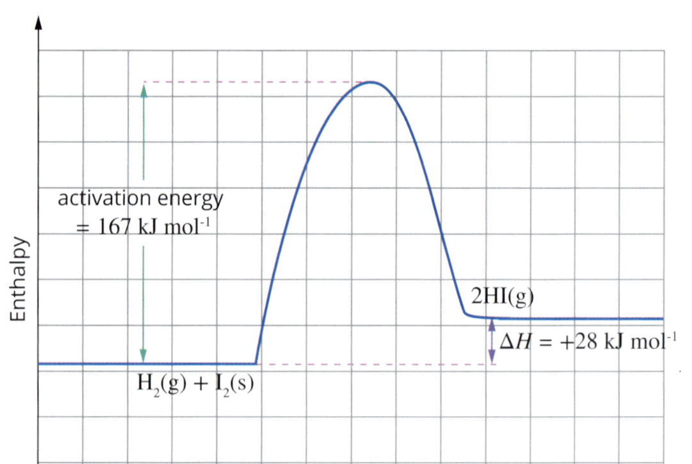

b Endothermic. ΔH is positive therefore the enthalpy of the products is greater than the enthalpy of the reactants.
c −28 kJ mol^{-1}
d 139 kJ mol^{-1}

18.3 Applying collision theory

WE 18.3.1 The surface area of the iron anchor is relatively small so the frequency of collisions with reacting particles would be low. The concentration of oxygen at greater depths is also low so the frequency of collisions is further reduced. Therefore, the rate of corrosion is reduced.

1 D **2** A, B and C **3** D, F, B, E, C and A
4 **a** At higher temperatures, the molecules that react to form fibreglass plastics have greater energy. They collide more frequently and are more likely to have a total energy exceeding the activation energy of the reaction involved, increasing the rate of reaction.
b Fine particles have a large surface area, resulting in a high frequency of collisions of aluminium particles with gas molecules (such as oxygen) in the air and hence rapid reaction rate. The aluminium can burn vigorously and release a large quantity of heat.
c At high altitude, such as in Nepal, air pressure is considerably lower than at any location in the Australian bush and so the water boils at a lower temperature in Nepal (up to 30°C lower). Thus, the average kinetic energy of the molecules in the potato is lower, so they are less likely to have a total energy exceeding the activation energy of the reactions involved in cooking a potato, so the potato cooks slower.
5 **a** A is activation energy, B is ΔH, C is total enthalpy of reaction.
b A catalyst would lower the value of A and C, while the value of B would remain unchanged.

Chapter 18 review

1 D   2 C   3 B   4 B

5 a A gas is produced so mass is lost from the mixture.
  b HNO_3 is in excess, so Cu is limiting.
  c

Mass of mixture (g)
$1.00\ mol\ L^{-1}$ nitric acid
Time (s)

Decreased rate of mass loss is due to lower nitric acid concentration.
Copper is still limiting, so final mass remains the same.

  d

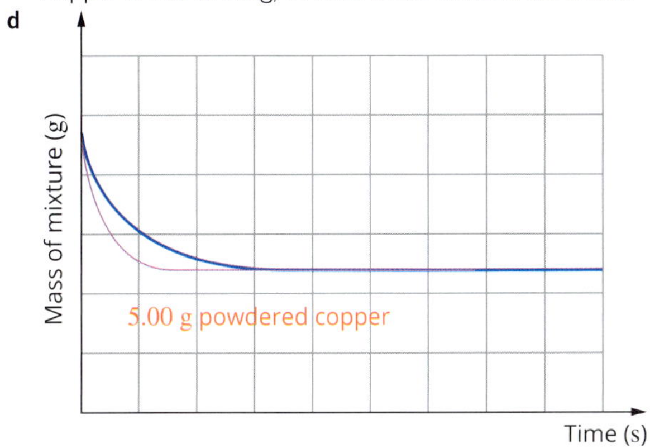

Increased rate of mass loss is due to increased copper surface area.
Copper is still limiting, so final mass remains the same.

6 Reactant particles must collide with each other, collide with sufficient energy to break the bonds within the reactants, collide with the correct orientation to break the bonds within the reactants and so allow the formation of new products.

7 Collision 1 has the correct collision orientation, allowing bonds to break within the reactants and bonds to form within the products.

8 a

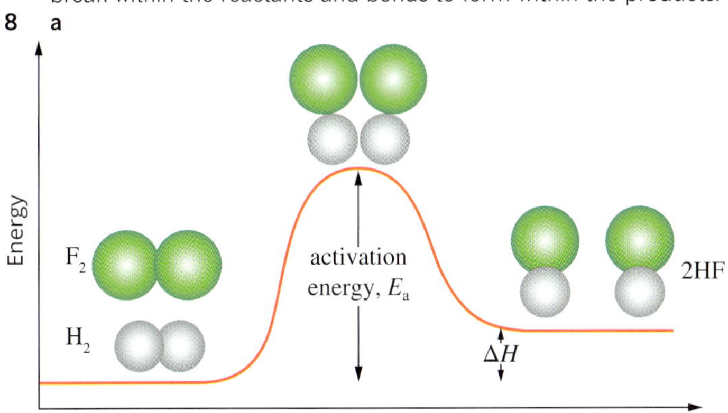

  b the minimum amount of energy required to break bonds in the reactants in order to form products in a reaction
  c endothermic
  d

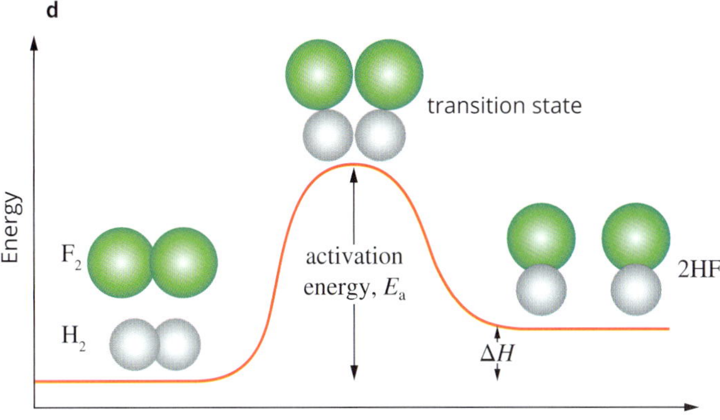

  e F–F and H–H bonds are being broken. H–F bonds are beginning to form.

9 a the single H–H bond in each hydrogen molecule and the double O=O bond in each oxygen molecule
  b two H–O bonds in each new water molecule during the reaction
  c The energy change for the reaction is the difference between the energy absorbed to break the bonds in the H_2 and O_2 reactants, and the energy released when the bonds in the H_2O product are made.
  d No reaction occurs until sufficient energy is supplied to overcome the activation energy.

10 B   11 D

12 a At lower temperatures, the molecules that react to cause the apple to brown have less energy. They collide less frequently and are less likely to have a total energy exceeding the activation energy of the reaction involved, decreasing the rate of reaction.
  b Using pure oxygen gas instead of air increases the concentration of oxygen. This results in a higher frequency of collisions between hydrogen and oxygen molecules and hence an increased reaction rate.

13 A   14 C

15 a surface area of a solid reactant, concentration of reactants in a solution, pressure of any gaseous reactants, temperature of the reaction, presence of a catalyst
  b i surface area of a solid reactant, concentration of reactants in a solution, pressure of any gaseous reactants
    ii temperature of the reaction (also increases collision frequency), presence of a catalyst

16 a $CaCO_3(s) + 2HCl(aq) \rightarrow CaCl_2(aq) + CO_2(g) + H_2O(l)$
  b $CaCO_3$ is in excess.
  c • a decrease in mass of reaction mixture as $CO_2(g)$ escapes to the atmosphere
    • an increase in pH with a pH probe as acid is consumed
  d The rate of reaction with the smaller lumps will be faster. The smaller lumps have a larger surface area so more collisions can occur per second.
  e increase temperature; increase concentration of hydrochloric acid. Both increase the frequency of collisions between reactants, and temperature also increases the proportion of successful collisions.

17 a Higher body temperature increases the rate of reactions. Increased pulse and breathing rate increases the concentration of reactants.
  b Lower body temperature decreases rate of metabolic reactions in the body, which protects organs from damage.

18 • Grind up the sugar crystals or use caster sugar.
  • Use a cup of hot water to dissolve the sugar.
  • Gently heat the sugar and water mixture while the sugar was dissolving.
  • Stir the sugar and water mixture while the sugar was dissolving.

Chapter 19 Catalysts

19.1 Catalysts

1 D

2 **a** a substance that increases the rate of a chemical reaction without itself undergoing permanent change

b the minimum amount of energy required to break the bonds in reactants in order to form products in a reaction

3 When salt is mixed with sugar, the salt acts as a catalyst and lowers the activation energy of the combustion reaction between sugar and oxygen.

4 **a** Reactions involving a heterogeneous catalyst take place at the surface of the catalyst. Reactants form bonds with the catalyst, lowering the activation energy of reactions and allowing them to proceed more rapidly.

b A porous pellet has a much larger surface area than a solid lump. More reactants may be in contact with the surface of a porous pellet at any instant, producing a faster rate of reaction.

Chapter 19 review

1 **a** Increasing the pressure of the gases would cause an increase in the number of collisions between the reactant molecules in a given time, so more collisions would occur with the correct orientation to react and with energy that is greater than or equal to the activation energy. As a result, the rate of reaction would increase.

b Adding a catalyst would allow the reaction to occur by a different pathway with a lower activation energy. The proportion of collisions with energy greater than the activation energy would thus be increased. As a result, the rate ofreaction would increase.

2 **a** $1370\,kJ\,mol^{-1}$ **b** $-572\,kJ\,mol^{-1}$

c

without catalyst

Energy

$2H_2(g) + O_2(g)$

$2H_2O(g)$

with catalyst

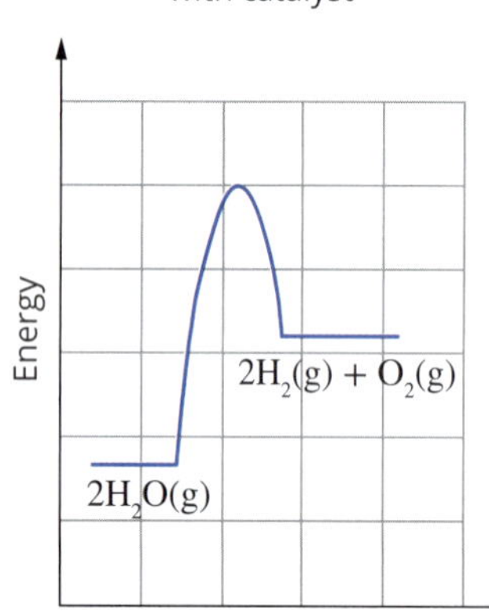

d $+572\,kJ\,mol^{-1}$

3 **a** Homogeneous. Hydrogen peroxide is a liquid and the dichromate ion catalyst is in an aqueous solution, thus they are the same phase.

b Heterogeneous. The reactants are gaseous while the catalyst is solid (V_2O_5).

c Heterogeneous. The reactants are gaseous while the catalyst is solid (nickel metal).

d Homogeneous. Both the reactant (ethanol) and catalyst (enzymes) are in aqueous solution.

4
- more easily separated from the products of a reaction
- much easier to reuse
- able to be used at high temperatures.

5 **a** false **b** true **c** false **d** false

6 C

7 The rapid effervescence on addition of the cobalt(II) ions is evidence that CO_2 is being rapidly formed in the reaction which suggests the cobalt(II) ions have increased the rate of reaction. The colour change from pink to green suggests that the cobalt(II) ions are involved in the reaction and are changed to an intermediate during the course of the reaction, but the intermediate is changed back to cobalt(II) ions at the end of the reaction.

8 A and B

9 The intricate three-dimensional structure of their active site is destroyed at pH values outside these ranges. The enzyme and substrate will be unable to form the enzyme–substrate complex that allows the reaction to proceed.

10 Catalytic converters are able to convert pollutants formed during combustion of hydrocarbon fuels in an internal combustion engine into CO_2, H_2O and N_2.

11 X, Z, W, Y

12 The platinum gauze is acting as a catalyst. It provides an alternative reaction pathway with a lower activation energy than the reaction pathway without the gauze. This allows hydrogen and oxygen gases to react at a lower temperature.

13 Nanomaterials have a very large surface area so there is much greater contact between reactant particles and catalyst, which means more catalytic reactions can occur in any given time.

The larger surface area of nanomaterials means it is possible to use less material to achieve the same level of catalytic activity.

14 C

15 Nitrogen molecules, N_2, consist of two nitrogen *atoms* held together by a triple covalent bond. This is a particularly *strong* covalent bond and so the reaction has a large activation energy in the absence of a catalyst. Iron is a suitable catalyst for this reaction, capable of providing a new reaction pathway and so *decreasing* the activation energy. Nitrogen and hydrogen molecules *adsorb* onto the iron surface and form bonds with the iron surface. The *covalent* bonds in the N_2 and H_2 molecules are *broken* and individual neighbouring nitrogen and hydrogen atoms *form covalent bonds* and become ammonia molecules. The ammonia molecules then leave the iron surface.

16 **a**

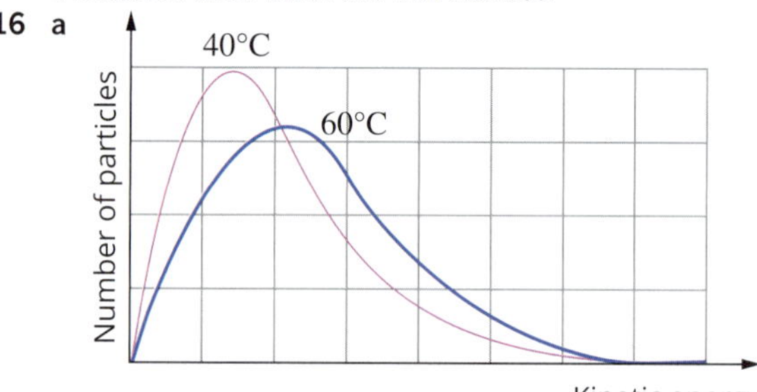

b

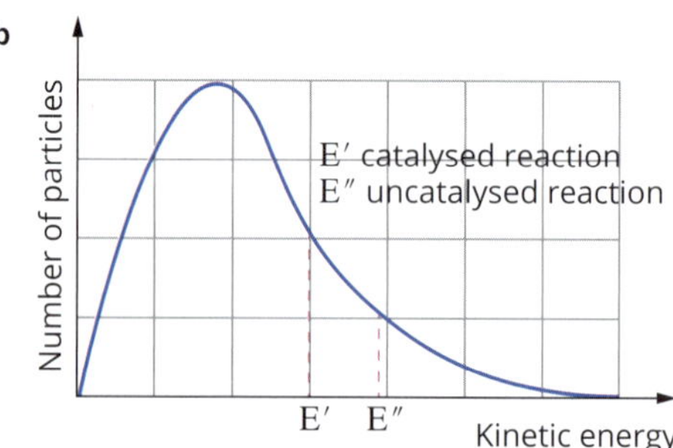

c When a catalyst is present, the reaction proceeds by an alternative reaction pathway with a lower activation energy (*E′*) than in the uncatalysed reaction (*E″*). At a given temperature, there will be a greater proportion of reacting particles that have a kinetic energy equal to or greater than the activation energies (*E′* than *E″*). As more reactants have sufficient energy to react, the rate of reaction increases.

17 a, c

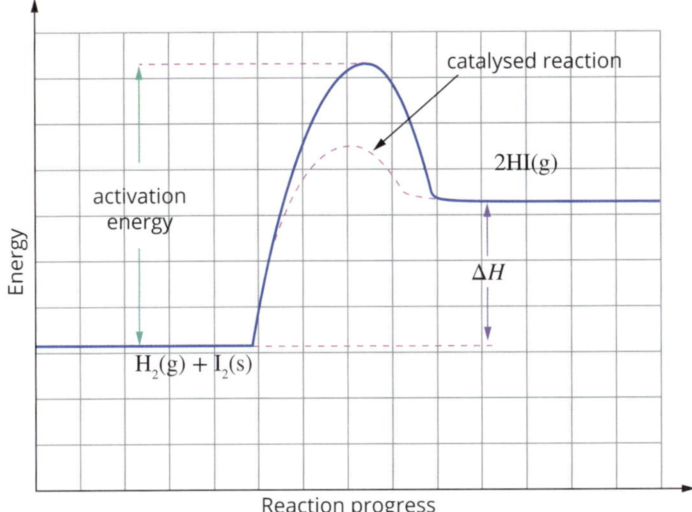

b endothermic

Chapter 20 Science inquiry skills in chemistry

20.1 Questioning

WE 20.1.1 The energy released per gram of the hydrocarbon increases as the molecular mass of the hydrocarbon fuel increases.

1 **a** A hypothesis is a statement that can be tested. This involves making a prediction based on previous observations.
b A theory is a hypothesis that is supported by a great deal of evidence from a wide variety of sources. A principle is a theory that is so strongly supported by evidence that it is unlikely to be shown to be untrue in the future.

2 A

3 **a** electrical conductivity **b** concentration of lead
c electrical conductivity **d** pH

4 qualitative **5** C **6** C

20.2 Planning investigations

1 B

2 **a** In a control experiment, two groups of subjects are tested; the groups, or the tests performed on them, are identical except for a single factor (the variable) which is not changed in the control experiment.
b The dependent variable is the variable that is measured to determine the effect of changes in the independent variable. The independent (experimental) variable is the variable that is changed in an experiment. For example, in an experiment testing the effect of soil pH on flower colour, the independent variable would be soil pH and the dependent variable would be flower colour.

3 **a** valid **b** reliable **c** accurate

4 **a** type of soft drink **b** pH
c temperature of solutions

5 **a** Litmus paper and universal indicator give qualitative information about pH through colour.
b A calibrated pH meter will give quantitative information and is more accurate than using litmus paper or universal indicator.

20.3 Uncertainty and error in data

1 **a** systematic errors **b** random errors

2 **a** systematic (evaporation will lead to a value that is always higher than the true value after evaporation takes place)
b random **c** mistake

3 There could be many reasons why the same experimental results cannot be obtained. The experimental design may be poor because of:

- a lack of objectivity
- a lack of clear and simple instructions
- a lack of appropriate equipment
- a failure to control variables.

Other problems not specifically related to the experiment could be:

- a poor hypothesis that could not be tested objectively
- conclusions that do not agree with the results
- interpretations that are subjective.

20.4 Processing data and information

1 22.7

2 Add a trend line or line of best fit.

3 **a**

Volume of carbon dioxide produced in first three seconds of reaction of sodium hydrogencarbonate and solutions of citric acid

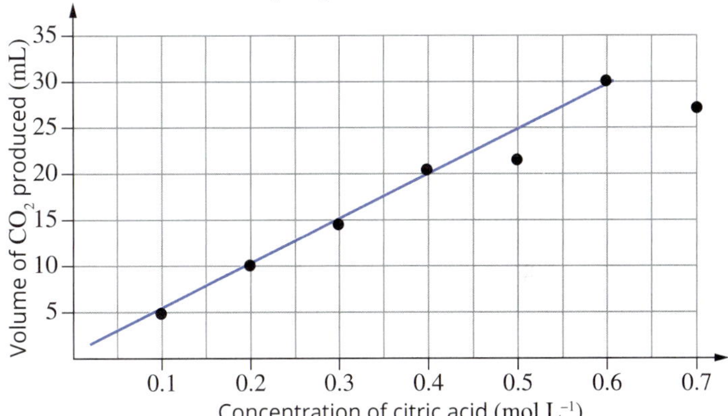

b Data point 4 (0.5, 21.5) is an outlier.
c An outlier is a point in the data that does not fit the trend.

20.5 Analysing data and information

1 A linear graph shows the proportional relationship between two variables.

2 an inversely proportional relationship

3 directly proportional

4 time restraints and limited resources

20.6 Conclusions

1 B

2 The statement 'Many repeats of the procedure were conducted' is unquantified. 'Thirty repeats of the procedure were conducted' is better because the number of trials is quantified.

20.7 Communicating

1 B **2** D

3 **a** g mol^{-1} **b** mol L^{-1}

4 Divide the value in g by 1000.

5 to reflect how we use quantities in the laboratory or to make the numbers easier to comprehend

Chapter 20 review

1 **a** dependent variable **b** controlled variable
c independent variable

2 **a** reflect **b** create **c** analyse **d** investigate
e apply **f** identify **g** describe

3 Independent variable = source of the water, dependent variable = heavy metal concentration, controlled variables = temperature, time of testing

4 **a** bar graph **b** line graph
c scatter graph (with line of best fit) **d** pie diagram

5 a It can dissolve or eat away at substances including tissues such as your skin or lungs.
b It is toxic (poisonous) if inhaled.
c It is a highly combustible liquid that could catch on fire.

6 Accuracy refers to the ability of the method to obtain the correct measurement close to a true or accepted value.

Validity refers to whether an experiment or investigation is in fact testing the set hypothesis and aims.

7 a mistake b random error c systematic error

8 an exponential/logarithmic relationship

9 a reliability b validity c accuracy d precision

10 Evaluate the method; identify issues that could affect validity, accuracy, precision and reliability of data; state systematic sources of error and uncertainty; and recommend improvements to the investigation if it is to be repeated.

11 A trend is a pattern or relationship that can be seen between the dependent and independent variables. It may be linear, in which the variables change in direct proportion to each other to produce a straight trend line. The relationship may be in proportion but non-linear, giving a curved trend line. The relationship may also be inverse, in which one variable decreases in response to the other variable increasing. This could be linear or non-linear.

12 any issues that could have affected the validity, accuracy, precision or reliability of the data plus any sources of error or uncertainty

13 a form of systematic error resulting from a researcher's personal preferences or motivations

14 C

15 to avoid plagiarism and ensure creators and sources are properly credited for their work

16 A

17 a 0.0300 L or 3.00×10^{-2} L b 34 500 g or 3.45×10^{4} g

18 a Aim: To determine the effect of increasing water temperature on the electrical conductivity of water.
b independent variable—water temperature; dependent variable—electrical conductivity of water; controlled variables—pH, water source, sampling container
c The data collected would be conductivity using a probe and, therefore, it would be quantitative.
d 10 mL measuring cylinder ±0.1 mL, pH probe ±0.02 alcohol-filled glass thermometer ±0.1°C, electrical conductivity probe ±2
e Raw data is data collected from experiments and recorded as measurements are taken. Processed data tabulates this into a form in which the reader can clearly see the temperature of the water and the conductivity at each separate temperature value. This can also then be processed and graphed with the independent variable on the *x*-axis (temperature) and dependent variable on the *y*-axis (conductivity). If the hypothesis is correct, the graph for this experiment would look like the one below:

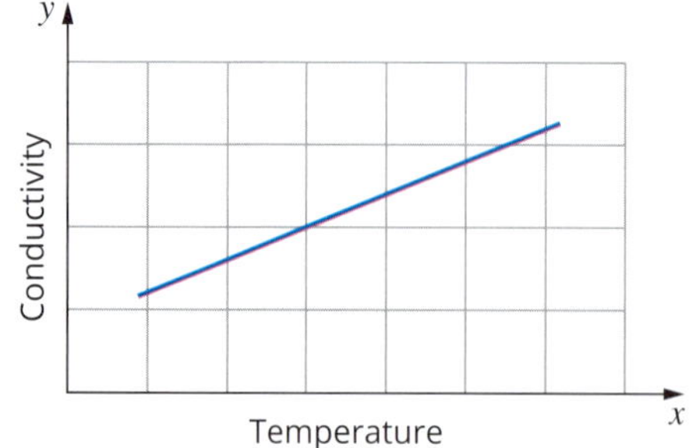

Unit 2 review: Molecular interactions and reactions

Section 1: Multiple choice

1	B	2	C	3	C	4	A
5	C	6	D	7	D	8	C
9	A	10	D	11	B	12	C

Section 2: Short answer

1 a

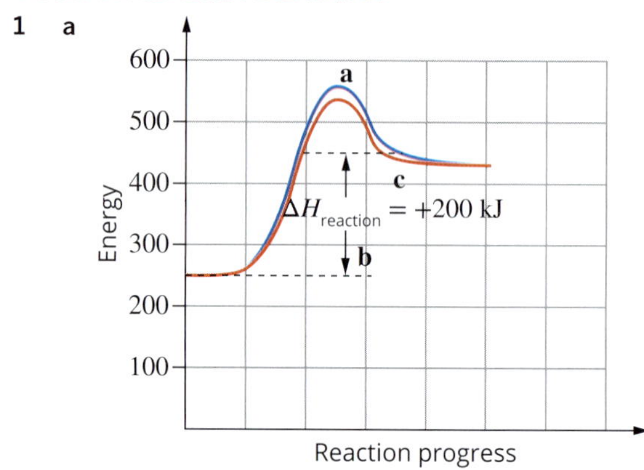

2 a $2H_2O_2(aq) \rightarrow 2H_2O(l) + O_2(g)$ + heat
b

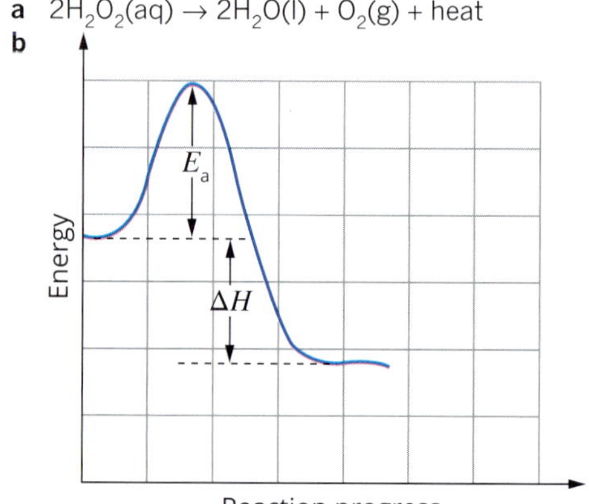

c

Energy

E_a

ΔH

Reaction progress

3 a $H_2SO_4(aq) + 2KOH(aq) \rightarrow H_2O(l) + K_2SO_4(aq)$
b $H^+(aq) + OH^-(aq) \rightarrow H_2O(l)$
c $6HCl(aq) + 2Al(s) \rightarrow 2AlCl_3(aq) + 3H_2(g)$
d $6H^+(aq) + 2Al(s) \rightarrow 2Al^{3+}(aq) + 3H_2(g)$

4 a 0.250 mol L^{-1}
b 3.51×10^{5} g (3 sig. figs)

5 The molecules are all polar and bond in the liquid state by dipole–dipole forces between the molecules. As molar mass increases, the force of attraction between the molecules increase therefore the boiling point increases.

Between the HF molecules there is a particularly strong dipole–dipole force called a hydrogen bond. This causes the liquid HF to have a higher BP than the other hydrogen halides despite it having a lower molar mass.

6 **a** Boyle's law: As pressure increases the gaseous particles (atoms or molecules) are forced closer together and therefore volume decreases as pressure increases.

b Charles' law. As temperature increases, the molecules (or atoms) gain more kinetic energy and will move faster and further causing a gas to expand when heated.

Section 3: Extended answer

1 $4NH_3(g) + 5O_2(g) \rightarrow 4NO(g) + 6H_2O(g)$

a **i** 2.35×10^6 g (3 sig. figs)

ii 1.33×10^6 L at STP (3 sig. figs)

b The rate of a chemical reaction depends on the frequency of successful collisions between the reaction particles (atoms, ions and molecules). The higher the number of collisions per second, the more chance of a successful collision and the faster the rate.

i High concentration of reacting molecules causes more frequent collisions per second, and so there is more chance of a successful collision and the reaction rate will increase.

ii High temperature increases the kinetic energy of the molecules and therefore collision frequency. Higher temperature also means that more reacting molecules have enough energy to overcome the activation energy.

iii High pressure causes more molecules to occupy a space or volume and so collide more frequently, giving more chance of a successful collision and so a higher rate of reaction.

iv A catalyst provides another reaction path that has a lower activation energy and reaction will continue more rapidly. The finely divided platinum provides a much greater surface area on which a reaction may occur and therefore speeds up the reaction.

2 **a** as a percentage (parts per 100) and most commonly as ppm

b 0.616 mol per 100.0 g

c 6.16 mol L^{-1}

d saturated

e The solubility changes as temperature changes so must be stated for comparison.

f The beaker would contain a clear, colourless liquid with a white solid on the bottom.

g When dissolved in water, the ionic crystal dissociates into freely moving ions, which are attracted to the positive and negative electrodes. This movement of ions allows the solution to conduct a current.

$$\underset{\text{ionic solid}}{NaCl(s)} \xrightarrow{H_2O} \underset{\text{free mobile ions}}{Na^+(aq) + Cl^-(aq)}$$

h When sugar dissolves in water, the weak molecular bonds (hydrogen bonds) between the molecules are broken by the interaction (bonding) with the water molecules. As there are no charged particles, the solution is a non-conductor. There will, however, be a very low conductivity due to the weak ionisation of water.

$$\underset{\text{covalent molecular solid}}{C_{12}H_{22}O_{11}(s)} \xrightarrow{H_2O} \underset{\text{neutral molecules}}{C_{12}H_{22}O_{11}(aq)}$$

Glossary

A

absorbance A measure of the capacity of a substance to absorb light of a specified wavelength.

absorption line The individual colours of light in a continuous spectrum that are absorbed by the hydrogen atoms.

absorption spectrum The collection of absorption lines.

accuracy The ability to obtain measurements that are very close to the true or accepted value of the quantity.

acid A substance capable of producing hydrogen ions in solution (Arrhenius model) or donating a hydrogen ion (Bronsted-Lowry model).

acid–base reaction A reaction in which an acid reacts with a base.

acid rain Rainwater that has reacted with acidic emissions and has a pH less than 5.5.

acidic solution An aqueous solution in which the concentration of hydronium ions (H^+) is greater than the concentration of hydroxide ions (OH^-). At 25°C, pH < 7.

acidification An increase in atmospheric carbon dioxide concentration causing an increase in the acidity of the oceans.

acidity The concentration of H^+ ions in an aqueous solution. Acidity is measured using the pH scale.

activation energy The minimum energy required by reactants for a reaction to occur; symbol E_a. This energy is needed to break the bonds between atoms in the reactants to allow products to form.

active site The specific part of the enzyme molecule with which a reactant can interact.

addition reaction A reaction in which a molecule binds to an unsaturated hydrocarbon, forming a single carbon–carbon bond. In this process two reactant molecules form one product.

adsorption The adhesion of molecules or substances to the surface of a solid or liquid.

alkali A soluble base or a solution of a soluble base.

alkali metal A group 1 metal—Li, Na, K, Rb, Cs and Fr.

alkane A saturated hydrocarbon; general formula C_nH_{2n+2}.

alkene An unsaturated hydrocarbon containing one carbon–carbon double bond; general formula C_nH_{2n}.

alkyl group A group obtained by removing a hydrogen atom from an alkane; general formula C_nH_{2n+1}, e.g. methyl ($-CH_3$).

allotrope Different forms of the same element in which the atoms combine in different ways.

alloy A substance formed when other materials (e.g. carbon, other metals) are mixed with a metal.

alpha particle A positively charged particle formed from ionising radiation.

amorphous A structure that has no consistent arrangement of particles.

amount A measure used by chemists for counting particles; the unit is the mole.

anion A negatively charged ion, e.g. a chloride ion, Cl^-.

annealing Heating a metal to a moderate temperature and then allowing it to cool slowly to make it softer and more ductile.

aqueous When a chemical species has been dissolved in water, the resulting solution is said to be aqueous. This can be shown by writing '(aq)' after the name or symbol of the chemical.

Arrhenius model A model that defines an acid as a substance that ionises water to produce H^+ ions and a base as a substance that dissociates in water to form OH^- ions.

artesian basin An underground area of porous rock surrounded by rock that is not permeable to water. Rain seeps into the rock and is stored underground.

asymmetrical molecule A molecule in which the polar bonds are unevenly (or asymmetrically) distributed. The bond dipoles do not cancel and an overall molecular dipole is created.

atom The basic building block of matter. It is made up of subatomic particles—protons, neutrons and electrons.

atomic absorption spectroscopy (AAS) An analytical technique that uses light absorption to measure the concentration of a metal in a sample.

atomic number The number of protons in the nucleus of an atom; identical to the charge number of the nucleus; symbol Z.

atomic radius A measurement used for the size of atoms; determined by measuring half the distance between two adjacent atoms in a molecule or structure.

atomic theory of matter A theory proposed by John Dalton in 1802 that states that all matter is made up of atoms. He said that atoms are indivisible, atoms of the same element are identical and compounds are made up of different types of atoms in fixed ratios.

Avogadro's law At the same temperature and pressure, equal volumes of all gases contain equal numbers of particles.

Avogadro's number The number of particles in a mole; symbol $N_A = 6.022 \times 10^{23}\,mol^{-1}$.

B

ball-and-stick model A model that displays both the three-dimensional position of the atoms and the bonds between them. The atoms are represented by coloured spheres that are connected by rods to represent the bonds.

bar Unit of pressure equal to 100 kPa.

base A substance capable of producing hydroxide ions in solution (Arrhenius model) or accepting a proton (Bronsted-Lowry model).

basic solution An aqueous solution in which $[H^+] < [OH^-]$. For a basic solution at 25°C, pH > 7.

bias Measured values consistently in one direction from the actual value; they may be too high or too low.

biodegradable Capable of being decomposed by bacteria or other living organisms.

biodiesel A renewable fuel usually produced from vegetable oil from sources such as soyabean, canola or palm oil.

bioethanol A renewable fuel produced from crops such as sugar cane.

biofuel Fuel derived from plant materials such as grains (maize, wheat, barley or sorghum), sugar cane and vegetable waste, and vegetable oils.

biogas A renewable fuel that can be used to generate electricity.

Bohr diagram A simple diagram that shows the arrangement of electrons around the nucleus.

Bohr model A theory of the atom proposed by Niels Bohr that states that electrons in an atom occupy fixed, circular orbits that correspond to specific energy levels.

bore water Water collected in aquifers (underground water-bearing rock) below the Earth's surface. Bore water may be accessed by drilling and sinking a bore pipe into the aquifer.

brittle Shatters when given a sharp tap.

buckyball A ball-like polyhedral molecule consisting of carbon atoms of the type found in fullerenes.

C

calibrate To determine, check or rectify the graduation of any instrument giving quantitative measurements.

calibration curve A plot of data involving two variables that is used to determine the values for one of the variables.

carbon-12 The isotope of carbon that has a mass number of 12. The isotope contains six protons and six neutrons. One atom of carbon-12 is taken as having a mass of exactly 12 atomic mass units. This is the standard from which all other relative masses are calculated.

carbon nanotube A tube-shaped nanoparticle, made of carbon atoms.

carbon neutral A description of a process that overall does not contribute to a change in atmospheric carbon.

carrier gas An unreactive gas, often nitrogen, used in gas chromatography.

catalysis The increase in the rate of a chemical reaction due to the presence of a catalyst.

catalyst A substance that increases the rate of a reaction but is not consumed in the reaction. The catalyst provides a new reaction pathway with a lower activation energy.

cation A positively charged ion, e.g. a sodium ion, Na^+.

centrifugation A separation process that uses the action of rapid rotation to accelerate the settling of particles in a solid–liquid mixture.

ceramic Material that is produced by the firing (heating followed by cooling) of clay.

chemical (HAZCHEM) code Classification of chemicals into hazardous categories.

chemical contaminant An element or a compound that may be harmful if consumed in drinking water. It can be naturally occurring or synthetic.

chemical energy Form of energy stored in the chemical bonds between atoms and molecules. This energy results from attractions between electrons and protons in atoms; repulsions between nuclei; repulsions between electrons; movement of electrons; vibrations and rotations around bonds.

chemical formula A representation of a substance using symbols for its constituent elements. It shows the ratio of atoms present in the substance.

chemical symbol A symbolic representation of an element, usually one or two letters, where the first letter is capitalised and the second letter is lower case, e.g. carbon's symbol is C and sodium's symbol is Na.

chemiluminescence The light emitted by a chemical reaction that does not produce significant amounts of heat.

chemiluminescent Emitting light as the result of a chemical reaction.

chlorination A step in the purification of drinking water. The clear water is treated with gaseous chlorine to destroy any bacteria; the main purpose of chlorination is to remove biological contaminants.

chromatogram The output of a chromatography procedure. In TLC and paper chromatography, it is the pattern of bands or spots formed on a plate or on the paper. In HPLC, it is the graph produced.

chromatography A technique for separating the components of a mixture. The components are carried by a mobile phase over the adsorbent surface of the stationary phase.

collision theory A theoretical model that accounts for rates of chemical reactions in terms of collisions between particles occurring during a chemical reaction.

colloid A mixture in which very small particles are spread throughout a liquid, solid or gas. The particles are bigger than single molecules, but so small that they do not settle on standing.

column chromatography A chromatographic technique in which the stationary phase is in a column, e.g. high performance liquid chromatography (HPLC).

combustion A rapid reaction with oxygen accompanied by the release of large amounts of heat; also called burning.

complete combustion A hydrocarbon undergoes complete combustion with oxygen at high temperatures when the only products are carbon dioxide and water.

components The chemicals in a mixture.

composite material A combination of two or more distinct materials with significantly different physical and chemical properties.

compound A pure substance made up of different types of atoms combined in a fixed ratio.

concentrated solution A solution that has a relatively high ratio of solute to solvent.

concentration A measure of how much solute is dissolved in a specified volume or mass of solution.

condensed structural formula A simple representation of the structural formula of an organic molecule. A condensed structural formula shows the atoms connected to each carbon atom, but not all the bonds, e.g. $CH_3CH_2CH_3$.

conductivity Permitting the flow of electric charges.

conductor An object or type of material that permits the flow of electric charges, e.g. a wire is an electrical conductor that can carry electricity along its length.

control experiment An identical experiment carried out at the same time, but the independent variable is not changed.

controlled variable A variable that must be kept constant during an investigation.

core charge (or effective nuclear charge) The effective nuclear charge experienced by the outer-shell electrons in an atom. It indicates the attractive force felt by the valence electrons towards the nucleus.

covalent bond The force of attraction formed when one or more pairs of electrons are shared between two nuclei.

covalent layer lattice An arrangement of atoms in a lattice in which there are strong covalent bonds between the atoms that have formed in a layer.

covalent network lattice An arrangement of atoms in a lattice in which there are strong covalent bonds between the atoms in all three dimensions.

cracking A chemical process during which carbon–carbon bonds in alkanes are broken to form smaller molecules and some unsaturated molecules.

credible Reliable and can be backed up with evidence; a credible source provides information that one can believe to be true.

crude oil A mixture of hydrocarbons that originates from the remains of prehistoric marine microorganisms. The organisms have been broken down by high temperatures and pressures over millions of years; also called petroleum.

crystal A solid made up of atoms or molecules arranged in a repeating three-dimensional pattern.

crystal lattice The symmetrical three-dimensional arrangement of atoms or ions inside a crystal.

crystallisation The process in which solid crystals are deposited when the concentration of a solute in a solution increases past the point of saturation.

crystallise Form solid crystals.

D

decantation A separation process in which a layer of liquid is separated from the solid that has settled at the bottom of the liquid.

decomposition A reaction in which a compound is broken down into smaller parts.

delocalise Spread out.

delocalised electron An electron that is not restricted to the region between two atoms.

density A measure of the amount of mass per unit volume. It has the SI units of $kg\,m^{-3}$, but is commonly quoted in $g\ cm^{-3}$ mL^{-1}.

dependent variable The variable that is measured or observed to determine the effect of changes in the independent variable.

desalinated seawater Fresh water made by removing the salt from seawater.

desalination The removal of salts from seawater to obtain fresh water.

desorption The breaking of the attraction between a substance and the surface to which the substance is adsorbed.

diamond A form of pure carbon that is the hardest naturally occurring substance.

diatomic molecule A molecule formed from two atoms only, e.g. Cl_2.

diffusion All the gases in a gas mixture spread to fill the available space. The speed at which the gases spread depends upon the mass of the gas particles.

dilute solution A solution that has a relatively low ratio of solute to solvent.

dilution The addition of a solvent to a solution to reduce its concentration.

dipole The separation of areas of positive and negative charge in a molecule.

dipole–dipole force A form of intermolecular force that occurs between polar molecules where the partially positively charged end of one molecule is attracted to the partially negatively charged end of another molecule.

diprotic acid An acid that can ionise in water to form two H^+ ions.

discrete A separate object or particle.

dispersion force The force of attraction between molecules due to temporary dipoles induced in the molecules. The temporary dipoles are the result of fluctuations in the electron density.

dissociate Break up.

dissociation A process in which molecules or ionic compounds separate or split into smaller particles such as atoms or ions. Examples of dissociation reactions include the solution of NaCl solid in water, forming $Na^+(aq)$ and $Cl^-(aq)$ ions, and the reaction of HCl gas with water, forming $H^+(aq)$ and $Cl^-(aq)$.

dissolution The process of dissolving a solute in a solvent to form a solution.

dissolve To incorporate a solid or gas into a liquid to form a solution.

distillation The process of separating a solution by evaporating, condensing and collecting the component of the solution with the lowest boiling point.

double covalent bond A covalent bond in which four electrons (two electron pairs) are shared.

ductile Able to be drawn into a wire.

E

elastic collision Collision between particles where kinetic energy is conserved.

electric dipole See *dipole*.

electrical conductivity The degree to which a specified material or solution conducts an electric current.

electrolyte A solution or molten substance that conducts electricity by means of the movement of ions, e.g. a solution of sodium chloride (NaCl(aq)).

electromagnetic radiation A form of energy that moves through space. Visible light, radio waves and X-rays are forms of electromagnetic radiation.

electromagnetic spectrum All possible frequencies of electromagnetic radiation shown in order of their wavelengths or frequencies.

electron A negatively charged, subatomic particle that occupies the region around the nucleus of an atom.

electron configuration In the shell model of an atom, the electronic configuration is a means of representing the number of electrons in each shell.

electron density The concentration of electrons that usually refer to the regions around an atom or molecule.

electron dot diagram See *Lewis structure diagram*.

electron shell In the shell model of an atom, an electron shell is the fixed energy level that corresponds to a circular orbit of the electrons. In the Schrödinger model, a shell contains subshells and orbitals of equal or similar energy.

electron transfer diagram A diagram that shows how electrons move from a metal atom to a non-metal to form ions.

electronegativity The ability of an atom to attract electrons in a covalent bond towards itself.

electrostatic attraction The force of attraction between a positively charged particle and a negatively charged particle.

element A substance made up of atoms with the same atomic number.

elemental analysis A process that determines the mass of each element in a sample of a compound.

eluent The solvent that carries the components and passes through a chromatography column.

emission line When an electron absorbs energy it jumps to a higher energy state. Shortly afterwards the electron returns to the lower energy level, releasing a fixed amount of energy as a particular colour of light.

emission spectrum A spectrum produced when an element is excited by heat or radiation. It appears as distinct lines characteristic of the element.

empirical formula A formula that shows the simplest whole number ratio of the elements in a compound, e.g. CH_2 is the empirical formula of propene (C_3H_6).

endothermic A reaction that absorbs energy from the surroundings; ΔH is positive.

energy content The chemical energy of a substance.

energy density The energy released per litre of fuel.

energy efficiency The percentage of energy from a source that is converted to useful energy

energy level One of the different shells of an atom in which an electron can be found.

energy profile diagram The energy changes that occur during the course of a reaction.

enhanced greenhouse effect The increased levels of greenhouse gases produced by our use of fossil fuels that cause global warming and trigger consequential shifts in weather patterns and climate.

enthalpy Heat content. The sum of the chemical potential and kinetic energies in a substance; symbol H.

enthalpy change The difference in the total enthalpy of the products and the total energy of the reactants; symbol ΔH. Also known as heat of reaction.

enzyme Biological catalyst.

excess reactant A reactant that is not completely consumed in a chemical reaction.

excited state A term used to describe an atom in which electrons occupy higher energy levels than the lowest possible energy levels.

exothermic A reaction that releases energy into the surroundings; ΔH is negative.

expertise Expert knowledge or skills in a field.

F

fermentation A process that uses other enzymes from yeasts to convert glucose and other small sugar molecules to ethanol and carbon dioxide.

filtrate The purified liquid that collects in the flask when a mixture is filtered.

filtration The process of removing solids from a liquid or gas by passing the mixture through a fine mesh or filter.

first ionisation energy The energy required to remove one electron from an atom of an element in the gas phase.

flame test Determination of the metallic elements present in a compound by inserting a sample of the compound into a non-luminous Bunsen burner flame. Some metals produce particular colours when they are heated.

floc Aluminium hydroxide precipitate produced in the flocculation process, which traps other fine particles and removes colour and some microorganisms from the water.

flocculation Process in water purification where small suspended particles come together to form larger, heavier particles, which are usually insoluble.

fluoridation Process in which fluoride is added to drinking water before it is released from storages. Fluoride in drinking water helps to reduce the incidence of tooth decay.

formula A representation of an element or compound using symbols for its constituent elements. It shows the ratio of atoms in a compound or the number of atoms in a molecule.

fossil fuel A fuel formed by the decomposition of plant and animal material over millions of years, including coal, oil and natural gas.

fracking Process of extraction of natural gas from coal or shale deposits.

fractional distillation A a form of distillation that separates solutions of liquids according to their boiling point.

fuel Substances that have chemical energy stored within them that can be released relatively easily.

full equation A representation of a reaction that uses the formulae of all reactants and products in a chemical reaction.

fullerene A molecule composed entirely of carbon, in the form of a hollow sphere or tube. Each carbon atom is bonded to three other carbon atoms.

G

gas chromatography (GC) A very sensitive form of chromatography in which the mobile phase is a gas.

geometric isomer A type of stereoisomer that can occur when there is restricted rotation in a molecule. Restricted rotation can occur about a carbon–carbon double bond or a ring. Stereoisomers of alkenes can be distinguished using the labels *cis*- and *trans*-.

graphene A form of carbon consisting of planar sheets one atom thick in which each carbon atom is bonded to three neighbouring carbon atoms.

graphite A form of carbon in which the carbon atoms are arranged in layers.

Great Artesian Basin A very large reservoir of groundwater in Australia. It is the largest basin of its kind in the world.

greenhouse effect The warming of the Earth's atmosphere due to the absorption of infrared radiation by gases such as carbon dioxide, water and methane.

greenhouse gas Gases in the atmosphere that absorb and re-radiate the enery as infra-red radiation. Carbon dioxide, methane, water vapour, nitrogen oxides and ozone are greenhouse gases.

ground state A term used to describe an atom in which the electrons occupy the lowest possible energy levels.

groundwater Water found below the Earth's surface in porous rock or fractures.

group (periodic table) A vertical column of elements in the periodic table.

H

heat capacity A measure of a substance's capacity to absorb and store heat energy. The heat capacity of 1 g of water is 4.18 $J\,°C^{-1}$. This tells you 1 g of water will absorb 4.18 J of heat energy to heat up by 1°C.

heat of combustion The enthalpy change that occurs when a specified amount (e.g. 1 g, 1 L, 1 mol) of a fuel burns completely in oxygen.

heat of reaction The exchange of heat between a system and its surroundings during a chemical reaction under constant pressure; symbol ΔH. Also known as enthalpy change.

heat treatment Heating a metal in different ways to alter its structure and physical properties.

heavy metal A metal with high density; usually used to describe a metal that poses a threat to health.

heterogeneous Diverse, different. A heterogeneous substance or solution possesses two or more different types of phases in the one sample, e.g. a suspension.

heterogeneous catalyst A catalyst that has a different physical state (phase) from the reactants and products.

high-performance (or high-pressure) liquid chromatography (HPLC) A very sensitive technique used to separate the components in a mixture, to identify each component, and to measure the concentrations of the components. It relies on pumps to pass a pressurised liquid solvent containing the sample mixture through a column filled with a solid adsorbent material.

homogeneous Uniform. The components of a homogeneous substance are uniformly distributed throughout the substance, e.g. a solution is homogeneous because the solute and the solvent cannot be distinguished from each other.

homogeneous catalyst A catalyst that has the same physical state (phase) as the reactants and products.

homologous series A series of compounds with similar properties and the same general formula, in which each member contains one CH_2 unit more than the previous member.

hydrated An ion surrounded by water molecules. Hydrated ions can be found in aqueous solutions or crystalline solids.

hydride A compound in which hydrogen is bonded to another element. HF, HCl and HI are hydrides of group 17 elements.

hydrocarbon A compound that contains carbon and hydrogen only, e.g. the alkanes, alkenes and alkynes.

hydrogen bond A type of intermolecular, dipole–dipole force where a hydrogen atom is covalently bonded to a highly electronegative atom such as oxygen, nitrogen or fluorine. Due to the disparity of electronegativity values between the atoms involved, the hydrogen develops a partial positive charge and bonds to lone pairs of electrons on neighbouring atoms of oxygen, nitrogen or fluorine.

hydronium ion The $H_3O^+(aq)$ ion.

hydroxide ion The $OH^-(aq)$ ion.

hypothesis A prediction based on previous knowledge; a possible outcome of the experiment.

I

ideal gas A gas that obeys the gas equation at all temperatures and pressures. An ideal gas is a theoretical gas in which there are no intermolecular attractive forces and the gas particles have no volume.

incomplete combustion Combustion that takes place when the oxygen supply is limited. Products are carbon monoxide (or carbon) and water.

independent variable A variable that is changed by the researcher.

indicator A substance that is different colours in its acid and base forms.

inert Not chemically reactive.

injection port Opening through which the sample is injected in gas chromography.

insecticide A substance used for killing insects.

instantaneous dipole A net dipole formed in a molecule due to temporary fluctuations in the electron density in the molecule.

intermolecular force An electrostatic force of attraction between molecules, including dipole–dipole forces, hydrogen bonds and dispersion forces.

interstitial alloy An alloy made by adding smaller atoms to a metal.

intramolecular bond A force that holds the atoms within a molecule together.

ion A positively or negatively charged atom or group of atoms.

ion–dipole attraction The attraction that forms between dissociated ions and polar water molecules when an ionic solid dissolves in water.

ionic bonding A type of chemical bonding that involves the electrostatic attraction between oppositely charged ions.

ionic compound A type of chemical compound that involves the electrostatic attraction between oppositely charged ions.

ionic equation An equation for a reaction that shows only the species involved in the reaction, e.g. $Cu^{2+}(aq) + 2OH^-(aq) \rightarrow Cu(OH)_2(s)$.

ionisation (i) The removal of one or more electrons from an atom or ion; (ii) the reaction of a molecular substance with a solvent to form ions in solution.

ionisation energy The energy required to remove one electron from an atom of an element in the gas phase.

ionise The reaction of a molecular substance with a solvent to form ions in solution. When some polar molecules dissolve in water they ionise to form a hydrogen ion, e.g. $HCl(g) \rightarrow H^+(aq) + Cl^-(aq)$.

isotope Each of two or more forms of the same element that contain equal numbers of protons but different numbers of neutrons in their nuclei, e.g. ^{12}C and ^{13}C are isotopes of carbon.

K

kinetic energy The energy that a particle or body has due to its motion ($KE = \frac{1}{2}mv^2$).

kinetic energy distribution diagram See *Maxwell–Boltzmann distribution graph.*

kinetic molecular theory A theory that aims to explain the behaviour of gases by assuming gases are composed of a large number of particles in random motion, these particles move in straight lines and have elastic collisions, the gas particles are very small, there is no attraction between the particles, and the average kinetic energy is related to the temperature of the gas.

L

latent heat The heat energy required to change the state of a substance at a constant temperature.

latent heat of fusion The energy required to change a fixed amount of solid to liquid at its melting temperature. The latent heat of fusion of water is $6.0\,kJ\,mol^{-1}$, meaning $6.0\,kJ$ of energy is needed to change 1 mole of water from a solid to a liquid at 0°C.

latent heat of vaporisation The energy required to change a fixed amount of liquid to a gas at its boiling temperature. The latent heat of vaporisation of water is $44.0\,kJ\,mol^{-1}$, meaning $44.0\,kJ$ is needed to change 1 mole of water from a liquid to a gas at 100°C.

lattice A regular arrangement of large numbers of atoms, ions or molecules.

law of conservation of energy A scientific law that states energy cannot be created or destroyed; it can only be transformed from type of energy to another.

Lewis structure diagram A representation of the electron arrangement in a molecule in which outer-shell electrons are represented by dots or crosses; also called electron dot diagram.

limewater test A test for carbon dioxide gas. The presence of carbon dioxide is detected by bubbling the gas through a calcium hydroxide solution ($Ca(OH)_2(aq)$). The limewater reacts with the carbon dioxide and turns milky.

limiting reagent A reactant that is completely consumed in a reaction and which determines the amount of products formed.

line of best fit See *trend line.*

liquefied petroleum gas (LPG) A fuel comprising a mixture of propane and butane, which are liquefied under pressure.

lone electron pair An outer-shell electron pair that does not form a bond with other atoms.

M

main group element An element in groups 1, 2 or 13–18 in the periodic table.

malleable Able to be bent or beaten into sheets.

mass number The number of protons and neutrons in the nucleus of an atom.

mass spectrometer An instrument that measures the mass-to-charge ratio of particles.

mass spectrum A plot of the isotopic mass, relative to the mass of carbon-12 taken as 12 units exactly, against the relative abundance of each isotope present in a sample.

material A substance that can be used to make objects.

matter Anything that has mass and occupies space.

Maxwell–Boltzmann distribution graph A graph showing the range of kinetic energies of particles in a sample; also called a kinetic energy distribution diagram.

mean The sum of all the values in a data set divided by the number of values in the data set. It is commonly known as the average of a set of numbers.

metallic bonding The electrostatic attractive forces between delocalised valence electrons and positively charged metal ions.

metallic bonding model A description that explains the properties and behaviour of metals in terms of the particles in metals.

metallic character Describes how closely an element exhibits the properties commonly associated with metals, namely, that it readily loses an electron to form a cation. This is closely related to ionisation energy.

metallic lattice Tightly packed arrangement of metal atoms in a 'sea' of electrons.

metallic nanomaterial A material with metal atoms arranged to make structures that have a least one dimension that is 1–100 nm ($1\,nm = 10^{-9}\,m$).

metalloid An element that displays both metallic and non-metallic properties, e.g. germanium, silicon, arsenic, tellurium.

mineral A naturally occurring inorganic substance that is solid and can be represented by a chemical formula, e.g. quartz.

miscible Liquids that can be mixed in any ratio to form a homogeneous solution.

mobile phase The phase that moves over the stationary phase in a chromatographic separation.

model A description that scientists use to represent the important features of a system or phenomenon.

molar mass The mass of 1 mole of a substance measured in $g\,mol^{-1}$; symbol *M*.

molarity The amount of solute, in moles, dissolved in 1 litre of solution ($mol\,L^{-1}$).

molar volume The volume occupied by 1 mole of gas at a specified set of conditions. At standard temperature (0°C or 273 K) and pressure 100 kPa), the molar volume of a gas is $22.7\,L\,mol^{-1}$.

mole The amount of substance that contains the same number of fundamental particles as there are atoms in 12 g of carbon-12; symbol *n*; unit mol.

mole ratio The ratio of species involved in a chemical reaction, based on the ratio of their coefficients in the reaction equation.

molecular formula A formula of a compound that gives the actual number and type of atoms present in a molecule. It may be the same as or different from the empirical formula.

molecule A group of two or more atoms covalently bonded together.

molten Materials that are normally found as solids but are liquid, melted, due to elevated temperature.

monoprotic acid An acid molecule that generates only one hydrogen ion when ionised in water.

N

nanomaterial A material with nanoscale features.

nanometre 10^{-9} of a metre.

nanoparticle A particle in the size range 1–100 nm.

nanorod A nanoscale rod in which length and width are in the range 1–100 nm. The length of the rod is 3–5 times its width.

nanoscale The scale used to classify objects 1–100 nm in size.

nanotechnology The use of technologies that manipulate and investigate the properties of nanomaterials.

nanotube An allotrope of carbon that consists of layers of carbon atoms formed into a long cylinder.

nanowire A wire that has a diameter measured on the nanoscale. Its length is unrestricted.

natural gas A fossil fuel mainly composed of methane (CH_4) together with small amounts of other hydrocarbons such as ethane (C_2H_6) and propane (C_3H_8). Water, sulfur, carbon dioxide and nitrogen may also be present in natural gas.

neutralisation reaction An acid reacts with a base in stoichiometric proportions to form a salt plus water.

neutralise To react an acid with a base in stoichiometric proportions to form a solution of a salt and water.

neutral solution A solution in which the concentrations of H^+ ions equals the concentration of OH^- ions; is neither acidic nor basic. At 25°C a neutral solution has a pH of 7.

neutron An uncharged subatomic particle found in the nucleus of an atom.

noble gas An unreactive gaseous element in group 18 of the periodic table. With the exception of helium, noble gases have eight electrons in their outer shells.

non-bonding electron An outer-shell electron that is not shared between atoms.

non-polar Bonds or molecules that do not have a permanent dipole. They have an even distribution of charge.

non-polar solvent A liquid or solvent that is a compound of two or more elements whose electronegativities are almost the same, e.g. oil.

non-renewable Resources that are used faster than they can be replaced, e.g. coal, oil and natural gas.

nucleon A particle that makes up the nucleus of an atom, i.e. protons and neutrons.

nucleus The positively charged core at the centre of an atom, consisting of protons and neutrons.

O

octet rule A rule used as part of the explanation for electron configuration and in bonding. The rule is that during a chemical reaction, atoms tend to lose, gain or share their valence electrons so that there are eight electrons in the outer shell.

ore A mineral or an aggregate of minerals that contains a valuable constituent, such as a metal, which is mined or extracted.

origin The point at which a small spot of a mixture is placed so that it can be separated by paper or thin-layer chromatography.

osmosis The tendency of a solvent to pass through a semipermeable membrane from an area of low solute concentration to an area of higher solute concentration.

outermost shell See *valence shell.*

outlier A value that lies outside most of the other values in a set of data.

P

paper chromatography An analytical technique for separating and identifying mixtures, which uses paper as the stationary phase.

partial pressure The pressure exerted by one component of a mixture of gases. The total pressure of a mixture of gases is equal to the sum of the individual partial pressures of each component in the mixture.

parts per million (ppm) A unit of concentration that states the number of grams of solute in 1 million grams of solution. It is equivalent to the number of milligrams of solute per kilogram of solution.

pascal Unit of pressure equal to 1 newton per square metre ($1\,N\,m^{-2}$), where: pressure = $\frac{\text{force}}{\text{area}}$.

percentage composition The proportion by mass of the different elements in a compound.
% by mass of an element in a compound
$= \frac{\text{mass of the element present}}{\text{total mass of the compound}} \times 100$

period (periodic table) A horizontal row of elements in the periodic table. The start of a new period corresponds to the outer electron of that element beginning a new shell.

periodic law The way properties of elements vary periodically with their atomic number.

periodic table A table that organises the elements by grouping them according to their electronic configurations.

personal protective equipment (PPE) Safety equipment, such as safety glasses, lab coats, gloves, that is worn in the laboratory.

petrodiesel A hydrocarbon fraction obtained from crude oil.

pH A measure of acidity and the concentration of hydronium ions, in solution. At 25°C, acidic solutions have a pH value less than 7 and bases have a pH value greater than 7. Mathematically, pH is defined as $pH = -\log_{10}[H^+]$.

pH scale See *pH.*

phenomenon Something that occurs and can be observed (or felt, heard etc.) to have occurred.

photochemical smog Atmospheric pollution produced through the action of sunlight on nitrogen oxides and unburned hydrocarbons to form ozone and other pollutants. The nitrogen oxides formed in high-temperature reactions such as those that occur in car engines and lightning strikes.

photosynthesis A reaction that occurs in the leaves of plants between carbon dioxide and water, in the presence of sunlight and chlorophyll, to form glucose and oxygen. The photosynthesis process can be represented by the equation:
$6CO_2(g) + 6H_2O(l) \rightarrow C_6H_{12}O_6(aq) + 6O_2(g)$

photovoltaic cell A device constructed from a specialised semiconductor that can produce a flow of electrons from light energy.

plastic A property of a material that can be reshaped by application of heat and pressure. In society, some polymers are often referred to as plastics.

polar Bonds or molecules with a permanent dipole. They have an uneven distribution of charge.

polarity The measure of how polar a molecule or bond is. The difference in charge between the positive and negative ends of an electric dipole. The difference in charge between the positive and negative ends of a polar molecule or covalent bond.

polar solvent A solvent that can dissolve substances consisting of polar molecules or charged ions, but will not dissolve solutes made up of non-polar molecules. For example, the polar nature of the ethanol molecule means it readily dissolves in water, which is also polar.

polyatomic ion An ion that is made up of more than one element, e.g. the carbonate ion (CO_3^{2-}).

polyatomic molecule A molecule that consists of more than two atoms, e.g. H_2O.

polymer A long-chain molecule that is formed by the reaction of large numbers of repeating units (monomers).

polyprotic acid An acid molecule that generates more than one hydrogen ion when ionised in water.

potable water Water that is suitable for drinking.

precipitate The solid formed during a reaction in which two or more solutions are mixed.

precipitation reaction A reaction between substances in solution in which one of the products is insoluble.

precision When repeated measurements of the same quantity give values that are in close agreement.

pressure The force exerted per unit area of a surface.

primary source A source that is a first-hand account.

protected catchment An area set aside to harvest and store water. There is no public access to protected catchments to ensure the quality of the water is high.

proton A positively charged, subatomic particle bound to neutrons in the nucleus of an atom.

Q

qualitative Relating to quality and not measured values.

qualitative analysis An analysis to determine the identity of the chemical(s) present in a substance.

quantised In specific quantities or chunks.

quantitative Relating to measured values.

quantitative analysis An analysis to determine the concentration of the chemicals present in a mixture.

quenching Heating a metal to a moderate temperature and then cooling it rapidly to make it harder and more brittle.

R

radioactive Spontaneously undergoing nuclear decay to produce radiation such as beta particles, alpha particles and gamma rays.

random error An error that is caused by measurements of a quantity being either above or below the actual quantity being measured. (The effects of random errors can be reduced by taking the average of many observations.)

range The spread of values in the data set. It is taken as the largest data value minus the smallest data value.

rate of reaction The change in concentration of a reactant or product per unit time.

raw data The information and results collected and recorded during an experiment.

reaction pathway A series of chemical reactions that converts a starting material into a product in a number of steps.

reactivity The ease with which a chemical can undergo reactions.

reactivity series of metals A ranking of metals in increasing order of their reactivity (ability to be oxidised) with the half-equations written as reduction equations of the corresponding ion. Least reactive metals are at the top and most reactive metals are at the bottom.

recycled water Water recovered by the purification of waste water.

relative atomic mass The weighted average of the relative isotopic masses of an element on the scale where ^{12}C is taken as 12 units exactly; symbol A_r.

relative formula mass The mass of a formula unit relative to the mass of an atom of ^{12}C taken as 12 units exactly. It is numerically equal to the sum of the relative atomic masses of the atoms making up the formula. Substances that contain atoms or ions bonded in lattice structures have a relative formula mass. Such compounds include ionic compounds and covalent network substances.

relative isotopic abundance The percentage abundance of a particular isotope in a sample of an element.

relative isotopic mass The mass of an atom of the isotope relative to the mass of an atom of ^{12}C taken as 12 units exactly.

relative molecular mass The mass of a molecule relative to the mass of an atom of ^{12}C taken as 12 units exactly; symbol M_r.

reliability A measure of how an experiment can be repeated several times with consistent results.

renewable Energy that can be obtained from natural resources that can be constantly replenished.

reputation The belief that someone or something has a particular characteristic.

research question A statement defining what is being investigated.

residue The solid collected in the filter paper when a mixture is filtered.

retardation factor (R_f) The ratio of the distance a component has moved from the origin to the distance the solvent has moved from the origin.

retention time (R_t) The time taken for a component to pass through a chromatography column.

reverse osmosis A process by which pure water can be obtained from salt water. Pressure is applied to the salt water, causing a net flow of water molecules away from the solution through a semipermeable membrane.

risk assessment A formal way of identifying risks and assessing potential harm from hazards in an experiment.

S

Safety Data Sheet (SDS) A summary of the risks of using a particular chemical, including measures to be followed to reduce risk.

salinity The presence of salt in water and soil that can damage plants or inhibit their growth.

salt A substance formed from a metal or ammonium cation and an anion. Salts are the products of reactions between acids and bases, metal oxides, carbonates and reactive metals.

saturated (i) A hydrocarbon that is composed of molecules with only carbon–carbon single bonds. (ii) Combined with or containing all the solute that can normally be dissolved at a particular temperature.

saturated solution A solution that cannot dissolve any more solute at the given temperature.

scanning tunnelling microscope (STM) A microscope that images atoms by using a sharp metallic tip to sense the atoms on the surface of a crystal.

secondary source A source derived from the original data or account.

sediment The solid material that settles to the bottom of a body of water.

sedimentation The process of solid particles settling out of a liquid.

seed crystal A small crystal from which a large crystal of the same material can typically be grown.

separation funnel A funnel with a large bulb and a tap in its output tube that is used to separate immiscible liquids.

settling A process in the purification of drinking water that utilises gravity to remove suspended solids from the water.

sieving The process of separating large solid particles from smaller solid particles by passing the mixture through a mesh.

SI units International System of Units based on the metre, kilogram, second, ampere, kelvin, candela and mole.

side group A group of atoms attached to a backbone chain of a long molecule.

significant figures The numbers that convey meaning and precision; usually depends on the scale of the measuring instrument.

single covalent bond A covalent bond in which two electrons are shared between two nuclei. It is depicted in a valence structure as a line between the two atoms involved.

solubility A measure of the amount of solute dissolved in a given amount of solvent at a given temperature.

solubility curve A graph of solubility versus temperature for a particular solute dissolved in a particular solvent.

solubility table A reference table that can be used to predict the solubility of ionic compounds.

solute A substance that dissolves in a solvent, e.g. sugar is the solute when it dissolves in water.

solution A homogeneous mixture of a solute dissolved in a solvent.

solvent A substance, usually a liquid, that is able to dissolve a solute to form a solution. Water is a very good solvent.

space-filling model A three-dimensional model in which the atoms are represented by spheres with radii proportional to the radii of the atoms involved. The distances between spheres are also proportional to the distances between the atomic nuclei.

specific heat capacity The amount of energy required to raise the temperature of an amount of a substance, usually 1 gram, by 1°C. The unit for specific heat capacity is usually $J\,g^{-1}\,°C^{-1}$, e.g. the specific heat capacity of water is $4.18\,J\,g^{-1}\,°C^{-1}$.

spectator ion An ion that remains in solution and is unchanged in the course of a reaction. Spectator ions are not included in ionic equations.

spectroscopy The study of the way that radiation, such as light and radio waves, interacts with matter.

standard atomic weight The relative atomic mass of an element based on the agreed proportions of isotopes in a 'normal' sample of the element on Earth.

standard solution A solution that has an accurately known concentration.

standard temperature and pressure (STP) Conditions of temperature and pressure relevant to a gas, where the temperature is 0°C (273 K) and pressure is 100 kPa.

stationary phase A solid, or a solid that is coated in a viscous liquid, used in chromatography. The components of a mixture undergo adsorption to this phase as they are carried along by the mobile phase.

steel A generally hard, strong, durable and not malleable alloy of iron and carbon, usually containing 0.2–1.5% carbon, often with other constituents such as manganese, chromium, nickel, molybdenum, copper, tungsten, cobalt or silicon, depending on the desired alloy properties.

stem name The name that corresponds to the prefix for the longest chain of carbons in the molecule.

sterile An environment that is totally clean and free from living microorganisms.

stoichiometric calculation The calculation of relative amounts of reactants and products in a chemical reaction.

stoichiometry The calculation of relative amounts of reactants and products in a chemical reaction. Chemical equations give the ratios of the amounts (moles) of the reactants and products.

strong acid An acid that completely ionises in aqueous solution.

strong base A base that completely dissociates in aqueous solution.

structural formula A formula that represents the three-dimensional arrangement of atoms in a molecule and shows all bonds as well as all atoms.

structural isomer A compound with the same molecular formula, but different structures.

subatomic particle A particle that makes up an atom—protons, neutrons and electrons.

sublimation The process by which a substance goes directly from the solid phase to the gaseous phase, without passing through a liquid phase.

substitutional alloy An alloy made from elements of similar chemical properties and size.

successive ionisation energy The energy required to achieve the sequential removal of electrons from the atom.

supersaturated solution An unstable solution that has more solute dissolved at a given temperature than a saturated solution.

surface area (chemistry**)** The area of all surfaces of the substance that are exposed to the other reactants. This is proportional to the amount of particles available at the surface to react.

surface tension The resistance of a liquid to increase its surface area.

surroundings (chemistry**)** The rest of the universe around a particular chemical reaction. The chemical reaction is the system. Energy moves between the system and surroundings in exothermic and endothermic reactions.

suspension A heterogeneous mixture containing solid particles that are large enough to sink to the bottom of the mixture if it is left to stand (sedimentation), e.g. sand in a container of water that has just been shaken.

sustainability The combination of meeting the long-term ongoing needs of society, while still reaching the immediate and short-term demands.

symmetrical molecule A molecule in which the polar bonds are evenly (or symmetrically) distributed. The bond dipoles cancel out and do not create an overall molecular dipole.

system In chemistry, the system is a chemical reaction. A system operates within its surroundings. Energy can move between the two. You always consider energy as absorbed or released from the perspective of the system. For example, energy is absorbed into the system from the surroundings, or energy is released by the system to the surroundings.

systematic error An error that produces a constant bias (measurements are either always above, or always below, the actual value) in measurement. (Systematic errors are eliminated or minimised through calibration of apparatus and the careful design of a procedure.)

T

tempering A process in which a metal that has been quenched is warmed again to a lower temperature to reduce its brittleness but to retain its hardness.

temporary dipole A net dipole formed in a molecule due to temporary fluctuations in the electron density in the molecule.

tensile strength The maximum resistance of a material to a force that is pulling it apart before breaking, measured as the maximum stress the material can withstand without tearing.

tetrahedral shape The shape of a molecule with a central atom surrounded by four other atoms. The bond angle between two outer atoms and the central atom is 109.5°.

thermochemical equation A chemical equation that includes the enthalpy change of the reaction, ΔH. For example, $2H_2(g) + O_2(g) \rightarrow 2H_2O(g)$ $\Delta H = -572\,kJ\,mol^{-1}$.

thin-layer chromatography An analytical technique for separating and identifying mixtures; it uses a thin layer of fine powder spread on a glass or plastic plate as the stationary phase.

transition metal An element in groups 3–12 in the periodic table.

transition state An arrangement of atoms in a reaction that occurs when sufficient energy is absorbed for the activation energy to be reached. It represents the stage of maximum potential energy for the reaction. Bond breaking and bond forming are both occurring at this stage, and the arrangement of atoms is unstable.

trend An observed pattern of data in a particular direction.

trend line Points can be joined with a single smooth straight or curved line; also known as a line of best fit.

triple covalent bond A covalent bond in which six electrons are shared between two nuclei. It is depicted in a valence structure as three lines between the two atoms involved.

triprotic acid An acid molecule that generates three hydronium ions when ionised in water.

U

uncertainty An error associated with measurements made during experimental work.

unsaturated A hydrocarbon composed of molecules with one or more carbon–carbon double or triple bonds.

unsaturated solution A solution that could dissolve more solute at the given temperature.

V

valence electron An electron found in the valence shell; an outermost electron in an atom or ion.

valence shell The highest energy shell (outer shell) of an atom that contains electrons.

valence shell electron pair repulsion theory (VSEPR) A model used to predict the shape of molecules. The basis of VSEPR is that the valence electron pairs surrounding an atom mutually repel each other, and therefore adopt an arrangement that minimises this repulsion, thus determining the molecular shape.

valence structure A representation of the electron arrangement in a molecule in which covalent bonds are represented by lines.

valency The charge of an ion is equal to its valency. It is a measure of the combining ability of an atom with other atoms.

validity Refers to whether the evidence supports the argument.

vapour pressure The pressure exerted by molecules that evaporate from a liquid in an enclosed vessel at the point where the rates of evaporation and condensation are in equilibrium.

variable Any factor that can be controlled, changed and measured in an experiment.

volume The amount of space that a substance occupies. It can be calculated by multiplying length by width by depth of a regular solid. Otherwise it can be determined by finding the volume of water displaced by the substance.

W

water cycle The continuous process by which water is circulated throughout the Earth and the atmosphere. Water passes into the atmosphere as water vapour, precipitates to Earth in liquid or solid form, and returns to the atmosphere through evaporation.

weak acid An acid that is partly ionised in water.

weak base A base that partially dissociates in water to produce hydroxide (OH^-) ions.

work hardening The increase in strength that results from processes such as hammering, rolling and drawing a metal.

Index

Attributions

The following abbreviations are used in this list: t = top, b = bottom, l = left, r = right, c = centre.

FRONT COVER: Shutterstock: Nik Merkulov.
123RF: alicenerr, p. 87tl; Elnur Amikishiyev, p. 306lt; Asaf Eliason, p. 91 (twenty-cent coin, two-dollar coin); Sergei Gorin, p. 222t; Anton Lebedev, p. 360b, 360tl, 360tr; Ivan Mikhaylov, p. 477b; Andriy Popov, p. 212b; Rido, p. 365tr; Gino Santa Maria, p. 89tt; Ron Sumners, p. 86tl; Kim Takhyzh-Sviridov, p. 374c; Tatyana Vyc, p. 442t; Aleksandrs Tihonovs, pp. 86br, 89tb; George Tsartsianidis, p. 394l; tykhyi, p. 86tr.
AAP: Tracey Nearmy, p. 32.
Age Fotostock: Arnold Mondadori Editore S.P., p. 285tr; Caspar Benson, p. 441.
Alamy Stock Photo: p. 335; Derrick Alderman, p. 63t (UV light, visible light); ASK Images, p. 78bl; John Cancalosi, p. 85cr; Paul Carstairs, p. 305; Phil Degginger, pp. 64t, 65c; DOE Photo, p. 167l; FLPA, p. 339r; fStop Images GmbH, p. 307b; Greenshoots Communications, p. 230b; ITAR-TASS Photo Agency, p. 47t; Sergey Krotov, p. 189; PNL, p. 387bl; Randsc, p. 126–7; R A Rayworth, p. 44t; Science History Images, p. 64c; Sciencephotos, p. 410; Kumar Sriskandan, p. 394r; Arthur Streeton, Golden summer, Eaglemont 1889 Oil on canvas/National Gallery of Australia, Canberra/Painting, p. 78tl; studiomode, p. 185r; The Print Collector, p. 22b; WorldFoto, p. 241b; World History Archive, p. 22c.
Australian Government: National Health and Medical Research Council and National Resource Management Ministerial Council 2004: Diagram from *Water Made Clear NHMRC, A consumer guide to accompany the Australian Drinking Water Guidelines 2004,* p. 334t.
Berrybank Farm: Jock Charles, p. 250tr.
Biochar.org: Christoph Steiner/UrbanFoodPlus Project/University of Kassel, p. 149b.
Book, Natalie: p. 184 (phenylethene).
BP: Courtesy of BP, p. 17b.
Bridgeman Images: Stained glass of St George, Southern German, 1400-10/State Hermitage Museum, St Petersburg, Russia, p. 93t.
California Institute of Technology: Dr Chi Ma, p. 8br (sem of opal).
Chevron Australia – Communications: Chevron Australia Pty Ltd, p. 241t.
Chillmaid, Martyn F.: p. 427.
CSIRO: Licensed under CC BY 3.0, p. 68b; scienceimage, p. 147t.
Derry, Lanna: Chemistry Education Association, p. 446t.
DK Images: Mike Dunning, p. 354.
Dorling Kindersley: Andrew Kerr, p. 250b.
Faber-Castell: p. 147b.
Fairfax Photo Sales: John Donegan/The Age, p. 250tl; Tony Ashby, p. 343br; Graham Tidy/Canberra Times/Fairfax, p. 48.
Fotolia: ASK-Fotografie, p. 377tr; ASP Inc, p. 399t; Barabas Attila, p. 198; Harald Biebel, p. 86bl; BillionPhotos.com, p. 91 (saxophone); bit24, p. 190t; blauviolette, p. 202b; bluesnote, p. 210tl; Joggie Botma, p. 208br; Greg Brave, p. 374r; Richard Carey, p. 306br; cherniyvg, p. 240cb; Copri, p. 148tl; dabjola, p. 268lb; Jeffrey Daly, p. 93ct; Mikhail Dudarev, p. 351; efired, p. 355tr; gmstockstudio, p. 222c; Michael Ireland, p. 208bl; irisphoto1, p. 367r; Craig Jewell, p. 208tl; KAR, p. 476; lexaarts, p. 94 (satellite); Lonely, p. 202t; Lustil, p. 148bl; Jan Matoska, p. 425tr; Dean Miller, p. 442bl; Mirco251280, p. 75; Melinda Nagy, p. 336; Tyler Olson, p. 395; osmar01, p. 94br; Pabijan, p. 192tl; Nicholas Piccillo, p. 192r (roses); piotr_roae, p. 105tr; rekemp, p. 310tl; Federico Rostagno, p. 480tl; sehbaer_nrw, p. 268lt; shaiith, p. 306tr; Tatiana Shepeleva, pp. 9tr, 156; Angel Simon, p. 298bl; tadamichi, p. 489tr; Ina van Hateren, p. 366; Maren Winter, p. 240t; Monika Wisniewska, p. 94 (dressing); WoGi, p. 503c, 503l, 503r; zavgsg, p. 311r.
Fundamental Photographs, NYC: 2007 Richard Megna, pp. 369l, 369r, 403t.
Getty Images: Andrew Lambert Photography, p. 479t; Dimas Ardian/Stringer, p. 200t; Olivier Lantzend Arffer, p. 92t; Ross Barnett, p. 109; Bloomberg, p. 224c; Heidi Gutman/NBC NewsWire, p. 11b; Ken Lucas, p. 85bl; milanfoto, p. 165bl; New York Daily News Archive, p. 133tr; Jochen Tack, p. 343bl; Andrew Unangst, p. 142br; Universal History Archive, p. 390l; Keith Wood, p. 224b.
Gislason, Kari: p. 338.
ImageFolk: David Butow, p. 339l; Lowell Georgia, p. 144t; Egmont Strigl/Westend61, p. 148tr.
Images of Elements: images-of-elements.com, p. 77r.
International Union of Pure and Applied Chemistry Periodic Table of the Elements (2016): p. 47b.
Ixom: Ixom Operations Pty Ltd, p. 481b, 481t.
Lindsay, Dan: © 2009 Porsche Ceramic Composite Brakes on Porsche Carrera S (997), p. 104cl.
Lochman Transparencies: p. 377b.
Malcolm Cross: pp. 10b, 284b.
Manildra Group: p. 252.
McBroom, Alice: p. 91 (statue).
NASA Images: MSFC/David Higginbotham/Emmett Given, p. 76; NASA Goddard Space Flight Center. Licensed under CC BY 2.0., p. 65bl.
Newspix: Brett Faulkner, p. 108.
NRM North: p. 341t.
Oxford Designers & Illustrators Ltd. Pearson Education Ltd: p. 62.
Perth Stadium: Courtesy of Perth Stadium, p. 23b.
Picture Media Pty Ltd: Sergei Karpukhin/Reuters, p. 144br.
Powerhouse Museum: Michael Myers, p. 104br.
Sanders, Robert: p. 68t.
SA Water: Courtesy of the Australian Water Quality Centre, p. 340t.
School Curriculum and Standards Authority, Western Australia (SCSA): © School Curriculum and Standards Authority, Government of Western Australia, 2014, pp. 1, 3, 21, 43, 73, 99, 127, 141, 159, 189, 207, 221, 261, 263, 283, 305, 331, 383, 409, 439, 457, 469.
Science Photo Library: pp. 43, 65br, 67b, 84t, 159, 233t, 285bl, 318r, 374lb, 396, 426; 452l; Steve Allen, p. 440c; AMMRF, University of Sydney, p. 73; Ramon Andrade/3Dciencia, p. 153; Andrew Lambert Photography, pp. 17tr, 84b, 84c, 85t, 145, 199t, 411, 415, 418, 444; ANIMATE4.COM, p. 21; Samuel Ashfield, p. 293b; Dr Timothy Baker/Visuals Unlimited, Inc., p. 8lc; Bill Beatty, Visuals Unlimited, p. 375l; J. Bernholc et al/North Carolina State University, p. 152tc; Massimo Brega, p. 284t; Massimo Brega/The Lighthouse, pp. 2–3; Dr John Brackenbury, pp. 262–3; British Antarctic Survey, p. 316r; Martyn F. Chillmaid, pp. 194t, 218tl, 309c, 377tl, 424l; Carlos Clarivan, pp. 190b, 346b; Chemical Heritage Foundation, p. 22t; Custom Medical Stock Photo, p. 74l; Colin Cuthbert, p. 373l; Stefan Diller, p. 91br; Dorling Kindersley/UIG, pp. 5tr, 146t; Peter Eaton-Requimte/University of Porto, Visuals Unlimited, p. 8ld; Emilio Segre Visual Archives/American Institute of Physics, p. 265tl; Clive Freeman/Biosym Technologies, p. 458lt; Garo/Phanie, p. 185l; Gustoimages, p. 409; Adam Hart-Davis, pp. 348cr, 440b; Mikkel Juul Jensen, p. 167r; Louise Murray, p. 18; Russell Kightley, pp. 9tl, 156l; Chris Knapton, p. 221; Laguna Design, p. 458lc; Lawrence Berkeley National Laboratory, p. 299; Claus Lunau, p. 332; Dr P. Marazzi, p. 424tr; David McCarthy, p. 94tl; Medi-mation, pp. 8br (sem of butterfly), 8br (butterfly), 14b, 15b, 15tr; Miriam and Ira D. Wallach

Division of Art, Prints and Photographs/New York Public Library, p. 231b; B. Murton/Southampton Oceanography Centre, p. 384t; NASA, p. 130; Natural History Museum, London, p. 424br; Susumu Nishinaga, p. 8la; Northwestern University, p. 8le; Ornl, p. 288; Alfred Pasieka, pp. 251, 348br; David Parker/IMI/University of Birmingham High TC Consortium, p. 6tl; P. Plailly/E. Daynes, p. 452br; Power and Syred, pp. 8lb, 149t; Alexis Rosenfeld, p. 314t; Science Picture Co, pp. 9b, 93cb, 156r; Sinclair Stammers, p. 285br; Volker Steger, p. 390br; TEK Image, p. 142bl; Sheila Terry, p. 387br; Geoff Tompkinson, p. 283; Trevor Clifford Photography, p. 96; U.S. Department of Energy, p. 16b; Victor Habbick Visions, p. 151; Dr Mark J. Winter, pp. 98–9; Peidong Yang/UC Berkeley, p. 93b.

Science Source – Photo Researchers, Inc: Charles D. Winters, p. 384b.

Shutterstock: p. 479b; Africa Studio, p. 355tl; Aleoks, pp. 260–1, 505–8; Anneka, pp. 92b, 375r; AS Food Studio, p. 440t; astudio, p. 16t; Marijus Auruskevicius, p. 104tl; AVprophoto, p. 10cr; Arvind Balaraman, pp. 90b, 210b; Andrey N. Bannov, p. 13t; Eva Blanda, p. 425b; Eugene Berman, p. 74b; Boykung, p. 78r; Natalia Bratslavsky, p. 11tl;Hung Chung Chih, p. 233b; Christian Delbert, p. 28; dio5050, p. 277l; Durden Images, p. 374lt; Elena Elisseeva, p. 247; Elovich, p. 141; Featureflash, p. 89tc; Fly_dragonfly, p. 355bl; Ryan R Fox, p. 105c; ggw, pp. 382–3; Richard Goldberg, p. 240ct; Tatiana Grozetskaya, p. 144bl; Iculig, p. 470; irin-k, p. 152tr; ItsAngela, p. 10l; Adam J, p. 457; jannoon028, p. 390tr; Szasz-Fabian Jozsef, p. 91 (teeth); kaetana, p. 6b; Kekyalyaynen, p. 15tl; Kichigin, p. 277br; LagunaticPhoto, pp. 206–7; JP Lavoie, p. 106tl; Frederic Legrand – COMEO, p. 155b; Looker_Studio, p. 469; Viacheslav Lopatin, p. 74c; magnetix, p. 268br; Alexandr Makarov, p. 87bl; MarKord, p. 11tr; mdmworks, p. 152tl; Nik Merkulov, p. 1; molekuul.be, pp. 136cl (ball-and-stick model), 136cl (space-filling model); Nuttapong, p. 192r (paper); Alon Othnay, p. 100; siamionau pavel, p. 192r (shoes); Plati Photography, p. 501; PomInOz, p. 229; prudkov, p. 240b; S-F, p. 363; Fouad A. Saad, p. 269; Gino Santa Maria, p. 439; sergeevspb, p. 148cl; John D Sirlin, p. 23cl; Stable, p. 192cl; Standard Studio, p. 480r; streetprince, p. 6cl; TFoxFoto, p. 1, 257-9; Stefano Tinti, p. 5b; tonyz20, p. 8br (opal); Tooykrub, p. 393; udaix, p. 309t; Chepko Danil Vitalevich, p. 331; Taras Vyshnya, p. 215t; Whiteaster, p. 359tr.

Tentotwo: p. 334b.

University of Idaho Coeur d'Alene: Marie Pengilly, p. 340b.

University of Queensland: p. 165br.

Volvo Car Corporation: Volvo Cars, p. 155t.